Novel Antibacterial Agents

Novel Antibacterial Agents

Editors

Fiorella Meneghetti
Daniela Barlocco

MDPI • Basel • Beijing • Wuhan • Barcelona • Belgrade • Manchester • Tokyo • Cluj • Tianjin

Editors
Fiorella Meneghetti
Department of Pharmaceutical Sciences
University of Milan
Milan
Italy

Daniela Barlocco
Department of Pharmaceutical Sciences
University of Milan
Milan
Italy

Editorial Office
MDPI
St. Alban-Anlage 66
4052 Basel, Switzerland

This is a reprint of articles from the Special Issue published online in the open access journal *Pharmaceuticals* (ISSN 1424-8247) (available at: www.mdpi.com/journal/pharmaceuticals/special_issues/Antibacterial_Agents).

For citation purposes, cite each article independently as indicated on the article page online and as indicated below:

LastName, A.A.; LastName, B.B.; LastName, C.C. Article Title. *Journal Name* **Year**, *Volume Number*, Page Range.

ISBN 978-3-0365-2861-8 (Hbk)
ISBN 978-3-0365-2860-1 (PDF)

Contents

About the Editors

Fiorella Meneghetti

Fiorella Meneghetti earned her degree with honors in Medicinal Chemistry and Technology at the University of Padova in 1998. In the same year, she obtained the qualification to practice as a pharmacist. In 2002, she received her Ph.D. in Bio-Chemical Sciences from the University of Torino. Then, she joined the University of Milan with a grant entitled "Structural investigation of redox proteins for the nanobiotechologies" (2002-2005). F.M. has been Assistant Professor since January 1st, 2006 (SSD CHIM/08) at the Department of Pharmaceutical Sciences of the University of Milan; she is responsible for the teaching of "Chemical and Toxicological Analysis" and "Drug Analysis". The main area of her research is the crystallization and structural determination of drugs and enzymes. F.M. is co-author of 97 scientific articles published on international journals of relevant impact factor and several presentations (oral/poster) at national and international conferences. Her research activity has been focused on the following fields: 1) structural analysis of enzymes; 2) crystal structures of pharmaceutical compounds; and 3) structural characterization of metal complexes.

Daniela Barlocco

Daniela Barlocco received her laurea degree in Pharmacy at the University of Pavia (Italy) in 1969 and her laurea degree in Chemistry in 1977 at the same University. In 1978, she joined the University of Milan, Faculty of Pharmacy. In 1992, she moved to the University of Modena for three years as associate professor. In 1995, she returned to the Faculty of Pharmacy of Milan, where she worked as professor of Medicinal Chemistry until her retirement in 2016. She is member of several Editorial Boards, including MRMC and Current Medicinal Chemistry. She spent several months as visiting professor at SmithKline Beecham (Harlow, UK) and at the University College (London). In 2005, she was a member of the Evaluation Panel for the Faculty of Pharmacy in Helsinki. Her research interests include synthetic and theoretical chemistry of pyridazines and medicinal chemistry of several enzyme inhibitors, e.g., AR, TS, ACAT, SSAO, STAT.

Preface to "Novel Antibacterial Agents"

This book was devoted to the latest advances achieved in the antibacterial field, with a focus on the recent efforts made to develop new antimicrobial agents with novel modes of action, and a perspective on future directions of this line of research.

Antimicrobial resistance has become a major threat to global health, and the twenty-two published articles here reported put in evidence that the discovery and development of new antibiotics are extremely challenging. The antimicrobial research covers a wide area, spanning from the design of new compounds, also supported by molecular modeling techniques, their synthesis and characterization, and biological tests.

In this context, the current crisis caused by the COVID-19 pandemic, but also older threats, such as the human immunodeficiency virus or the hepatitis C virus, require greater attention than ever.

The research works described in this book provide an extremely useful example of the results achieved in the field of antibacterial drug development. The search for new chemical entities was approached starting from both natural and synthetic compounds and addressing different targets. In addition, recent findings were presented and discussed highlighting the strategies to fight bacterial resistance. Detailed references to the state-of-the-art can be found in this book.

We strongly encourage the wide group of readers to explore the book that we are presenting, to get inspired to develop new approaches for the diagnosis and treatment of antibacterial diseases, and to circumvent resistance issues.

We are extremely grateful to all the authors for the hard work they have done in producing the chapters. We thank MDPI for the decision to publish this book and Ms. Fendy Fan for the kind assistance and technical support.

Fiorella Meneghetti, Daniela Barlocco
Editors

pharmaceuticals

MDPI

Editorial

Special Issue "Novel Antibacterial Agents"

Fiorella Meneghetti * and Daniela Barlocco

Department of Pharmaceutical Sciences, University of Milan, Via L. Mangiagalli 25, 20133 Milan, Italy; daniela.barlocco@unimi.it

* Correspondence: fiorella.meneghetti@unimi.it

This Special Issue of *Pharmaceuticals* is devoted to significant advances achieved in the field of antibacterial agents. Here, we report recent efforts made to develop new antimicrobials with novel modes of action/resistance, and offer perspectives on the future directions of antibacterial agents.

Antimicrobial resistance has become a major threat to global health and the twenty-two published articles, included here, evidence that the discovery and development of new antibiotics is extremely challenging.

This Special Issue is focused on the search for new chemical entities, starting from both natural and synthetic compounds and addressing different targets. In addition, recent findings are presented and discussed, highlighting strategies of fighting bacterial resistance.

Investigation into antimicrobials covers a wide research area, as emphasized in this Special Issue, spanning from the design, synthesis, and characterization of new compounds, supported by molecular modeling techniques, to the development of biological tests. This is all possible thanks to the contributions of experts.

Natural products, as rich sources of chemical diversity, offer excellent possibilities to identify novel leads in medicinal chemistry. In this direction, S. Garzoli et al. [1], via Headspace-Gas Chromatography/Mass Spectrometry (HS-GC/MS), identified 28 components, mainly belonging to the monoterpenes family, in the essential oils from needles (EOs) of four *Pinaceae*. Both the liquid and vapour phases were evaluated for their antibacterial activity against three Gram-negative (*Escherichia coli*, *Pseudomonas fluorescens*, and *Acinetobacter bohemicus*) and two Gram-positive (*Kocuria marina* and *Bacillus cereus*) bacteria using different assays. Better results were obtained with the vapour phase. In addition, a concentration-dependent antioxidant activity was evidenced for all the EOs. The Authors highlighted the importance of α-pinene as a novel natural antibacterial agent.

Other interesting antioxidant compounds were identified by L.C. Nascimento da Silva et al. [2], in *B. tetraphylla* leaf methanolic extracts (BTME) which, when tested on *Tenebrio molitor* larvae inoculated with heat-inactivated *E. coli*, proved to be able to protect the larvae from the stress. A mixture of aliphatic (terpenes, fatty acids, carbohydrates) and aromatic compounds (phenolic derivatives) were evidenced in BTME using NMR analysis.

Antioxidant properties, in addition to antimicrobial and cytotoxic activity, were reported by A.J.L. Pombeiro et al. [3] in a series of saccharin-tetrazolyl and -thiadiazolyl derivatives. The best antioxidant results were shown for (*N*-(1-methyl-2*H*-tetrazol-5-yl)-*N*-(1,1-dioxo-1,2-benzisothiazol-3-yl) amine, while all the compounds had insignificant toxicity when tested in an *Artemia salina* model.

Several EOs, derived from *Origanum majorana*, *Rosmarinus officinalis*, and *Thymus zygis* medicinal plants were investigated by A. Gaber et al. [4] for their ability to inhibit biofilm formation and eradicate methicillin-resistant *Staphylococcus aureus* (MRSA) isolates. The best activity was found in *T. Zygis*, but all the studied EOs showed interesting properties (the percentage of inhibition ranging from 10.20 to 95.91%, and percentage of eradication ranging from 12.65 to 98.01%), thus, representing potential alternatives to antibiotics.

Citation: Meneghetti, F.; Barlocco, D. Special Issue "Novel Antibacterial Agents". *Pharmaceuticals* **2021**, *14*, 382. https://doi.org/10.3390/ph14040382

Received: 14 April 2021
Accepted: 16 April 2021
Published: 19 April 2021

The inhibition of biofilm formation was also studied also by B. Citterio et al. [5], who investigated the properties of the marine bisindole alkaloid 2,2-bis(6-bromo-1*H*-indol-3-yl)ethanamine and its fluorinated analogue which were tested both for their potential use as antibiotic adjuvants and antibiofilm agents against *S. aureus* CH 10850 (MRSA) and *S. aureus* ATCC 29213 (MSSA). The fluorinated derivative showed a higher potency in eradicating a preformed biofilm. Both compounds showed a safe profile and were indicated for in vivo application as adjuvants to restore antibiotic treatment against MRSA.

Biofilm formation is an urgent problem in dentistry, due to the high porosity and absorptiveness of the polymers commonly used for the 3D-printed dental prostheses. D.-H. Lee et al. [6] studied the effects of a new chlorhexidine (CHX)-loaded polydimethylsiloxane (PDMS)-based coating material on the surface microstructure, surface wettability and antibacterial activity of 3D-printing dental polymer. The antibacterial was first encapsulated in mesoporous silica nanoparticles (MSN). The significant experimental results supported this approach.

Functionalization of oxidized multi-walled carbon nanotubes (oxCNTs) was studied by Z. Sideratou et al. [7], who used a simple non-covalent modification procedure, using hyperbranched poly(ethyleneimine) derivatives (QPEIs), with various quaternization degrees. The aqueous dispersion of this material was found to be stable over 12 months and to be provided with significant antibacterial and anti-cyanobacterial activity. It was suggested by the Authors for application in the disinfection industry.

W.Y. Bang et al. [8] proposed the use of the cell-free supernatant from *Lactobacillus plantarum* NIBR97 as an alternative to chemical disinfectants. By scanning electron microscopy (SEM), this was shown to cause cellular lysis in bacterial membranes, thus, indicating the involvement of peptides or proteins in its mechanism of action and suggesting the use of proteinase K treatment to support its antibacterial activity. In addition, it showed good antiviral properties, possibly through a different mechanism of action.

A study by A. Baycal et al. [9] on *Moringa oleifera* reported the correlation between the capability of laser-induced breakdown spectroscopy (LIBS) to monitor the elemental compositions of plants and their biological effects. In particular, the bioactive components of the seed (MOS) alcoholic extract were tested against *Escherichia coli* and *Staphylococcus aureus* via an agar well diffusion (AWD) assay and scanning electron microscopy (SEM). An interesting activity against Gram-positive bacteria was evidenced. In addition, the authors reported that MOS extract exhibited significant inhibitory properties on HCT116 cell growth, whereas no effects were noticed in a parallel assay on HEK-293 cells. According to the authors, the antibacterial and anticancer potency of the MOS extract could be attributed to several complexes, e.g., ethyl ester and D-allose and hexadecenoic, oleic and palmitic acids.

J.-H Moon et al. [10] reported their results on 3-pentylcatechol (PC), a synthetic urushiol derivative, able to inhibit the growth of *Helicobacter pylori* in an in vitro assay, in comparison with triple therapy (omeprazole, metronidazole, and clarithromycin). The authors report that PC displayed better effects than triple therapy at all doses. Interestingly, synergism was shown when PC and triple therapy were co-administered.

The composition and biological properties of *Thymus mastichina*, a diffuse semishrub of jungles and the rocky landscapes of the Iberian Peninsula, were reviewed by A.R.T.S. Araujo et al. [11]. Several properties were reported both for the extracts and the essential oils. In particular, activity against methicillin resistant bacteria and antifungal activity should be highlighted, though several other properties (e.g., anticancer, antiviral, insecticidal, repellent, anti-Alzheimer's, and anti-inflammatory) have been investigated. The authors suggest that *Thymus mastichina* should be considered for use in food and cosmetic applications other than in the pharmaceutical field.

Antibiotic resistance was also the target of the paper by M. Sessevmez et al. [12], who indicated bacteriophages as an alternative method, which was first proposed in the early 20th century by d'Herelle, Bruynoghe and Maisin to treat bacterial infections. Different administration methods are possible, including topical application, inhalation, oral or parenteral delivery. *Pseudomonas aeruginosa*, *Mycobacterium tuberculosis* and *Acinetobacter baumannii*, responsible for the main drug resistant infections, are potential targets of phages, which are developed under a strict quality control regime.

Tuberculosis was also the target of F. Meneghetti et al. [13], who described novel furan derivatives as inhibitors of salicylate synthase MbtI, an essential enzyme of mycobacterium, absent in human cells. The best compound, which provided comparable inhibitory properties to the previous leads but endowed a better antitubercular activity, was 5-(3-cyano-5-(trifluoromethyl)phenyl)furan-2-carboxylic acid, which will form the basis of future studies.

Covalent inhibitors of another bacterial enzyme for which there is no human orthologue, namely UDP-*N*-acetylglucosamine enolpyruvyl transferase (MurA), have been developed by G.M. Keserü et al. [14], who indicated bromo-cyclobutenaminones as novel electrophilic probes by screening a small library of cyclobutenone derivatives. The bromine atom has been recognized as an essential requirement, and MS/MS experiments have led to the suggestion that Cys115 is also involved. The stability and bioavailability of these compounds was also assessed.

Another alternative to the use of antibiotics in bacterial infections was proposed by F.M. Goycoolea et al. [15] who studied a library of 23 pure compounds of different chemical structures, assessing their quorum sensing (QS) inhibition activity. The best results were obtained with phenazine carboxylic acid (PCA), 2-heptyl-3-hydroxy-4-quinolone (PQS), 1*H*-2-methyl-4-quinolone (MOQ) and genipin, which exhibited QS inhibition activity without compromising bacterial growth.

A specific disease, characterized by complex aetiological mechanisms, namely atopic dermatitis (AD), was the topic of a review by R. Di Marco et al. [16]. Loss of the skin barrier, linked to dysbiosis and immunological dysfunction, which causes an imbalance in the ratio between the pathogen *Staphylococcus aureus* and/or other microorganisms residing in the skin, is considered as a crucial factor contributing to AD. Though the review suggests several treatments for the disease, including the use of bacteria and/or microbiota transplantation, together with different drug delivery systems, the authors conclude that a standardized process is necessary in order to obtain reliable data.

F. Sparatore et al. [17] addressed the topic of Leishmaniases, the therapy of which is presently based on expensive drugs which are associated with severe side-effects and the treat of resistance. They tested sixteen lucanthone and four amitriptyline analogues in vitro, all of which were characterized by a bulky quinolizidinylalkyl moiety, against *Leishmania tropica* and *L. infantum* promastigotes. All compounds displayed significant activity (IC_{50} in the low μM range) and low cytotoxicity. The authors suggest that these analogues could act through trypanothione reductase (TryR) inhibition.

In the search for selective agents that are capable of treating bacterial diarrhoea without affecting the host intestinal microbiota, L. Kokoska et al. [18] investigated ten phytochemicals and their synthetic analogues (berberine, bismuth subsalicylate, ferron, 8-hydroxyquinoline, chloroxine, nitroxoline, salicylic acid, sanguinarine, tannic acid, and zinc pyrithione) in vitro and compared the results with six commercial antibiotics (ceftriaxone, ciprofloxacin, chloramphenicol, metronidazole, tetracycline, and vancomycin) against 21 intestinal pathogenic/probiotic (e.g., *Salmonella* spp. and *bifidobacteria*) bacterial strains and three intestinal cancer/normal (Caco-2, HT29, and FHs 74 Int) cell lines. Several compounds, e.g., chloroxine, ciprofloxacin, nitroxoline, tetracycline, and zinc pyrithione exhibited potent selective growth-inhibitory activity against pathogens, while 8-hydroxyquinoline and sanguinarine provided the best activity towards cancer cells. It should also be noted that none of the compounds were found to be cytotoxic to normal cells. Once more, plants have been suggested as a promising source for novel drugs.

N. D'Amelio et al. [19] investigated the mechanism of action of a synthetic peptide (K11), whose antibacterial properties have been previously reported. Via a liquid and solid-state NMR technique, they studied the interaction of K11 with different biomimetic membranes and reported that this can destabilize them. In addition, via molecular dynamic simulations, they suggested that K11 could penetrate the membranes in four steps (anchoring, twisting, helix flip, and internalization) involving several lysine residues.

Additionally, also within the field of synthetic compounds, B. Altava et al. [20] investigated how they could vary the antibacterial properties of imidazole and its imidazolium salts derived from *L*-valine and *L*-phenylalanine, by modulating their lipophilicity. When tested on *E. Coli* and *B. Subtilis* bacterial strains, very encouraging results were obtained. In particular, the minimum bactericidal concentration (MBC) of one compound towards *B. subtilis* was found to be lower than the IC_{50} cytotoxicity value for the control cell line, HEK-293. Moreover, the capability of these structures to aggregate in different media was investigated to establish the importance of the monomeric species.

The effects of substituent variation on a different substrate, namely *tris*(1*H*-indol-3-yl)methylium, was investigated by A.S. Trenin et al. [21]. The synthesized compounds were tested on 12 bacterial strains that were either sensitive or resistant to meticillin. The results indicated that antibacterial properties depended on the chain length, with the best activity residing in compounds with C5–C6 chains. The most interesting compounds showed better antibacterial properties than levofloxacin on MRSA and had a very safe profile.

Antibacterial and antifungal activity was investigated in vitro by P. Eleftheriou et al. [22] on a series of 3-amino-5-(indol-3-yl) methylene-4-oxo-2-thioxothiazolidine derivatives. Compounds exhibited significant activity both on Gram-positive and Gram-negative bacteria, demonstrating a potency greater than ampicillin. Similarly, their antifungal activity was superior to that of ketoconazole. Docking studies have suggested that their antibacterial activity could be derived from *E. coli* Mur B inhibition, while CYP51 inhibition would be responsible for the antifungal activity.

The research described in the articles constituting this Special Issue collectively provides extremely useful examples of the results that have been recently achieved in the field of antibacterial drug development.

We hope the readers enjoy this Special Issue and are inspired to develop new approaches for antibacterial disease diagnosis, treatment and to circumvent resistance issues.

Author Contributions: F.M. and D.B. contributed equally to this Editorial. All authors have read and agreed to the published version of the manuscript.

Funding: This research was funded by University of Milan, Linea B.

Institutional Review Board Statement: Not applicable.

Informed Consent Statement: Not applicable.

Acknowledgments: We are extremely grateful to all authors for their hard work to produce an updated and comprehensive issue on antibacterial agents in a timely fashion. We would also like to thank the Reviewers who carefully evaluated the submitted manuscripts. We thank the Editor-in-Chief of Pharmaceuticals, J.J. Vanden Eynde, for giving us the opportunity to serve as Guest Editors and F. Fan for the kind assistance and technical support.

Conflicts of Interest: The authors declare no conflict of interest.

References

1. Garzoli, S.; Masci, V.L.; Caradonna, V.; Tiezzi, A.; Giacomello, P.; Ovidi, E. Liquid and Vapor Phase of Four Conifer-Derived Essential Oils: Comparison of Chemical Compositions and Antimicrobial and Antioxidant Properties. *Pharmaceuticals* **2021**, *14*, 134. [CrossRef]
2. Silva, T.F.; Cavalcanti Filho, J.R.N.; Barreto Fonsêca, M.M.L.; Santos, N.M.; Barbosa da Silva, A.C.; Zagmignan, A.; Abreu, A.G.; Sant'Anna da Silva, A.P.; Lima, V.L.M.; Silva, N.H.; et al. Products Derived from *Buchenavia tetraphylla* Leaves Have In Vitro Antioxidant Activity and Protect *Tenebrio molitor* Larvae against *Escherichia coli*-Induced Injury. *Pharmaceuticals* **2020**, *13*, 46. [CrossRef]
3. Frija, L.M.T.; Ntungwe, E.; Sitarek, P.; Andrade, J.M.; Toma, M.; Śliwiński, T.; Cabral, L.S.; Cristiano, M.L.; Rijo, P.; Pombeiro, A.J.L. In Vitro Assessment of Antimicrobial, Antioxidant, and Cytotoxic Properties of Saccharin–Tetrazolyl and–Thiadiazolyl Derivatives: The Simple Dependence of the pH Value on Antimicrobial Activity. *Pharmaceuticals* **2019**, *12*, 167. [CrossRef]
4. Ben Abdallah, F.; Lagha, R.; Gaber, A. Biofilm Inhibition and Eradication Properties of Medicinal Plant Essential Oils against Methicillin-Resistant *Staphylococcus aureus* Clinical Isolates. *Pharmaceuticals* **2020**, *13*, 369. [CrossRef] [PubMed]

5. Campana, R.; Mangiaterra, G.; Tiboni, M.; Frangipani, E.; Biavasco, F.; Lucarini, S.; Citterio, B. A Fluorinated Analogue of Marine Bisindole Alkaloid 2,2-Bis(6-bromo-1*H*-indol-3-yl)ethanamine as Potential Anti-Biofilm Agent and Antibiotic Adjuvant Against *Staphylococcus aureus*. *Pharmaceuticals* **2020**, *13*, 210. [CrossRef]
6. Mai, H.-N.; Hyun, D.C.; Park, J.H.; Kim, D.-Y.; Lee, S.M.; Lee, D.-H. Antibacterial Drug-Release Polydimethylsiloxane Coating for 3D-Printing Dental Polymer: Surface Alterations and Antimicrobial Effects. *Pharmaceuticals* **2020**, *13*, 304. [CrossRef]
7. Heliopoulos, N.S.; Kythreoti, G.; Lyra, K.M.; Panagiotaki, K.N.; Papavasiliou, A.; Sakellis, E.; Papageorgiou, S.; Kouloumpis, A.; Gournis, D.; Katsaros, F.K.; et al. Cytotoxicity Effects of Water-Soluble Multi-Walled Carbon Nanotubes Decorated with Quaternized Hyperbranched Poly(ethyleneimine) Derivatives on Autotrophic and Heterotrophic Gram-Negative Bacteria. *Pharmaceuticals* **2020**, *13*, 293. [CrossRef]
8. Kim, S.W.; Kang, S.I.; Shin, D.H.; Oh, S.Y.; Lee, C.W.; Yang, Y.; Son, Y.K.; Yang, H.-S.; Lee, B.-H.; An, H.-J.; et al. Potential of Cell-Free Supernatant from *Lactobacillus plantarum* NIBR97, Including Novel Bacteriocins, as a Natural Alternative to Chemical Disinfectants. *Pharmaceuticals* **2020**, *13*, 266. [CrossRef] [PubMed]
9. Aldakheel, R.K.; Rehman, S.; Almessiere, M.A.; Khan, F.A.; Gondal, M.A.; Mostafa, A.; Baykal, A. Bactericidal and In Vitro Cytotoxicity of *Moringa oleifera* Seed Extract and Its Elemental Analysis Using Laser-Induced Breakdown Spectroscopy. *Pharmaceuticals* **2020**, *13*, 193. [CrossRef] [PubMed]
10. Jeong, H.Y.; Lee, T.H.; Kim, J.G.; Lee, S.; Moon, C.; Truong, X.T.; Jeon, T.-I.; Moon, J.-H. 3-Pentylcatechol, a Non-Allergenic Urushiol Derivative, Displays Anti-*Helicobacter pylori* Activity In Vivo. *Pharmaceuticals* **2020**, *13*, 384. [CrossRef]
11. Rodrigues, M.; Lopes, A.C.; Vaz, F.; Filipe, M.; Alves, G.; Ribeiro, M.P.; Coutinho, P.; Araujo, A.R.T.S. *Thymus mastichina*: Composition and Biological Properties with a Focus on Antimicrobial Activity. *Pharmaceuticals* **2020**, *13*, 479. [CrossRef]
12. Düzgüneş, N.; Sessevmez, M.; Yildirim, M. Bacteriophage Therapy of Bacterial Infections: The Rediscovered Frontier. *Pharmaceuticals* **2021**, *14*, 34. [CrossRef]
13. Mori, M.; Stelitano, G.; Chiarelli, L.R.; Cazzaniga, G.; Gelain, A.; Barlocco, D.; Pini, E.; Meneghetti, F.; Villa, S. Synthesis, Characterization, and Biological Evaluation of New Derivatives Targeting MbtI as Antitubercular Agents. *Pharmaceuticals* **2021**, *14*, 155. [CrossRef] [PubMed]
14. Hamilton, D.J.; Ábrányi-Balogh, P.; Keeley, A.; Petri, L.; Hrast, M.; Imre, T.; Wijtmans, M.; Gobec, S.; Esch, I.J.P.; Keserű, G.M. Bromo-Cyclobutenaminones as New Covalent UDP-*N*-Acetylglucosamine Enolpyruvyl Transferase (MurA) Inhibitors. *Pharmaceuticals* **2020**, *13*, 362. [CrossRef] [PubMed]
15. Qin, X.; Vila-Sanjurjo, C.; Singh, R.; Philipp, B.; Goycoolea, F.M. Screening of Bacterial Quorum Sensing Inhibitors in a *Vibrio fischeri* LuxR-Based Synthetic Fluorescent *E. coli* Biosensor. *Pharmaceuticals* **2020**, *13*, 263. [CrossRef]
16. Magnifico, I.; Petronio Petronio, G.; Venditti, N.; Cutuli, M.A.; Pietrangelo, L.; Vergalito, F.; Mangano, K.; Zella, D.; Di Marco, R. Atopic Dermatitis as a Multifactorial Skin Disorder. Can the Analysis of Pathophysiological Targets Represent the Winning Therapeutic Strategy? *Pharmaceuticals* **2020**, *13*, 411. [CrossRef] [PubMed]
17. Tonelli, M.; Sparatore, A.; Basilico, N.; Cavicchini, L.; Parapini, S.; Tasso, B.; Laurini, E.; Pricl, S.; Boido, V.; Sparatore, F. Quinolizidine-Derived Lucanthone and Amitriptyline Analogues Endowed with Potent Antileishmanial Activity. *Pharmaceuticals* **2020**, *13*, 339. [CrossRef]
18. Kudera, T.; Doskocil, I.; Salmonova, H.; Petrtyl, M.; Skrivanova, E.; Kokoska, L. In Vitro Selective Growth-Inhibitory Activities of Phytochemicals, Synthetic Phytochemical Analogs, and Antibiotics against Diarrheagenic/Probiotic Bacteria and Cancer/Normal Intestinal Cells. *Pharmaceuticals* **2020**, *13*, 233. [CrossRef] [PubMed]
19. Ramos-Martín, F.; Herrera-León, C.; Antonietti, V.; Sonnet, P.; Sarazin, C.; D'Amelio, N. Antimicrobial Peptide K11 Selectively Recognizes Bacterial Biomimetic Membranes and Acts by Twisting Their Bilayers. *Pharmaceuticals* **2021**, *14*, 1. [CrossRef] [PubMed]
20. Valls, A.; Andreu, J.J.; Falomir, E.; Luis, S.V.; Atrián-Blasco, E.; Mitchell, S.G.; Altava, B. Imidazole and Imidazolium Antibacterial Drugs Derived from Amino Acids. *Pharmaceuticals* **2020**, *13*, 482. [CrossRef]
21. Lavrenov, S.N.; Isakova, E.B.; Panov, A.A.; Simonov, A.Y.; Tatarskiy, V.V.; Trenin, A.S. *N*-(Hydroxyalkyl) Derivatives of *tris*(1*H*-indol-3-yl)methylium Salts as Promising Antibacterial Agents: Synthesis and Biological Evaluation. *Pharmaceuticals* **2020**, *13*, 469. [CrossRef] [PubMed]
22. Horishny, V.; Kartsev, V.; Matiychuk, V.; Geronikaki, A.; Anthi, P.; Pogodin, P.; Poroikov, V.; Ivanov, M.; Kostic, M.; Soković, M.D.; et al. 3-Amino-5-(indol-3-yl)methylene-4-oxo-2-thioxothiazolidine Derivatives as Antimicrobial Agents: Synthesis, Computational and Biological Evaluation. *Pharmaceuticals* **2020**, *13*, 229. [CrossRef] [PubMed]

 *pharmaceuticals*

Article

Synthesis, Characterization, and Biological Evaluation of New Derivatives Targeting MbtI as Antitubercular Agents

Matteo Mori [1,†], Giovanni Stelitano [2,†], Laurent R. Chiarelli [2], Giulia Cazzaniga [1], Arianna Gelain [1], Daniela Barlocco [1], Elena Pini [1], Fiorella Meneghetti [1,*] and Stefania Villa [1]

1 Department of Pharmaceutical Sciences, University of Milan, Via L. Mangiagalli 25, 20133 Milano, Italy; matteo.mori@unimi.it (M.M.); giulia.cazzaniga@unimi.it (G.C.); arianna.gelain@unimi.it (A.G.); daniela.barlocco@unimi.it (D.B.); elena.pini@unimi.it (E.P.); stefania.villa@unimi.it (S.V.)
2 Department of Biology and Biotechnology "Lazzaro Spallanzani", University of Pavia, via A. Ferrata 9, 27100 Pavia, Italy; giovanni.stelitano01@universitadipavia.it (G.S.); laurent.chiarelli@unipv.it (L.R.C.)
* Correspondence: fiorella.meneghetti@unimi.it; Tel.: +39-02-503-19306
† These authors contributed equally to this work.

Abstract: Tuberculosis (TB) causes millions of deaths every year, ranking as one of the most dangerous infectious diseases worldwide. Because several pathogenic strains of *Mycobacterium tuberculosis* (Mtb) have developed resistance against most of the established anti-TB drugs, new therapeutic options are urgently needed. An attractive target for the development of new antitubercular agents is the salicylate synthase MbtI, an essential enzyme for the mycobacterial siderophore biochemical machinery, absent in human cells. A set of analogues of **I** and **II**, two of the most potent MbtI inhibitors identified to date, was synthesized, characterized, and tested to elucidate the structural requirements for achieving an efficient MbtI inhibition and a potent antitubercular activity with this class of compounds. The structure-activity relationships (SAR) here discussed evidenced the importance of the furan as part of the pharmacophore and led to the preparation of six new compounds (**IV–IX**), which gave us the opportunity to examine a hitherto unexplored position of the phenyl ring. Among them emerged 5-(3-cyano-5-(trifluoromethyl)phenyl)furan-2-carboxylic acid (**IV**), endowed with comparable inhibitory properties to the previous leads, but a better antitubercular activity, which is a key issue in MbtI inhibitor research. Therefore, compound **IV** offers promising prospects for future studies on the development of novel agents against mycobacterial infections.

Keywords: tuberculosis; mycobactins; furan; siderophores; drug design; bioisosterism; drug resistance

Citation: Mori, M.; Stelitano, G.; Chiarelli, L.R.; Cazzaniga, G.; Gelain, A.; Barlocco, D.; Pini, E.; Meneghetti, F.; Villa, S. Synthesis, Characterization, and Biological Evaluation of New Derivatives Targeting MbtI as Antitubercular Agents. *Pharmaceuticals* **2021**, *14*, 155. https://doi.org/10.3390/ph14020155

Academic Editor: Pascal Sonnet
Received: 12 January 2021
Accepted: 9 February 2021
Published: 13 February 2021

1. Introduction

Tuberculosis (TB), the infectious disease caused by *Mycobacterium tuberculosis* (Mtb), represents a global emergency requiring new therapeutic options, mainly because of the rapid spread of drug-resistant strains, which are causing an alarming rise in clinical cases.

According to the 2020 World Health Organization (WHO) report [1], TB was responsible for around 1.4 million deaths and over 10 million new infections worldwide in 2019; these numbers are expected to rise significantly in 2020 as a consequence of the coronavirus disease 2019 (COVID-19) pandemic. Additionally, it is estimated that Mtb exists in its latent form in approximately one-quarter of the global population [1].

Although the investigation of new pharmaceutical forms for the delivery of current antitubercular drugs may contribute to enhance patient compliance and limit the spread of the disease [2,3], the development of new therapeutic options represents an even more pressing need. While drug-susceptible TB can be cured within 6–8 months with the current standard treatment regimen [4], multi- and extensively drug-resistant (MDR/XDR) infections are treated for at least 20 months with poor outcomes [5], posing a serious threat to human health. The continuous genetic adaptation and rapid propagation of drug-resistant pathogens have led to an expected drop in the therapeutic efficacy of the current anti-TB drugs, forcing the scientists to face new challenges in the discovery of novel molecular entities to address this issue. Hence, the development of innovative compounds targeting both replicating and dormant Mtb bacilli is critical for the design of more effective and shorter therapies.

To address the urgent need of selective antitubercular drugs with novel mechanisms of action, new drug targets have been recently explored and validated [6–8]. Among them, the mycobactin biosynthetic pathway, which leads to the synthesis of siderophores capable of sequestering host iron, has been identified as a source of promising candidates [9,10]. Indeed, the siderophore biochemical machinery is significantly upregulated under iron-deficient conditions, common in infected macrophages, constituting one of the major pathogenic determinants of TB. Moreover, it is absent in humans, thus minimizing the risk of off-target effects.

Among the four enzymes involved in mycobactin biosynthesis and currently under investigation as potential drug targets (i.e., MbtI, MbtA, MbtM, and PPTase), we focused our attention on the Mg^{2+}-dependent bifunctional salicylate synthase MbtI, which catalyzes the first two steps in the production of all mycobacterial siderophores. This enzyme belongs to the family of the chorismate-utilizing enzymes (CUEs) [11] and it catalyzes the reactions shown in Figure 1.

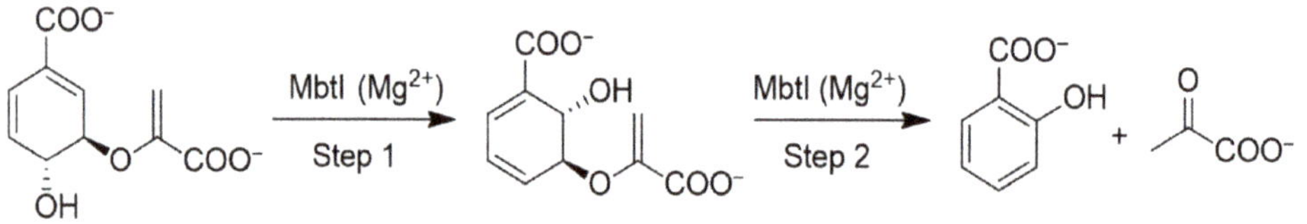

Figure 1. Reactions catalyzed by MbtI.

In this context, we developed in recent years a series of furan-based carboxylic acids as MbtI inhibitors [12–15]. Among this class of compounds, **I** and **II** (Figure 2) emerged as the best leads, exhibiting a strong MbtI inhibition, conceivably related to their antitubercular activity, and a negligible cytotoxicity towards eukaryotic cells. When analyzing the structure-activity relationships (SAR) of these compounds, we observed that the activity of the substances was closely related to the presence of an electron withdrawing moiety in a suitable position of the phenyl ring. The removal of the substituent from the phenyl of our furan-based leads (**III**, Figure 2) resulted in a complete loss of activity [14].

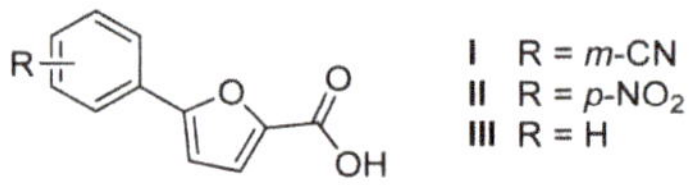

Figure 2. Chemical structure of the lead compounds **I**, **II**, and **III**.

Encouraged by these studies, and with the aim of enriching our arsenal of MbtI inhibitors with compounds exhibiting enhanced antitubercular activities, we enlarged our set of derivatives to include compounds bearing different heterocyclic scaffolds.

In some literature cases, the furan core was successfully replaced by other heterocycles to improve the cellular activity; indeed, according to Hinsberg's "ring equivalence" theory, the concept of isosterism and bioisosterism can be extended to heterocycles [16]. Here, we investigated whether the furan moiety could be successfully replaced by any of the heterocycles shown in Figure 3, also considering that extensive research efforts have been devoted to the exploration of heterocyclic compounds as antimycobacterial agents [17].

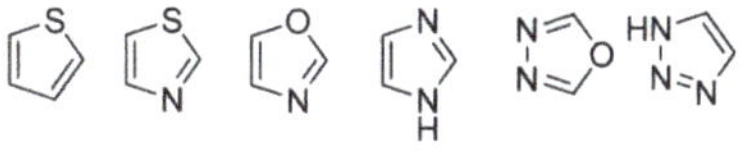

Figure 3. Chemical structure of the heterocyclic cores tested in this study: **1** (thiophene), **2** (thiazole), **3** (oxazole), **4** (imidazole), **5** (1,3,4-oxadiazole), **6** (1,2,3-triazole).

We considered the introduction of a thiophene (**1**), because, in several literature examples, the use of this ring has resulted in an improvement of the antimycobacterial activity [18–20]. The furan was then substituted by a thiazole (**2**); aside from being the most common heterocycle in drug design [21], this ring is part of the chemical structure of many compounds endowed with antitubercular activity [22]. To expand our investigations, we synthesized two derivatives bearing an oxazole (**3**), where the sulfur atom of the thiazole ring is substituted by an oxygen, a sulfur isostere [23]. The imidazole was then selected as an attractive isostere of thiazole and oxazole; notably, nitroimidazopyran PA-824 [24] has recently moved to advanced-stage clinical trials, inspiring the development of anti-TB agents featuring this moiety [25]. Finally, we explored the 1,3,4-oxadiazole (**5**), as it was reported to interact with some of the newest anti-TB targets [26], and the 1,2,3-triazole (**6**), whose importance is demonstrated by the antitubercular agent I-A09, which is under preclinical trials [27].

In the first part of this work, we designed, synthesized, and evaluated the biological activity of novel heterocyclic compounds belonging to two homologous series (Table 1), bearing the *m*-CN (series A, compounds **1a–6a**) and *p*-NO_2 (series B, **1b–6b**) substituent, respectively.

Table 1. In vitro activity of compounds **1a,b–6a,b**.

Structure	**Series A** R = *m*-CNPh		**Series B** R = *p*-NO_2Ph	
	%RA *	**IC_{50} ** (μM)**	**%RA ***	**IC_{50} ** (μM)**
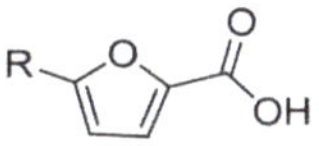	**I**		**II**	
	3.1 ± 1.0	6.3 ± 0.9	18.2 ± 5.1	7.6 ± 1.6
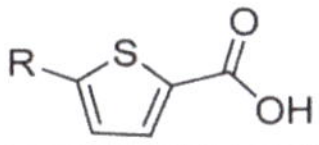	**1a**		**1b**	
	23.7 ± 5.0	35.5 ± 1.9	22.5 ±10.8	18.6 ± 1.7
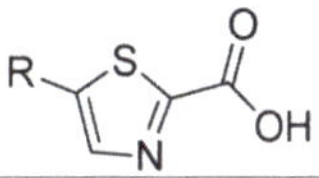	**2a**		**2b**	
	24.8 ± 1.8	41.9 ± 7.3	32.8 ± 2.3	-
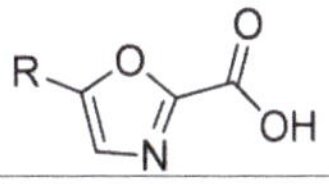	**3a**		**3b**	
	38.0 ± 7.6	-	23.0 ± 7.8	21.1 ± 2.7
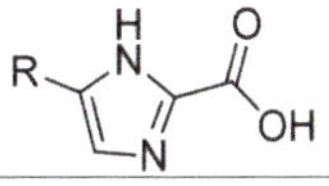	**4a**		**4b**	
	21.2 ± 1.0	39.3 ± 3.0	18.2 ± 4.9	27.5 ± 2.3
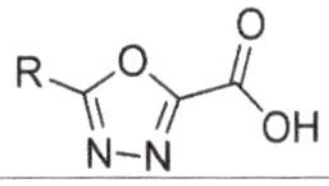	**5a**		**5b**	
	75.0 ± 9.8	-	59.0 ± 4.3	-
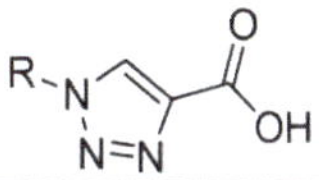	**6a**		**6b**	
	67.9 ± 6.6	-	32.7 ± 7.8	-

* % residual enzymatic activity at 100 μM; ** only for compounds with %RA ≤ 25%.

The modest biological activity of the new derivatives prompted us to reconsider the furan as the best heterocyclic core, suggesting the critical nature of an appropriately substituted furan moiety to maintain a significant enzymatic inhibition and to achieve antitubercular activity.

On this basis, and considering our previous results on disubstituted derivatives, we decided to design and synthesize six new furan-based analogues (**IV–IX**, Figure 4). In particular, **IV** was investigated as the isomer of 5-(2-cyano-4-(trifluoromethyl)phenyl)furan-2-carboxylic acid, which exhibited an interesting inhibitory effect (IC_{50} of about 18 μM) [13]. The new analogue **IV** bears the CN and CF_3 moieties in the relative *meta* positions to avoid the steric interactions between the two adjacent groups, which had proven to be detrimental for the activity [13]. Compound **V** was synthesized to evaluate the role of the 3-CN moiety, as our past works had shown that this group was superior to other substituents in terms of enzymatic activity [12,13].

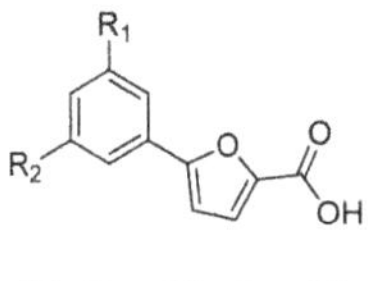

IV R_1 = CN; R_2 = CF_3
V R_1 = CF_3; R_2 = CF_3
VI R_1 = CN; R_2 = F
VII R_1 = CN; R_2 = OCH_3
VIII R_1 = CN; R_2 = CH_3
IX R_1 = CN; R_2 = OH

Figure 4. Chemical structure of compounds **IV–IX**.

Finally, compounds **VI–IX** were prepared to examine the influence of the substituent at position 5 of the phenyl ring on the biological activity of this class of compounds.

This strategy gave us the opportunity to examine a hitherto unexplored position of the phenyl ring (**IV–IX**), leading to the discovery of novel MbtI inhibitors endowed with antimycobacterial activity.

2. Results

2.1. Chemistry

The synthetic procedures adopted for the preparation of the compounds are heterogeneous and reflect the diverse approaches needed for the obtainment of the various heterocyclic derivatives. Where possible, the synthetic strategies were designed and optimized to afford the desired compounds, starting from the same commercially available reagents. All the compounds were characterized by means of mono- and bi-dimensional NMR techniques, FT-IR, ESI-MS, and elemental analysis. The procedures for the synthesis of series A and B (compounds **1a,b–6a,b**) are depicted in Schemes 1–6; all details regarding procedures and analytical data are reported in the Supplementary Materials.

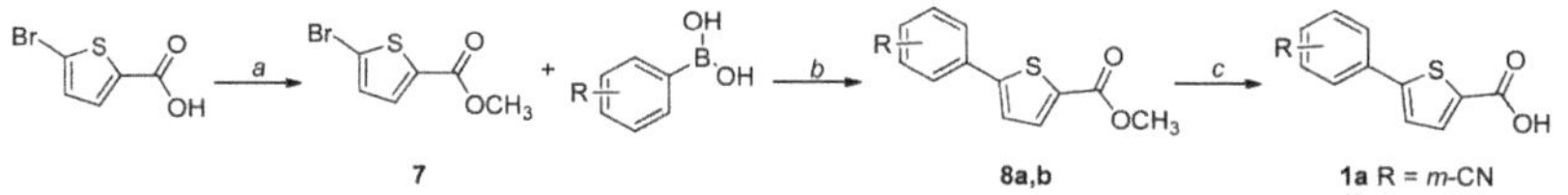

1a R = *m*-CN
1b R = *p*-NO_2

Scheme 1. Synthetic procedure for the preparation of **1a,b**. Reagents and conditions: (**a**) MeOH, conc. H_2SO_4, reflux, overnight; (**b**) $Pd(PPh_3)_2Cl_2$, 2 M Na_2CO_3, dry 1,4-dioxane, 90 °C, overnight, N_2 atm; (**c**) *1.* LiOH, THF-H_2O 2:1, r.t., 2 h for **1a**; 1 M NaOH, EtOH-THF 1:1, reflux, 5 h for **1b**; *2.* 1 M HCl, 0 °C.

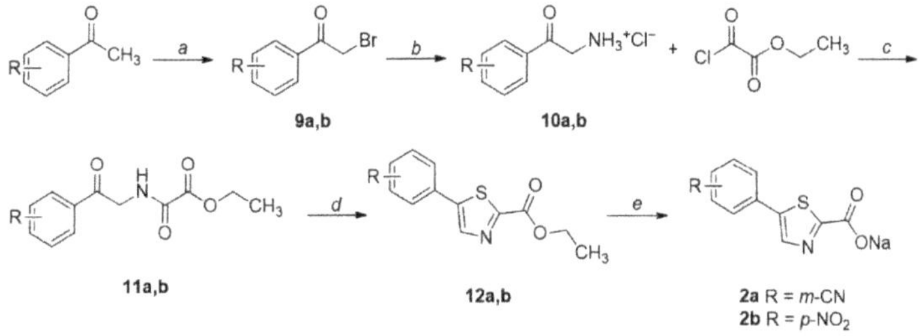

Scheme 2. Synthetic procedure for the preparation of **2a,b**. Reagents and conditions: (**a**) NBS, *p*-TsOH, DCM, overnight, r.t, N_2 atm.; (**b**) *1.* hexamine, DCM, 8 h, r.t.; *2.* conc. HCl, EtOH, overnight, r.t.; (**c**) TEA, EtOAc, 3 h, reflux; (**d**) Lawesson's reagent, 1,4-dioxane, 2 h, reflux; (**e**) NaOH, THF-H_2O 1:1, 1.5 h, r.t.

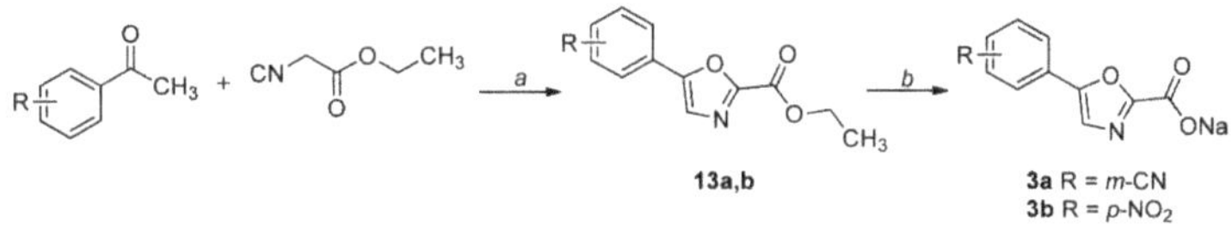

Scheme 3. Synthetic procedure for the preparation of **3a,b**. Reagents and conditions: (**a**) I_2, DMSO, 3 h, 130 °C; (**b**) NaOH, THF-H_2O 1:1, 1.5 h, r.t.

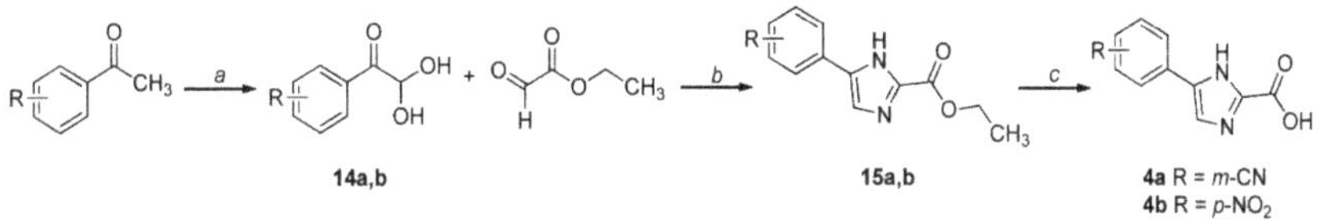

Scheme 4. Synthetic procedure for the preparation of **4a,b**. Reagents and conditions: (**a**) SeO_2, 1,4-dioxane/H_2O, reflux, 7 h, N_2 atm; (**b**) NH_4OAc, CH_3CN, H_2O, r.t., 2 h; (**c**) *1.* LiOH, THF-H_2O 2:1, r.t., overnight for **4a**; NaOH, THF-H_2O 1:1, reflux, 6 h for **4b**; *2.* 3 M HCl, 0 °C.

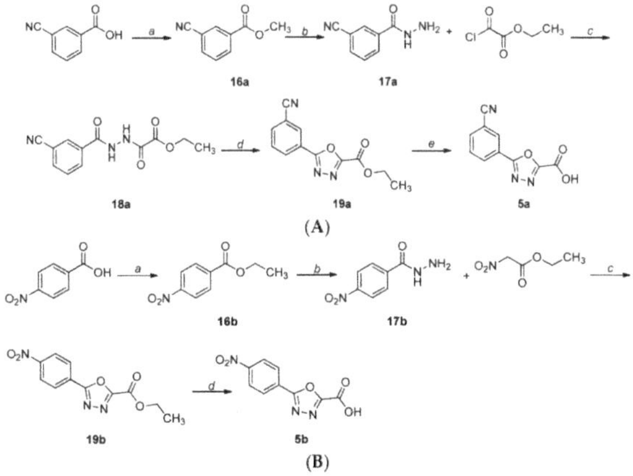

Scheme 5. (**A**) Synthetic procedure for the preparation of **5a**. Reagents and conditions: (**a**) dry MeOH, conc. H_2SO_4, reflux, 3 h, N_2 atm; (**b**) $NH_2NH_2 \cdot H_2O$, MeOH, r.t., overnight; (**c**) TEA, DCM, r.t., 2 h; (**d**) TEA, DCM, TsCl, r.t., 2 h; (**e**) *1.* NaOH, THF-H_2O 1:1, r.t., 1 h; *2.* Amberlite IR120, 0 °C. (**B**) Synthetic procedure for the preparation of **5b**. Reagents and conditions: (**a**) EtOH, conc. H_2SO_4, reflux, overnight; (**b**) $NH_2NH_2 \cdot H_2O$, EtOH, reflux, overnight; (**c**) 86% PPA, 120-130 °C, 1.5 h; (**d**) *1.* LiOH, THF-H_2O 1:1, r.t., 1 h; *2.* Amberlite IR120, 0 °C.

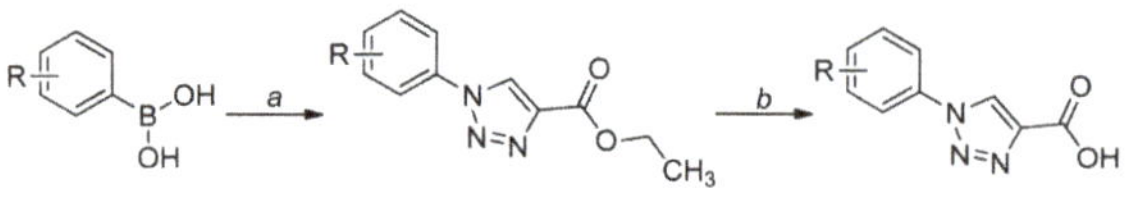

20a,b

6a R = *m*-CN
6b R = *p*-NO_2

Scheme 6. Synthetic procedure for the preparation of **6a,b**. Reagents and conditions: (**a**) *1.* NaN_3, $Cu(OAc)_2$, MeOH, 55 °C, 1.5-4 h, N_2 atm; *2.* ethyl propiolate, (+)-sodium L-ascorbate, r.t., overnight-24 h; (**b**) *1.* LiOH, THF-H_2O 2:1, r.t., 1 h for **6a**; NaOH, THF-H_2O 1:1, reflux, 5 h for **6b**; *2.* 3 M HCl, 0 °C for **6a**; 1 M HCl, 0 °C for **6b**.

The key intermediate (**7**) for the synthesis of **1a,b** was obtained through a Fischer–Speier esterification of the commercially available 5-bromo-2-thiophenecarboxylic acid. Then, a palladium-catalyzed Suzuki-Miyaura coupling led to **8a,b**, which were hydrolyzed to the corresponding acids (**1a,b**) under basic conditions (Scheme 1).

The synthesis of the thiazole-based derivatives (**2a,b**) started from the bromination of the appropriate acetophenone, leading to **9a,b**. The hydrochloride salts of the corresponding amines (**10a,b**), obtained through the Delépine reaction, were *N*-acylated with ethyl chlorooxoacetate to afford the corresponding amides (**11a,b**). The formation of the thiazole ring was performed using the Lawesson's reagent, leading to the esters **12a,b**, which were hydrolyzed under basic conditions and isolated as sodium salts (Scheme 2) [28].

Oxazole-based derivatives were obtained through an iodine-promoted formal [3+2] cycloaddition of methyl ketones to α-methylenyl isocyanides: in particular, ethyl isocyanoacetate was reacted with the suitable acetophenone to afford 2,5-disubstituted oxazole esters (**13a,b**). The intermediates were then hydrolyzed under basic conditions and isolated as sodium salts (Scheme 3) [29].

For the synthesis of **4a,b**, the suitably substituted geminal diols **14a,b** were obtained from the commercially available acetophenone in the presence of selenium dioxide. Then, these intermediates were reacted with ethyl 2-oxoacetate and ammonium acetate, affording the imidazole esters **15a,b**, which were finally hydrolyzed to the corresponding acids (**4a,b**) under basic conditions (Scheme 4) [30].

For the synthesis of the *m*-CN-substituted 1,3,4-oxadiazole derivative (**5a**), the commercially available 3-cyanobenzoic acid was converted to the corresponding methyl ester (**16a**) and reacted with hydrazine hydrate to afford the hydrazide **17a**. This intermediate was acylated with ethyl-chlorooxoacetate to **18a**, which was cyclized to **19a** using *p*-toluensulfonyl chloride in the presence of triethylamine. The oxadiazole ester was then hydrolyzed under basic conditions to give **5a** (Scheme 5A) [31].

Concerning the *p*-NO_2-substituted 1,3,4-oxadiazole analogue (**5b**), the hydrazide **17b**, obtained as described above, was reacted with ethyl 2-nitroacetate in polyphosphoric acid to afford the ester **19b**, which was hydrolyzed to the corresponding acid (**5b**) under basic conditions (Scheme 5B) [32].

The 1,4-substituted triazole esters **20a,b** were obtained through a one-pot Huisgen cycloaddition, starting from the appropriate phenylboronic acid: the synthesis of the azide was followed by the addition of ethyl propiolate, leading to the desired intermediates. The final acids (**6a,b**) were obtained through a base-catalyzed hydrolysis of the ester function (Scheme 6).

The new furan derivatives **IV–IX** were synthesized according to previously published procedures [13]. **V** was obtained by the same approach adopted for **1a,b**, employing (3,5-bis(trifluoromethyl)phenyl)boronic acid and methyl 5-bromofuran-2-carboxylate in a traditional Suzuki–Miyaura reaction, followed by a hydrolysis of the ester function [13]. **IV** was obtained by reacting 3-bromo-5-(trifluoromethyl)benzonitrile with (5-(methoxycarbonyl)furan-2-yl)boronic acid in a microwave-assisted Suzuki-Miyaura coupling, followed by a base-catalyzed hydrolysis of the ester function [13]. Similarly, 3-bromo-5-fluorobenzonitrile, 3-bromo-5-methoxybenzonitrile, 3-bromo-5-methylbenzonitrile, and

3-bromo-5-hydroxybenzonitrile were used as starting compounds for **VI**, **VII**, **VIII**, and **IX**, respectively.

2.2. Biological Studies

To pursue our aim of investigating the role of the heterocyclic core in MbtI inhibition, we decided to compare two sets of data, derived from our previous leads: **I**, characterized by the presence of the *m*-CN group (series A), and **II**, bearing the less druggable *p*-NO_2 group, but capable of potently inhibiting MbtI (series B) [12]. Therefore, keeping the cyano and the nitro group in their original positions, we explored the effects of the variation of the five-membered ring on the activity against the enzyme. The results of the in vitro assays on compounds **1a,b–6a,b**, calculated as previously reported [12], are listed in Table 1.

As for the thiophene analogues, the biological tests showed that **1a** and **1b** are approximately equipotent, with 23% residual enzymatic activity at 100 μM (%RA). The corresponding thiazole derivatives **2a** and **2b** are weaker inhibitors, especially in the presence of the *p*-NO_2 substitution. In some cases, thiazoles have been identified as pan-assay interference compounds (PAINS) [33]. To exclude this possibility, we tested **2a** and **2b**, along with all the other compounds published herein, against the PAINS filters of four online-based services, namely FAF-Drugs4 [34], SmartsFilter (https://chiltepin.health.unm.edu/tomcat/biocomp/smartsfilter accessed on 12 January 2021), SwissADME [35], and Zinc Patterns (http://zinc15.docking.org/patterns/home/ accessed on 12 January 2021). Notably, none of the molecules were identified as potential PAINS. As for the oxazole derivatives, **3a** showed a negligible activity, while **3b** evidenced a modest activity. Although we had envisioned that the structural features of the imidazole ring could be beneficial to form interactions within the MbtI active site, derivatives **4a** and **4b** displayed only a weak activity. Finally, the replacement of the furan with the oxadiazole and triazole cores in **5a–b** and **6a–b**, respectively, afforded compounds devoid of any significant effect against MbtI.

The minimal inhibitory concentration (MIC^{99}) of the derivatives exhibiting an IC_{50} lower than 30 μM was determined against the nonpathogenic *M. bovis* BCG, in iron-limiting conditions (chelated Sauton's medium), using the resazurin reduction assay method (REMA). All of them displayed MIC^{99} values greater than 250 μM, which did not represent a significant improvement with respect to the previous leads.

In light of these findings, and considering previous SAR data, we designed the new derivatives **IV–IX**. The furan ring was chosen as the central core of these compounds, because it proved to be the best option and an important portion of our pharmacophore model. This additional investigation was undertaken to explore a new position of the phenyl ring, with the final goal of identifying the structural requirements needed to improve the antitubercular potency of these compounds.

The disubstituted derivatives **IV–IX** were tested for their effects against the recombinant MbtI, prepared, and assayed as previously reported [12]; their in vitro activities are shown in Table 2.

Table 2. In vitro activity of compounds **IV–IX**.

Entry	% RA *	IC_{50} ** (μM)
IV	0.7 ± 2.7	15.5 ± 3.1
V	8.9 ± 1.6	18.8 ± 6.8
VI	8.7 ± 1.4	17.3 ± 3.1
VII	10.5 ± 3.9	14.5 ± 2.1
VIII	16.7 ± 3.8	29.1 ± 1.8
IX	15.3 ± 2.6	33.5 ± 3.4

* % residual enzymatic activity at 100 μM; ** only for compounds with %RA $\leq$ 25%.

Compound **IV** was designed with the cyano group in the *meta* position because our past works had shown that this group was superior to other options in terms of enzymatic activity; moreover, it features a trifluoromethyl moiety in 5, which allowed us to explore a hitherto unconsidered substitution pattern on the phenyl ring. **IV** was found to be a potent MbtI inhibitor (≈ 1% RA at 100 μM), with an IC_{50} of ≈ 15 μM. Compound **V**, carrying two trifluoromethyl moieties, was also effective against MbtI, though displaying a slightly higher IC_{50} with respect to **IV** (≈ 19 μM vs. 15 μM). Then, we tested compounds **VI–IX**, maintaining the original cyano group in 3 and featuring different moieties in 5. Firstly, we assayed the fluorine-substituted compound **VI**, which showed an IC_{50} value similar to that of **V** (IC_{50} ≈17 μM vs. 15 μM). Compound **VII**, bearing a methoxy moiety in 5, displayed a comparable IC_{50} with respect to **IV**. Derivatives **VIII** and **IX**, carrying the CH_3 and the OH groups respectively, showed weaker inhibitory properties compared to compound **IV** (IC_{50} ≈ 29 μM and 33 μM) (see Table 2).

The four candidates exhibiting promising inhibitory properties (IC_{50} < 30 μM) were tested for their antimycobacterial activities against the nonpathogenic *M. bovis* BCG, in iron-limiting conditions (chelated Sauton's medium), using the REMA method. In this assay, compound **IV** showed the best antimycobacterial activity, with a MIC^{99} value of 125 μM.

Due to its better bactericidal activity, **IV** emerged as the best inhibitor out of this furan series: its halved MIC^{99} compared to **I** (125 μM vs. 250 μM) highlighted the better drugability of this compound with respect to our previous candidates.

Therefore, we submitted compound **IV** to a kinetic analysis, which demonstrated the competitive nature of its inhibition against MbtI, with a K_i value of 9.2 ± 0.7 μM (Figure 5).

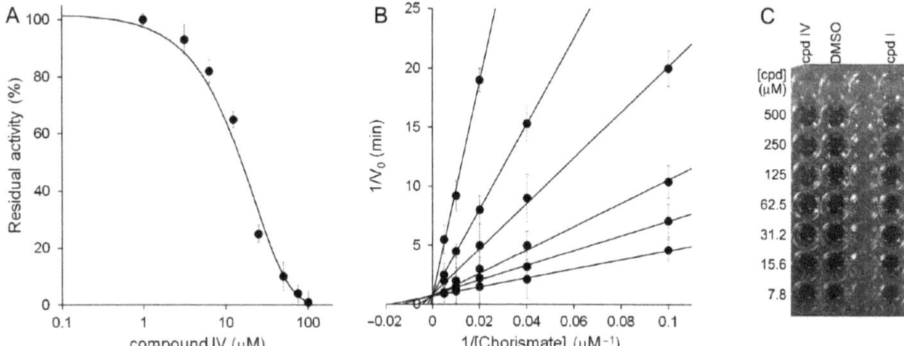

Figure 5. Biological characterization of **IV**. (**A**) IC_{50} determination of **IV** against MbtI activity. (**B**) Global reciprocal plot of data from MbtI steady-state kinetics analysis towards chorismic acid, in the presence of different concentrations of **IV** (50, 20, 10, 5, 1, and 0 μM). (**C**) MIC^{99} determination of **IV** against *M. bovis* BCG growth.

3. Discussion

In this work, we first applied a bioisosteric replacement strategy introducing in our leads **I** and **II** structural modifications in the five-membered core to alter the compound's electronic distribution and lipophilicity, with the aim of improving the target engagement and the antimycobacterial activity.

Contrary to traditional bioisosteric principles, the biological profiles of the thiophene analogues were not comparable to those of the furan derivatives. Conversely, their inhibitory effect followed the general trend exhibited by the thiazole, oxazole, and imidazole derivatives, suggesting that other factors prevail over the bioisosterism of the two nuclei. Interestingly, a significant decline in the activity was observed in the oxadiazole- and triazole-based compounds. Regarding the substitution of the phenyl ring, the presence of the *m*-cyano or *p*-nitro groups did not seem to impact significantly on the variations of the activity.

In previous work, we reported the cocrystal structure of MbtI in complex with **I** and described the key interactions of the compound within the active site of the enzyme. Briefly, **I** forms H-bonds through its carboxylic group with Tyr385, Arg405, and an ordered water molecule; the oxygen of the furan interacts with Arg405, while the phenyl ring forms a cation-π interaction with Lys438 and a Van der Waals contact with Thr361. Finally, the cyano group interacts with Lys205, a key amino acid involved in the first step of the catalytic reaction. While the diminished activity of the triazole ring may be justified by the absence of a heteroatom capable of accepting a H bond from Arg405, the formulation of a hypothesis to explain the superiority of the furan with respect to the remaining cores is more arduous. A computational analysis of the binding modes of the tested compounds did not reveal significant disparities with respect to that of the lead molecule (unpublished data), suggesting that other influencing elements must be involved. Similarly, an in-silico comparison of the physicochemical characteristics of the compounds did not lead to meaningful results. Despite the different heterocyclic nuclei impart modifications to the overall properties of the molecules, a correspondence between the alteration of a parameter and the biological activity could not be unequivocally established. Therefore, it is reasonable to assume that the superior activity of the furan derivatives cannot be merely linked to the variation of one single parameter, but rather it is the result of a much more complex intertwinement of unrelated minimal modifications. The inherent multifactorial nature of these processes makes it hard, and potentially misleading, to seek a simplistic, univocal interpretation of these results. Hence, it is our opinion that the biological activity, empirically detected with our assays, is the only meaningful parameter that should be considered while determining the best heterocyclic core for this class of compounds.

In addition, when working with mycobacterial enzymes, in vitro activity may not necessarily correlate with the efficacy against the mycobacteria; therefore, compounds showing a weaker inhibitory effect against the purified target could display a better activity against bacterial growth, for instance due to an improved membrane permeability. On these bases, the MIC of the most potent molecules (IC_{50} < 30 μM) was determined against *M. bovis* BCG. Although, in some literature cases, the furan core was successfully replaced by other heterocycles to improve the cellular activity [36], this was not our case. None of the compounds belonging to series A and B exhibited improved antitubercular action, all of them having MIC^{99} values greater than 250 μM.

Overall, these biological results confirmed the furan as the best heterocyclic moiety among the options explored in this study and prompted us to reconsider this ring as the best scaffold to gain MbtI inhibition and antimycobacterial activity. In this regard, new modifications to the phenyl ring were attempted to improve the biological profile of our compounds. In our previous work, we discovered that the *m*-CN substitution offered the possibility to achieve better results in terms of enzyme inhibition compared to the other groups [12]. Moreover, in the context of a previously published disubstituted series, we took into account 5-(2-cyano-4-(trifluoromethyl)phenyl)furan-2-carboxylic acid, which proved to possess an interesting inhibitory effect (IC_{50} of about 18 μM). Therefore, we decided to explore the activity of its isomer 5-(3-cyano-5-(trifluoromethyl)phenyl)furan-2-carboxylic acid (**IV**, Figure 4), bearing the cyano group in the preferred position 3. Additionally, the relocation of the trifluoromethyl moiety to position 5 to avoid the steric interactions between adjacent groups allowed us to examine a hitherto unexplored substitution site on the phenyl ring. The new isomer revealed an interesting IC_{50} value of about 15 μM. Following the same strategy, we synthesized compound **V**, an isomer of 5-(2,4-bis(trifluoromethyl)phenyl)furan-2-carboxylic acid (IC_{50} of about 13 μM) [13]. The new analogue exhibited slightly lower inhibitory properties ($IC_{50} \approx 19$ μM) with respect to the parent compound, while maintaining a promising activity. Subsequently, to further explore the SAR of the 5-substituent, we synthesized and tested compounds **VI–IX**, bearing the CN group in 3.

The presence of different substituents at position 5 of compounds **IV–IX** did not seem to significantly affect their inhibitory activity against MbtI, with IC_{50} ranging from 15 μM to 33 μM. By contrast, the 5-CF_3 group of **IV** was able to significantly ameliorate its antimycobacterial properties with respect to the lead **I**. Indeed, **IV** displayed a MIC^{99} value of 125 μM, far better than that of **I** (250 μM).

Further biological studies demonstrated that **IV** was a competitive inhibitor of MbtI, exhibiting a K*i* of about 9 μM, roughly comparable to that of **I**.

Overall, these SAR observations demonstrated the essentiality of the furan core and the advantages of a 3,5-disubstitution of the phenyl ring to achieve a potent in vitro activity against MbtI and a significant antimycobacterial effect.

The improvement of the MIC of new compounds is a common goal in antitubercular drug discovery, as reported in a very recent review, which also supported the importance of the mycobactin biosynthetic pathway for the development of anti-TB agents [37]. The increased antitubercular activity of **IV** with respect to our previous leads opens new avenues for structural modifications towards improved candidates.

4. Material and Methods

4.1. Chemistry

All starting materials, chemicals, and solvents were purchased from commercial suppliers (Sigma-Aldrich, St. Louis, MI, USA; FluoroChem, Hadfield, UK; Carlo Erba, Cornaredo, Italy) and used as received. Anhydrous solvents were utilized without further drying. Aluminum-backed Silica Gel 60 plates (0.2 mm; Merck, Darmstadt, Germany) were used for analytical thin-layer chromatography (TLC), to follow the course of the reactions. Microwave-assisted reactions were carried out with a Biotage® Initiator Classic (Biotage, Uppsala, Sweden). Silica gel 60 (40–63 μM; Merck) was used for the purification of intermediates and final compounds, through flash column chromatography. Melting points were determined in open capillary tubes with a Stuart SMP30 Melting Point Apparatus (Cole-Parmer Stuart, Stone, UK). All tested compounds were characterized by means of mono- and bi-dimensional NMR techniques, FT-IR, and ESI-MS. ^1H and ^{13}C NMR spectra were acquired at ambient temperature with a Varian Oxford 300 MHz instrument (Varian, Palo Alto, CA, USA) or a Bruker Avance 300 MHz instrument (Bruker, Billerica, MA, USA), operating at 300 MHz for ^1H and 75 MHz for ^{13}C. Chemical shifts are expressed in ppm (δ), while *J*-couplings are given in Hertz. The full decoupling mode was employed for ^{13}C spectra when the relaxation times of the carbons did not allow for a sufficient resolution using the APT sequence. The 2D-NOESY sequence was employed to unambiguously assign the hydrogen signals, when appropriate. HMBC and HSQC analyses were performed to aid the assignment of ^{13}C NMR signals, when necessary. ATR-FT-IR spectra were acquired with a Perkin Elmer Spectrum One FT-IR (Perkin Elmer, Waltham, MA, USA), equipped with a Perkin Elmer Universal ATR sampling accessory consisting of a diamond crystal. Analyses were performed in a spectral region between 4000 and 650 cm^{-1} and analyzed by transmittance technique with 28 scansions and 4 cm^{-1} resolution. MS analyses were carried out with a Thermo Fisher (Waltham, MA, USA) LCQ Fleet system, equipped with an ESI electrospray ionization source and an Ion Trap mass analyzer; ionization: ESI positive or ESI negative; capillary temperature: 250 °C; source voltage: 5.50 kV; source current: 4.00 μA; multipole 1 and 2 offset: −5.50 V and −7.50 V, respectively; intermultipole lens voltage: −16.00 V; trap DC offset voltage: −10.00 V. The purity of the tested compounds was assessed by means of elemental analysis using a EuroVector EA 3000 CHNS-O analyzer (EuroVector, Pavia, Italy). All experimental values are within ± 0.40% of the theoretical predictions, indicating a ≥ 95% purity.

All synthetic procedures are reported in the Supplementary Materials (SM).

5-(3-Cyanophenyl)thiophene-2-carboxylic acid (1a). The compound was synthesized by a specific procedure, reported in SM. Aspect: white solid. Mp: 207 °C. TLC (DCM-MeOH 9:1): R_f = 0.20. The following analytical data are referred to the sodium salt. ^1H NMR (300 MHz, DMSO-d_6) δ 8.09 (t, *J* = 1.7 Hz, 1H, H_6), 7.91 (ddd, *J* = 7.9, 2.0, 1.2 Hz, 1H,

H_8), 7.70 (dt, J = 7.7, 1.4 Hz, 1H, H_{10}), 7.57 (t, J = 7.8 Hz, 1H, H_9), 7.48 (d, J = 3.7 Hz, 1H, H_3), 7.21 (d, J = 3.7 Hz, 1H, H_4) ppm. ^{13}C NMR (75 MHz, DMSO-d_6) δ 164.80 (COO^-), 149.77 (C_2), 141.67 (C_5), 136.20 ($C_{5'}$), 131.04 (C_8), 130.74 (C_9), 130.17 (C_{10}), 128.71 (C_6, C_3), 125.45 (C_4), 119.01 (CN), 112.70 (C_7) ppm. FT-IR (ATR) ν = 2235, 1578, 1537, 1450, 1397, 1335, 811, 789, 767, 682, 675 cm^{-1}. Anal. calcd. for $C_{12}H_6NNaO_2S$: C, 57.37; H, 2.41; N, 5.58; S, 12.76. Found: C, 57.48; H, 2.45; N, 5.61; S, 12.83.

5-(4-Nitrophenyl)thiophene-2-carboxylic acid (1b). The compound was obtained according to Procedure A (SM). Starting compound: methyl 5-(4-nitrophenyl)thiophene-2-carboxylate. Yield: 98%. Aspect: yellow solid. Mp: 189 °C. TLC (DCM-MeOH 9:1): R_f = 0.20. The following analytical data are referred to the sodium salt. ^{1}H NMR (300 MHz, DMSO-d_6) δ 8.21 (d, J = 9.0 Hz, 2H, $H_{7,7'}$), 7.97 (d, J = 9.0 Hz, 2H, $H_{6,6'}$), 7.57 (d, J = 3.7 Hz, 1H, H_3), 7.22 (d, J = 3.7 Hz, 1H, H_4) ppm. ^{13}C NMR (75 MHz, DMSO-d_6) δ 164.05 (COO^-), 152.14 (C_2), 146.32 (C_8), 141.46 (C_5), 141.24 ($C_{5'}$), 128.78 (C_3), 126.96 (C_4), 126.15 ($C_{7,7'}$), 124.86 ($C_{6,6'}$) ppm. FT-IR (KBr) ν = 3435, 2920, 2550, 1927, 1664, 1622, 1514, 1450, 1365, 995, 833, 704 cm^{-1}. ESI-MS (m/z) calcd for $C_{11}H_6NNaO_4S$ 270.99, found 204.71 $[M\text{-}CO_2Na]^-$. Anal. calcd. for $C_{11}H_6NNaO_4S$: C, 48.71; H, 2.23; N, 5.16; S, 11.82. Found: C, 48.75; H, 2.27; N, 5.14; S, 11.71.

Sodium 5-(3-cyanophenyl)thiazole-2-carboxylate (2a). The compound was obtained according to Procedure B (SM). Starting compound: ethyl 5-(3-cyanophenyl)thiazole-2-carboxylate. Yield: 86%. Aspect: white solid. Mp: >300 °C (dec.). ^{1}H NMR (300 MHz, DMSO-d_6) δ 8.20 (s, 1H, H_4), 8.19 (t, J = 1.4 Hz, 1H, H_6), 7.95 (ddd, J = 7.9, 1.9, 1.2, Hz, 1H, H_{10}), 7.77 (dt, J = 7.7, 1.2, H_8), 7.60 (dt, J = 7.9, 0.5 Hz, 1H, H_9) ppm. ^{13}C NMR (75 MHz, DMSO-d_6) δ 173.60 (C_2), 161.61 (COO^-), 140.73 (C_4), 138.39 (C_5), 133.51 ($C_{5'}$), 131.81 (C_8), 131.51 (C_{10}), 130.85 (C_9), 129.86 (C_6), 118.85 (CN), 112.81 (C_7) ppm. FT-IR (ATR) ν = 3354, 2235, 1663, 1641, 1578, 1440, 1407, 1366, 1110, 866, 806, 796 cm^{-1}. Anal. calcd. for $C_{11}H_5N_2NaO_2S$: C, 52.38; H, 2.00; N, 11.11; S, 12.71. Found: C, 52.51; H, 2.02; N, 11.09; S, 12.75.

Sodium 5-(4-nitrophenyl)thiazole-2-carboxylate (2b). The compound was obtained according to Procedure A (SM). Starting compound: ethyl 5-(4-nitrophenyl)thiazole-2-carboxylate. Yield: 85%. Aspect: dark green solid. Mp: >300 °C (dec.). ^{1}H NMR (300 MHz, DMSO-d_6) δ 8.29 (s, 1H, H_4), 8.23 (d, J = 6.0 Hz, 2H, $H_{7,7'}$), 7.93 (d, J = 6.0 Hz, 2H, $H_{6,6'}$) ppm. ^{13}C NMR (75 MHz, DMSO-d_6) δ 174.68 (C_2), 161.40 (COO^-), 146.99 (C_8), 141.97 (C_4), 138.75 ($C_{5'}$), 138.37 (C_5), 127.62 ($C_{7,7'}$), 124.86 ($C_{6,6'}$) ppm. FT-IR (ATR) ν = 3648, 3297, 3100, 2963, 1675, 1645, 1621, 1595, 1514, 1424, 1408, 1365, 1343, 1108, 847 cm^{-1}. ESI-MS (m/z) calcd for $C_{10}H_5N_2NaO_4S$ 272.21, found 205.78 $[M\text{-}CO_2Na]^-$. Anal. calcd. for $C_{10}H_5N_2NaO_4S$: C, 44.12; H, 1.85; N, 10.29; S, 11.78. Found: C, 44.31; H, 1.87; N, 10.34; S, 11.67.

Sodium 5-(3-cyanophenyl)oxazole-2-carboxylate (3a). The compound was obtained according to Procedure B (SM). Starting compound: ethyl 5-(3-cyanophenyl)oxazole-2-carboxylate. Yield: 89%. Aspect: grey solid. Mp: >300 °C (dec.). ^{1}H NMR (300 MHz, DMSO-d_6) δ 8.17 (t, J = 1.6 Hz, H_6), 7.99 (dt, J = 8.0, 1.6 Hz, 1H, H_{10}), 7.74 (dt, J = 8.0, 1.6 Hz, 1H, H_8), 7.66 (t, J = 8.0 Hz, H_9) ppm. ^{13}C NMR (75 MHz, DMSO-d_6) δ 161.87 (COO^-), 157.88 (C_2), 148.06 (C_5), 131.96 (C_9), 130.85 (C_8), 129.69 ($C_{5'}$), 128.69 (C_{10}), 127.90 (C_6), 125.07 (C_4), 118.78 (CN), 112.74 (C_7) ppm. FT-IR (ATR) ν = 3522, 3383, 2234, 1650, 1616, 1520, 1422, 1389, 1318, 1273, 1216, 965, 825, 817, 795 cm^{-1}. ESI-MS (m/z) calcd. for $C_{11}H_5N_2NaO_3$ 236.16, found 169.67 $[M\text{-}CO_2Na]^-$. Anal. calcd. for $C_{11}H_5N_2NaO_3$: C, 55.94; H, 2.13; N, 11.86. Found: C, 56.03; H, 2.15; N, 11.93.

Sodium 5-(4-nitrophenyl)oxazole-2-carboxylate (3b). The compound was obtained according to Procedure B (SM). Starting compound: ethyl 5-(4-nitrophenyl)oxazole-2-carboxylate. Yield: 80%. Aspect: pale yellow solid. Mp: >300 °C (dec.). ^{1}H NMR (300 MHz, DMSO-d_6) δ 8.30 (d, J = 8.9 Hz, 2H, $H_{7,7'}$), 7.96 (d, J = 6.0 Hz, 2H, $H_{6,6'}$), 7.87 (s, 1H, H_4) ppm. ^{13}C APT NMR (75 MHz, DMSO-d_6) δ 162.28 (COO^-), 157.95 (C_2), 148.37 (C_5), 147.06 (C_8), 134.32 ($C_{5'}$), 126.99 (C_4), 125.31 ($C_{7,7'}$), 124.92 ($C_{6,6'}$) ppm. FT-IR (ATR) ν = 3436, 2964, 1645, 1607, 1512, 1388, 1346, 1261, 1108, 854, 818 cm^{-1}. ESI-MS (m/z) calcd. for $C_{10}H_5N_2NaO_5$

256.15, found 189.94 [M-CO_2Na]$^-$. Anal. calcd. for $C_{10}H_5N_2NaO_5$: C, 46.89; H, 1.97; N, 10.94. Found: C, 46.59; H, 1.98; N, 10.92.

5-(3-Cyanophenyl)-1H-imidazole-2-carboxylic acid (4a). The compound was synthesized through a specific procedure, reported in SM. Aspect: white solid. Mp: 165 °C. TLC (DCM-MeOH 7:3): R_f = 0.42. ^{1}H NMR (300 MHz, DMSO-d_6) δ 12.0-9.0 (broad s exch. D_2O, 2H, NH_2^+), 8.25 (s, 1H, H_6), 8.16 (d, J = 7.8 Hz, 1H, H_8), 7.99 (s, 1H, H_4), 7.68 (d, J = 7.8 Hz, 1H, H_{10}), 7.58 (t, J = 7.8 Hz, 1H, H_9) ppm. The compound degrades in solution at room temperature, during the acquisition of the ^{13}C NMR spectrum. FT-IR (ATR) ν = 3205, 2228, 1666, 1601, 1514, 1473, 1426, 1403, 1334, 1130, 1089, 811, 780, 680 cm^{-1}. ESI-MS (m/z) calcd. for $C_{11}H_7N_3O_2$ 213.19, found 212.42 [M-H]$^-$. Anal. calcd. for $C_{11}H_7N_3O_2$: C, 61.97; H, 3.31; N, 19.71. Found: C, 62.03; H, 3.35; N, 19.82.

5-(4-Nitrophenyl)-1H-imidazole-2-carboxylic acid (4b). The compound was obtained according to Procedure C (SM). Starting compound: ethyl 5-(4-nitrophenyl)-1*H*-imidazole-2-carboxylate. Yield: 66%. Aspect: red solid. Mp: 137 °C. TLC (DCM-MeOH 7:3): R_f = 0.40. The following analytical data are referred to the sodium salt. ^{1}H NMR (300 MHz, DMSO-d_6) δ 8.15-8.09 (m, 4H, $H_{6,6'}$, $H_{7,7'}$), 7.76 (s, 1H, H_4) ppm. ^{13}C NMR (75 MHz, DMSO-d_6) δ 162.09 (COO$^-$), 149.48 (C_2), 145.46 (C_8), 142.17 (C_5), 137.58 ($C_{5'}$), 125.45 ($C_{7,7'}$), 124.19 ($C_{6,6'}$), 117.67 (C_4) ppm. FT-IR (ATR) ν = 3607, 3156, 1652, 1600, 1494, 1472, 1415, 1342, 1135, 1112, 992, 849, 751 cm^{-1}. ESI-MS (m/z) calcd for $C_{10}H_7N_3O_4$ 233.18, found 232.32 [M-H]$^-$. Anal. calcd. for $C_{10}H_6N_3NaO_4$: C, 47.07; H, 2.37; N, 16.47. Found: C, 47.31; H, 2.39; N, 16.36.

5-(3-Cyanophenyl)-1,3,4-oxadiazole-2-carboxylic acid (5a). The compound was obtained according to Procedure D (SM). Starting compound: ethyl 5-(3-cyanophenyl)-1,3,4-oxadiazole-2-carboxylate. Yield: quantitative. Aspect: white solid. Mp: 222 °C (dec.). TLC (DCM-MeOH 7:3): R_f = 0.42. The following analytical data are referred to the sodium salt. ^{1}H NMR (300 MHz, DMSO-d_6) δ 8.35 (t, J = 1.8 Hz, 1H, H_6), 8.28 (dt, J = 8.0, 1.4 Hz, 1H, H_8), 8.06 (dt, J = 7.8, 1.4 Hz, 1H, H_{10}), 7.79 (dt, J = 7.9, 0.7 Hz, 1H, H_9) ppm. The compound degrades in solution at room temperature, during the acquisition of the ^{13}C NMR spectrum. FT-IR (ATR) ν = 3543, 3384, 2232, 1651, 1614, 1549, 1400, 1343, 1228, 1183, 1086, 812, 807, 679 cm^{-1}. Anal. calcd. for $C_{10}H_4N_3NaO_3$: C, 50.65; H, 1.70; N, 17.72. Found: C, 50.71; H, 1.81; N, 17.87.

5-(4-Nitrophenyl)-1,3,4-oxadiazole-2-carboxylic acid (5b). The compound was obtained according to Procedure D (SM). Starting compound: ethyl 5-(4-nitrophenyl)-1,3,4-oxadiazole-2-carboxylate. Yield: quantitative. Aspect: pale yellow solid. Mp: 220 °C (dec.). TLC (DCM-MeOH 7:3): R_f = 0.44. ^{1}H NMR (300 MHz, DMSO-d_6) δ 13.7 (broad s exch. D_2O, 1H, COOH), 8.30 (d, J = 8.9 Hz, 2H, $H_{7,7'}$), 8.15 (d, J = 8.9 Hz, 2H, $H_{6,6'}$) ppm. ^{13}C NMR (75 MHz, DMSO-d_6) δ 166.24 (COOH), 150.51 (C_5), 136.98 ($C_{5'}$), 131.14 ($C_{7,7'}$), 124.15 ($C_{6,6'}$) ppm. FT-IR (ATR) ν = 2962, 2924, 2853, 1691, 1603, 1520, 1258, 1080, 1013, 789 cm^{-1}. Anal. calcd. for $C_9H_5N_3O_5$: C, 45.97; H, 2.14; N, 17.87. Found: C, 46.02; H, 2.17; N, 17.96.

1-(3-Cyanophenyl)-1H-1,2,3-triazole-4-carboxylic acid (6a). The compound was synthesized through a specific procedure, reported in SM. Aspect: white solid. TLC (DCM-MeOH 9:1): R_f = 0.14. ^{1}H NMR (300 MHz, DMSO-d_6) δ 8.90 (s, 1H, H_1), 8.44 (t, J = 2.0 Hz, 1H, H_6), 8.30 (ddd, J = 1.2, 2.0, 8.1 Hz, 1H, H_8), 7.91 (dt, J = 1.2, 8.1 Hz, 1H, H_{10}), 7.77 (t, J = 8.1 Hz, 1H, H_9) ppm. ^{13}C NMR (75 MHz, DMSO-d_6) δ 177.11 (C_2), 163.73 (COOH), 137.81 ($C_{5'}$), 132.35 (C_8), 131.66 (C_9), 124.99 (C_{10}), 124.28 (C_6), 123.68 (C_1), 118.30 (CN), 113.25 (C_7) ppm. FT-IR (ATR) ν = 3389, 3091, 2235, 1589, 1557, 1536, 1403, 1343, 1312, 1021, 794 cm^{-1}. Anal. calcd. for $C_{10}H_6N_4O_2$: C, 56.08; H, 2.82; N, 26.16. Found: C, 56.27; H, 2.91; N, 26.35.

1-(4-Nitrophenyl)-1H-1,2,3-triazole-4-carboxylic acid (6b). The compound was obtained according to Procedure A (SM). Starting compound: ethyl 1-(4-nitrophenyl)-1*H*-1,2,3-triazole-4-carboxylate. Yield: 91%. Aspect: white solid. Mp: 175 °C. TLC (DCM-MeOH 8:2): R_f = 0.12. ^{1}H NMR (300 MHz, DMSO-d_6) δ 13.4 (broad s exch. D_2O, 1 H, COOH), 9.59 (s, 1H, H_1), 8.46 (d, J = 7.0 Hz, 2H, $H_{7,7'}$), 8.30 (d, J = 7.0 Hz, 2H, $H_{6,6'}$) ppm. ^{13}C NMR (75 MHz, DMSO-d_6) δ 161.73 (COOH), 147.60 (C_8), 141.69 (C_2), 140.96 ($C_{5'}$), 128.02 (C_1), 125.90 ($C_{7,7'}$), 121.72 ($C_{6,6'}$) ppm. FT-IR (ATR) ν = 3249, 3142, 3102, 3061, 2962, 2913, 2866,

1729, 1704, 1596, 1516, 1342, 1268, 1219, 1151, 1032, 982, 869, 854 cm^{-1}. ESI-MS (m/z) calcd for $C_9H_6N_4O_4$ 234.17, found 161.30 $[M-CO_2-N_2]^-$. Anal. calcd. for $C_9H_6N_4O_4$: C, 46.16; H, 2.58; N, 23.93. Found: C, 46.25; H, 2.60; N, 23.97.

5-(3-Cyano-5-(trifluoromethyl)phenyl)furan-2-carboxylic acid (IV). The compound was synthesized according to a previously published procedure [13]. Yield: 70%. Aspect: white solid. Mp: 210 °C. TLC (DCM-MeOH 7:3): R_f = 0.33. ^{1}H NMR (300 MHz, DMSO-d_6) δ 13.59-13.10 (broad s. exch. D_2O, 1H, COOH), 8.62-8.57 (m, 1H, H_7), 8.41-8.33 (m, 2H, H_{11}, H_9), 7.52 (d, J = 3.7 Hz, 1H, H_4), 7.38 (d, J = 3.7 Hz, 1H, H_3) ppm. ^{13}C APT NMR (75 MHz, DMSO-d_6) δ 159.45 (COOH), 152.84 (C_5), 146.18 (C_2), 132.17 (C_7), 132.17-131.78-131.34-130.90 (q, C_{10}), 131.99 (C_6), 129.14 (C_9), 128.80-125.19-121.56-117.94 (q, CF_3), 125.09 (C_{11}), 120.09 (C_3), 117.53 (CN), 114.28 (C_8), 111.74 (C_4) ppm. FT-IR (ATR): ν = 3130, 3082, 2960, 2917, 2849, 2237, 1688, 1578, 1519, 1455, 1438, 1410, 1344, 1274, 1256, 1163, 1133, 1114, 1085, 1032, 900 cm^{-1}. ESI-MS (m/z) calcd for $C_{13}H_6F_3NO_3$ 281.03, found 280.14 $[M-H]^-$. Anal. calcd. for $C_{13}H_6F_3NO_3$: C, 55.53; H, 2.15; N, 4.98. Found: C, 55.47; H, 2.19; N, 5.01.

5-(3,5-Bis(trifluoromethyl)phenyl)furan-2-carboxylic acid (V). The compound was synthesized according to a previously published procedure [13]. Yield: 91%. Aspect: white solid. Mp: 168 °C. TLC (DCM-MeOH 7:3): R_f = 0.39. ^{1}H NMR (300 MHz, DMSO-d_6) δ 13.65-13.20 (broad s. exch. D_2O, 1H, COOH), 8.41-8.36 (m, 2H, H_7, H_{11}), 8.12-8.07 (m, 1H, H_9), 7.59 (d, J = 3.7 Hz, 1H, H_3), 7.38 (d, J = 3.7 Hz, 1H, H_4) ppm. ^{13}C APT NMR (75 MHz, DMSO-d_6) δ 159.48 (COOH), 153.19 (C_5), 146.03 (C_2), 132.34-131.90-131.46-131.03 (q, C_8), 132.00 (C_6), 128.98-125.36-121.74-118.12 (q, CF_3), 125.00 (C_{11}), 122.14 (C_9), 120.14 (C_3), 111.73 (C_4) ppm. FT-IR (ATR): ν = 2960, 2925, 2855, 1689, 1621, 1591, 1526, 1455, 1420, 1366, 1278, 1161, 1124, 1081, 1027, 896 cm^{-1}. ESI-MS (m/z) calcd for $C_{13}H_6F_6O_6$ 324.18, found 323.16 $[M-H]^-$. Anal. calcd. for $C_{13}H_6F_6O_3$: C, 48.17; H, 1.87. Found: C, 48.02; H, 1.91.

5-(3-cyano-5-fluorophenyl)furan-2-carboxylic acid (VI). The compound was synthesized according to a previously published procedure [13]. Yield: quantitative. Aspect: white solid. Mp: 247 °C. TLC (DCM-MeOH 7:3): R_f = 0.26. ^{1}H NMR (300 MHz, DMSO-d_6) δ 13.38 (broad s exch D_2O, 1H, COOH), 8.14 (t, J = 1.6, 1H, H_7), 7.96 (dd, J = 7.8, 1.6 Hz, 1H, H_{11}), 7.88 (dd, J = 6.0, 1.6 Hz, 1H, H_9), 7.43 (d, J = 3.7 Hz, 1H, H_4), 7.37 (d, J = 3.7 Hz, 1H, H_3) ppm. ^{13}C APT NMR (75 MHz, DMSO-d_6) δ 164.18-160.91 (d, CF), 159.53 (COOH), 153.20-153.16 (d, C_5), 145.89 (C_2), 133.05-132.93 (d, C_6) 124.92-124.87 (d, C_7), 120.19 (C_3), 119.75-119.41 (d, C_{11}), 117.72-117.68 (d, CN), 116.57-116.25 (d, C_9), 114.46-114.32 (d, C_8), 111.43 (C_4) ppm. FT-IR (ATR): ν = 3113, 2916, 2849, 2663, 2575, 2231, 1675, 1594, 1519, 1435, 1308, 1214, 1173, 1028, 866, 809, 760 cm^{-1}. ESI-MS (m/z) calcd for $C_{12}H_6FNO_3$ 231.18, found 230.50 $[M-H]^-$. Anal. calcd. for $C_{12}H_6F NO_3$: C, 62.34; H, 2.62. Found: C, 62.53; H, 2.51.

5-(3-Cyano-5-methoxyphenyl)furan-2-carboxylic acid (VII). The compound was synthesized according to a previously published procedure [13]. Yield: 80%. Aspect: white solid. Mp: 226 °C. TLC (DCM-MeOH 7:3): R_f = 0.46. ^{1}H NMR (300 MHz, DMSO-d_6) δ 13.42-13.19 (broad s. exch. D_2O, 1H, COOH), 7.82 (t, J = 1.4 Hz, 1H, H_7), 7.59 (dd, J = 2.5, 1.5 Hz, 1H, H_{11}), 7.45 (dd, J = 2.5, 1.3 Hz, 1H, H_9), 7.35 (d, J = 3.7 Hz, 1H, H_3), 7.33 (d, J = 3.7 Hz, 1H, H_4), 3.87 (s, 3H, CH_3) ppm. ^{13}C NMR (75 MHz, DMSO-d_6) δ 160.43 (C_{10}) 159.61 (COOH), 154.2 (C_5), 145.46 (C_2), 132.08 (C_6), 120.73 (C_3), 120.20 (C_7), 118.66 (CN), 117.65 (C_9), 115.02 (C_{11}), 113.72 (C_8), 110.60 (C_4), 56.49 (CH_3) ppm. FT-IR (ATR): ν = 3116, 3086, 2926, 2574, 2229, 1693, 1608, 1594, 1572, 1515, 1461, 1427, 1305, 1215, 1167, 1033, 960 cm^{-1}. ESI-MS (*m/z*) calcd for $C_{13}H_9O_4$ 243.05, found 242.28 $[M-H]^-$. Anal. calcd. for $C_{13}H_9F_3O_4$: C, 54.56; H, 3.17. Found: C, 54.71; H, 3.23.

5-(3-cyano-5-methylphenyl)furan-2-carboxylic acid (VIII). The compound was synthesized according to a previously published procedure [13]. Yield: 90%. Aspect: white solid. Mp: 238 °C. TLC (DCM-MeOH 7:3): R_f = 0.31 ^{1}H NMR (300 MHz, DMSO-d_6) δ 13.20 (broad s exch D_2O, 1H, COOH), 8.08-8.02 (m, 1H, H_7), 7.93-7.88 (m, 1H, H_{11}), 7.67-7.62 (m, 1H, H_9), 7.33 (d, J = 3.6 Hz, 1H, H_3), 7.28 (d, J = 3.6 Hz, 1H, H_4), 2.39 (s, 3H, CH_3) ppm. ^{13}C NMR (75 MHz, DMSO-d_6) δ 159.56 (COOH), 154.49 (C_2), 145.39 (C_5), 140.84 (C_{10}), 132.82 (C_9), 130.66 (C_6), 129.51 (C_{11}), 125.61 (C_7), 120.13 (C_3), 118.82 (CN), 112.64 (C_8), 110.06 (C_4), 20.89 (CH_3) ppm. FT-IR (ATR): ν = 3119, 2926, 2849, 2692, 2579, 2228, 1726, 1682, 1584, 1520, 1422, 1299,

1169, 1029, 858, 810, 760 cm^{-1}. ESI-MS (m/z) calcd for $C_{13}H_9NO_3$ 227.063, found 226.35 $[M-H]^-$. Anal. calcd. for $C_{13}H_9$ NO_3: C, 68.72; H, 3.99. Found: C, 68.93; H, 4.01.

5-(3-cyano-5-hydroxyphenyl)furan-2-carboxylic acid (IX). The compound was synthesized according to a previously published procedure [13]. Yield: 95%. Aspect: white solid. Mp: 280 °C (dec.). TLC (DCM-MeOH 7:3 and 3 drops of CH_3COOH): R_f = 0.44. ^1H NMR (300 MHz, DMSO-d_6) δ 13.22 (broad s exch. D_2O, 1H, COOH), 10.54 (broad s exch. D_2O, 1H, OH), 7.71 (t, J = 1.5, 1H, H_7), 7.48 (dd, J = 2.4, 1.5 Hz, 1H, H_{11}), 7.32 (d, J = 3.6 Hz, 1H, H_3), 7.26 (d, J = 3.6 Hz, 1H, H_4), 7.14 (dd, J = 2.4, 1.5, 1H, H_9) ppm. ^{13}C APT NMR (75 MHz, DMSO-d_6) δ 159.50 (C_{10}), 158.90 (COOH), 154.42 (C_5), 145.27 (C_2), 132.09 (C_6), 120.10 (C_3), 119.32 (C_7), 118.93 (C_9), 118.74 (CN), 115.92 (C_{11}), 113.50 (C_8), 110.08 (C_4) ppm. FT-IR (ATR): ν = 3400, 3108, 2602, 2228, 1652, 1595, 1516, 1487, 1439, 1310, 1241, 1213, 1174, 1152, 1035, 963, 881, 816, 668 cm^{-1}. ESI-MS (m/z) calcd. for $C_{12}H_7NO_4$ 229.19, found 228.29 $[M-H]^-$. Anal. calcd. for $C_{12}H_7$ NO_4: C, 62.89; H, 3.08. Found: C, 62.53; H, 3.05.

4.2. Biological Activities

4.2.1. MbtI Enzymatic Assays

Recombinant *M. tuberculosis* MbtI was produced and purified as previously reported [14]. Enzyme activity was determined at 37 °C, measuring the formation of salicylic acid by a fluorimetric assay, slightly modified from Vasan et al. [38]. Briefly, the reactions were performed in a final volume of 400 µL of 50 mM Hepes pH 7.5, 5 mM $MgCl_2$, containing 1-2 µM MbtI, by the addition of chorismic acid, and monitored using a Perkin-Elmer LS3 fluorimeter (Ex. λ = 305 nm, Em. λ = 420 nm). Inhibition assays were performed in the presence of the compound at 100 µM (stock solution 20 mM in DMSO) and 50 µM chorismic acid. Where possible, compounds were tested both as free acids and sodium salts, providing analogous results. For compounds inhibiting by more than 75% the initial activity, IC_{50} values were determined. To this end, the activity was measured at different compound concentrations, and values were calculated according to Equation (1), with Origin 8 software:

$$A_{[I]} = A_{[0]} \times \left(1 - \frac{[I]}{[I] + IC_{50}}\right) \quad (1)$$

where $A_{[I]}$ is the activity at inhibitor concentration [I] and A[0] is the activity in the absence of the inhibitor.

The K_i was determined at different substrate [S] and compound concentrations, using Equation (2):

$$v = \frac{V_{max}[S]}{[S] + K_m\left(1 + \frac{[I]}{K_i}\right)} \quad (2)$$

4.2.2. MIC Determination

The minimal inhibitory concentration (MIC^{99}) of the most active compounds was determined against *M. bovis* BCG in low-iron chelated Sauton's medium, by the 2-fold microdilution method in U-bottom 96-well microtiter plates, as previously reported [13]. To this purpose, cells were grown in 7H9 medium, sub-cultured in chelated Sauton's medium, and then diluted to an OD_{600} of 0.01 in chelated Sauton's containing different concentrations of the test compound. After 15 days of incubation at 37 °C, the growth was evaluated by the resazurin reduction assay method (REMA). Thirty microliters of a 0.01% solution of filter-sterilized resazurin sodium salt were added to each well, and the microtiters were re-incubated at the same temperature for 24 h. The MIC was defined as the lowest concentration of the drug that prevented a change in color from blue to pink, which indicates bacterial growth.

5. Conclusions

In this paper, a SAR study on our series of MbtI inhibitors led to the identification of new candidates, endowed with a potent activity against the enzyme and encouraging bactericidal properties. For the new products, we described the design, synthesis, analytical characterization, and biological activity.

Firstly, two sets of compounds, series A and B, incorporating a variety of heterocyclic motifs were biologically evaluated against MbtI and in whole-cell assays against *M. bovis* BCG. This approach led to the disclosure of **1b, 3b**, and **4b** provided with a moderate activity (IC_{50} in the range 18–27 μM), but low bactericidal effects. Overall, the obtained results confirmed that the furan core was a better scaffold to gain MbtI inhibition in comparison with several other heterocycles.

These findings provided the basis for the design of the new furan-based derivatives **IV–IX**, which were synthesized, characterized, and tested. The best compound of this series, **IV**, bearing the preferred cyano group at position 3 and a trifluoromethyl moiety in 5, showed a potent MbtI inhibitory effect (%RA $\approx$ 1%, $IC_{50} \approx 15$ μM, $Ki \approx 9$ μM), comparable to the previous lead **I**, but exhibiting an enhanced antimycobacterial action ($MIC^{99} = 125$ μM), thus becoming one of the few potent MbtI inhibitors endowed with a promising antitubercular activity.

These observations justify the selection of **IV** as the new lead of our next optimization campaign, which will contribute to strengthen the perspectives of anti-TB drug discovery.

Supplementary Materials: The Supplementary Materials are available online at https://www.mdpi.com/1424-8247/14/2/155/s1.

Author Contributions: Conceptualization: F.M., M.M., and S.V. Synthesis and characterization of the compounds: E.P., M.M., G.C., A.G., and S.V. Biological experiments: L.R.C. and G.S. Financial resources: S.V., F.M., A.G., and E.P. F.M. supervised the whole study and wrote the paper. F.M., S.V., L.R.C., M.M., and D.B. reviewed and edited the manuscript. All authors have read and agreed to the published version of the manuscript.

Funding: This research received no external funding.

Institutional Review Board Statement: Not applicable.

Informed Consent Statement: Not applicable.

Data Availability Statement: The data presented in this study are available within the article.

Acknowledgments: All authors would like to acknowledge the University of Milan for funding this work (Linea B). Moreover, they gratefully thank Giulia Gwen Ballabio and Matteo Catalano for their helpful contribution.

Conflicts of Interest: The authors declare no conflict of interest.

References

1. *Global Tuberculosis Report 2020*; World Health Organization: Geneva, Switzerland, 2020; ISBN 978-92-4-001313-1.
2. Xu, K.; Liang, Z.C.; Ding, X.; Hu, H.; Liu, S.; Nurmik, M.; Bi, S.; Hu, F.; Ji, Z.; Ren, J.; et al. Nanomaterials in the Prevention, Diagnosis, and Treatment of *Mycobacterium Tuberculosis* Infections. *Adv. Healthc. Mater.* **2018**, *7*, 1700509. [CrossRef] [PubMed]
3. Truzzi, E.; Meneghetti, F.; Mori, M.; Costantino, L.; Iannuccelli, V.; Maretti, E.; Domenici, F.; Castellano, C.; Rogers, S.; Capocefalo, A.; et al. Drugs/Lamellae Interface Influences the Inner Structure of Double-Loaded Liposomes for Inhaled Anti-TB Therapy: An In-Depth Small-Angle Neutron Scattering Investigation. *J. Colloid Interface Sci.* **2019**, *541*, 399–406. [CrossRef]
4. Nahid, P.; Dorman, S.E.; Alipanah, N.; Barry, P.M.; Brozek, J.L.; Cattamanchi, A.; Chaisson, L.H.; Chaisson, R.E.; Daley, C.L.; Grzemska, M.; et al. Official American Thoracic Society/Centers for Disease Control and Prevention/Infectious Diseases Society of America Clinical Practice Guidelines: Treatment of Drug-Susceptible Tuberculosis. *Clin. Infect. Dis.* **2016**, *63*, e147–e195. [CrossRef]
5. Falzon, D.; Jaramillo, E.; Schünemann, H.J.; Arentz, M.; Bauer, M.; Bayona, J.; Blanc, L.; Caminero, J.A.; Daley, C.L.; Duncombe, C.; et al. WHO Guidelines for the Programmatic Management of Drug-Resistant Tuberculosis: 2011 Update. *Eur. Respir. J.* **2011**, *38*, 516–528. [CrossRef]
6. Fanzani, L.; Porta, F.; Meneghetti, F.; Villa, S.; Gelain, A.; Lucarelli, A.P.; Parisini, E. *Mycobacterium tuberculosis* Low Molecular Weight Phosphatases (MPtpA and MPtpB): From Biological Insight to Inhibitors. *Curr. Med. Chem.* **2015**, *22*, 3110–3132. [CrossRef] [PubMed]
7. Stelitano, G.; Sammartino, J.C.; Chiarelli, L.R. Multitargeting Compounds: A Promising Strategy to Overcome Multi-Drug Resistant Tuberculosis. *Molecules* **2020**, *25*, 1239. [CrossRef] [PubMed]

8. Mori, M.; Sammartino, J.C.; Costantino, L.; Gelain, A.; Meneghetti, F.; Villa, S.; Chiarelli, L.R. An Overview on the Potential Antimycobacterial Agents Targeting Serine/Threonine Protein Kinases from *Mycobacterium tuberculosis*. *Curr. Top. Med. Chem.* **2019**, *19*, 646–661. [CrossRef] [PubMed]
9. Meneghetti, F.; Villa, S.; Gelain, A.; Barlocco, D.; Chiarelli, L.R.; Pasca, M.R.; Costantino, L. Iron Acquisition Pathways as Targets for Antitubercular Drugs. *Curr. Med. Chem.* **2016**, *23*, 4009–4026. [CrossRef] [PubMed]
10. Chao, A.; Sieminski, P.J.; Owens, C.P.; Goulding, C.W. Iron Acquisition in Mycobacterium tuberculosis. *Chem. Rev.* **2019**, *119*, 1193–1220. [CrossRef] [PubMed]
11. Zhang, X.-K.; Liu, F.; Fiers, W.D.; Sun, W.-M.; Guo, J.; Liu, Z.; Aldrich, C.C. Synthesis of Transition-State Inhibitors of Chorismate Utilizing Enzymes from Bromobenzene *cis* -1,2-Dihydrodiol. *J. Org. Chem.* **2017**, *82*, 3432–3440. [CrossRef]
12. Mori, M.; Stelitano, G.; Gelain, A.; Pini, E.; Chiarelli, L.R.; Sammartino, J.C.; Poli, G.; Tuccinardi, T.; Beretta, G.; Porta, A.; et al. Shedding X-ray Light on the Role of Magnesium in the Activity of *M. tuberculosis* Salicylate Synthase (MbtI) for Drug Design. *J. Med. Chem.* **2020**, *63*, 7066–7080. [CrossRef]
13. Chiarelli, L.R.; Mori, M.; Beretta, G.; Gelain, A.; Pini, E.; Sammartino, J.C.; Stelitano, G.; Barlocco, D.; Costantino, L.; Lapillo, M.; et al. New Insight into Structure-Activity of Furan-based Salicylate Synthase (MbtI) Inhibitors as Potential Antitubercular Agents. *J. Enzym. Inhib. Med. Chem.* **2019**, *34*, 823–828. [CrossRef]
14. Chiarelli, L.R.; Mori, M.; Barlocco, D.; Beretta, G.; Gelain, A.; Pini, E.; Porcino, M.; Mori, G.; Stelitano, G.; Costantino, L.; et al. Discovery and Development of Novel Salicylate Synthase (MbtI) Furanic Inhibitors as Antitubercular Agents. *Eur. J. Med. Chem.* **2018**, *155*, 754–763. [CrossRef] [PubMed]
15. Pini, E.; Poli, G.; Tuccinardi, T.; Chiarelli, L.; Mori, M.; Gelain, A.; Costantino, L.; Villa, S.; Meneghetti, F.; Barlocco, D. New Chromane-Based Derivatives as Inhibitors of *Mycobacterium tuberculosis* Salicylate Synthase (MbtI): Preliminary Biological Evaluation and Molecular Modeling Studies. *Molecules* **2018**, *23*, 1506. [CrossRef] [PubMed]
16. Hinsberg, O. Über β-Naphtolsulfid und Iso-β-naphtolsulfid. *J. Prakt. Chem.* **1916**, *93*, 277–301. [CrossRef]
17. Chauhan, P.M.S.; Sunduru, N.; Sharma, M. Recent Advances in the Design and Synthesis of Heterocycles as Anti-Tubercular Agents. *Future Med. Chem.* **2010**, *2*, 1469–1500. [CrossRef] [PubMed]
18. Singh, P.; Kumar, S.K.; Maurya, V.K.; Mehta, B.K.; Ahmad, H.; Dwivedi, A.K.; Chaturvedi, V.; Thakur, T.S.; Sinha, S. S-Enantiomer of the Antitubercular Compound S006-830 Complements Activity of Frontline TB Drugs and Targets Biogenesis of *Mycobacterium tuberculosis* Cell Envelope. *ACS Omega* **2017**, *2*, 8453–8465. [CrossRef]
19. Aggarwal, A.; Parai, M.K.; Shetty, N.; Wallis, D.; Woolhiser, L.; Hastings, C.; Dutta, N.K.; Galaviz, S.; Dhakal, R.C.; Shrestha, R.; et al. Development of a Novel Lead that Targets *M. tuberculosis* Polyketide Synthase 13. *Cell* **2017**, *170*, 249–259.e25. [CrossRef]
20. Wilson, R.; Kumar, P.; Parashar, V.; Vilchèze, C.; Veyron-Churlet, R.; Freundlich, J.S.; Barnes, S.W.; Walker, J.R.; Szymonifka, M.J.; Marchiano, E.; et al. Antituberculosis Thiophenes Define a Requirement for Pks13 in Mycolic Acid Biosynthesis. *Nat. Chem. Biol.* **2013**, *9*, 499–506. [CrossRef]
21. Nayak, S.; Gaonkar, S.L. A Review on Recent Synthetic Strategies and Pharmacological Importance of 1,3-Thiazole Derivatives. *Mini Rev. Med. Chem.* **2018**, *19*, 215–238. [CrossRef] [PubMed]
22. Machado, D.; Azzali, E.; Couto, I.; Costantino, G.; Pieroni, M.; Viveiros, M. Adjuvant Therapies Against Tuberculosis: Discovery of a 2-Aminothiazole Targeting Mycobacterium tuberculosis Energetics. *Future Microbiol.* **2018**, *13*, 1383–1402. [CrossRef]
23. Azzali, E.; Machado, D.; Kaushik, A.; Vacondio, F.; Flisi, S.; Cabassi, C.S.; Lamichhane, G.; Viveiros, M.; Costantino, G.; Pieroni, M. Substituted N-Phenyl-5-(2-(phenylamino)thiazol-4-yl)isoxazole-3-carboxamides Are Valuable Antitubercular Candidates that Evade Innate Efflux Machinery. *J. Med. Chem.* **2017**, *60*, 7108–7122. [CrossRef]
24. Tyagi, S.; Nuermberger, E.; Yoshimatsu, T.; Williams, K.; Rosenthal, I.; Lounis, N.; Bishai, W.; Grosset, J. Bactericidal Activity of the Nitroimidazopyran PA-824 in a Murine Model of Tuberculosis. *Antimicrob. Agents Chemother.* **2005**, *49*, 2289–2293. [CrossRef] [PubMed]
25. Pieroni, M.; Wan, B.; Zuliani, V.; Franzblau, S.G.; Costantino, G.; Rivara, M. Discovery of Antitubercular 2,4-diphenyl-1H-imidazoles from Chemical Library Repositioning and Rational Design. *Eur. J. Med. Chem.* **2015**, *100*, 44–49. [CrossRef] [PubMed]
26. De, S.S.; Khambete, M.P.; Degani, M.S. Oxadiazole Scaffolds in Anti-Tuberculosis Drug Discovery. *Bioorg. Med. Chem. Lett.* **2019**, *29*, 1999–2007. [CrossRef] [PubMed]
27. Zhou, B.; He, Y.; Zhang, X.; Xu, J.; Luo, Y.; Wang, Y.; Franzblau, S.G.; Yang, Z.; Chan, R.J.; Liu, Y.; et al. Targeting Mycobacterium Protein Tyrosine Phosphatase B for Antituberculosis Agents. *Proc. Natl. Acad. Sci. USA* **2010**, *107*, 4573–4578. [CrossRef] [PubMed]
28. Kadam, K.S.; Jadhav, R.D.; Kandre, S.; Guha, T.; Reddy, M.M.K.; Brahma, M.K.; Deshmukh, N.J.; Dixit, A.; Doshi, L.; Srinivasan, S.; et al. Evaluation of Thiazole Containing Biaryl Analogs as Diacylglycerol Acyltransferase 1 (DGAT1) Inhibitors. *Eur. J. Med. Chem.* **2013**, *65*, 337–347. [CrossRef]
29. Wu, X.; Geng, X.; Zhao, P.; Zhang, J.; Wu, Y.D.; Wu, A.X. I2-Promoted Formal [3+2] Cycloaddition of α-Methylenyl Isocyanides with Methyl Ketones: A Route to 2,5-Disubstituted Oxazoles. *Chem. Commun.* **2017**, *53*, 3438–3441. [CrossRef]
30. Curreli, F.; Kwon, Y.D.; Belov, D.S.; Ramesh, R.R.; Kurkin, A.V.; Altieri, A.; Kwong, P.D.; Debnath, A.K. Synthesis, Antiviral Potency, in Vitro ADMET, and X-ray Structure of Potent CD4 Mimics as Entry Inhibitors That Target the Phe43 Cavity of HIV-1 gp120. *J. Med. Chem.* **2017**, *60*, 3124–3153. [CrossRef]

31. Leung, D.; Du, W.; Hardouin, C.; Cheng, H.; Hwang, I.; Cravatt, B.F.; Boger, D.L. Discovery of an Exceptionally Potent and Selective Class of Fatty Acid Amide Hydrolase Inhibitors Enlisting Proteome-Wide Selectivity Screening: Concurrent Optimization of Enzyme Inhibitor Potency and Selectivity. *Bioorg. Med. Chem. Lett.* **2005**, *15*, 1423–1428. [CrossRef]
32. Aksenov, A.V.; Khamraev, V.; Aksenov, N.A.; Kirilov, N.K.; Domenyuk, D.A.; Zelensky, V.A.; Rubin, M. Electrophilic Activation of Nitroalkanes in Efficient Synthesis of 1,3,4-Oxadiazoles. *RSC Adv.* **2019**, *9*, 6636–6642. [CrossRef]
33. Devine, S.M.; Mulcair, M.D.; Debono, C.O.; Leung, E.W.W.; Nissink, J.W.M.; Lim, S.S.; Chandrashekaran, I.R.; Vazirani, M.; Mohanty, B.; Simpson, J.S.; et al. Promiscuous 2-Aminothiazoles (PrATs): A Frequent Hitting Scaffold. *J. Med. Chem.* **2015**, *58*, 1205–1214. [CrossRef] [PubMed]
34. Lagorce, D.; Bouslama, L.; Becot, J.; Miteva, M.A.; Villoutreix, B.O. FAF-Drugs4: Free ADME-tox Filtering Computations for Chemical Biology and Early Stages Drug Discovery. *Bioinformatics* **2017**, *33*, 3658–3660. [CrossRef] [PubMed]
35. Daina, A.; Michielin, O.; Zoete, V. SwissADME: A Free WebTool to Evaluate Pharmacokinetics, Drug-Likeness and Medicinal Chemistry Friendliness of Small Molecules. *Sci. Rep.* **2017**, *7*, 42717. [CrossRef] [PubMed]
36. Chapman, T.M.; Bouloc, N.; Buxton, R.S.; Chugh, J.; Lougheed, K.E.A.; Osborne, S.A.; Saxty, B.; Smerdon, S.J.; Taylor, D.L.; Whalley, D. Substituted Aminopyrimidine Protein Kinase B (PknB) Inhibitors Show Activity Against *Mycobacterium tuberculosis*. *Bioorg. Med. Chem. Lett.* **2012**, *22*, 3349–3353. [CrossRef] [PubMed]
37. Shyam, M.; Shilkar, D.; Verma, H.; Dev, A.; Sinha, B.N.; Brucoli, F.; Bhakta, S.; Jayaprakash, V. The Mycobactin Biosynthesis Pathway: A Prospective Therapeutic Target in the Battle against Tuberculosis. *J. Med. Chem.* **2020**. [CrossRef]
38. Vasan, M.; Neres, J.; Williams, J.; Wilson, D.J.; Teitelbaum, A.M.; Remmel, R.P.; Aldrich, C.C. Inhibitors of the Salicylate Synthase (MbtI) from *Mycobacterium tuberculosis* Discovered by High-Throughput Screening. *ChemMedChem* **2010**, *5*, 2079–2087. [CrossRef]

 *pharmaceuticals* 

Article

Liquid and Vapor Phase of Four Conifer-Derived Essential Oils: Comparison of Chemical Compositions and Antimicrobial and Antioxidant Properties

Stefania Garzoli [1,*,†], Valentina Laghezza Masci [2,†], Valentina Caradonna [2], Antonio Tiezzi [2], Pierluigi Giacomello [1] and Elisa Ovidi [2]

1 Department of Drug Chemistry and Technology, Sapienza University, 00185 Rome, Italy; pierluigi.giacomello@uniroma1.it
2 Department for the Innovation in Biological, Agrofood and Forestal Systems, Tuscia University, 01100 Viterbo, Italy; laghezzamasci@unitus.it (V.L.M.); valecarad@gmail.com (V.C.); antoniot@unitus.it (A.T.); eovidi@unitus.it (E.O.)
* Correspondence: stefania.garzoli@uniroma1.it
† These authors contributed equally.

Abstract: In this study, the chemical composition of the vapor and liquid phase of *Pinus cembra* L., *Pinus mugo* Turra, *Picea abies* L., and *Abies Alba* M. needles essential oils (EOs) was investigated by Headspace-Gas Chromatography/Mass Spectrometry (HS-GC/MS). In the examined EOs, a total of twenty-eight components were identified, most of which belong to the monoterpenes family. α-Pinene (16.6–44.0%), β-pinene (7.5–44.7%), limonene (9.5–32.5%), and γ-terpinene (0.3–19.7%) were the most abundant components of the liquid phase. Such major compounds were also detected in the vapor phase of all EOs, and α-pinene reached higher relative percentages than in the liquid phase. Then, both the liquid and vapor phases were evaluated in terms of antibacterial activity against three Gram-negative bacteria (*Escherichia coli*, *Pseudomonas fluorescens*, and *Acinetobacter bohemicus*) and two Gram-positive bacteria (*Kocuria marina* and *Bacillus cereus*) using a microwell dilution assay, disc diffusion assay, and vapor phase test. The lowest Minimum Inhibitory Concentration (MIC) (13.28 mg/mL) and Minimal Bactericidal Concentration (MBC) (26.56 mg/mL) values, which correspond to the highest antibacterial activities, were reported for *P. abies* EO against *A. bohemicus* and for *A. alba* EO against *A. bohemicus* and *B. cereus*. The vapor phase of all the tested EOs was more active than liquid phase, showing the inhibition halos from 41.00 ± 10.15 mm to 80.00 ± 0.00 mm for three bacterial strains (*A. bohemicus*, *K. marina*, and *B. cereus*). Furthermore, antioxidant activities were also investigated by 2,2-diphenyl-1-picrylhydrazyl (DPPH) and 2,2′-azinobis (3- ethylbenzothiazoline-6-sulfonic acid) diammonium salt (ABTS) assays, and a concentration-dependent antioxidant capacity for all EOs was found. *P. mugo* EO showed the best antioxidant activity than the other Pinaceae EOs. The four Pinaceae EOs could be further investigated for their promising antibacterial and antioxidant properties, and, in particular, α-pinene seems to have interesting possibilities for use as a novel natural antibacterial agent.

Keywords: *Pinus cembra* L.; *Pinus mugo* Turra; *Picea abies* L.; *Abies alba* M.; essential oil; chemical investigation; HS-GC/MS; antibacterial activity; antioxidant activity

Citation: Garzoli, S.; Masci, V.L.; Caradonna, V.; Tiezzi, A.; Giacomello, P.; Ovidi, E. Liquid and Vapor Phase of Four Conifer-Derived Essential Oils: Comparison of Chemical Compositions and Antimicrobial and Antioxidant Properties. *Pharmaceuticals* **2021**, *14*, 134. https://doi.org/10.3390/ph14020134

Academic Editor: Fiorella Meneghetti; Daniela Barlocco

Received: 30 December 2020
Accepted: 4 February 2021
Published: 8 February 2021

1. Introduction

Since ancient times, plants are a source of different kinds of compounds that humans used for their numerous biological activities and as a source for drug development [1]. Nowadays, the studies on antioxidant and antimicrobial activities of natural products are of considerable interest due to the importance of identifying and characterizing new bioactive molecules for applications in different fields as food preservation and packaging, antibiotically resistance phenomenon, and plant diseases.

Among plant secondary metabolites, essential oils (EOs), biosynthesized by glandular trichomes and other secretory structures in plants, are liquids particularly rich in volatile molecules such as monoterpene and sesquiterpene hydrocarbons, oxygenated monoterpenes and sesquiterpenes, esters, aldehydes, ketones, alcohols, phenols, and oxides [2–4]. The chemical composition of EOs can vary from plant to plant and even in the same species and depends on several factors such as post-harvest conservation conditions [5], extraction methods [6–8] and times [9,10], microclimate, and site in which the plant is growing [11,12]. EOs contribute to the plant relations with environment and with other organisms, and humans and animals take advantage of the abundance of such bioactive molecules from the plant kingdom [4]. Numerous papers deal with the biological activities of the EOs such as antioxidant and anti-inflammatory properties, antibacterial and antifungal activities, immunomodulatory effects, and cytotoxic activities against different cancer cell lines [13–19]. Gymnosperms, the Cupressaceae, and the Pinaceae families produce economically important EOs [20]. The Pinaceae family is the largest family of non-flowering seed plants and comprises 11 genera and approximately 230 species of trees, rarely shrubs, which are widely distributed in the Northern Hemisphere [21,22]. The biological activities of Pinaceae EOs reflect the richness in their chemical composition. Antioxidant, antibacterial, antifungal, insect larvicidal, anti-inflammatory, and antiproliferative activities are reported for different genus of the Pinaceae family [16,23–30].

In our searching and studying of natural compounds, in the present paper, we investigated and compared the chemical composition and the antimicrobial and antioxidant properties of the vapor and liquid phase of four Pinaceae EOs from *Pinus cembra* L. and *Pinus mugo* Turra, which belong to the Pinus genus, and *Picea abies* L. and *Abies alba* M., which belong to the Picea and Abies genus, respectively.

2. Results

2.1. Liquid and Vapor Phases EOs Chemical Composition

By Gas Chromatography-Mass Spectrometry (GC/MS) and Headspace (HS)-GC/MS analysis, the composition of the vapor and liquid phase of all EOs was described. Twenty components were identified in *P. cembra* and *P. mugo* EOs, and they are listed in Table 1. The most abundant component was α-pinene (44.0% when using GC/MS and 65.6% when using HS/GC-MS) followed by γ-terpinene (19.7% GC/MS; 11.0% HS/GC-MS), limonene (14.8% GC/MS and 8.2%; HS/GC-MS) and β-pinene (12.5% GC/MS; 12.4% HS/GC-MS) in *P. cembra* EO. On the contrary, β-pinene (43.3% GC/MS; 42.3% HS/GC-MS) was the major compound in *P. mugo*. EO followed by α-pinene (16.6% GC/MS; 31.6% HS/GC-MS) and limonene (9.5% GC/MS; 7.8% HS/GC-MS). β-Phellandrene (16.0%) as well as other minor compounds such as p-cymenene (0.1%), copaene (0.1%), and bornyl acetate (3.0%) appeared only in the liquid phase of *P. mugo* EO. On the other hand, α-phellandrene (0.7%) was detected only in *P. mugo* vapor phase EO. In particular, in the vapor phase of both EOs, the components from N° 11 to N° 20 were missing except for β-caryophyllene (0.1%), which was detected in *P. mugo* EO.

Twenty-one components were identified in *P. abies* and *A. alba* EOs, and they are listed in Table 2. β-Pinene was the principal compound in *P. abies* EO (20.2% when using GC/MS and 34.5% when using HS/GC-MS), while α-pinene (30.8% GC/MS; 51.3% HS/GC-MS) was the principal compound in *P. abies* EO. The second most abundant component was α-pinene (20.2% and 34.5%) in *P. abies* EO and limonene (32.5% and 19.0%) in *A. alba* EO when using GC/MS and HS/GC-MS, respectively. p-Cymene (0.2%; 0.1%), camphor (1.2%; 0.2%), and borneol (2.1%; 0.2%) were detected only in the liquid and vapor phase, respectively of *P. abies* EOs. α-Himachalene (0.3%), citronellol acetate (0.4%), humulene (1.6%), and caryophyllene oxide (0.1) appeared only in the liquid phase of *A. alba* EO. Lastly, in the vapor phase of both EOs, the components from N° 12 to N° 21 were missing except for borneol (0.2%), which was detected in *P. abies* EO.

Table 1. Chemical composition (%) of liquid and vapor phases of *P. cembra* and *P. mugo* EOs.

N°	COMPONENT [1]	LRI [2]	LRI [3]	*Pc*(%) [4]	*Pc*(%) [5]	*Pm*(%) [6]	*Pm*(%) [7]
1	α-pinene	1019	1021	44.0	65.6	16.6	31.6
2	camphene	1062	1065	1.6	2.3	1.6	2.9
3	β-pinene	1098	1099	12.5	12.4	43.3	42.3
4	β-thujene	1120	1118	0.3	0.2	0.3	-
5	α-phellandrene	1158	1160	-	-	-	0.7
6	limonene	1197	1198	14.8	8.2	9.5	7.8
7	β-phellandrene	1210	1207	-	-	16.0	-
8	γ-terpinene	1270	1241	19.7	11.0	0.3	13.3
9	p-cymene	1270	1268	0.1	Tr	0.2	0.2
10	terpinolene	1285	1282	0.5	0.2	2.1	1.1
11	p-cymenene	1431	1435	-	-	0.1	-
12	copaene	1491	1487	-	-	0.1	-
13	bornyl acetate	1571	1567	-	-	3.0	-
14	thymol methyl ether	1576	1575	1.2	tr	-	-
15	β-caryophyllene	1620	1619	0.4	-	3.6	0.1
16	α-terpineol	1655	1655	0.2	-	0.2	tr
17	humulene	1670	1667	0.5	-	1.1	-
18	γ-muurolene	1674	1673	0.4	-	0.3	-
19	α-muurolene	1691	1690	2.4	-	0.9	-
20	δ-cadinene	1760	1758	1.4	-	0.5	-
	SUM			100.0	99.9	99.7	100.0
	Monoterpenes			93.7	99.9	93.2	100.0
	Sesquiterpenes			5.1	-	6.5	-
	Other			1.2	-	-	-

[1] The components are reported according their elution order on a polar column; [2] Linear retention indices measured on polar column; [3] Linear retention indices from the literature; [4] Percentage values of *P. cembra* EO components (%); [5] Percentage values of *P. cembra* EO components (vapor phase); [6] Percentage mean values of *P. mugo* EO components (%); [7] Percentage mean values of *P. mugo* EO components (vapor phase); -Not detected; tr: traces (mean value < 0.1%).

Table 2. Chemical composition (%) of liquid and vapor phases of *P. abies* and *A. alba* essential oils (EOs).

N°	COMPONENT [1]	LRI [2]	LRI [3]	*Pa*(%) [4]	*Pa*(%) [5]	*Aa*(%) [6]	*Aa*(%) [7]
1	santene	980	984	0.7	1.9	1.4	4.2
2	α-pinene	1019	1021	20.2	34.5	30.8	51.3
3	camphene	1062	1065	7.2	10.5	11.2	16.5
4	β-pinene	1098	1099	44.7	43.8	7.5	7.3
5	limonene	1197	1198	14.2	8.0	32.5	19.0
6	γ-terpinene	1240	1241	0.3	0.1	1.1	1.2
7	p-cymene	1270	1268	0.2	0.1	-	-
8	terpinolene	1285	1282	0.6	0.2	0.5	0.1
9	α-longipinene	1480	1477	0.3	-	0.9	-
10	camphor	1506	1507	1.2	0.2	-	-
11	bornyl acetate	1571	1567	3.7	0.3	4.2	0.4
12	longifolene	1585	1583	1.1	-	0.6	-
13	β-caryophyllene	1620	1619	1.3	-	5.8	-
14	α-himachalene	1641	1637	-	-	0.3	-
15	citronellol acetate	1646	1644	-	-	0.4	-
16	α-terpineol	1655	1655	0.4	tr	0.2	-
17	humulene	1670	1667	-	-	1.6	-
18	γ-muurolene	1674	1673	0.4	-	0.3	-
19	borneol	1677	1675	2.1	0.2	-	-
20	δ-cadinene	1760	1758	0.9	-	0.6	-
21	caryophyllene oxide	1895	1892	-	-	0.1	-
	SUM			99.5	99.8	100.0	100.0
	Monoterpenes			94.8	97.9	88.4	95.8
	Sesquiterpenes			4.0	0.2	9.9	-
	Other			0.7	1.9	1.7	4.2

[1] The components are reported according their elution order on polar column; [2] Linear retention indices measured on polar column; [3] Linear retention indices from the literature; [4] Percentage mean values of *P. abies* EO components (%); [5] Percentage mean values of *P. abies* EO components (vapor phase); [6] Percentage mean values of *A. alba* EO components (%); [7] Percentage mean values of *A. alba* EO components (vapor phase); -Not detected; tr: traces (mean value < 0.1%).

Among the most abundant compounds, particular attention was paid to α-pinene, as it always reached higher percentages in the vapor phase than in the liquid phase of the investigated EOs. The compared values are as follows: (44.0% vs. 65.6%), (16.6% vs. 31.6%),

(20.2% vs. 35.5%), and (30.8% vs. 51.3%) liquid and vapor phase in *P. cembra*, *P. mugo*, *P. abies*, and *A. alba* EOs, respectively (Figure 1).

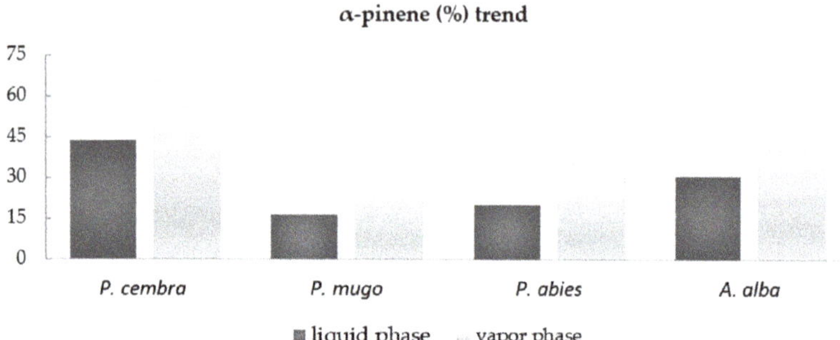

Figure 1. Bar graph of α-pinene percentage trend in liquid and vapor phase EOs.

2.2. Antibacterial Activities of P. Cembra, P. Mughus, P. Abies, and A. Alba EOs

The antibacterial activities of the Pinaceae EOs were evaluated for three Gram-negative (*Escherichia coli*, *Pseudomonas fluorescens*, and *Acinetobacter bohemicus*) and two Gram-positive bacteria (*Kocuria marina* and *Bacillus cereus* using micro dilution assay to determine Minimum Inhibitory Concentration (MIC) and the Minimal Bactericidal Concentration (MBC), and the MBC/MIC ratio defines an agent as bacteriostatic when the MBC/MIC ratio > 4 and as bactericidal when the MBC/MIC ratio ≤ 4 [31]. Furthermore, the disc diffusion assay by contact with the essential oil determined the diameter of bacterial growth inhibition zone (IZ), and the vapor phase test determined the antibacterial growth inhibition zone (Vapor IZ) by more volatile molecules of the EO in a preservative atmosphere. The antibacterial results of the tested EOs are summarized in Tables 3–6 reporting the MIC, MBC, MBC/MIC ratio, IZ, and vapor IZ following the treatments for each bacterial strain. In Table 3, the treatment with *P. cembra* EO showed MIC and MBC values of 53.12 mg/mL for *E. coli*, *P. fluorescens*, and *K. marina*, while MIC values were 26.56 mg/mL for *A. bohemicus* and *B. cereus*, and MBC values were 26.56 mg/mL and 53.12 mg/mL, respectively. MBC/MIC ratio defined the *P. cembra* EO as bactericidal against all bacterial strains. No effects were observed with the disc diffusion assay and with the vapor phase test for *P. cembra* EO against *E. coli* and *P. fluorescens*. The IZ and vapor IZ values were 17.67 ± 0.58 mm and 67.33 ± 2.52 mm for *A. bohemicus*, 9.33 ± 0.58 mm and 80.00 ± 0.00 mm for *K. marina*, and 11.67 ± 1.15 mm and 80.00 ± 0.00 mm for *B. cereus*, respectively.

Table 3. Antibacterial activity of *P. cembra* EO.

Strains	*Pinus cembra*				
	MIC [1]	MBC [2]	MBC/MIC Ratio	IZ [3]	Vapor IZ [4]
E. coli	53.12	53.12	1	0.00	0.00
P. fluorescens	53.12	53.12	1	0.00	0.00
A. bohemicus	26.56	26.56	1	17.67 ± 0.58	67.33 ± 2.52
K. marina	53.12	53.12	1	9.33 ± 0.58	80.00 ± 0.00
B. cereus	26.56	53.12	2	11.67 ± 1.15	80.00 ± 0.00

[1] Minimal Inhibitory Concentration expressed in mg/mL of EO treatment; [2] Minimal Bactericidal Concentration expressed in mg/mL of EO treatment; [3] Growth inhibition zone by disc diffusion assay expressed in mm; [4] Growth inhibition zone by vapor phase test expressed in mm. Values are expressed as means ± SD. $p < 0.05$ from one-way analysis of variance test (ANOVA).

Table 4 summarizes the antibacterial tests for *P. mugo* EO. MIC and MBC values were 52.16 mg/mL for *E. coli*, *P. fluorescens*, and *K. marina*, while MIC and MBC values were 26.08 mg/mL for *A. bohemicus* and MIC and MBC values were 26.08 mg/mL and

52.16 mg/mL for *B. cereus*, respectively. The MBC/MIC ratio defined as bactericidal the *P. mugo* EO against all bacterial strains. No effects were observed by the disc diffusion assay and by the vapor phase test for *P. cembra* EO on *P. fluorescens*. *P. mugo* EO was not highly active against *E. coli* with an IZ value of 9.67 ± 0.58 mm, while no growth inhibition zone was observed by the vapor phase test. Higher antibacterial activity was observed for the other bacterial strains: IZ and vapor IZ values were 25.33 ± 4.51 mm and 41.00 ± 10.15 mm for *A. bohemicus*, 11.33 ± 1.15 mm and 80.00 ± 0.00 mm for *K. marina*, and 15.67 ± 1.15 mm and 76.67 ± 5.77 mm for *B. cereus*, respectively. The *P. mugo* EO vapor phase was more active than the liquid phase against *A. bohemicus*, *K. marina*, and *B. cereus*.

Table 4. Antibacterial activity of *P. mugo* EO.

Strains	*Pinus mugo*				
	MIC [1]	MBC [2]	MBC/MIC Ratio	IZ [3]	Vapor IZ [4]
E. coli	52.16	52.16	1	9.67 ± 0.58	0.00
P. fluorescens	52.16	52.16	1	0.00	0.00
A. bohemicus	26.56	26.56	1	25.33 ± 4.51	41 ± 10.15
K. marina	52.16	52.16	1	11.33 ± 1.15	80.00 ± 0.00
B. cereus	26.56	52.16	2	15.67 ± 1.15	76.67 ± 5.77

[1] Minimal Inhibitory Concentration expressed in mg/mL of EO treatment; [2] Minimal Bactericidal Concentration expressed in mg/mL of EO treatment; [3] Growth inhibition zone by disc diffusion assay expressed in mm; [4] Growth inhibition zone by vapor phase test expressed in mm. Values are expressed as means ± SD. $p < 0.05$ from one-way analysis of variance test (ANOVA).

Table 5. Antibacterial activity of *P. abies* EO.

Strains/Origin	*Picea abies*				
	MIC [1]	MBC [2]	MBC/MIC Ratio	IZ [3]	Vapor IZ [4]
E. coli	53.12	53.12	1	0.00	0.00
P. fluorescens	53.12	53.12	1	0.00	0.00
A. bohemicus	13.28	26.56	2	18.67 ± 1.53	76.67 ± 5.77
K. marina	53.12	53.12	1	9.67 ± 1.15	80.00 ± 0.00
B. cereus	26.56	53.12	2	11.67 ± 1.53	80.00 ± 0.00

[1] Minimal Inhibitory Concentration expressed in mg/mL of EO treatment; [2] Minimal Bactericidal Concentration expressed in mg/mL of EO treatment; [3] Growth inhibition zone by disc diffusion assay expressed in mm; [4] Growth inhibition zone by vapor phase test expressed in mm. Values are expressed as means ± SD. $p < 0.05$ from one-way analysis of variance test (ANOVA).

Table 6. Antibacterial activity of *A. alba* EO.

Strains/Origin	*Abies alba*				
	MIC [1]	MBC [2]	MBC/MIC Ratio	IZ [3]	Vapor IZ [4]
E. coli	51.28	51.28	1	0.00	0.00
P. fluorescens	51.28	51.28	1	0.00	0.00
A. bohemicus	12.82	25.64	2	19.67 ± 0.58	80.00 ±00
K. marina	51.28	51.28	1	7.67 ± 1.15	80.00 ± 00
B. cereus	12.82	25.64	2	15.00 ± 2.65	66.67 ± 11.55

[1] Minimal Inhibitory Concentration expressed in mg/mL of EO treatment; [2] Minimal Bactericidal Concentration expressed in mg/mL of EO treatment; [3] Growth inhibition zone by disc diffusion assay expressed in mm; [4] Growth inhibition zone by vapor phase test expressed in mm. Values are expressed as means ± SD. $p < 0.05$. $p < 0.05$ from one-way analysis of variance test (ANOVA).

P. abies EO antibacterial activity is reported in Table 5. MIC and MBC values were 53.12 mg/mL for *E. coli*, *P. fluorescens*, and *K. marina*. For *A. bohemicus*, a lower MIC value was observed (13.28 mg/mL), whereas MBC was 26.56 mg/mL. The antibacterial activity for *B. cereus* was 26.56 mg/mL and 53.12 mg/mL for the MIC and MBC values, respectively. As obtained by the MBC/MIC ratio, *P. abies* EO was bactericidal against all bacterial strains. No effects were observed by the disc diffusion assay and by the vapor phase test for *P. abies* EO on *P. fluorescens* and *E. coli*. IZ and vapor IZ values were 18.67 ± 1.53 mm and 76.67 ± 5.77 mm for *A. bohemicus*, 9.67 ± 1.15 mm and 80.00 ± 0.00 mm for *K. marina* and 11.67 ± 1.53 mm and 80.00 ± 0.00 mm for *B. cereus*, respectively. The *P. abies* EO vapor phase was more active rather than the liquid phase against *A. bohemicus*, *K. marina*, and *B. cereus*.

The results for *A. alba* antibacterial activity are reported in Table 6. MIC and MBC values were 51.28 mg/mL for *E. coli*, *P. fluorescens*, and *K. marina*. Lower MIC and MBC values were found for *A. bohemicus* and for *B. cereus* (12.82 mg/mL and 25.64 mg/mL, respectively). The MBC/MIC ratio defined as bactericidal the *A. alba* EO against all bacterial strains. No effects were observed in the disc diffusion assay and in the vapor phase test for *A. alba* EO on *E. coli* and *P. fluorescens*. Higher antibacterial activity was detected for the other bacterial strains: IZ and vapor IZ values were 19.67 ± 0.58 mm and 80.00 ± 00 mm for *A. bohemicus*, 7.67 ± 1.15 and 80.00 ± 0.00 for *K. marina*, and 15.00 ± 2.65 and 66.67 ± 11.55 for *B. cereus*, respectively. The vapor phase test revealed that the activities of the *A. alba* EO against *A. bohemicus*, *K. marina*, and *B. cereus* were higher than those of the liquid phase.

2.3. Antioxidant Activity

To determine the antioxidant activity of the four Pinaceae EOs, 2,2-Diphenyl-1-picrylhydrazyl (DPPH) scavenging activity and 2,2′-azinobis (3- ethylbenzothiazoline-6-sulfonic acid) diammonium salt (ABTS) radical scavenging assay, based on the reaction of the potential antioxidant with colored radicals, were carried out. The antioxidant activity results are reported in Table 7. In all EOs, a concentration-dependent antioxidant capacity was found. In both tests, *P. mugo* EO showed the highest antioxidant activity than the other Pinaceae EOs. This EO exhibits lower IC_{50} values (3.08 µg/mL and 43.08 µg/mL for DPPH and ABTS assays, respectively) and higher TEAC values (7.65 mol/mg and 14.01 mol/mg for DPPH and ABTS assays, respectively). The second effective essential oil was the *A. alba* EO with IC_{50} values of 7.84 µg/mL and 44.23 µg/mL and TEAC values of 1.63 mol/mg and 13.26 mol/mg in the DPPH method and ABTS method, respectively. The TEAC values of *P. cembra* and *P. abies* EOs were almost identical: 1.63 mol/mg and 1.68 mol/mg in the DPPH assay, respectively and 13.26 mol/mg for both the EOs in the ABTS assay. Taking into account the ABTS test, the IC_{50} amount was 44.90 µg/mL in *P. cembra* EO and 45.00 µg/mL in *A. alba* EO. In the DPPH test, both EOs showed similar IC_{50} values, too (13.01 µg/mL for *P. cembra* and 13.05 µg/mL for *P. abies*). *A. alba* EO remains the least effective in antioxidant capacity of the analyzed Pinaceae EOs.

Table 7. Effects of *P. mugo*, *P. cembra*, *P. abies*, and *A. alba* EOs in different antioxidant assays.

Assay	Values Expressed as	*P. cembra*	*P. mugo*	*P. abies*	*A. alba*
DPPH	IC_{50} *	13.01 ± 0.86	3.08 ± 0.65	13.05 ± 3.09	7.84 ± 1.70
	TEAC **	1.63 ± 0.46	7.65 ± 1.33	1.68 ± 0.64	3.01 ± 0.48
ABTS	IC_{50} *	44.90 ± 2.06	43.08 ±6.95	45.00 ± 6.26	44.23 ± 1.10
	TEAC **	13.26 ± 1.45	14.01 ± 2.01	13.26 ± 0.52	13.65 ± 0.49

* µg/mL of essential oil; ** µM of Trolox equivalent/mg of essential oil. Values are expressed as means ± SD. $p < 0.05$.

3. Discussion

The chemical profile of both the vapor and liquid phase and the antibacterial and antioxidant activities of four Pinaceae EOs, obtained from needles by steam distillation, were investigated using different kinds of techniques and assays. In the literature, a few papers reporting the Pinaceae EOs chemical composition are present, and no report describes the volatile composition of the vapor phase of the conifer-derived EOs by HS-GC/MS, as we applied for our investigation. In our investigations, the chemical constituents resulted primarily monoterpenoids and their contents were higher in the vapor phases of *P. cembra* (99.9%) and *P. mugo* (100.0%) EOs than the vapor phases of the *P. abies* (97.9%) and *A. alba* (95.8%) EOs. The major compounds of the *P. cembra* EO were α-pinene (44.0%), γ-terpinene (19.7%), limonene (14.8%), and β-pinene (12.5%). Similar composition was described by Lis et al. [32], where the needle oil was dominated by α-pinene (48.4%), limonene (7.5%), and β-phellanderene (3.1%); α-pinene was also the major component (69.14%) in needle EO of *P. cembra* growing in Romania [33]. The composition of the EO from twig tips with needles of the *P. cembra* L. growing in Salzburg Alps was represented by α-pinene (43.9–48.3%), β-phellandrene (13.1–17.2%), and β-pinene (6.6–9.3%) [34]. *Pinus cembra* needles EO from Slovakia consisted of α-pinene (53.2%), limonene (11.4%), and β-phellandrene (9.4%) [35].

The main components of *P. mugo* EO were β-pinene (43.3%), α-pinene (16.6%), β-phellandrene (16.0%), and limonene (9.5%) with a low percentage of β-caryophyllene (3.6%). A different composition was reported for *P. mugo* EO from needles growing in Poland where 3-carene (23.8 %), myrcene (22.3 %), and α-pinene (10.3 %) resulted as the main components [36]. 3-Carene (31.73%) was also the major compound in EO of *P. mugo* from Kosovo [37], followed by α-pinene (19.95%) and β-phellandrene (13.49%). *P.mugo* needles EOs from Macedonian [38] and Serbia [39] mainly consisted of Δ^3-carene (amount up to 35% and 23.9%), α- and β-pinene (up to 20% and 17.9%) and β-phellandrene (amount about 15% and 7.2%), respectively.

In *P. abies* EO, we found β-pinene (44.7%) as the most abundant component followed by α-pinene (20.2%), limonene (14.2%), and camphene (7.2%). A different composition has been described for the EOs from shoots of *P. abies* that grow wild in different locations of Romania, which are characterized by limonene (from 6.27% up to 12.98%), camphene (from 3.89% up to 14.07%), α-pinene (from 2.44% up to 10.42%), and β-myrcene (from 0.44% up to 10.12%) [40].

The chemical composition of *A. alba* EO showed two components such as limonene and α-pinene with a similar percentage (32.5% and 30.8%) followed by camphene (11.2%) and β-pinene (7.5%). The same compounds were listed with an inverted trend in *A. alba* EO from Montenegro where β-pinene (32.8%) was the major component followed by α-pinene (17.3%) and camphene (16.7%) [41]. On the contrary, α-limonene (about 70%) and α-pinene (57%) were the major compounds in *A. alba* EO from seeds and cones respectively [42]. In *A. alba* EO from Poland, limonene was the component with the higher percentage (82.9%) detected in seed EO, whereas α-pinene (50.0%) was the main component in cone EO [43]. According to the literature [44] and on the basis of the reported data, it becomes evident that the chemical composition of the EOs from species belonging to the Pinaceae family can depend by multiple factors such as part of the plant examined, its geographic origin, and also, extraction methods and storage [45].

MIC and MBC values defined by the microwell dilution method were tested against *E. coli*, *P. fluorescens* and *A. bohemicus* Gram-negative bacteria strains and *K. marina* and *B. cereus* Gram-positive bacteria strains. The lowest MIC (13.28 mg/mL) and MBC (26.56 mg/mL) values, which correspond to the highest antibacterial activities, were reported for *P. abies* EO against *A. bohemicus* and for *A. alba* EO against *A. bohemicus* and *B. cereus* with 12.82 mg/mL (MIC) and 25.64 mg/mL (MBC).

The increase of antibiotically resistance phenomenon in human and animal pathologies has determined the intensification of research on new natural antimicrobial substances [19,46,47], and in this view, several studies were carried out to investigate the biological activities of Pinaceae EOs and the roles of their molecules. *P. abies* EO extracted by supercritical carbon dioxide was investigated for antimicrobial properties on *E. coli* using the isothermal calorimetry technique, and it inhibited the growth and interfered with the metabolic activity of the microorganism [48]. Kartnig et al. [49] determined the antibacterial activities of the essential oils of young pine shoots on different bacterial strains also from human patients, and significant activities were revealed against G+ bacteria strains and *Candida* species tested. Apetrei et al. [25] reported that needles and twigs essential oils of *Pinus cembra* showed high activity against *Sarcina lutea* and *Staphylococcus aureus* and no activity against *B. cereus*, *E. coli* and *Pseudomonas aeruginosa*.

The antibacterial activities of the four Pinaceae EOs were also confirmed by agar diffusion and disk volatilization methods by which the IZ and vapor IZ were measured in mm of inhibition halos. For all the tested EOs, the vapor phases were more active than the liquid phases, showing the inhibition halos from 41.00 ± 10.15 mm to 80.00 ± 0.00 mm for three bacterial strains (*A. bohemicus*, *K. marina*, and *B. cereus*). Concerning *E. coli* and *P. fluorescens*, a very low or null activity was reported. The results showed high activities of the EOs against *A. bohemicus*, *K. marina*, and *B. cereus* and a scarce or null activity against *E. coli* and *P. fluorescens*. The highest activities obtained by vapor phases of all EOs against *A. bohemicus*, *K. marina*, and *B. cereus* could be related with the presence of α-pinene. In the graph bar (Figure 1), the relative percentages of α-pinene were reported. It reached higher percentages in the vapor phase than in the liquid phase of all investigated EOs. In particular, liquid and vapor phase values were as follows: (44.0% vs. 65.6%), (16.6% vs. 31.6%), (20.2% vs. 35.5%), and (30.8% vs. 51.3%), in *P. cembra*, *P. mugo*, *P. abies*, and *A. alba* EOs respectively. These results suggest that α-pinene could play an important role for the detected antibacterial activity. Some papers reported α-pinene from Pinaceae EOs as the main compound showing good biological activity; it was the principal constituent (5.2–37.0%) in five Moroccan Pinus species EOs [50] and in *Pinus peuce* Griseb. EOs (12.89–27.34%) growing on three different locations in R. Macedonia [51].

Different studies confirmed the antibacterial properties of α-pinene [52]. Freitas et al. [53] reported that α-pinene has antibacterial and antibiotic-modulating activities against *S. aureus;* it also increases the activity of norfloxacin against *E. coli* and norfloxacin and gentamicin against *S. aureus*. Furthermore, Hippeli et al. [54] described an anti-inflammatory potential of *P. mugo* EOs and its main compound α-pinene, while Cole et al. [55] showed anti-proliferative activity on the MCF-7 cell line. On the other hand, Kurti et al. [37] attributed the antimicrobial activities of some Pinus species EOs from Kosovo to the hexane/diethyl ether fractions, which were mainly composed by oxygenated monoterpenes.

In the present study, the susceptibility of bacteria does not seem to be related with the features of the cell Gram-positive and Gram-negative bacteria wall structure, since the more sensitive bacteria strains, *A. bohemicus*, *K. marina*, and *B. cereus* do not belong to the same group. Generally, Gram-negative are more resistant than Gram-positive bacteria, because the cell wall does not allow the entrance into the cell of hydrophobic molecules present in the essential oils [56,57], although some exceptions have been shown [58,59]. In a comparative study of the essential oils from four Pinus species [30], it was found that the sensitivity of the tested bacterial pathogens cannot be related with the cell wall structure. Different mechanisms of action can explain the EOs antimicrobial activities, and their wide variety of molecular components can act at multiple levels [60].

The DPPH and ABTS assays demonstrated a significant antioxidant activity for all Pinaceae EOs. *P. mugo* EO was the more active with an IC_{50} 3.08 ± 0.65 and 43.08 ± 6.95 µg/mL for DPPH and ABTS assays, respectively. The values expressed in Trolox equivalent (TEAC) confirmed identical results. A comparative investigation has been carried out on *P. halepensis* EO chemical composition and antioxidant activities, with respect to the impact of geographic variation and environmental conditions [61].

The variety of compounds that are present in the investigated EOs confers them numerous biological properties, and their antioxidant activities could be related to the presence of monoterpenes. Wang et al. [62] studied the antioxidant activities of seven terpenoids found in wine, and among the tested compounds, α-pinene and limonene had the highest DPPH free radical scavenging and the highest reducing power. Wojtunik-Kulesza [63] reviewed the monoterpenes biological properties and antioxidant activities of α-pinene were also reported.

4. Materials and Methods

4.1. Materials

P. cembra L., *P. mugo* Turra, *A. alba* M., and *P. abies* L. bio essential oils (IT BIO 013 n° BZ-43509-AB) from needles growing in Alto Adige, Italy were obtained by steam distillation for 6 h extraction time and were directly provided by Bergila GmbH Srl (Falzes/Issengo-Bolzano). Methanol, 2,2-Diphenyl-1-picrylhydrazyl (DPPH), 6-hydroxy-2,5,7,8-tetramethylchroman-2-carboxylic acid (Trolox), 2,2′-azinobis (3-ethylbenzothiazoline-6-sulfonic acid) diammonium salt (ABTS), potassium persulfate ($K_2S_2O_8$), LB Broth with Agar and Thiazolyl Blue Tetrazolium Bromide (MTT) were from Sigma-Aldrich (Darmstadt, Germany). Gentamicin sulfate was purchased from Biochrom PAN-Bio-Tech GmbH (Aidenbach, Germany).

4.2. Gas Chromatography–Mass Spectrometry (GC–MS) Analysis

To describe the chemical composition of the EOs, a gas chromatograph with a flame ionization detector (FID) directly coupled to a mass spectrometer (MS) Perkin Elmer Clarus 500 model (Waltham, MA, USA) was used. The GC was equipped with a Restek Stabilwax (fused-silica) polar capillary column. Helium was used as carrier gas at a flow rate of 1 mL/min. The injector was set to a 280 °C, and the oven temperature program was as follows: isothermal at 60 °C for 5 min, then ramped to 220 °C at a rate of 6 °C min^{-1}, and finally isothermal at 220 °C for 20 min. One uL of EO was diluted in 1 mL of methanol, and the injection volume was 1 μL. The Electron Impact-Mass Spectrometer (EI-MS) mass spectra were recorded at 70 eV (EI) and were scanned in the range of 40–500 m/z. The ion source and the connection parts temperature was 220 °C. The injector split ratio was 1:20. The GC-TIC mass spectra were obtained by the TurboMass data analysis software (Perkin Elmer). The identification of components was performed by matching their mass spectra with those stored in the Wiley and NIST 02 mass spectra libraries database. Furthermore, the linear retention indices (LRIs) (relative to C8–C30 aliphatic hydrocarbons, injected in the column at the same operating conditions described above) were calculated and compared with available retention data present in the literature. The relative percentages of all identified components were obtained by peak area normalization from GC-FID chromatograms without the use of an internal standard or correction factors and expressed in percentages. All analyses were repeated twice.

4.3. Headspace GC-MS Analysis

The volatile chemical profile of essential oils was carried out with a Perkin Elmer Headspace Turbomatrix 40 (Waltham, MA, USA) autosampler connected to GC-MS [64,65]. One mL of the each EO was placed in 20 mL vials sealed with headspace PTFE-coated silicone rubber septa and caps. To optimize the headspace procedure for the determination of volatile organic compounds (VOCs), more operative parameters were optimized. The gas phase of the sealed vials was equilibrated for 20 min at 60 °C and was followed immediately by compound desorption into GC injector in splitless mode. Quantification of compounds was performed by GC-FID in the same conditions described in the previous paragraph.

4.4. Antibacterial Activities of the Pinaceae Essential Oils

The antibacterial activities were investigated by using different methods, the Minimal Inhibitory Concentration (MIC), the Minimal Bactericidal Concentration (MBC), the agar diffusion method, and Vapor Phase Test (VPT).

4.4.1. Bacterial Strains

Five bacterial strains from the culture collections of the Plant Cytology and Biotechnology Laboratory of Tuscia University were tested to evaluate the antibacterial activities of *P. cembra* L., *P. mugo* Turra, *A. alba* M., and *P. abies* L. essential oils: *Escherichia coli* ATCC 25922, *Pseudomonas fluorescens* ATCC 13525, and *Acinetobacter bohemicus* DSM 102855 among Gram-negative and *Kocuria marina* DSM 16420 and *Bacillus cereus* ATCC 10876 among Gram-positive. All tested bacterial strains were maintained on LB broth (10 g tryptone, 5 g yeast extract, 10 g NaCl per liter, autoclaved at 121 °C for 20 min) with agar. Bacteria cultures were maintained at two different temperatures: 26 °C for *P. fluorescens*, *A. bohemicus*, and *B. cereus* and 37 °C for *K. marina* and *E. coli*. All inocula were prepared with fresh cultures plated the day before the test.

4.4.2. Minimum Inhibitory Concentration (MIC)

The MIC is defined as the lowest concentration of antimicrobial agent that completely inhibits the growth of the microorganism as detected by the unaided eye and was carried out according to the microwell dilution method. Briefly, 12 dilutions of the four essential oils in LB broth (for *P. cembra* from 53.12 to 0.01 mg/mL; for *P. mugo* from 52.16 to 0.01 mg/mL; for *A. alba* form 51.28 to 0.01 mg/mL and for *P. abies* from 53.12 to 0.01 mg/mL), a control with the same percentage of DMSO (from 6.25% to 0.003%) in Lysogeny broth, a growth control without treatments, a positive control with gentamicin diluted from 100 to 0.05 μg/mL, and a sterility control without bacteria were plated on 96 microwell plates. Then, 50 μL of bacterial inoculum, 10^6 CFU/mL, were added in each well, except for the sterility control, and the plates were incubated for 24 h at the corresponding temperature. The visualization of the inhibition activity was obtained by 20 μL of a solution of 3-(4,5-dimethylthiazol-2-yl)-2,5-diphenyltetrazolium bromide (200 μg/mL, MTT) added to each well. The assay was carried out in triplicate. The MBC/MIC ratio was reported to interpret the activity of the essential oil, and an antimicrobial agent is considered bacteriostatic when the ratio MBC/MIC > 4 and bactericidal when the ratio MBC/MIC is ≤4 [31].

4.4.3. Minimum Bactericidal Concentration (MBC)

To verify the lowest concentration at which the tested essential oils kill the bacterial cells, which is defined the Minimum Bactericidal Concentration (MBC), 10 μL of the last four dilutions from microwell dilution method in which no bacteria growth was observed were plated on a Petri plate with LB agar and incubated for 24 h. The concentration at which no growth on agar was observed defined MBC values. The assay was carried out in triplicate.

4.4.4. Agar Diffusion Method

To determine the diameter of the halo inhibition of the bacteria growth induced by *P. abies*, *A. alba*, *P. cembra*, and *P. mugo* essential oils, the bacterial strains were suspended in LB broth to obtain a turbidity of 0.5 McFarland (approximately 10^8 Colony-Forming Unit/mL—CFU/mL) and then plated on LB broth with agar in a Petri plate. Sterile disks (6 mm diameter, Oxoid) were placed on the agar and impregnated with 10 μL of samples. Two μL of gentamicin from a stock solution (10 mg/mL) was used as a positive control After 24 h, the inhibitory activities of each essential oil were recorded as mm of halo diameter without growth [58] using a vernier caliper rule. The mean and the respective standard deviation (SD) of the measured halo in three independent experiments were recorded.

4.4.5. Vapor Phase Test (VPT)

The antibacterial activity of the Pinaceae essential oils in the vapor phase was evaluated by the modified disk volatilization method [66,67]. LB agar were poured into an 80 mm plastic Petri dish and a lower amount into its cover. Each bacterial suspension containing 10^8 CFU/mL was plated on the LB agar medium. Then, 10 μL of tested essential oils were added to a 6 mm sterile disk and placed on agar in the covered Petri plate. Liquid LB agar was put in the space between the cover and the base of the Petri dishes to facilitate the sealing and to prevent any vapor leakage. The Petri plates were incubated for 24 h in an inverted position, and afterwards, the inhibition halos were measured. Negative controls were carried out without the essential. All VPTs were carried out in triplicate.

4.5. Antioxidant Activity

To assess the antioxidant activity of the four Pinaceae essential oils, DPPH radical scavenging activity and ABTS radical scavenging assay, which are based on the reaction of the potential antioxidant with colored radicals, were carried out.

4.5.1. DPPH Scavenging Activity Assay

In DPPH radical scavenging assay, the Pinaceae essential oils antioxidant activities were calculated against the 1,1-diphenyl-2-picrilidrazil radical (DPPH•) using the method described by Sanchez-Moreno et al. [68]. First, 100 μL of fresh solution of a solid crystalline DPPH• (0.2 mM) in methanol were added to 100 μL of 12 geometric dilutions in methanol of each essential oil inside a 96-well plate. Geometric dilutions of the samples in methanol were used as sample blanks. In blank DPPH samples, essential oils were omitted. As a positive control, dilutions were prepared starting from Trolox solution (1 mM) in methanol. The samples were incubated for 30 min in the dark at room temperature, and the absorbances decreases were measured at 517 nm using a Tecan Sunrise^{TM} UV-vis spectrophotometer. The assay was repeated three times.

4.5.2. ABTS Radical Scavenging Assay

The radical scavenging activities of the Pinaceae essential oils were also calculated using the ABTS (2,2′-azino-bis (3-ethylbenzothiazoline-6-sulfonic acid) diammonium salt) assay described by Re et al. [69] with some modifications. The radical cation ABTS+• was produced by reacting ABTS aqueous solution (7 mM) with $K_2S_2O_8$ (140 mM) following an incubation for 16 h in the dark at room temperature before use. The ABTS+• solution was diluted with ethanol to reach an absorbance of 0.70 ± 0.02 at 734 nm, and 1980 μL was mixed with 20 μL of the essential oil dilutions in ethanol. The resulting solutions were incubated for 5 min at room temperature. Afterwards, the absorbances were measured at 734 nm using a Jasco V-630 UV-Visible spectrophotometer and using Spectra Manager^{TM} software. Furthermore, the absorbance of the ABTS+• blank, consisting of 20 μL of ethanol dissolved in 1980 μL of ABTS+• solution, was measured. The assay was repeated three times.

4.5.3. IC_{50} and TEAC Calculation

Trolox and samples calibration curves were obtained by plotting the inhibition ratio against sample concentrations. The inhibition ratio was calculated using the following formula:

$$IR\% = \frac{A\ blank - A\ sample}{A\ blank} \times 100. \quad (1)$$

The IC50 parameter was calculated using the sample calibration curve. A lower value of the IC50 parameter correspond to a lower concentration of the EO that can scavenge 50% of DPPH· molecules; therefore, it indicates a higher antioxidant activity.

The Trolox equivalent antioxidant capacity (TEAC) index was obtained from the ratio between the Trolox IC50 (μM) and the sample IC50 (mg/L):

$$TEAC = \frac{IC50_{trolox}}{IC50_{sample}}. \quad (2)$$

4.6. Statistical Analysis

The results were expressed as means ± standard deviation (SD). The one-way analysis of variance test (ANOVA) using GraphPad Prism software (GraphPad Prism 5.0, GraphPad Software, Inc., San Diego, CA, USA) was used to evaluate statistical discrepancies between the groups (p values < 0.05).

5. Conclusions

In this study, for the first time, the chemical composition of the liquid and vapor phase of four Pinaceae EOs was investigated by the HS-GC/MS technique. The results of analyses showed that these EOs are rich in monoterpenoids and highlight that α-pinene, one of the main compounds, is more abundant in the vapor phase of each oil than in the liquid phase. The antimicrobial and antioxidant activities were also reported and compared. The vapor phase of each EO resulted more active against the investigated bacterial strains.

The biological effects of the Pinaceae EOs combined with their bioavailability makes them promising sources for possible application in different fields such as pharmacology, pharmacognosy, and phytochemistry.

Author Contributions: Conceptualization, S.G.; V.L.M.; E.O.; investigation, S.G., V.L.M.; V.C.; data curation, S.G.; V.L.M.; E.O.; writing—original draft preparation, S.G.; V.L.M.; E.O.; writing—review and editing, S.G.; A.T.; E.O.; supervision, P.G; A.T.; funding acquisition, P.G.; A.T. All the authors critically edited the manuscript before submission. All authors have read and agreed to the published version of the manuscript.

Funding: This research received no external funding.

Institutional Review Board Statement: Not applicable.

Informed Consent Statement: Not applicable.

Data Availability Statement: Not applicable.

Acknowledgments: The authors are thankful to Bergila, GmbH Srl (Falzes/Issengo-Bolzano) Italy, for providing *Pinus cembra* L., *Pinus mugo* Turra, *Picea abies* L. and *Abies Alba* M. essential oils.

Conflicts of Interest: The authors declare no conflict of interest.

References

1. Bernardini, S.; Tiezzi, A.; Masci, V.L.; Ovidi, E. Natural products for human health: An historical overview of the drug discovery approaches. *Nat. Prod. Res.* **2018**, *32*, 1926–1950. [CrossRef]
2. Tongnuanchan, P.; Benjakul, S. Essential Oils: Extraction, Bioactivities, and Their Uses for Food Preservation. *J. Food Sci.* **2014**, *79*, R1231–R1249. [CrossRef]
3. El Asbahani, A.; Miladi, K.; Badri, W.; Sala, M.; Addi, E.H.A.; Casabianca, H.; El Mousadik, A.; Hartmann, D.; Jilale, A.; Renaud, F.N.R.; et al. Essential oils: From extraction to encapsulation. *Int. J. Pharm.* **2015**, *483*, 220–243. [CrossRef]
4. Sharifi-Rad, J.; Sureda, A.; Tenore, G.C.; Daglia, M.; Sharifi-Rad, M.; Valussi, M.; Tundis, R.; Sharifi-Rad, M.; Loizzo, M.R.; Ademiluyi, A.O.; et al. Biological Activities of Essential Oils: From Plant Chemoecology to Traditional Healing Systems. *Molecules* **2017**, *22*, 70. [CrossRef]
5. Turek, C.; Stintzing, F.C. Stability of Essential Oils: A Review. *Compr. Rev. Food Sci. Food Saf.* **2013**, *12*, 40–53. [CrossRef]
6. Kamiie, Y.; Sagisaka, M.; Nagaki, M. Essential oil composition of Lavandula angustifolia "Hidcote": Comparison of hydrodistillation and supercritical fluid extraction methods. *Trans. Mater. Res. Soc. Jpn.* **2014**, *39*, 485–489. [CrossRef]
7. Chenni, M.; El Abed, D.; Rakotomanomana, N.; Fernandez, X.; Chemat, F. Comparative Study of Essential Oils Extracted from Egyptian Basil Leaves (*Ocimum basilicum* L.) Using Hydro-Distillation and Solvent-Free Microwave Extraction. *Molecules* **2016**, *21*, 113. [CrossRef] [PubMed]
8. Elyemni, M.; Louaste, B.; Nechad, I.; Elkamli, T.; Bouia, A.; Taleb, M.; Chaouch, M.; Eloutassi, N. Extraction of Essential Oils of *Rosmarinus officinalis* L. by Two Different Methods: Hydrodistillation and Microwave Assisted Hydrodistillation. *Sci. World J.* **2019**, *2019*, 3659432. [CrossRef]
9. Garzoli, S.; Božović, M.; Baldisserotto, A.; Sabatino, M.; Cesa, S.; Pepi, F.; Vicentini, C.B.; Manfredini, S.; Ragno, R. Essential oil extraction, chemical analysis and anti-Candida activity of Foeniculum vulgare Miller—new approaches. *Nat. Prod. Res.* **2017**, *32*, 1254–1259. [CrossRef]
10. Božović, M.; Garzoli, S.; Baldisserotto, A.; Romagnoli, C.; Pepi, F.; Cesa, S.; Vertuani, S.; Manfredini, S.; Ragno, R. *Melissa officinalis* L. subsp. altissima (Sibth. & Sm.) Arcang. essential oil: Chemical composition and preliminary antimicrobial investigation of samples obtained at different harvesting periods and by fractionated extractions. *Ind. Crop. Prod.* **2018**, *117*, 317–321. [CrossRef]
11. Hendawy, S.; El Gendy, A.; Omer, E.; Pistelli, L.; Pistelli, L. Growth, Yield and Chemical Composition of Essential Oil of Mentha piperita var. multimentha Grown Under Different Agro-ecological Locations in Egypt. *J. Essent. Oil Bear. Plants* **2018**, *21*, 23–39. [CrossRef]
12. Chen, S.-X.; Yang, K.; Xiang, J.-Y.; Raymond Kwaku, O.; Han, J.-X.; Zhu, X.-A.; Huang, Y.-T.; Liu, L.-J.; Shen, S.-B.; Li, H.-Z.; et al. Comparison of Chemical Compositions of the Pepper EOs From Different Cultivars and Their AChE Inhibitory Activity. *Nat. Prod. Commun.* **2020**, *15*, 1934578X20971469.
13. Edris, A.E. Pharmaceutical and therapeutic Potentials of essential oils and their individual volatile constituents: A review. *Phytother. Res.* **2007**, *21*, 308–323. [CrossRef]

14. Bakkali, F.; Averbeck, S.; Averbeck, D.; Idaomar, M. Biological effects of essential oils-a review. *Food Chem. Toxicol.* **2008**, *46*, 446–475. [CrossRef]
15. Elshafie, H.S.; Camele, I. An Overview of the Biological Effects of Some Mediterranean Essential Oils on Human Health. *BioMed Res. Int.* **2017**, *2017*, 9268468. [CrossRef]
16. Basholli-Salihu, M.; Schuster, R.; Hajdari, A.; Mulla, D.; Viernstein, H.; Mustafa, B.; Mueller, M. Phytochemical composition, anti-inflammatory activity and cytotoxic effects of essential oils from three Pinus spp. *Pharm. Biol.* **2017**, *55*, 1553–1560. [CrossRef]
17. de Lavor, É.M.; Fernandes, A.W.C.; de Andrade Teles, R.B.; Leal, A.E.B.P.; de Oliveira, R.G., Jr.; Silva, M.G.; de Oliveira, A.P.; Silva, J.C.; de Moura Fontes Araújo, M.T.; Coutinho, H.D.M.; et al. Essential Oils and Their Major Compounds in the Treatment of Chronic Inflammation: A Review of Antioxidant Potential in Preclinical Studies and Molecular Mechanisms. *Oxidative Med. Cell. Longev.* **2018**, *2018*, 1–23. [CrossRef] [PubMed]
18. Peterfalvi, A.; Miko, E.; Nagy, T.; Reger, B.; Simon, D.; Miseta, A.; Czéh, B.; Szereday, L. Much More Than a Pleasant Scent: A Review on Essential Oils Supporting the Immune System. *Molecules* **2019**, *24*, 4530. [CrossRef]
19. Wińska, K.; Mączka, W.; Łyczko, J.; Grabarczyk, M.; Czubaszek, A.; Szumny, A. Essential Oils as Antimicrobial Agents—Myth or Real Alternative? *Molecules* **2019**, *24*, 2130. [CrossRef]
20. Husnu Can Baser, K.; Buchbauer, G. *Handbook of Essential Oils*; Routledge: Milton Park, UK, 2015.
21. Ran, J.-H.; Shen, T.-T.; Wu, H.; Gong, X.; Wang, X.-Q. Phylogeny and evolutionary history of Pinaceae updated by transcriptomic analysis. *Mol. Phylogenetics Evol.* **2018**, *129*, 106–116. [CrossRef] [PubMed]
22. Simpson, M.G. Evolution and Diversity of Woody and Seed Plants. In *Plant Systematics*; Elsevier: New York, NY, USA, 2019; pp. 131–165.
23. Bağci, E.; Diğrak, M. Antimicrobial Activity of Essential Oils of someAbies (Fir) Species from Turkey. *Flavour Fragr. J.* **1996**, *11*, 251–256. [CrossRef]
24. Hong, E.-J.; Na, K.-J.; Choi, I.-G.; Choi, K.-C.; Jeung, E.-B. Antibacterial and Antifungal Effects of Essential Oils from Coniferous Trees. *Biol. Pharm. Bull.* **2004**, *27*, 863–866. [CrossRef] [PubMed]
25. Lungu, C.; Tuchilus, C.; Aprotosoaie, A.C.; Oprea, A.; Malterud, K.E.; Miron, A. Chemical, Antioxidant and Antimicrobial Investigations of *Pinus cembra* L. Bark and Needles. *Molecules* **2011**, *16*, 7773–7788. [CrossRef]
26. Politeo, O.; Skočibušić, M.; Maravić, A.; Ruscic, M.; Milos, M. Chemical Composition and Antimicrobial Activity of the Essential Oil of Endemic Dalmatian Black Pine (*Pinus nigra* ssp. dalmatica). *Chem. Biodivers.* **2011**, *8*, 540–547. [CrossRef]
27. Zeng, W.; Zhang, Z.; Gao, H.; Jia, L.-R.; He, Q. Chemical Composition, Antioxidant, and Antimicrobial Activities of Essential Oil from Pine Needle (Cedrus deodara). *J. Food Sci.* **2012**, *77*, C824–C829. [CrossRef]
28. Ismail, A.; Hanana, M.; Gargouri, S.; Jamoussi, B.; Hamrouni, L.; Amri, I. Comparative study of two coniferous species (*Pinus pinaster* Aiton and *Cupressus sempervirens* L. var. *dupreziana* [A. Camus] Silba) essential oils: Chemical composition and biological activity. *Chil. J. Agric. Res.* **2013**, *73*, 259–266. [CrossRef]
29. Xie, Q.; Liu, Z.; Li, Z. Chemical Composition and Antioxidant Activity of Essential Oil of Six Pinus Taxa Native to China. *Molecules* **2015**, *20*, 9380–9392. [CrossRef]
30. Mitić, Z.S.; Jovanović, B.; Jovanović, S.Č.; Mihajilov-Krstev, T.; Stojanović-Radić, Z.Z.; Cvetković, V.J.; Mitrović, T.L.; Marin, P.D.; Zlatković, B.; Stojanović, G.S. Comparative study of the essential oils of four Pinus species: Chemical composition, antimicrobial and insect larvicidal activity. *Ind. Crop. Prod.* **2018**, *111*, 55–62. [CrossRef]
31. Gatsing, D.; Tchakoute, V.; Ngamga, D.; Kuiate, J.; Tamokou, J.; Nji-Nkah, B.; Tchouanguep, F.; Fodouop, S. In vitro antibacterial activity of Crinum purpurascens Herb. leaf extract against the Salmonella species causing typhoid fever and its toxicological evaluation. *Iran. J. Med. Sci.* **2009**, *34*, 126–136.
32. Lis, A.; Kalinowska, A.; Krajewska, A.; Mellor, K. Chemical Composition of the Essential Oils from Different Morphological Parts of *Pinus cembra* L. *Chem. Biodivers.* **2017**, *14*, e1600345. [CrossRef] [PubMed]
33. Apetrei, C.L.; Spac, A.; Brebu, C.M.; Miron, A.; Popa, G. Composition and antioxidant and antimicrobial activities of the essential oils of a full-grown *Pinus cembra* L. tree from the Calimani Mountains (Romania). *J. Serb. Chem. Soc.* **2013**, *78*, 27–37. [CrossRef]
34. Chizzola, R.; Müllner, K. Variability of volatiles in *Pinus cembra* L. within and between trees from a stand in the Salzburg Alps (Austria) as assessed by essential oil and SPME analysis. *Genet. Resour. Crop. Evol.* **2020**, *68*, 567–579. [CrossRef]
35. Nikolić, B.; Todosijevićb, M.; Ratknića, M.; Đorđevićc, I.; Stankovićd, I.; Cvetkovićd, M.D.; Marine, P.; Teševićb, V. Terpenes and n-alkanes in needles of Pinus cembra. *NPC* **2018**, *13*, 1035–1037.
36. Lis, A.; Lukas, M.; Mellor, K. Comparison of Chemical Composition of the Essential Oils from Different Botanical Organs of Pinus mugo Growing in Poland. *Chem. Biodivers.* **2019**, *16*, e1900397. [CrossRef] [PubMed]
37. Kurti, F.; Giorgi, A.; Beretta, G.; Mustafa, B.; Gelmini, F.; Testa, C.; Angioletti, S.; Giupponi, L.; Zilio, E.; Pentimalli, D.; et al. Chemical composition, antioxidant and antimicrobial activities of essential oils of different Pinus species from Kosovo. *J. Essent. Oil Res.* **2019**, *31*, 263–275. [CrossRef]
38. Karapandzova, M.; Stefkov, G.; Karanfilova, I.C.; Panovska, T.K.; Stanoeva, I.P.; Stefova, M.; Kulevanova, S. Chemical Characterization and Antioxidant Activity of Mountain Pine (*Pinus mugo* Turra, Pinaceae) from Republic of Macedonia. *Rec. Nat. Prod.* **2019**, *13*, 50–63. [CrossRef]
39. Stevanovic, T.; Garneau, F.-X.; Jean, F.-I.; Vilotic, D.; Petrović, S.; Ruzic, N. The essential oil composition of *Pinus mugo* Turra from Serbia. *Flavour Fragr. J.* **2004**, *20*, 96–97. [CrossRef]

40. Radulescu, V.; Saviuc, C.; Chifiriuc, C.; Oprea, E.; Ilies, D.C.; Marutescu, L.; Lazar, V. Chemical composition and antimicrobial activity of essential oil from shoots spruce (*Picea abies* L.). *Rev. Chim.* **2011**, *62*, 69–72.
41. Chalchat, J.-C.; Sidibé, L.; Maksimovic, Z.A.; Petrovic, S.D.; Gorunovic, M.S. Essential Oil of Abies alba Mill., Pinaceae, from the Pilot Production in Montenegro. *J. Essent. Oil Res.* **2001**, *13*, 288–289. [CrossRef]
42. Wajs, A.; Urbańska, J.; Zaleśkiewicz, E.; Bonikowski, R. Composition of Essential Oil from Seeds and Cones of Abies alba. *Nat. Prod. Commun.* **2010**, *5*, 1291–1294. [CrossRef]
43. Wajs-Bonikowska, A.; Sienkiewiczb, M.; Stobieckaa, A.; Macia, A.; Szokac, L.; Karna, E. Chemical Composition and Biological Activity of Abies alba and A. Koreana Seed and Cone Essential Oils and Characterization of Their Seed Hydrolates. *Chem. Biodivers.* **2015**, *12*, 407–418. [CrossRef]
44. Tümen, I.; Hafızoğlu, H.; Kilic, A.; Dönmez, I.E.; Sivrikaya, H.; Reunanen, M. Yields and Constituents of Essential Oil from Cones of Pinaceae spp. Natively Grown in Turkey. *Molecules* **2010**, *15*, 5797–5806. [CrossRef]
45. Sedláková, J.; Lojková, L.; Kubáň, V. Gas Chromatographic determination of monoterpenes in spruce needles (*Picea abies*, *P. omorica*, and *P. pungens*) after supercritical fluid extraction. *Chem. Pap.* **2003**, *57*, 359–363.
46. Pereira, R.L.; Pereira, A.M.; Ferreira, M.O.S.; Fontenelle, R.; Saker-Sampaio, S.S.; Santos, H.N.; Bandeira, P.A.; Vasconcelos, M.A.N.; Queiroz, J.; Braz-Filho, R.H.; et al. Evaluation of the antimicrobial and antioxidant activity of 7-hydroxy-4′,6-dimethoxy-isoflavone and essential oil from *Myroxylon peruiferum* L.f. *An. Acad. Bras. Ciências* **2019**, *91*, e20180204. [CrossRef] [PubMed]
47. Donadu, M.G.; Le, N.T.; Ho, D.V.; Doan, T.Q.; Le, A.T.; Raal, A.; Usai, M.; Marchetti, M.; Sanna, G.; Madeddu, S.; et al. Phytochemical Compositions and Biological Activities of Essential Oils from the Leaves, Rhizomes and Whole Plant of Hornstedtia bella Škorničk. *Antibiotics* **2020**, *9*, 334. [CrossRef] [PubMed]
48. Haman, N.; Morozova, K.; Tonon, G.; Scampicchio, M.; Ferrentino, G. Antimicrobial Effect of Picea abies Extracts on E. coli Growth. *Molecules* **2019**, *24*, 4053. [CrossRef] [PubMed]
49. Kartnig, T.; Still, F.; Reinthaler, F. Antimicrobial activity of the essential oil of young pine shoots (*Picea abies* L.). *J. Ethnopharmacol.* **1991**, *35*, 155–157. [CrossRef]
50. Hmamouchi, M.; Hamamouchi, J.; Zouhdi, M.; Bessiere, J.M. Chemical and Antimicrobial Properties of Essential Oils of Five Moroccan Pinaceae. *J. Essent. Oil Res.* **2001**, *13*, 298–302. [CrossRef]
51. Karapandzova, M.; Stefkov, G.; Trajkovska-Dokic, E.; Kaftandzieva, A.; Kulevanova, S. Antimicrobial activity of needle essential oil of Pinus peuce Griseb. (Pinaceae) from Macedonian flora. *Maced. Pharm. Bull.* **2012**, *57*, 25–36. [CrossRef]
52. Da Silva, A.C.R.; Lopes, P.M.; De Azevedo, M.M.B.; Costa, D.C.M.; Alviano, C.S.; Alviano, D.S. Biological Activities of a-Pinene and β-Pinene Enantiomers. *Molecules* **2012**, *17*, 6305–6316. [CrossRef]
53. Freitas, P.R.; De Araújo, A.C.J.; Barbosa, C.R.D.S.; Muniz, D.F.; Da Silva, A.C.A.; Rocha, J.E.; Oliveira-Tintino, C.D.D.M.; Ribeiro-Filho, J.; Da Silva, L.E.; Confortin, C.; et al. GC-MS-FID and potentiation of the antibiotic activity of the essential oil of Baccharis reticulata (ruiz & pav.) pers. and α-pinene. *Ind. Crop. Prod.* **2020**, *145*, 112106. [CrossRef]
54. Hippeli, S.; Graßmann, J.; Vollmann, R.; Elstner, E.F. Latschenkieferöl: Möglicher Wirkmechanismus für antientzündliche Wirkung. *Pharm. Zeitung.* **2004**, *149*, 35–36.
55. Cole, R.A.; Bansal, A.; Moriarity, D.M.; Haber, W.A.; Setzer, W.N. Chemical composition and cytotoxic activity of the leaf essential oil of Eugenia zuchowskiae from Monteverde, Costa Rica. *J. Nat. Med.* **2007**, *61*, 414–417. [CrossRef]
56. Ait-Ouazzou, A.; Cherrat, L.; Espina, L.; Lorán, S.; Rota, C.; Pagán, R. The antimicrobial activity of hydrophobic essential oil constituents acting alone or in combined processes of food preservation. *Innov. Food Sci. Emerg. Technol.* **2011**, *12*, 320–329. [CrossRef]
57. Trombetta, D.; Castelli, F.; Sarpietro, M.G.; Venuti, V.; Cristani, M.; Daniele, C.; Saija, A.; Mazzanti, G.; Bisignano, G. Mechanisms of Antibacterial Action of Three Monoterpenes. *Antimicrob. Agents Chemother.* **2005**, *49*, 2474–2478. [CrossRef]
58. Burt, S. Essential oils: Their antibacterial properties and potential applications in foods—a review. *Int. J. Food Microbiol.* **2004**, *94*, 223–253. [CrossRef] [PubMed]
59. Imelouane, B.; Amhamdi, H.; Wathelet, J.P.; Ankit, M.; Khedid, K.; El Bachiri, A. Chemical composition of the essential oil of thyme (Thymus vulgaris) from Eastern Morocco. *Int. J. Agric. Biol.* **2009**, *11*, 205–208.
60. Nazzaro, F.; Fratianni, F.; De Martino, L.; Coppola, R.; De Feo, V. Effect of Essential Oils on Pathogenic Bacteria. *Pharmaceuticals* **2013**, *6*, 1451–1474. [CrossRef] [PubMed]
61. Djerrad, Z.; Kadik, L.; Djouahri, A. Chemical variability and antioxidant activities among Pinus halepensis Mill. essential oils provenances, depending on geographic variation and environmental conditions. *Ind. Crop. Prod.* **2015**, *74*, 440–449. [CrossRef]
62. Wang, C.-Y.; Chen, Y.-W.; Hou, C.-Y. Antioxidant and antibacterial activity of seven predominant terpenoids. *Int. J. Food Prop.* **2019**, *22*, 230–238. [CrossRef]
63. Wojtunik-Kulesza, K.A.; Kasprzak, K.; Oniszczuk, T.; Oniszczuk, A. Natural Monoterpenes: Much More than Only a Scent. *Chem. Biodivers.* **2019**, *16*, e1900434. [CrossRef] [PubMed]
64. Garzoli, S.; Turchetti, G.; Giacomello, P.; Tiezzi, A.; Masci, V.L.; Ovidi, E. Liquid and Vapour Phase of Lavandin (Lavandula × intermedia) Essential Oil: Chemical Composition and Antimicrobial Activity. *Molecules* **2019**, *24*, 2701. [CrossRef] [PubMed]
65. Oliva, A.; Garzoli, S.; Sabatino, M.; Tadić, V.; Costantini, S.; Ragno, R.; Božović, M. Chemical composition and antimicrobial activity of essential oil of Helichrysum italicum (Roth) G. Don fil. (Asteraceae) from Montenegro. *Nat. Prod. Res.* **2020**, *34*, 445–448. [CrossRef] [PubMed]

66. Balouiri, M.; Sadiki, M.; Ibnsouda, S.K. Methods for in vitro evaluating antimicrobial activity: A review. *J. Pharm. Anal.* **2016**, *6*, 71–79. [CrossRef]
67. Wang, T.-H.; Hsia, S.-M.; Wu, C.-H.; Ko, S.-Y.; Chen, M.Y.; Shih, Y.-H.; Shieh, T.-M.; Chuang, L.-C.; Wu, C.-Y. Evaluation of the Antibacterial Potential of Liquid and Vapor Phase Phenolic Essential Oil Compounds against Oral Microorganisms. *PLoS ONE* **2016**, *11*, e0163147. [CrossRef]
68. Sanchez-Moreno, C.; Larrauri, J.A.; Saura-Calixto, F. A procedure to measure the antiradical efficiency of polyphenols. *J. Sci. Food Agric.* **1998**, *76*, 270–276. [CrossRef]
69. Re, R.; Pellegrini, N.; Proteggente, A.; Pannala, A.; Yang, M.; Rice-Evans, C. Antioxidant activity applying an improved ABTS radical cation decolorization assay. *Free Radic. Biol. Med.* **1999**, *26*, 1231–1237. [CrossRef]

pharmaceuticals

MDPI

Review

Bacteriophage Therapy of Bacterial Infections: The Rediscovered Frontier

Nejat Düzgüneş [1,*], Melike Sessevmez [2] and Metin Yildirim [3]

[1] Department of Biomedical Sciences, Arthur A. Dugoni School of Dentistry, University of the Pacific, San Francisco, CA 94103, USA

[2] Department of Pharmaceutical Technology, Faculty of Pharmacy, Istanbul University, Istanbul 34116, Turkey; melikesessevmez@gmail.com

[3] Department of Pharmacy Services, Vocational School of Health Services, Tarsus University, Mersin 33400, Turkey; metinyildirim4@gmail.com

* Correspondence: nduzgunes@pacific.edu

Abstract: Antibiotic-resistant infections present a serious health concern worldwide. It is estimated that there are 2.8 million antibiotic-resistant infections and 35,000 deaths in the United States every year. Such microorganisms include *Acinetobacter*, Enterobacterioceae, *Pseudomonas*, *Staphylococcus* and *Mycobacterium*. Alternative treatment methods are, thus, necessary to treat such infections. Bacteriophages are viruses of bacteria. In a lytic infection, the newly formed phage particles lyse the bacterium and continue to infect other bacteria. In the early 20th century, d'Herelle, Bruynoghe and Maisin used bacterium-specific phages to treat bacterial infections. Bacteriophages are being identified, purified and developed as pharmaceutically acceptable macromolecular "drugs," undergoing strict quality control. Phages can be applied topically or delivered by inhalation, orally or parenterally. Some of the major drug-resistant infections that are potential targets of pharmaceutically prepared phages are *Pseudomonas aeruginosa*, *Mycobacterium tuberculosis* and *Acinetobacter baumannii*.

Keywords: lytic infection; antibiotic-resistance; *Mycobacterium tuberculosis*; *Acinetobacter baumannii*; *Pseudomonas aeruginosa*; phage production; magistral phage; pulmonary delivery; oral administration; topical delivery

Citation: Düzgüneş, N.; Sessevmez, M.; Yildirim, M. Bacteriophage Therapy of Bacterial Infections: The Rediscovered Frontier. *Pharmaceuticals* **2021**, *14*, 34. https://doi.org/10.3390/ph14010034

Received: 22 November 2020
Accepted: 29 December 2020
Published: 5 January 2021

1. Introduction: Bacteriophage Treatment of a Serious Infection

A 68-year-old man with diabetes developed necrotizing pancreatitis that was complicated by a pancreatic pseudocyst infected with a multi-drug-resistant strain of *Acinetobacter baumannii* [1]. *A. baumannii* is a Gram-negative nosocomial pathogen involved in bacteremia, meningitis and pulmonary infections with a high mortality rate. It is one of the "ESKAPE" microorganisms that are grouped together because of the common occurrence of multi-drug-resistance in the group. These microorganisms include *Enterococcus faecium*, *Staphylococcus aureus*, *Klebsiella pneumoniae*, *Acinetobacter baumannii*, *Pseudomonas aeruginosa* and *Enterobacter* species. The condition of the patient was deteriorating rapidly despite antibiotic treatment, to which he was obviously not responding. Bacteriophage therapy was initiated as part of an emergency investigational new drug protocol.

Bacteriophages are viruses of bacteria. Phages can cause either lytic or lysogenic infections in bacteria after attaching to a receptor or receptors on the bacterial surface and delivering their genome into the bacteria. In a lytic infection, the phage replicates and the new phage particles lyse the bacterium and continue to infect other bacteria (Figure 1). In a lysogenic infection, a DNA phage inserts its genetic material into the bacterial chromosome, and the genome is passed on to daughter cells as the bacterium divides. The integrated DNA may be activated by changes in environmental conditions to excise itself from the chromosome, producing phage particles that become lytic [2–4].

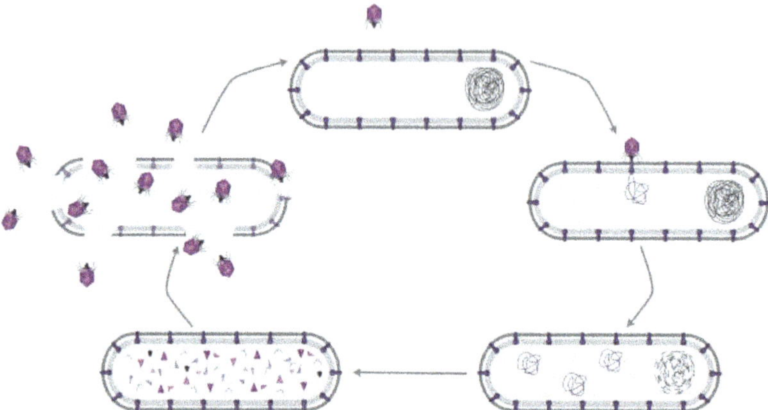

Figure 1. The lytic infection cycle of a bacteriophage. A phage particle attaches to a receptor on the surface of a host bacterium and delivers its genome into the cytoplasm. Phage proteins and replicate genomes are synthesized and self-assemble into new phage particles that eventually lyse the bacterium. The phages then infect other bacteria with the particular receptor (reproduced with permission from Kortright et al., 2019 [4]).

A large number of phage types that could specifically lyse *A. baumannii* were tested against three strains of the bacterium obtained previously from the patient. The phages had been collected by the Biological Defense Research Directorate of the Naval Medical Research Center. Most of the phages, however, were not lytic towards the first clinical bacterial sample obtained from the patient. Six phage types inhibited bacterial growth for 20 h. Four of these phages were pooled and, when tested against the bacterial isolate from the patient, had superior activity compared to each of the phages by themselves. The patient showed clear clinical improvement within 2 days of the administration of the phage cocktail containing a total of 5×10^9 particles. Initially, the phage mixture was delivered through the percutaneous catheters draining the pseudocyst cavity, the gallbladder and the intra-abdominal cavity and repeated every 6 to 12 h. Thirty-six hours after the initiation of treatment, the phage cocktail was given intravenously [1]. This treatment reversed the patient's clinical decline, cleared the *A. baumannii* infection and returned the patient to health.

The British bacteriologist Ernest Hankin (1896) noted the anti-*Vibrio cholerae* activity of water from two rivers in India and suggested that a substance that could pass through porcelain filters caused this observation and possibly limited cholera epidemics.

The Russian bacteriologist Gamaleya reported a similar phenomenon with *Bacillus subtilis* [5]. The English bacteriologist Frederick Twort hypothesized in 1915 that the antibacterial effect could be mediated by a virus [6]. The French-Canadian microbiologist Felix d'Herelle, at the Pasteur Institute in Paris, identified non-bacterial microorganisms from the stools of patients suffering from severe hemorrhagic dysentery. These microorganisms formed plaques in cultures of *Shigella* isolated from the patients [7]. In 1919, d'Herelle used a phage preparation to treat a boy with dysentery who recovered within a day and then three additional patients who started to recover within a day [8,9]. The first report of phage therapy was published in 1921 by Richard Bruynoghe and Joseph Maisin [10]. They injected bacteriophages into and around surgically opened staphylococcal skin lesions, which regressed within 1–2 days.

2. Phages as Pharmaceuticals

2.1. Phage Isolation and Enrichment

The key processes in phage therapy protocols are phage selection and isolation. The wrong choices can have fatal consequences [11]. Generally, two methods are used when choosing the appropriate phage for therapy: (1) A phage cocktail, such as Pyophage and Intestiphage. These preparations have a broader spectrum of activity than a single phage component and do not allow resistance to develop within a short time. (2) A pathogen-specific phage. Bacteria are isolated from the infection and tested for susceptibility to particular phages isolated previously [11].

Samples for phage isolation are taken from environments where the bacterial host can often be found, including soil, plant residues, fecal matter, wastewater and sewage (Figure 2). Phages against *Shigella dysenteriae* 2308 were isolated from the New York City sewage by Dubos et al. [12]. B_VpS_BA3 and vB_VpS_CA8 phages against *Vibrio parahaemolyticus* were isolated from sewage collected in China [13]. The vB_KpnP_IME337 phage against carbapenem-resistant *Klebsiella pneumoniae* was isolated from hospital sewage in China [14]. Li et al. [15] isolated 54 novel phages against the same organism from medical and domestic sewage wastewater. The newly isolated phage P545 had a relatively wide host range and strong antibacterial activity.

Although there are differences in phage isolation, the basic principle of the methods is the same as that developed by d'Herelle, and they are generally characterized as enrichment procedures [16]. First, the presence of phages is detected in the collected sample. Selection of the bacterial host is vital for the isolation of a phage in a recently acquired sample from the environment. Solid samples are mixed with sterile broth or buffer and then subjected to centrifugation and filtration [17]. Bacteria of interest are incubated overnight with the environmental sample. The bacteria that have survived the attack of the lytic phages are removed from the mixture by centrifugation or filtration, or both. The presence of phages in the filtrate is then assessed by plaque assay or by qPCR. The isolated phages have to be analyzed for their virulence, i.e., their ability to lyse target bacteria and the range of bacterial types they are able to infect. In an alternate method, samples from the environment are plated directly onto a lawn of particular bacteria and the presence of plaques resulting from bacterial lysis is detected. The latter method has been used to discover phages that lyse *Escherichia coli* and various bacteria from dental plaque and the oral cavity [17,18].

In the procedure described in detail by Luong et al. [19], the target bacterial strain is isolated and incubated with the phages. Then, after several agar plaque isolations, a single plaque is cultivated overnight. The isolated phage genome is sequenced to screen and identify lysogenic and deleterious genes. Phages are grown at liter scale, and the lysate is purified to eliminate any bacteria and cellular debris by pressure-driven filtration through filters of 0.8-, 0.45- and 0.22-µm pore size, followed by cross-flow ultra-filtration to eliminate debris smaller than 100 kD. This process eliminates endotoxin, exotoxins, peptidoglycan, nucleic acids and flagella. The phage particles are purified by CsCl density gradient centrifugation and dialysis to eliminate the CsCl. Any residual endotoxin molecules are removed by lipopolysaccharide-affinity chromatography. The last step ensures that the phage preparations do not cause inflammation or endotoxic shock when administered to patients [19].

Phages are expected to be found where the host bacteria reside. For example, phages that infect intestinal bacteria can be isolated from fecal material, and phages against epidermal bacteria such as *Staphylococcus aureus* are most likely isolated from skin samples or wound exudates. Identifying a phage against a particular bacterium is not straightforward, however. Whereas phages that lyse antibiotic-resistant *Klebsiella pneumoniae* and *Pseudomonas aeruginosa* were readily isolated from sewage samples, phages against antibiotic-resistant *Acinetobacter baumannii* were not found as frequently [20]. Furthermore, phages against methicillin-resistant *Staphylococcus aureus* were identified only rarely.

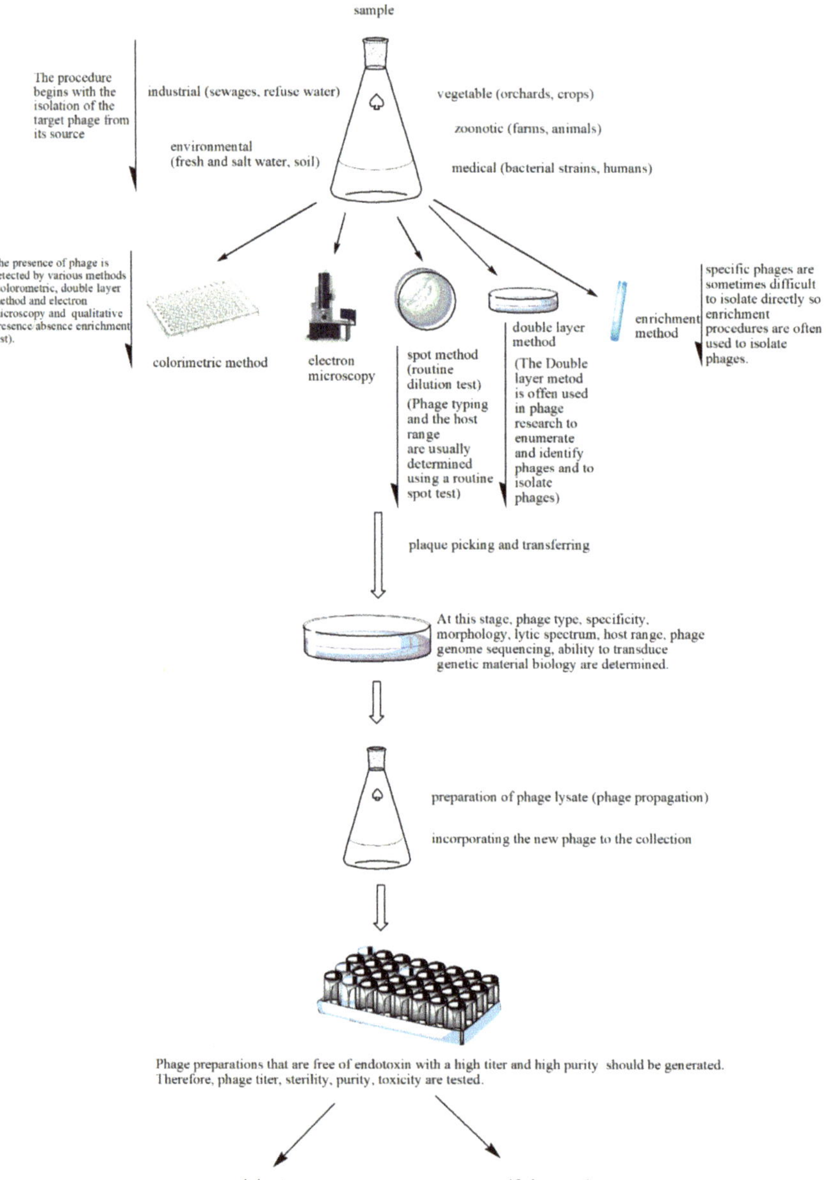

Figure 2. Stages of phage preparation. The environment (e.g., wastewater, farms and soil) is a source for all types of phages. The presence of phages in the tested sample is determined by different methods such as the double layer agar method, spot assays, the colorimetric method, the enrichment method or electron microscopy. The plaques that indicate lytic activity are picked up and transferred for the determination of phage type, specificity, etc. A phage lysate is prepared. At this stage, multiple procedures are performed to check for sterility (microbial contamination), toxicity (bacterial endotoxin or lipopolysaccharide (LPS) quantification), bacterial DNA contamination and phage titer. The purified phage preparation is stored.

The choice of a host for phage isolation may also depend on the ease of culturing the bacteria, as in the use of *Mycobacterium smegmatis* to isolate phages that will infect other *Mycobacterium* species. *M. smegmatis* grows much faster than *M. tuberculosis* and thus can produce a lawn on the appropriate agar surface for testing phage activity [17,20,21]. The isolated phages would then be tested further on the specific target *Mycobacterium* species.

Swanstrom et al. [22] investigated the variables contributing to the generation of high-titer phage stocks, using the agar layer method and coli phage T4r. The numbers of virus particles and bacteria per plate, the incubation period, the amount of soft agar in the agar layer as well as the broth volume used for virus extraction from the agar were found to be significant factors. When these factors were optimized, stock concentrations in the range of 10^{11}–10^{12} infectious particles/mL could be obtained [22].

Echeverría-Vega et al. [23] used a straightforward protocol for the isolation of bacteriophages from coastal organisms. They also validated the protocol for the isolation of lytic bacteriophages for the fish pathogen bacterium *Vibrio ordalii*. This method has particular utility for the recovery of bacteriophages for use as natural antimicrobial agents in aquacultures. In the enrichment method, samples are added to the host produced in a suitable medium, incubated and then centrifuged. The suspension containing the phage is filtered and applied at different concentrations onto an agar medium with target bacteria. The formed plaques are then counted. Thanks to the enrichment method, phages at low concentration can reach the desired level in culture. Enrichment is an advantageous method in cases where the amount of phages is low [24]. Numerous lytic phages were isolated against *Caulobacter* and *Asticcacaulis* bacteria using the enrichment method [25]. Methods such as spot testing, plaque testing, culture lysis and the calorimetric method are used in the detection of newly isolated bacteriophages (Figure 2) [26–31]. In the spot assay [26], bacteria are grown in Luria-Bertani broth, and after they are in the early log phase, they are mixed with soft agar and poured onto a Petri plate with previously poured agar. A phage filtrate is then placed on the soft agar and the plates are incubated overnight at 37 °C, after which bacterial lysis zones are counted [28]. In the double layer agar method, a bacterial culture in the log phase is mixed with a purified phage preparation and incubated briefly to allow for phage adsorption. This mixture is combined with soft agar and poured onto a previously solidified agar layer to form a homogeneous layer. After incubation at 37 °C for 24 h, plaque formation is observed, indicating phage activity. The plaques are resuspended in Mg (SM) buffer [28].

2.2. Phage Production

Bacteriophages need a host cell to reproduce. Understanding the interactions between host bacteria and bacteriophages is a crucial step in estimating the risks in production, including possible mutations in either microorganism [32]. The production process may also be affected by the nutrient composition, oxygenation, temperature and pH [33].

The substrate and temperature chosen for phage infection and bacterial growth are important factors. Fermentation is an important stage for host bacteria to multiply and produce bacteriophages. The sterilization step is performed to destroy undesired microorganisms. The bioreactor can be sterilized with heat, medium or a combination of these. During the fermentation process, the injected air is filtered through an in-line membrane. The air released after fermentation is filtered after it condenses [34].

Phages are grown basically in shaker flasks or stirred tank bioreactors. The latter are used to carry out industrial-scale production of bacteriophages, which has been divided into three different systems: batch, semi-continuous and continuous [35]. Each system has brought about its distinct benefits and drawbacks, discussed earlier by Merabishvili et al. [36]. Mancuso et al. [37] developed a production process that makes it possible to obtain high titers of *E. coli* T3 phages at high concentrations (10^{11} PFU mL^{-1}) using two continuous stirred tank bioreactors. The first bioreactor is just for propagation of the host bacteria at a steady-state growth rate by using controllable dilution rates and growth-limiting substrate (glucose). The second bioreactor is used for bacteriophage production and is fed from the host bacteria of the first bioreactor. Besides achieving high phage productivity of bacteriophages via the production process, the mutation risk of the host bacteria potentially caused by bacteriophages is suppressed.

2.3. Phage Purification and Quality Control

For the pharmaceutical application of phages, it is necessary to first carry out the purification process. The bacteriophage of interest is separated from host bacteria cells and debris by centrifugation, microfiltration or by using these methods together. The potential presence of any toxins in the preparations would be detrimental to the final pharmaceutical product. A Chamberland filter of 0.1–1 μm was used for bacteriophage preparations to be used in human trials [38]. It was recently clarified with a 0.2-μm filter pore size. Purification procedures of phages should follow the Critical Quality Attributes (CQA) specification [34]. The process of removing endotoxins from phages is complex because lipopolysaccharide forms micelles that have approximately the same size as phages. Therefore, extra purification methods such as ion exchange, affinity chromatography and solvent extraction are needed for lysates of phage-infected Gram-negative bacteria [33].

Endotoxin. In bacteriophage products, endotoxin measurement is critical. Gel clot, turbidimetric and chromatic methods are used for endotoxin determination in bacteriophage products. The *Limulus* amoebocyte lysate (LAL) assay is the most commonly used method [39].

Transmission electron microscopy (TEM). The specific morphology of phages in a final product can be viewed by transmission electron microscopy. Merabishvili et al. [40] used TEM for confirmation of the presence of the expected virion morphologic particles as well as their specific interaction with the target bacteria.

Titer. The process of determining phage concentration by dilution and plating with susceptible cells is called titering or the plaque assay. A bacteriophage capable of productively infecting a cell is named a plaque-forming unit (PFU/mL) [41].

pH. In a therapeutic formulation, the pH value is very important. According to the European Pharmacopoeia, the pH should be in the range 6.0–8.0 [42].

Nucleic Acid Contaminants. Because phages break down bacterial DNA, the presence and concentration of nucleic acid residues in final products should be determined. qPCR can be used for this purpose [33].

2.4. Phage Stability and Storage Conditions

Once solutions of phages are prepared, the biological properties of the phages have to be preserved during storage. Freeze-drying, spray-drying or encapsulation methods can be used to increase phage stability, as well as adding stabilizing additives to their solutions [43–46]. The quality, safety and storage conditions of phages to be prepared for use in treatment should be validated [47]. González-Menéndez et al. [48] investigated different preservation techniques for the storage of *Staphylococcus* phages (phiIPLA88, phiIPLA35, phiIPLA-RODI and phiIPLA-C1C). They evaluated the stability of phages at different temperatures (−20, −80 and −196 °C) and time periods (1, 6, 12 and 24 months). They also investigated various stabilization enhancing agents, including disaccharides, glycerol, sorbitol and skim milk. They showed that at −80 and −196 °C, all phages showed good viability after 24 months, regardless of the stabilizer [48].

2.5. Therapeutic Phages

Hyman et al. [17] proposed the following characteristics of phages to be used for therapeutic purposes: (a) The phage should be virulent and be able to cause complete cytotoxicity to the target bacterium. (b) It should be exclusively lytic and should not become temperate (i.e., lysogenic). (c) The phage should have the potential to transduce the host bacteria. (d) It should have the desired host range. (e) It should be screened for toxin genes that can affect the patient. *Myoviridae, Siphoviridae and Podoviridae* families are used commonly for phage therapy [49,50]. There are approximately 800 phages against pathogens such as *Escherichia, Morganella, Klebsiella, Enterococcus, Pseudomonas, Staphylococcus* and *Salmonella* [31].

2.6. Magistral Phage

A "magistral preparation" or a "compounded prescription drug product", in Europe and the U.S., respectively, is defined as "any medicinal product prepared in a pharmacy in accordance with a medical prescription for an individual patient" [33,34,48,51,52]. Such preparations for a particular patient are mixed by a pharmacist from their individual ingredients based on a prescription from a physician. The magistral formula is a practical way for a medical doctor to personalize patient treatments to specific needs and to make medications available that do not exist commercially. Some medicines, including natural hormone combinations, are made as magistral preparations. It is expected that magistral preparations will become more readily available as novel medicines are developed to treat rare conditions.

A magistral phage preparation is prepared from a phage bank, which is a repository of well-characterized microorganisms. A phage as an active pharmaceutical ingredient (API) is produced using a suitable bacterial host. An approved laboratory then carries out External Quality Assessments to test the API's properties and quality. Active phage APIs are evaluated for activity against the target. Finally, phage APIs are mixed with a suitable carrier system. There are currently no guidelines on the preparation, formulation and use of magistral phages [52].

2.7. Topical Administration of Phages

Several studies have shown that local and topical phage applications are successful. In the treatment of infections caused by Staphylococci, *Klebsiella, Pseudomonas, Proteus* and *Escherichia* such as conjunctivitis, otitis, gingivitis, furunculosis, decubitis ulcer, open wound infection, burns, osthitis (caused by fractures) and chronic suppurative fistulae, phage cocktails have been applied locally [53–56]. A commercial product called PhagoBioDerm, which targets *P. aeruginosa, S. aureus* and *Streptococcus* spp. and contains phages as well as cipro-floxacin, can be applied directly over infected wounds. Goode et al. [57] eliminated *Salmonella* contamination on chicken skin by using a lytic bacteriophage. Vieira et al. [58] performed phage therapy against multidrug-resistant *P. aeruginosa* that had caused skin infections. Thanks to phage therapy, the amount of *P. aeruginosa* 709 present in human skin decreased by four orders of magnitude.

2.8. Pulmonary Phage Delivery

The first studies of inhaled phage therapy were carried out in the early 1960s. Such treatments at a more advanced level were performed in Russia, Poland and Georgia. Although there have been many successful trials, some treatments have not had a positive outcome because of a lack of phage variety, quality control and technical knowledge [59]. Phages may be encapsulated in polymers, nanoparticles and liposomes for stability during storage, including storage as a freeze-dried preparation [60]. Liposome encapsulation was found to facilitate phage entry into macrophages. Treatment of experimental *K. pneumoniae*-induced lobar pneumonia was more effective with liposome-entrapped phages administered intraperitoneally as late as 3 days post-infection, whereas free phages provided a therapeutic effect only if they were administered at 1 day after infection [61]. Systemic side effects were reduced by the use of liposomal phage.

Liquid formulations using intranasal instillation and nebulization in phage studies against respiratory infections on animal models are quite popular. Liquid phage formulations are stable, easily aerosolizable and easy to formulate compared to other carrier systems [59]. Carrigy et al. [62] tested pre-exposure prophylactic aerosol delivery of the anti-tuberculosis bacteriophage D29 as an option for protection against *Mycobacterium tuberculosis* infection and proposed that mycobacteriophage aerosols at sufficient doses may be protective against *M. tuberculosis* infection. The same group studied the titer reduction and phage delivery rate of three inhalation devices (Vibrating Mesh Nebulizer, Jet Nebulizer and Soft Mist Inhaler) with the mycobacteriophage D29 and showed that this method of administration is suitable for phage delivery to lung tissue [63]. Golshahi et al. [64] determined that the inhaled formulation of bacteriophages gives successful results in the treatment of cystic fibrosis pulmonary infections.

2.9. Parenteral Phage Application

Phages are rapidly eliminated by the immune system when administered intravenously. Lin et al. [65] investigated the intravenous administration of the anti-pseudomonal phage øPEV20 in *P. aeruginosa*-infected rats and demonstrated dose-dependent pharmacodynamics. Intravenous administration of øPEV20 at a dose of $>10^4$ PFU/mouse resulted in rapid bacterial killing and >8-$\log_{10}$ CFU/mL reduction in bacterial load compared with the initial inoculum and untreated controls at 2.5 h. However, treatment at a dose of $<10^4$ øPEV20 PFU/mouse was ineffective against pan-drug-resistant *P. aeruginosa*. McVay et al. [66] injected a *P. aeruginosa* phage cocktail with three different administration methods (subcutaneous (s.c.), intramuscular (i.m.) and intraperitoneal (i.p.)) to *P. aeruginosa*-infected mice. Without treatment, the survival rate was 6%, and i.p. administration of phage resulted in the highest rate of survival (87%). According to the results of pharmacokinetic studies on phages, compared to other administration routes, phages reached the target at higher concentrations and faster when given via the i.p. route.

2.10. Oral Phage Therapy

Oral formulations of bacteriophages are generally used to target acute gastrointestinal infections. However, there are a number of factors for the treatment to be successful, including stability and effective phage dose at the site of infection. A significant decrease in phage titers occurs before the phages reach the site of infection. Phage viability and activity decrease as a result of gastric acidity and digestive enzymes such as pepsin and pancreatin. Thus, it is necessary to prepare new dosage forms. Vinner et al. [67] encapsulated enteric bacteriophage K1F against *E. coli* in a pH-responsive solid formulation and examined the viability of these bacteriophages at different pH values. They found that the microencapsulation process preserved phages for an extended period in the gastric acid environment. The encapsulated phages were active in killing *E. coli* co-incubated with human epithelial cells, which are normally stressed in the presence of the bacteria alone. There were no stability problems for the encapsulated phages that were refrigerated for 4 weeks. Stanford et al. [68] used polymer-encapsulated wV8, rV5, wV7 and wV11 phages, which are targeted to *E. coli*. Then, the phages were exposed to pH 3.2 for 20 min. The unencapsulated phages lost their activity while the encapsulated phages recovered 13.6% of their activity. Vinner et al. [69] prepared the encapsulated bacteriophage Felix O1, which is specific to *Salmonella*, by spray-drying, employing a commercially available pH-sensitive copolymer of methyl methacrylate and methacrylic acid. The inclusion of trehalose in the formulation protected the phages from the effects of spray-drying, maintaining the original phage titer. In a different approach, Colom et al. [70] encapsulated the phages UAB_Phi20, UAB_Phi78 and UAB_Ph87 individually in a complex mixture of lipids that produced a net positive charge on the ensuing liposomes. The encapsulation efficiencies were relatively high, in the range of 47–49%, which is most likely the result of phage binding to the net cationic lipid mixture. The encapsulated phages were more effective than plain phages in *S. enterica* ser. Typhimurium-infected chickens at only 8 days following treatment, with a 3.9 $\log_{10}$ reduction [70].

3. Mycobacteriophage Therapy of *Mycobacterium tuberculosis*

There are more than 170 *Mycobacterium* species that have great variety in terms of their pathogenicity in humans [71]. In addition to *M. tuberculosis*, *M. ulcerans* and *M. leprae* cause Buruli ulcer and leprosy, respectively [72]. *M. tuberculosis* is a well-known example of an intracellular bacterium that localizes inside phagosomes of macrophages of the host and causes tuberculosis (TB), which primarily affects the lungs [73]. Multi-drug-resistant (MDR) TB cases have emerged in the late 1980s and early 1990s. These strains are resistant to the first-line drugs against TB, rifampicin and isoniazid. In 2018, the World Health Organization (WHO) reported that 484,000 new TB patients failed to respond to rifampicin. Seventy-eight percent of these patients were infected with MDR-TB [74].

Alternative treatment approaches for MDR-TB have become crucial to managing the disease. One of these approaches is mycobacteriophage therapy. More than 70 years have passed since mycobacteriophages were isolated for the first time [75]. So far, 11,282 mycobacteriophages have been isolated [76].

Bacteriophages can enter macrophages by four main routes [77] (Figure 3): (1) Endocytotic uptake of the bacteriophage alone; (2) entry into the macrophage via pathogenic bacteria together with the bound phage; (3) uptake of the bacteriophage and non-pathogenic bacteria; (4) internalization of the bacteriophage that has been encapsulated by poly-mers or liposomes. Relatively non-pathogenic vectors, such as *M. smegmatis,* can be used to deliver phages to the same intracellular compartments where *M. tuberculosis* is found [78]. The lytic mycobacteriophage TM4 was delivered in this manner to *M. tuberculosis*-infected RAW264.7 macrophages and reduced the bacterial counts. By contrast, the phage alone was ineffective. The administration of the *M. smegmatis*-TM4 complex to *M. avium*-infected mice significantly decreased the bacterial counts in the spleen, whereas TM4 or *M. smegmatis* alone had no effect [79]. The authors suggested that phage resistance (which was observed in their study) could be overcome by the use of phage cocktails.

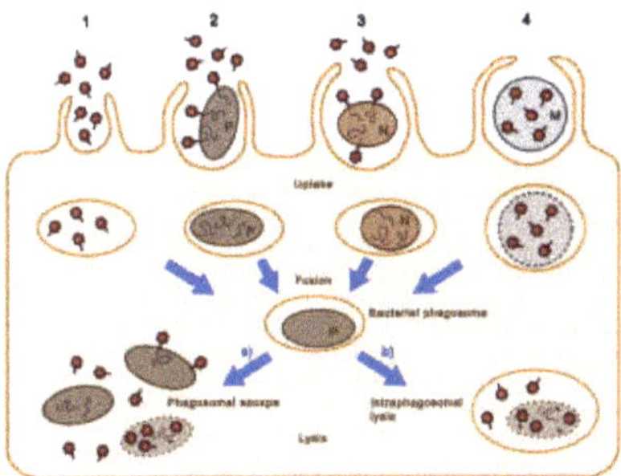

Figure 3. Possible cellular entry pathways of bacteriophages. The pathways 1–4 are described in the text. Dark red hexagons, bacteriophage; brown-filled ovals, bacteria (pathogenic bacteria, P; non-pathogenic bacteria, N); curly lines within bacteria, bacteriophage nucleic acids; orange ovals, endosomal vesicles; blue-gray circles and M, microcapsules; dotted lines, degrading bacterial membrane (reproduced with permission from Nieth et al., 2015 [77]).

The mycobacteriophage D29 was used to treat *M. tuberculosis* H37Rv inside RAW 264.7 macrophages [80]. The phage, administered twice over a 24-h period, caused an eight-fold reduction in the CFUs, indicating that it was able to access the intracellular compartment occupied by the bacteria. The phage was encapsulated in (or associated with) liposomes comprising phosphatidylcholine, cholesterol and Tween-80 and which were sized by extrusion through membranes of 400-nm diameter. This formulation applied to infected macrophages resulted in a two-fold improvement of the antimycobacterial effect over that of the free phage. In an in vitro model of tuberculous granuloma developed from peripheral blood mononuclear cells of patients with TB, liposomal phage was about nine-fold more effective than free D29 [81].

Aerosolized bacteriophage D29 was used to investigate the possibility of protecting mice against *M. tuberculosis* infection [62]. This treatment significantly decreased the *M. tuberculosis* counts in the lungs 1 day and 3 weeks after challenge. The authors suggested that aerosolized mycobacteriophages may be useful in conferring additional protection to healthcare workers who may be at risk of exposure to tuberculosis. D29 was also employed in a murine footpad model in treating Buruli ulcer, which is caused by *Mycobacterium ulcerans* [82]. In infected patients, the bacterium causes necrosis of the skin, subcutaneous tissue as well as bone. If the disease reaches advanced stages, surgical resection of the skin may be necessary. The subcutaneous injection of D29 resulted in a decrease in pathology and mycobacterial counts. It also caused increased production of cytokines, including IFN-γ, in the footpads and draining lymph nodes. Endolysins are bacteriophage-encoded peptidoglycan-disrupting enzymes synthesized at the last stage of the phage life cycle in the infected bacteria [83]. One endolysin, lysine B, was found to lyse *M. ulcerans* infecting the footpad of experimental mice [84].

Developing mycobacteriophages into efficient therapeutic pharmaceuticals has focused on improving their uptake into macrophages and co-localization with the intracellular mycobacteria. In this process, however, it is essential to maintain the stability of the formulation and the vitality of the mycobacteriophages. In the next step, well-established in vitro and in vivo studies of effective and stable mycobacteriophage formulations are expected to translate into clinical studies with successful outcomes.

4. Bacteriophage Therapy of *Pseudomonas aeruginosa*

P. aeruginosa are Gram-negative aerobic bacteria classified as Gammaproteobacteria that can cause severe necrotizing bronchopneumonia, burn wound infections, urinary tract infections, otitis externa, eye infections and bacteremia [85].

In a murine model of sepsis caused by *P. aeruginosa* via the gut, the lytic phage KPP10 administered orally increased the survival rate from 0% in the controls to 67% [86]. The number of viable bacteria in the liver, spleen and blood were reduced in the phage-treated group, as were the levels of inflammatory cytokines in the liver and blood. Imipenem-resistant *P. aeruginosa* delivered i.p. resulted in bacteremia and killed 100% of experimental mice within 24 h [87]; the i.p. administration of the phage ØA392 within 1 h of infection was able to rescue all the animals. The phages were found in blood within 2 h. However, delivery of the phage at 3 h post-infection resulted in only 50% survival. In a murine burn-wound model, fatal infection by *P. aeruginosa* could be reduced to 87% survival when a three-phage cocktail was given i.p. [67]. The phages rapidly distributed to the blood, liver and spleen. In a similar study, i.p. delivery of multi-drug-resistant *P. aeruginosa* caused fatal bacteremia in mice within 2 d [88]. A phage strain that had lytic activity against numerous multi-drug-resistant *P. aeruginosa* given 45 min after bacterial infection resulted in 100% survival. Fifty percent of the animals could be saved even when the therapy was applied at a point where the animals were sick. The therapeutic effect of the phage was also shown not to be the result of a non-specific immune response.

When mice with acute lung infection with intranasally administered bioluminescent *P. aeruginosa*, which resulted in the death of all the animals within 2 days, were treated with bacteriophage PAK-P1-to-bacterium ratios of 1:1 and 10:1 via the same route, they survived until the end of the 12-day experiment [89]. Bacteriophage treatment also prevented lung infection when administered 24 h before inoculation of bacteria. Two phages were isolated from wastewater, the myovirus φNH-4 and the podovirus φMR299-2, and used to treat *P. aeruginosa* infection in murine lungs [90]. The pathogen was reduced by three to four orders of magnitude in 6 h. A mixture of the two phages could kill biofilms of mucoid and nonmucoid strains of *P. aeruginosa* on CFBE41o-cystic fibrosis bronchial epithelial cells, and the phages were shown to multiply over 24 h.

Phage GNCP treatment of multi-drug-resistant *P. aeruginosa* infection in diabetic and non-diabetic mice, which caused fatal bacteremia within 2 d, at a 10:1 ratio of phage:bacteria resulted in protection of 90% of diabetic animals and 100% of non-diabetic animals [91]. Bacteriophages were also effective in reducing inflammation in a murine acute infection model of *P. aeruginosa* [92]. The titer of phage PEV31 delivered intratracheally to mice without bacterial infection decreased with a $t_{1/2}$ of about 8 h. In mice infected with *P. aeruginosa*, the phage titer increased by about two orders of magnitude in 16 h, and bacterial growth was suppressed, whereas it increased exponentially in the untreated animals [93].

5. Clinical Cases Treated with Bacteriophages

Lung transplant recipients with life-threatening multi-drug-resistant *P. aeruginosa* or *Burkholderia dolosa* infections were treated with lytic bacteriophages targeting the bacterial strains, together with antibiotics [94]. Two patients with *P. aeruginosa* infection responded to the treatment and could leave the hospital. The patient with recurrent *B. dolosa* infection did not respond to bacteriophage therapy. The safety and feasibility of phage treatment of patients with various infections at a single center in the U.S. was established, although two of the 10 patients described did not respond to therapy [95].

P. aeruginosa can infect prosthetic vascular grafts that often do not respond to antibiotic therapy [96]. The bacteriophage OMKO1, together with ceftazidime, was used to treat infection of an aortic graft, which was resolved and did not recur.

Bacteriophage therapy was applied to chronic non-healing wounds that were infected with *E. coli*, *S. aureus* and *P. aeruginosa* and that did not respond to antibiotic therapy [97]. The application of a cocktail of bacteriophages over the wounds every other day resulted in the resolution of the infection after 3–5 doses. The wounds healed completely in seven out of 20 patients and formed healthy granulation tissue and margins in the other patients.

In a trial that included 48 patients with non-healing wounds, a single phage against a particular bacterial infection or multiple phages targeting multiple bacteria were applied every other day for 5 to 7 days [98]. The cure rate was 81%, with diabetic patients having a lower rate (74%).

A 65-year-old woman with a post-operative left-eye corneal abscess and interstitial keratitis was treated for many years with various antibiotics but remained positive for vancomycin-intermediate sensitivity *S. aureus* in the nasal cavity, skin and eye [99]. She then underwent topical and intravenous phage therapy with the bacteriophage SATA-8505. This phage strain is active against the methicillin-resistant *S. aureus* strain USA300 and has been patented. Ocular and nasal cultures from the patient 3 and 6 months after therapy showed no infection.

6. Conclusions

The importance of phage therapy for bacterial infections has been recognized by both academic institutions and the pharmaceutical industry. The Center for Phage Applications and Therapeutics at the University of California San Diego, the Center for Phage Technology at Texas A&M University at College Station and the Pittsburgh Bacteriophage Institute at the University of Pittsburgh are examples of academic institutions. Companies focusing on phage therapy include the Eliava Institute and affiliated companies in Tblisi, InnoPhage in Porto, Adaptive Phage Therapeutics in Gaithersburg, Intralytics in Columbia, Maryland, and Armata Pharmaceuticals in Marina Del Ray. Thus, it appears that phage therapy will be widely available, next to newly developed antibiotics, to teat multi-drug-resistant infections.

Although small-scale studies have demonstrated the potential of phage therapy for bacterial infections, especially in cases of severe antibiotic resistance, the widespread applicability of this therapy has not been shown in clinical trials [100]. In a clinical trial involving patients with urinary tract infections, phages administered directly into the bladder were no more effective than placebo or antibiotics [101]. Burn-wound infections with *P. aeruginosa* were treated with either the phage PP1131 or 1% sulfadiazine silver emulsion cream, the standard of care, in a multi-center clinical trial. Phage treatment at the relatively low dose of 10^6 plaque-forming units per mL was not as effective as the standard of care [102].

To supplement phage therapy, it may be possible to utilize antibiotics and phages simultaneously in some circumstances. In an in vitro study of *P. aeruginosa* and *S. aureus* biofilms, either alone or in combination, the phage EPA1 that infects *P. aeruginosa* and different antibiotics, the simultaneous application of the two agents drastically increased the cytotoxicity against the bacteria [103]. The addition of gentamicin or ciprofloxacin after a 6-h treatment with the phage appeared to eradicate the bacterial biofilms, with higher gentamicin concentrations being necessary for treating combined biofilms.

Clinical trials of bacteriophage therapy of bacterial infections are still at an early stage. Optimal conditions of phage use, including their concentration, the time and sequence of administration and their combination with the appropriate antibiotics, are likely to establish the effectiveness and reliability of this medicine. Even until such standards are established, their ability to save the patient described in the Introduction is a most welcome addition to the practice of medicine.

Funding: This research received no external funding.

Conflicts of Interest: The authors declare no conflict of interest.

References

1. Schooley, R.T.; Biswas, B.; Gill, J.J.; Hernandez-Morales, A.; Lancaster, J.; Lessor, L.; Barr, J.J.; Reed, S.L.; Rohwer, F.; Benler, S.; et al. Development and use of personalized bacteriophage-based therapeutic cocktails to treat a patient with a disseminated resistant *Acinetobacter baumannii* infection. *Antimicrob. Agents Chemother.* **2017**, *61*, e00954-17. [CrossRef]
2. Stone, E.; Campbell, K.; Grant, I.; McAuliffe, O. Understanding and exploiting phage-host interactions. *Viruses* **2019**, *11*, 567. [CrossRef] [PubMed]
3. Dunne, M.; Hupfeld, M.; Klumpp, J.; Loessner, M.J. Molecular basis of bacterial host interactions by gram-positive targeting bacteriophages. *Viruses* **2018**, *10*, 397. [CrossRef] [PubMed]
4. Kortright, K.E.; Chan, B.K.; Koff, J.L.; Turner, P.E. Phage therapy: A renewed approach to combat antibiotic-resistant bacteria. *Cell Host Microbe* **2019**, *25*, 219–232. [CrossRef] [PubMed]
5. Samsygina, G.A.; Boni, E.G. Bacteriophages and phage therapy in pediatric practice. *Pediatriia* **1984**, *4*, 67–70.
6. Twort, F.W. An investigation on the nature of ultra-microscopic viruses. *Lancet* **1915**, *186*, 1241–1243. [CrossRef]
7. D'Herelle, F. Sur un microbe invisible antagoniste des bacilles dysenteriques. *C. R. Acad. Sci.* **1917**, *165*, 373–375.
8. Summers, W.C. *Felix d'Herelle and the Origins of Molecular Biology*; Yale University Press: New Haven, CT, USA, 1999.
9. Sulakvelidze, A.; Alavidze, Z.; Morris, J.G. Bacteriophage therapy. *Antimicrob. Agents Chemother.* **2001**, *45*, 649–659. [CrossRef]
10. Bruynoghe, R.A.J.M.; Maisin, J. Essais de thérapeutique au moyen du bacteriophage. *C. R. Soc. Biol.* **1921**, *85*, 1120–1121.
11. Abedon, S.T.; Kuhl, S.J.; Blasdel, B.G.; Kutter, E.M. Phage treatment of human infections. *Bacteriophage* **2011**, *1*, 66–85. [CrossRef]
12. Dubos, R.J.; Straus, J.H.; Pierce, C. The multiplication of bacteriophage in vivo and its protective effect against an experimental infection with *Shigella dysenteriae*. *J. Exp. Med.* **1943**, *78*, 161. [CrossRef] [PubMed]
13. Yang, M.; Liang, Y.; Huang, S.; Zhang, J.; Wang, J.; Chen, H.; Ye, Y.; Gao, X.; Wu, Q.; Tan, Z. Isolation and characterization of the novel phages vB_VpS_BA3 and vB_VpS_CA8 for lysing *Vibrio parahaemolyticus*. *Front. Microbiol.* **2020**, *11*, 259. [CrossRef] [PubMed]
14. Gao, M.; Wang, C.; Qiang, X.; Liu, H.; Li, P.; Pei, G.; Zhang, X.; Mi, Z.; Huang, Y.; Tong, Y.; et al. Isolation and characterization of a novel bacteriophage infecting carbapenem-resistant Klebsiella pneumoniae. *Curr. Microbiol.* **2020**, *77*, 722–729. [CrossRef]
15. Li, M.; Guo, M.; Chen, L.; Zhu, C.; Xiao, Y.; Li, P.; Guo, H.; Chen, L.; Zhang, W.; Du, H. Isolation and characterization of novel lytic bacteriophages infecting epidemic carbapenem-resistant *Klebsiella pneumoniae* strains. *Front. Microbiol.* **2020**, *11*, 1554. [CrossRef] [PubMed]
16. Van Twest, R.; Kropinski, A.M. Bacteriophage enrichment from water and soil. In *Bacteriophages*; Clokie, M.R., Kropinski, A.M., Eds.; Humana Press: Totowa, NJ, USA, 2009; Volume 501, pp. 15–21.
17. Hyman, P. Phages for phage therapy: Isolation, characterization, and host range breadth. *Pharmaceuticals* **2019**, *12*, 35. [CrossRef]
18. Tylenda, C.A.; Calvert, C.; Kolenbrander, P.E.; Tylenda, A. Isolation of Actinomyces bacteriophage from human dental plaque. *Infect. Immun.* **1985**, *49*, 1–6. [CrossRef]
19. Luong, T.; Salabarria, A.C.; Edwards, R.A.; Roach, D.R. Standardized bacteriophage purification for personalized phage therapy. *Nat. Protoc.* **2020**, *15*, 2867–2890. [CrossRef] [PubMed]
20. Mattila, S.; Ruotsalainen, P.; Jalasvuori, M. On-demand isolation of bacteriophages against drug-resistant bacteria for personalized phage therapy. *Front. Microbiol.* **2015**, *6*, 1271. [CrossRef]
21. Hatfull, G.F. The secret lives of mycobacteriophages. *Adv. Virus Res.* **2012**, *82*, 179–288. [PubMed]
22. Swanstrom, M.; Adams, M.H. Agar layer method for production of high titer phage stocks. *Proc. Soc. Exp. Biol. Med.* **1951**, *78*, 372–375. [CrossRef]
23. Echeverría-Vega, A.; Morales-Vicencio, P.; Saez-Saavedra, C.; Gordillo-Fuenzalida, F.; Araya, R. A rapid and simple protocol for the isolation of bacteriophages from coastal organisms. *MethodsX* **2019**, *6*, 2614–2619. [CrossRef] [PubMed]
24. Gill, J.J.; Hyman, P. Phage choice, isolation, and preparation for phage therapy. *Curr. Pharm. Biotechnol.* **2010**, *11*, 2–14. [CrossRef] [PubMed]

25. Schmidt, J.M.; Stanier, R.Y. Isolation and characterization of bacteriophages active against stalked bacteria. *Microbiology* **1965**, *39*, 95–107. [CrossRef] [PubMed]
26. Armon, R.; Kott, Y. A simple, rapid and sensitive presence/absence detection test for bacteriophage in drinking water. *J. Appl. Bacteriol.* **1993**, *74*, 490–496. [CrossRef]
27. Mirzaei, M.K.; Nilsson, A.S. Isolation of phages for phage therapy: A comparison of spot tests and efficiency of plating analyses for determination of host range and efficacy. *PLoS ONE* **2015**, *10*, e0118557. [CrossRef]
28. Pallavali, R.R.; Degati, V.L.; Lomada, D.; Reddy, M.C.; Durbaka, V.R.P. Isolation and in vitro evaluation of bacteriophages against MDR-bacterial isolates from septic wound infections. *PLoS ONE* **2017**, *12*, e0179245. [CrossRef]
29. Cerca, N.; Oliveria, R.; Azeredo, J. Susceptibility of Staphylococcus epidermidis planktonic cells and biofilms to the lytic action of staphylococcus bacteriophage *K. Lett. Appl. Microbiol.* **2007**, *45*, 313–317. [CrossRef]
30. Maskow, T.; Kiesel, B.; Schubert, T.; Yong, Z.; Harms, H.; Yao, J. Calorimetric real time monitoring of lambda prophage induction. *J. Virol. Methods* **2010**, *168*, 126–132. [CrossRef]
31. Weber-Dąbrowska, B.; Jończyk-Matysiak, E.; Żaczek, M.; Łobocka, M.; Łusiak-Szelachowska, M.; Górski, A. Bacteriophage procurement for therapeutic purposes. *Front. Microbiol.* **2016**, *7*, 1177. [CrossRef]
32. García, R.; Latz, S.; Romero, J.; Higuera, G.; García, K.; Bastías, R. Bacteriophage Production Models: An Overview. *Front. Microbiol.* **2019**, *10*, 1187. [CrossRef]
33. Mutti, M.; Corsini, L. Robust approaches for the production of active ingredient and drug product for human phage therapy. *Front. Microbiol.* **2019**, *10*, 2289. [CrossRef] [PubMed]
34. Regulski, K.; Champion-Arnaud, P.; Gabard, J. Bacteriophage manufacturing: From early twentieth-century processes to current GMP. In *Bacteriophages*; Harper, D.R., Abedon, S.T., Burrowes, B.J., McConville, M.L., Eds.; Springer: Cham, Switzerland, 2018; pp. 1–31.
35. Jurač, K.; Nabergoj, D.; Podgornik, A. Bacteriophage production processes. *Appl. Microbiol. Biotechnol.* **2019**, *103*, 685–694. [CrossRef] [PubMed]
36. Merabishvili, M.; Pirnay, J.P.; Vogele, K.; Malik, D.J. Production of phage therapeutics and formulations: Innovative approaches. In *Phage Therapy: A Practical Approach*; Górski, A., Międzybrodzki, R., Borysowski, J., Eds.; Springer: Cham, Switzerland, 2019; pp. 3–41.
37. Mancuso, F.; Shi, J.; Malik, D.J. High throughput manufacturing of bacteriophages using continuous stirred tank bioreactors connected in series to ensure optimum host bacteria physiology for phage production. *Viruses* **2018**, *10*, 537. [CrossRef] [PubMed]
38. De Vicente Jordana, R. Study on adsorption of bacteriophage by filters. *Appl. Microbiol.* **1959**, *7*, 239–247. [CrossRef] [PubMed]
39. Szermer-Olearnik, B.; Boratyński, J. Removal of endotoxins from bacteriophage preparations by extraction with organic solvents. *PLoS ONE* **2015**, *10*, e0122672. [CrossRef]
40. Merabishvili, M.; Pirnay, J.P.; Verbeken, G.; Chanishvili, N.; Tediashvili, M.; Lashkhi, N.; Glonti, T.; Krylov, V.; Mast, J.; Van Parys, L.; et al. Quality-controlled small-scale production of a well-defined bacteriophage cocktail for use in human clinical trials. *PLoS ONE* **2009**, *4*, e4944. [CrossRef]
41. Stephenson, F.H. *Calculations for Molecular Biology and Biotechnology*, 3rd ed.; Academic Press: Cambridge, MA, USA, 2016.
42. Council of Europe. *European Pharmacopoeia*, 4th ed; Council of Europe: Strasbourg, France, 2002.
43. Manohar, P.; Ramesh, N. Improved lyophilization conditions for long-term storage of bacteriophages. *Sci. Rep.* **2019**, *9*, 15242. [CrossRef]
44. Leung, S.S.Y.; Parumasivam, T.; Gao, F.G.; Carter, E.A.; Carrigy, N.; Vehring, R.; Finlay, W.H.; Morales, S.; Britton, W.J.; Kutter, E.; et al. Effect of storage conditions on the stability of spray dried, inhalable bacteriophage powders. *Int. J. Pharm.* **2017**, *521*, 141–149. [CrossRef]
45. Dini, C.; Islan, G.A.; de Urraza, P.J.; Castro, G.R. Novel biopolymer matrices for microencapsulation of phages: Enhanced protection against acidity and protease activity. *Macromol. Biosci.* **2012**, *12*, 1200–1208. [CrossRef]
46. Merabishvili, M.; Vervaet, C.; Pirnay, J.-P.; De Vos, D.; Verbeken, G.; Mast, J.; Chanishvili, N.; Vaneechoutte, M. Stability of Staphylococcus aureus phage ISP after freeze-drying (lyophilization). *PLoS ONE* **2013**, *8*, 17559. [CrossRef]
47. Pirnay, J.P.; Blasdel, B.G.; Bretaudeau, L.; Buckling, A.; Chanishvili, N.; Clark, J.R.; Corte-Real, S.; Debarbieux, L.; Dublanchet, A.; De Vos, D.; et al. Quality and safety requirements for sustainable phage therapy products. *Pharm. Res.* **2015**, *32*, 2173–2179. [CrossRef] [PubMed]
48. Gonzalez-Menendez, E.; Fernandez, L.; Gutierrez, D.; Rodriguez, A.; Martinez, B.; Garcia, P. Comparative analysis of different preservation techniques for the storage of *Staphylococcus* phages aimed for the industrial development of phage-based antimicrobial products. *PLoS ONE* **2018**, *13*, e0205728. [CrossRef] [PubMed]
49. Kornienko, M.; Kuptsov, N.; Gorodnichev, R.; Bespiatykh, D.; Guliaev, A.; Letarova, M.; Kulikov, E.; Veselovsky, V.; Malakhova, M.; Ilina, E.; et al. Contribution of Podoviridae and Myoviridae bacteriophages to the effectiveness of anti-staphylococcal therapeutic cocktails. *Sci. Rep.* **2020**, *10*, 18612. [CrossRef] [PubMed]
50. Chen, Y.; Sun, E.; Song, J.; Tong, Y.; Wu, B. Three Salmonella enterica serovar Enteritidis bacteriophages from the Siphoviridae family are promising candidates for phage therapy. *Can. J. Microbiol.* **2018**, *6*, 865–875. [CrossRef] [PubMed]
51. Drulis-Kawa, Z.; Majkowska-Skrobek, G.; Maciejewska, B. Bacteriophages and phage-derived proteins—Application approaches. *Curr. Med. Chem.* **2015**, *22*, 1757–1773. [CrossRef]

52. Pirnay, J.P.; Verbeken, G.; Ceyssens, P.J.; Huys, I.; De Vos, D.; Ameloot, C.; Fauconnier, A. The magistral phage. *Viruses* **2018**, *10*, 64. [CrossRef]
53. Chang, R.Y.K.; Morales, S.; Okamoto, Y.; Chan, H.K. Topical application of bacteriophages for treatment of wound infections. *Trans. Res.* **2020**, *220*, 153–166. [CrossRef]
54. Hawkins, C.; Harper, D.; Burch, D.; Änggård, E.; Soothill, J. Topical treatment of Pseudomonas aeruginosa otitis of dogs with a bacteriophage mixture: A before/after clinical trial. *Vet. Microbiol.* **2010**, *146*, 309–313. [CrossRef]
55. Chadha, P.; Katare, O.P.; Chibber, S. In vivo efficacy of single phage versus phage cocktail in resolving burn wound infection in BALB/c mice. *Microb. Pathog.* **2016**, *99*, 68–77. [CrossRef]
56. Morozova, V.V.; Vlassov, V.V.; Tikunova, N.V. Applications of bacteriophages in the treatment of localized infections in humans. *Front. Microbiol.* **2018**, *9*, 1696. [CrossRef]
57. Goode, D.; Allen, V.M.; Barrow, P.A. Reduction of experimental *Salmonella* and *Campylobacter* contamination of chicken skin by application of lytic bacteriophages. *Appl. Environ. Microbiol.* **2003**, *69*, 5032–5036. [CrossRef] [PubMed]
58. Vieira, A.; Silva, Y.J.; Cunha, A.; Gomes, N.C.M.; Ackermann, H.W.; Almeida, A. Phage therapy to control multidrug-resistant *Pseudomonas aeruginosa* skin infections: In vitro and ex vivo experiments. *Eur. J. Clin. Microbiol. Infect. Dis.* **2012**, *31*, 3241–3249. [CrossRef] [PubMed]
59. Chang, R.Y.K.; Wallin, M.; Lin, Y.; Leung, S.S.Y.; Wang, H.; Morales, S.; Chan, H.K. Phage therapy for respiratory infections. *Adv. Drug Deliv. Rev.* **2018**, *133*, 76–86. [CrossRef] [PubMed]
60. Malik, D.J.; Sokolov, I.J.; Vinner, G.K.; Mancuso, F.; Cinquerrui, S.; Vladisavljevic, G.T.; Clokie, M.R.J.; Stapley, A.G.F.; Kirpichnikova, A. Formulation, stabilisation and encapsulation of bacteriophage for phage therapy. *Adv. Colloid Interface Sci.* **2017**, *249*, 100–133. [CrossRef] [PubMed]
61. Singla, S.; Harjai, K.; Katare, O.P.; Chhibber, S. Bacteriophage-loaded nanostructured lipid carrier: Improved pharmacokinetics mediates effective resolution of *Klebsiella pneumoniae*–induced lobar pneumonia. *J. Infect. Dis.* **2015**, *212*, 325–334. [CrossRef]
62. Carrigy, N.B.; Larsen, S.E.; Reese, V.; Pecor, T.; Harrison, M.; Kuehl, P.J.; Hatfull, G.F.; Sauvageau, D.; Baldwin, S.L.; Finlay, W.H.; et al. Prophylaxis of *Mycobacterium tuberculosis* H37Rv infection in a preclinical mouse model via inhalation of nebulized bacteriophage D29. *Antimicrob. Agents Chemother.* **2019**, *63*, e00871-19. [CrossRef]
63. Carrigy, N.B.; Chang, R.Y.; Leung, S.S.; Harrison, M.; Petrova, Z.; Pope, W.H.; Hatfull, G.F.; Britton, W.J.; Chan, H.K.; Sauvageau, D.; et al. Anti-tuberculosis bacteriophage D29 delivery with a vibrating mesh nebulizer, jet nebulizer, and soft mist inhaler. *Pharm. Res.* **2017**, *34*, 2084–2096. [CrossRef]
64. Golshahi, L.; Lynch, K.H.; Dennis, J.J.; Finlay, W.H. In vitro lung delivery of bacteriophages KS4-M and ΦKZ using dry powder inhalers for treatment of *Burkholderia cepacia* complex and *Pseudomonas aeruginosa* infections in cystic fibrosis. *J. Appl. Microbiol.* **2011**, *110*, 106–117. [CrossRef]
65. Lin, Y.W.; Chang, R.Y.; Rao, G.G.; Jermain, B.; Han, M.L.; Zhao, J.X.; Chen, K.; Wang, J.P.; Barr, J.J.; Schooley, R.T.; et al. Pharmacokinetics/pharmacodynamics of antipseudomonal bacteriophage therapy in rats: A proof-of-concept study. *Clin. Microbiol. Infect.* **2020**. [CrossRef]
66. McVay, C.S.; Velásquez, M.; Fralick, J.A. Phage therapy of Pseudomonas aeruginosa infection in a mouse burn wound model. *Antimicrob. Agents Chemother.* **2007**, *51*, 1934–1938. [CrossRef]
67. Vinner, G.K.; Richards, K.; Leppanen, M.; Sagona, A.P.; Malik, D.J. Microencapsulation of enteric bacteriophages in a pH-responsive solid oral dosage formulation using a scalable membrane emulsification process. *Pharmaceutics* **2019**, *11*, 475. [CrossRef] [PubMed]
68. Stanford, K.; McAllister, T.A.; Niu, Y.D.; Stephens, T.P.; Mazzocco, A.; Waddell, T.E.; Johnson, R.P. Oral delivery systems for encapsulated bacteriophages targeted at *Escherichia coli* O157: H7 in feedlot cattle. *J. Food Prot.* **2010**, *73*, 1304–1312. [CrossRef] [PubMed]
69. Vinner, G.K.; Rezaie-Yazdi, Z.; Leppanen, M.; Stapley, A.G.; Leaper, M.C.; Malik, D.J. Microencapsulation of salmonella-specific bacteriophage Felix O1 using spray-drying in a pH-responsive formulation and direct compression tableting of powders into a solid oral dosage form. *Pharmaceuticals* **2019**, *12*, 43. [CrossRef] [PubMed]
70. Colom, J.; Cano-Sarabia, M.; Otero, J.; Cortés, P.; Maspoch, D.; Llagostera, M. Liposome-encapsulated bacteriophages for enhanced oral phage therapy against Salmonella spp. *Appl. Environ. Microbiol.* **2015**, *81*, 4841–4849. [CrossRef]
71. Forbes, B.A. Mycobacterial taxonomy. *J. Clin. Microbiol.* **2017**, *55*, 380–383. [CrossRef]
72. Heifets, L. Mycobacterial infections caused by nontuberculous mycobacteria. *Semin. Respir. Crit. Care Med.* **2004**, *25*, 283–295. [CrossRef]
73. Russell, D.G. *Mycobacterium tuberculosis*: Here today, and here tomorrow. *Nat. Rev. Mol. Cell Biol.* **2001**, *2*, 569–578. [CrossRef]
74. Tuberculosis (TB). Available online: https://www.who.int/news-room/fact-sheets/detail/tuberculosis (accessed on 11 October 2019).
75. McNerney, R. TB: The return of the phage. A review of fifty years of mycobacteriophage research. *Int. J. Tuberc. Lung Dis.* **1999**, *3*, 179–184.
76. Host Genus: Mycobacterium. Available online: https://phagesdb.org/hosts/genera/1/ (accessed on 9 September 2020).
77. Nieth, A.; Verseux, C.; Römer, W. A question of attire: Dressing up bacteriophage therapy for the battle against antibiotic-resistant intracellular bacteria. *Springer Sci. Rev.* **2015**, *3*, 1–11. [CrossRef]

78. Broxmeyer, L.; Sosnowska, D.; Miltner, E.; Chacón, O.; Wagner, D.; McGarvey, J.; Barletta, R.G.; Bermudez, L.E. Killing of *Mycobacterium avium* and *Mycobacterium tuberculosis* by a mycobacteriophage delivered by a nonvirulent mycobacterium: A model for phage therapy of intracellular bacterial pathogens. *J. Infect. Dis.* **2002**, *186*, 1155–1160. [CrossRef]
79. Danelishvili, L.; Young, L.S.; Bermudez, L.E. In vivo efficacy of phage therapy for *Mycobacterium avium* infection as delivered by a nonvirulent mycobacterium. *Microb. Drug Resist.* **2006**, *12*, 1–6. [CrossRef] [PubMed]
80. Lapenkova, M.B.; Smirnova, N.S.; Rutkevich, P.N.; Vladimirsky, M.A. Evaluation of the efficiency of lytic Mycobacteriophage D29 on the model of M. tuberculosis-infected macrophage RAW 264 cell line. *Bull. Exp. Biol. Med.* **2018**, *164*, 344–346. [CrossRef] [PubMed]
81. Lapenkova, M.B.; Alyapkina, Y.S.; Vladimirsky, M.A. Bactericidal activity of liposomal form of lytic Mycobacteriophage D29 in cell models of tuberculosis infection in vitro. *Bull. Exp. Biol. Med.* **2020**, *169*, 361–364. [CrossRef] [PubMed]
82. Trigo, G.; Martins, T.G.; Fraga, A.G.; Longatto-Filho, A.; Castro, A.G.; Azeredo, J.; Pedrosa, J. Phage therapy is effective against infection by *Mycobacterium ulcerans* in a murine footpad model. *PLoS Negl. Trop. Dis.* **2013**, *7*, e2183. [CrossRef] [PubMed]
83. Loessner, M.J. Bacteriophage endolysins—Current state of research and applications. *Curr. Opin. Microbiol.* **2005**, *8*, 480–487. [CrossRef] [PubMed]
84. Fraga, A.G.; Trigo, G.; Murthy, R.K.; Akhtar, S.; Hebbur, M.; Pacheco, A.R.; Dominguez, J.; Silva-Gomes, R.; Gonçalves, C.M.; Oliveira, H.; et al. Antimicrobial activity of Mycobacteriophage D29 Lysin B during *Mycobacterium ulcerans* infection. *PLoS Negl. Trop. Dis.* **2019**, *13*, e0007113. [CrossRef]
85. Düzgüneş, N. *Medical Microbiology and Immunology for Dentistry*, 1st ed.; Quintessence Publishing Company: Hanover Park, IL, USA, 2016; p. xi+290.
86. Watanabe, R.; Matsumoto, T.; Sano, G.; Ishii, Y.; Tateda, K.; Sumiyama, Y.; Uchiyama, J.; Sakurai, S.; Matsuzaki, S.; Imai, S.; et al. Efficacy of bacteriophage therapy against gut-derived sepsis caused by *Pseudomonas aeruginosa* in mice. *Antimicrob. Agents Chemother.* **2007**, *51*, 446–452. [CrossRef]
87. Wang, J.; Hu, B.; Xu, M.; Yan, Q.; Liu, S.; Zhu, X.; Sun, Z.; Reed, E.; Ding, L.; Gong, J.; et al. Use of bacteriophage in the treatment of experimental animal bacteremia from imipenem-resistant *Pseudomonas aeruginosa*. *Int. J. Mol. Med.* **2006**, *17*, 309–317. [CrossRef]
88. Vinodkumar, C.S.; Kalsurmath, S.; Neelagund, Y.F. Utility of lytic bacteriophage in the treatment of multidrug-resistant *Pseudomonas aeruginosa* septicemia in mice. *Indian J. Pathol. Microbiol.* **2008**, *51*, 360–366. [CrossRef]
89. Debarbieux, L.; Leduc, D.; Maura, D.; Morello, E.; Criscuolo, A.; Grossi, O.; Balloy, V.; Touqui, L. Bacteriophages can treat and prevent *Pseudomonas aeruginosa* lung infections. *J. Infect. Dis.* **2010**, *201*, 1096–1104. [CrossRef] [PubMed]
90. Alemayehu, D.; Casey, P.G.; McAuliffe, O.; Guinane, C.M.; Martin, J.G.; Shanahan, F.; Coffey, A.; Ross, R.P.; Hill, C. Bacteriophages φMR299-2 and φNH-4 can eliminate *Pseudomonas aeruginosa* in the murine lung and on cystic fibrosis lung airway cells. *mBio* **2012**, *3*, e00029-12. [CrossRef] [PubMed]
91. Shivshetty, N.; Hosamani, R.; Ahmed, L.; Oli, A.K.; Sannauallah, S.; Sharanbassappa, S.; Patil, S.A.; Kelmani, C.R. Experimental protection of diabetic mice against lethal *P. aeruginosa* infection by bacteriophage. *Biomed. Res. Int.* **2014**, *2014*, 793242. [CrossRef] [PubMed]
92. Pabary, R.; Singh, C.; Morales, S.; Bush, A.; Alshafi, K.; Bilton, D.; Alton, E.W.; Smithyman, A.; Davies, J.C. Antipseudomonal bacteriophage reduces infective burden and inflammatory response in murine lung. *Antimicrob. Agents Chemother.* **2015**, *60*, 744–751. [CrossRef] [PubMed]
93. Chow, M.Y.T.; Chang, R.Y.K.; Li, M.; Wang, Y.; Lin, Y.; Morales, S.; McLachlan, A.J.; Kutter, E.; Li, J.; Chan, H.K. Pharmacokinetics and time-kill of inhaled antipseudomonal bacteriophage therapy in mice. *Antimicrob. Agents Chemother.* **2020**. [CrossRef]
94. Law, N.; Yung, G.; Turowski, J.; Anesi, J.; Strathdee, S.A.; Schooley, R.T. Early clinical experience of bacteriophage therapy in 3 lung transplant recipients. *Am. J. Transplant.* **2019**, *19*, 2631–2639.
95. Aslam, S.; Lampley, E.; Wooten, D.; Karris, M.; Benson, C.; Strathdee, S.; Schooley, R.T. Lessons learned from the first 10 consecutive cases of intravenous bacteriophage therapy to treat multidrug-resistant bacterial infections at a single center in the United States. *Open Forum Infect. Dis.* **2020**, *7*, ofaa389. [CrossRef]
96. Chan, B.K.; Turner, P.E.; Kim, S.; Mojibian, H.R.; Elefteriades, J.A.; Narayan, D. Phage treatment of an aortic graft infected with *Pseudomonas aeruginosa*. *Evol. Med. Public Health* **2018**, *2018*, 60–66. [CrossRef]
97. Gupta, P.; Singh, H.S.; Shukla, V.K.; Nath, G.; Bhartiya, S.K. Bacteriophage therapy of chronic nonhealing wound: Clinical study. *Int. J. Low. Extrem. Wounds* **2019**, *18*, 171–175. [CrossRef]
98. Patel, D.R.; Bhartiya, S.K.; Kumar, R.; Shukla, V.K.; Nath, G. Use of customized bacteriophages in the treatment of chronic nonhealing wounds: A prospective study. *Int. J. Low. Extrem. Wounds* **2019**. [CrossRef]
99. Fadlallah, A.; Chelala, E.; Legeais, J.M. Corneal infection therapy with topical bacteriophage administration. *Open J. Ophthalmol.* **2015**, *9*, 167. [CrossRef]
100. Górski, A.; Borysowski, J.; Międzybrodzki, R. Phage therapy: Towards a successful clinical trial. *Antibiotics* **2020**, *9*, 827. [CrossRef] [PubMed]
101. Leitner, L.; Ujmajuridze, A.; Chanishvili, N.; Goderdzishvili, M.; Chkonia, I.; Rigvava, S.; Chkhotua, A.; Changashvili, G.; McCallin, S.; Schneider, M.P.; et al. Intravesical bacteriophages for treating urinary tract infections in patients undergoing transurethral resection of the prostate: A randomised, placebo-controlled, double-blind clinical trial. *Lancet Infect. Dis.* **2020**. [CrossRef]

102. Jault, P.; Leclerc, T.; Jennes, S.; Pirnay, J.-P.; Que, Y.A.; Resch, G.; Rousseau, A.F.; Ravat, F.; Carsin, H.; Le Floch, R.; et al. Efficacy and tolerability of a cocktail of bacteriophages to treat burn wounds infected by Pseudomonas aeruginosa (PhagoBurn): A randomised, controlled, double-blind phase 1/2 trial. *Lancet Infect. Dis.* **2019**, *19*, 35–45. [CrossRef]
103. Aktürk, E.; Oliveira, H.; Santos, S.B.; Costa, S.; Kuyumcu, S.; Melo, L.D.R.; Azeredo, J. Synergistic action of phage and antibiotics: Parameters to enhance the killing efficacy against mono and dual-species biofilms. *Antibiotics* **2019**, *8*, 103. [CrossRef] [PubMed]

pharmaceuticals

MDPI

Article

Antimicrobial Peptide K11 Selectively Recognizes Bacterial Biomimetic Membranes and Acts by Twisting Their Bilayers

Francisco Ramos-Martín [1,*], Claudia Herrera-León [1], Viviane Antonietti [2], Pascal Sonnet [2], Catherine Sarazin [1] and Nicola D'Amelio [1,*]

[1] Unité de Génie Enzymatique et Cellulaire UMR 7025 CNRS, Université de Picardie Jules Verne, 80039 Amiens, France; claudia.herrera@u-picardie.fr (C.H.-L.); catherine.sarazin@u-picardie.fr (C.S.)

[2] Agents Infectieux, Résistance et Chimiothérapie, AGIR UR 4294, Université de Picardie Jules Verne, UFR de Pharmacie, 80037 Amiens, France; viviane.silva-pires@u-picardie.fr (V.A.); pascal.sonnet@u-picardie.fr (P.S.)

* Correspondence: francisco.ramos@u-picardie.fr (F.R.-M.); nicola.damelio@u-picardie.fr (N.D.); Tel.: +33-3-22-82-74-73 (F.R.-M. & N.D.); Fax: +33-3-22-82-75-95 (F.R.-M. & N.D.)

Abstract: K11 is a synthetic peptide originating from the introduction of a lysine residue in position 11 within the sequence of a rationally designed antibacterial scaffold. Despite its remarkable antibacterial properties towards many ESKAPE bacteria and its optimal therapeutic index (320), a detailed description of its mechanism of action is missing. As most antimicrobial peptides act by destabilizing the membranes of the target organisms, we investigated the interaction of K11 with biomimetic membranes of various phospholipid compositions by liquid and solid-state NMR. Our data show that K11 can selectively destabilize bacterial biomimetic membranes and torque the surface of their bilayers. The same is observed for membranes containing other negatively charged phospholipids which might suggest additional biological activities. Molecular dynamic simulations reveal that K11 can penetrate the membrane in four steps: after binding to phosphate groups by means of the lysine residue at the N-terminus (anchoring), three couples of lysine residues act subsequently to exert a torque in the membrane (twisting) which allows the insertion of aromatic side chains at both termini (insertion) eventually leading to the flip of the amphipathic helix inside the bilayer core (helix flip and internalization).

Keywords: antimicrobial peptide; biomembranes; ESKAPE; antibiotic resistance; NMR; molecular dynamics; biophysics; sequence alignment

Citation: Ramos-Martín, F.; Herrera-León, C.; Antonietti, V.; Sonnet, P.; Sarazin, C.; D'Amelio, N. Antimicrobial Peptide K11 Selectively Recognizes Bacterial Biomimetic Membranes and Acts by Twisting Their Bilayers. *Pharmaceuticals* **2021**, *14*, 1. https://dx.doi.org/10.3390/ph14010001

Received: 18 November 2020
Accepted: 19 December 2020
Published: 22 December 2020

1. Introduction

The persistent use of antibiotics, self-medication and exposure to nosocomial infections has provoked the emergence of multidrug resistant (MDR) bacteria worldwide [1–4]. The term "ESKAPE" was adopted to refer to some of the most relevant pathogens associated with the highest risk of mortality by the World Health Organization (WHO) [5], namely *Enterococcus faecium*, *Staphylococcus aureus*, *Klebsiella pneumoniae*, *Acinetobacter baumannii*, *Pseudomonas aeruginosa* and *Enterobacter* spp.

In the quest for new molecules able to overcome this major health issue, antimicrobial peptides (AMPs) are promising alternatives to classical antibiotics, due to their low tendency to resistance [6]. AMPs are natural peptides found in all life kingdoms which can be considered components of the innate immunity against bacteria but also fungi, parasites, virus and cancer. Their reduced tendency to resistance is intrinsically due to their mechanism of action causing the selective disruption of bacterial membranes by acting on the lipidic organization of membranes whose lipid composition cannot be changed by a simple point mutation. While exceptions exist [7], their efficacy is proven by the fact that they have been evolutionarily optimized over millions of years, their fast killing rate discourages the rise of drug-resistant mutants [8] and horizontal transfer of resistance genes against AMPs is infrequent [9]. As opposed to standard antibiotics, many AMPs are able to rapidly permeate bacteria and cause irreversible damage to their cell membranes, leading to the death of microorganisms [10,11]. In some cases, their action is also intracellular [12,13].

Several AMPs have been rationally optimized and in this work we focus on K11, a synthetic AMP which was reported to exert antimicrobial action against many of the mentioned ESKAPE bacteria such as *Acinetobacter baumannii*, methicillin-resistant *Staphylococcus aureus*, *Pseudomonas aeruginosa*, *Staphylococcus epidermidis*, and *Klebsiella pneumoniae* [14,15]. K11 has also been successfully used in-vivo as a topic hydrogel solution against *A. baumannii*-infected wounds [15]. Its mechanism of action deserves special attention considering that many of its bacterial targets [14,15] cause complex infections because of their ability to form biofilms [16–18] or change their membrane composition. For example, *A. baumannii* is not only able to form biofilms on biotic and abiotic surfaces but it can also develop resistance to colistin by incorporating phospholipids such as phosphatidylethanolamine (PE), cardiolipin (CL) and monolysocardiolipin to remodel its lipid composition [18,19].

From the point of view of the sequence, K11 (KWKSFIKKLTKKFLHSAKKF-NH2) is an example of synthetic peptide inspired by natural AMPs (cecropin A1, melittin and magainin) [14,15]. More specifically, K11 is one member of a group of peptides synthesized from the CP-P designed antibacterial scaffold (KWKSFIKKLTSKFLHLAKKF). This template was created [14] from the N-terminus of CP26 peptide (inspired by cecropin A1 and melittin) and C-terminus from P18 peptide (inspired by cecropin A1 and magainin) [20]. While CP26 has been reported to target bacterial lipopolysaccharides (LPS) [21], P18 also displays anticancer activity [20]. Most importantly, both CP26 and P18 display antimicrobial activity and negligible toxicity. The introduction of a lysine in position 11 in the CP-P template (hence the name) led to the K11, a peptide with improved values of the therapeutic index (320) [14]. It is believed that the introduction of lysine 11, besides changing the net positive charge, would also alter its amphipathic structure. However, more structural studies are needed to elucidate its mode of action [14].

The interesting properties of K11 prompted us to investigate its interaction with biomimetic membranes by liquid and solid-state NMR spectroscopy (ssNMR) and Molecular Dynamic simulations (MD). Nowadays many different lipidic systems have been optimized for such kinds of studies, going from dodecylphosphocholine (DPC) micelles to bicelles and liposomes with variable phospholipid and sterol compositions reproducing those of the target organisms. The membrane of K11 bacterial targets contains PE, phosphatidylglycerol (PG) and CL in various amounts, as most bacteria. In particular, *Pseudomonas aeruginosa* [22], *Escherichia coli* [23], *Salmonella paratyphi* [24], *Acinetobacter baumannii*, and *Klebsiella pneumoniae* are rich in PE, as expected for the outer membrane of many gram-negative bacteria [25]. Some of its gram-positive targets such as *Bacillus subtilis* and *Bacillus pumilus*, contain PE and PG (although the distribution of phospholipids is unclear) [26–28], while PG or CL clearly prevail in others, such as *Staphylococcus epidermidis* [29,30], *Staphylococcus aureus* [25] and *Micrococcus luteus* [31]. Independently of the relative composition of PG and PE, a special network of H-bond or water-bridged interactions can be established between the two phospholipids [25,32,33], whose ratio can be modulated by bacteria in response to external agents or conditions [22,28,34]. For example, *S. aureus* and *S. epidermidis* can increase their amount of CL under high salt conditions [29,30].

In this work, we show that K11 is able to penetrate biomimetic membranes reproducing the phospholipid composition found in bacteria. Most intriguing, we show that the peptide might act by twisting the membrane using couples of lysine residues. According with this mechanism, the introduction of lysine 11 (whose introduction in the related CP-P peptide significantly improved the therapeutic index) would act in couple with lysine 12 and synergically with all other lysines to torque the membrane, thus facilitating the insertion of aromatic residues at both termini (phenylalanine or tryptophan) and eventually the full peptide in the innermost part of bacterial bilayers. Additionally, for the first time we have observed an interaction with phosphatidylserine (PS), a phospholipid often involved in a wide range of biological processes including viral infection and carcinogenesis [35–37].

2. Results and Discussion

The work was organized as follows. First, property-alignment [38] was used to highlight important motifs along the sequence and to explore further possible activities of K11. Second, we studied the structure of the peptide in solution and in the presence of simple biomimetic models (micelles and isotropic bicelles). Third, ssNMR was used to characterize the effect of K11 on the lipid assembly in vesicles containing various compositions of PE, PG and CL, due to their importance in bacterial membranes. In the attempt to understand the selectivity and low toxicity of K11, membranes containing phosphatidylcholine (PC) were used to mimic the outer leaflet of eukaryotic cells. PS was also considered to explain further predicted activities. Finally, the studied systems including an even larger variety of phospholipids were studied by molecular dynamic simulations for a deeper understanding of the experimental results and of the mechanism of action.

2.1. Property-Sequence Alignment of K11 Highlight Antibacterial Motifs and Predicts Further Activities

In order to highlight structure-function relations, we performed property-alignment in the ADAPTABLE web server [38]. It is important to point out that property alignment clusters sequences with specific activities (antibacterial in this case). The K11 peptide is one of a series of synthetic peptides obtained by designed mutations of the CP-P template [14]. All of them are present in the ADAPTABLE database but they are not meaningful when evaluating the importance of conserved residues among evolutionary-distant sequences. Excluding these entries (peptides 2–22), the sequence-related family (SR family) (Figure S1A) shows that the KWK motif at the N terminus and a large portion of the C-terminus seem to be recurrent in peptides with antibacterial activity. Interestingly, eight out of nine meaningful sequences exhibit activity towards a large variety of cancers [39] (Figure S1B). Such predicted activity could be explained by the fact that one of K11 precursors, P18 has also anticancer properties [20]. One related peptide exhibits antifungal activity against *Candida albicans* and *Trichosporon beigelii* [40]. It should be noted that PS plays a relevant role in both cancer development [35,41–43] and *Candida albicans* virulence [44–48] for which PG [49] and PI [47] are also important.

2.2. K11 Peptide Is Unstructured in Aqueous Solution

The ^{1}H and ^{13}C NMR assignment of K11 is reported in Table S1 and Figure 1A,B. The deviations from random coil values [50–52] indicate that the peptide is mainly unstructured in solution (Figure 1C,D), as also confirmed by circular dichroism (CD) (Figure 1E). The formation of a stable helix (theoretical helical wheel in Figure 1F), which would approach many positive charges arising from eight lysine residues, is probably disfavored in the absence of charge-compensating molecular partners.

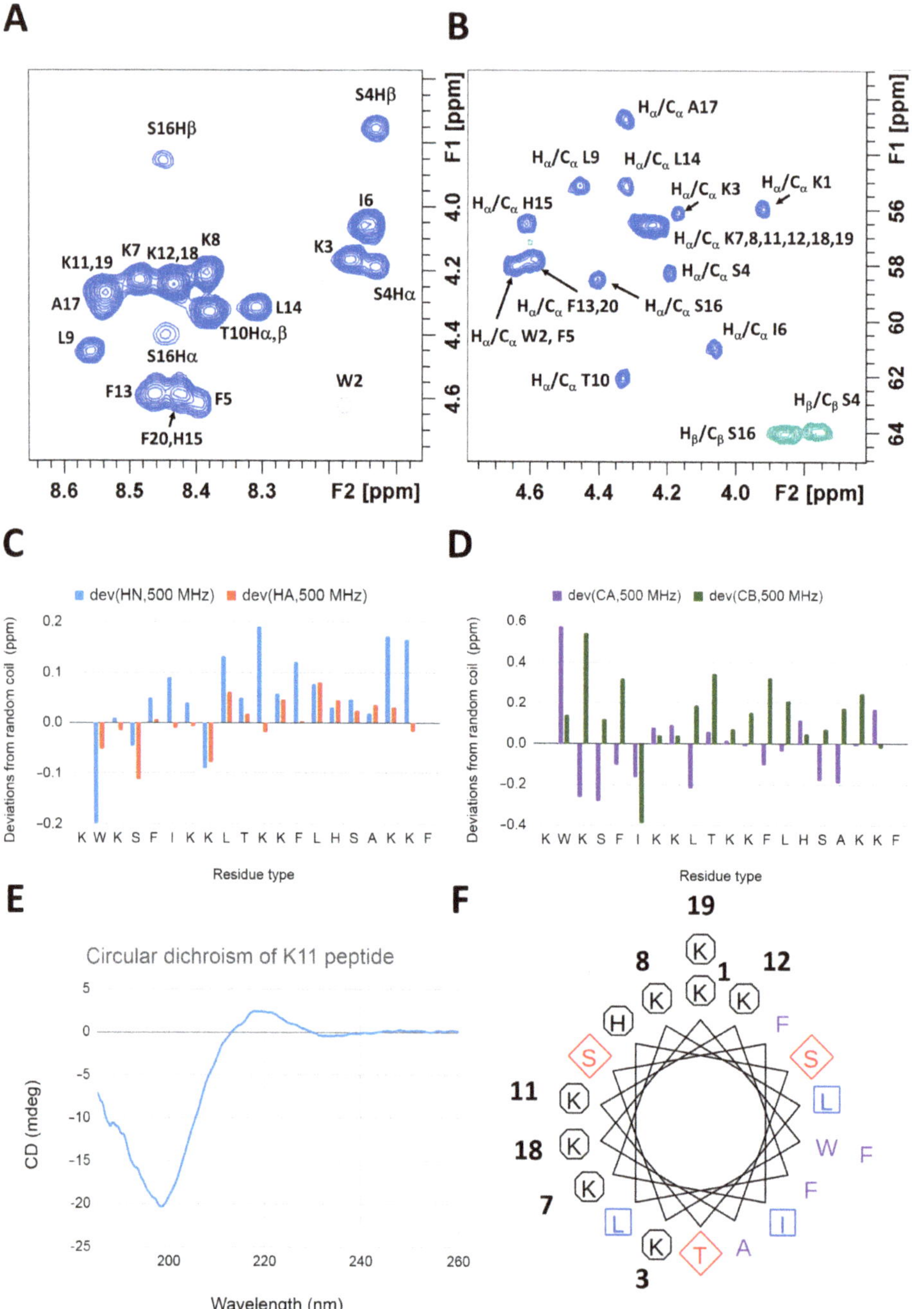

Figure 1. (**A**,**B**) The ^{1}H and ^{13}C NMR assignment of K11 on the amide/Hα (**A**) and side chain (**B**) regions of ^{1}H, ^{1}H-TOCSY and ^{1}H,^{13}C-HSQC spectrum, respectively. (**C**,**D**) Chemical shift deviations from random coil values of amide HN and Hα protons (**C**) and Cα and Cβ carbons (**D**); (**E**) CD spectrum of K11 in solution; (**F**) helical wheel showing the disposition of side chains in alpha helical conformation.

2.2.1. K11 Peptide Assumes Alpha Helical Conformation in a Lipidic Environment

The titration of K11 with a concentrated solution of DPC induces drastic changes in the ^{1}H NMR spectrum as shown in Figure 2A. New peaks appear in the spectrum while the signals originating from the unbound peptide gradually disappear as the concentration of DPC increases. The slow exchange regime is clearly exemplified by the isolated signal of W2 Hδ1, disappearing at its original shift of 10.2 ppm and reappearing at 10.8 ppm (see Figure 2A). A closer analysis of ^{1}H,^{13}C-HSQC spectrum (Figure S2A–D), reveals that all aromatic residues (Figure S2A,B) are deeply affected by the presence of the micelles with the exception of H15, whose signals only slightly shift and partially lose intensity (see Figure S2A). A significant shift is also observed for aliphatic signals of F5, I6, L9, T10, F13, L14, A17, F20 (Figure S2C,D) which would be located on the same molecular face in case an alpha helix is formed. A clear proximity of the aromatic (Figure S2E) and aliphatic (Figure S2F) side chains to the DPC acyl chains is demonstrated by the NOE cross peaks in the NOESY spectrum. NOEs with the acyl chain of DPC (whose assignment was based on the literature [53]) but not with its choline headgroup testify a rather deep insertion of the peptide into the micelle and the absence of a specific interaction with the headgroup. While most ^{13}C backbone signals are lost in the ^{1}H,^{13}C-HSQC spectrum, we were able to assign all Hα protons. Their values were used to predict the secondary structure [50–52] of the peptide bound to DPC micelles. The negative deviations from theoretical random coil values unequivocally indicate that the peptide assumes an alpha helical conformation (see Figures 2B and 1C for comparison). Accordingly, all the weak $HN_i/HN_{i-1/i+1}$ NOEs observed in the free peptide gain in intensity (data not shown). All-atom MD simulations are in perfect agreement with NMR data (Figure 2C), showing how the peptide conserves its alpha helical conformation along the full trajectory of 500 ns.

The radial distribution function [54] of each oxygen and nitrogen atom of the membrane from all O/N atoms of the peptide can be used to highlight key interatomic interactions. By focusing on its maximum value in the distance range of H-bond and salt bridges, we obtain a quantification of the frequency at which these interactions occur all along the last 250 ns of the simulations. Results are shown in Figure 2D, revealing that lysine side chains are able to recognize the phosphate groups of the phospholipids. Interestingly, only K1, 3, 8 and 12 interact frequently while K7, 11, 18 and 19 seem slightly less involved (Figure S3). While K7 and 11 might not be at the optimal distance, the absence of interaction of K19 might be due to the high curvature of the micelle, not allowing all lysines of the helix to interact at the same time.

2.2.2. In the Presence of Biomimetic Bicelles K11 Peptide Possibly Assumes a Conformation Similar to That Found with Micelles

In order to better understand the influence of the curvature, we studied the interaction of K11 peptide with isotropic bicelles, better representing a more extended surface in solution [55]. Isotropic bicelles can be formed by a mixture of DMPC (1,2-dimyristoyl-sn-glycero-3-phosphocholine) and DHPC (1,2-dihexanoyl-sn-glycero-3-phosphocholine) [56]. The short acyl chain of DHPC is able to stabilize the bilayer formed by DMPC, whose myristoyl hydrophobic chains would be otherwise exposed to the solvent [57]. Fast tumbling isotropic bicelles can be obtained at a DMPC/DHPC ratio 1:2 (q = 0.5). As in the case of micelles, the ^{1}H NMR spectrum of K11 peptide drastically changes in the presence of bicelles (70 mM) but NMR signals become too broad for a new assignment. However, the NMR spectrum resembles the one observed in the presence of micelles (see Figure S4) suggesting that the same helical conformation is formed.

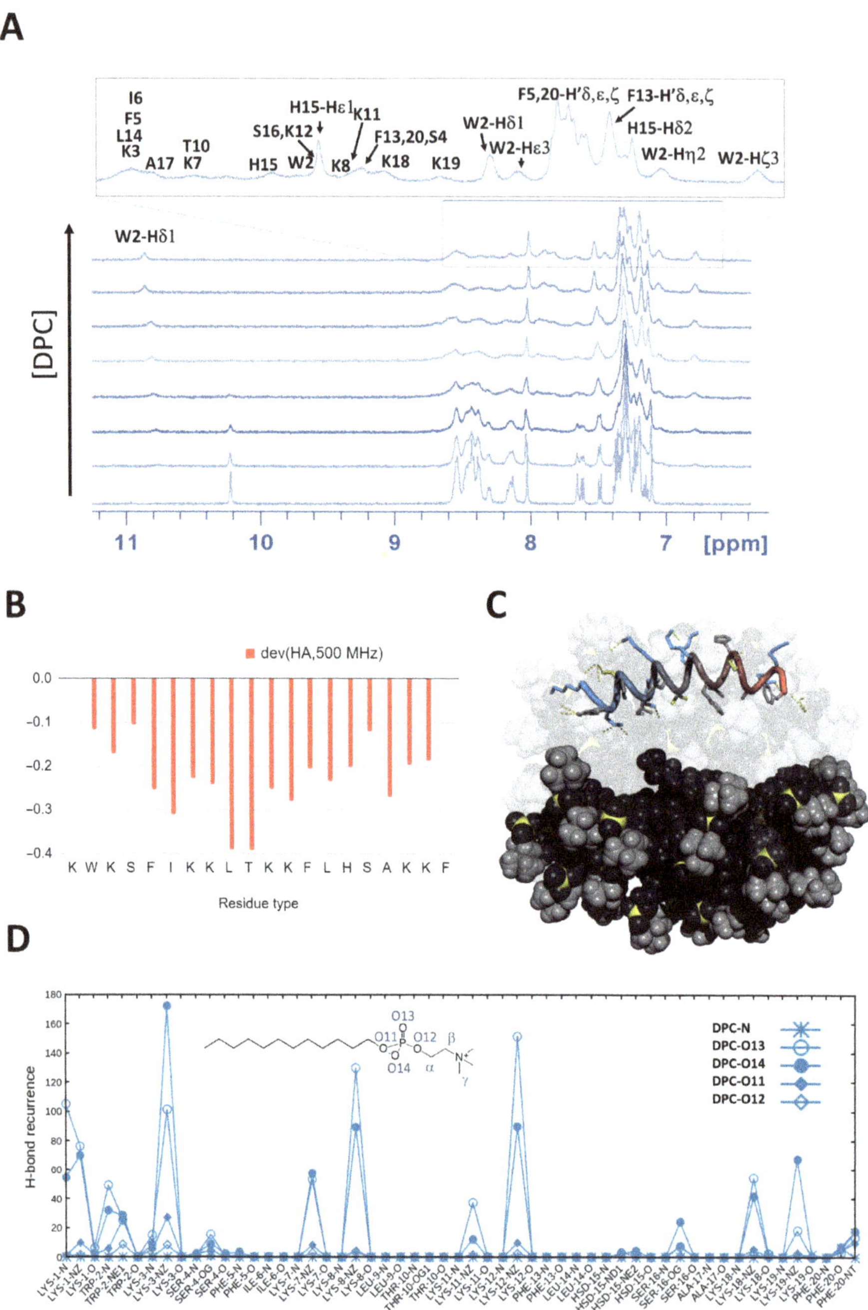

Figure 2. (**A**) ^{1}H NMR spectra of K11 1 mM in the presence of DPC at concentrations 1, 2, 4, 10, 20, 30, 60 mM. The NMR assignment in the presence of micelles is also shown. (**B**) Chemical shift deviations from random coil values of Hα protons whose negative deviations indicate an alpha helical conformation; (**C**) MD snapshot of K11 interacting with DPC micelles; (**D**) Polar contacts recurrence (H-bonds and salt bridges) along MD simulation.

In DMPC/DHPC bicelles part of DMPC lipids can be substituted by phospholipids with different headgroups to mimic different biological membranes [57]. In this way, bicelles containing PE, PG, and PS were formed and tested in their interaction with K11. Figure S4 shows that, although the spectra are qualitatively similar, the linewidth is significantly larger in the case of DMPG (1,2-dimyristoyl-sn-glycero-3-phospho-(1′-rac-glycerol)) and DMPS (1,2-dimyristoyl-sn-glycero-3-phospho-L-serine), probably indicating that the peptide is less mobile inside these bilayers because of a stronger interaction. Interestingly, both DMPG and DMPS introduce negative charges in the bicelles which might stabilize the structure of the positively charged K11 peptide.

2.2.3. K11 Selectivity Perturbs the Core of Liposomes with Bacterial Phospholipid Compositions

In order to ascertain the effect of K11 on the lipid assembly of the membrane, we studied the interaction of K11 with multilamellar vesicles (MLVs) by ssNMR.

MLVs are more suitable to mimic lipid bilayers of biological membranes because of their hydration state and the lower curvature than bicelles. They also allow to vary the phospholipid composition more freely [55] while bicelles always require a large part of DMPC and DHPC. Moreover, MLVs can be prepared by using commercially available phospholipids bearing deuterated palmitic chains allowing to sense the organization of the hydrophobic core of the lipid bilayer by ^{2}H NMR. Thus, the order parameter for each C-^{2}H bond of the chain can be measured by means of ^{2}H quadrupolar splitting [58–60]. Figure 3 shows ^{2}H spectra of MLVs with various phospholipid compositions and the effect of the presence of K11. Each spectrum results from a superposition of the quadrupolar doublet arising from different C-^{2}H bonds. Since the mobility of unsaturated chains in bilayers increases as we move away from the headgroup, the quadrupolar splitting also decreases, with the consequence that methyl groups appear at the center of the envelope while the carbon in position 2 appears at the extremities of the spectrum, with all the remaining C-^{2}H in-between. Even though we could not measure the order parameters for each C-H moiety because of the low resolution, the overall behavior is very clear when observing the width of the superimposed signals. The presence of K11 does not perturb the ^{2}H spectra of POPC (1-palmitoyl-2-oleoyl-glycero-3-phosphocholine) membranes and POPE (30% POPC) (1-palmitoyl-2-oleoyl-sn-glycero-3-phosphoethanolamine), suggesting that the peptide is not able to penetrate deeply into these bilayers. This observation stresses the importance of the curvature in biomimetic models. Despite the presence of the same headgroup, K11 deeply penetrates DPC micelles but not POPC MLVs, where lipids are more closely compacted because of the locally almost planar surface as opposed to the high curvature of micelles [55].

Quite interestingly, K11 deeply affects the ^{2}H spectrum of POPG (1-palmitoyl-2-oleoyl-sn-glycero-3-phospho-(1′-rac-glycerol)) and POPS (1-palmitoyl-2-oleoyl-sn-glycero-3-phospho-L-serine) MLVs. These headgroups are commonly found in bacteria and cancer cells, respectively, but also fungi, as hypothesized in the activity prediction based on property-alignment by ADAPTABLE. Our data reproduce qualitatively what was found with bicelles, where the increased linewidth observed with PG and PS headgroups suggested a stronger binding (see Section 2.2.2). For both POPG and POPS MLVs, the apparent reduction of the quadrupolar splitting and the loss of resolution reflects a drastic increase in the phospholipid acyl chain mobility, most probably due to the internalization of the peptide in the bilayers. Encouraged by these results, we prepared MLVs using a mixture of PE and PG headgroups, typically found in bacteria [33]. As shown in Figure 3, K11 is able to perturb the fluidity of such bilayers even more and the effect becomes really important in the presence of CL, also found in bacteria [61]. Figure 3 shows how K11 affects also membranes mainly constituted by CL (CL 50% / POPC 50%). It should be noted that in the cases of CL and POPE, the addition of POPC was necessary for the formation of a MLV, due to their intrinsic shape and different T_m [32,62–68].

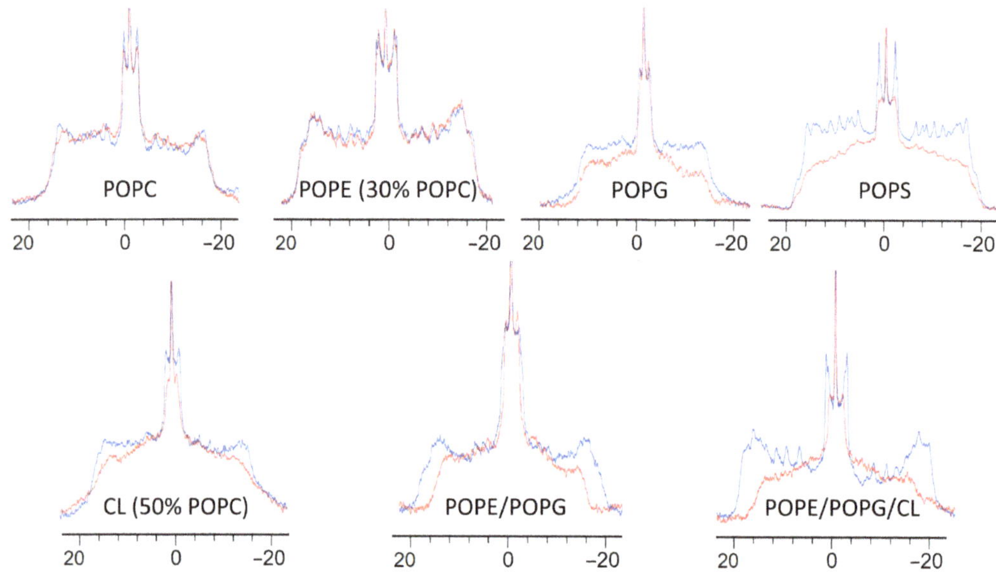

Figure 3. Static ^{2}H NMR spectra of various multilamellar vesicles (MLVs) in the absence (blue) and in the presence (red) of K11 peptide.

2.3. MD Simulations Provide a Molecular Picture of the Interaction

In order to get insight into the details of the interaction between K11 peptide and biomembranes, we performed MD simulations using a variety of phospholipid combinations involved in bacteria, cancer and fungi. Figure 4 shows the most significant snapshot for each run, and in particular the frames where the peptide comes close to the membranes.

In the case of POPC, K11 stays away from the bilayer during most of the simulation and even when it approaches the membrane, there is no evidence of a specific interaction. For POPE (containing 30% of POPC to reproduce the experimental conditions), the situation is only slightly different. In this case, the peptide can lay on the surface sporadically and very few interactions are established.

The situation is radically different for POPG and in general for all negatively charged phospholipids: POPS, POPI (1-palmitoyl-2-oleoyl-sn-glycero-3-phosphoinositol), and CL. A clearer picture comes from the analysis of polar interactions shown in Figures S5 and S6. As in Section 2.2.1, we calculated the recurrence of H-bond and salt bridges (Figures S5 and S6). In the case of POPG we observe an important interaction of the N-terminal lysine with the oxygen atoms of POPG phosphate groups by means of both the backbone and side chain amines. Such an interaction is consistently observed when the peptide significantly interacts, suggesting that K11 peptide approaches the membranes with the first lysine residue. Furthermore, also the side chains of lysine residues in position 3, 7 and 8 establish similar contacts. The selectivity for negatively charged phospholipids like POPG is probably due to the electrostatic attraction leading the positively charged K11 peptide (+8 at physiological pH) towards the negative charges introduced by POPG headgroup but also to the availability of multiple oxygen atoms provided by its glycerol moiety (inositol and carboxylate in the case of POPI and POPS, respectively), which are available for hydrogen bonding or the formation of extra salt bridges in the case of POPS. Not surprisingly, a strong interaction is also observed with CL due to the structural similarity to POPG, the exposition of phosphate groups at the membrane surface and the doubly negative charge. When PG or CL are in mixtures, they are involved in the large majority of the peptide-membrane interactions (see Figure S6).

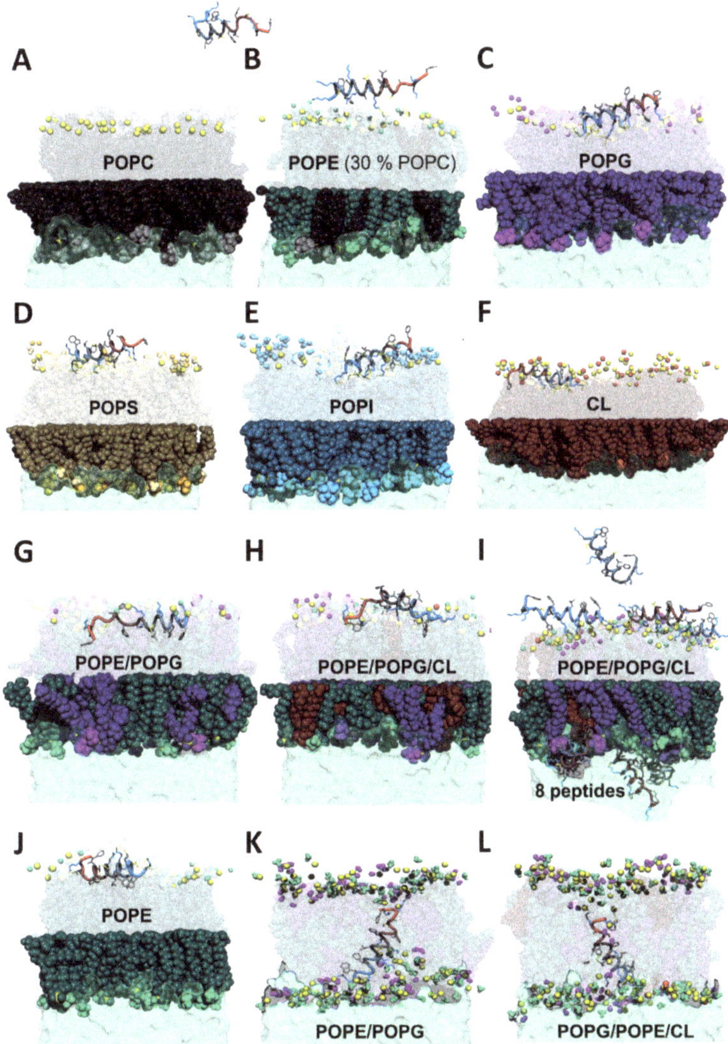

Figure 4. MD snapshots representative of K11 peptide interacting with several membranes of variable phospholipid compositions. Color code: (**A**) POPC black (body) and light gray (choline group); (**B**) POPE dark green (body), turquoise (headgroup) and light green (amine of the headgroup); (**C**) POPG dark violet (body), violet (headgroup) and light violet (hydroxyls of the headgroup); (**D**) POPS brown (body), gold (headgroup), light yellow (amine of the headgroup) and orange (carboxyl of the headgroup); (**E**) POPI blue (body), light blue (headgroup) and cyan (hydroxyls of the headgroup); (**F**) CL dark red (body) and light red (headgroup). Panels G-H show lipid mixtures typically found in bacteria: (**G**) POPE/POPG and (**H**) POPE/POPG/CL. Panel (**I**) represents an example of calculation with eight peptides while panel (**J**) refer to a calculation of one peptide interacting with a pure POPE membrane, differing from that in panel B for the absence of POPC. Snapshots in panels (**K**), (**L**) refer to examples of simulations where K11 is purposely placed inside the membrane at the start of the calculation. In all panels the phosphorus atom of phospholipids is shown as a yellow sphere; for clarity, only functional moieties of headgroups are represented as spheres either in the upper leaflet, or in both leaflets (panels K, L). K11 peptide is shown as a "tube" colored from blue (N-terminus) to red (C-terminus) except in panel I where each of the eight peptide has a different color. Side chains are shown as sticks with the following color code: positively charged (blue), negatively charged (red), non-polar (light gray), polar (yellow).

The Coulomb attraction is clearly a key factor to understand such selectivity but also other factors play a role: the inter-lipid spacing (modulated by the curvature [69,70]), the steric hindrance of the headgroups and the amount of inter-lipid interactions [34,35]. The first factor explains why K11 can penetrate DPC micelles (whose PC headgroup is not found in bacteria) but doesn't seem to affect POPC liposomes significantly. In this case, the high curvature makes the inter-lipid spacing wider, facilitating the access to the micelle core. The second factor explains the preference for CL over PE and PG when the three are present (see Figure S6); in the case of CL phosphate moieties are directly accessible as its headgroup doesn't present steric hindrance. The third factor is also important, because the insertion of the peptide implies breaking a number of favorable interactions (like the ones present in PE/PG membranes [25,32,33]). Besides these effects, the presence of negative charges on the membranes does incentivize the binding not only in the case of PS but also with PG and PI (Figures S5 and S6). In similar conditions of curvature, we can expect that the accessibility to phosphate moieties is modulated by the steric hindrance of headgroups but also the network of inter-lipid interactions. Last but not least, the mobility of the peptide in the complex and the degree of lipid-order destabilization might contribute to the overall process as entropic contribution [71].

2.3.1. K11 Exerts a Twisting Effect of Its Target Membranes

Quite interestingly, the membrane planarity is heavily perturbed and almost twisted when K11 peptide interacts. A closer analysis of the helix wheel reveals that the peptide could act as a screw twisting the membrane by means of couples of lysine residues (Figure 5). In all MD simulations in the presence of membranes containing PG, PS, PI or CL we observe the same behavior (see Figure 5A): the peptide approaches the membrane with the N-terminal lysine (step 1, anchoring), grabs onto the available oxygen atoms (most frequently phosphate oxygen O13 and O14 but also headgroup oxygens) and deforms the membrane (ste p2, twisting). The deformation allows the insertion of terminal aromatic groups (F20 but in some cases also W2) inside the bilayer (step 3, insertion). F20 and W2 are the only amino acids whose hydrophobic side chains are readily available and this is due to the fact that the amphipathic helix formed by K11 approaches the membrane with its hydrophilic face. In particular, in the simulation containing POPE/POPG, we observe a further important step. The insertion of the aromatic ring of F20 eventually determines the flip of the full hydrophobic face of the helix into the bilayer thus reaching the hydrophobic core (step 4, helix flip and internalization). The peptide remains inserted even prolonging the simulation up to 2 μs.

A closer look to the relative disposition of lysine residues in an alpha helix conformation can explain the twisting effect on the membrane (Figure 5B). We believe that K11 peptide actually works as a screw. By landing on the surface with the first lysine, the peptide anchors to the available oxygen atoms. These may arise from the phospholipid phosphate groups or oxygens in the headgroups. Such an anchoring is quite effective because lysine 1 bears two amine moieties that can bind in a bidentate fashion. Figures S5 and S6 shows how such an interaction is present in almost all simulations involving charged phospholipids with high occurrence. We can imagine dissecting the helix of the peptide with three planes almost orthogonal to its long axis, each containing two lysine residues (K7 and 8, K11 and 12, K18 and 19). The K1 anchoring step is followed by the establishment of interactions involving lysines 7 and 8 with available nearby membrane oxygen atoms. These bindings have the synergic effect of rotating the membrane in their plane (see Figure 5B). Such rotation is subsequently reproduced in the plane of lysines 11 and 12 and in that of lysines 18 and 19. As the couples of lysines 7–8, 11–12 and 18–19 are located with different phases in the helix wheel, these subsequent rotations have the effect of a twist. In particular, while lysines 7–8 and 11–12 would determine a clockwise rotation, lysines 18–19 would act in the opposite sense because of their intermediate phase in the wheel.

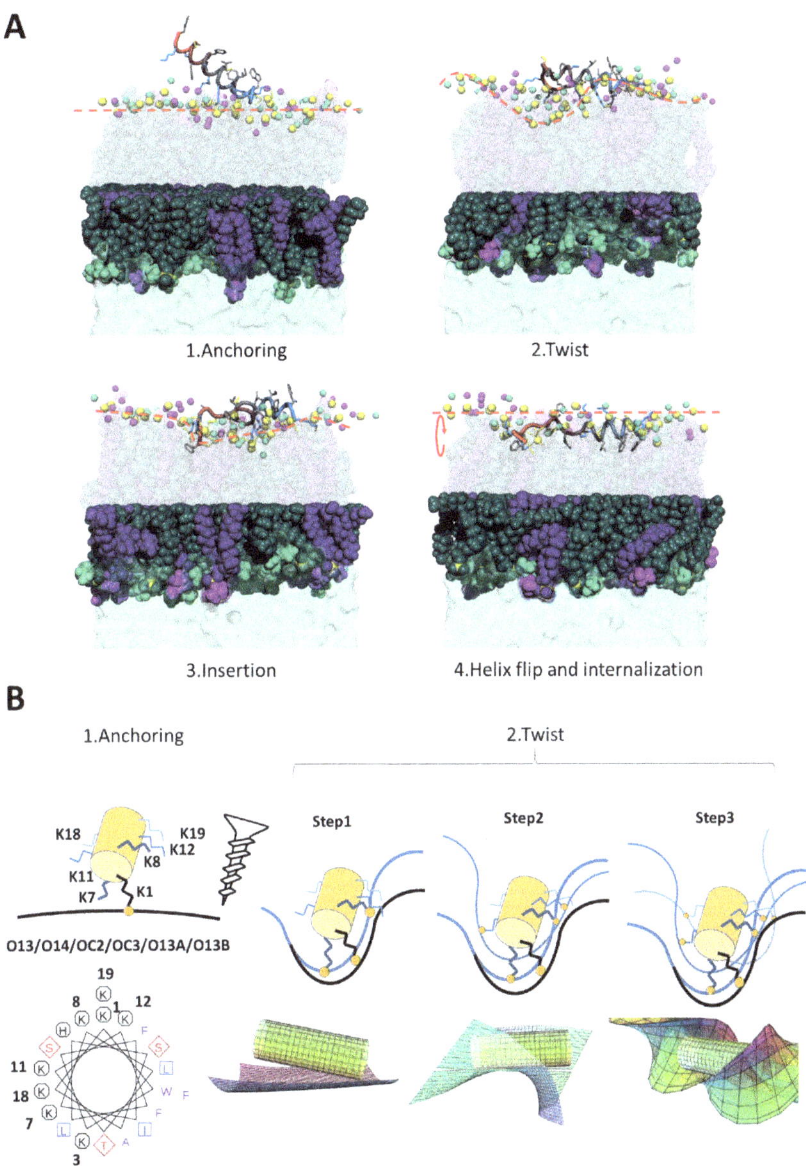

Figure 5. (**A**) Proposed mechanism of action of K11 peptide in four steps. The peptide first anchors to the membrane by the lysine residue in position 1 (anchoring) which can bind membrane oxygen atoms in a bidentate fashion. The peptide then twists the membrane (as described in panel B) thus allowing the insertion of terminal aromatic side chains. Finally, the peptide flips inside the bilayer. For color codes refer to the caption of Figure 4. (**B**) Mechanism by which K11 might exert a torque on the target membrane. Yellow circles represent oxygen atoms on the surface of the membrane and available for H-bonding or salt bridges with lysine side chains. The helix formed by K11 is represented as a cylinder. Lysine residues and membrane planes are represented in blue color whose intensity degrades with the distance from the observer. The torque is achieved in three subsequent steps, each rotating the membrane in the plane described by each couple of lysine residue. A geometrical representation of the effect on the membrane for each step is exemplified under each step. Image generated with the help of CalcPlot3D software [72].

Significant perturbation of the membrane can be more easily visualized by monitoring the area per lipid along the MD trajectories. Although the perturbation can be detected in simulations with one peptide, the effect is amplified with the introduction of several peptides (see Figure S7). With the exception of POPC and POPE, we observe a decrease in area per lipid of the upper leaflet and an increase in that of the lower leaflet, indicating that the peptide exerts a pressure causing the membrane to invaginate (negative curvature).

2.3.2. K11 First Rigidifies the Membrane and Subsequently Makes It More Fluid

The last step of the mechanism proposed in Figure 5 (step 4, helix flip and internalization) is essential because it explains the reduction of the lipid chain order experimentally demonstrated for the phospholipid acyl chains by the perturbation of ^{2}H NMR spectra (Figure 3). It should be stressed that peptide anchoring (step 1 in Figure 5A) and membrane twisting (step 2 in Figure 5A) actually increase the order parameters of acyl side chains while the internalization (step 4 in Figure 5A) of the peptide in the hydrophobic core reduces it (Figure 6), as experimentally observed (Figure 3). The description of the proposed mechanism of action in four steps may require extending our 500 ns simulations further. The complete helix flip (step 4 in Figure 5A) is only observed in one of the three repetitions of the simulation with POPE/POPG membranes. We can hypothesize that in POPE/POPG mixtures (which better represent the bacterial membrane with respect to pure POPG) the activation energy for the penetration of the peptide is lower, thus allowing its detection in our 500 ns-long simulations. This is a reasonable hypothesis when considering that the PE headgroup has a smaller steric hindrance than that of PG and could facilitate the entrance of the peptide, as can also be rarely observed in simulations with pure POPE (Figure 4J). For this reason, we have extended the POPE/POPG calculation up to 2 μs.

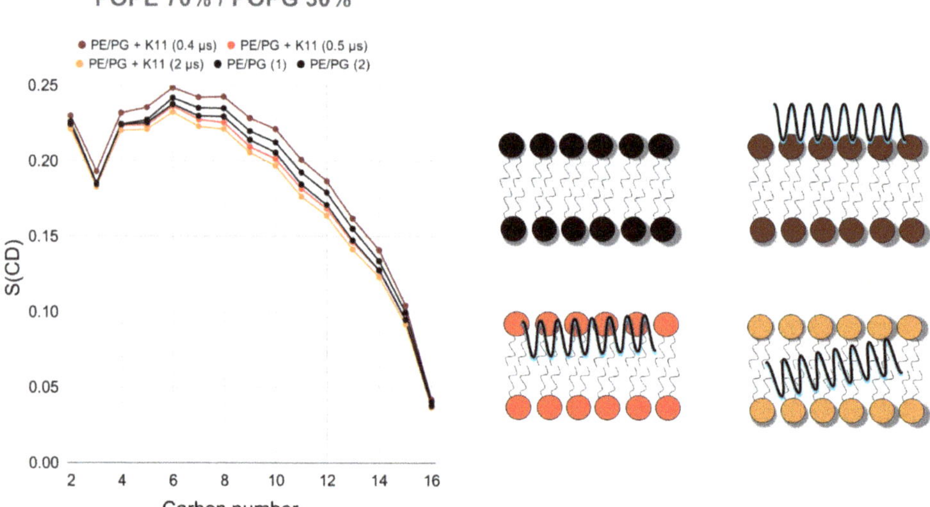

Figure 6. Order parameter of C-H moieties in palmitoyl side chains in membranes containing POPE (70%) and POPG (30%) as calculated from MD simulations in the absence (2 repetitions in black labeled as 1 and 2) and in the presence of K11 peptide. The order parameter varies in different ways along the 2 μs trajectory. The membrane is rigidified upon interaction (brown curve) but becomes more fluid as the peptide penetrates (red) and becomes internalized (orange curve).

It should be noted that the formation of the complex takes place in slow exchange in the NMR time scale (Figure 2A), meaning that $k_{ex} << |\Delta\omega|$. k_{ex} is the exchange rate constant ($k_{ex} = k_{on}[L] + k_{off}$ where k_{on} and k_{off} are the on and off-rate constants for the formation of the complex between the peptide P and the membrane M according to the equation P + M → PL) and $|\Delta\omega|$ is the chemical shift difference between the free and the bound form of the peptide [73]. Our deviations are on the order of 0.2 ppm (Figure 2B), meaning that at 500 MHz $|\Delta\omega|$ is ~600 s^{-1} ($|\Delta\omega| = 2\pi|\Delta\nu|$, where $|\Delta\nu|$ is the chemical shift difference in Hz). Our slow exchange conditions therefore limit the value of k_{ex} and k_{off} to a maximum of 600 s^{-1} and the lifetime of the complex to a minimum of 1.7 milliseconds or much more, including the case of irreversible binding (the lifetime is the inverse of the off-rate constant). The detection of such long processes [74,75] would require more advanced sampling algorithms including dual-resolution MD [76], coarse-grain simulations, steered MD [77], umbrella-sampling [78,79], metadynamics [80,81], or replica exchange, among others [75,82,83]. This is beyond the scope of this work that aims at characterizing the steps at the very beginning of the interaction, in order to unravel the mode of action. The choice of all-atom MD allows us to directly compare the calculation with NMR data providing specific information on hydrogen and carbon atoms.

Peptide concentration plays an important role in the mechanism of action of AMPs because antimicrobials can act synergically to destabilize the target membrane using different strategies (carpet, pore formation by toroidal or barrel-stave models [76,84–86]). In order to confirm the hypothesis of an initial rigidification we calculated the order parameter for all membranes (Figure S8) increasing the number of peptides to simulate a higher concentration (one snapshot example of such calculation is shown in Figure 4I). Figure S8 shows a rigidification clearly visible for PG and PS membranes, as expected. The increase of order observed upon peptide binding is not so uncommon, and depends on factors like lipid composition of the membrane, temperature, and charge [58,60,87–89]. Sometimes, peptides that attach to the surface of the bilayer can increase acyl chain packing [90,91], especially when a strong electrostatic attraction is established [91]. The rigidification effect upon binding is also consistent with the observed hydrophobic thickness (Figure S9), that greatly increases for POPG, POPG containing membranes and CL as compared to POPC. Furthermore, the observed decrease in the electron density (Figure S9) can be a consequence of more water molecules being located near the polar head groups due to more loosely packaging caused by the presence of the peptides [88,92,93].

The effect of fluidification following the internalization was confirmed (Figure S10) by placing the peptide inside the bilayers at the beginning of the simulations (see example snapshots in Figure 4K,L).

2.3.3. K11 Approaches Phospholipids Head Groups from Opposite Leaflets Possibly Leading to Membrane Disassembly after Entering the Bilayer

The simulations of the fully internalized peptide can be thought of as a "prolongation" for those in which the peptide is able to access the membrane core. These simulations allow us to bypass the longer time scales needed to observe the full process. Two snapshots are shown in Figure 4K,L and they testify to a quite interesting phenomenon. The length of K11 helix is slightly shorter than the membrane thickness with the result that both the N-terminus and the C-terminus of K11 tend to recall polar head groups in the membrane core by binding with their oxygen atoms. Polar head groups on opposite leaflets almost come in close proximity. This is possible because polar head groups are initially grabbed by peripheral lysines residues and subsequently "walk" by detaching and attaching to the ones in the center of the helix. This is particularly evident in bacterial biomimetic membranes (Figure 4K,L). Once inside the bilayer, this mechanism would allow K11 to disassemble the membrane.

2.3.4. PS Targeting Opens the Way to Possible New Biological Activities

The data presented in this work indicates that K11 destabilizes PS containing membranes (Figures 3 and 4 but also Figures S5, S7 and S8), and this could be an indication of a possible anticancer activity, as already shown for some members [39] of its SR family (see Section 2.1 and Figure S1). It has to be noted that K11 was created as a combination of CP26 peptide (inspired by cecropin A1 and melittin) and a C-terminus from P18 peptide (inspired by cecropin A1 and magainin), which displays anticancer activity [20]. Similarly to what happens in apoptotic cells, cancerous cells tend to expose PS, a phospholipid normally found in the inner leaflet of the membrane [35]. A specific interaction with PS is probably the reason why a considerable number of antimicrobial peptides produced in eukaryotes (or inspired by them like K11) display anticancer activity while displaying low hemolytic activity and toxicity to healthy ones. Their eukaryotic origin explains their selectivity [94–96] As PS-targeting has proved to be effective as anti-cancer [36] or antiviral [97] therapies, the selective recognition of PS by K11 should not be undervalued.

One member of the K11 SR family also displays activity against fungi like *Candida albicans*. As shown in this work, K11 targets PS and PI, both being relevant for *Candida* virulence [44–48], together with PG [49].

3. Materials and Methods

3.1. Synthesis of K11 Peptide

Fmoc(9-fluorophenylmethoxy)-amino acids, Fmoc-Tyr(tBu)-AC TentaGel® resin (0.22 mmol/g, particle size: 90 μm) and Fmoc-TentaGel®-S RAM resin (0.24 mmol/g, particle size: 90 μm) were purchased from Iris Biotech (Germany). The other chemical compounds were purchased from VWR Chemicals, Iris Biotech or Acros and used without further purification. The peptides were synthesized on an CEM Liberty 1 Microwave Peptide Synthesizer, using standard automated continuous-flow microwave solid-phase peptide synthesis methods. Five-fold molar excess of the above amino acids was used in a typical coupling reaction. Fmoc-deprotection was accomplished by treatment with 20% (*v*/*v*) piperidine in *N*-methyl-2-pyrrolidone (NMP) at 75 °C. The coupling reaction was achieved by treatment with 2-(1*H*-benzotriazol-1-yl)-1,1,3,3-tetramethyluronium hexafluorophosphate (HBTU) and *N*,*N*-diisopropylethylamine (DIEA) in NMP using a standard microwave protocol (75 °C). The peptide was cleaved and side-chain deprotected by treatment of the peptide resin with a mixture of 1.85 mL of trifluoroacetic acid (TFA), 50 μL of triisopropylsilane, 50 μL H_2O and 50 mg of DL-dithiothreitol, in respective percent proportions, 92.5/2.5/2.5/2.5, during 4 h at room temperature. The solid support was removed by filtration, the filtrate concentrated under reduced pressure, and the peptide precipitated from diethyl ether. The precipitate was washed several times with diethyl ether and dried under reduced pressure. The peptides were purified on an RP-HPLC C18 column (Phenomenex® C18, Jupiter 4μ Proteo, 90 Å, 250 × 21.20 mm) using a mixture of aqueous 0.1% (*v*/*v*) TFA (A) and 0.1% (*v*/*v*) TFA in acetonitrile (B) as the mobile phase (flow rate of 3 mL/min) and employing UV detection at 210 and 254 nm. The purity of all peptides was found to be >95%.

K11 peptide was obtained as a white powder, with a total yield of 21.7%, after purification by reverse-phase HPLC (96% analytical purity) (see Figure S11). The concentration of the sample was determined by dissolving a precise amount of the powder in a precise volume of the buffer. The concentrated solution was subsequently divided in aliquots and lyophilized. Once redissolved in buffer, the concentration was confirmed by the absorbance at 280 nm, using a molar extinction coefficient of 5500 cm^{-1}, M-1 (only one tryptophan is present) estimated by the ProtParam tool [98] of Expasy server (https://web.expasy.org/protparam/).

3.2. Sequence Alignment by ADAPTABLE Web Server

The family of peptides sequence-related to K11 (KWKSFIKKLTKKFLHSAKKF) was created by the family generator page of ADAPTABLE webserver (http://gec.u-picardie.fr/adaptable/) using "Create the family of a specific peptide" option with the following parameters: "antibacterial = y"; "activity (μM) = 1"; "Substitution matrix = Blosum45"; "Minimum % of similarity = 51". As ADAPTABLE continuously updates with new entries sequence-related families might change slightly with the time [38].

3.3. Sample Preparation, NMR Experiments and Analysis

Backbone and sequential resonance assignments were achieved by ^{1}H,^{13}C-HSQC, ^{1}H,^{1}H-TOCSY (mixing of 60 ms), and ^{1}H,^{1}H-NOESY (mixing of 200 ms) recorded on a 500 MHz Bruker spectrometer equipped with a 5 mm Broadband Inverse (BBI) probe. Deuterated sodium 3-(trimethylsilyl)propionate-d4 (TSP-d4) at a concentration of 100 μM was used as internal reference for chemical shift. Reference random coil values in our experimental conditions (T = 278 K, pH 6.6 and ionic strength 0.02 M) were calculated by POTENCY web server (https://st-protein02.chem.au.dk/potenci/) [99].

CD spectra were obtained in the far-UV (260–185 nm) on a J-815 Jasco spectropolarimeter (Tokyo, Japan). The CD measurement was performed at 5 °C, using a 1 mm path cell, with 5 accumulations for a 216.0 mg/mL sample in 10 mM sodium phosphate buffer, pH 6.6. All CD spectra measured were baseline corrected by subtracting the buffer spectrum.

A 1 mM sample of K11 (90% 10 mM phosphate buffer/10% D_2O, pH 6.6) was titrated with a 1 M stock solution of DPC to a final DPC concentration of 60 mM. Titration was followed by 1D ^{1}H-NMR at 278 K. For the assignment of the interacting form of the peptide 2D ^{1}H,^{1}H-NOESY and ^{1}H,^{13}C-HSQC were recorded at total DPC concentrations of 60 mM.

Bicelles were prepared as follows. A mixture of 33.3% DMPC and 66.7% DHPC in chloroform was used to obtain isotropic bicelles at a molar (q) ratio of 0.5. The solvent was evaporated under a nitrogen flow and the samples were then lyophilized and resuspended in a 10 mM phosphate buffer (pH 6.6) to reach a final concentration of 1 M (stock solution). DMPG, DMPS and DMPE containing bicelles were prepared as described above, except part of DMPC was replaced by DMPG (25%), DMPS (25%) or DMPE (10%) reproducing previous experiments [57]. A 1 mM sample of K11 (90% 10 mM phosphate buffer/10% D_2O, pH 6.6) was titrated with bicelles up to a final lipid concentration of 70 mM and monitored at 278 K by a 1D ^{1}H-NMR spectrum recorded after each addition.

MLVs containing deuterated palmitoyl chains were prepared according to the conventional protocol [100–103] using the following proportions: 50%:50% POPC/POPC:d31, 50%:50% POPG/POPG:d31, 50%:50% POPS/POPS:d31; 70%:30% POPE:d31/POPG, 50%:50% CL/POPC:d31, 70%:30% POPE:d31/POPC and 67%:27%:6% POPE:d31/POPG/CL. Lipids were solubilized in chloroform and solutions were mixed in order to obtain the right proportions in a total lipid amount of 60 mM. The resulting solution was evaporated under nitrogen gas flow. The sample was hydrated with ultrapure water, well-vortexed to promote a total hydration and lyophilized overnight to remove the traces of solvents. The resulting powder containing lipids was hydrated by 80 μL of ultra-pure water (for non-charged lipids) or 10 mM phosphate buffer pH 6.6 100 mM NaCl (for charged lipids), vortexed and homogenized using four free-thaw cycles involving one step of freezing (−80 °C, 15 min) followed by thawing (40 °C, 15 min) and shaking. Finally, the MLV samples were placed in a 7-mm ssNMR rotor to perform the experiments. 2.4 mM of peptide were added for interaction studies.

ssNMR experiments were recorded at 310 K on a Bruker Avance Biospin 300 WB (7.05 T) equipped with a CP-MAS 7-mm probe (Bruker Biospin, Karlsruhe, Germany). Static ^{2}H NMR was carried out applying a phase cycled quadrupolar echo pulse sequence (90°x-τ-90°y-τ-acq) [104]. The parameters used are listed below: spectral width of 150 kHz, $\pi/2$ pulse of 5.25 μs, an interpulse delay of 40 μs, a recycled delay of 1.5 s, and a number of acquisitions ranging from 8 k to 14 k depending on samples. For all spectra, an exponential line broadening of 100 Hz was applied before Fourier-transform from the top of the echo signal.

3.4. Molecular Dynamics Simulations

Systems for simulations were prepared using CHARMM-GUI [105–107]. A total of 128 lipid molecules were placed in each lipid bilayer (i.e., 64 lipids in each leaflet) and peptide molecules were placed over the upper leaflet at non-interacting distance (>10 Å). Lysine residues were protonated while histidine residue was protonated only on nitrogen in position δ. Initial peptide structure was obtained via I-TASSER [108] prediction tool, that produced a similar construct as the one produced by PEPFOLD [109,110] software. This structure was almost completely helical. Amidation of the C-terminus was achieved via the CHARMM terminal group patching functionality which is fully integrated in the CHARMM-GUI workflow. In case of calculations with eight peptides, they were placed next to each other but not in contact. A water layer of 50-Å thickness was added above and below the lipid bilayer which resulted in about 15,000 water molecules (30,000 in the case of CL) with small variations depending on the nature of the membrane. Systems were neutralized with Na^+ or Cl^- counterions.

MD simulations were performed using GROMACS software [111] and CHARMM36 force field [112] under semi-isotropic (for bilayers) and isotropic (for micelles) NPT conditions [113,114]. The TIP3P model [54] was used to describe water molecules. Each system was energy-minimized with a steepest-descent algorithm for 5000 steps. Systems were equilibrated with the Berendsen barostat [115] and Parrinello-Rahman barostat [116,117] was used to maintain pressure (1 bar) semi-isotropically with a time constant of 5 ps and a compressibility of 4.5×10^{-5} bar^{-1}. The Nose-Hoover thermostat [118,119] was chosen to maintain the systems at 310 K with a time constant of 1 ps. All bonds were constrained using the LINear Constraint Solver (LINCS) algorithm, which allowed an integration step of 2 fs. PBC (periodic boundary conditions) were employed for all simulations, and the particle mesh Ewald (PME) method [120] was used for long-range electrostatic interactions. After the standard CHARMM-GUI minimization and equilibration steps [113], the production run was performed for 500 ns (except when mentioned explicitly) and the whole process (minimization, equilibration and production run) was repeated once in the absence of peptide and twice in its presence. Convergence was assessed using RMSD and polar contacts analysis (see Figure S12).

All MD trajectories were analyzed using GROMACS tools [121,122] and Fatslim [123]. MOLMOL [124] and VMD [125] were used for visualization. Graphs and images were produced with GNUplot [126] and PyMol [127].

4. Conclusions

In this work, we have shown how the K11 peptide, largely unstructured in solution, assumes alpha helical conformation in the presence of biomimetic membranes. The interaction has very different consequences on the stability of the membrane depending on its nature. While PC and PE/PC bilayers are largely unaffected, PG, PS, PI and CL strongly interact with lysine residues. When examining bacterial-like mixtures containing both PG and PE, the large majority of the peptide-membrane interactions takes place with PG and the structurally related CL, if present. However, the same mechanism might well be active in the presence of PS, often exposed on the outer leaflet of cancer cells, which would suggest a potential anticancer activity of K11, as already described for its related peptides. The analysis of polar contacts reveals that lysine side chains tend to interact with oxygen atoms of the phosphate moiety (or the carboxylate of the serine in PS) rather than the OH of the glycerol or inositol head group, indicating that the recognition is based on the formation of salt bridges rather than H-bonds. This explains the lower affinity for PC and PE where the negative charge is neutralized by the choline and ethanolamine moieties, respectively. Once the salt bridges are formed, the peptide might penetrate as a screw, anchoring to the target with its N-terminus and twisting the membrane by further subsequent salt bridges involving pairs of lysine residues. The torque allows then the insertion of terminal hydrophobic side chains and eventually the internalization of the full peptide. Once inside, K11 can approach phospholipid head groups on opposite leaflets causing an effective disruption of the membrane potentially leading to the bacterial death.

Supplementary Materials: The following are available online at https://www.mdpi.com/1424-8247/14/1/1/s1, Figure S1: Sequence alignment of K11 peptide used as a bait in the ADAPTABLE web server; Figure S2: ^{1}H,^{13}C-HSQC spectral regions and assignment of K11 in solution and in the presence of DPC micelles, Figure S3: Minimum distance of each lysine side chain amine (atom name NZ) from membrane phosphorus atoms along the simulation trajectory of K11 interacting with DPC micelles, Figure S4: ^{1}H NMR normalized spectra of K11 in the presence of DMPC/DHPC, DMPC/DHPC/DMPE, DMPC/DHPC/DMPG, and DMPC/DHPC/DMPS bicelles, Figures S5 and S6: Occurrence of polar atom contacts between K11 peptide and various membrane bilayers, Figure S7: Area per lipid in bilayers containing various phospholipids compositions as calculated from MD simulations in the presence of eight K11 peptides, Figure S8: Order parameter of C-H moieties in palmitoyl side chains in membranes containing various phospholipids compositions as calculated from multiple repetitions of MD simulations in the absence (2 repetitions in black labeled as 1 and 2) and in the presence (3 repetitions in red labeled from 1 to 3) of eight K11 peptides. The panel in the right bottom corner is an example of MD snapshot with POPE/POPG bilayer (color code in the caption of Figure 4). TOCL2 refers to CL, Figure S9: Electron density profiles for POPC, POPG and POPE/POPG/CL in presence of eight K11 peptides, Figure S10. Order parameter of C-H moieties in palmitoyl side chains in membranes containing various phospholipids compositions as calculated from multiple repetitions of MD simulations in the absence (2 repetitions in black labeled as 1 and 2) and in the presence (3 repetitions in red labeled from 1 to 3) of K11 peptide initially placed inside the bilayer. The panel in the right bottom corner is an example of MD snapshot with POPE/POPG/CL bilayer (color code in the caption of Figure 4). TOCL2 refers to C, Figure S11: Analytical purity of K11 peptide; Table S1: ^{1}H and ^{13}C NMR assignment of K11 peptide, Figure S12: Convergence analysis of the simulation of K11 peptide in the presence of POPE/POPG membrane. (A) Peptide RMSD (Cα carbon); (B) Polar contact block analysis in time intervals.

Author Contributions: Conceptualization, F.R.-M. and N.D.; Data curation, N.D.; Formal analysis, F.R.-M. and C.H.-L.; Funding acquisition, C.S. and N.D.; Investigation, F.R.-M., C.H.-L. and V.A.; Methodology, F.R.-M. and N.D.; Project administration, N.D.; Resources, P.S. and C.S.; Software, F.R.-M. and N.D.; Supervision, N.D.; Validation, N.D.; Visualization, F.R.-M. and N.D.; Writing—original draft, F.R.-M. and N.D.; Writing—review & editing, F.R.-M., C.H.-L., P.S., C.S. and N.D. All authors have read and agreed to the published version of the manuscript.

Funding: Francisco Ramos-Martín's PhD scholarship was co-funded by Conseil régional des Hauts-de-France and by European Fund for Economic and Regional Development (FEDER); Claudia Herrera-León's PhD scholarship was funded by the National Council for Science and Technology (CONACYT). This work was partly supported through the ANR Natural-Arsenal project. Publication fees were partly funded by the University of Picardie Jules Verne.

Acknowledgments: We would like to thank Zakaria Bouchouireb, Professor Manuel Dauchez, University of Reims Champagne-Ardenne (URCA), UMR CNRS 7369 MEDyC for useful discussion and Dominique Cailleu for his competence in the setting up of NMR experiments. We also thank the Matrics platform at the University "Picardie Jules Verne" and the "Méso-centre de Calcul Scientifique Intensif" at the University of Lille for providing computing resources.

Conflicts of Interest: The authors declare that they have no competing interest.

References

1. Allegranzi, B.; Bagheri Nejad, S.; Combescure, C.; Graafmans, W.; Attar, H.; Donaldson, L.; Pittet, D. Burden of endemic health-care-associated infection in developing countries: Systematic review and meta-analysis. *Lancet* **2011**, *377*, 228–241. [CrossRef]
2. Ibrahim, M.E.; Bilal, N.E.; Hamid, M.E. Increased multi-drug resistant Escherichia coli from hospitals in Khartoum state, Sudan. *Afr. Health Sci.* **2012**, *12*, 368–375. [CrossRef] [PubMed]
3. Pendleton, J.N.; Gorman, S.P.; Gilmore, B.F. Clinical relevance of the ESKAPE pathogens. *Expert Rev. Anti. Infect. Ther.* **2013**, *11*, 297–308. [CrossRef]
4. Mulani, M.S.; Kamble, E.E.; Kumkar, S.N.; Tawre, M.S.; Pardesi, K.R. Emerging Strategies to Combat ESKAPE Pathogens in the Era of Antimicrobial Resistance: A Review. *Front. Microbiol.* **2019**, *10*, 539. [CrossRef]
5. Tacconelli, E.; Carrara, E.; Savoldi, A.; Harbarth, S.; Mendelson, M.; Monnet, D.L.; Pulcini, C.; Kahlmeter, G.; Kluytmans, J.; Carmeli, Y.; et al. Discovery, research, and development of new antibiotics: The WHO priority list of antibiotic-resistant bacteria and tuberculosis. *Lancet Infect. Dis.* **2018**, *18*, 318–327. [CrossRef]
6. Wang, G. *Antimicrobial Peptides: Discovery, Design and Novel Therapeutic Strategies*, 2nd ed.; CABI: Wallingford, UK, 2017; ISBN 9781786390394.
7. Joo, H.-S.; Fu, C.-I.; Otto, M. Bacterial strategies of resistance to antimicrobial peptides. *Philos. Trans. R. Soc. Lond. B Biol. Sci.* **2016**, *371*, 20150292. [CrossRef]

8. Yu, G.; Baeder, D.Y.; Regoes, R.R.; Rolff, J. Predicting drug resistance evolution: Insights from antimicrobial peptides and antibiotics. *Proc. Biol. Sci.* **2018**, *285*, 20172687. [CrossRef]
9. Kintses, B.; Méhi, O.; Ari, E.; Számel, M.; Györkei, Á.; Jangir, P.K.; Nagy, I.; Pál, F.; Fekete, G.; Tengölics, R.; et al. Phylogenetic barriers to horizontal transfer of antimicrobial peptide resistance genes in the human gut microbiota. *Nat. Microbiol.* **2019**, *4*, 447–458. [CrossRef] [PubMed]
10. Berglund, N.A.; Piggot, T.J.; Jefferies, D.; Sessions, R.B.; Bond, P.J.; Khalid, S. Interaction of the Antimicrobial Peptide Polymyxin B1 with Both Membranes of E. coli: A Molecular Dynamics Study. *PLoS Comput. Biol.* **2015**, *11*, e1004180. [CrossRef] [PubMed]
11. Pfalzgraff, A.; Brandenburg, K.; Weindl, G. Antimicrobial Peptides and Their Therapeutic Potential for Bacterial Skin Infections and Wounds. *Front. Pharmacol.* **2018**, *9*, 281. [CrossRef] [PubMed]
12. Zhu, S.; Sani, M.-A.; Separovic, F. Interaction of cationic antimicrobial peptides from Australian frogs with lipid membranes. *Pept. Sci.* **2018**, *110*, e24061. [CrossRef]
13. Giuliani, A.; Pirri, G.; Bozzi, A.; Di Giulio, A.; Aschi, M.; Rinaldi, A.C. Antimicrobial peptides: Natural templates for synthetic membrane-active compounds. *Cell. Mol. Life Sci.* **2008**, *65*, 2450–2460. [CrossRef] [PubMed]
14. Jin-Jiang, H.; Jin-Chun, L.; Min, L.; Qing-Shan, H.; Guo-Dong, L. The Design and Construction of K11: A Novel α-Helical Antimicrobial Peptide. *Int. J. Microbiol.* **2012**, *2012*, 764834. [CrossRef] [PubMed]
15. Rishi, P.; Vashist, T.; Sharma, A.; Kaur, A.; Kaur, A.; Kaur, N.; Kaur, I.P.; Tewari, R. Efficacy of designer K11 antimicrobial peptide (a hybrid of melittin, cecropin A1 and magainin 2) against Acinetobacter baumannii-infected wounds. *Pathog. Dis.* **2018**, *76*, fty072. [CrossRef] [PubMed]
16. Hobby, C.R.; Herndon, J.L.; Morrow, C.A.; Peters, R.E.; Symes, S.J.K.; Giles, D.K. Exogenous fatty acids alter phospholipid composition, membrane permeability, capacity for biofilm formation, and antimicrobial peptide susceptibility in Klebsiella pneumoniae. *MicrobiologyOpen* **2019**, *8*, e00635. [CrossRef]
17. Benamara, H.; Rihouey, C.; Jouenne, T.; Alexandre, S. Impact of the biofilm mode of growth on the inner membrane phospholipid composition and lipid domains in Pseudomonas aeruginosa. *Biochim. Biophys. Acta BBA Biomembr.* **2011**, *1808*, 98–105. [CrossRef] [PubMed]
18. Harding, C.M.; Hennon, S.W.; Feldman, M.F. Uncovering the mechanisms of Acinetobacter baumannii virulence. *Nat. Rev. Microbiol.* **2018**, *16*, 91–102. [CrossRef]
19. Lopalco, P.; Stahl, J.; Annese, C.; Averhoff, B.; Corcelli, A. Identification of unique cardiolipin and monolysocardiolipin species in Acinetobacter baumannii. *Sci. Rep.* **2017**, *7*, 2972. [CrossRef]
20. Shin, S.Y.; Lee, S.H.; Yang, S.T.; Park, E.J.; Lee, D.G.; Lee, M.K.; Eom, S.H.; Song, W.K.; Kim, Y.; Hahm, K.S.; et al. Antibacterial, antitumor and hemolytic activities of alpha-helical antibiotic peptide, P18 and its analogs. *J. Pept. Res.* **2001**, *58*, 504–514. [CrossRef]
21. Scott, M.G.; Yan, H.; Hancock, R.E.W. Biological Properties of Structurally Related α-Helical Cationic Antimicrobial Peptides. *Infect. Immun.* **1999**, *67*, 2005–2009. [CrossRef]
22. Randle, C.L.; Albro, P.W.; Dittmer, J.C. The phosphoglyceride composition of gram-negative bacteria and the changes in composition during growth. *Biochim. Biophys. Acta BBA Lipids Lipid Metab.* **1969**, *187*, 214–220. [CrossRef]
23. Shokri, A.; Larsson, G. Characterisation of the Escherichia coli membrane structure and function during fedbatch cultivation. *Microb. Cell Fact.* **2004**, *3*, 9. [CrossRef] [PubMed]
24. Modak, M.J.; Nair, S.; Venkataraman, A. Studies on the Fatty Acid Composition of some Salmonellas. *J. Gen. Microbiol.* **1970**, *60*, 151–157. [CrossRef] [PubMed]
25. Galanth, C.; Abbassi, F.; Lequin, O.; Ayala-Sanmartin, J.; Ladram, A.; Nicolas, P.; Amiche, M. Mechanism of antibacterial action of dermaseptin B2: Interplay between helix-hinge-helix structure and membrane curvature strain. *Biochemistry* **2009**, *48*, 313–327. [CrossRef] [PubMed]
26. Shivaji, S.; Chaturvedi, P.; Suresh, K.; Reddy, G.S.N.; Dutt, C.B.S.; Wainwright, M.; Narlikar, J.V.; Bhargava, P.M. Bacillus aerius sp. nov., Bacillus aerophilus sp. nov., Bacillus stratosphericus sp. nov. and Bacillus altitudinis sp. nov., isolated from cryogenic tubes used for collecting air samples from high altitudes. *Int. J. Syst. Evol. Microbiol.* **2006**, *56*, 1465–1473. [CrossRef] [PubMed]
27. Kamp, J.A.F.; Houtsmuller, U.M.T.; Van Deenen, L.L.M. On the phospholipids of Bacillus megaterium. *Biochim. Biophys. Acta BBA Lipids Lipid Metab.* **1965**, *106*, 438–441. [CrossRef]
28. Bishop, D.G.; Op den Kamp, J.A.; van Deenen, L.L. The distribution of lipids in the protoplast membranes of Bacillus subtilis. A study with phospholipase C and trinitrobenzenesulphonic acid. *Eur. J. Biochem.* **1977**, *80*, 381–391. [CrossRef]
29. Komaratat, P.; Kates, M. The lipid composition of a halotolerant species of Staphylococcus epidermidis. *Biochim. Biophys. Acta* **1975**, *398*, 464–484. [CrossRef]
30. Kanemasa, Y.; Yoshioka, T.; Hayashi, H. Alteration of the phospholipid composition of Staphylococcus aureus cultured in medium containing NaCl. *Biochim. Biophys. Acta* **1972**, *280*, 444–450.
31. De Bony, J.; Lopez, A.; Gilleron, M.; Welby, M.; Lanéelle, G.; Rousseau, B.; Beaucourt, J.P.; Tocanne, J.F. Transverse and lateral distribution of phospholipids and glycolipids in the membrane of the bacterium Micrococcus luteus. *Biochemistry* **1989**, *28*, 3728–3737. [CrossRef]
32. Tari, A.; Huang, L. Structure and function relationship of phosphatidylglycerol in the stabilization of phosphatidylethanolamine bilayer. *Biochemistry* **1989**, *28*, 7708–7712. [CrossRef] [PubMed]

33. Murzyn, K.; Róg, T.; Pasenkiewicz-Gierula, M. Phosphatidylethanolamine-phosphatidylglycerol bilayer as a model of the inner bacterial membrane. *Biophys. J.* **2005**, *88*, 1091–1103. [CrossRef] [PubMed]
34. Reddy, P.H.; Burra, S.S.; Murthy, P.S. Correlation between calmodulin-like protein, phospholipids, and growth in glucose-grown Mycobacterium phlei. *Can. J. Microbiol.* **1992**, *38*, 339–342. [CrossRef] [PubMed]
35. Bevers, E.M.; Williamson, P.L. Getting to the Outer Leaflet: Physiology of Phosphatidylserine Exposure at the Plasma Membrane. *Physiol. Rev.* **2016**, *96*, 605–645. [CrossRef]
36. Kenis, H.; Reutelingsperger, C. Targeting phosphatidylserine in anti-cancer therapy. *Curr. Pharm. Des.* **2009**, *15*, 2719–2723. [CrossRef]
37. Zwaal, R.F.A.; Comfurius, P.; Bevers, E.M. Surface exposure of phosphatidylserine in pathological cells. *CMLS Cell. Mol. Life Sci.* **2005**, *62*, 971–988. [CrossRef] [PubMed]
38. Ramos-Martín, F.; Annaval, T.; Buchoux, S.; Sarazin, C.; D'Amelio, N. ADAPTABLE: A comprehensive web platform of antimicrobial peptides tailored to the user's research. *Life Sci. Alliance* **2019**, *2*, e201900512. [CrossRef]
39. Shin, S.Y.; Lee, M.K.; Kim, K.L.; Hahm, K.S. Structure-antitumor and hemolytic activity relationships of synthetic peptides derived from cecropin A-magainin 2 and cecropin A-melittin hybrid peptides. *J. Pept. Res.* **1997**, *50*, 279–285. [CrossRef]
40. Park, Y.; Lee, D.G.; Hahm, K.-S. Antibiotic activity of Leu-Lys rich model peptides. *Biotechnol. Lett.* **2003**, *25*, 1305–1310. [CrossRef]
41. Gray, M.; Gong, J.; Nguyen, V.; Osada, T.; Hartman, Z.; Hutchins, J.; Freimark, B.; Lyerly, K. Targeting of phosphatidylserine by monoclonal antibodies augments the activity of paclitaxel and anti-PD1/PD-L1 therapy in the murine breast model E0771. *J. ImmunoTher. Cancer* **2015**, *3*, P357. [CrossRef]
42. Riedl, S.; Rinner, B.; Asslaber, M.; Schaider, H.; Walzer, S.; Novak, A.; Lohner, K.; Zweytick, D. In search of a novel target-phosphatidylserine exposed by non-apoptotic tumor cells and metastases of malignancies with poor treatment efficacy. *Biochim. Biophys. Acta* **2011**, *1808*, 2638–2645. [CrossRef] [PubMed]
43. Ran, S.; Downes, A.; Thorpe, P.E. Increased exposure of anionic phospholipids on the surface of tumor blood vessels. *Cancer Res.* **2002**, *62*, 6132–6140. [PubMed]
44. Hasim, S.; Vaughn, E.N.; Donohoe, D.; Gordon, D.M.; Pfiffner, S.; Reynolds, T.B. Influence of phosphatidylserine and phosphatidylethanolamine on farnesol tolerance in Candida albicans. *Yeast* **2018**, *35*, 343–351. [CrossRef] [PubMed]
45. Khandelwal, N.K.; Sarkar, P.; Gaur, N.A.; Chattopadhyay, A.; Prasad, R. Phosphatidylserine decarboxylase governs plasma membrane fluidity and impacts drug susceptibilities of Candida albicans cells. *Biochim. Biophys. Acta Biomembr.* **2018**, *1860*, 2308–2319. [CrossRef] [PubMed]
46. Cassilly, C.; Reynolds, T. PS, It's Complicated: The Roles of Phosphatidylserine and Phosphatidylethanolamine in the Pathogenesis of Candida albicans and Other Microbial Pathogens. *J. Fungi* **2018**, *4*, 28. [CrossRef] [PubMed]
47. Makovitzki, A.; Avrahami, D.; Shai, Y. Ultrashort antibacterial and antifungal lipopeptides. *Proc. Natl. Acad. Sci. USA* **2006**, *103*, 15997–16002. [CrossRef]
48. Lösel, D.M. Lipids in the Structure and Function of Fungal Membranes. In *Biochemistry of Cell Walls and Membranes in Fungi*; Kuhn, P.J., Ed.; Springer: Berlin/Heidelberg, Germany, 1990; pp. 119–133.
49. Mahto, K.K.; Singh, A.; Khandelwal, N.K.; Bhardwaj, N.; Jha, J.; Prasad, R. An assessment of growth media enrichment on lipid metabolome and the concurrent phenotypic properties of Candida albicans. *PLoS ONE* **2014**, *9*, e113664. [CrossRef]
50. Wishart, D.S.; Sykes, B.D.; Richards, F.M. The chemical shift index: A fast and simple method for the assignment of protein secondary structure through NMR spectroscopy. *Biochemistry* **1992**, *31*, 1647–1651. [CrossRef]
51. Wishart, D.S.; Sykes, B.D. The 13 C Chemical-Shift Index: A simple method for the identification of protein secondary structure using 13 C chemical-shift data. *J. Biomol. NMR* **1994**, *4*, 171–180. [CrossRef]
52. Wishart, D.S. Interpreting protein chemical shift data. *Prog. Nucl. Magn. Reson. Spectrosc.* **2011**, *58*, 62–87. [CrossRef]
53. Beswick, V.; Guerois, R.; Cordier-Ochsenbein, F.; Coïc, Y.M.; Tam, H.D.; Tostain, J.; Noël, J.P.; Sanson, A.; Neumann, J.M. Dodecylphosphocholine micelles as a membrane-like environment: New results from NMR relaxation and paramagnetic relaxation enhancement analysis. *Eur. Biophys. J.* **1999**, *28*, 48–58. [CrossRef] [PubMed]
54. Berendsen, H.J.C.; Postma, J.P.M.; van Gunsteren, W.F.; Hermans, J. Interaction Models for Water in Relation to Protein Hydration. In *The Jerusalem Symposia on Quantum Chemistry and Biochemistry*; Springer Publishing: New York, NY, USA, 1981; pp. 331–342.
55. Porcelli, F.; Ramamoorthy, A.; Barany, G.; Veglia, G. On the role of NMR spectroscopy for characterization of antimicrobial peptides. *Methods Mol. Biol.* **2013**, *1063*, 159–180. [PubMed]
56. van Dam, L.; Karlsson, G.; Edwards, K. Direct observation and characterization of DMPC/DHPC aggregates under conditions relevant for biological solution NMR. *Biochim. Biophys. Acta* **2004**, *1664*, 241–256. [CrossRef] [PubMed]
57. Marcotte, I.; Auger, M. Bicelles as model membranes for solid- and solution-state NMR studies of membrane peptides and proteins. *Concepts Magn. Reson. Part A* **2005**, *24A*, 17–37. [CrossRef]
58. Davis, J.H. The description of membrane lipid conformation, order and dynamics by 2H-NMR. *Biochim. Biophys. Acta BBA Rev. Biomembr.* **1983**, *737*, 117–171. [CrossRef]
59. Molugu, T.R.; Lee, S.; Brown, M.F. Concepts and Methods of Solid-State NMR Spectroscopy Applied to Biomembranes. *Chem. Rev.* **2017**, *117*, 12087–12132. [CrossRef]
60. Salnikov, E.S.; Mason, A.J.; Bechinger, B. Membrane order perturbation in the presence of antimicrobial peptides by (2)H solid-state NMR spectroscopy. *Biochimie* **2009**, *91*, 734–743. [CrossRef]

61. Romantsov, T.; Guan, Z.; Wood, J.M. Cardiolipin and the osmotic stress responses of bacteria. *Biochim. Biophys. Acta* **2009**, *1788*, 2092–2100. [CrossRef]
62. Sendecki, A.M.; Poyton, M.F.; Baxter, A.J.; Yang, T.; Cremer, P.S. Supported Lipid Bilayers with Phosphatidylethanolamine as the Major Component. *Langmuir* **2017**, *33*, 13423–13429. [CrossRef]
63. Lewis, R.N.A.H.; McElhaney, R.N. The physicochemical properties of cardiolipin bilayers and cardiolipin-containing lipid membranes. *Biochim. Biophys. Acta* **2009**, *1788*, 2069–2079. [CrossRef]
64. Szoka, F., Jr.; Papahadjopoulos, D. Comparative properties and methods of preparation of lipid vesicles (liposomes). *Annu. Rev. Biophys. Bioeng.* **1980**, *9*, 467–508. [CrossRef] [PubMed]
65. Papahadjopoulos, D.; Miller, N. Phospholipid model membranes. I. Structural characteristics of hydrated liquid crystals. *Biochim. Biophys. Acta* **1967**, *135*, 624–638. [CrossRef]
66. Litman, B.J. Lipid model membranes. Characterization of mixed phospholipid vesicles. *Biochemistry* **1973**, *12*, 2545–2554. [CrossRef] [PubMed]
67. Tinker, D.O.; Pinteric, L. On the identification of lamellar and hexagonal phases in negatively stained phospholipid-water systems. *Biochemistry* **1971**, *10*, 860–865. [CrossRef] [PubMed]
68. Junger, E.; Reinauer, H. Liquid crystalline phases of hydrated phosphatidylethanolamine. *Biochim. Biophys. Acta* **1969**, *183*, 304–308. [CrossRef]
69. Yesylevskyy, S.O.; Rivel, T.; Ramseyer, C. The influence of curvature on the properties of the plasma membrane. Insights from atomistic molecular dynamics simulations. *Sci. Rep.* **2017**, *7*, 1–13. [CrossRef]
70. Brown, M.F. Curvature forces in membrane lipid-protein interactions. *Biochemistry* **2012**, *51*, 9782–9795. [CrossRef]
71. Klebe, G. Protein–Ligand Interactions as the Basis for Drug Action. In *Drug Design*; Springer: Berlin/Heidelberg, Germany, 2013; pp. 61–88.
72. Seeburger, P. *Calcplot3d, an Exploration Environment for Multivariable Calculus-Taylor Polynomials of a Function of Two Variables (1st and 2nd Degree)*; Convergence; MAA: Washington, DC, USA, 2011.
73. Teilum, K.; Kunze, M.B.A.; Erlendsson, S.; Kragelund, B.B. (S)Pinning down protein interactions by NMR. *Protein Sci.* **2017**, *26*, 436–451. [CrossRef]
74. Yang, Z.; Choi, H.; Weisshaar, J.C. Melittin-Induced Permeabilization, Re-sealing, and Re-permeabilization of E. coli Membranes. *Biophys. J.* **2018**, *114*, 368–379. [CrossRef]
75. Avci, F.G.; Akbulut, B.S.; Ozkirimli, E. Membrane Active Peptides and Their Biophysical Characterization. *Biomolecules* **2018**, *8*, 77. [CrossRef]
76. Orsi, M.; Noro, M.G.; Essex, J.W. Dual-resolution molecular dynamics simulation of antimicrobials in biomembranes. *J. R. Soc. Interface* **2011**, *8*, 826–841. [CrossRef] [PubMed]
77. Isralewitz, B.; Baudry, J.; Gullingsrud, J.; Kosztin, D.; Schulten, K. Steered molecular dynamics investigations of protein function. *J. Mol. Graph. Model.* **2001**, *19*, 13–25. [CrossRef]
78. Kästner, J. Umbrella sampling. *WIREs Comput. Mol. Sci.* **2011**, *1*, 932–942. [CrossRef]
79. Domański, J.; Hedger, G.; Best, R.B.; Stansfeld, P.J.; Sansom, M.S.P. Convergence and Sampling in Determining Free Energy Landscapes for Membrane Protein Association. *J. Phys. Chem. B* **2017**, *121*, 3364–3375. [CrossRef]
80. Barducci, A.; Bonomi, M.; Parrinello, M. Metadynamics. *WIREs Comput. Mol. Sci.* **2011**, *1*, 826–843. [CrossRef]
81. Bussi, G.; Laio, A. Using metadynamics to explore complex free-energy landscapes. *Nat. Rev. Phys.* **2020**, *2*, 200–212. [CrossRef]
82. Marrink, S.J.; Corradi, V.; Souza, P.C.T.; Ingólfsson, H.I.; Tieleman, D.P.; Sansom, M.S.P. Computational Modeling of Realistic Cell Membranes. *Chem. Rev.* **2019**, *119*, 6184–6226. [CrossRef]
83. Mori, T.; Miyashita, N.; Im, W.; Feig, M.; Sugita, Y. Molecular dynamics simulations of biological membranes and membrane proteins using enhanced conformational sampling algorithms. *Biochim. Biophys. Acta* **2016**, *1858*, 1635–1651. [CrossRef]
84. Nguyen, L.T.; Haney, E.F.; Vogel, H.J. The expanding scope of antimicrobial peptide structures and their modes of action. *Trends Biotechnol.* **2011**, *29*, 464–472. [CrossRef]
85. Aisenbrey, C.; Marquette, A.; Bechinger, B. The Mechanisms of Action of Cationic Antimicrobial Peptides Refined by Novel Concepts from Biophysical Investigations. *Adv. Exp. Med. Biol.* **2019**, *1117*, 33–64.
86. Marquette, A.; Bechinger, B. Biophysical Investigations Elucidating the Mechanisms of Action of Antimicrobial Peptides and Their Synergism. *Biomolecules* **2018**, *8*, 18. [CrossRef] [PubMed]
87. Ouellet, M.; Bernard, G.; Voyer, N.; Auger, M. Insights on the interactions of synthetic amphipathic peptides with model membranes as revealed by 31P and 2H solid-state NMR and infrared spectroscopies. *Biophys. J.* **2006**, *90*, 4071–4084. [CrossRef] [PubMed]
88. Shahane, G.; Ding, W.; Palaiokostas, M.; Azevedo, H.S.; Orsi, M. Interaction of Antimicrobial Lipopeptides with Bacterial Lipid Bilayers. *J. Membr. Biol.* **2019**, *252*, 317–329. [CrossRef] [PubMed]
89. Dufourc, E.J.; Dufourcq, J.; Birkbeck, T.H.; Freer, J.H. Delta-haemolysin from Staphylococcus aureus and model membranes. A solid-state 2H-NMR and 31P-NMR study. *Eur. J. Biochem.* **1990**, *187*, 581–587. [CrossRef]
90. Dufourc, E.J.; Smith, I.C.; Dufourcq, J. Molecular details of melittin-induced lysis of phospholipid membranes as revealed by deuterium and phosphorus NMR. *Biochemistry* **1986**, *25*, 6448–6455. [CrossRef] [PubMed]
91. Henzler-Wildman, K.A.; Martinez, G.V.; Brown, M.F.; Ramamoorthy, A. Perturbation of the hydrophobic core of lipid bilayers by the human antimicrobial peptide LL-37. *Biochemistry* **2004**, *43*, 8459–8469. [CrossRef]

92. Zhuang, X.; Makover, J.R.; Im, W.; Klauda, J.B. A systematic molecular dynamics simulation study of temperature dependent bilayer structural properties. *Biochim. Biophys. Acta* **2014**, *1838*, 2520–2529. [CrossRef]
93. Smondyrev, A.M.; Berkowitz, M.L. Structure of Dipalmitoylphosphatidylcholine/Cholesterol Bilayer at Low and High Cholesterol Concentrations: Molecular Dynamics Simulation. *Biophys. J.* **1999**, *77*, 2075–2089. [CrossRef]
94. Xia, L.; Zhang, F.; Liu, Z.; Ma, J.; Yang, J. Expression and characterization of cecropinXJ, a bioactive antimicrobial peptide from (Bombycidae, Lepidoptera) in Escherichia coli. *Exp. Ther. Med.* **2013**, *5*, 1745–1751. [CrossRef]
95. Romoli, O.; Mukherjee, S.; Mohid, S.A.; Dutta, A.; Montali, A.; Franzolin, E.; Brady, D.; Zito, F.; Bergantino, E.; Rampazzo, C.; et al. Enhanced Silkworm Cecropin B Antimicrobial Activity against from Single Amino Acid Variation. *ACS Infect. Dis.* **2019**, *5*, 1200–1213. [CrossRef]
96. Liu, D.; Liu, J.; Li, J.; Xia, L.; Yang, J.; Sun, S.; Ma, J.; Zhang, F. A potential food biopreservative, CecXJ-37N, non-covalently intercalates into the nucleotides of bacterial genomic DNA beyond membrane attack. *Food Chem.* **2017**, *217*, 576–584. [CrossRef] [PubMed]
97. Chen, Y.-H.; Du, W.; Hagemeijer, M.C.; Takvorian, P.M.; Pau, C.; Cali, A.; Brantner, C.A.; Stempinski, E.S.; Connelly, P.S.; Ma, H.-C.; et al. Phosphatidylserine vesicles enable efficient en bloc transmission of enteroviruses. *Cell* **2015**, *160*, 619–630. [CrossRef] [PubMed]
98. Gasteiger, E.; Hoogland, C.; Gattiker, A.; Duvaud, S.; Wilkins, M.R.; Appel, R.D.; Bairoch, A. Protein Identification and Analysis Tools on the ExPASy Server. In *The Proteomics Protocols Handbook*; Springer Publishing: New York, NY, USA, 2005; pp. 571–607.
99. Nielsen, J.T.; Mulder, F.A.A. POTENCI: Prediction of temperature, neighbor and pH-corrected chemical shifts for intrinsically disordered proteins. *J. Biomol. NMR* **2018**, *70*, 141–165. [CrossRef] [PubMed]
100. Monnier, N.; Furlan, A.L.; Buchoux, S.; Deleu, M.; Dauchez, M.; Rippa, S.; Sarazin, C. Exploring the Dual Interaction of Natural Rhamnolipids with Plant and Fungal Biomimetic Plasma Membranes through Biophysical Studies. *Int. J. Mol. Sci.* **2019**, *20*, 1009. [CrossRef] [PubMed]
101. Furlan, A.L.; Castets, A.; Nallet, F.; Pianet, I.; Grélard, A.; Dufourc, E.J.; Géan, J. Red wine tannins fluidify and precipitate lipid liposomes and bicelles. A role for lipids in wine tasting? *Langmuir* **2014**, *30*, 5518–5526. [CrossRef] [PubMed]
102. Furlan, A.L.; Jobin, M.-L.; Pianet, I.; Dufourc, E.J.; Géan, J. Flavanol/lipid interaction: A novel molecular perspective in the description of wine astringency & bitterness and antioxidant action. *Tetrahedron* **2015**, *71*, 3143–3147.
103. Grélard, A.; Guichard, P.; Bonnafous, P.; Marco, S.; Lambert, O.; Manin, C.; Ronzon, F.; Dufourc, E.J. Hepatitis B subvirus particles display both a fluid bilayer membrane and a strong resistance to freeze drying: A study by solid-state NMR, light scattering, and cryo-electron microscopy/tomography. *FASEB J.* **2013**, *27*, 4316–4326. [CrossRef]
104. Davis, J.H.; Jeffrey, K.R.; Bloom, M.; Valic, M.I.; Higgs, T.P. Quadrupolar echo deuteron magnetic resonance spectroscopy in ordered hydrocarbon chains. *Chem. Phys. Lett.* **1976**, *42*, 390–394. [CrossRef]
105. Jo, S.; Lim, J.B.; Klauda, J.B.; Im, W. CHARMM-GUI Membrane Builder for mixed bilayers and its application to yeast membranes. *Biophys. J.* **2009**, *97*, 50–58. [CrossRef]
106. Lee, J.; Cheng, X.; Swails, J.M.; Yeom, M.S.; Eastman, P.K.; Lemkul, J.A.; Wei, S.; Buckner, J.; Jeong, J.C.; Qi, Y.; et al. CHARMM-GUI Input Generator for NAMD, GROMACS, AMBER, OpenMM, and CHARMM/OpenMM Simulations Using the CHARMM36 Additive Force Field. *J. Chem. Theory Comput.* **2016**, *12*, 405–413. [CrossRef]
107. Wu, E.L.; Cheng, X.; Jo, S.; Rui, H.; Song, K.C.; Dávila-Contreras, E.M.; Qi, Y.; Lee, J.; Monje-Galvan, V.; Venable, R.M.; et al. CHARMM-GUI Membrane Builder toward realistic biological membrane simulations. *J. Comput. Chem.* **2014**, *35*, 1997–2004. [CrossRef] [PubMed]
108. Zhang, Y. I-TASSER server for protein 3D structure prediction. *BMC Bioinform.* **2008**, *9*, 40. [CrossRef] [PubMed]
109. Thévenet, P.; Shen, Y.; Maupetit, J.; Guyon, F.; Derreumaux, P.; Tufféry, P. PEP-FOLD: An updated de novo structure prediction server for both linear and disulfide bonded cyclic peptides. *Nucleic Acids Res.* **2012**, *40*, W288–W293. [CrossRef]
110. Shen, Y.; Maupetit, J.; Derreumaux, P.; Tufféry, P. Improved PEP-FOLD Approach for Peptide and Miniprotein Structure Prediction. *J. Chem. Theory Comput.* **2014**, *10*, 4745–4758. [CrossRef] [PubMed]
111. Abraham, M.J.; Murtola, T.; Schulz, R.; Páll, S.; Smith, J.C.; Hess, B.; Lindahl, E. GROMACS: High performance molecular simulations through multi-level parallelism from laptops to supercomputers. *SoftwareX* **2015**, *1–2*, 19–25. [CrossRef]
112. Klauda, J.B.; Venable, R.M.; Freites, J.A.; O'Connor, J.W.; Tobias, D.J.; Mondragon-Ramirez, C.; Vorobyov, I.; MacKerell, A.D., Jr.; Pastor, R.W. Update of the CHARMM all-atom additive force field for lipids: Validation on six lipid types. *J. Phys. Chem. B* **2010**, *114*, 7830–7843. [CrossRef]
113. Jo, S.; Kim, T.; Im, W. Automated builder and database of protein/membrane complexes for molecular dynamics simulations. *PLoS ONE* **2007**, *2*, e880. [CrossRef]
114. Cheng, X.; Jo, S.; Lee, H.S.; Klauda, J.B.; Im, W. CHARMM-GUI micelle builder for pure/mixed micelle and protein/micelle complex systems. *J. Chem. Inf. Model.* **2013**, *53*, 2171–2180. [CrossRef]
115. Berendsen, H.J.C.; Postma, J.P.M.; van Gunsteren, W.F.; DiNola, A.; Haak, J.R. Molecular dynamics with coupling to an external bath. *J. Chem. Phys.* **1984**, *81*, 3684–3690. [CrossRef]
116. Parrinello, M.; Rahman, A. Polymorphic transitions in single crystals: A new molecular dynamics method. *J. Appl. Phys.* **1981**, *52*, 7182–7190. [CrossRef]
117. Nosé, S.; Klein, M.L. Constant pressure molecular dynamics for molecular systems. *Mol. Phys.* **1983**, *50*, 1055–1076. [CrossRef]

118. Nosé, S. A unified formulation of the constant temperature molecular dynamics methods. *J. Chem. Phys.* **1984**, *81*, 511–519. [CrossRef]
119. Hoover, W.G. Canonical dynamics: Equilibrium phase-space distributions. *Phys. Rev. A Gen. Phys.* **1985**, *31*, 1695–1697. [CrossRef] [PubMed]
120. Essmann, U.; Perera, L.; Berkowitz, M.L.; Darden, T.; Lee, H.; Pedersen, L.G. A smooth particle mesh Ewald method. *J. Chem. Phys.* **1995**, *103*, 8577–8593. [CrossRef]
121. Smith, D.J.; Klauda, J.B.; Sodt, A.J. Simulation Best Practices for Lipid Membranes [Article v1.0]. *Living J. Comput. Mol. Sci.* **2019**, *1*, 5966. [CrossRef]
122. Lemkul, J. From Proteins to Perturbed Hamiltonians: A Suite of Tutorials for the GROMACS-2018 Molecular Simulation Package [Article v1.0]. *Living J. Comput. Mol. Sci.* **2019**, *1*, 5068. [CrossRef]
123. Buchoux, S. FATSLiM: A fast and robust software to analyze MD simulations of membranes. *Bioinformatics* **2017**, *33*, 133–134. [CrossRef]
124. Koradi, R.; Billeter, M.; Wüthrich, K. MOLMOL: A program for display and analysis of macromolecular structures. *J. Mol. Graph.* **1996**, *14*, 51–55. [CrossRef]
125. Humphrey, W.; Dalke, A.; Schulten, K. VMD: Visual molecular dynamics. *J. Mol. Graph.* **1996**, *14*, 33–38. [CrossRef]
126. Janert, P.K. *Gnuplot in Action: Understanding Data with Graphs*; Manning Publications: Shelter Island, NY, USA, 2016; ISBN 9781933988399.
127. DeLano, W.L. Pymol: An open-source molecular graphics tool. *CCP4 Newsl. Protein Crystallogr.* **2002**, *40*, 82–92.

Article

Imidazole and Imidazolium Antibacterial Drugs Derived from Amino Acids

Adriana Valls [1], Jose J. Andreu [1], Eva Falomir [1], Santiago V. Luis [1], Elena Atrián-Blasco [2,3,*], Scott G. Mitchell [2,3,*] and Belén Altava [1,*]

[1] Departamento de Química Inorgánica y Orgánica, Universitat Jaume I, Av. Sos Baynat s/n, 12071 Castellón, Spain; avalls@uji.es (A.V.); al314093@uji.es (J.J.A.); efalomir@uji.es (E.F.); luiss@uji.es (S.V.L.)

[2] Instituto de Nanociencia y Materiales de Aragón (INMA), Consejo Superior de Investigaciones Científicas-Universidad de Zaragoza, 50009 Zaragoza, Spain

[3] CIBER de Bioingeniería, Biomateriales y Nanomedicina, Instituto de Salud Carlos III, 28029 Madrid, Spain

* Correspondence: elenaab@unizar.es (E.A.-B.); scott.mitchell@csic.es (S.G.M.); altava@uji.es (B.A.)

Received: 18 November 2020; Accepted: 17 December 2020; Published: 21 December 2020

Abstract: The antibacterial activity of imidazole and imidazolium salts is highly dependent upon their lipophilicity, which can be tuned through the introduction of different hydrophobic substituents on the nitrogen atoms of the imidazole or imidazolium ring of the molecule. Taking this into consideration, we have synthesized and characterized a series of imidazole and imidazolium salts derived from *L*-valine and *L*-phenylalanine containing different hydrophobic groups and tested their antibacterial activity against two model bacterial strains, Gram-negative *E. coli* and Gram-positive *B. subtilis*. Importantly, the results demonstrate that the minimum bactericidal concentration (MBC) of these derivatives can be tuned to fall close to the cytotoxicity values in eukaryotic cell lines. The MBC value of one of these compounds toward *B. subtilis* was found to be lower than the IC_{50} cytotoxicity value for the control cell line, HEK-293. Furthermore, the aggregation behavior of these compounds has been studied in pure water, in cell culture media, and in mixtures thereof, in order to determine if the compounds formed self-assembled aggregates at their bioactive concentrations with the aim of determining whether the monomeric species were in fact responsible for the observed antibacterial activity. Overall, these results indicate that imidazole and imidazolium compounds derived from *L*-valine and *L*-phenylalanine—with different alkyl lengths in the amide substitution—can serve as potent antibacterial agents with low cytotoxicity to human cell lines.

Keywords: imidazole and imidazolium salts; amino acid; antibacterial agents; aggregation; lipophilicity

1. Introduction

Aromatic heterocycles, particularly the imidazole ring, have been used in the last decades as structural skeletons to obtain different types of bioactive compounds with antibacterial, antifungal, anticancer, antiviral, antidiabetic, and other properties [1–4]. The search for new potent drug molecules derived from imidazole continues to be an intense area of investigation in medicinal chemistry [5–7]. Moreover, pharmaceutical research, manufacture, and regulation are enhancing the development of solid active ingredients, delivered as powders or tablets; however, many solid drugs which perform well in in vitro evaluation remain too insoluble for the body to absorb [8,9]. Most of the bioactive agents sold for pharmaceutical or food industries are salts [10,11], and in this context, ionic liquids (ILs) represent a promising class of drug candidates whose physicochemical and pharmaceutical properties can be easily tuned [12–17]. In this regard, the imidazolium skeleton can be transformed into ionic liquids with promisingly potent pharmacological properties [18–21]. Consequently, monoimidazolium [22–24] and bisimidazolium [25–28] salts have been explored as a new generation of antibacterial agents. In this context, amino acid-based monoimidazolium salts with good bacterial toxicity have been reported in the literature [29].

Our research group has an ongoing interest in the biomimetic and bioactive capacity of imidazole or imidazolium amino acid derivatives [30–33]. Herein, imidazole, monotopic, and ditopic imidazolium salts derived from *L*-valine and *L*-phenylalanine with different alkyl lengths in the amide substitution were synthesized and characterized comprehensively. The antibacterial activity of these amino acid-based imidazolium salts against Gram-negative *Escherichia coli* DH5-α (herein *E. coli*) and Gram-positive *Bacillus subtilis* 1904-E (herein *B. subtilis*) were evaluated and their cytotoxicity was also studied using a human embryonic kidney cell line (HEK-293). Finally, due to the amphiphilic character of these compounds and the strong tendency towards self-aggregation of ionic liquid-related surfactants based on imidazolium salts [34–37], we investigated the spontaneous aggregation behavior of these compounds in water and in bacterial cell culture medium. Through optical and scanning electron microscopy, as well as UV-vis and fluorescence spectroscopy, we have extracted structure-property relationships between the degree of aggregation/self-assembly of the *L*-valine and *L*-phenylalanine derivatives and their corresponding antibacterial activity and cytotoxicity [38].

2. Results

2.1. Synthesis

The imidazole-amino acid derivatives **1a**, **2a**, and **3a** were obtained from the corresponding α-amino amide as previously described [31]. Monotopic and ditopic -imidazolium salts **1b–3b** and **1c–3c** were obtained in high yield by treatment of the corresponding imidazole with benzyl bromide or 1,3-bromomethylbenzene, respectively, as described in our previous publications (Scheme 1) [30,32].

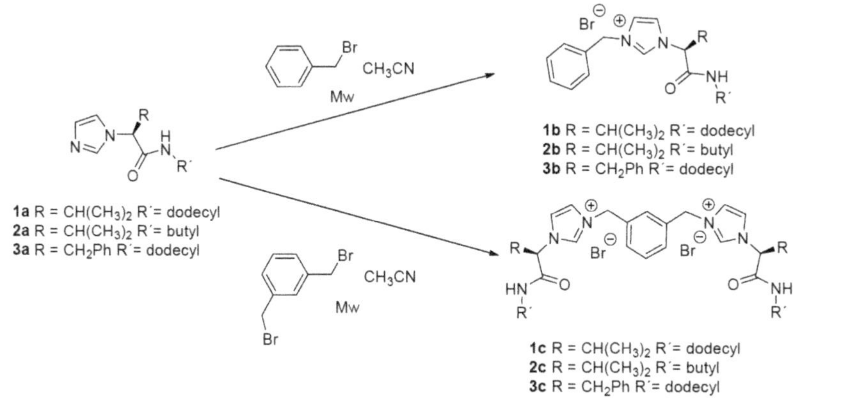

1a R = $CH(CH_3)_2$ R′= dodecyl
2a R = $CH(CH_3)_2$ R′= butyl
3a R = CH_2Ph R′= dodecyl

1b R = $CH(CH_3)_2$ R′= dodecyl
2b R = $CH(CH_3)_2$ R′= butyl
3b R = CH_2Ph R′= dodecyl

1c R = $CH(CH_3)_2$ R′= dodecyl
2c R = $CH(CH_3)_2$ R′= butyl
3c R = CH_2Ph R′= dodecyl

Scheme 1. Synthesis of the amino acid-based imidazolium salts in this report.

2.2. Antibacterial and Cytotoxicity Studies

The in vitro antibacterial activities of the synthesized compounds were examined against *E. coli* and *B. subtilis*. Bacteria were incubated in culture media with varying concentrations of the examined compound and the antibacterial properties were determined by observation of the optical density at 560 nm (bacteriostatic activity) and by the Resazurin cell viability assay (bactericidal activity). The corresponding minimal bactericidal concentration (MBC) and minimum inhibitory concentration (MIC) values that were obtained are summarized in Table 1 (please refer to Table S1 for MIC and MBC values in μM).

The outer membrane of Gram-negative bacteria such as *E. coli* includes porins, which allow the passage of small hydrophilic molecules across the membrane, and lipopolysaccharide molecules that extend into extracellular space. Thus, the observed trend in the activity results could be explained by the relative lipophilicity of the compounds combined with their capacity to disrupt the cell membrane [39,40]. The relative lipophilicity of the compounds was determined theoretically using VCCLab and Molinspiration softwares (LogP values Table 1) and experimentally (retention time values from HPLC, Table 1, Figures S3–S8). The HPLC method used was first validated using different lipophilic commercial compounds with LogP values from 1 to 4.5 (Figure S3a). The structure-activity relationships of the compounds will be further discussed in Section 3.

Table 1. Minimum inhibitory concentration (MIC, μg/mL), minimal bactericidal concentration (MBC, μg/mL), the half maximal inhibitory concentration IC_{50} (μg/mL) values and the partition coefficient log P values and HPLC retention times for the different compounds.

Entry	Compound	LogP [a]	Retention Time (min) [b]	*E. coli*		*B. subtilis*		HEK-293 IC_{50} μg/mL
				MIC μg/mL [e]	MBC μg/mL [e]	MIC μg/mL [e]	MBC μg/mL [e]	
1	**1a**	3.01 [c]	6.1	>2000	>2000	16	16	3.2 ± 0.5
2	**1b**	3.63 [d]	6.2	128	256	4	8	0.8 ± 0.2
3	**1c**	4.83 [d]	7.7	256	256	16	32	18 ± 4
4	**2a**	−0.67 [c]	3.4	>2000	>2000	>2000	>2000	>45
5	**2b**	−0.20 [d]	3.7	>2000	>2000	>1000	>1000	>79
6	**2c**	−1.34 [d]	3.1	1000	>2000	128	256	>142
7	**3a**	3.50 [c]	6.7	1000	>2000	16	16	7.3 ± 1.2
8	**3b**	4.26 [d]	8.8	32	128	4	4	61 ± 6
9	**3c**	5.96 [d]	10.0	2000	2000	64	64	37 ± 5
10	Alamethicin [41]	–	–			16	–	62.5 [f] [42]

[a] Average of values calculated using VCCLab and Molinspiration software. [b] Retention time in HPLC C18 reverse phase, CH_3CN/H_2O 70/30 (0.1% HCO_2H). [c] Value for the protonated forms. [d] Calculated for the imidazolium cations. [e] MIC/MBC values were obtained from a minimum of three separate experiments. Please refer to Supporting Information for MIC/MBC/IC_{50} values in μM. [f] Cytotoxicity against MRC-5 cells.

The MBC values obtained were plotted against the logP values of the corresponding compounds to study the possible correlation between activity and lipophilicity (Figure 1).

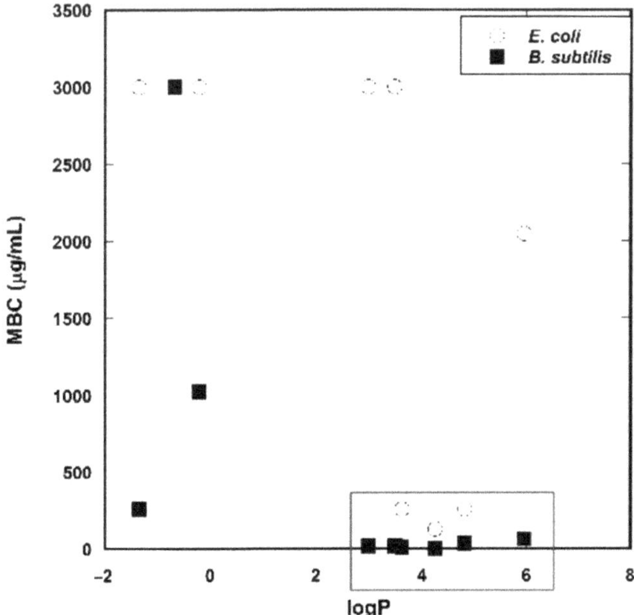

Figure 1. Plot of Log P vs. MBC. * For the sake of clarity, MBC values > 2000 μg/mL are given the value of 3000 μg/mL.

Electron microscopy is a powerful tool to further assess the effect of the imidazole derivatives and imidazolium salts on bacterial cell growth, inhibition, and death. Two compounds which showed from moderate to good antibacterial activity—the bisimidazolium salt **1c** and the monoimidazolium salt **3b**, respectively—were chosen for electron microscopy characterization. For these studies, *E. coli* and *B. subtilis* were inoculated for 20 h with each compound at their corresponding MIC (Figure 2) and $\frac{1}{2}$ MIC (Figures S1 and S2) concentrations and fixed with glutaraldehyde.

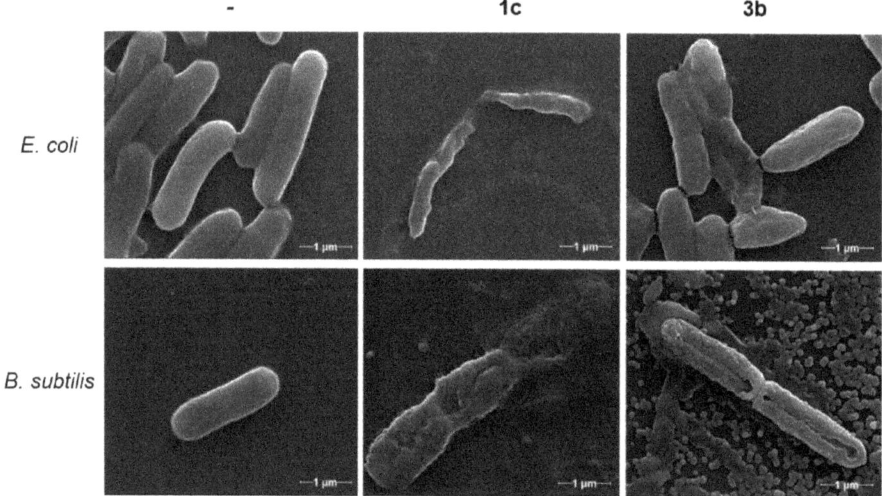

Figure 2. Scanning electron microscopy (SEM) images of *E. coli* and *B. subtilis* without treatment (-) and after incubation with compounds **1c** and **3b** at their corresponding MIC (60,000×). See Supporting Information for additional SEM images.

2.3. *Aggregation Studies*

The self-assembly of the compounds in aqueous medium and in the bacterial cell culture medium (LB broth) was investigated by optical and scanning electron microscopy, as well as UV-vis and fluorescence spectroscopy.

2.3.1. Fluorescence Spectroscopy

To investigate the microenvironment of the critical aggregation concentration (CAC) for self-assembly in water and in the bacterial cell culture medium by fluorescence, the intensity ratio of two of the peaks (I_1/I_3) of the pyrene fluorescence spectrum was used [43–45]. Plots of the pyrene I_1/I_3 ratio as a function of the total surfactant concentration show a typical sigmoidal decrease in the region where self-assembly takes place. At low concentrations, this ratio is larger as it corresponds to a polar environment for pyrene. When the surfactant concentration increases this ratio decreases rapidly, as the self-assembly favors the location of pyrene in a more hydrophobic environment, until reaching a roughly constant value because of the full incorporation of the probe into the hydrophobic region of the aggregates. Different approaches have been used to estimate CAC values from I_1/I_3 ratios [46]. The most common approach is the use of the break points, either directly or by extrapolating the values from the intersection of the two straight lines defined at the constant and variable regions of the I_1/I_3 sigmoidal curve [43,46–48]. As CAC represents the threshold of concentration at which self-aggregation starts, the corresponding value can be estimated from the break point at lower concentration (see Figures S9–S14) [49–52].

Fluorescence studies were carried out using MilliQ® water and 1/1 MilliQ® water/bacterial cell culture medium, because with the pure cell culture medium, a strong broad fluorescence emission band was observed precluding an accurate analysis. Furthermore, compound **3c** could not be studied due to solubility problems. The corresponding CACs obtained in water and in the 1/1 mixture of water/bacterial cell culture medium by fluorescence are shown in Table 2.

Furthermore, the MBC values of the compounds against *B. subtilis* were compared to their CAC (Figure 3). This comparison can shed light on the active form of the molecules exerting the antibacterial action, i.e., monomeric or aggregated structures.

Table 2. Estimated critical aggregation concentration (CAC) values obtained in aqueous and bacterial cell culture medium using fluorescence spectroscopy at 25 °C.

Entry	Amphiphilic Compound	CAC Fluorescence (mM) [a]		
		W [b]	CCM:W [c]	
			CAC1	CAC2
1	**1a**	0.085	0.004	0.045
2	**1b**	0.084	0.006	0.048
3	**1c**	0.010	0.016	0.21
4	**2a**	4.31	2.89	5.4
5	**2b**	4.57	3.46	8.4
6	**2c**	2.34	2.25	3.54
7	**3a**	0.033	0.018	0.325
8	**3b**	0.098	0.063	0.33
9	**3c**	nd [d]	nd [d]	

[a] CAC values from the break point. [b] In water. [c] In the 1/1 bacterial cell culture medium/water. [d] Low solubility.

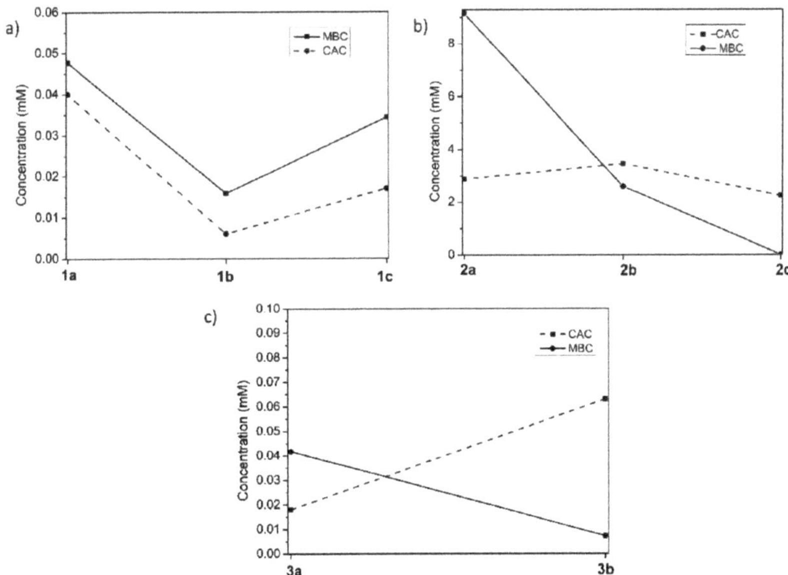

Figure 3. Correlation of MBC for *B. subtilis* and CAC for the different series of compounds in the bacterial cell culture medium: (**a**) valine derivatives with long alkyl chain; (**b**) valine derivatives with short chain; and (**c**) phenylalanine derivatives with long alkyl chain.

2.3.2. Optical Microscopy and Scanning Electron Microscopy (SEM)

The morphology of the aggregates in water, in 1/1 water/bacterial cell culture medium, and in the cell culture medium at concentrations above the CAC, were studied by optical microscopy (Figures S15–S17) and SEM (Figure 4 and Figure S18).

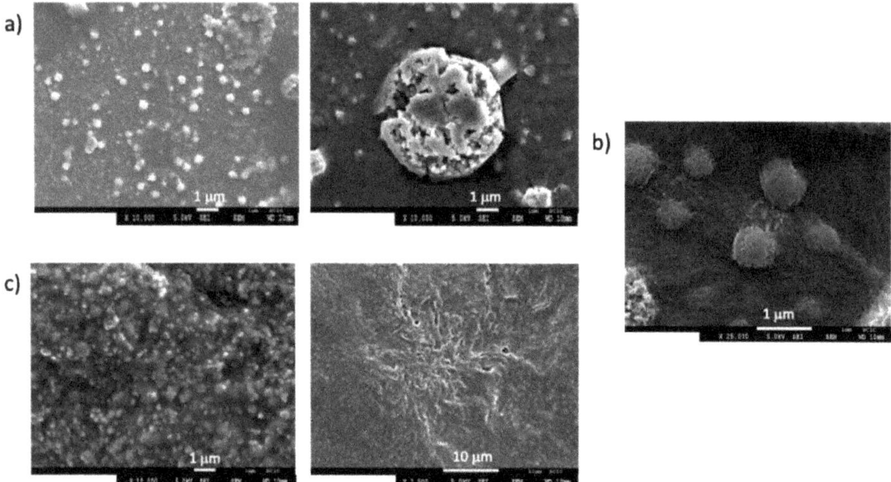

Figure 4. SEM images for **3b** (**a**) 0.5 mM in water; (**b**) 0.7 mM in 1/1 water/bacterial cell culture medium; and (**c**) 0.7 mM in the bacterial cell culture medium.

2.3.3. UV-Vis Spectroscopy

The aggregation and stability of the aggregates in the different solvents, water, 1/1 water/bacterial cell culture, and bacterial cell culture medium for **1a–c** were studied by UV-vis at 25 °C measuring the absorbance at 600 nm, for 1 mM colloidal solutions (Figure 5, Figures S19 and S20) [53,54].

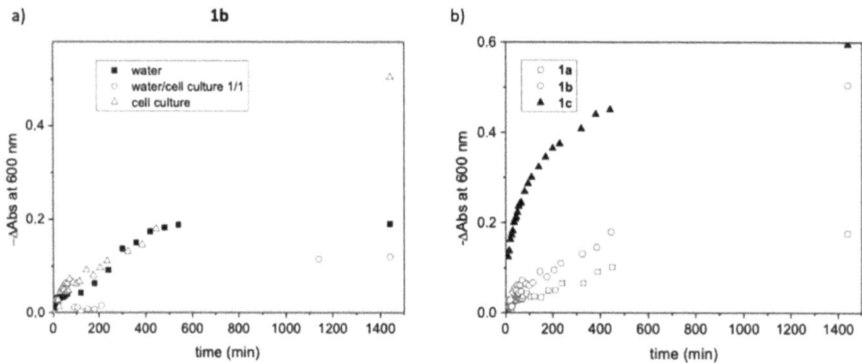

Figure 5. (**a**) Change in absorbance at 600 nm with respect to time for **1b** at 1 mM in different media. (**b**) Change in absorbance at 600 nm with respect to time for **1a–1c** at 1 mM in the bacterial cell culture medium.

3. Discussion

The Gram-positive cell wall is composed of a thick, multilayered peptidoglycan sheath outside of the cytoplasmic membrane, while the Gram-negative cell wall is composed of an outer membrane linked by lipoproteins to thin, mainly single-layered peptidoglycan. The peptidoglycan is located within the periplasmic space that is created between the outer and inner membranes [55]. It therefore follows that all the tested compounds were more active against *B. subtilis* (Gram-positive) than *E. coli* (Gram-negative) (i.e., see Table 1, entries 1 and 7). Compound **3b** (entry 8) proved to be the most active antibacterial agent possessing an MBC as low as 4 and 128 μg/mL against *B. subtilis* and *E. coli* respectively; while those compounds with shorter alkyl chains presented the lowest activity. Some of the compounds (Table 1, entries 2, 3, 6, 7, and 8) presented MIC values lower than MBC values. An antimicrobial compound is considered to be bactericidal whenever its MBC to MIC ratio is less than or equal to four. Compounds with MBC/MIC >4 are considered to be bacteriostatic [56]. In all the cases in which we could obtain exact MBC and MIC values (Table 1, entries 2, 3, 8 and 9), their ratio was equal to or minor than four. Therefore, compounds **1b**, **1c**, **3b**, and **3c** can be considered to possess true bactericidal activity.

The structural element which clearly relates to a better activity is the use of longer alkyl chains (compounds **1** and **3**) in contrast to shorter alkyl chains (compounds **2**), probably due to an increased lipophilicity. The toxicity towards miscellaneous bacterial strains of alkyl imidazolium salts has been reported to increase with the length of the alkyl chain [57]. The monoimidazolium salts with these longer alkyl chains (**1a-b** and **3a-b**) were more active against *B. subtilis* and *E. coli* than the bisimidazolium counterparts, with the monotopic salt **3b** showing the lowest MIC and MBC values (entries 2 and 8, Table 1), this indicates that the introduction of two hydrophobic alkyl chains contributes greatly to decrease the activity, opposite to the trends observed in the literature [27].

Regarding amino acid nature, the phenylalanine monotopic salt with long alkyl chain (**3b**) presented lower MIC and MBC values against *E. coli* than the analogous valine compound (**1b**) (entries 2 and 8, Table 1), however for ditopic salts, the behavior is the opposite, **3c** presented higher MIC and MBC values than **1c** (entries 3 and 9, Table 1).

Although a similar biological activity for the Gram-positive and Gram-negative organisms is preferred, it is not always the case with different strains of microorganisms. Some authors have described that Gram-positive organisms preferred a more lipophilic molecule than the Gram-negative ones [58]. This has been attributed to the difference in the cell outer membrane between bacterial types and strains: while Gram-positive bacteria have a very simple cell wall, the outer membrane of Gram-negative bacteria contains lipopolysaccharides which are cross-bridged by divalent cations, adding strength to the membrane and impermeability to lipophilic molecules. This agreed with the results obtained in Table 1 where the more lipophilic compounds showed less activity against *E. coli*.

A good correlation was obtained between the theoretical logP, calculated using the average values from VCCLab software and Molinspiration, and the retention time observed from HPLC (Figure S1). The activity observed against *B. subtilis* increases for compounds with logP > 3 with MBC ≤ 64 mg/mL. However, MBC values greater than 2000 μg/mL were obtained for both lipophilic and lipophobic compounds, with the exception of the lipophilic compounds **1b**, **1c**, and **3b** which present lower MBC values (Figure 1).

From the SEM images in Figure 2, both bacteria strains incubated with compounds **1c** and **3b** show clear signs of damage: from morphological changes to disruption of the cell membrane and leakage of cytoplasmatic material, ending with the disintegration of the bacteria into small fragments. Most images, especially of *B. subtilis*, show an "implosion" of bacteria, with a marked depression in the middle of the cell (also refer to Figures S1 and S2). In some of the images, aggregates of the compounds can be seen surrounding the bacteria, many of which are attached to the cell membrane.

Human embryonic kidney cells, HEK-293, were chosen as a cell model to evaluate the cytotoxicity of the compounds. The HEK-293 cells were incubated in the cell culture medium with varying concentrations of the examined compound and the impact of treatment was measured using the MTT cell viability assay [46]. The results indicated that compounds **1a–1c**, **3a**, and **3c** were considerably toxic with IC_{50} values lower than 36 μM (Table 1 and Table S1). Surprisingly, compound **3b** derived from phenylalanine was less toxic to the HEK-293 cell line at concentrations 15 times higher than the MIC and MBC for the *B. subtilis* strain (Table 1, entry 8) [59].

Comparing results with the commercial antibiotic alamethicin, the corresponding MIC value of **3b** against *B. subtilis* is lower than alamethicin, (Table 1, entry 10) [41], whereas the toxicity of alamethicin against MRC-5 human cells is similar to the cytotoxicity of the imidazolium salt **3b** against HEK-293 line [42].

To gain a more detailed understanding on the mechanism of cytotoxicity of these imidazole and imidazolium salts on bacteria, we studied the possible structure–activity relationship between their antibacterial activities and their aggregation behavior in bacterial cell medium [23,38].

From pyrene fluorescence studies in pure water, the plot of the I_1/I_3 ratio for the corresponding emission spectra vs. concentrations showed one single break point (Figure S9, Figure S11, and Figure S13) reaching in all cases values of $I_1/I_3 \approx 1.3$ or lower after the break point. However, in the bacterial cell culture medium, the plot of the I_1/I_3 ratios for the corresponding emission spectra vs. concentration presented two single break point and, in some cases, even three points. The two break points observed for all the compounds suggest the presence of two different processes. The first one takes place at I_1/I_3 values observed for pyrene in the absence of compounds ($I_1/I_3 = 1.16$ in water/1/1 bacterial cell culture medium) and reveals that pyrene is fully exposed to the polar solvent mixture in this first aggregation step. At the second break point, at higher concentrations, this ratio reaches lower values suggesting the formation in this region of aggregates in which the probe molecule is less solvent exposed [60]. For example, for compound **1b**, the first break point at 6 μM leads to aggregates with an appreciable solvent exposed probe ($I_1/I_3 \approx 1.1$) and the second process starts at *ca.* 48 μM affording aggregates, providing a low polarity microenvironment to pyrene, reaching I_1/I_3 values ≈ 0.8 for 1 mM concentration.

The CAC values obtained for compounds with long alkyl chain were in the μM range while for compounds **2a-c** with short alkyl chains, the CAC values obtained were in the mM range (i.e., entries 1 and 4, Table 2), where the lowest CAC values for the ditopic salts were found in water. In general, for the imidazole and monotopic salts, changing the medium from water to water/bacterial cell culture medium led to a decrease in the CAC values (i.e., Table 2, entries 1 and 2), however for ditopic salts, the CAC values did change significantly (Table 2, entries 3 and 6).

Comparing the first CAC values obtained in 1/1 water/bacterial cell culture medium and the MBC for *B. subtilis* for the different series of compounds in Figure 3, it can be observed as compounds **1a–1c** presented the CAC below the MBC, implying that these compounds exist in an aggregated form at the MBC concentration, meaning fewer imidazolium monomers will be present at these concentrations, less than is needed to produce a significant biologic effect, thus increased overall concentrations are needed to obtain the desired bactericidal effects if the monomeric form is the responsibility of the corresponding bioactivity. However, different behavior was observed for the series **3a–3b** derived from phenylalanine. Figure 3c shows how the CAC line intersects the MBC line, indicating that compound **3b** is not aggregated at the corresponding MBC value and exerts a high bactericidal effect, as observed in Table 2 (entry 8). Finally, for the **2a–2c** series, the CAC is below the MBC for **2a** but above for **2b** and **2c**, illustrating that the monomeric form is responsible for the corresponding antibacterial activity (Table 2, entry 6).

In addition to the results above, the compounds containing dodecyl chains can easily align with lipids and hence, accumulate within the bacterial cell membrane. In this regard, compounds with longer alkyl tails have CACs in the μM range in the bacterial cell culture medium and thus easily self-assemble, leading to an easy accumulation within the cell membrane. Therefore, resulting in a lower effective concentration at the site of action within the cellular cytoplasm lower. This accumulation could lead to a biocidal mechanism based on [38], as is observed in Figure 2. Furthermore, it appears that the shorter chain length results in reduced membrane interaction and an energetically unfavorable micelle formation, meaning low self-assembling capability, which leads to a lower overall bacterial cytotoxicity as seen from the corresponding MIC and MBC values (Table 1).

Consequently, by comparing the MBC for *B. subtilis* and CAC of the imidazole and imidazolium series, this study has provided a better understanding of the relationship between the biological activities of these compounds correlated with their aggregation capabilities. The results demonstrate that the compounds with longer alkyl chains provide excellent antimicrobial activity although most of them are aggregated at the antimicrobial response concentration, with only compound **3b** existing in its monomeric form at its corresponding MBC value.

Optical microscopy confirmed the formation of spherical aggregates between 0.5–20 μm diameter in size in the three different media (Figures S15–S17). Regarding the medium, in general, in the culture medium, the dispersity of the aggregates decreased for compounds with longer alkyl chains (see Figures S15 and S17). Furthermore, in the culture medium, compounds **2a–c** and **3b** were able to form worm-like aggregates at the studied concentrations (Figures S16 and S17).

SEM images for **1c** and **3b** in water, in 1/1 water/bacterial cell culture medium and in the bacterial cell culture medium revealed the formation of different aggregates morphologies. Compound **1c** produced spherical aggregates with <3 μm diameter size in all three media, with the aggregates in water being more distorted (Figure S18). Compound **3b** was able to form spherical aggregates in water and water/bacterial cell culture medium (Figure 4a,b), while in the pure culture medium, different morphologies were observed, such as spherical aggregates <1 μm in diameter coexisting with fibrillary aggregates (Figure 4c). When viewed at higher magnification, it is observed that the larger spherical aggregates consisted of several smaller aggregates or dendritic fibrillary aggregates for **3b** and **1c,** respectively (Figure 4 and Figure S18).

Regarding the stability of the aggregates formed, studies by UV-vis spectroscopy for **1a–c** are gathered in Figure 5. Figure 5a shows the change in absorbance ($-\Delta A_{600}$) of compound **1b** (1 mM) at 600 nm with respect to time in the different media. The initial rate for the change in absorbance associated to the destabilization of the aggregates is defined by the slope of the linear region of the initial $-\Delta A_{600}$ versus time plot. The slope and the total change in the absorbance was the smallest when using the water/bacterial cell culture medium, being the rate at the initial region for pure water and bacterial cell culture medium in the same ranges. However, it must be highlighted that the absorbance decreases until reaching a zero value after 500 min for pure water. A different behavior was obtained for **1a** and **1c** (see Figures S19 and S20). For compound **1a**, the absorbance at 600 nm decreased with time in the three media with similar rates, while for **1c**, the rate followed the order water > bacterial cell culture medium > 1/1 water/bacterial cell culture medium, reaching almost zero values in the tree media after 24 h.

Overall, the results obtained show that in the pure culture medium, the stability of the aggregates follows the order **1a** > **1b** > **1c** (Figure 5b), indicating that the introduction of the two headgroups and hydrophobic alkyl chains in **1c** contributes to a minor stabilization of the aggregates in this medium.

4. Materials and Methods

4.1. Materials

4.1.1. Reagents and Culture Media

Resazurin sodium salt and dimethyl sulfoxide (DMSO) were bought from Sigma-Aldrich. Luria-Bertani (LB) liquid broth (Miller's formulation) and nutrient broth (NB) were freshly prepared and sterilized by autoclave. Broth powders were bought from Scharlab. Tryptone soy agar plates were purchased from Thermo Scientific. Glutaraldehyde was purchased in solution at 25% in H_2O and Grade II from Sigma Aldrich and used as provided. Phosphate buffer was prepared from the solid salts NaH_2PO_4 and Na_2HPO_4, both purchased from Aldrich at qualities 99% and 99.5% respectively, by dissolving them in MilliQ® water and adjusting pH with NaOH and HCl solutions.

Cell culture media for cytotoxicity studies were purchased from Gibco (Grand Island, NY, USA). Fetal bovine serum (FBS) was obtained from HyClone (UT, USA). Supplements and other chemicals not listed in this section were obtained from Sigma Chemical Co. (St. Louis, MO, USA). Plastics for cell culture were supplied by Thermo Scientific BioLite (Madrid, Spain). All tested compounds were dissolved in DMSO at a concentration of 10 mM and stored at −20 °C until use. HEK-293 cell lines were maintained in Dulbecco's modified Eagle's medium (DMEM) containing glucose (1 g/L), glutamine (2 mM), penicillin (50 μg/mL), streptomycin (50 μg/mL), and amphotericin B (1.25 μg/mL) supplemented with 10% FBS.

Reagents and solvents, including NMR solvents, were purchased from commercial suppliers and were used without further purification except for pyrene, used for fluorescence studies, that was crystallized twice from methanol. Deionized water was obtained from a MilliQ® equipment (Burlington, MA, USA). Imidazoles **1a** and **2a** and imidazolium salts **1b** and **1c** were prepared as previously described [30,32].

4.1.2. Synthesis and Characterization

Imidazole **3a** and compounds **2b–c** and **3b–c** were prepared following the synthetic protocols.

*General procedure for compound **3a**:* To a mixture of glyoxal (40% aq., 1.1 equiv, 2.6 mL) and formaldehyde (37% aq., 5.0 equiv., 7.7 mL), the (*S*)-2- amino-N-dodecyl-3- phenylpropanamide compound (1.0 equiv., 6.9 g, 20.8 mmol) and ammonium acetate (1.1 equiv, 1.8 g, 23.4 mmol) were dissolved previously in methanol and added. The reaction mixture was stirred at room temperature for 48 h. The solvent was evaporated under reduced pressure and the resulting crude residue was treated with saturated Na_2CO_3 solution, extracted with CH_2Cl_2 (3×), dried with anhydrous $MgSO_4$, filtered, and concentrated.

*General procedure for compounds **2b–3b**:* To a mixture of compound **2a–3a** (1.1 equiv) and bromomethylbenzene (1.0 equiv) were dissolved in acetonitrile (5 mL). The reaction was carried out under microwave irradiation using 120 W, 1.72×10^6 Pa, 150 °C, and 1 h. After solvent evaporation, the remaining solid was washed with diethyl ether (×3) to afford the desired compound.

*General procedure for compounds **2c–3c**:* To a mixture of compound **2a–3a** (2.2 equiv) and 1, 3-(bis-bromomethyl)benzene (1.0 equiv) were dissolved in acetonitrile (5 mL). The reaction was carried out under microwave irradiation using 120 W, 1.72×10^6 Pa, 150 °C, and 1 h. After solvent evaporation, the remaining solid was washed with diethyl ether (×3) to afford the desired compound.

*Compound **3a**:* yellow liquid (7 g, 88%), $[\alpha]^{25}{}_D = -7.11$ (c = 0.01, MeOH); m.p= 56.1 °C. ^{1}H NMR (400 MHz, $CDCl_3$ and CD_3OD) δ 7.26–7.07 (m, 4H), 7.00 (m, 1H), 6.97–6.89 (m, 3H), 6.48 (t, J = 5.7 Hz, NH), 4.67 (dd, J = 9.1, 6.0 Hz, 1H), 3.46 (dd, J = 14.0, 6.0 Hz, 1H), 3.19–3.00 (m, 3H), 1.30 (m, 2H), 1.26–1.04 (m, 18H), 0.85–0.77 (m, 3H). ^{13}C NMR (101 MHz, $CDCl_3$) δ 168.3, 134.0, 136.3, 129.6, 128.8, 128.7, 127.2, 118.0, 62.9, 39.8, 39.3, 31.9, 29.6, 29.6, 29.5, 29.3, 29.2, 26.8, 22.7, 14.1. MS (ESI) (m/z) calcd. for $C_{24}H_{37}N_3O$ $[M+H]^+$ = 384.3; found 383.4 (100%), 767.7 (35%, $[M+M+H]^+$). IR (ATR) = 3309, 2953, 2919, 2850, 1656, 1549, 1493, 1469, 1454 cm^{-1}. Calculated for $C_{24}H_{37}N_3O{\cdot}4H_2O$: C 63.27, H 9.96, N 9.22; found C 62.97, H 9.74, N 9.58.

*Compound **2b**:* yellow oil (150 mg, 93%); $[\alpha]^{25}{}_D = 41.07$ (c = 0.01, MeOH); m.p. 33 °C. ^{1}H NMR (400 MHz, $CDCl_3$) δ 9.93 (s, 1H), 8.62 (t, J = 5.7 Hz, 1H), 7.75 (t, J = 1.8 Hz, 1H), 7.38–7.21 (m, 5H), 7.11 (s, 1H), 5.66 (d, J = 10.6 Hz, 1H), 5.47–5.27 (dd, J = 10.6, 2.2 Hz, 2H), 3.34–3.17 (m, 1H), 3.12–2.96 (m, 1H), 2.45–2.37 (m, 1H), 1.54–1.40 (m, 2H), 1.32–1.21 (m, 2H), 1.03 (d, J = 6.5 Hz, 3H), 0.81 (t, J = 7.3 Hz, 3H), 0.76 (d, J = 6.6 Hz, 3H). ^{13}C NMR (101 MHz, $CDCl_3$) δ 167.0, 136.1, 131.8, 130.1, 129.8, 128.6, 121.8, 120.9, 67.7, 53.9, 39.6, 31.2, 31.0, 20.2, 18.8, 18.3, 13.7. MS (ESI) (m/z) calcd. for $C_{19}H_{28}N_3O$ $[M]^+$ = 314.2; found 314.5 (100%); IR (ATR)= 3220, 3063, 2962, 2933, 2873, 1672, 1550, 1497, 1456, 1327, 1225, 1152 cm^{-1}. Calculated for $C_{19}H_{28}N_3OBr$: C 57.87, H 7.16, N 10.66; found C 57.80, H 6.98, N 11.01.

*Compound **3b**:* yellow viscous solid (129 mg, 90%); $[\alpha]^{25}{}_D = -31.07$ (c = 0.01, MeOH); m.p = 16 °C. ^{1}H NMR (300 MHz, $CDCl_3$) δ 9.53 (s, 1H), 8.59 (t, J = 5.5 Hz, NH), 7.74 (s, 1H), 7.41–7.14 (m, 7H), 7.04–6.95 (m, 2H), 6.92 (s, 1H), 6.55 (m, 1H), 5.14 (q, J = 14.7 Hz, 2H), 3.51–2.95 (m, 4H), 1.83 (s, 3H), 1.44 (m, 2H), 1.17 (s, 16H), 0.92–0.70 (m, 3H). ^{13}C NMR (101 MHz, $CDCl_3$) δ 166.6, 136.1, 134.3, 131.9, 129.8, 129.7, 129.1, 129.0, 128.2, 127.5, 121.8, 120.9, 62.3, 53.6, 40.0, 39.0, 31.9, 29.7, 29.6, 29.5, 29.4,

29.2, 28.9, 27.0, 22.7, 14.1. MS (ESI) (m/z) calcd. for $C_{31}H_{44}N_3O$ $[M]^+$ = 474.4; found 474.7 (100%); IR (ATR) = 3297, 3061, 2966, 2922, 2851, 1654, 1557, 1495, 1453 cm^{-1}. Calculated for $C_{31}H_{44}N_3OBr{\cdot}H_2O$: C 65.02, H 8.10, N 7.34; found C 65.46, H 8.24, N 7.68.

Compound **2c**: yellow oil (104 mg, 90%); $[\alpha]^{25}{}_D$ = 6.93 (c = 0.01, MeOH); m.p.= 85 °C. ^1H NMR (500 MHz, $CDCl_3$) δ 10.01 (s, 2H), 8.40 (t, J = 5.6 Hz, 2H), 8.12 (s, 2H), 7.68 (d, J = 1.3 Hz, 2H), 7.47–7.36 (m, 2H), 7.34–7.23 (m, 2H), 5.73 (d, J = 14.4 Hz, 2H), 5.55–5.38 (dd, J = 10.7 Hz, J = 3.5 Hz, 4H), 3.37–3.25 (m, 2H), 3.15–3.00 (m, 2H), 2.53–2.38 (m, 2H), 1.64–1.43 (m, 4H), 1.39–1.20 (m, 4H), 1.08 (d, J = 6.6 Hz, 6H), 0.87 (t, J = 7.3 Hz, 6H), 0.82 (d, J = 6.7 Hz, 6H). 13C NMR (126 MHz, $CDCl_3$) δ 206.8, 166.7, 136.5, 134.1, 130.6, 130.5, 129.9, 122.4, 120.8, 77.3, 77.0, 76.8, 68.1, 53.2, 39.5, 31.1, 30.9, 30.9, 20.1, 18.8, 18.4, 13.6. MS (ESI) (m/z) calcd. for $C_{32}H_{50}N_6O_2$ $[M]^{2+}$ = 275.2; found 275.3 (100%); IR (ATR)= 3412, 3228, 3125, 3066, 2962, 2933, 2873, 1671, 1550, 1465, 1360, 1298, 1226, 1152 cm^{-1}. Calculated for $C_{32}H_{50}N_6O_2Br_2{\cdot}2H_2O$: C 51.48, H 7.29, N 11.26; found C 50.84, H 7.32, N 11.46.

Compound **3c**: yellow viscous solid (126 mg, 82%); $[\alpha]^{25}{}_D$ = 17.33 (c = 0.01, MeOH); m.p. = 60 °C. ^1H NMR (300 MHz, $CDCl_3$) δ 9.47 (s, 2H), 8.21 (t, J = 5.7 Hz, NH), 7.81–7.51 (m, 6H), 7.29–7.06 (m, 16H), 6.13 (t, J = 8.0 Hz, 2H), 5.31 (s, 4H), 3.37 (dd, J = 13.6, 7.2 Hz, 2H), 3.25–3.10 (m, 4H), 3.00–2.84 (m, 2H), 1.41–1.29 (m, 4H), 1.23–1.05 (m, 32H), 0.80 (m, 6H). ^{13}C NMR (101 MHz, $CDCl_3$) δ 166.3, 136.3, 134.3, 133.6, 130.6, 130.3, 129.7, 129.2, 128.9, 127.5, 122.3, 121.0, 62.6, 53.0, 39.9, 38.9, 31.9, 29.7, 29.7, 29.6, 29.5, 29.4, 29.2, 28.9, 26.9, 22.7, 14.1. MS (ESI) (m/z) calcd. for $C_{56}H_{82}N_6O_2$ $[M]^{2+}$ = 435.3; found 435.7 (100%); IR (ATR)= 3294, 3063, 2923, 2852, 1656, 1554, 1495, 1454, 1362 cm^{-1}. Calculated for $C_{56}H_{82}N_6O_2Br_2$: C 65.23, H 8.02, N 8.15; found C 65.76, H 8.57, N 8.34.

4.1.3. Microorganisms and Growth Conditions

Two bacterial strains were used in the antibacterial assays: *Escherichia coli* DH5α as a Gram-negative model and *Bacillus subtilis* 1904-E as a Gram-positive model. Both bacterial strains were donated to our laboratory and can be provided on request by contacting the corresponding authors. Both bacterial strains were incubated at 37 °C and the pre-inoculum incubation time was of 24 h. Liquid Luria-Bertani (LB) medium was used for *E. coli* DH5α and nutrient broth (NB) for *B. subtilis*.

4.2. Methods

4.2.1. Bacterial Proliferation Assay in Presence of Imidazole Derivatives

The bacteria cell bank suspensions were thawed and inoculated in the appropriate liquid broth for 24 h at 37 °C with mild agitation. A dilution from these culture solutions was used for the following tests, corresponding to an inoculum of 1×10^7 CFU/mL. Stock solutions of all the tested compounds were prepared in DMSO at a concentration of 100 mg/mL, aliquoted, and stored at −20 °C.

(A) Bacterial growth inhibition assay: Conditions here described are for testing 6 different concentrations of the compounds, with triplicates of each condition. Therefore, 4 compounds were tested per plate. An adapted version of the microdilution method was used. Firstly, the imidazole derivatives were dissolved in the corresponding broth at 2× the highest tested concentration. Then, 100 μL of the 2× solutions were added to the first (A) and second (B) row wells of a 96-well plate. In addition, 100 μL of liquid medium had been previously added to rows B to F. Subsequent dilutions at 1:2 are prepared in rows B to F, by withdrawal of 100 μL from the previous row (more concentrated) to the next row (half diluted), mixing well. Then, 100 μL were discarded from the last row (F). By now, there are 100 μL in each well, and 100 μL of bacterial suspension at 10^7 CFU/mL were added to each well. Then, the 96-well plates were incubated for 24 h at 37 °C under mild agitation. Bacterial growth was controlled both by visual observation of the turbidity in each well and by measuring the optical density (OD) at 560 nm at time 0 h and 24 h. Results are recorded as the lowest concentration of antimicrobial agent that inhibits visible growth of the bacteria, and were compared with the OD variation of a control culture containing *E. coli* or *B. subtilis* (+ control) and of solution of the tested compounds without bacteria (- control).

(B) Bacterial cell viability assay: Cell viability was analyzed using a Resazurin (7-Hydroxy-3H-phenoxazin-3-one 10-oxide) assay in a 96-well plate. Once the bacterial cultures of growth inhibition assay had been grown for a total of 24 h, 25 µL of a 0.1 mg/mL resazurin (prepared in LB or NB medium) were added to each well and incubated in the dark at 37 °C for 1 h under stirring. Resazurin has a blue color at the testing pH and turns pink when reduced by the viable bacteria to resorufin. Therefore, pink wells indicate metabolizing bacteria, while blue wells are indicative of bacteria that have lost their ability to convert resazurin to resorufin. Different controls were made in order to corroborate the MBC value obtained by the resazurin assay. The change of color was confirmed at 1, 4, and 24 h after its addition. The viability of bacteria was verified (either confirmed or rejected) by the colony plate-counting method, by seeding 10 µL from the cell culture onto tryptone soy agar plates and observing the presence or absence of bacterial growth after 24 h at 37 °C.

4.2.2. Log P Calculation and Retention Time Determination

LogP values for the different compounds were calculated using VCCLab (ALOGPS 2.1) and Molinspiration (miLogP2.2) softwares. We used LogP as the average of these values. The protonated forms for the imidazole derivatives were considered. Reverse phase HPLC (equipment: Agilent technologies 1100 series, column: Xterra MS C18 4.6 × 150 mm (5 µmol/L)) was also used for measuring the relative lipophilicity of these compounds, since the retention time of each molecule on the reverse phase column is related to its lipophilicity. All the products were dissolved in MeOH at 2 mmol/L concentration and eluted using 70/30 acetonitrile/water and 0.1% of formic acid for 15 min and flow rate 0.2 mL/min at 25 °C. λ used was 254, 280, and 220 nm taking the corresponding chromatogram with higher mAU (see Supporting Information).

4.2.3. H NMR Studies

NMR experiments were carried out on a Varian INOVA 500 spectrometer (500 MHz for ^{1}H and 125 MHz for ^{13}C), on a Bruker Avance III HD 400 spectrometer (400 MHz for ^{1}H and 100 MHz for ^{13}C) or on a Bruker Avance III HD 300 spectrometer (300 MHz for ^{1}H and 75 MHz for ^{13}C) at 25 °C. Chemical shifts are reported in ppm using TMS as the reference.

4.2.4. Fluorescence Spectroscopy Measurements

Pyrene was used as a fluorescence probe to determine the CAC of the compounds in water and 1/1 water/bacterial cell culture medium at 25 ± 1 °C. Fluorescence measurements were performed with a Spex Fluorolog 3-11 instrument equipped with a 450 W xenon lamp (right angle mode). Firstly, a stock pyrene solution of 1.98×10^{-4} mol/L was prepared in ultrapure methanol. Then, solutions of the imidazole and imidazolium salt compounds (ranging from 6 to 3×10^{-3} mmol/L) were prepared in different vials and 5 µL of pyrene solution was transferred into the vials, reaching a final pyrene concentration of 9.89×10^{-7} mol/L in each vial. Fluorescence spectra of pyrene were recorded from 200 to 650 nm after excitation at 337 nm, and the spectra were not corrected for the Xe lamp spectral response. The slit width was set at 5 nm for both excitation and emission. The peak intensities at 373 and 385 nm were determined as I_1 and I_3, respectively. The ratios of the peak intensities at 373 and 385 nm (I_1/I_3) for the emission spectra were recorded as a function of the logarithm of concentration. The CAC values were taken from the break point. Samples were excited with a 337 nm NanoLED.

4.2.5. Optical Images

Images were recorded with OLYMPUS COVER-018 microscopy, BX51TF model, at 25 °C. Experiments were carried out in water, 1/1 water/bacterial cell culture medium, and in bacterial cell culture medium.

4.2.6. Scanning Electron Microscopy (SEM)

SEM images of the compounds were obtained using a JEOL 7001F microscope with a digital camera; while SEM images of the incubated bacteria were obtained using an Inspect F50 microscope, at 10 kV and spot size of 3.0, with a digital camera. Bacteria solutions at ca. 0.5×10^7 CFU/mL were incubated overnight without and with compounds **1c** and **3b** at their $\frac{1}{2}$ MIC and MIC. After this, bacteria were washed with sterile PBS and fixed by incubation for 2 h in a 2.5% glutaraldehyde solution in phosphate buffer 10 mmol/L at pH 7.2. The fixed bacteria were subsequently washed once with phosphate buffer saline solution and four times with MilliQ water to remove any residual salts and glutaraldehyde. Finally, bacteria were resuspended in MilliQ water and 10 μL of these solutions were placed on silicon wafers and allowed to dry by evaporation overnight. Samples were coated with platinum using the sputtering technique in which microscopic particles of platinum are rejected from the surface after the material is itself bombarded by energetic particles of a plasma or gas. Experiments were carried out in water, 1/1 water/bacterial cell culture medium, and in bacterial cell culture medium.

4.2.7. UV-Vis Spectroscopy

UV-Vis absorption spectra of the colloidal solutions were recorded on a Hewlett-Packard 8453 spectrophotometer at 25 °C. Experiments were carried out in water, 1/1 water/bacterial cell culture medium, and in bacterial cell culture medium.

4.2.8. Cell Proliferation Assay for Cytotoxicity Studies

In 96-well plates, 3×10^3 HEK-293 cells per well were seeded and incubated with serial dilutions of the tested compounds (from 200 to 0.2 μM) to a total volume of 100 μL of their growth media. The 3-(4,5-dimethylthiazol-2-yl)-2,5-diphenyltetrazolium bromide (MTT; Sigma Chemical Co.) dye reduction assay in 96-well microplates was used, as previously described [18]. After 2 days of incubation (37 °C, 5% CO_2 in a humid atmosphere), 10 μL of MTT (5 mg/mL in phosphate-buffered saline, PBS) was added to each well, and the plate was incubated for a further 3 h (37 °C). The supernatant was discarded and replaced by 100 μL of DMSO to dissolve formazan crystals. The absorbance was then read at 540 nm by Multiskan™ FC microplate reader. For all concentrations of compound, cell viability was expressed as the percentage of the ratio between the mean absorbance of treated cells and the mean absorbance of untreated cells. Three independent experiments were performed, and the IC_{50} values (i.e., concentration half inhibiting cell proliferation) were graphically determined using GraphPad Prism 4 software.

Statistical analysis: GraphPad Prism v4.0 software (GraphPad Software Inc., La Jolla, CA, USA) was used for statistical analysis. For all experiments, the obtained results of the triplicates were represented as means with standard deviation (SD).

5. Conclusions

A series of novel imidazole and imidazolium salts derived from *L*-valine and *L*-phenylalanine containing different hydrophobic groups have been synthesized and their antibacterial activity studied against *E. coli* and *B. subtilis*. The results demonstrate that an optimum lipophilicity of the alkyl chain and the amino acid side chain is needed to achieve antibacterial activity. The compounds presented better antibacterial activity against *B. subtilis* than *E. coli*, where compound **1a–1b** and **3a–3b** were the most active against *B. subtilis*, showing MBC values corresponding to 16 μg/mL or lower. Monotopic compound **3b** was 15 times less active against human embryonic kidney cells HEK-293 than toward *B. subtilis*, thus demonstrating its potential as an effective antibacterial agent with good biocompatibility. Aqueous aggregation studies revealed CAC values for compounds **1a–1c** and **3a–3c** in the μM range in water alone, however these CAC values decreased for imidazole and monotopic species when water was replaced with bacterial cell culture medium. Optical microscopy and SEM images confirmed the formation of these spherical aggregates. It is important to note that most of the bioactive compounds were aggregated to some extent at their MIC/MBC concentrations, however the monotopic compound **3b** was not aggregated at its corresponding MBC, suggesting that the monomeric species was responsible for the observed antibacterial activity.

Supplementary Materials: The following are available online at http://www.mdpi.com/1424-8247/13/12/482/s1.

Author Contributions: Conceptualization, B.A. and S.G.M.; methodology, B.A. and S.G.M.; software, A.V. and J.J.A.; validation, B.A., S.G.M. and E.A.-B.; formal analysis, B.A.; investigation, A.V., J.J.A., E.A.-B., and E.F.; resources, B.A.; data curation, S.G.M.; writing—original draft preparation, B.A.; writing—review and editing, B.A., S.G.M. and E.A.-B.; supervision, B.A., S.G.M. and S.V.L.; funding acquisition, S.V.L., E.A.-B., and S.G.M. All authors have read and agreed to the published version of the manuscript.

Funding: E.A.B. and S.G.M acknowledge funding from the Ministerio de Ciencia e Innovación (Spain) (PID2019-109333RB-I00) and the European Union's Horizon 2020 research and innovation program (Marie Skłodowska-Curie grant agreement No 845427). S.V.L and B.A. acknowledge funding from Ministerio de Ciencia e Innovación, RTI2018-098233-B-C22 and Pla de Promoció de la Investigació de la Universitat Jaume I, UJI-B2019-40. A.V. was funded by Ministerio de Ciencia e Innovación within the predoctoral fellowship program, grant FPU15/01191.

Acknowledgments: The electron microscopy characterization was conducted at the Laboratorio de Microscopias Avanzadas (LMA) at Universidad de Zaragoza. Authors acknowledge the LMA for offering access to their instruments and expertise. Technical support from the SECIC of the UJI is acknowledged.

Conflicts of Interest: The authors declare no conflict of interest.

References

1. Rani, N.; Sharma, A.; Singh, R. Imidazoles as Promising Scaffolds for Antibacterial Activity: A Review. *Mini-Rev. Med. Chem.* **2013**, *13*, 1812–1835. [CrossRef] [PubMed]
2. Duan, Y.T.; Wang, Z.C.; Sang, Y.L.; Tao, X.X.; Zhu, H.L. Exploration of Structure-Based on Imidazole Core as Antibacterial Agents. *Curr. Top. Med. Chem.* **2013**, *13*, 3118–3130. [CrossRef] [PubMed]
3. Li, W.J.; Li, Q.; Liu, D.; Ding, M.W. Synthesis, Fungicidal Activity, and Sterol 14α-Demethylase Binding Interaction of 2-Azolyl-3,4-dihydroquinazolines on Penicillium digitatum. *J. Agric. Food Chem.* **2013**, *61*, 1419–1426. [CrossRef] [PubMed]
4. Chen, L.; Zhao, B.; Fan, Z.J.; Liu, X.M.; Wu, Q.F.; Li, H.P.; Wang, H.X. Synthesis of Novel 3,4-Chloroisothiazole-Based Imidazoles as Fungicides and Evaluation of Their Mode of Action. *J. Agric. Food Chem.* **2018**, *66*, 7319–7327. [CrossRef]
5. Hu, Y.; Shen, Y.F.; Wu, X.H.; Tu, X.; Wang, G.X. Synthesis and biological evaluation of coumarin derivatives containing imidazole skeleton as potential antibacterial agents. *Eur. J. Med. Chem.* **2018**, *143*, 958–969. [CrossRef]
6. Wang, P.-Y.; Wang, M.-W.; Zeng, D.; Xiang, M.; Rao, J.-R.; Liu, Q.-Q.; Liu, L.-W.; Wu, Z.-B.; Li, Z.; Song, B.A.; et al. Rational Optimization and Action Mechanism of Novel Imidazole (or Imidazolium)-Labeled 1,3,4 Oxadiazole Thioethers as Promising Antibacterial Agents against Plant Bacterial Diseases. *J. Agric. Food Chem.* **2019**, *67*, 3535–3545. [CrossRef]
7. Rossi, R.; Ciofalo, M. An Updated Review on the Synthesis and Antibacterial Activity of Molecular Hybrids and Conjugates Bearing Imidazole Moiety. *Molecules* **2020**, *25*, 5133. [CrossRef]
8. Shamshina, J.L.; Kelley, S.P.; Gurau, G.; Rogers, R.D. Chemistry: Develop ionic liquid drugs. *Nature* **2015**, *528*, 188–189. [CrossRef]
9. Hauss, D. Oral lipid-based formulations. *J. Adv. Drug Deliv. Rev.* **2007**, *59*, 667–676. [CrossRef]
10. Guillory, J.K. *Pharmaceutical Salts: Properties, Selection, and Use*, 2nd ed.; Stahl, P.H., Wermuth, C.G., Eds.; Wiley-VCH: Weinheim, Germany, 2002. [CrossRef]
11. Becerril, R.; Nerín, C.; Silva, F. Encapsulation Systems for Antimicrobial Food Packaging Components: An Update. *Molecules* **2020**, *25*, 1134. [CrossRef]
12. Stoimenovski, J.; MacFarlane, D.R.; Bica, K.; Rogers, R.D. Crystalline vs. Ionic Liquid Salt Forms of Active Pharmaceutical Ingredients: A Position Paper. *Pharm. Res.* **2010**, *27*, 521–526. [CrossRef] [PubMed]
13. Shadid, M.; Gurau, G.; Shamshina, J.L.; Chuang, B.-C.; Hailu, S.; Guan, E.; Chowdhury, S.K.; Wu, J.T.; Rizvi, S.A.A.; Griffin, R.J.; et al. Sulfasalazine in ionic liquid form with improved solubility and exposure. *Med. Chem. Comm.* **2015**, *6*, 1837–1841. [CrossRef]

14. Egorova, K.S.; Gordeev, E.G.; Ananikov, V.P. Biological activity of ionic liquids and their application in pharmaceutics and medicine. *Chem. Rev.* **2017**, *117*, 7132–7189. [CrossRef]
15. Ferraz, R.; Branco, L.C.; Prudencio, C.; Noronha, J.P.; Petrovski, Z. Ionic liquids as active pharmaceutical ingredients. *ChemMedChem* **2011**, *6*, 975–985. [CrossRef]
16. Miskiewicz, A.; Ceranowicz, P.; Szymczak, M.; Bartus, K.; Kowalczyk, P. The Use of Liquids Ionic Fluids as Pharmaceutically Active Substances Helpful in Combating Nosocomial Infections Induced by Klebsiella Pneumoniae New Delhi Strain, Acinetobacter Baumannii and Enterococcus Species. *Int. J. Mol. Sci.* **2018**, *19*, 2779. [CrossRef]
17. Cuervo-Rodríguez, R.; Muñoz-Bonilla, A.; López-Fabal, F.; Fernández-García, M. Hemolytic and Antimicrobial Activities of a Series of Cationic Amphiphilic Copolymers Comprised of Same Centered Comonomers with Thiazole Moieties and Polyethylene Glycol Derivatives. *Polymers* **2020**, *12*, 972. [CrossRef]
18. Messali, M.; Moussa, Z.; Alzahrani, A.Y.; El-Naggar, M.Y.; ElDouhaibi, A.S.; Judeh, Z.M.A.; Hammouti, B. Synthesis, characterization and the antimicrobial activity of new eco-friendly ionic liquids. *Chemosphere* **2013**, *91*, 1627–1634. [CrossRef]
19. Wang, D.; Richter, C.; Rühling, A.; Drücker, P.; Siegmund, D.; Metzler-Nolte, N.; Glorius, F.; Galla, H.-J. A Remarkably Simple Class of Imidazolium-Based Lipids and Their Biological Properties. *Chem. Eur. J.* **2015**, *21*, 15123–15126. [CrossRef]
20. Chen, H.-L.; Kao, H.-F.; Wang, J.-Y.; Wei, G.-T. Cytotoxicity of Imidazole Ionic Liquids in Human Lung Carcinoma A549 Cell Line. *J. Chin. Chem. Soc.* **2014**, *61*, 763–769. [CrossRef]
21. Malhotra, S.V.; Kumar, V.; Velez, C.; Zayas, B. Imidazolium-Derived Ionic Salts Induce Inhibition of Cancerous Cell Growth through Apoptosis. *MedChemComm* **2014**, *5*, 1404–1409. [CrossRef]
22. Pernak, J.; Sobaszkiewicz, K.; Mirska, I. Anti-microbial activities of ionic liquids. *Green Chem.* **2003**, *5*, 52–56. [CrossRef]
23. Kuznetsova, D.A.; Gabdrakhmanov, D.R.; Lukashenko, S.S.; Voloshina, A.D.; Sapunova, A.S.; Kulik, N.V.; Nizameev, I.R.; Kadirov, M.K.; Kashapov, R.R.; Zakharova, Y.L. Supramolecular systems based on cationic imidazole-containing amphiphiles bearing hydroxyethyl fragment: Aggregation properties and functional activity. *J. Mol. Liq.* **2019**, *289*, 111058. [CrossRef]
24. Garcia, M.T.; Ribosa, I.; Perez, L.; Manresa, A. Micellization and antimicrobial properties of surface-active ionic liquids containing cleavable carbonate linkages. *Langmuir* **2017**, *33*, 6511–6520. [CrossRef]
25. Gindri, I.M.; Siddiqui, D.A.; Bhardwaj, P.; Rodriguez, L.C.; Palmer, K.L.; Frizzo, C.P.; Martinsc, M.A.P.; Rodrigues, D.C. Dicationic imidazolium-based ionic liquids: A new strategy for non-toxic and antimicrobial materials. *RSC Adv.* **2014**, *4*, 62594–62602. [CrossRef]
26. Pałkowski, Ł.; Błaszczyński, J.; Skrzypczak, A.; Błaszczak, J.; Kozakowska, K.; Wróblewska, J.; Kożuszko, S.; Gospodarek, E.; Krysiński, J.; Słowiński, R. Antimicrobial activity and SAR study of new gemini imidazolium-based chlorides. *J. Chem. Biol. Drug Des.* **2014**, *83*, 278–288. [CrossRef]
27. Voloshina, A.D.; Gumerova, S.K.; Sapunova, A.S.; Kulik, N.V.; Mirgorodskaya, A.B.; Kotenko, A.A.; Prokopyeva, T.M.; Mikhailov, V.A.; Zakharova, L.Y.; Sinyashin, O.G. The structure—Activity correlation in the family of dicationic imidazolium surfactants: Antimicrobial properties and cytotoxic effect. *BBA Gen. Subj.* **2020**, *1864*, 129728. [CrossRef] [PubMed]
28. Wang, L.; Qin, H.; Ding, L.; Huo, S.; Deng, Q.; Zhao, B.; Meng, L.; Yan, T. Preparation of a novel class of cationic gemini imidazolium surfactants containing amide groups as the spacer: Their surface properties and antimicrobial activity. *J. Surfactant Deterg.* **2014**, *17*, 1099–1106. [CrossRef]
29. Kapitanov, I.V.; Jordan, A.; Karpichev, Y.; Spulak, M.; Perez, L.; Kellett, A.; Kümmerer, K.; Gathergood, N. Synthesis, self-assembly, bacterial and fungal toxicity, and preliminary biodegradation studies of a series of L-phenylalanine-derived surface-active ionic liquids. *Green Chem.* **2019**, *21*, 1777–1794. [CrossRef]
30. González, L.; Escorihuela, J.; Altava, B.; Burguete, M.I.; Luis, S.V. Chiral Room Temperature Ionic Liquids as Enantioselective Promoters for the Asymmetric Aldol Reaction. *Eur. J. Org. Chem.* **2014**, *2014*, 5356–5363. [CrossRef]
31. González, L.; Altava, B.; Bolte, M.; Burguete, M.I.; García-Verdugo, E.; Luis, S.V. Synthesis of Chiral Room Temperature Ionic Liquids from Amino Acids–Application in Chiral Molecular Recognition. *Eur. J. Org. Chem.* **2012**, *2012*, 4996–5009. [CrossRef]
32. González-Mendoza, L.; Escorihuela, J.; Altava, B.; Burguete, M.I.; Luis, S.V. Application of optically active chiral bis(imidazolium) salts as potential receptors of chiral dicarboxylate salts of biological relevance. *Org. Biomol. Chem.* **2015**, *13*, 5450–5459. [CrossRef] [PubMed]

33. González-Mendoza, L.; Escorihuela, J.; Altava, B.; Burguete, M.I.; Hernando, E.; Luis, S.V.; Quesada, R.; Vicent, C. Bis(imidazolium) salts derived from amino acids as receptors and transport agents for chloride anions. *RSC Adv.* **2015**, *5*, 34415–34423. [CrossRef]
34. Baltazar, Q.Q.; Chandawalla, J.; Sawyer, K.; Anderson, J.L. Interfacial and micellar properties of imidazolium-based monocationic and dicationic ionic liquids. *Colloids Surf. A* **2007**, *302*, 150–156. [CrossRef]
35. Kamboj, R.; Singh, S.; Bhadani, A.; Kataria, H.; Kaur, G. Gemini Imidazolium Surfactants: Synthesis and Their Biophysiochemical Study. *Langmuir* **2012**, *28*, 11969–11978. [CrossRef]
36. Zhuang, L.-H. Synthesis and properties of novel ester-containing gemini imidazolium surfactants. *J. Colloid Interface Sci.* **2013**, *408*, 94–100. [CrossRef] [PubMed]
37. Bhadani, A.; Singh, T.M.S.; Sakai, K.; Sakai, H.; Abe, M. Structural diversity, physicochemical properties and application of imidazolium surfactants: Recent advances. *Adv. Colloid Interface Sci.* **2016**, *231*, 36–58. [CrossRef] [PubMed]
38. Wang, D.; Galla, H.-J.; Drücker, P. Membrane interactions of ionic liquids and imidazolium salts. *Biophys. Rev.* **2018**, *10*, 735–746. [CrossRef]
39. Knight, N.J.; Hernando, E.; Haynes, C.J.E.; Busschaert, M.; Clarke, H.J.; Takimoto, K.; García-Valverde, M.; Frey, J.G.; Quesada, R.; Gale, P.A. QSAR analysis of substituent effects on tambjamine anion transporters. *Chem. Sci.* **2016**, *7*, 1600–1608. [CrossRef]
40. Gorczyca, M.; Korchowiec, B.; Korchowiec, J.; Trojan, S.; Rubio-Magnieto, J.; Luis, S.V.; Rogalska, E. A Study of the Interaction between a Family of Gemini Amphiphilic Pseudopeptides and Model Monomolecular Film Membranes Formed with a Cardiolipin. *J. Phys. Chem. B* **2015**, *119*, 6668–6679. [CrossRef]
41. Barns, K.J.; Weisshaar, J.C. Single-cell, time-resolved study of the effects of the antimicrobial peptide alamethicin on Bacillus subtilis. *Biochim. Biophys. Acta* **2016**, *1858*, 725–732. [CrossRef]
42. Ishiyama, A.; Otoguro, K.; Iwatsuki, M.; Namatame, M.; Nishihara, A.; Nonaka, K.; Kinoshita, Y.; Takahashi, Y.; Masuma, R.; Shiomi, K.; et al. In vitro and in vivo antitrypanosomal activities of three peptide antibiotics: Leucinostatin A and B, alamethicin I and tsushimycin. *J. Antibiot.* **2009**, *62*, 303–308. [CrossRef] [PubMed]
43. Kalyanasundaram, K.; Thomas, J.K. Environmental effects on vibronic band intensities in pyrene monomer fluorescence and their application in studies of micellar Systems. *J. Am. Chem. Soc.* **1977**, *99*, 2039–2044. [CrossRef]
44. Kalyanasundaram, K. *Photochemistry in Microheterogeneous Systems*, 1st ed.; Academic Press: New York, NY, USA, 1987.
45. Aguiar, J.; Carpena, P.; Molina-Bolívar, J.A.; Carnero Ruiz, C. On the determination of the critical micelle concentrationby the pyrene 1:3 ratio method. *J. Colloid Interface Sci.* **2003**, *258*, 116–122. [CrossRef]
46. Stockert, J.C.; Horobin, R.W.; Colombo, L.L.; Blázquez-Castro, A. Tetrazolium salts and formazan products in Cell Biology: Viability assessment, fluorescence imaging, and labeling perspectives. *Acta Histochem.* **2018**, *120*, 159–167. [CrossRef] [PubMed]
47. Frindi, M.; Michels, B.; Zana, R. Ultrasonic Absorption Studies of Surfactant Exchange between Micelles and Bulk Phase In Aqueous Micellar Solutions of Nonionic Surfactants with Short Alkyl Chains. 1,2-Hexanedlol and 1,2,3-Octanetrlol. *J. Phys. Chem.* **1991**, *95*, 4832–4837. [CrossRef]
48. Regev, O.; Zana, R. Aggregation Behavior of Tyloxapol, a Nonionic Surfactant Oligomer, in Aqueous Solution. *J. Colloid Interface Sci.* **1999**, *210*, 8–17. [CrossRef]
49. Ananthapadmanabhan, K.P.; Goddard, E.D.; Turro, N.J.; Kuo, P.L. Fluorescence Probes for Critical Micelle Concentration. *Langmuir* **1985**, *2*, 352–355. [CrossRef]
50. Liu, C.G.; Desai, K.G.H.; Chen, X.G.; Park, H.J. Linolenic acid-modified chitosan for formation of selfassembled nanoparticles. *J. Agric. Food Chem.* **2005**, *53*, 437–441. [CrossRef]
51. Dong, X.; Liu, C. Preparation and Characterization of Self-Assembled Nanoparticles of Hyaluronic Acid-Deoxycholic Acid Conjugates. *J. Nanomat.* **2010**, *2010*, 1–9. [CrossRef]
52. Yoshimura, T.; Ichinokawa, T.; Kaji, M.; Esumi, K. Synthesis and surface-active properties of sulfobetaine-type zwitterionic gemini surfactants. *Colloids Surf. A Physicochem. Eng. Asp.* **2006**, *273*, 208–212. [CrossRef]
53. Gregory, J. Monitoring particle aggregation processes. *Adv. Colloid Interface Sci.* **2009**, *147–148*, 109–123. [CrossRef] [PubMed]
54. Aslan, K.; Luhrs, C.C.; Pérez-Luna, V.H. Controlled and Reversible Aggregation of Biotinylated Gold Nanoparticles with Streptavidin. *J. Phys. Chem. B* **2004**, *108*, 15631–15639. [CrossRef]
55. Brown, L.; Wolf, J.M.; Prados-Rosales, R.; Casadevall, A. Through the wall: Extracellular vesicles in gram-positive bacteria, mycobacteria and fungi. *Nat. Rev. Microbiol.* **2015**, *13*, 620–630. [CrossRef] [PubMed]

56. Bury-Moné, S. Antibacterial Therapeutic Agents: Antibiotics and Bacteriophages. In *Reference Module in Biomedical Sciences*, 3rd ed.; Elsevier: Amsterdam, The Netherlands, 2014; pp. 1–13. ISBN 9780128012383.
57. Ghanema, O.B.; Mutaliba, M.J.A.; El-Harbawi, M.; Gonfaa, G.; Kait, C.F.; Alitheend, N.B.M.; Leveque, J.M. Effect of imidazolium-based ionic liquids on bacterial growth inhibition investigated via experimental and QSAR modelling studies. *J. Hazard. Mater.* **2015**, *297*, 198–206. [CrossRef] [PubMed]
58. Lien, E.; Hansch, C.; Anderson, S. Structure-activity correlations for antibacterial agents on gram-positive and gram-negative cells. *J. Med. Chem.* **1968**, *11*, 430–441. [CrossRef] [PubMed]
59. Coleman, D.; Špulák, S.; Garcia, M.T.; Gathergood, N. Antimicrobial toxicity studies of ionic liquids leading to a 'hit' MRSA selective antibacterial imidazolium salt. *Green Chem.* **2012**, *14*, 1350–1356. [CrossRef]
60. Roy, S.; Dey, J. Spontaneously Formed Vesicles of Sodium N-(11-Acrylamidoundecanoyl)-glycinate and l-Alaninate in Water. *Langmuir* **2005**, *21*, 10362–10369. [CrossRef]

Publisher's Note: MDPI stays neutral with regard to jurisdictional claims in published maps and institutional affiliations.

Review

Thymus mastichina: Composition and Biological Properties with a Focus on Antimicrobial Activity

Márcio Rodrigues [1,2,3,*], Ana Clara Lopes [1], Filipa Vaz [1], Melanie Filipe [1], Gilberto Alves [3], Maximiano P. Ribeiro [1,2,3], Paula Coutinho [1,2,3,*] and André R. T. S. Araujo [1,2,4,*]

1 School of Health Sciences, Polytechnic Institute of Guarda, Rua da Cadeia, 6300-035 Guarda, Portugal; claralopes28@gmail.com (A.C.L.); filipa.a.c.vaz@hotmail.com (F.V.); melaniemfilipe@gmail.com (M.F.); mribeiro@ipg.pt (M.P.R.)

2 Research Unit for Inland Development (UDI), Polytechnic Institute of Guarda, Av. Dr. Francisco Sá Carneiro, 50, 6300-559 Guarda, Portugal

3 CICS-UBI—Health Sciences Research Centre, University of Beira Interior, Av. Infante D. Henrique, 6200-506 Covilhã, Portugal; gilberto@fcsaude.ubi.pt

4 LAQV/REQUIMTE, Department of Chemical Sciences, Faculty of Pharmacy, University of Porto, Rua Jorge Viterbo Ferreira, 228, 4050-313 Porto, Portugal

* Correspondence: marciorodrigues@ipg.pt (M.R.); coutinho@ipg.pt (P.C.); andrearaujo@ipg.pt (A.R.T.S.A.); Tel.: +351-271-220-191 (M.R. & P.C. & A.R.T.S.A.)

Received: 1 October 2020; Accepted: 17 December 2020; Published: 19 December 2020

Abstract: *Thymus mastichina* has the appearance of a semishrub and can be found in jungles and rocky lands of the Iberian Peninsula. This work aimed to review and gather available scientific information on the composition and biological properties of *T. mastichina*. The main constituents of *T. mastichina* essential oil are 1,8-cineole (or eucalyptol) and linalool, while the extracts are characterized by the presence of flavonoids, phenolic acids, and terpenes. The essential oil and extracts of *T. mastichina* have demonstrated a wide diversity of biological activities. They showed antibacterial activity against several bacteria such as *Escherichia coli*, *Proteus mirabilis*, *Salmonella* subsp., methicillin-resistant and methicillin-sensitive *Staphylococcus aureus*, *Listeria monocytogenes EGD*, *Bacillus cereus*, and *Pseudomonas*, among others, and antifungal activity against *Candida* spp. and *Fusarium* spp. Additionally, it has antioxidant activity, which has been evaluated through different methods. Furthermore, other activities have also been studied, such as anticancer, antiviral, insecticidal, repellent, anti-Alzheimer, and anti-inflammatory activity. In conclusion, considering the biological activities reported for the essential oil and extracts of *T. mastichina*, its potential as a preservative agent could be explored to be used in the food, cosmetic, or pharmaceutical industries.

Keywords: antimicrobial; biological activities; essential oil; extract; *Thymus mastichina*

1. Introduction

Thymus mastichina L. (Figure 1) is an endemic species of the Iberian Peninsula, commonly known as "Bela-Luz", "Sal-Puro", "Tomilho-alvadio-do-Algarve", "Mastic thyme", and "Spanish marjoram" and belongs to the Lamiaceae family [1–4]. *T. mastichina* species can be classified into two subspecies: *T. mastichina* subsp. *donyanae* and *T. mastichina* subsp. *mastichina*; the first of which is present in Algarve (Portugal) and Huelva and Seville (Spain) and the latter extends throughout the Iberian Peninsula [5,6]. This aromatic plant is a semiwoody shrub that grows up to 50 cm tall and is characterized by simple and opposite leaves and bilabiate flower groups in a flower head or capitula, which blossom from April to June [2,7,8]. *T. mastichina* can be found in jungles, uncultivated, ruderal, and rupicolous lands and in dry stony open places, except in calcareous regions [1,8], being very resistant to frost, diseases, and pests. *T. mastichina* is known for its strong eucalyptus odor and it has been used for various health conditions due to its antiseptic, digestive, antirheumatic, and antitussive effects [2,6,9,10].

Figure 1. *Thymus mastichina* plant (source Planalto Dourado™ Essential Oils Enterprise, from Freixedas, Guarda, Portugal).

T. mastichina can be used in fresh or dry form and its leaves are traditionally used as a condiment/spices flavoring, in seasoning traditional dishes and salads, to preserve olives, to aromatize olive oil, and as a substitute for salt [10,11]. This medicinal and aromatic plant is also used as a source of essential oil in the cosmetic and perfume industries [10,12]. Infusions with dry parts of the plant have been used to relieve colds, cough, throat irritations, and abdominal pain, while infusions with fresh parts of the plant have been used for indigestion and stomach pain [13]. Thus, there are different products based on *T. mastichina* that are commercially available in Portugal and Spain (e.g., "Bela-Luz" essential oil, "Marjoram Spanish" essential oil, "Tomilho Bela Luz" herbs). In this work, we compiled the available information regarding the chemical composition of the essential oil and extracts of *T. mastichina* to review published studies in which its biological activities have been evaluated.

2. Search Strategy

The search was performed on the following databases: PubMed and Web of Science. Various combinations of the following terms were queried: *T. mastichina*, thyme, antibacterial, antifungal, antimicrobial, anti-inflammatory, anti-Alzheimer, antioxidant, anticancer, antiviral, insecticidal, repellent, essential oil, plant extract, biological activity, and composition. In addition, references cited in related publications were followed up. The selection of articles was performed by its relevance to the purpose of this review. It should be noted that no date or language criteria were defined as filters in the search strategy implemented.

3. Chemical Composition of *T. mastichina* Essential Oils and Extracts

The essential oil of *T. mastichina* is usually obtained mainly by hydrodistillation for 2–4 h, with low yields (from 0.4% to 6.90% (*v/w*)) (Table 1). Other extraction methods were also used, such as microdistillation, that showed higher amounts of 1,8-cineole plus limonene in comparison with hydrodistillation. However, in general, few differences were found between the different methodologies. The yield may vary depending on several factors, namely the part of the plant used, place of harvest, period of the year, storage time, extraction time, and type of fertilization, among others [14].

Table 1. Obtention features and characterization of the *Thymus mastichina* essential oil and its main constituents (equal or higher than 5%).

Plant Material (Growth Phase)	Period of Year	Source	Yield	Major Constituents	Reference
Flowering branches	May–July	Trás-os-Montes; Beira Alta; Beira Baixa; Estremadura; Ribatejo; Alto Alentejo; Algarve (Portugal)	2.2% (*v*/*w*)	1,8-cineole (53.3%); linalool (5.5%)	[15]
Leaves	December; May	Nave do Barão, Algarve (Portugal)	-	1,8-cineole (46.29%); camphor (10.77%); camphene (6.31%); α-pinene (5.23%)	[16]
Leaves	-	Vadofresno, Córdoba (Spain)	-	1,8-cineole (24.81%) 1,8-cineole (18.87%)	[17]
Leaves and flowers	December; May	Algarve (Portugal)	-	1,8-cineole (46.3–50.4%); camphor (9.6–10.8%); camphene (5.0–6.3%); α-pinene (4.0–5.3%);	[18]
Leaves	-	S. Brás de Alportel, Algarve (Portugal)	0.4–0.9% (*v*/*w*)	1,8-cineole (50.2–61.0%); camphor (7.6–10,1%); δ-terpineol (6.5–9.7%); camphene (4.4–6.1%)	[19]
Flowers			1.6–2.2% (*v*/*w*)	1,8-cineole (46.7–50.2%); δ-terpineol (5.9–8.2%)	
Aerial parts	December; May; June; October January; May, June; October (2 years)	S. Brás de Alportel, Algarve (Portugal)	-	1,8-cineole (42.1–50.43%); camphor (7.4–11.5%); camphene (3.1–6.3%); α-terpineol (3.4–5.7%); *trans*-sabinene hydrate (0.2–5.6%); α-pinene (3.1–5.3%)	[20]

Table 1. *Cont.*

Plant Material (Growth Phase)	Period of Year	Source	Yield	Major Constituents	Reference
Flowers (full flowering phase)	May	S. Brás de Alportel, Algarve, Sotavento (Portugal)	-	1,8-cineole (46.9%); camphor (6.7%); α-terpineol (5.2%)	[21]
Leaves (full flowering phase)				1,8-cineole (42.4%); camphor (7.7%); borneol (6.8%); α-terpineol (6.1%)	
Aerial parts (beginning of flowering phase)		Sesimbra, Estremadura (Portugal)	-	Chemotype A (aerial parts): linalool (44.4%); 1,8-cineole (37.4%) Chemotype B (aerial parts): linalool (61.4%); camphor (5.3%)	
Aerial parts (flowering phase)	-	Algarve (Portugal)	-	1,8-cineole (45.3%); camphor (8.5%); camphene (6.6%); α-pinene (5.4%); limonene (5.2%); borneol (5.0%)	[22]
		Estremadura (Portugal)		linalool (52.3%); 1,8-cineole (9.6%); limonene (6.4%); p-cymene (6.2%)	
Aerial parts	October; January; April; June	Nave do Barão, Algarve (Portugal)	0.7–3.6% (*v/w*)	1,8-cineole (45.1–58.6%); camphor (5.5–8.9%); α-pinene (4.6–6.8%); camphene (4.3–6.0%)	[23]
Aerial parts (vegetative phase and flowering phase)	October; May	Sesimbra (Portugal)	0.7–2.7% (*v/w*)	linalool (58.7–69%); 1,8-cineole (1.1–10.8%); elemol (0.9–6.6%); camphor (2.4–5.3%)	[24]

Table 1. *Cont.*

Plant Material (Growth Phase)	Period of Year	Source	Yield	Major Constituents	Reference
Aerial parts	January	Direção Regional de Agricultura de Trás-os-Montes (Portugal)	1.3% (*v/w*)	1,8-cineole (57.8%); limonene (10.8%)	[25]
Aerial parts	-	Direção Regional de Agricultura de Trás-os-Montes, Mirandela (Portugal)	-	1,8-cineole (67.4%)	[26]
Leaves	June	Sesimbra (Portugal)	2.2%	linalool (68.5%); 1,8-cineole (9.4%)	[8]
Flowers			2.6%	linalool (73.5%); 1,8-cineole (10.2%)	
Leaves		Arrábida (Portugal)	1.7%	1,8-cineole (69.2%); linalool (6.3%)	
Flowers			3.5%	1,8-cineole (54.6%); linalool (13.7%)	
Leaves		Mértola (Portugal)	2.0%	1,8-cineole (44.2%)	
Flowers			3.0%	1,8-cineole (39.4%); linalool (8.1%)	
Leaves		S. Brás de Alportel (Portugal)	1.4%	1,8-cineole (49.7%)	
Flowers			2.0%	1,8-cineole (48.5%)	
Aerial parts (flowering phase)		Mirandela (Portugal)	2.4% (*v/w*)	1,8-cineole (64.1%); α-terpineol (5.6%)	[27]
Aerial parts (vegetative phase)	January	Algarve (Portugal)	2.3% (*v/w*)	1,8-cineole (49.4%); limonene (9.3%)	[28]
Aerial parts	-	Córdoba (Spain)	-	1,8-cineole (45.67%); linalool (27.88%)	[29]
Aerial parts	-	Mértola (Portugal)	1.0–1.3% (*v/w*)	1,8-cineole (61.0%)	[30]
		Vila Real de Santo António (Portugal)		1,8-cineole (49.4%)	
		Sesimbra (Portugal)		linalool (39.7%); 1,8-cineole (9.6%)	

Table 1. *Cont.*

Plant Material (Growth Phase)	Period of Year	Source	Yield	Major Constituents	Reference
Plants (flowering phase)	-	Direção Regional de Agricultura e Pescas do Algarve (Portugal)	-	1,8-cineole (41.0%); β-pinene + trans-sabinene (7.0%); camphor (6.9%); borneol (6.5%); α-pinene (6.0%); camphene (5.5%)	[31]
Aerial parts (flowering phase)	June	Direção Regional de Agricultura e Pescas do Algarve (Portugal)	4% (*w*/*w*)	1,8-cineole (44%); camphor (10%); borneol (7%); camphene (7%); α-pinene (6%); α-terpineol (5%)	[32]
Plant (flowering phase)	Summer	Tordesillas, Valladolid; Truchas, Peradoones, Carrocera, Boñar, León; Almazán, Soria; Riaza, Villacastín, Segovia; Serranilos, Avila; Saldaña, Palencia (Spain)	3.40–6.90%	-	[33]
Aerial parts	-	Direção Regional de Agricultura e Pescas do Algarve (Portugal)	6.3% (*w*/*w*)	1,8-cineole (49.4%); α-pinene (7.0%); camphene (6.9%); camphor (5.8%); β-pinene (5.3%)	[14]
Whole plants	-	Barcelona (Spain)	-	1,8-cineole (52.57%); linalool (12.78%)	[34]
Commercial samples: leaves, stem, and flowers	-	Esencias Martinez Lozano, Murcia (Spain)	-	1,8-cineole (51.94%); linalool (19.90%)	[35]
Aerial parts (vegetative phase)	-	Coimbra (Portugal)	1.17% (*v*/*w*)	1,8-cineole (46%); limonene (23%)	[36]

Table 1. *Cont.*

Plant Material (Growth Phase)	Period of Year	Source	Yield	Major Constituents	Reference
Aerial parts (flowering phase)	June–July	Béjar, Valdemierque, Mozarbez, Golpejas, Salamanca; Carrocera, Boñar, Truchas, Peranzanes, León; Salas de los Infante, Lerma, Oña, Burgos; Villacastin, Riaza, Coca, Prádena, Segovia; Vinuesa, Aldealpozo, Almazán, Langa de Duero, Soria (Spain)	2.27–6.48% (*v*/*w*)	1,8-cineole (56.80–69.60%); linalool (0.62–15.7%); α-terpineol (2.07–5.99%); β-pinene (1.72–5.63%); limonene (1.07–5.10%)	[37]
Aerial parts	-	Vila Chã (Portugal)	-	1,8-cineole (47.4%); thymol (13.7%); p-cymene (9.7%); γ-terpinene (7.3%)	[38]
Flowers and leaves	June	Carrocera, Léon (Spain)	-	1,8-cineole + limonene (61.6%); linalool (6%); β-pinene (5.7%) 1,8-cineole + limonene (69.3%) 1,8-cineole + limonene (64%)	[39]
Plants (flowering phase)	-	Algarve (Portugal)	1.2% (*v*/*w*)	1,8-cineole (52.8%); α-pinene (7.2%); camphene (7.2%); camphor (7.2%)	[40]
Plants grown in vitro (all parts except roots)	-	Urbino (Italy)	0.56% (*v*/*w*)	1,8-cineole (55.6%); linalool (24.5%); β-pinene (5.9%)	[41]
Commercial samples		Planalto Dourado™; Freixedas (Portugal)	-	1.8-cineole (49.94%); linalool (5.66%); α-terpineol (5.59%); β-pinene (5.54%)	[42]

Table 1. *Cont.*

Plant Material (Growth Phase)	Period of Year	Source	Yield	Major Constituents	Reference
Aerial parts	-	Évora, Alentejo (Portugal)	1.1% (*v*/*w*)	1,8-cineole (72.0%); α-terpineol (9.0%)	[43]
Flowers		Badajoz (Spain)	-	limonene + 1,8-cineole (71.82%); β-myrcene (9.81%); α-terpineol (5.32%); camphene (5.15%)	[44]
Fruits				limonene + 1,8-cineole (78.37%); β-myrcene (5.69%); α-terpineol (5.05%)	
Leaves	Summer	Alentejo (Portugal)	-	1,8-cineole (74.2%); α-terpenyl acetate (7.9%)	[45]
Leaves	-	UNIQ F&F Co., Ltd. (Seoul, Korea)	-	β-pinene (5.81%); 1,8-cineole (64.61%); linalool (15.28%)	[46]
Aerial parts	July	Murcia (Spain)	1.8–3.6% (*v*/*w*)	1,8-cineole (38.8–74.0%); linalool (13.3–42.7%)	[4]
Leaves and stem	-	Ciudad Real (Spain)	-	1,8-cineole (43.26%); linalool (36.72%); linalyl acetate (5.58%)	[47]
Commercial samples	-	Ervitas Catitas (Portugal)		1,8-cineole (55.9%); β-pinene (10.8%)	[48]
Aerial parts		Évora, Alentejo (Portugal)	1.06% (*v*/*w*)	1.8-cineole (71.2%); α-terpineol (9.7%)	[2]

As shown in Figure 2 and Table 1, the composition of the essential oil of *T. mastichina* is quite diverse, being constituted by hydrocarbon and oxygenated monoterpenes and sesquiterpenes, presenting main constituents of 1,8-cineole (also known as eucalyptol) and linalool. The major constituents of the essential oil of *T. mastichina* have been determined by gas chromatography [2,4,8,14–20,23–32,34–45,47,48]. Table 1 summarizes the major constituents present and it can be seen that their composition varies according to origin. In fact, there are three main subtypes of essential oils, depending on the main compounds present: 1,8-cineole, linalool, and 1,8-cineole/linalool. In Portugal, the chemotypes of linalool and 1,8-cineole/linalool are only found in Estremadura, mainly at Arrábida and Sesimbra, while the 1,8-cineole chemotype is distributed throughout the country [8,21]. As expected, there are differences in the phytochemical composition among different species of Thymus; for instance, the main compound presented in *Thymus vulgaris* is thymol, which is present in low percentages in *T. mastichina* [49].

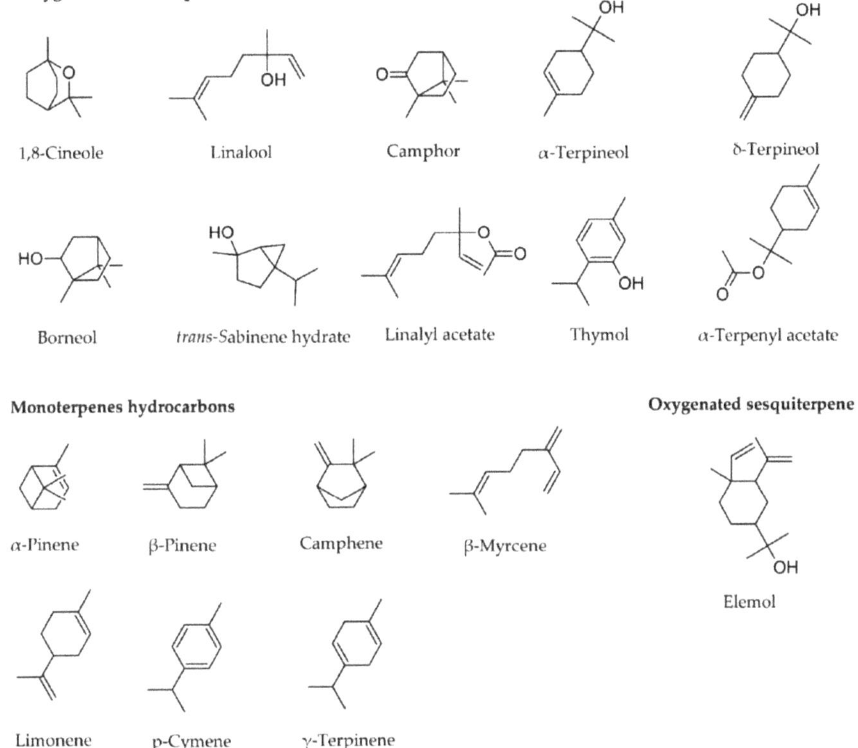

Figure 2. Chemical structures of main constituents of the *Thymus mastichina* essential oil categorized in oxygenated and hydrocarbons monoterpenes and hydrocarbon sesquiterpene.

As presented in Table S1, in addition to *T. mastichina* essential oil, in some cases after hydrodistillation, the decoction water (the remaining hydrodistillation aqueous phase) was also collected [12,14,32,43,50]. Conversely, in some studies, extraction from the aerial parts of the plant was performed using different solvents (hexane, dichloromethane, ethanol, diethyl ether, ethyl acetate, *n*-butanol, and water) [10,13,32,37,43,50,51]. Furthermore, in some cases, the extracts were obtained by ultrasound [9,51]. The extracts obtained from the aerial parts of the plant were characterized by the presence of different polyphenol classes, in particular, flavonoids (apigenin, kaempferol, luteolin, naringenin, quercetin, sakuranetin, sterubin), phenolic acids (caffeic acid, chlorogenic acid, 2-methoxysalicylic acid, 3-methoxysalicylic acid, rosmarinic acid, salvianolic acids I and K and derivatives), phenolic terpene (carnosol) and hexoside and glycoside derivatives; other compounds identified were steroid (β-sitosterol), triterpenoids (oleanolic acid, ursolic acid), and xanthophyll lutein [9,10,12,37,51,52]. In Figure S1 the chemical structures of several compounds present in the extract are presented. In opposition to *T. mastichina* essential oil, in which the composition is extensively characterized, the extract phenolic profile has been less investigated and diverse chromatographic peaks detected during the phytochemical characterization remain unidentified.

4. Biological Properties

The diverse bioactivities of *T. mastichina* essential oil and extracts are related to the chemical composition. Essential oil and extracts from the aerial parts of *T. mastichina* have been mainly described for their antibacterial and antifungal activities, but also for antioxidant, anticancer, antiviral, insecticidal, insect repellent, and anti-enzymatic activities (anti-Alzheimer, anti-inflammatory, α-amylase, and α-glucosidase).

4.1. Antibacterial and Antifungal Activities

The antibacterial and antifungal activities of the essential oils and their main compounds were evaluated by several researchers. The effect of antibacterial activity of essential oils may inhibit the growth of bacteria (bacteriostatic) or destroy bacterial cells (bactericidal). Nevertheless, it is difficult to distinguish these actions, as the antibacterial activity evaluation if frequently based on the most known and basic methods such as disk-diffusion and broth microdilution for the determination of the diameter of the zone of inhibition, minimum inhibitory concentration (MIC), and minimum lethal concentration (MLC), and mixtures of methods, as can be seen in Table 2.

The antimicrobial activity of *T. mastichina* essential oil from the chemotype of Algarve and two chemotypes of Sesimbra (Estremadura) in Portugal, measured by disc agar diffusion method, was confirmed by Faleiro et al. [21]. The tested microorganisms (*Escherichia coli*, *Proteus mirabilis*, *Salmonella* subsp., *Staphylococcus aureus*, *Listeria monocytogenes EGD*) presented different sensitivities. In particular, *T. mastichina* essential oil (3 μL) from Algarve showed the highest activity against *S. aureus* showing a diameter of the zone of inhibition of 13.7 and 15.7 mm for the flower and leaf, respectively, while *T. mastichina* essential oil from Sesimbra (Estremadura) had the highest activity for *S. aureus* showing a diameter of the zone of inhibition of 13.3 mm for chemotype A. Additionally, the antimicrobial activity observed with *T. mastichina* essential oil was explored to determine if it was due to the main constituents present in the different oil chemotypes (linalool, 1,8-cineole and linalool/1,8-cineole (1:1) mixture); for this purpose, the antimicrobial activity of these constituents alone was tested. It was concluded a higher antimicrobial activity of linalool compared with 1,8-cineole. Moreover, possible antagonist and synergistic effects of the various constituents of the essential oil were registered, once *E. coli* was susceptible to linalool but not to the mixture of linalool plus 1,8-cineole, and for *C. albicans* the mixture of linalool plus 1,8-cineole produced a slight increase in the antimicrobial activity whereas was not susceptible to 1,8-cineole. In a previous work of the same authors, and using the same method, the results suggested that the higher antimicrobial activity of the *T. mastichina* essential oil against *Salmonella* was associated to higher amounts of camphor present in *T. mastichina* essential oil comparatively to the essential oils of other plants [16].

Table 2. Antibacterial and antifungal activity of *Thymus mastichina* essential oils and extracts.

Origin	Micro-Organisms	Species	Measured Response and Results Obtained		References
S. Brás de Alportel, Algarve, Sotavento (Portugal)			Diameter of the zone of inhibition (mm), including the diameter of the disc (6 mm)		[21]
			Flower	Leaf	
	Gram-negative bacteria	*Escherichia coli*	8.0	14.0	
		Proteus mirabilis	7.0	7.3	
		Salmonella subsp.	8.0	8.7	
	Gram-positive bacteria	*Staphylococcus aureus*	13.7	15.7	
		Listeria monocytogenes EGD	9.7	12.3	
	Fungus	*Candida albicans*	10.0	11.0	
Sesimbra, Estremadura (Portugal)			Chemotype A	Chemotype B	
	Gram-negative bacteria	*Escherichia coli*	7.5	10.6	
		Proteus mirabilis	7.5	10.0	
		Salmonella subsp.	6.3	7.0	
	Gram-positive bacteria	*Staphylococcus aureus*	13.3	9.6	
		Listeria monocytogenes EGD	ND	11.0	
	Fungus	*Candida albicans*	10.6	13.6	
Direção Regional de Agricultura de Trás-os-Montes, Mirandela (Portugal)	Fungi		MIC (µL/mL)	MLC (µL/mL)	[26]
		Candida albicans	2.5	2.5	
		Candida albicans	1.25–2.5	2.5	
		Candida albicans	2.5	5.0	
		Candida tropicalis	2.5–5.0	5.0	
		Candida tropicalis	5.0–10.0	5.0	
		Candida glabrata	1.25–2.5	5.0	
		Candida glabrata	2.5	5.0	
		Candida krusei	1.25–2.5	2.5	
		Candida guilhermondii	1.25	1.25	
		Candida parapsilosis	2.5–5.0	5.0	
Córdoba (Spain)	Gram-negative bacteria		MIC (%, *v/v*)		[29]
		Escherichia coli—origin in poultry	4		
		Salmonella enteritidis—origin in poultry	4		
		Salmonella essen—origin in poultry	4		
		Escherichia coli (ETEC)—origin in pig	4		
		Salmonella choleraesuis—origin in pig	4		
		Salmonella typhimurium—origin in pig	4		

Table 2. *Cont.*

Origin	Micro-Organisms	Species	Measured Response and Results Obtained						References
			Area of the inhibition zone (mm^2) excluding the film area						
			6%	7%		8%		9%	
Barcelona (Spain)	Gram-positive bacteria	*Listeria innocua*	NA [a]	0.79 [a]		0.79 [a]		NF [a]	[34]
		Methicillin-resistant *Staphylococcus aureus*	NA [a]	NA [a]		NA [a]		NF [a]	
	Gram-negative bacteria	*Salmonella enteritidis*	NA [a]	NA [a]		NA [a]		NF [a]	
		Pseudomona fragi	NA [a]	NA [a]		NA [a]		NF [a]	
			MIC microdilution technique (µg/mL)		MIC dilution technique (µg/mL)		MBC broth dilution techniques (µg/mL)		
			After 24 h	After 48 h	After 24 h	After 48 h	Microdilution	Tube dilution	
Monteloeder, SL (Elche, Spain)	Gram-negative bacteria	*Escherichia coli*	12,800 [b]	25,600 [b]	12,800 [b]	25,600 [b]	51,200 [b]	51,200 [b]	[53]
		Salmonella enterica	6400 [b]	12,800 [b]	12,800 [b]	25,600 [b]	25,600 [b]	51,200 [b]	
		Enterobacter aerogenes	12,800 [b]	51,200 [b]	51,200 [b]	102,400 [b]	51,200 [b]	102,400 [b]	
	Gram-positive bacteria	*Bacillus cereus*	1600 [b]	3200 [b]	3200 [b]	3200 [b]	6400 [b]	6400 [b]	
		Methicillin-resistant *Staphylococcus aureus*	400 [b]	800 [b]	800 [b]	1600 [b]	1600 [b]	3200 [b]	
			Diameter of the inhibition zone (mm) including disc diameter (9 mm)				MIC (µL/mL)		
Esencias Martinez Lozano (Murcia, Spain)	Gram-positive bacteria	*Listeria innocua*	26.83				3.75		[35]
	Gram-negative bacteria	*Serratia marcescens*	12.36				7.5		
		Pseudomonas fragi	11.68				3.75		
		Pseudomonas fluorescens	9.0				3.75		
		Aeromonas hydrophila	11.29				3.75		
		Shewanella putrefaciens	9.0				3.75		
		Achromobacter denitrificans	10.69				3.75		
		Enterobacter amnigenus	12.51				7.5		
		Enterobacter gergoviae	12.14				7.5		
		Alcaligenes faecalis	23.50				3.75		
			Diameter of the inhibition zone (mm) including disc diameter (10 mm)						
			1%				2%		
Esencias Martinez Lozano (Murcia, Spain)	Gram-positive bacteria	*Listeria innocua*	17.92 [c]				25.51 [c]		[54]
	Gram-negative bacteria	*Serratia marcescens*	21.15 [c]				32.36 [c]		
		Enterobacter amnigenus	NA [c]				NA [c]		
		Alcaligenes faecalis	18.42 [c]				28.29 [c]		

Table 2. *Cont.*

Origin	Micro-Organisms	Species	Measured Response and Results Obtained			References
			Diameter of the inhibition zone (mm) including disc diameter (9 mm)			
			Minced beef	Cooked ham	Dry-cured sausage	
Esencias Martinez Lozano (Murcia, Spain)	Gram-positive bacteria	*Listeria innocua*	34.98	15.23	19.45	[55]
	Gram-negative bacteria	*Achromobacter denitrificans*	11.29	13.29	15.87	
		Alcaligenes faecalis	16.91	15.34	16.03	
		Aeromonas hydrophila	14.7	12.13	24.94	
		Enterobacter amnigenus	10.97	10.69	17.31	
		Enterobacter gergoviae	10.82	13.81	9	
		Pseudomonas fluorescens	12.07	12.86	16.7	
		Pseudomonas fragi	11.61	11.78	14.19	
		Serratia marcescens	11.84	12.69	11.49	
		Shewanella putrefaciens	13.09	14.34	15.82	
			MIC (μg/mL)	MFC (mg/mL)		
Urbino (Italy)	Fungi	*Fusarium culmorum*	1500	2		[41]
		Fusarium graminearum	1500	2		
		Fusarium poae	1500	2		
		Fusarium avenaceum	1500	2		
		Fusarium equiseti	2100	2.4		
		Fusarium semitectum	2000	2.4		
		Fusarium sporotrichoides	2000	2.4		
		Fusarium nivale	2000	2.4		
			MIC (mg/mL)	MBC (mg/mL)		
Alentejo (Portugal)	Gram-positive bacteria	Methicillin-sensitive *Staphylococcus aureus*	20	40		[45]
		Bacillus subtilis	15	30		
	Gram-negative bacteria	*Escherichia coli*	15	30		
		Pseudomonas aeruginosa	20	70		
				ED_{50}		
Ciudad Real (Spain)	Fungi	*Botrytis cinerea*		-		[47]
		Sclerotinia sclerotiorum		14.87		
		Fusarium oxysporum		58.0		
		Phytophthora parasitica		22.0		
		Alternaria brassicae		>100		
		Cladobotryum mycophilum		14.1		
		Trichoderma agressivum		-		

Table 2. *Cont.*

Origin	Micro-Organisms	Species	Measured Response and Results Obtained		References
			MIC (mg/mL)	MBC (mg/mL)	
Murcia (Spain)	Gram-negative bacteria	*Escherichia coli*	2.3–9.4	2.3–9.4	[4]
	Gram-positive bacteria	Methicillin-sensitive *Staphylococcus aureus*	2.3–4.7	4.6–4.7	
	Fungus	*Candida albicans*	2.3–4.7	2.3–4.7	
			Inhibition growth zone (mm)	MIC (μg/mL)	
Ervitas Catitas (Portugal)	Gram-positive bacteria	Methicillin-sensitive *Staphylococcus aureus* (isolates)	9.0–11.8	500–4000 (or higher)	[48]
		Staphylococcus epidermidis (isolates)	ND; 9.0–13.8	4000–4000 (or higher)	
			Inhibition growth zone (mm)	MIC (μL/mL)	
Évora, Alentejo (Portugal)	Gram-positive bacteria	*Methicillin-sensitive Staphylococcus aureus*	19	>2	[2]
		Staphylococcus epidermidis	21	>2	
		Enterococcus faecalis	21	>2	
	Gram-negative bacteria	*Escherichia coli*	11	>2	
		Morganella morganii	17	1.1	
		Proteus mirabilis	9	0.5	
		Salmonella enteritidis	11	0.1	
		Salmonella typhimurium	8	>2	
		Pseudomonas aeruginosa	17	1.1	

ED_{50} (effective dose 50), concentration that inhibits mycelial growth by 50%; MBC, minimum bactericidal concentration; MFC, minimal fungicidal concentration; MIC, minimum inhibitory concentration; MLC, minimum lethal concentration; NA, not active; ND, not determined; NF, no film formed. [a] Whey protein isolate films incorporated with *T. mastichina* essential oil [b] Mixture of *Rosmarinus officinalis*, *Salvia lavandulifolia*, and *T. mastichina* and chitosan [c] Chitosan edible film disks incorporated with *T. mastichina* essential oil.

T. mastichina essential oils with origin in different bioclimatic zones from Murcia (Spain) showed activity (growth inhibition) against Gram-positive (methicillin-sensitive *S. aureus*), Gram-negative (*E. coli*), and fungi (*C. albicans*). However, some differences were identified among them, probably due to the influence of the clime in the composition of the essential oil. In particular, *T. mastichina* essential oil from the Supra-Mediterranean bioclimatic zone (Moratalla, Spain) produced higher inhibition against *C. albicans* than the *T. mastichina* essential oils from other bioclimatic zones (Caravaca de la Cruz and Lorca, Spain) due to the high concentration of linalool [4]. However, other studies reported lower antibacterial activities of *T. mastichina* essential oil than those found in this study [29,45]. Recently Arantes et al. [2] described the *T. mastichina* essential oil broad spectrum of antibacterial activity against several strains. It was observed higher susceptibility (lower MICs) observed in Gram-negative bacteria (*E. coli*, *Morganella morganii*, *P. mirabilis*, *Salmonella enteritidis*, *Salmonella typhimurium*, and *Pseudomonas aeruginosa*) than in Gram-positive ones (methicillin-sensitive *S. aureus*, *Staphylococcus epidermidis*, and *Enterococcus faecalis*) for both agar disc diffusion assay and MICs determination by broth microdilution assay. The higher antibacterial activity against Gram-negative is suggested to be correlated with the presence of monoterpene and phenolic compounds capable of disintegrating the outer membrane of Gram-negative strains. In fact, essential oils are characterized by unique antibacterial potential due to the high number of chemical compounds present in their composition, which act simultaneously, preventing resistance mechanisms in bacteria. Furthermore, synergistic interactions between compounds of essential oils can potentiate their natural antimicrobial effect. Thus, antimicrobial potential cannot be associated with only one component or mechanism of action. Nevertheless, due to the lipophilic character of essential oils, the mechanism of action could be related to the alteration of cell membrane properties.

The antifungal activity of *T. mastichina* essential oils from Sesimbra (Estremadura, Portugal) was observed against *C. albicans* [21]. The antifungal capacity of the *T. mastichina* essential oil against *Candida* spp. have also been evaluated through the macrodilution method that enables the determination of MIC and MLC [26]. Flow cytometry was also used as a complementary method for the study of the mechanisms responsible for antifungal activity. *T. mastichina* essential oil showed higher inhibition compared to the other *Thymus* species tested, with a lower MIC concentration obtained for *T. mastichina* against *Candida* spp. varied from 1.25 to 10.00 μL/mL, depending on the species of *Candida*, while the MLC remained at 5 μL/mL for almost all species. This study described the potent antifungal activity of *T. mastichina* essential oil against *Candida* spp., warranting future therapeutic trials on mucocutaneous candidosis. Compared with other studies, similar MIC values for *Candida* were found using *T. mastichina* essential oil from Portugal [4]. A remarkable increase in antifungal activity of the mixture of extracts of *Rosmarinus officinalis*, *Salvia lavandulifolia*, *T. mastichina*, and chitosan against different yeasts (*C. albicans*, *Pichia anomala*, *Pichia membranaefaciens*, and *Saccharomyces cerevisiae*) and filamentous fungi (*Aspergillus niger* and *Penicillium digitatum* strains belonging to the collection of fungi isolated from citrus) was observed [53]. However, the results obtained did not enable the determination of MICs or minimum fungicidal concentrations (MFCs) because the values were higher than the maximum concentrations tested. Additionally, the fungicide activity of *T. mastichina* extracts (at 20–25 mg/mL) from plants micropropagated in vitro against *Aspergillus fumigatus* was demonstrated for the first time [56]. In another study, the antifungal capacity of *T. mastichina* essential oil was determined against *Fusarium* spp. using the agar dilution method. *T. mastichina* essential oil showed antifungal activity against pathogenic fungi strains of the genus *Fusarium* with MICs and MFCs ranging from 1500 to 2100 μg/mL and from 2.0 to 2.4 mg/mL, respectively. In this study, the antifungal activity of the two main constituents, 1,8-cineole and linalool, was also evaluated and considering the obtained MICs and MFCs, the antifungal activity of the essential oil seemed to be due to the presence of major constituents [41].

On the other hand, the potential use of *T. mastichina* essential oil as an active and functional ingredient in food products, and the antimicrobial activity against zoonotic and food spoilers and foodborne microorganisms was evaluated in several studies.

The effect of *T. mastichina* essential oil was evaluated on several bacteria of the Enterobacteriaceae family (*E. coli, Salmonella* spp.) with an origin in poultry and pigs species, was registered with MIC values with 4% (*v*/*v*) [29], as well as in other studies against methicillin-sensitive *S. aureus* and *S. epidermidis* isolates from ovine mastitic milk [48].

Gram-negative bacteria (*E. coli, Salmonella enterica*, and *Enterobacter aerogenes*) and Gram-positive bacteria (*Bacillus cereus* and methicillin-resistant *S. aureus*) were used for determination of MIC, through dilution and microdilution techniques, of a mixture of extracts of *Rosmarinus officinalis, Salvia lavandulifolia, T. mastichina*, and chitosan [53]. All tested extracts demonstrated noticeable antimicrobial activities against spoilage and foodborne pathogens such as *B. cereus*, methicillin-resistant *S. aureus, E. coli, E. aerogenes, S. enterica*, and yeast-like fungi, without interference in sensory properties. Vieira et al. [45] also observed *T. mastichina* essential oil activity, with MIC values of 15 mg/mL (*Bacillus subtilis* and *E. coli*) and 20 mg/mL (methicillin-sensitive *S. aureus* and *P. aeruginosa*); and the same pattern for MLC, with a value of 40 mg/mL for *S. aureus*, 30 mg/mL for *B. subtilis* and *E. coli*, and 70 mg/mL for *P. aeruginosa*, and suggesting this aromatic plant to be used in control pathogenic microorganisms in deteriorating foods. In another study, the antibacterial activity of *T. mastichina* essential oil was assayed in vitro by a microdilution method against both Gram-positive (*Listeria innocua*, methicillin-resistant *S. aureus*, and *B. cereus*) and Gram-negative bacteria (*S. enterica* and *E. coli*). In this study, the inhibition percentage increased with the *T. mastichina* essential oil concentration and higher inhibition was observed for *L. innocua* [44]. Furthermore, the antimicrobial activity of whey protein isolate-based edible films incorporated with *T. mastichina* essential oil were tested against Gram-positive bacteria (*L. innocua* and methicillin-resistant *S. aureus*) and Gram-negative bacteria (*S. enteritidis* and *Pseudomonas fragi*) to be useful as a coating in the food industry. In this context, it should be highlighted that the antimicrobial activity was only observed against *L. innocua* and using the whey protein films containing 7% and 8% of *T. mastichina* essential oil [34]. This work suggests the possibility of using films incorporating essential oils on food systems.

The active antimicrobial activity of *T. mastichina* essential oil (30 μL) applied through the disk-diffusion method was confirmed by Ballester-Costa et al. [35] against *L. innocua* and *Alcaligenes faecalis, Serratia marcescens, Enterobacter amnigenus*, and *Enterobacter gergoviae*, but not active against *P. fragi, Pseudomonas fluorescens, Aeromonas hydrophila, Shewanella putrefaciens, Achromobacter denitrificans*, and *E. gergoviae*. Additionally, MIC of *T. matichina* essential oil determined by microdilution assay was between 3.75 and 7.5 μL/mL for all strains. The differences obtained in both methods are related to the lower dispersion of the essential oils on a solid medium and consequently reduced ability to access to the microorganism in the disk diffusion method, confirming the unreliability of this method for essential oil evaluation. Posteriorly, the same authors tested the effect of chitosan edible film disks incorporated with the essential oil of *T. mastichina* at concentrations of 1% and 2% against *L. innocua, S. marcescens, E. amnigenus*, and *A. faecalis*. At both concentrations, antibacterial activity was observed for *S. marcescens, L. innocua*, and *A. faecalis*, with activity against *S. marcescens* being higher. However, antibacterial activity against *E. amnigenus* was not registered [54]. Besides, in the following year, they evaluated the activity of *T. mastichina* essential oil (30 μL), applied through the disk-diffusion method, against the bacteria *A. denitrificans, A. faecalis, A. hydrophila, E. amnigenus, E. gergoviae, L. innocua, P. fluorescens, P. fragi, S. marcescens*, and *S. putrefaciens* using, as culture medium, extracts from meat homogenates (minced beef, cooked Ham, or dry-cured sausage). *T. mastichina* essential oils were extremely active against *L. innocua* in minced beef and active for *A. hydrophila* in dry-cured sausage, while for the remaining bacteria only moderate activity or absence of activity was found [55]. In this way, it was suggested the use of *T. mastichina* essential oil as a "green" preservative agent in the food industry, per se or incorporated in edible films. In addition, it should be highlighted that its efficacy as an antibacterial agent has been demonstrated in model systems that closely simulate food composition.

Due to the serious damage caused by fungal pathogens of agricultural interest, the possible future application of the essential oils as alternative antifungal agents was evaluated by [47]. In this study, *T. mastichina* essential oil showed either partial or complete antifungal activity against plant and mushroom pathogenic fungi (*Sclerotinia sclerotiorum*, *Fusarium oxysporum*, *Phytophthora parasitica*, *Alternaria brassicae*, and *Cladobotryum mycophilum*) by the disk-diffusion assay; the inhibitory effect of essential oils was dose-dependent on the eight tested fungi enabling the determination of ED_{50} values for most of them.

Rapid antibacterial screening of essential oils using the agar diffusion technique is usually conducted. However, the lack of standardized methods makes direct comparison of results between studies difficult [57]. The problems related to oils dispersion and lipophilic constituents in aqueous media, and varying methods for determining numbers of viable bacteria remaining after the addition of the oil were the main causes of unreliability and inconsistent results obtained from disc diffusion, well diffusion, and agar dilution methods. Nevertheless, the broth dilution method, using emulsifier, seems to be the most accurate method for testing the antimicrobial activity of the hydrophobic and viscous essential oils.

4.2. Antioxidant Activity

The use of antioxidants is useful in the food industry to avoid rancidity and/or deterioration of foods [58] and also to prevent reactive oxygen species (ROS) formation, such as superoxide anion, hydrogen peroxide, and hydroxyl radicals. The ROS are capable of inducing lipid peroxidation, which may result in damage of membranes, lipids, lipoproteins, and induce DNA mutations that are linked to several diseases such as rheumatoid arthritis, atherosclerosis, ischemia, carcinogenesis, and aging [31]. The antioxidant properties of different *T. mastichina* plant extracts, essential oils, and pure compounds have been evaluated using a quite diverse number of in vitro assays, namely, those that evaluate lipid peroxidation, free radical scavenging ability, and chelating metal ions. In a study conducted by Miguel et al. [22], *T. mastichina* essential oil, as well as its main constituents (1,8-cineole and linalool), evaluated through the peroxide values expressed as percentage of inhibition, showed antioxidant activity higher than that shown by the synthetic antioxidant butylated hydroxytoluene (BHT). Nevertheless, the results cannot be explained only by some of its constituents because there may be synergistic or antagonistic effects among them. Due to the demonstrated antioxidant activity, essential oils of this species seem to be a good alternative to some synthetic antioxidants. Miguel et al. [25] conducted another study, in which *T. mastichina* essential oil was tested by a modified thiobarbituric acid-reactive substances (TBARS) assay in which the antioxidant capacity was evaluated measuring the ability to inhibit lipid peroxidation. In this modified TBARS assay egg yolk was used (as a lipid-rich medium) in the presence or absence of the radical inducer of lipid peroxidation, 2,2′-azobis (2 amidinopropane) dihydrochloride (ABAP). At concentrations of 62.5–500 mg/L of essential oil, the antioxidant capacity in the absence of the peroxyl radical inducer ABAP, presented values between 9.6% and 38.9% and in the presence of the peroxyl radical inducer, ABAP lower values were achieved (−19.5–16.0%). In 2007, the same research group compared the antioxidant activity of *T. mastichina* essential oils, over a concentration range (160–1000 mg/L), isolated from different populations. The highest differences in the antioxidant activities of these essential oils were observed at the lowest concentration tested (160 mg/L), in which the *T. mastichina* essential oil from Mértola showed the lowest activity (20%), whereas *T. mastichina* essential oil from Sesimbra exhibited the highest activity (42%). Similarly, for the highest concentration tested (1000 mg/L), *T. mastichina* essential oil from Mértola showed an antioxidant index of 59% and *T. mastichina* essential oil from Sesimbra presented an inhibition percentage of 79%. At a concentration of 1000 mg/L, the *T. mastichina* essential oils from Vila Real de Santo António and Sesimbra showed a higher ability to inhibit lipid oxidation than α-tocopherol and were within the same range of activity of BHA. In comparison with assays without ABAP, the presence of the radical inducer reduced the ability of *T. mastichina* essential oils to prevent oxidation, particularly at concentrations of 160, 800, and 1000 mg/L [30].

In a study conducted by Galego et al. [31], the antioxidant activity of *T. mastichina* essential oil was determined using different methods, such as TBARS, measuring the scavenging effect of the substances on the 2,2-diphenyl-1-picrylhydrazyl (DPPH) radicals, determining the ferric reducing antioxidant power (FRAP) based on the principle that substances, which have reduction potential, react with potassium ferricyanide (Fe^{3+}) to form potassium ferrocyanide (Fe^{2+}), which then reacts with ferric chloride to form ferric–ferrous complex, and also monitoring the chelating effect on ferrous ions (Fe^{2+}). The results showed that even at higher concentrations of *T. mastichina* essential oil (1000 mg/L), the antioxidant index was around 50% as observed with the TBARS method, while for the synthetic antioxidants, BHA and BHT, the antioxidant index was approximately 100%. In other methods, the difference between the activity of the essential oil and synthetic antioxidants BHA and BHT was more accentuated showing lesser antioxidant activity.

According to Bentes et al. [32], the antioxidant activities of *T. mastichina* essential oil were screened by five different methods: DPPH free radical scavenging, modified TBARS (using egg yolk as a lipid-rich medium) for measuring the inhibition of lipid peroxidation, FRAP assay based on the reduction of ferric iron (Fe^{3+}) to ferrous iron (Fe^{2+}) by antioxidants present in the samples, chelating activity on ferrous ions (Fe^{2+}), and superoxide anion radical scavenging. With the DPPH free radical scavenging assay, *T. mastichina* essential oil was almost totally ineffective as an antioxidant, in contrast to the remaining water-soluble hydrodistillation-aqueous phase extract that showed similar activity to that of α-tocopherol, particularly at higher concentrations (75 and 100 mg/L). This result suggests that a considerable portion of the antioxidant compounds were retained in the remaining hydrodistillation-aqueous water. In this study, the FRAP assay of *T. mastichina* essential oil was also compared. However, the correlation between DPPH and the FRAP assay activities was not as clear in the evaluation of different extracts. Whereas the hydrodistillation-aqueous phase extracts were the best antioxidants when evaluated by DPPH, the *T. mastichina* methanolic extract had a greater capacity for reducing Fe^{3+} than the hydrodistillation-aqueous phase extract. In addition, in the superoxide anion assay, only the hydrodistillation-aqueous phase and methanolic extracts were able to scavenge superoxide radicals, but the first ones were the most effective. The superoxide radical scavenging capacity of these extracts was even higher than that of the positive control (ascorbic acid, particularly at 500 mg/L). *T. mastichina* methanol extracts were powerful superoxide radical scavengers with 40–60% activity.

To determine the antioxidant activity in plant extracts, four different methods were used: the DPPH radical scavenging assay, the TBARS assay in brain homogenates, the FRAP assay of ferric iron (Fe^{3+}) to ferrous iron (Fe^{2+}), and the system β-carotene/linoleic acid based on the inhibition of β-carotene bleaching in the presence of linoleic acid radicals. The IC_{50} in the β-carotene/linoleic acid assay was achieved at a value of 0.90 mg/mL, in the DPPH radical scavenging assay, the value obtained was 0.69 mg/mL, the TBARS assay yielded a value of 0.43 mg/mL, and in the FRAP assay, the lowest value was obtained at 0.23 mg/mL [13].

As Asensio-S.-Manzanera et al. [33] reported, the DPPH and FRAP assays were used to evaluate the antioxidant activity on dry plant extracts of *T. mastichina* and dry residues after hydrodistillation. In the dry plant extracts, a scavenging effect was observed through the assays of DPPH (EC_{50} 0.59–1.78 mg/mL) and FRAP (EC_{50} 0.77–2.05 mg/mL). While for the hydrodistilled residue, the EC_{50} values were higher, showing less antioxidant activity. These results could be related to the higher phenol content in the dry extracts compared with the hydrodistilled residue, indicating that a considerable amount of antioxidants were retained in the remaining hydrodistilled water and *T. mastichina* essential oil.

Albano et al. [14], evaluated the antioxidant activity of *T. mastichina* essential oil and decoction water extract. The DPPH method was used with the essential oil and decoction water extract, while the scavenging activity of the superoxide anion radical was only used successfully with decoction water extract. For the DPPH method, the IC_{50} of the essential oil was much higher (6706.8 μg/mL) than that of the extract (4.2 μg/mL). The evaluation of the antioxidant activity by the superoxide anion scavenging activity method in the extract revealed that the IC_{50} of *T. mastichina* was 14.8 μg/mL. In this study, no correlation was detected between total phenols and removal of superoxide anion radicals.

The antioxidant activity of *T. mastichina* ethanolic extracts, obtained by the *Soxhlet* system or ultrasound method, was also evaluated through the β-carotene/linoleate model system, FRAP, DPPH radical scavenging, and iron and copper ion chelation. In general, good antioxidant activities were obtained; in particular, for the *T. mastichina* extract obtained by Soxhlet extraction, which was even comparable to the antioxidant standard red grape pomace [51].

In a study conducted by Delgado et al. [37], the DPPH and FRAP assays of the methanolic extracts and essential oils from 20 different populations were examined. All methanolic extracts and essential oils presented a similar concentration-dependent pattern, increasing the scavenging activity against DPPH with the increase in concentration. However, much higher DPPH radical scavenging abilities were obtained with the methanolic extracts than with the essential oils. The results of the inhibition of DPPH radicals ranged between 86.6% and 93.9% when the highest methanolic extract concentration was used. In contrast, the essential oils of *T. mastichina* presented a much lower DPPH radical scavenging activity, varying between 30.8% and 57.7%, even when the highest concentration was used. Relative to the reductive power assay, it was only determined for the methanolic extracts because it was not possible to perform the assay correctly with the essential oils. Interestingly, in this study, the authors tried to relate the antioxidant activity to their chemical composition. In fact, *T. mastichina* methanolic extracts were rich in rosmarinic acid, which is a polyphenol carboxylic acid, known for possessing antioxidant activity, whereas the essential oils had low contents of thymol, among others, that could explain the low antioxidant activity.

The antioxidant activity of 14 populations of *T. mastichina* methanolic extracts grown in an experimental plot was analyzed by DPPH and FRAP assays to determine antioxidant activity. Population means for DPPH activity ranges were 44–98 mg TE/g dry weight (DW), while FRAP antioxidant capacity was 52–115 mg of Trolox equivalents (TE)/g DW. In general, populations with high DPPH free radical scavenging assay also showed high FRAP antioxidant activity and vice-versa. In this study, the rosmarinic acid contributed mainly to the FRAP antioxidant capacity and total phenolic content, while the unknown compound (peak 3) contributed mainly to the DPPH assay. This study showed high intrapopulation variability and, above all, high interpopulation variability [9].

Chitosan edible films incorporated with *T. mastichina* essential oil (concentrations of 1% and 2%) were tested by DPPH and FRAP methods. The DPPH method demonstrated lower values (0.29 and 0.44 mg/g, respectively) compared to the FRAP method (2.21 and 3.99 mg/g, respectively) [54]. These investigators also evaluated the antioxidant activity but under different conditions. The concentrations used ranged from 0.23 to 30 mg/mL of *T. mastichina* essential oil for DPPH methods, 2,2′-azino-bis(3-ethylbenzothiazoline-6-sulfonic acid) (ABTS) radical cation scavenging activity assay, and FRAP assay, while the concentrations used ranged from 0.15 to 20 mg/mL of *T. mastichina* essential oil for the ferrous ion-chelating ability assay. The DPPH method showed the lowest IC_{50} values (3.11 mg/mL), followed by the ABTS radical scavenging method (3.73 mg/mL), followed by the ferrous ion-chelating activity (9.61 mg/mL). Relative to the FRAP assay, the results were 19.26 mg TE/mL. *T. mastichina* essential oil can be used in general by the food and pharmaceutical industry as a potential natural additive, replacing or reducing the use of chemical substances, since it has antioxidant properties [55].

In the study conducted by Delgado-Adámez et al. [44], the antioxidant activity of *T. mastichina* essential oil showed an antioxidant activity lower than 4 g Trolox/L determined by the ABTS radical method.

In another study, the antioxidant activities of *T. mastichina* essential oil and aqueous extract were evaluated. The essential oil in the β-carotene/linoleic acid method presented an IC_{50} of 0.622 mg/mL, while the aqueous extract presented values of 0.017 mg/mL activity. In the radical DPPH method, the IC_{50} values were 9.052 and 0.104 mg/mL for the essential oil and aqueous extract, respectively. Finally, for the FRAP test, the IC_{50} values presented the same pattern (18.687 and 0.109 mg/mL for the essential oil and aqueous extract, respectively). Thus, it was clear that the extract showed higher antioxidant activity than the essential oil due to the high content of phenolic compounds in extracts [43]. These results are in accordance with the study of Albano et al. [14], in which the extracts also showed higher scavenging ability. In a recent study from the same investigators, the antioxidant activity of *T. mastichina* essential oil was corroborated through the DPPH radical method, FRAP assay, and β-carotene/linoleic acid system. This activity could be related to the high content of oxygenated monoterpenes (85.9%), which act as radical scavengers and ferric reducers with high activity to protect the lipid substrate [2].

The antioxidant activities of *T. mastichina* essential oils have been evaluated using several complementary methods: oxygen radical absorbance capacity (ORAC) that measures the antioxidant capacity against peroxyl radicals, ABTS, DPPH, TBARS, and chelating power. The results obtained in the different methods were the following: ORAC method (163.5–735.1 mgTE/g), ABTS method (0.8–4.3 mg TE/g), DPPH assay (53.5–76.1 mg TE/kg), TBARS method (0.9–1.2 mg BHT equivalents (BHTE)/g), and chelating power (0.6–1.6 mg ethylenediaminetetraacetic acid equivalents (EDTAE)/g). In general, the different antioxidant activities were related to the individual constituents that were also tested in these assays [4].

Taghouti et al. [10] showed that the *T. mastichina* hydroethanolic extract presented a significantly higher scavenging activity of the ABTS radical cation (≈1.48 mmol TE/g extract) than the aqueous decoction extract (≈0.96 mmol TE/g extract). For the hydroxyl radical and nitric oxide radical scavenging assays, both extracts showed a similar capacity for scavenging. These screening assays showed that *T. mastichina* extracts may be a potential source of phenolic compounds with relevant antioxidant activities, inclusively using the decoction that is the traditional method of consumption.

4.3. Anticancer Activity

Dichloromethane and ethanol extracts from the aerial parts of *T. mastichina* were evaluated regarding anticancer activity on the colon cancer cell line HCT, presenting IC_{50} values of 2.8 and 12 μg/mL, respectively. Additionally, one constituent of the extract, ursolic acid, was found to have an IC_{50} value of 6.8 μg/mL, while the other compounds in the extract were inactive with an $IC_{50} >$ 20 μg/mL. A mixture of oleanolic and ursolic acid showed higher cytotoxicity than pure ursolic acid (IC_{50} of 2.8 μg/mL). The presence of these constituents identified by colon cancer cytotoxicity-guided activity indicates that *T. mastichina* extracts may have a protective effect against colon cancer [52].

The cytotoxic effects of *T. mastichina* essential oil were evaluated on human epithelioid cervix carcinoma (HeLa) and human histiocytic leukemia (U937) cell lines. A dose-dependent decrease in the survival of both tumor cell lines was observed after treatment with *T. mastichina* essential oil; this decrease was statistically significant at a concentration of 0.1% (*v*/*v*) *T. mastichina* essential oil [44].

The antiproliferative effect of *T. mastichina* essential oil against human breast carcinoma cell line MDA-MB-231 was also evaluated, showing an IC_{50} value of 108.5 μg/mL. From the literature, studies reported that antiproliferative activity could be related to 1,8-cineole content and are dependent on the monoterpenes content and their ability to affect oxidative stress [2]. It has been appointed that the preventive effect of essential oils on cancer disorders could be related to the promotion of cell cycle arrest, stimulating cell apoptosis and DNA repair mechanisms, inhibiting cancer cell proliferation, metastasis formation, and multidrug resistance, which makes them potential candidates for adjuvant anticancer therapeutic agents.

In a recent study, a *T. mastichina* aqueous decoction and hydroethanolic extract presented a dose and time-dependent inhibitory effect on the cell viability of a human colon adenocarcinoma (Caco-2) cell line and human hepatocellular carcinoma (HepG2) cell line. The hydroethanolic extract presented higher antiproliferative activity/cytotoxicity on Caco-2 cells (IC_{50}: 71.18 and 51.30 μg/mL after 24 and 48 h of incubation, respectively) than the aqueous decoction extracts (IC_{50}: 220.68 and 95.65 μg/mL after 24 and 48 h of incubation, respectively), which was correlated with its higher phenolic content. In addition, it should be noted that the Caco-2 cells were more sensitive than HepG2 cells, as it presented lower IC_{50} values for both extracts [10].

4.4. Antiviral Activity

T. mastichina essential oil was evaluated for its ability to reduce or eliminate the most emergent foodborne viruses in the food industry, evaluating its potential to inactivate two model nonenveloped viruses, a human norovirus surrogate, murine norovirus (MNV-1) with RNA genome, and a human adenovirus serotype 2 (HAdV-2) with DNA genome. However, no significant reduction of virus titers was observed when *T. mastichina* essential oil was used at different temperatures and times [59].

The influenza virus is associated with respiratory tract complications. Thus, in a study conducted by Choi [46], *T. mastichina* essential oil demonstrated interesting anti-influenza activity of reducing visible cytopathic effects of the A/WS/33 virus. This anti-influenza A/WS/33 activity of *T. mastichina* essential oil appeared to be associated with the constituent linalool.

4.5. Insecticidal and Insect Repellent Activity

In recent years, plants have been identified for their insecticidal or larvicidal properties and used to control insect vectors offering an economically viable and ecofriendly approach. One of the studies was based on exposing *Spodoptera littoralis* larvae to *T. mastichina* essential oil and verifying larval mortality. The essential oil was highly toxic with a lethal concentration at 50% (LC_{50}) of 19.3 mL/m^3 when applied by fumigation. After topical application, *T. mastichina* essential oil was also highly toxic with a LC_{50} of 0.034 µL/larvae [60]. The same investigator also evaluated the topical and fumigant activity of essentials oils on the adult house fly (*Musca domestica* L.) and determined a topical LD_{50} of 33 µg/fly and fumigant LD_{50} of 7.3 µg/cm^3 [61].

4.6. Anti-Alzheimer Activity

Alzheimer's disease is characterized by the loss of cholinergic neurons, leading to the progressive reduction of acetylcholine in the brain and cognitive impairment. Inhibition of acetylcholinesterase (AChE) and butyrylcholinesterase (BChE) has great potential in the treatment of Alzheimer's disease and special focus has been directed to these targets.

Albano et al. [14] reported the AChE inhibition activity of *T. mastichina* essential oil for the first time, with IC_{50} values of 45.8 µg/mL related to the 1,8-cineole constituent. However, the assessment of decoction water extracts was not possible for AChE inhibition capacity. Aazza et al. [40] also reported AChE inhibition activity of *T. mastichina* essential oil but poor activity was observed (IC_{50} of 0.1 mg/mL), despite the fact that the main constituent was also 1,8-cineole.

In another study, *T. mastichina* essential oil revealed a high ability to inhibit cholinesterase activity with an IC_{50} of 78.8 and 217.1 µg/mL for AChE and BChE, respectively. The aqueous extract also showed AChE activity, with IC_{50} values of 1003.6 and 779.1 µg/mL for AChE and BChE, respectively. These results suggest that the essential oils and extracts of this aromatic plant could be useful in the treatment of Alzheimer's disease [43].

Finally, *T. mastichina* essential oil was also reported to have AChE inhibition activity with an IC_{50} of 57.5–117.2 µg/mL. These results support the possible use of *T. mastichina* essential oils as an aid in the treatment of Alzheimer's disease or in its prevention for people with family precedents [4].

4.7. Anti-Inflammatory Activity

The 5-lipoxygenase (5-LOX) activity is an assay used to evaluate both anti-inflammatory and antioxidant activities.

Albano and Miguel [50] evaluated the 5-LOX activity of deodorized (divided into three fractions: the first one suspended in methanol; the second fractionated with water and chloroform; the third with chloroform), organic (diethyl ether, ethyl ether, and *n*-butanol), and aqueous extracts of *T. mastichina*. This study reported that lower IC_{50} values (12.2 µg/mL) were obtained for the water-insoluble deodorized chloroformic fraction of deodorized extract. In contrast, a higher IC_{50} was obtained using diethyl ether, with an IC_{50} of 62.5 µg/mL. Finally, for the deodorized fraction suspended in methanol and the aqueous extract, it was not possible to determine the IC_{50}. For most of these extracts in the 5-LOX assay, there was a correlation between phenol content and IC_{50} values, meaning that a higher phenolic content in the extract resulted in a lower 5-LOX activity.

The *T. mastichina* essential oil revealed anti-inflammatory activity, being able to inhibit 5-LOX, with an IC_{50} of 1084.5 μg/mL, whereas the extracts showed an IC_{50} of 66.7 μg/mL. The constituents of the *T. mastichina* essential oil have also been shown to be as effective as 5-LOX inhibitors, such as 1,8-cineole [14]. In another study, a similar IC_{50} of 0.73 mg/mL for the *T. mastichina* essential oil was also reported [40].

T. mastichina essential oil was reported to present 5-LOX inhibition activity that was expressed as the degree of inhibition (DI (%)). *T. mastichina* essential oil, at a concentration of 150 μg/mL, showed a DI between 40.8% and 56.7% [4].

4.8. α-Amylase and α-Glucosidase Activity

Inhibition of α-amylase and α-glucosidase activity was also studied for *T. mastichina* essential oil, with reported IC_{50} values of 4.6 and 0.1 mg/mL, respectively [40]. The inhibition of these enzymes could result in the slow and prolonged release of glucose into circulation and, consequently, the retardation of sudden hyperglycemia after the consumption of a meal.

5. Conclusions

T. mastichina essential oil was obtained mainly by hydrodistillation, consisting mainly of 1,8-cineole (eucalyptol), linalool, limonene, camphor, borneol, and α-terpineol, as well as other volatile compounds. Conversely, despite being lesser studied, *T. mastichina* extracts using different solvents were also characterized, being composed of 2-methoxysalicylic acid, 3-methoxysalicylic acid apigenin, caffeic acid, chlorogenic acid, kaempferol, luteolin, quercetin, rosmarinic acid, sakuranetin, sterubin, salvianolic acid derivatives, and hexoside and glycoside derivatives, among other constituents.

T. mastichina has been traditionally used as a flavoring for food and in the treatment of health conditions due to its antiseptic, digestive, antirheumatic, and antitussive effects. Regarding the biological activities reported in different studies, *T. mastichina* essential oil and/or extracts also have antibacterial, antifungal, antioxidant, insecticide, repellent, antiviral, anti-Alzheimer, and anti-inflammatory activities. The antibacterial and antifungal activities of *T. mastichina* are an important characteristic for the use of these plants for production as natural antimicrobial agents that could be used as preservatives against diverse Gram-positive and negative bacteria and fungi. In addition, the antioxidant activity of *T. mastichina* was also largely explored through different assays, representing an interesting alternative to synthetic antioxidants. Although little attention has been paid to other activities, such as insecticide, repellent, antiviral, anti-Alzheimer, and anti-inflammatory activities, *T. mastichina* also showed interesting potential for these activities. In some studies, these effects were related to the composition and were tested to understand if some compounds were primarily responsible for the observed activity.

In conclusion, attending to its traditional use and reported biological activities, *T. mastichina* essential oil and/or extracts could present a noteworthy role as preservatives and salt substitutes in food industries, as perfumes in cosmetic industry, and as sources of bioactive compounds for pharmaceutical industries.

Supplementary Materials: The following are available online at http://www.mdpi.com/1424-8247/13/12/479/s1, Table S1: Obtention features and characterization of *Thymus mastichina* extracts and its main constituents and total phenolic and flavonoids contents, Figure S1: Chemical structures of majority of the identified constituents of the *Thymus mastichina* extracts (flavonoids, phenolic acids, phenolic terpene, steroid, triterpenoids and xanthophyll).

Author Contributions: Conceptualization, M.R., A.R.T.S.A.; writing—original draft preparation, M.R., A.C.L., F.V., M.F.; writing—review and editing, G.A., M.P.R., P.C., A.R.T.S.A.; funding acquisition, M.R., M.P.R., P.C., A.R.T.S.A. All authors have read and agreed to the published version of the manuscript.

Funding: This research was funded by Fundação para a Ciência e Tecnologia (FCT), Fundo Europeu de Desenvolvimento Regional (FEDER) and COMPETE 2020 for financial support under the research project "The development of dermo-biotechnological applications using natural resources in the Beira and Serra da Estrela regions—DermoBio" ref. SAICT-POL/23925/2016 presented in the Notice for the Presentation of Applications No. 02/SAICT/2016—Scientific Research and Technological Development Projects (IC & DT) in Copromotion.

Acknowledgments: The authors would like to thank to Conceição Matos from Planalto Dourado^{TM} Essential Oils Enterprise for kindly providing the photos of *Thymus mastichina* from Freixedas, Guarda, Portugal.

Conflicts of Interest: The authors declare no conflict of interest.

References

1. Utad, J.B. Ficha da Espécie *Thymus mastichina*. Available online: https://jb.utad.pt/especie/Thymus_mastichina (accessed on 23 July 2020).
2. Arantes, S.M.; Piçarra, A.; Guerreiro, M.; Salvador, C.; Candeias, F.; Caldeira, A.T.; Martins, M.R. Toxicological and pharmacological properties of essential oils of *Calamintha nepeta, Origanum virens* and *Thymus mastichina* of Alentejo (Portugal). *Food Chem. Toxicol.* **2019**, *133*, 110747. [CrossRef] [PubMed]
3. Póvoa, O.; Delgado, F. Tipos e Espécies de PAM. Guia Para a Produção de Plantas Aromáticas e Medicinais em Portugal. 2015. Available online: http://epam.pt/guia/tipos-e-especies-de-pam/ (accessed on 18 September 2020).
4. Cutillas, A.-B.B.; Carrasco, A.; Martinez-Gutierrez, R.; Tomas, V.; Tudela, J. *Thymus mastichina* L. essential oils from Murcia (Spain): Composition and antioxidant, antienzymatic and antimicrobial bioactivities. *PLoS ONE* **2018**, *13*, e0190790. [CrossRef] [PubMed]
5. Girón, V.; Garnatje, T.; Vallès, J.; Pérez-Collazos, E.; Catalán, P.; Valdés, B. Geographical distribution of diploid and tetraploid cytotypes of *Thymus* sect. *mastichina* (Lamiaceae) in the Iberian Peninsula, genome size and evolutionary implications. *Folia Geobot.* **2012**, *47*, 441–460.
6. Méndez-Tovar, I.; Martín, H.; Santiago, Y.; Ibeas, A.; Herrero, B.; Asensio-S-Manzanera, M.C. Variation in morphological traits among *Thymus mastichina* (L.) L. populations. *Genet. Resour. Crop Evol.* **2015**, *62*, 1257–1267. [CrossRef]
7. Miguel, G.; Guerrero, C.; Rodrigues, H.; Brito, J.; Venâncio, F.; Tavares, R.; Duarte, F. Effect of substrate on the essential oils composition of *Thymus mastichina* (L.) L. subsp. *mastichina* collected in Sesimbra region (Portugal). In *Natural Products in the New Millennium: Prospects and Industrial Application*; Springer: Dordrecht, The Netherlands, 2002; pp. 143–148.
8. Miguel, M.G.; Duarte, F.L.; Venâncio, F.; Tavares, R. Comparison of the main components of the essential oils from flowers and leaves of *Thymus mastichina* (L.) L. ssp. *mastichina* collected at different regions of portugal. *J. Essent. Oil Res.* **2004**, *16*, 323–327.
9. Méndez-Tovar, I.; Sponza, S.; Asensio-S-Manzanera, M.C.; Novak, J. Contribution of the main polyphenols of *Thymus mastichina* subsp: *Mastichina* to its antioxidant properties. *Ind. Crops Prod.* **2015**, *66*, 291–298. [CrossRef]
10. Taghouti, M.; Martins-Gomes, C.; Schäfer, J.; Santos, J.A.; Bunzel, M.; Nunes, F.M.; Silva, A.M. Chemical characterization and bioactivity of extracts from *Thymus mastichina*: A *Thymus* with a distinct salvianolic acid composition. *Antioxidants* **2019**, *9*, 34. [CrossRef]
11. Barros, L.; Carvalho, A.M.; Ferreira, I.C.F.R. From famine plants to tasty and fragrant spices: Three Lamiaceae of general dietary relevance in traditional cuisine of Trás-os-Montes (Portugal). *LWT Food Sci. Technol.* **2011**, *44*, 543–548. [CrossRef]
12. Asensio-S-Manzanera, M.C.; Mendez, I.; Santiago, Y.; Martin, H.; Herrero, B. Phenolic compounds variability in hydrodistilled residue of Thymus mastichina. In *VII Congreso Iberico de Agroingeniería y Ciencias Hortícolas: Innovar y Producir Para el Futuro*; UPM: Madrid, Spain, 2014; pp. 2086–2091.
13. Barros, L.; Heleno, S.A.; Carvalho, A.M.; Ferreira, I.C.F.R. Lamiaceae often used in Portuguese folk medicine as a source of powerful antioxidants: Vitamins and phenolics. *LWT Food Sci. Technol.* **2010**, *43*, 544–550. [CrossRef]
14. Albano, S.M.; Sofia Lima, A.; Graça Miguel, M.; Pedro, L.G.; Barroso, J.G.; Figueiredo, A.C. Antioxidant, anti-5-lipoxygenase and antiacetylcholinesterase activities of essential oils and decoction waters of some aromatic plants. *Rec. Nat. Prod.* **2012**, *6*, 35–48.

15. Salgueiro, L.R.; Vila, R.; Tomàs, X.; Cañigueral, S.; Da Cunha, A.P.; Adzet, T. Composition and variability of the essential oils of *Thymus* species from section *mastichina* from Portugal. *Biochem. Syst. Ecol.* **1997**, *25*, 659–672. [CrossRef]
16. Faleiro, L.; Miguel, G.M.; Guerrero, C.A.C.; Brito, J.M.C. Antimicrobial activity of essential oils of *Rosmarinus officinalis* L., *Thymus mastichina* (L) L. ssp *Mastichina* and *Thymus albicans* Hofmanns & link. *Acta Hortic.* **1999**, *501*, 45–48.
17. Jiménez-Carmona, M.M.; Ubera, J.L.; Luque De Castro, M.D. Comparison of continuous subcritical water extraction and hydrodistillation of marjoram essential oil. *J. Chromatogr. A* **1999**, *855*, 625–632. [CrossRef]
18. Miguel, M.G.; Guerrero, C.A.C.; Brito, J.M.C.; Venâncio, F.; Tavares, R.; Martins, A.; Duarte, F. Essential oils from *Thymus mastichina* (L.) L. ssp. *mastichina* and *Thymus albicans* Hoffmanns & link. *Acta Hortic.* **1999**, *500*, 59–63.
19. Miguel, M.G.; Duarte, F.; Venâncio, F.; Tavares, R. Chemical composition of the essential oils from *Thymus mastichina* over a day period. *Acta Hortic.* **2002**, *597*, 87–90. [CrossRef]
20. Miguel, M.G.; Duarte, F.; Venâncio, F.; Tavares, R.; Guerrero, C.; Martins, H.; Carrasco, J. Changes of the chemical composition of the essential oil of portuguese *Thymus mastichina* in the course of two vegetation cycles. *Acta Hortic.* **2002**, *576*, 83–86. [CrossRef]
21. Faleiro, M.L.; Miguel, M.G.; Ladeiro, F.; Venâncio, F.; Tavares, R.; Brito, J.C.; Figueiredo, A.C.; Barroso, J.G.; Pedro, L.G. Antimicrobial activity of essential oils isolated from Portuguese endemic species of *Thymus*. *Lett. Appl. Microbiol.* **2003**, *36*, 35–40. [CrossRef]
22. Miguel, M.G.; Figueiredo, A.C.; Costa, M.M.; Martins, D.; Duarte, J.; Barroso, J.G.; Pedro, L.G. Effect of the volatile constituents isolated from *Thymus albicans*, *Th. mastichina*, *Th. carnosus* and *Thymbra capitata* in sunflower oil. *Nahr. Food* **2003**, *47*, 397–402. [CrossRef]
23. Miguel, M.G.; Guerrero, C.; Rodrigues, H.; Brito, J.; Duarte, F.; Venâncio, F.; Tavares, R. Essential oils of Portuguese *Thymus mastichina* (L.) L. subsp. *mastichina* grown on different substrates and harvested on different dates. *J. Hortic. Sci. Biotechnol.* **2003**, *78*, 355–358.
24. Miguel, M.G.; Guerrero, C.; Rodrigues, H.; Brito, J.C.; Duarte, F.; Venâncio, F.; Tavares, R. Main components of the essential oils from wild portuguese *Thymus mastichina* (L.) L. ssp. *mastichina* in different developmental stages or under culture conditions. *J. Essent. Oil Res.* **2004**, *16*, 111–114. [CrossRef]
25. Miguel, G.; Simões, M.; Figueiredo, A.C.; Barroso, J.G.; Pedro, L.G.; Carvalho, L. Composition and antioxidant activities of the essential oils of *Thymus caespititius*, *Thymus camphoratus* and *Thymus mastichina*. *Food Chem.* **2004**, *86*, 183–188. [CrossRef]
26. Pina-Vaz, C.; Rodrigues, A.G.; Pinto, E.; Costa-de-Oliveira, S.; Tavares, C.; Salgueiro, L.; Cavaleiro, C.; Gonçalves, M.J.; Martinez-de-Oliveira, J. Antifungal activity of *Thymus* oils and their major compounds. *J. Eur. Acad. Dermatol. Venereol.* **2004**, *18*, 73–78. [CrossRef] [PubMed]
27. Moldão-Martins, M.; Beirão-da-Costa, S.; Neves, C.; Cavaleiro, C.; Salgueiro, L.; Luísa Beirão-da-Costa, M. Olive oil flavoured by the essential oils of *Mentha* × *piperita* and *Thymus mastichina* L. *Food Qual. Prefer.* **2004**, *15*, 447–452. [CrossRef]
28. Miguel, M.G.; Falcato-Simões, M.; Figueiredo, A.C.; Barroso, J.M.G.; Pedro, L.G.; Carvalho, L.M. Evaluation of the antioxidant activity of *Thymbra capitata*, *Thymus mastichina* and *Thymus camphoratus* essential oils. *J. Food Lipids* **2005**, *12*, 181–197.
29. Peñalver, P.; Huerta, B.; Borge, C.; Astorga, R.; Romero, R.; Perea, A. Antimicrobial activity of five essential oils against origin strains of the Enterobacteriaceae family. *APMIS* **2005**, *113*, 1–6. [CrossRef]
30. Miguel, M.G.; Costa, L.A.; Figueiredo, A.C.; Barroso, J.G.; Pedro, L.G. Assessment of the antioxidant ability of *Thymus albicans*, *Th. mastichina*, *Th. camphoratus* and *Th. carnosus* essential oils by TBARS and Micellar Model systems. *Nat. Prod. Commun.* **2007**, *2*, 399–406.
31. Galego, L.; Almeida, V.; Gonçalves, V.; Costa, M.; Monteiro, I.; Matos, F.; Miguel, G. Antioxidant activity of the essential oils of *Thymbra capitata*, *Origanum vulgare*, *Thymus mastichina*, and *Calamintha baetica*. *Acta Hortic.* **2008**, *765*, 325–334. [CrossRef]
32. Bentes, J.; Miguel, M.G.; Monteiro, I.; Costa, M.; Figueiredo, A.C.; Barroso, J.G.; Pedro, L.G. Antioxidant activities of essential oils and extracts of Portuguese *Thymbra capitata* and *Thymus mastichina*. *Ital. J. Food Sci.* **2009**, *XVIII*, 1–125.
33. Asensio-S-Manzanera, M.C.; Martín, H.; Sanz, M.A.; Herrero, B. Antioxidant activity of *Lavandula latifolia*, *Salvia lavandulifolia* and *Thymus mastichina* collected in Spain. *Acta Hortic.* **2011**, *925*, 281–290. [CrossRef]

34. Fernández-Pan, I.; Royo, M.; Ignacio Maté, J. Antimicrobial activity of whey protein isolate edible films with essential oils against food spoilers and foodborne pathogens. *J. Food Sci.* **2012**, *77*, M383–M390.
35. Ballester-Costa, C.; Sendra, E.; Fernández-López, J.; Pérez-Álvarez, J.A.; Viuda-Martos, M. Chemical composition and in vitro antibacterial properties of essential oils of four *Thymus* species from organic growth. *Ind. Crops Prod.* **2013**, *50*, 304–311. [CrossRef]
36. Faria, J.M.S.; Barbosa, P.; Bennett, R.N.; Mota, M.; Figueiredo, A.C. Bioactivity against *Bursaphelenchus xylophilus*: Nematotoxics from essential oils, essential oils fractions and decoction waters. *Phytochemistry* **2013**, *94*, 220–228. [CrossRef] [PubMed]
37. Delgado, T.; Marinero, P.; Asensio-S-Manzanera, M.C.; Asensio, C.; Herrero, B.; Pereira, J.A.; Ramalhosa, E. Antioxidant activity of twenty wild Spanish *Thymus mastichina* L. populations and its relation with their chemical composition. *LWT Food Sci. Technol.* **2014**, *57*, 412–418. [CrossRef]
38. Miguel, M.G.; Gago, C.; Antunes, M.D.; Megías, C.; Cortés-Giraldo, I.; Vioque, J.; Lima, A.S.; Figueiredo, A.C. Antioxidant and antiproliferative activities of the essential oils from *Thymbra capitata* and *Thymus* species grown in Portugal. *Evid. Based Complement. Altern. Med.* **2015**, *2015*, 1–8.
39. Méndez-Tovar, I.; Sponza, S.; Asensio-S-Manzanera, C.; Schmiderer, C.; Novak, J. Volatile fraction differences for Lamiaceae species using different extraction methodologies. *J. Essent. Oil Res.* **2015**, *27*, 497–505. [CrossRef]
40. Aazza, S.; El-Guendouz, S.; Miguel, M.G.; Dulce Antunes, M.; Leonor Faleiro, M.; Isabel Correia, A.; Cristina Figueiredo, A. Antioxidant, anti-inflammatory and anti-hyperglycaemic activities of essential oils from *Thymbra capitata, Thymus albicans, Thymus caespititius, Thymus carnosus, Thymus lotocephalus* and *Thymus mastichina* from Portugal. *Nat. Prod. Commun.* **2016**, *11*, 1029–1038.
41. Fraternale, D.; Giamperi, L.; Ricci, D. Chemical Composition and antifungal activity of essential oil obtained from in vitro plants of *Thymus mastichina* L. *J. Essent. Oil Res.* **2003**, *15*, 278–281.
42. Ibáñez, M.D.; Blázquez, M.A. Herbicidal value of essential oils from oregano-like flavour species. *Food Agric. Immunol.* **2017**, *28*, 1168–1180. [CrossRef]
43. Arantes, S.; Piçarra, A.; Candeias, F.; Caldeira, A.T.; Martins, M.R.; Teixeira, D. Antioxidant activity and cholinesterase inhibition studies of four flavouring herbs from Alentejo. *Nat. Prod. Res.* **2017**, *31*, 2183–2187. [CrossRef]
44. Delgado-Adámez, J.; Garrido, M.; Bote, M.E.; Fuentes-Pérez, M.C.; Espino, J.; Martín-Vertedor, D. Chemical composition and bioactivity of essential oils from flower and fruit of *Thymbra capitata* and *Thymus* species. *J. Food Sci. Technol.* **2017**, *54*, 1857–1865.
45. Vieira, M.; Bessa, L.J.; Martins, M.R.; Arantes, S.; Teixeira, A.P.S.; Mendes, Â.; Martins da Costa, P.; Belo, A.D.F. Chemical composition, antibacterial, antibiofilm and synergistic properties of essential oils from *Eucalyptus globulus* Labill. and seven mediterranean aromatic plants. *Chem. Biodivers.* **2017**, *14*, e1700006. [CrossRef] [PubMed]
46. Choi, H.-J. Chemical constituents of essential oils possessing anti-influenza A/WS/33 virus activity. *Osong Public Health Res. Perspect.* **2018**, *9*, 348–353. [CrossRef] [PubMed]
47. Diánez, F.; Santos, M.; Parra, C.; Navarro, M.J.; Blanco, R.; Gea, F.J. Screening of antifungal activity of 12 essential oils against eight pathogenic fungi of vegetables and mushroom. *Lett. Appl. Microbiol.* **2018**, *67*, 400–410. [CrossRef] [PubMed]
48. Queiroga, M.C.; Pinto Coelho, M.; Arantes, S.M.; Potes, M.E.; Martins, M.R. Antimicrobial activity of essential oils of Lamiaceae aromatic spices towards sheep mastitis-causing *Staphylococcus aureus* and *Staphylococcus epidermidis*. *J. Essent. Oil Bear. Plants* **2018**, *21*, 1155–1165. [CrossRef]
49. Salehi, B.; Mishra, A.P.; Shukla, I.; Sharifi-Rad, M.; del Mar Contreras, M.; Segura-Carretero, A.; Fathi, H.; Nasrabadi, N.N.; Kobarfard, F.; Sharifi-Rad, J. Thymol, thyme, and other plant sources: Health and potential uses. *Phyther. Res.* **2018**, *32*, 1688–1706. [CrossRef]
50. Albano, S.M.; Miguel, M.G. Biological activities of extracts of plants grown in Portugal. *Ind. Crops Prod.* **2011**, *33*, 338–343. [CrossRef]
51. Sánchez-Vioque, R.; Polissiou, M.; Astraka, K.; de los Mozos-Pascual, M.; Tarantilis, P.; Herraiz-Peñalver, D.; Santana-Méridas, O. Polyphenol composition and antioxidant and metal chelating activities of the solid residues from the essential oil industry. *Ind. Crops Prod.* **2013**, *49*, 150–159. [CrossRef]

52. Gordo, J.; Máximo, P.; Cabrita, E.; Lourenço, A.; Oliva, A.; Almeida, J.; Filipe, M.; Cruz, P.; Barcia, R.; Santos, M.; et al. *Thymus mastichina*: Chemical constituents and their anti-cancer activity. *Nat. Prod. Commun.* **2012**, *7*, 1491–1494.
53. Giner, M.J.; Vegara, S.; Funes, L.; Martí, N.; Saura, D.; Micol, V.; Valero, M. Antimicrobial activity of food-compatible plant extracts and chitosan against naturally occurring micro-organisms in tomato juice. *J. Sci. Food Agric.* **2012**, *92*, 1917–1923. [CrossRef]
54. Ballester-Costa, C.; Sendra, E.; Fernández-López, J.; Viuda-Martos, M. Evaluation of the antibacterial and antioxidant activities of chitosan edible films incorporated with organic essential oils obtained from four *Thymus* species. *J. Food Sci. Technol.* **2016**, *53*, 3374–3379. [CrossRef]
55. Ballester-Costa, C.; Sendra, E.; Viuda-Martos, M. Assessment of antioxidant and antibacterial properties on meat homogenates of essential oils obtained from four *Thymus* species achieved from organic growth. *Foods* **2017**, *6*, 59. [CrossRef] [PubMed]
56. Leal, F.; Coelho, A.C.; Soriano, T.; Alves, C.; Matos, M. Fungicide activity of *Thymus mastichina* and *Mentha rotundifolia* in plants in vitro. *J. Med. Food* **2013**, *16*, 273. [CrossRef]
57. Hood, J.R.; Wilkinson, J.M.; Cavanagh, H.M.A. Evaluation of common antibacterial screening methods utilized in essential oil research. *J. Essent. Oil Res.* **2003**, *15*, 428–433. [CrossRef]
58. Figueiredo, A.; Barroso, J.G.; Pedro, L.G.; Salgueiro, L.; Miguel, M.G.; Faleiro, M.L. Portuguese *Thymbra* and *Thymus* species volatiles: Chemical composition and biological activities. *Curr. Pharm. Des.* **2008**, *14*, 3120–3140. [CrossRef]
59. Kovač, K.; Diez-Valcarce, M.; Raspor, P.; Hernández, M.; Rodríguez-Lázaro, D. Natural plant essential oils do not inactivate non-enveloped enteric viruses. *Food Environ. Virol.* **2012**, *4*, 209–212.
60. Pavela, R. Insecticidal activity of some essential oils against larvae of *Spodoptera littoralis*. *Fitoterapia* **2005**, *76*, 691–696. [CrossRef] [PubMed]
61. Pavela, R. Insecticidal properties of several essential oils on the house fly (*Musca domestica* L.). *Phyther. Res.* **2008**, *22*, 274–278. [CrossRef] [PubMed]

Article

N-(Hydroxyalkyl) Derivatives of *tris*(1*H*-indol-3-yl)methylium Salts as Promising Antibacterial Agents: Synthesis and Biological Evaluation

Sergey N. Lavrenov [1,*], Elena B. Isakova [1], Alexey A. Panov [1], Alexander Y. Simonov [1], Viktor V. Tatarskiy [2,3] and Alexey S. Trenin [1]

1 Gause Institute of New Antibiotics, 11 B. Pirogovskaya Street, 119021 Moscow, Russia; ebisakova@yandex.ru (E.B.I.); 7745243@mail.ru (A.A.P.); simonov-live@inbox.ru (A.Y.S.); as-trenin@mail.ru (A.S.T.)

2 N. N. Blokhin Russian Cancer Research Center, 24 Kashirskoe Shosse, 115478 Moscow, Russia; tatarskii@gmail.com

3 National University of Science and Technology MISiS, 4 Leninsky Ave., 119049 Moscow, Russia

* Correspondence: lavrenov@mail.ru; Tel.: +7-9262-249-777

Received: 26 November 2020; Accepted: 15 December 2020; Published: 16 December 2020

Abstract: The wide spread of pathogens resistance requires the development of new antimicrobial agents capable of overcoming drug resistance. The main objective of the study is to elucidate the effect of substitutions in *tris*(1*H*-indol-3-yl)methylium derivatives on their antibacterial activity and toxicity to human cells. A series of new compounds were synthesized and tested. Their antibacterial activity in vitro was performed on 12 bacterial strains, including drug resistant strains, that were clinical isolates or collection strains. The cytotoxic effect of the compounds was determined using an test with HPF-hTERT (human postnatal fibroblasts, immortalized with hTERT) cells. The activity of the obtained compounds depended on the carbon chain length. Derivatives with C5–C6 chains were more active. The minimum inhibitory concentration (MIC) of the most active compound on Gram-positive bacteria, including MRSA, was 0.5 µg/mL. Compounds with C5–C6 chains also revealed high activity against *Staphylococcus epidermidis* (1.0 and 0.5 µg/mL, respectively) and moderate activity against Gram-negative bacteria *Escherichia coli* (8 µg/mL) and *Klebsiella pneumonia* (2 and 8 µg/mL, respectively). However, they have no activity against *Salmonella cholerasuis* and *Pseudomonas aeruginosa*. The most active compounds revealed higher antibacterial activity on MRSA than the reference drug levofloxacin, and their ratio between antibacterial and cytotoxic activity exceeded 10 times. The data obtained provide a basis for further study of this promising group of substances.

Keywords: *tris*(1*H*-indol-3-yl)methylium; turbomycin; indole derivatives; antibacterial action; overcoming of drug resistance

1. Introduction

In recent years, the situation in the field of therapy of infectious diseases has been significantly complicated due to the wide spread of pathogens resistant to known antibiotic drugs [1–3]. To solve this problem, it is proposed to try and prevent the drug resistance by rational use of antibiotics, and by development of new antimicrobial agents capable of overcoming drug resistance. The latter direction is one of the most important ones of modern medicinal chemistry [4]. The most promising compounds are the ones that have low toxicity for human cells and retain high activity on resistant strains of pathogens. The main objective of this study was to elucidate the effect of substituents on the activity of compounds that contain *tris*(1-alkylindol-3-yl)methylium core with respect to various test microorganisms, as well as their toxicity for human cells.

Recently, compounds containing triphenylmethyl or triindolylmethyl fragments have attracted the interest of researchers. This is due to the fact that some compounds of this type exhibit useful biological properties, such as antimicrobial and antiproliferative [5–13].

Earlier, we studied the structure-antimicrobial and cytotoxic activity relationship among a new class of compounds—*tris*(1-alkylindol-3-yl)methylium salts, structurally similar to the natural antibiotic turbomycin A [5]. Among them, a number of substances with a high (submicromolar) activity, even on multidrug-resistant strains of *Staphylococcus aureus*, were found. The first symmetrical alkyl derivatives we obtained were highly active against bacteria, but their toxicity was higher than the antibacterial activity [11,12]. Their structure needed to be improved. One of the successful attempts were chimeric structures [14] combining fragments of *tris*(1-alkylindol-3-yl)methylium and 3,4-disubstituted pyrrole-2,5-diones (maleinimides) [15,16]. They had a relatively low toxicity to human cells, and at the same time, a high activity on resistant strains of Gram-positive bacteria [14], acting by disrupting the functioning of their membranes [17].

When analyzing the structure-activity relationship of these compounds, we observed that the activity of these substances is closely related to their lipophilicity. The most promising substances had a $LogP_{ow}$ (partition between octanol and water) in the region of 2–5, and if the $LogP_{ow}$ was lower than that, the substance had neither pronounced antibacterial activity nor cytotoxicity. If the $LogP_{ow}$ was higher then 5, the cytotoxicity was higher than the antibacterial activity. Thus, the assumption was made about the optimal region of LogPow, where one can expect to find substances with a good ratio of antibacterial activity to cytotoxicity. Symmetrical *N*-(hydroxyalkyl) derivatives of *tris*(1*H*-indol-3-yl)methylium seem very promising in this regard, since by changing the length of the hydrocarbon part of the substituent, it is possible to adjust the lipophilicity of the molecule. In addition, the presence of hydroxyl groups improves the solubility of substances in aqueous media. This paper presents the synthesis and study of antibacterial and cytotoxic activity in the homologous series of *N*-(hydroxyalkyl) derivatives of *tris*(1*H*-indol-3-yl)methylium with a hydrocarbon chain length from C2 to C6.

2. Results and Discussion

2.1. Chemistry

The title compounds were synthesized according to Scheme 1. To start, reagents 2-(1*H*-indol-1-yl) acetic acid (**1a**) and 3-(1*H*-indol-1-yl)propanoic acid (**1b**) were used, which were converted to the corresponding (1*H*-indol-1-yl)alkanols **3a** and **3b** by $LiAlH_4$ reduction. To obtain (1*H*-indol-1-yl)alkanols **3c–e**, another synthesis method was used—alkylation of indole by ω-bromoalkanols **2c–e**. The hydroxyl group of compounds **3a–e** was then protected by acetylation to obtain substances **4a–e**, which were then subjected to formylation by the Wilsmeier–Haack method, with further deacetylation to obtain 1-(ω-hydroxyalkyl)-1*H*-indole-3-carbaldehydes **6a–e**. Without the protection of the hydroxyl group, formylation of **3a–e** is not possible, the reaction leading to formation of a complex mixture of products. To obtain symmetric *tris*(1-[ω-hydroxyalkyl]-1*H*-indol-3-yl)methanes **7a–e**, we condensed (1*H*-indol-1-yl)alkanoles **3c–e** with the corresponding aldehydes **6a–e** in 2:1 molar ratio in boiling methanol using dysprosium triflate $Dy(OTf)_3$ as a catalyst. As a final stage, compounds **7a–e** were oxidized by DDQ (2,3-dichloro-5,6-dicyano-1,4-benzoquinone) in THF with subsequent HCl treatment to yield the target *tris*(1-[ω-hydroxyalkyl]-1H-indol-3-yl)methylium salts **8a–e**. All synthesized compounds were characterized by NMR spectroscopy, high-resolution mass spectrometry (HRMS), and HPLC.

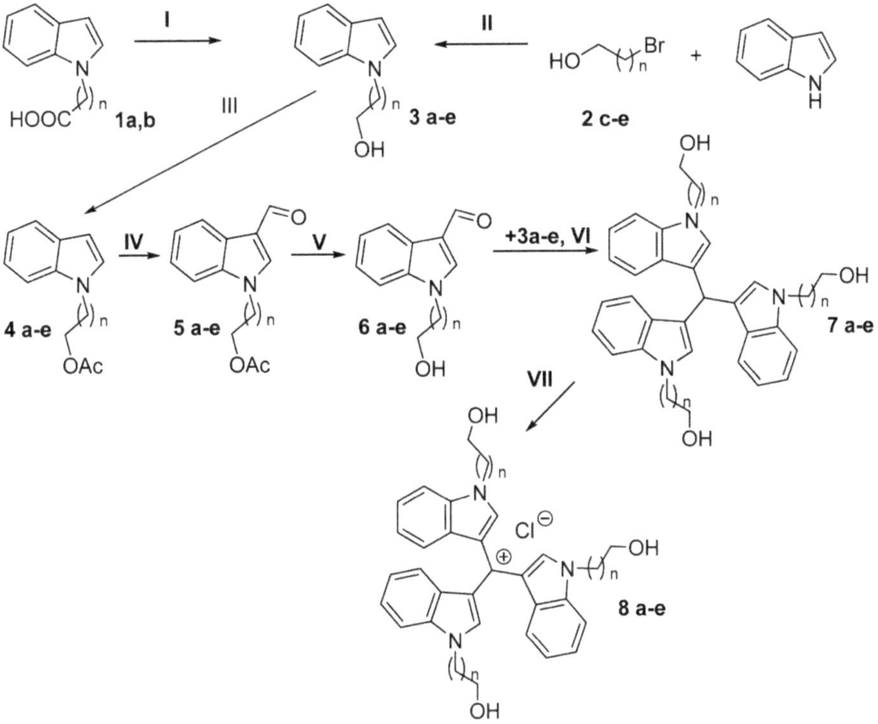

Scheme 1. Synthesis of all compounds. For compounds: **1a**: $n = 1$; **1b**: $n = 2$; **2–8a**: $n = 1$; **2–8b**: $n = 2$; **2–8c**: $n = 3$; **2–8d**: $n = 4$; **2–8e**: $n = 5$. Reagents and conditions: (I) $LiAlH_4$, THF, reflux; (II) KOH, DMSO, rt; (III) Ac_2O, pyridine, DMAP, rt; (IV) $POCl_3$, DMF, rt; (V) NaOMe, MeOH, rt; (VI) MeOH, $Dy(OTf)_3$, reflux; (VII) DDQ, THF, HCl rt.

2.2. Biological Evaluation

2.2.1. Antibacterial Activity

As reference compounds, we used levofloxacin, a common broad-spectrum antibiotic, and Brilliant Green, a common antiseptic with a structure somewhat similar to substances **8a–e,** being triarylmethylium salt. The compounds **7a–e** were insoluble in water, so for a test for antibacterial activity, a solubilizer had to be used. Addition of Kolliphor EL (5× by weight) provided enough solubility for the compounds. The compounds **8a–e** were soluble enough in water by themselves.

Trisidolylmethanes **7a–e** showed practically no antibacterial activity (MICs >64 μg/mL). This is in good agreement with our earlier data that triindolylmethanes are biologically inactive until they are oxidized to trisindolylmethylium salts [11,12,14]. Trisindolylmethylium salts **8a–e** exhibited significant antibacterial activity, as shown in Table 1. The data obtained show that **8a–b** substituted with C2 and C3 hydroxyalkyl substituents have practically no antibacterial activity, however, with further growth of the chain length starting from C4, pronounced activity appears, reaching 0.5 μg/mL in the C6 derivative (compound **8e**). The compounds which are active on bacteria but are low-toxic should balance between pore-forming activity on the lipid bilayers and non-selective detergent activity [17,18].

Compounds **8d** and **8e** were highly active against Gram-positive bacteria, including strains with antibiotic resistance. For example, activity of compound **8e** against *S. aureus* ATCC 25,923 and clinical isolate *S. aureus* 10, that were sensitive to all antibiotics, was 0.5 μg/mL. At the same time, activity of this compound against two methicillin resistant strains (MRSA) (*S. aureus* 5 and *S. aureus* 100 KC) was just the same high (0.5 μg/mL). The same activity (0.5 μg/mL) was revealed against *S. aureus* ATCC 3798, which was resistant not only to ampicillin, oxacillin, cefuroxime, and carbenicillin (antibiotics of penicillin and cefalosporine group), but also to clindamycin, erythromycin, rifampicin, ciprofloxacin, and levofloxacin. Compounds **8d** and **8e** were active against *S. aureus* ATCC 700699, which possess resistance to levofloxacin.

Compounds **8d** and **8e** were also active (1, 0.5 μg/mL) against *Staphylococcus epidermidis* 533, which is resistant to gentamicin, but, in contrast to **8d** (8 μg/mL), compound **8e** was almost inactive (>64 μg/mL) to *Enterococcus faecium* 569, which possess resistance to cefuroxime, clindamycin, gentamycin, vancomycin, and doxycycline.

A moderate level of activity of **8d** and **8e** was found against *Escherichia coli* ATCC 25,922 and *Klebsiella pneumoniae* ATCC 13883. Against *K. pneumoniae* compound **8d** was slightly more active (2 μg/mL) than compound **8e** (8 μg/mL). However, it should be noted that these two compounds, in general, were significantly less active against Gram-negative bacteria than against Gram-positive ones.

Table 1. Antibacterial and cytotoxic activity of **8a–e**.

Bacterial Strains	**MIC, μg/mL**						
	Lf	**BG**	**8a**	**8b**	**8c**	**8d**	**8e**
S. aureus ATCC 25923	0.25	0.06	32	32	16	2	0.5
S. aureus 10	0.13	0.13	16	8	8	1	0.5
S. aureus 5	0.25	0.06	64	16	8	4	0.5
S. aureus 100 KC	32	0.06	64	16	16	2	0.5
S. aureus ATCC 3798	32	0.06	32	16	16	2	0.5
S. aureus ATCC 700699	16	1	32	>64	8	2	0.5
S. epidermidis 533	0.5	0.06	32	32	16	1	0.5
E. faecium 569	1	0.25	>64	32	>64	8	>64
E. coli ATCC 25922	0.06	4	16	>64	32	8	8
K. pneumoniae ATCC 13883	0.25	8	>64	>64	8	2	8
S. cholerasuis ATCC 14028	0.13	16	>64	>64	>64	>64	>64
P. aeruginosa ATCC 27853	1	>64	>64	>64	>64	>64	>64
Test Cells	**IC_{50}, μg/mL**						
HPF-hTERT	>50	0.03	>50	>50	>50	13	2

Lf—levofloxacin; BG—brilliant green; IC_{50}—concentration of compound inhibiting the growth of cells by 50%.

2.2.2. Cytotoxic Activity

Compounds **7a–e** all showed similar cytotoxicity with IC_{50} higher than 50 μg/mL. Cytotoxicity of compounds **8a–e** (Table 1) depended on the length of the hydroxyalkyl chain and for C2–C4 substituents IC_{50}, were higher than 50 μg/mL. For C5 compound **8d**, it was 13 μg/mL, and then in C6, there is a sharp increase in cytotoxicity, which reached 2 μg/mL. Apparently, this rapid increase in cytotoxicity is associated with an increase in the total detergent activity of the molecule, which leads to a decrease in the selectivity of the action on the lipid layers of the membrane [17,18].

2.3. Study of the Relationship between Lipophility and Biological Activity

Analyzing the structure-activity relationship for compounds **8a–e**, it was observed that the activity of these substances is closely related to their lipophilicity. There are two possible tautomeric variations of methylium (Figure 1): form 1, where the positive charge is on the central carbon atom, and closer to the real form 2, where the positive charge is on one of the nitrogen atoms of the indole cycles.

It was shown that the calculated values in this case are very close to the real ones when the contribution of both forms is taken into account. In other words, the arithmetic mean ($miLogP_{ow\ form1}$ + $miLogP_{ow\ form2}$)/2 will be closer to the real $LogP_{ow}$ obtained experimentally (Table 2). Data analysis shows that molecules with a good ratio of antibacterial activity to cytotoxicity are more likely to be in the LogP range from 2 to 5. Above 5, a sharp increase in cytotoxicity begins, and below 2, there is an almost complete absence of antimicrobial activity. The best result is likely to be expected with a LogP in the region of 4. Computer calculations of lipophilicity in the Molinspiration package fairly adequately predict LogP for molecules and can be used to select potentially promising compounds of this group.

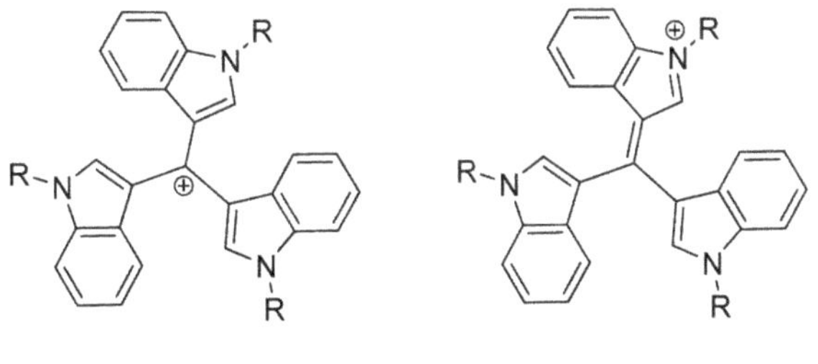

Tautomeric form 1 Tautomeric form 2

Figure 1. Tautomeric forms of compounds **8a–e**.

Table 2. Experimental and calculated values $LogP_{OW}$ for **8a–e**.

	8a	8b	8c	8d	8e
$LogP_{OW}$	1.15	1.2	2.9	4	5.4
$miLogP_{OW\ form1}$	1.65	2.46	3.27	4.79	6.30
$miLogP_{OW\ form2}$	0.22	1.03	1.84	3.36	4.87
$(miLogP_{OW\ form1} + miLogP_{OW\ form2})/2$	1.76	1.75	2.55	4.08	5.59

3. Materials and Methods

3.1. Chemistry

All the reagents were obtained commercially and used without further purification. Indole and all ω-bromoalkanols, all solvents, $LiAlH_4$, $Dy(OTf)_3$, DMAP, DDQ were purchased from Sigma-Aldrich; 2-(1*H*-indol-1-yl)acetic acid and 3-(1*H*-indol-1-yl)propanoic acid were purchased from Alinda company (www.alinda.ru). Purity of the compounds was checked by thin layer chromatography using silica-gel 60 F_{254}-coated Al plates (Merck) and spots were observed under UV light (254 nm). Column chromatography was performed on Kieselgel 60 (Merck). Proton nuclear magnetic resonance (^1H NMR) and carbon-13 nuclear magnetic resonance (^{13}C NMR) spectra (in DMSO-d6) were recorded on a Varian VXR-400 spectrometer at 400 and 100 MHz respectively, the chemical shift values are expressed in ppm (δ scale) using DMSO as an internal standard, the coupling constants expressed in Hz. The NMR spectra of the compounds **8a–e** were recorded at 80 °C to avoid peaks broadening. The mass spectral measurements were carried out by ESI method on microTOF-QII (Brucker Daltonics GmbH). Analytical high-performance liquid chromatography (HPLC) was performed on a Shimadzu LC-20AD system using Kromasil-100-5-C18 (Akzo-Nobel) column, 4.6 × 250 mm, 20 °C temperature, UV detection, mobile phase A—0.2% $HCOONH_4$), mobile phase B-MeCN, (pH 7.4), fl-1 mL/min., loop 20 mkl. The NMR spectra of the compounds **3–10** are presented in Supplementary file NMR_spectra.pdf.

3.2. Antibacterial Activity

Compounds were tested against Gram-positive and Gram-negative bacteria, including sensitive or drug resistant strains from American Type Culture Collection (ATCC), as well as resistant clinical isolates from the culture collection of the Laboratory for Control of Hospital Infections (Sechenov University, Moscow, Russia). Collection cultures of Gram-positive bacteria: *Staphylococcus aureus* ATCC 25923, *Staphylococcus aureus* ATCC 3798, *Staphylococcus aureus* ATCC 700,699 and clinical isolates of Gram-positive bacteria: *Staphylococcus aureus* 5, *Staphylococcus aureus* 10, *Staphylococcus aureus* 100 KS, *Staphylococcus epidermidis* 533, *Enterococcus faecium* 569 and collection cultures of Gram-negative bacteria: *Escherichia coli* ATCC 25922, *Klebsiella pneumoniae* ATCC 13883, *Salmonella cholerasuis* ATCC 14,028 were used.

For the cultivation of the strains, various nutrient media were used: Trypticase Soy Agar BBL for *Staphylococcus* sp., *E. coli*, *K. pneumoniae*, *S. cholerasuis* and Columbia Agar Base BBL for the cultivation of *Enterococcus* sp. Cultures grown on appropriate nutrient media at 35 °C for 1 day were used to set up experiments. For determination of the antibacterial action Mueller-Hinton (Acumedia, Baltimore, MD, USA) liquid medium was used. The minimum inhibitory concentrations (MIC) were determined by the microdilution method in 96 well sterile plates in a cation-adjusted Müller-Hinton medium in accordance with the requirements of the Institute of Clinical and Laboratory Standards (CLSI/NCCLS) [19]. MIC was defined as the minimum drug concentration that completely prevents the growth of the test organism.

3.3. Cytotoxic Activity

The cytotoxic properties of the compounds obtained were tested using the MTT assay as described previously [11] on the healthy donor (postnatal) human fibroblasts immortalized by transfection of the hTERT gene of the catalytic component of telomerase (hereinafter, FB).

Cells were grown at 37 °C and 5% CO_2. Human donor fibroblasts were cultivated in DMEM medium (Paneko, Russia) with addition of 10% FBS (Hyclone, Austria), 2 mM L-glutamine (Paneko, Russia), and 1% penicillin-streptomycin (Paneko, Russia). Cells were seeded at concentration 2500 cells/well in 96 well plates (Corning, NY, USA), and left overnight to attach. The next day, the cells were treated with compounds, with indicated concentrations (ten two-fold dilutions, starting from 50 uM) for 72 h. After incubation, MTT reagent (3-(4,5-dimethylthiazol-2-yl)-2,5-diphenyl tetrazolium bromide) (Sigma-Aldrich, Saint-Louis, MO, USA) was added to a final concentration of 0.5 ug/mL and the cells were incubated for 2 h at 37 °C and 5% CO_2. After incubation, the medium was discarded, and 100 uL of DMSO was added. The optical densities were read at 570 nm wavelength on Multiskan FC (ThermoFisher, Waltham, MA, USA). The OD values for controls were taken as 100%. The IC_{50} values were calculated in GraphPadRrism 6.0.

3.4. Determination of Lipophilicity

We used the partition coefficient as an indicator of lipophilicity. The partition coefficient (P) is defined as the ratio of the equilibrium solute concentrations in a two-phase system of immiscible solvents. The most common in practice is the octanol-water (P_{ow}) system. The partition coefficient is usually represented as a decimal logarithm ($LogP_{ow}$). It can be measured in several ways. A most advanced, accurate, and less time-consuming is the HPLC method for determining P_{ow} using high-performance liquid chromatography [20]. HPLC is performed on an analytical column with a solid phase containing long hydrocarbon chains chemically bound to silica gel. The retention time on such a column (R_t) is directly related to the partition coefficient P_{ow}. The most informative in our case was the partition coefficient at a physiological pH value of 7.4. It was at this pH that the main biological experiments with the studied substances were carried out, namely tests of antimicrobial activity and cytotoxicity. HPLC was performed on a Shimadzu LC-20AD system using Kromasil-100-5-C18 (Akzo-Nobel) column, 4.6 × 250 mm, 20 °C temperature, UV detection, mobile phase A—0.2% $HCOONH_4$), mobile phase B-MeCN, (pH 7.4), fl-1 mL/min., loop 20 mkl. After calibration using substances with a known $LogP_{ow}$, it is possible to recalculate R_t to $LogP_{ow}$. For calibration, we used aniline ($LogP_{ow}$ 0.9), p-chloroaniline ($LogP_{ow}$ 1.8), diphenylamine ($LogP_{ow}$ 3.4), triphenylamine ($LogP_{ow}$ 5.7).

To study the possibility of using computer models to calculate $LogP_{ow}$ [21], $miLogP_{ow}$ values were calculated for the same substances by the Molinspiration package [22], which is an online tool available at www.molinspiration.com. The method for logP prediction developed at Molinspiration ($miLogP_{ow}$) is based on group contributions. These have been obtained by fitting calculated logP with experimental logP for a training set more than twelve thousand, mostly drug-like molecules. In this way, hydrophobicity values for 35 small simple "basic" fragments were obtained, as well as values for 185 larger fragments, characterizing intramolecular hydrogen bonding contribution to logP and charge interactions. Molinspiration methodology for logP calculation is very robust and is able to process practically all organic molecules. For 50.5% of molecules, logP is predicted with error <0.25, for 80.2% with error <0.5 and for 96.5% with error <1.0. Only for 3.5% of structures, logP is predicted with error >1.0. The statistical parameters listed above rank Molinspiration miLogP as one of the best methods available for logP prediction. MiLogP is used due to its robustness and good prediction quality in the popular ZINC database for virtual screening. A report by the National Institute of Standards documenting excellent agreement between experimental logP and Molinspiration calculated logP for some industrial chemicals [23].

3.5. Chemical Experimental Data

2-(1*H*-Indol-1-yl)-ethanol **(3a)**.

To the boiling suspension of $LiAlH_4$ (15.2 g, 0.4 mol) in THF (500 mL), the solution of 2-(1*H*-indol-1-yl)acetic acid **1a** (17.56 g, 0.1 mol) in THF (100 mL) was gradually added, then the reaction mixture was refluxed for 5 h. After cooling to RT, the reaction mixture was quenched with KOH (20% aqueous solution), then was filtered and diluted with EtOAc (300 mL) and aqueous solution of citric acid (10.0 g in 100 mL) was added. The organic layer was separated, washed with water and brine, and evaporated in vacuo. The residue was purified by flash chromatography (50 g of silica gel) using EtOAc-hexane (1:10 to 1:1) as an eluent, to give **3a** (13.2 g, 82%) as a colorless oil.

^{1}H NMR: δ 7.52 (dt, 1H, J = 7.8, 1.1 Hz), 7.48–7.41 (m, 1H), 7.33 (d, 1H, J = 3.1 Hz), 7.10 (ddd, 1H, J = 8.2, 7.0, 1.3 Hz), 6.99 (ddd, 1H, J = 8.0, 7.0, 1.0 Hz), 6.40 (dd, 1H, J = 3.1, 0.9 Hz), 4.93 (t, 1H, J = 5.3 Hz), 4.19 (t, 2H, J = 5.7 Hz), 3.70 (q, 2H, J = 5.5 Hz). ^{13}C NMR: δ 136.33, 129.59, 128.53, 121.26, 120.73, 119.24, 110.32, 100.67, 60.76, 48.66, 48.64. HRMS (EI) m/z $[M + H]^+$ Calcd for $C_{10}H_{12}NO^+$ 162.0913; Found 162.0923.

3-(1*H*-Indol-1-yl)propan-1-ol **(3b)**.

The same procedure as above was carried out using 3-(1*H*-indol-1-yl)propanoic acid (18.9 g, 0.1 mol), to give **3b** (13.8 g, 79%) as a colorless oil.

^{1}H NMR: δ 7.53 (dt, 1H, J = 7.8, 1.0 Hz), 7.44 (dd, 1H, J = 8.3, 1.0 Hz), 7.32 (d, 1H, J = 3.1 Hz), 7.14–7.07 (m, 1H), 7.04–6.97 (m, 1H), 6.41 (dd, 1H, J = 3.2, 0.9 Hz), 4.69 (s, 1H), 4.21 (t, 2H, J = 6.9 Hz), 3.37 (d, 2H, J = 5.2 Hz), 1.88 (t, 2H, J = 6.6 Hz). ^{13}C NMR: δ 136.08, 129.12, 128.53, 121.38, 120.85, 119.27, 110.17, 100.82, 58.27, 42.84, 33.43. HRMS (EI) m/z $[M + H]^+$ Calcd for $C_{11}H_{14}NO^+$ 176.1070; Found 176.1073.

4-(1*H*-Indol-1-yl)butan-1-ol **(3c)**.

To the suspension of KOH (20 g, 0.35 mol) in DMSO (100 mL), indole (11.7 g, 0.1 mol) and 4-bromobutan-1-ol (16.8 g, 0.11 mol) were added. After intensive stirring at RT for 5 h, the reaction mixture was filtered, diluted with EtOAc (300 mL), and washed with an aqueous solution of citric acid (10.0 g, 100 mL). The organic layer was separated, washed with water and brine, and evaporated in vacuo. The residue was purified by flash chromatography (50 g of silica gel) using EtOAc-Hexane (1:10 to 1:1) as an eluent, to give **3c** (17.5 g, 93%) as a colorless oil.

^{1}H NMR: δ 7.53 (dt, 1H, J = 7.8, 1.0 Hz), 7.44 (dd, 1H, J = 8.3, 1.1 Hz), 7.33 (d, 1H, J = 3.1 Hz), 7.11 (ddd, 1H, J = 8.2, 6.9, 1.2 Hz), 7.00 (ddd, 1H, J = 8.0, 7.0, 1.0 Hz), 6.41 (dd, 1H, J = 3.1, 0.9 Hz), 4.47 (t, 1H, J = 5.1 Hz), 4.15 (t, 2H, J = 7.1 Hz), 3.39 (td, 2H, J = 6.4, 4.9 Hz), 1.77 (dq, 2H, J = 9.6, 7.2 Hz), 1.43–1.32 (m, 2H). ^{13}C NMR: δ 136.11, 129.00, 128.57, 121.34, 120.85, 119.23, 110.20, 100.79, 60.80, 45.84, 30.21, 27.09. HRMS (EI) m/z $[M + H]^+$ Calcd for $C_{12}H_{18}NO^+$ 190.1226; Found 190.1228.

5-(1*H*-Indol-1-yl)pentan-1-ol **(3d)**.

The same procedure as above was carried out using indole (11.7 g, 0.1 mol) and 5-bromopentan-1-ol (18.3 g, 0.11 mol), to give **3d** (18.1 g, 89%) as a colorless oil.

^{1}H NMR: δ 7.54 (dt, 1H, J = 7.9, 1.0 Hz), 7.42 (dd, 1H, J = 8.2, 1.1 Hz), 7.31 (d, 1H, J = 3.2 Hz), 7.12 (ddd, 1H, J = 8.2, 7.0, 1.3 Hz), 7.01 (ddd, 1H, J = 8.0, 6.9, 1.0 Hz), 6.41 (dd, 1H, J = 3.1, 0.9 Hz), 4.43 (s, 1H), 4.11 (t, 2H, J = 7.0 Hz), 3.37 (t, 2H, J = 6.5 Hz), 1.73 (p, 2H, J = 7.2 Hz), 1.49–1.37 (m, 2H), 1.33–1.19 (m, 2H). ^{13}C NMR: δ 136.10, 128.97, 128.57, 121.35, 120.86, 119.23, 110.14, 100.80, 61.06, 45.95, 32.54, 30.23, 23.37. HRMS (EI) m/z $[M + H]^+$ Calcd for $C_{13}H_{18}NO^+$ 204.1383; Found 204.1381.

6-(1*H*-Indol-1-yl)hexan-1-ol **(3e)**.

The same procedure as above was carried out using indole (11.7 g, 0.1 mol) and 6-bromohexan-1-ol (20.0 g, 0.11 mol), to give **3e** (19.5 g, 90%) as a colorless oil.

^{1}H NMR: δ 7.22–7.15 (m, 1H), 7.09–6.98 (m, 3H), 3.52 (t, 1H, J = 6.8 Hz), 2.85 (t, 1H, J = 7.2 Hz), 2.20 (d, 3H, J = 11.3 Hz), 1.77 (p, 1H, J = 6.9 Hz). ^{13}C NMR: δ 170.04, 165.67, 151.95, 136.37, 135.88, 134.80, 132.56, 130.15, 129.63, 125.82, 42.19, 37.21, 30.97, 28.14, 20.96, 20.84. HRMS (EI) m/z $[M + H]^+$ Calcd for $C_{14}H_{20}NO^+$ 218.1539; Found 218.1540.

2-(1*H*-Indol-1-yl)ethyl acetate **(4a)**.

To the solution of 2-indol-1-yl-ethanol **3a** (12 g, 75 mmol) in pyridine (100 mL), Ac_2O (8 mL, 10 mmol) and DMAP (100 mg, 0.08 mmol) were added. After stirring at rt for 5 h, the reaction mixture was evaporated in vacuo, diluted with EtOAc (300 mL) and washed with an aqueous solution of citric acid (1.0 g in 100 mL). The organic layer was separated, washed with water and brine and evaporated in vacuo. The residue was purified by flash chromatography (100 g of silica gel) using EtOAc-Hexane (1:10 to 1:3) as an eluent, to give **4a** (14.7 g, 97%) as a colorless oil. ^{1}H NMR: δ 7.63 (d, 1H, J = 7.8 Hz), 7.50 (dd, 1H, J = 8.3, 0.6 Hz), 7.35 (d, 1H, J = 3.2 Hz), 7.26–7.16 (m, 1H), 7.11 (td, 1H, J = 7.5, 0.9 Hz), 6.52 (dd, 1H, J = 3.1, 0.7 Hz), 4.37 (qd, 4H, J = 6.2, 1.5 Hz), 1.93 (s, 3H). ^{13}C NMR: δ 170.53, 136.43, 129.16, 128.75, 121.63, 120.96, 119.58, 110.09, 101.50, 63.40, 44.93, 20.86.

3-(1*H*-Indol-1-yl)propyl acetate **(4b)**.

The same procedure as above was carried out using 3-indol-1-yl-propan-1-ol **3b** (7.0 g, 40 mmol), to give **4b** (8.1 g, 94%) as a colorless oil. ^{1}H NMR: δ 7.62 (d, 1H, J = 7.8 Hz), 7.46 (d, 1H, J = 8.2 Hz), 7.31 (d, 1H, J = 3.1 Hz), 7.24–7.15 (m, 1H), 7.15–7.06 (m, 1H), 6.55 – 6.47 (m, 1H), 4.22 (t, 2H, J = 6.8 Hz), 3.96 (t, 2H, J = 6.4 Hz), 2.14–1.96 (m, 5H). ^{13}C NMR: δ 170.75, 136.20, 128.81, 128.74, 121.54, 120.97, 119.43, 110.00, 101.23, 61.71, 42.73, 29.29, 20.92.

4-(1*H*-Indol-1-yl)butyl acetate **(4c)**.

The same procedure as above was carried out using 4-(1*H*-indol-1-yl)butan-1-ol **3c** (10.0 g, 53 mmol), to give **4c** (11.7 g, 96%) as a colorless oil. ^{1}H NMR: δ 7.53 (d, 1H, J = 7.8 Hz), 7.47–7.39 (m, 1H), 7.32 (d, 1H, J = 3.1 Hz), 7.15–7.05 (m, 1H), 7.04–6.96 (m, 1H), 6.41 (dd, 1H, J = 3.1, 0.7 Hz), 4.15 (t, 2H, J = 7.0 Hz), 3.95 (t, 2H, J = 6.6 Hz), 1.94 (s, 3H), 1.92–1.69 (m, 2H), 1.69–1.42 (m, 2H). ^{13}C NMR: δ 170.85, 136.06, 128.93, 128.55, 121.40, 120.86, 119.28, 110.12, 100.92, 100.89, 85.49, 63.84, 45.44, 26.84, 25.99, 21.06, 21.05.

5-(1*H*-Indol-1-yl)pentyl acetate **(4d)**.

The same procedure as above was carried out using 5-(1*H*-indol-1-yl)pentan-1-ol **3d** (10.0 g, 49 mmol), to give **4d** (11.4 g, 95%) as a colorless oil. ^{1}H NMR: δ 7.52 (d, J = 7.8 Hz, 2H), 7.47–7.38 (m, 2H), 7.32 (d, J = 3.1 Hz, 2H), 7.15–7.05 (m, 2H), 7.04–6.94 (m, 2H), 6.40 (dd, J = 3.0, 0.6 Hz, 2H), 4.12 (t, J = 7.0 Hz, 4H), 3.92 (t, J = 6.6 Hz, 4H), 2.48 (s, 1H), 1.94 (s, 6H), 1.82–1.64 (m, 4H), 1.62–1.47 (m, 4H), 1.23 (dd, J = 9.2, 6.2 Hz, 4H). ^{13}C NMR: δ 170.84, 136.03, 128.97, 128.51, 121.33, 120.83, 119.22, 110.14, 100.77, 64.07, 45.69, 29.86, 28.10, 23.15, 21.10.

6-(1*H*-Indol-1-yl)hexyl acetate **(4e)**.

The same procedure as above was carried out using 6-(1*H*-indol-1-yl)hexan-1-ol **3e** (10.0 g, 46 mmol), to give **4e** (10.97 g, 93%) as a colorless amorphous solid. ^{1}H NMR: δ 7.51 (d, 2H, J = 7.8 Hz), 7.42 (d, 2H, J = 8.2 Hz), 7.31 (d, 2H, J = 3.0 Hz), 7.09 (d, 2H, J = 7.3 Hz), 6.99 (d, 2H, J = 7.3 Hz), 6.39 (d, 2H, J = 2.7 Hz), 4.12 (t, 4H, J = 7.0 Hz), 3.92 (t, 4H, J = 6.6 Hz), 1.95 (s, 6H), 1.72 (d, 3H, J = 7.2 Hz), 1.69–1.25 (m, 10H), 1.25–0.98 (m, 5H). ^{13}C NMR: δ 170.86, 136.05, 128.95, 128.51, 121.31, 120.82, 119.20, 110.11, 100.75, 64.16, 45.76, 30.15, 28.44, 26.34, 25.45, 21.11.

2-(3-Formyl-1*H*-indol-1-yl)ethyl acetate **(5a)**.

2-(1*H*-Indol-1-yl)ethyl acetate **4a** (14.7 g, 72 mmol) was dissolved in the solution of $POCl_3$ (0.9 mL, 10 mmol) in DMF (50 mL) and intensively stirred at 5 °C for 5 h. The reaction mixture was quenched with Na_2CO_3 (10% aqueous solution), diluted with EtOAc (100 mL) and water (200 mL). The organic layer was separated and the water layer was re-extracted with EtOAc (100 mL). The combined extracts were washed with water and brine and evaporated in vacuo. The residue was purified by flash chromatography (100 g of silica gel) using EtOAc-Hexane (1:5 to 1:1) as an eluent to give **5a** (11.9 g, 71%) as a colorless oil. ^{1}H NMR: δ 9.94 (s, 1H), 8.27 (s, 1H), 8.16 (d, 1H, J = 7.3 Hz), 7.61 (d, 1H, J = 7.9 Hz), 7.36–7.21 (m, 2H), 4.51 (t, 2H, J = 5.0 Hz), 4.38 (t, 2H, J = 5.1 Hz), 1.88 (s, 3H). ^{13}C NMR: δ 185.18, 170.48, 141.46, 137.60, 125.09, 124.04, 122.98, 121.56, 117.98, 111.35, 62.73, 45.81, 20.84.

3-(3-Formyl-1*H*-indol-1-yl)propyl acetate **(5b)**.

The same procedure as above was carried out using 3-(1*H*-indol-1-yl)propylacetate **4b** (8.0 g, 36.8 mmol), to give **5b** (6.1 g, 68%) as a colorless amorphous solid. 1H NMR δ 9.91 (s, 1H), 8.27 (s, 1H), 8.19–8.11 (m, 1H), 7.58 (d, J = 8.0 Hz, 1H), 7.33–7.21 (m, 2H), 4.32 (t, J = 6.9 Hz, 2H), 3.96 (t, J = 6.2 Hz, 2H), 3.53 (s, 2H), 2.18–2.02 (m, 2H), 1.91 (s, 3H). ^{13}C NMR: δ 184.98, 170.77, 170.76, 141.11, 137.43, 125.13, 124.00, 122.92, 121.56, 117.73, 111.29, 61.61, 43.80, 28.75, 20.91.

4-(3-Formyl-1*H*-indol-1-yl)butyl acetate **(5c)**.

The same procedure as above was carried out using 4-(1*H*-indol-1-yl)butyl acetate **4c** (11.7 g, 50 mmol), to give **5c** (9.7 g, 74%) as a colorless amorphous solid. ^{1}H NMR: δ 9.89 (s, 1H), 8.30 (s, 1H), 8.10 (d, 1H, J = 7.5 Hz), 7.61 (d, 1H, J = 8.1 Hz), 7.33–7.20 (m, 2H), 4.28 (t, 2H, J = 7.1 Hz), 3.98 (t, 2H, J = 6.6 Hz), 1.94 (s, 3H), 1.91–1.75 (m, 2H), 1.75–1.47 (m, 2H). ^{13}C NMR: δ 184.99, 170.84, 141.12, 137.41, 125.11, 123.97, 122.91, 121.51, 117.55, 111.46, 85.48, 63.71, 46.32, 26.39, 25.83, 21.08.

5-(3-Formyl-1*H*-indol-1-yl)pentyl acetate **(5d)**.

The same procedure as above was carried out using 5-(1*H*-indol-1-yl)pentyl acetate **4d** (11.4 g, 46 mmol), to give **5d** (9.78 g, 77%) as a colorless amorphous solid. ^{1}H NMR: δ 9.90 (s, 1H), 8.30 (s, 1H), 8.12 (d, 1H, J = 7.4 Hz), 7.61 (d, 1H, J = 8.1 Hz), 7.34–7.22 (m, 2H), 4.26 (t, 2H, J = 7.1 Hz), 3.95 (t, 2H, J = 6.6 Hz), 1.94 (s, 3H), 1.88–1.73 (m, 2H), 1.64–1.51 (m, 2H), 1.34–1.06 (m, 2H). ^{13}C NMR: δ 184.57, 170.46, 140.76, 137.06, 124.74, 123.57, 122.50, 121.13, 117.13, 111.09, 63.61, 46.18, 28.94, 27.63, 22.60, 20.70.

6-(3-Formyl-1*H*-indol-1-yl)hexyl acetate **(5e)**.

The same procedure as above was carried out using 6-(1*H*-indol-1-yl)hexyl acetate **4e** (10.9 g, 42 mmol), to give **5e** (8.8 g, 73%) as a colorless amorphous solid. ^{1}H NMR: δ 9.87 (s, 1H), 8.30 (s, 1H), 8.09 (d, 1H, J = 7.6 Hz), 7.61 (d, 1H, J = 8.2 Hz), 7.29 (dt, 2H, J = 8.2, 1.3 Hz), 7.23 (dt, 2H, J = 7.0, 1.0 Hz), 4.29 (t, 2H, J = 7.1 Hz), 3.92 (t, 2H, J = 7.0 Hz), 1.94 (s, 3H), 1.82–1.75 (m, 2H), 1.53–1.46 (m, 2H) 1.34–1.22 (m, 4H). ^{13}C NMR: δ 184.89, 170.55, 140.86, 137.15, 124.88, 123.53, 122.59, 121.19, 117.20, 111.01, 64.14, 46.64, 29.59, 28.37, 26.14, 25.38, 21.13.

1-(2-Hydroxyethyl)-1*H*-indole-3-carbaldehyde **(6a)**.

2-(3-Formyl-1*H*-indol-1-yl)ethyl acetate **5a** (11.9 g, 51 mmol) was dissolved in the solution Na (200 mg, 0.8 mmol) in MeOH (50 mL) and stirred at 10 min. Then, the reaction mixture was evaporated in vacuo, quenched with an aqueous solution of citric acid (0.5 g, 50 mL), and EtOAc (200 mL). The organic layer was separated and the water layer was re-extracted with EtOAc (100 mL). The extracts were combined, washed with water and brine and evaporated in vacuo. The residue was purified by flash chromatography (50 g of silica gel) using EtOAc-Hexane (1:5 to 1:0) as an eluent, to give **6a** (8.8 g, 91%) as a colorless amorphous solid. ^{1}H NMR: δ 9.90 (s, 1H), 8.24 (s, 1H), 8.14–8.08 (m, 1H), 7.60 (d, 1H, J = 8.0 Hz), 7.33–7.20 (m, 2H), 5.03 (t, 1H, J = 5.2 Hz), 4.30 (t, 2H, J = 5.3 Hz), 3.76 (q, 2H, J = 5.2 Hz). ^{13}C NMR: δ 185.08, 141.97, 137.70, 125.14, 123.83, 122.84, 121.42, 117.41, 111.63, 60.04, 49.51. HRMS (EI) m/z $[M + H]^+$ Calcd for $C_{11}H_{12}NO_2^+$ 190.0863; Found 190.0872.

1-(3-Hydroxypropyl)-1*H*-indole-3-carbaldehyde **(6b)**.

The same procedure as above was carried out using 3-(3-Formyl-1*H*-indol-1-yl)propyl acetate **(5b)** (6.0 g, 24 mmol), to give **6b** (4.5 g, 92%) as a colorless amorphous solid. ^{1}H NMR: δ 9.89 (s, 1H), 8.26 (s, 1H), 8.10 (d, 1H, J = 7.5 Hz), 7.59 (d, 1H, J = 8.1 Hz), 7.35–7.21 (m, 2H), 4.73 (t, 1H, J = 5.0), 4.32 (t, 2H, J = 7.0 Hz), 3.39 (dd, 2H, J = 11.2, 5.9 Hz), 1.94 (p, 2H, J = 6.5 Hz). ^{13}C NMR: δ 184.99, 141.29, 137.45, 125.11, 123.96, 122.90, 121.50, 117.48, 111.44, 57.99, 43.82, 32.79. HRMS (EI) m/z $[M + H]^+$ Calcd for $C_{12}H_{14}NO_2^+$ 204.1019; Found 204.1030.

1-(4-Hydroxybutyl)-1*H*-indole-3-carbaldehyde **(6c)**.

The same procedure as above was carried out using 4-(3-formyl-1*H*-indol-1-yl)butyl acetate **5c** (5.0 g, 19 mmol), to give **6c** (3.7 g, 90%) as a colorless amorphous solid. ^{1}H NMR: δ 9.89 (s, 1H), 8.29 (s,

1H), 8.10 (d, 1H, J = 7.6 Hz), 7.60 (d, 1H, J = 8.1 Hz), 7.33–7.21 (m, 2H), 4.47 (t, 1H, J = 5.1 Hz), 4.27 (t, 2H, J = 7.1 Hz), 1.90–1.76 (m, 2H), 1.47–1.33 (m, 2H). ^{13}C NMR: δ 184.93, 141.11, 137.46, 125.14, 123.93, 122.86, 121.49, 117.49, 111.49, 60.64, 46.69, 29.94, 26.60. HRMS (EI) m/z [M + H]$^+$ Calcd for $C_{13}H_{16}NO_2^+$ 218.1176; Found 218.1186.

1-(5-Hydroxypentyl)-1*H*-indole-3-carbaldehyde **(6d)**.

The same procedure as above was carried out using 5-(3-formyl-1*H*-indol-1-yl)pentyl acetate **5d** (5.0 g, 18 mmol), to give **6d** (3.9 g, 93%) as a colorless amorphous solid. ^{1}H NMR: δ 9.88 (s, 1H), 8.29 (s, 1H), 8.13–8.06 (m, 1H), 7.59 (d, 1H, J = 8.2 Hz), 7.32–7.20 (m, 2H), 4.41 (br, 1H), 4.24 (t, 2H, J = 7.0 Hz), 3.34 (t, 2H, J = 6.4 Hz), 1.78 (p, 2H, J = 7.2 Hz), 1.41 (p, 2H, J = 6.7 Hz), 1.31–1.20 (m, 2H). ^{13}C NMR (100 MHz, dmso) δ 185.00, 141.21, 137.46, 125.12, 123.96, 122.89, 121.51, 117.47, 111.49, 106.48, 60.90, 46.79, 32.36, 29.66, 23.18. HRMS (EI) m/z [M + H]$^+$ Calcd for $C_{14}H_{18}NO_2^+$ 232.1332; Found 232.1338.

1-(6-Hydroxyhexyl)-1*H*-indole-3-carbaldehyde **(6e)**.

The same procedure as above was carried out using 6-(3-formyl-1*H*-indol-1-yl)hexyl acetate **5e** (6.0 g, 20 mmol), to give **6e** (4.8 g, 94%) as a colorless amorphous solid. ^{1}H NMR: δ 9.88 (s, 1H), 8.29 (s, 1H), 8.09 (d, 1H, J = 7.7 Hz), 7.59 (d, 1H, J = 8.1 Hz), 7.33–7.19 (m, 2H), 4.38 (s, 1H), 4.24 (t, 2H J = 7.1 Hz), 3.33 (t, 2H, J = 6.3 Hz), 1.77 (p, 2H, J = 7.1 Hz), 1.41–1.19 (m, 6H). ^{13}C NMR: δ 185.11, 140.95, 137.09, 125.33, 123.99, 122.77, 121.82, 117.40, 111.85, 61.02, 45.91, 32.65, 30.34, 26.75, 26.77. HRMS (EI) m/z [M + H]$^+$ Calcd for $C_{15}H_{20}NO_2^+$ 246.1489; Found 246.1483.

tris(1-[2-Hydroxyethyl]-1*H*-indol-3-yl)methane **(7a)**.

To the solution of 2-indol-1-yl-ethanol **3a** (3.0 g, 18.7 mmol) in MeOH (100 mL), 1-(2-hydroxyethyl)-1*H*-indole-3-carbaldehyde **6a** (1.7 g, 9.1 mmol), AcOH (1 mL), and $Dy(OTf)_3$ (10 mg, 16.4 μmol) were added. After refluxing for 12 h, the reaction mixture was cooled to RT. The resulting suspension was filtered, the precipitate washed with MeOH (2 × 50 mL) and Et_2O (50 mL). The residue was dried in vacuo, to give **7a** (3.9 g, 89%) as a colorless amorphous solid.

^{1}H NMR: δ 7.52–7.32 (m, 6H), 7.05 (t, 3H, J = 7.5 Hz), 7.01 (s, 3H), 6.88 (t, 3H, J = 7.3 Hz), 6.03 (s, 1H), 4.80 (t, 3H, J = 5.0 Hz), 4.11 (t, 6H, J = 5.1 Hz), 3.64 (d, 6H, J = 5.3 Hz). ^{13}C NMR: δ 136.48, 127.25, 127.11, 120.59, 119.39, 118.0, 117.33, 109.77, 60.36, 48.12. HRMS (EI) m/z [M − H]+ Calcd for $C_{34}H_{37}N_3O_3^+$ 493.2365; Found 492.2282.

tris(1-[3-Hydroxypropyl]-1*H*-indol-3-yl)methane **(7b)**.

The same procedure as above was carried out using 3-indol-1-yl-propan-1-ol **3b** (3.0 g, 17.1 mmol) and 1-(3-hydroxypropyl)-1*H*-indole-3-carbaldehyde **6b** (1.7 g, 8.5 mmol), to give **7b** (3.8 g, 85%) as a colorless amorphous solid. ^{1}H NMR: δ 7.39 (t, 6H, J = 7.5 Hz), 7.05 (d, 3H, J = 7.9 Hz), 6.98 (s, 3H), 6.88 (d, 3H, J = 7.7 Hz), 6.03 (s, 1H), 4.52 (t, 3H, J = 5.0 Hz), 4.13 (t, 6H, J = 6.8 Hz), 3.42–3.23 (m, 6H), 1.89–1.67 (m, 6H). ^{13}C NMR: δ 136.2, 127.05, 126.69, 120.71, 119.55, 117.99, 117.24, 109.63, 57.79, 42.2, 33.03. HRMS (EI) m/z [M − H] + Calcd for $C_{34}H_{37}N_3O_3^+$ 535.2835; Found 534.2747.

tris(1-[4-Hydroxybutyl]-1*H*-indol-3-yl)methane **(7c)**.

The same procedure as above was carried out using 4-indol-1-yl-butan-1-ol **3c** (3.0 g, 15.8 mmol) and 1-(4-Hydroxybutyl)-1*H*-indole-3-carbaldehyde **6c** (1.6 g, 7.8 mmol), to give **7c** (3.7 g, 82%) as a colorless amorphous solid. ^{1}H NMR: δ 7.40 (d, J = 2.8 Hz, 3H), 7.38 (d, J = 3.3 Hz, 3H), 7.05 (t, J = 7.5 Hz, 3H), 6.96 (s, 3H), 6.86 (t, J = 7.4 Hz, 3H), 6.03 (s, 1H), 4.44 (t, J = 5.0 Hz, 3H), 4.07 (t, J = 6.8 Hz, 6H), 3.35 (q, J = 6.1 Hz, 6H), 1.90–1.48 (m, 6H), 1.48–1.11 (m, 6H). ^{13}C NMR: δ 157.88, 144.03, 138.55, 124.44, 123.13, 120.55, 112.03, 59.90, 46.90, 29.14, 25.85. HRMS (EI) m/z [M − H] + Calcd for $C_{37}H_{43}N_3O_3^+$ 577.3304; Found 576.3213.

tris(1-[5-Hydroxypentyl]-1*H*-indol-3-yl)methane **(7d)**.

The same procedure as above was carried out using 5-indol-1-yl-pentan-1-ol **3d** (3.0 g, 15.8 mmol) and 1-(5-hydroxypentyl)-1*H*-indole-3-carbaldehyde **5d** (1.6 g, 7.8 mmol), to give **7d** (4.2 g, 88%) as a colorless amorphous solid. ^{1}H NMR: δ 7.38 (d, 6H, J = 5.3 Hz), 7.05 (t, 3H, J = 7.3 Hz), 6.95 (s, 3H), 6.86 (t, 3H, J = 7.3 Hz), 6.02 (s, 1H), 4.33 (t, 3H, J = 4.5 Hz), 4.06 (br, 6H), 1.65 (t, 6H, J = 6.8 Hz), 1.39 (t, 6H, J = 6.7 Hz), 1.37–0.95 (m, 6H). ^{13}C NMR: δ 136.2, 127.05, 126.67, 120.68, 119.59, 117.91,

117.14, 109.64, 60.56, 45.24, 32.07, 29.69, 22.76. HRMS (EI) m/z [M] + Calcd for $C_{40}H_{49}N_3O_3^+$ 619.3774; Found 618.3677.

tris(1-[6-Hydroxyhexyl]-1*H*-indol-3-yl)methane (**7e**).

The same procedure as above was carried out using 6-indol-1-yl-hexan-1-ol **3e** (3.0 g, 13.8 mmol) and 1-(6-hydroxyhexyl)-1*H*-indole-3-carbaldehyde **6e** (1.7 g, 6.9 mmol), to give **7e** (3.7 g, 83%) as a colorless amorphous solid. ^{1}H NMR: δ 7.38 (d, 6H, J = 8.6 Hz), 7.05 (t, 3H, J = 7.6 Hz), 6.93 (s, 3H), 6.85 (t, 3H, J = 7.7 Hz), 6.02 (s, 1H), 4.30 (t, 3H, J = 5.1 Hz), 4.06 (t, 6H, J = 6.7 Hz), 3.40–3.20 (m, 6H), 1.73–1.52 (m, 6H), 1.40–1.05 (m, 18H). ^{13}C NMR (100 MHz, DMSO) δ = 136.18, 127.03, 126.67, 120.67, 119.56, 117.89, 117.09, 109.63, 60.54, 45.15, 32.43, 29.80, 26.07, 25.08. HRMS (EI) m/z [M]+ Calcd for $C_{43}H_{55}N_3O_3^+$ 661.4243; Found 660.4160.

tris(1-(2-Hydroxyethyl)-1*H*-indol-3-yl)methylium chloride (**8a**).

To the solution of *tris*(1-[2-hydroxyethyl]-1*H*-indol-3-yl)methane **7a** (2.0 g, 4.0 mmol) in THF (50 mL), DDQ (0.9 g, 4.0 mmol) was added. After stirring at rt for 1 h, the reaction mixture was quenched with conc. HCl (0.4 mL, 5 mmol) and evaporated in vacuo. The residue was purified by flash chromatography (100 g of silica gel) using CH_2Cl_2-MeOH (100:1 to 10:1) as an eluent to give **8a** (1.6 g, 78%) as a red amorphous solid. R_t = 3.39 min. ^{1}H NMR: δ 8.35 (s, 3H), 7.84 (d, 3H, J = 8.3 Hz), 7.40 (t, 3H, J = 7.9 Hz), 7.15–7.05 (m, 6H), 4.92 (s, 3H), 4.51 (t, 6H, J = 5.2 Hz), 3.93 (t, 6H). ^{13}C NMR: δ 144.8, 138.69, 126.74, 124.22, 122.98, 120.60, 117.42, 112.00, 59.12, 49.57. HRMS (EI) m/z [M]+ Calcd for $C_{31}H_{30}N_3O_3^+$ 492.2282; Found 492.2270.

tris(1-(3-Hydroxypropyl)-1*H*-indol-3-yl)methylium chloride (**8b**).

The same procedure as above was carried out using *tris*(1-[3-hydroxypropyl]-1*H*-indol-3-yl) methane **7b** (2.0 g, 3.7 mmol), to give **8b** (1.6 g, 73%) as a red amorphous solid. R_t = 3.55 min. ^{1}H NMR: δ 8.38 (s, 3H), 7.83 (d, 3H, J = 8.3 Hz), 7.41 (t, 3H, J = 7.6 Hz), 7.13 (t, 3H, J = 7.6 Hz), 7.02 (s, 3H), 4.52 (t, 6H, J = 6.9 Hz), 4.46 (br, 3H), 3.56 (t, 6H, J = 5.9 Hz), 2.32–1.92 (m, 6H). ^{13}C NMR: δ 157.77, 144.25, 138.49, 126.7, 124.35, 123.06, 120.49, 111.89, 57.47, 44.22, 31.89. HRMS (EI) m/z [M]+ Calcd for $C_{34}H_{36}N_3O_3^+$ 534.2751; Found 534.2745.

tris(1-(4-Hydroxybutyl)-1*H*-indol-3-yl)methylium chloride (**8c**).

The same procedure as above was carried out using *tris*(1-[4-hydroxybutyl]-1*H*-indol-3- yl)methane **7c** (2.0 g, 3.4 mmol), to give **8c** (1.7 g, 81%) as a red amorphous solid. R_t = 8.76 min. ^{1}H NMR: δ 8.42 (s, 3H), 7.83 (d, J = 8.3 Hz, 3H), 7.40 (t, J = 7.6 Hz, 3H), 7.12 (t, J = 7.5 Hz, 3H), 6.99 (d, J = 7.4 Hz, 3H), 4.49 (t, J = 7.1 Hz, 6H), 4.37 (s, 3H), 3.49 (t, J = 6.2 Hz, 6H), 2.15–1.90 (m, 6H), 1.68–1.47 (m, 6H). ^{13}C NMR: δ 157.88, 144.03, 138.55, 126.78, 124.44, 123.13, 120.55, 117.4, 112.03, 59.90, 46.90, 29.14, 25.85. HRMS (EI) m/z [M]+ Calcd for $C_{37}H_{42}N_3O_3^+$ 576.3221; Found 576.3227.

tris(1-(5-Hydroxypentyl)-1*H*-indol-3-yl)methylium chloride (**8d**).

The same procedure as above was carried out using *tris*(1-[5-hydroxypentyl]-1*H*-indol-3-yl) methane **7d** (2.0 g, 3.2 mmol), to give **8d** (1.8 g, 86%) as a red amorphous solid. R_t = 11.94 min. ^{1}H NMR: δ 8.42 (s, 3H), 7.84 (d, 3H, J = 8.3 Hz), 7.41 (t, 3H, J = 7.6 Hz), 7.13 (t, 3H, J = 7.5 Hz), 7.01 (s, 3H), 4.46 (t, 6H, J = 7.0 Hz), 4.17 (s, 3H), 3.44 (br, 6H), 2.20–1.76 (m, 6H), 1.57–1.40 (m, 12H). ^{13}C NMR: δ 138.47, 124.35, 123.06, 120.48, 111.94, 60.12, 46.93, 31.52, 28.73, 22.38. HRMS (EI) m/z [M]+ Calcd for $C_{40}H_{48}N_3O_3^+$ 618.3690; Found 618.3682.

tris(1-(6-hydroxyhexyl)-1*H*-indol-3-yl)methylium chloride (**8e**).

The same procedure as above was carried out using *tris*(1-[6-hydroxyhexyl]-1*H*-indol-3- yl)methane **7e** (2.0 g, 3.0 mmol), to give **8e** (1.7 g, 81%) as a red amorphous solid. R_t = 16.08 min. ^{1}H NMR: δ 8.42 (s, 3H), 7.84 (d, J = 8.3 Hz, 3H), 7.41 (t, J = 7.7 Hz, 3H), 7.12 (t, J = 7.5 Hz, 3H), 7.00 (d, J = 7.1 Hz, 3H), 4.45 (t, J = 7.1 Hz, 6H), 4.09 (br, 3H), 3.41 (t, J = 6.1 Hz, 6H), 2.17–1.75 (m, 6H), 1.50–1.35 (m, 18H). ^{13}C NMR: δ 157.82, 143.94, 138.49, 126.76, 124.36, 123.03, 120.44, 111.93, 60.24, 46.89, 31.87, 28.87, 25.61, 24.68. HRMS (EI) m/z [M]+ Calcd for $C_{43}H_{54}N_3O_3^+$ 660.4160; Found 660.4150.

4. Conclusions

The first 5 representatives of *N*-(hydroxyalkyl) derivatives of *tris*(1*H*-indol-3-yl)methylium salts were synthesized and tested. Substances **8d** and **8e** showed high activity on Gram-positive bacteria, including resistant strains, and slightly less on Gram-negative ones. At the same time, the cytotoxicity of **8d** was 13 times lower than the antibacterial activity, which indicates the possible prospects for further search among this group of substances. Despite the fact that the exact target of these substances has not yet been established, it is known that the mechanism of their action is associated with a disruption of the membrane. Analysis of the structure-activity relationship showed an empirical dependence of the ratio of antibacterial/cytotoxic activity on the lipophilicity of the molecule. It is found that the best ratio is most likely achieved with $LogP_{ow}$ close to 4. The possibility of theoretical calculation of $LogP_{ow}$ for predicting the activity of new molecules using the Molinspiration package is shown.

Supplementary Materials: The following are available online at http://www.mdpi.com/1424-8247/13/12/469/s1, Supplementary file manuscript-supplementary. pdf: NMR spectra of compounds **3–8**.

Author Contributions: Conceptualization, S.N.L.; methodology, S.N.L.; investigation, S.N.L., E.B.I., A.A.P., A.Y.S., V.V.T.; data curation, S.N.L., E.B.I., A.A.P., A.Y.S., V.V.T.; writing—original draft preparation, S.N.L.; writing—review and editing, S.N.L., A.A.P., A.Y.S., A.S.T.; supervision, S.N.L., A.S.T.; project administration, S.N.L., A.S.T.; All authors have read and agreed to the published version of the manuscript.

Funding: The research was supported by a grant from the Russian Science Foundation (project no. 16-15-10300P).

Conflicts of Interest: The authors declare no conflict of interest. The funders had no role in the design of the study; in the collection, analyses, or interpretation of data; in the writing of the manuscript, or in the decision to publish the results.

References

1. Shallcross, L.J.; Davies, S.C. The World Health Assembly resolution on antimicrobial resistance. *J. Antimicrob. Chemother.* **2014**, *69*, 2883–2885. [CrossRef] [PubMed]
2. Arias, C.A.; Murray, M.D. Antibiotic-resistant bugs in the 21st century-a clinical super-challenge. *N. Engl. J. Med.* **2009**, *360*, 439–443. [CrossRef] [PubMed]
3. Tsakou, F.; Jersie-Christensen, R.; Jenssen, H.; Mojsoska, B. The Role of Proteomics in Bacterial Response to Antibiotics. *Pharmaceuticals* **2020**, *13*, 214. [CrossRef] [PubMed]
4. MacKenzie, F.M.; Struelens, M.J.; Towner, K.J.; Gould, I.M. Report of the Consensus Conference on Antibiotic Resistance; Prevention and Control (ARPAC). *Clin. Microbiol. Infect.* **2005**, *11*, 938–954. [CrossRef] [PubMed]
5. Gillespie, D.E.; Brady, S.F.; Bettermann, A.D.; Cianciotto, N.P.; Liles, M.R.; Rondon, M.R.; Clardy, J.; Goodman, R.M.; Handelsman, J. Isolation of Antibiotics Turbomycin A and B from a Metagenomic Library of Soil Microbial DNA. *Appl. Environ. Microbiol.* **2002**, *8*, 4301–4306. [CrossRef] [PubMed]
6. Dothager, R.S.; Putt, K.S.; Allen, B.J.; Leslie, B.J.; Nesterenko, V.; Hergenrother, P.J. Synthesis and Identification of Small Molecules that Potently Induce Apoptosis in Melanoma Cells through G1 Cell Cycle Arrest. *J. Am. Chem. Soc.* **2005**, *127*, 8686. [CrossRef] [PubMed]
7. Palchaudhuri, R.; Nesterenko, V.; Hergenrother, P.J. The Complex Role of the Triphenylmethyl Motif in Anticancer Compounds. *J. Am. Chem. Soc.* **2008**, *130*, 10274. [CrossRef] [PubMed]
8. Palchaudhuri, R.; Hergenrother, P.J. Triphenylmethylamides (TPMAs): Structure–activity relationship of compounds that induce apoptosis in melanoma cells. *Bioorg. Med. Chem. Lett.* **2008**, *18*, 5888. [CrossRef] [PubMed]
9. Lavrenov, S.N.; Bychkova, O.P.; Dezhenkova, L.G.; Mkrtchyan, A.S.; Tatarskiy, V.V.; Tsvigun, E.A.; Trenin, A.S. Synthesis and study of cytotoxic activity of novel 3,3-bis(indol-3-yl)-1,3-dihydroindol-2-ones. *Chem. Heterocycl. Compd.* **2020**, *56*, 741–746. [CrossRef]
10. Al-Qawasmeh, R.A.; Lee, Y.; Cao, M.-Y.; Gu, X.; Vassilakos, A.; Wright, J.A.; Young, A. Triaryl methane derivatives as antiproliferative agents. *Bioorg. Med. Chem. Lett.* **2004**, *14*, 347. [CrossRef] [PubMed]

11. Lavrenov, S.N.; Luzikov, Y.N.; Bykov, E.E.; Reznikova, M.I.; Stepanova, E.V.; Glazunova, V.A.; Volodina, Y.L.; Tatarsky, V.V., Jr.; Shtil, A.A.; Preobrazhenskaya, M.N. Synthesis and cytotoxic potency of novel tris(1-alkylindol-3-yl)methylium salts: Role of N-alkyl substituents. *Bioorg. Med. Chem.* **2010**, *18*, 6905. [CrossRef] [PubMed]
12. Stepanova, E.V.; Shtil, A.A.; Lavrenov, S.N.; Bukhman, V.M.; Inshakov, A.N.; Mirchink, E.P.; Trenin, A.S.; Galatenko, O.A.; Isakova, E.B.; Glazunova, V.A.; et al. Tris(1-alkylindol-3-yl)methylium salts as a novel class of antitumor agents. *Russ. Chem. Bull. Int. Ed.* **2010**, *59*, 2259. [CrossRef]
13. Durandin, N.A.; Tsvetkov, V.B.; Bykov, E.E.; Kaluzhny, D.N.; Lavrenov, S.N.; Tevyashova, A.N.; Preobrazhenskaya, M.N. Quantitative parameters of complexes of tris(1-alkylindol-3-yl)methylium salts with serum albumin: Relevance for the design of drug candidates J. *Photochem. Photobiol. B Biol.* **2016**, *162*, 570–576. [CrossRef]
14. Lavrenov, S.N.; Simonov, A.Y.; Panov, A.A.; Lakatosh, S.A.; Isakova, E.B.; Tsvigun, E.A.; Bychkova, O.P.; Tatarskiy, V.V.; Ivanova, E.S.; Mirchink, E.P.; et al. Synthesis and biological activity of new antimicrobial agents-Hybrid derivatives of maleimides and triindolylmethanes. *Antibiot. Chemother.* **2018**, *63*, 4–10. (In Russian)
15. Panov, A.A.; Simonov, A.Y.; Lavrenov, S.N.; Lakatosh, S.A.; Trenin, A.S. 3,4-Disubstituted maleimides: Synthesis and biological activity (Review). *Chem. Heterocycl. Compd.* **2018**, *54*, 103–113. [CrossRef]
16. Panov, A.; Lavrenov, S.; Simonov, A.; Mirchink, E.; Isakova, E.; Trenin, A. Synthesis and antimicrobial activity of 3,4-bis(arylthio)maleimides. *J. Antibiot.* **2019**, *72*, 122–124. [CrossRef] [PubMed]
17. Efimova, S.S.; Tertychnaya, T.E.; Lavrenov, S.N.; Ostroumova, O.S. The mechanisms of action of triindolylmethane derivatives on lipid membranes. *Acta Nat.* **2019**, *11*, 38–45. [CrossRef]
18. Seddon, A.M.; Curnow, P.; Booth, P.J. Membrane proteins, lipids and detergents: Not just a soap opera. *Biochim. Biophys. Acta Rev. Biomembr.* **2004**, *1666*, 105–117. [CrossRef]
19. NCCLS. *Reference Method for Broth Dilution Antibacterial Susceptibility Testing*; Clinical and Laboratory Standards Institute: Wayne, PS, USA, 2000.
20. OECD. Test Guideline, Test № 117; OECD iLibrary is the Online Library of the Organisation for Economic Cooperation and Development (OECD). Available online: https://www.oecd-ilibrary.org/ (accessed on 15 December 2020).
21. Ciura, K.; Fedorowicz, J.; Andri, F.; Greber, K.E.; Gurgielewicz, A.; Sawicki, W.; Saczewski, J. Lipophilicity Determination of Quaternary (Fluoro)Quinolones by Chromatographic and Theoretical Approaches. *Int. J. Mol. Sci.* **2019**, *20*, 5288. [CrossRef]
22. Online Version v 2018 10. Available online: https://www.molinspiration.com (accessed on 15 December 2020).
23. logP–Octanol-Water Partition Coefficient Calculation. Available online: https://www.molinspiration.com/services/logp.pdf (accessed on 15 December 2020).

Review

Atopic Dermatitis as a Multifactorial Skin Disorder. Can the Analysis of Pathophysiological Targets Represent the Winning Therapeutic Strategy?

Irene Magnifico [1], Giulio Petronio Petronio [1,*], Noemi Venditti [1], Marco Alfio Cutuli [1], Laura Pietrangelo [1], Franca Vergalito [2], Katia Mangano [3], Davide Zella [4] and Roberto Di Marco [1]

1 Department of Health and Medical Sciences "V. Tiberio" Università degli Studi del Molise, 8600 Campobasso, Italy; i.magnifico@studenti.unimol.it (I.M.); n.venditti@studenti.unimol.it (N.V.); m.cutuli@studenti.unimol.it (M.A.C.); laura.pietrangelo@unimol.it (L.P.); roberto.dimarco@unimol.it (R.D.M.)

2 Department of Agricultural, Environmental and Food Sciences (DiAAA), Università degli Studi del Molise, 86100 Campobasso, Italy; franca.vergalito@unimol.it

3 Department of Biomedical and Biotechnological Sciences, Universitá degli Studi di Catania, 95123 Catania, Italy; kmangano@unict.it

4 Department of Biochemistry and Molecular Biology, School of Medicine, Institute of Human Virology, University of Maryland, Baltimore, MD 21201, USA; DZella@ihv.umaryland.edu

* Correspondence: giulio.petroniopetronio@unimol.it; Tel.: +39-0874-404688

Received: 16 October 2020; Accepted: 19 November 2020; Published: 22 November 2020

Abstract: Atopic dermatitis (AD) is a pathological skin condition with complex aetiological mechanisms that are difficult to fully understand. Scientific evidence suggests that of all the causes, the impairment of the skin barrier and cutaneous dysbiosis together with immunological dysfunction can be considered as the two main factors involved in this pathological skin condition. The loss of the skin barrier function is often linked to dysbiosis and immunological dysfunction, with an imbalance in the ratio between the pathogen *Staphylococcus aureus* and/or other microorganisms residing in the skin. The bibliographic research was conducted on PubMed, using the following keywords: 'atopic dermatitis', 'bacterial therapy', 'drug delivery system' and 'alternative therapy'. The main studies concerning microbial therapy, such as the use of bacteria and/or part thereof with microbiota transplantation, and drug delivery systems to recover skin barrier function have been summarized. The studies examined show great potential in the development of effective therapeutic strategies for AD and AD-like symptoms. Despite this promise, however, future investigative efforts should focus both on the replication of some of these studies on a larger scale, with clinical and demographic characteristics that reflect the general AD population, and on the process of standardisation, in order to produce reliable data.

Keywords: atopic dermatitis; skin barrier; cutaneous dysbiosis; *Staphylococcus aureus*; microbial therapy; drug delivery systems

1. Introduction

Atopic dermatitis (AD) is a chronic relapsing inflammatory skin disorder, affecting 7–10% of the adult population and 15–30% of children, and is associated with significant morbidity and decreased quality of life [1]. Although AD can occur at any age, the incidence peaks in infancy with approximately 45% of all cases beginning within the first six months of life, 60% during the first year, and 80–90% by an individual's fifth birthday [2]. The general term 'eczema' was initially used to describe the condition.

Subsequently, the correlation between eczema and other atopic disorders led to the coining of the term 'atopic dermatitis' in 1933 by Wise and Sulzberger [3]. The AD clinical pattern includes both pruritic and eczematous lesions and the pathophysiology is complex and multifactorial [3–6]. Current knowledge indicates that the main pathogenetic factors of AD are skin barrier dysfunction and dysbiosis of resident microbiota [7]. To these main factors, immunological dysregulation must be added. Skin barrier dysfunction induces immune dysregulation and immune dysregulation alters skin barrier function. Skin microbial dysbiosis also alters immune responses in AD ([8–10]). Therefore, the interaction between barrier dysfunction, microbial dysbiosis and immune dysregulation is at the basis of the worsening of the disease [8]. The skin barrier is localised to the uppermost area of the epidermis, which is the cornified layer (*stratum corneum*) forming by the migration of keratinocytes from the basal to the upper layers. Keratinocytes produce lipids, cyclic adenosine monophosphate (cyclic AMP), cathelicidin and beta-defensins, which form extracellular lipid-enriched layers, kill pathogens and play essential roles in maintaining skin homeostasis [11]. Epidermal barrier proteins, including filaggrin (FLG), keratins, loricrin, involucrin and intercellular proteins, are cross-linked to form an impermeable skin barrier [12]. The alteration in the protein and lipid content of the skin contributes to skin barrier dysfunction. The loss of the function of FLG and other proteins is strongly associated with the development of AD [13]. The overexpression of Th2 and Th22 cytokines altering the protein and lipid content of the skin contributes to skin barrier dysfunction [14]. When developing drug delivery systems (DDSs) for dermatological disorders such as AD, different features of the compromised skin should be considered. In infected, broken or damaged skin where the integrity of the *stratum corneum* is compromised, DDSs improve the efficiency of the formulation [15]. Numerous studies have shown how these systems can aid the delivery of payloads to target sites in dermatological disorder treatment. In particular, the potential for nanocarriers to serve as DDSs for effective AD management has been investigated [15,16].

In addition, an imbalance between *Staphylococcus aureus* (*S. aureus*) and the resident skin microbiota can generate a dysbiosis state that induces an alteration in the immune response and compromises the skin barrier [17]. The skin microbiota plays a role in protecting against infection and inflammation because they guarantee the normal function of the skin barrier. Indeed, viruses, fungi, and bacteria residing on the skin metabolise host proteins and lipids and produce bioactive molecules. These include free fatty acids, cAMP, phenol-soluble modulins (PSMs), microbial cell wall components and antibiotics like bacteriocins that can act on other microbes to inhibit pathogen invasion. All these substances target the host epithelium and stimulate keratinocyte-derived immune mediators such as complement and IL-1, or immune cells in the epidermis and dermis [18–20]. For instance, *Staphylococcus epidermidis (S. epidermidis)* suppresses inflammation by inducing the secretion of interleukin-10, an anti-inflammatory cytokine, from antigen-presenting cells [21,22]. In addition, is able to secrete a unique lipoteic acid that suppress both keratinocytes' inflammatory cytokines and inflammation through a TLR2-dependent mechanism [22,23].

The skin dysbiosis that occurs through an increase in the pathogen *S. aureus* and a variation in the composition and number of skin commensal bacteria also contributes to skin barrier defects and can be a trigger for AD [24]. Indeed, a recent analysis highlighted a prevalence of *S. aureus* on the skin of subjects with AD, with an abundance rate of 70% compared to 39% in the control group [25]. We now have a better understanding of the pathogenetic mechanism of *S. aureus*. This pathogen has numerous virulence factors that contribute to its pathogenesis.

Among these, those most commonly involved in the etiopathogenesis of AD are δ-toxin, phenol-soluble modulins, superantigens, protein A, pro-inflammatory lipoproteins and proteases [26].

In addition to *S. aureus*, skin dysbiosis may occur through an increase in the relative abundance of other species of the genus *Staphylococcus*, such as *S. haemolyticus*. Furthermore, reductions in microorganisms belonging to the genera *Streptococcus* spp., *Propionibacterium* spp., *Acinetobacter* spp., *Corynebacterium* spp. and *Prevotella* spp. have also been observed, which cannot be attributed to an increase in *S. aureus* [27]; on the other hand, *Propionibacterium acnes* was found less frequently on the skin of AD and it was inversely correlated to disease severity [28,29]. After a flare, the species that saw a reduction in their levels then saw an increase in relative abundance [27,29]. An important role is also played by fungal microbiota, which lead to a reduction in the relative abundance of *Malassezia* spp. and an increase in the enrichment of the *M. dermatis* and fungi not belonging to the genus *Malassezia, Aspergillus, Candida* and *Cryptococcus* [29–31]. The reconstitution of healthy microbial diversity, presumably by removing *S. aureus* and allowing the skin to repopulate with physiological microbiota, can restore the protective function of the skin and promote the healing process [7,32]. Within the scientific literature, clinical severity has been evaluated using the objective SCORAD index (scoring AD), which was developed by the European Task Force on Atopic Dermatitis (ETFAD) to create a consensus on assessment methods for AD. This system considers both objective signs (severity and extension) and subjective signs (pruritus and loss of sleep). The SCORAD (AD SCORing) allows a unique classification of the disease: mild, moderate or severe. In addition, a complete diagnosis also includes the evaluation of the intensity of the itching [33]. The European guidelines for the management of AD in adults and children are different for the each level of severity: baseline—emollients and bath oils; mild topical glucocorticosteroids; moderate topical tacrolimus or glucocorticosteroids; and severe systemic immunosuppression [34].

In this case, the new treatment options with antibodies (Ab), especially with the Ab Dupilumab, against interleukin-4 receptor revealed great potential without serious side effects [35–38]

Currently, available drugs are influenced by bioavailability and may give rise to severe adverse events. For example, the use of topical corticosteroids can improve the condition of AD patients, but over-use of corticosteroids during a long bout of sickness can cause some side effects such as hypertension, atrophy and tachyphylaxis result in cumulative toxicity [39]. Although the use of corticosteroids, supported by the use of emollient creams, are widely used in combination to improve symptoms, they do not ensure the complete elimination of AD [40]. The lack of a curative treatment has led to the search for alternative and/or complementary therapies. Microbial therapy and DDSs can help to restore healthy skin microbiota, which have been altered due to skin dysbiosis, and efficiently deliver drugs to skin compromised by AD in order to re-establish the normal function of the skin barrier [41].

This review aims to provide, for the first time, a broad view of AD in light of the newest scientific evidence correlating the two most relevant aspects of this pathology: restoration of healthy skin microbiota and DDSs.

2. Results

2.1. Microbial Therapy: Restoration of Healthy Skin Microbiota

The use of live/heat-killed or inactivated microorganism, the substances with microorganism-derivatives, and the rebalancing of the physiological skin microbiota through skin bacterial transplant may be considered the therapeutic landscape for AD, since they promote the correct functioning of the skin barrier [7,32,42]. Current scientific evidence shows the role of probiotics in improving the clinical course of AD by restoring skin microbiota homeostasis, maintaining lipid barrier functions and modulating the immune system [43]. In addition, some bacterial compounds such as cell wall fragments and their metabolites demonstrate greater stability than viable cells when kept at room temperature, making them more suitable for the formulation of topical preparations. For example, microbe free cultures are still able to exert antimicrobial and immunomodulatory activity in the same way as vital forms [44]. Lastly, studies on the effects of bacterial skin transplant (SBT), an intriguing treatment for the restoration of a healthy skin microbiome in AD patients, have yielded promising results in human and animal models [45]. Together, these approaches have low costs, few side effects, a more relaxed therapy (no daily application necessary) and a more lasting effect.

2.1.1. Live Microorganisms

The use of living microorganisms as food supplements or in medical practices for the treatment of bacterial vaginosis, vaginitis, childhood colic, obesity, type 2 diabetes and pharingotonsillitis is already well known [46,47]. Clinical and experimental research extensively documents the capacity for probiotics to go beyond positively influencing the intestinal functions, and to exert their benefits at the skin level thanks to their peculiar properties [43]. The topical administration of probiotics can increase skin ceramides, improve erythema, scaling and pruritus, and decrease the concentration of the pathogenic *S. aureus* [48].

There have been several studies into the use of live microorganisms for the treatment of AD, using both human and animal models. Seven of these studies are reviewed: three employed animal models; four involved clinical trials, of which three involved children and one adults (see Table S1A in the Supplementary Electronic Material for details).

Firstly, an in vivo study using Sprague-Dawley rats and ddY mice, and the oral administration of *Lactobacillus plantarum*. It has been proven that food supplementation of β-1,3/1,6-glucan and/or *L. plantarum LM1004* can reduce vasodilation, itching, oedema and regulates the immune response [49].

In a double-blind clinical trial on 50 children with moderate AD, the oral administration of a mixture of the probiotics *Bifidobacterium lactis*, *Bifidobacterium longum* and *Lactobacillus casei* was effective in reducing SCORAD index scores and reducing the use of topical steroids to treat flares when compared to the control arm. These findings suggest that such a mixture of probiotics can be used for the treatment of AD [50].

An in vivo study on SKH-1 hairless mice aimed to test a probiotic mixture of five bacterial strains, *Bifidobacterium longum*, *Lactobacillus helveticus*, *Lactococcus lactis*, *Streptococcus thermophilus* and *Lactobacillus rhamnosus*, in preserving skin integrity and homeostasis. It has been observed that daily oral treatment with the probiotic mixture, through modulation of the immune response, has significantly limited chronic skin inflammation, demonstrating its use in pathological dermatological conditions such as AD and psoriasis [51].

The oral administration of *Weissella cibaria* WIKIM28 in a mouse model of AD induced in BALB/c mice has shown that this bacterial strain can be a good candidate as a probiotic for AD prevention and improvement. Thus, the intake of this live microorganism improved AD-like skin lesions and exhibited excellent immunomodulatory activity [52].

A randomised, double-blind study carried out on 220 children affected by moderate/severe AD, showed that the oral administration of *Lactobacillus paracasei* and *Lactobacillus fermentum*, for 3 weeks led to decreased IgE, TNF-α, urine eosinophilic protein X and SCORAD scores. Thus indicating that supplementation of a probiotic mixture of *L. plantarum* and *L. fermentum* is associated with clinical improvement of AD [53].

Another trial on 43 children tested *Lactobacillus salivarius*, which, when orally administered, showed a significant improvement in clinical parameters, SCORAD scores and itch values [54].

In a prospective controlled pilot trial on 25 adults, the oral administration of the probiotic strain *L. salivarius* LS01 in association with *Streptococcus thermophilus*, significantly improved both SCORAD scores and the *S. aureus* count. Moreover, the combination of *S. thermophilus* ST10 with *L. salivarius* LS01 improved the overall effectiveness of the formulation by reducing the recovery time [55].

2.1.2. Heat-Killed or Inactivated Microorganisms.

The growing interest in the biological effects of heat-killed or inactivated microorganisms is already well documented. In particular, the use of heat-treated probiotic bacteria (lactic and bifidobacteria), together with their cell-free supernatants or selected purified cellular components in immunomodulation and maintaining the integrity of the intestinal barrier against enteropathogens is well known to the scientific community. Only recently, numerous scientific studies have investigated the role of these non-viable microorganisms in the management of dermatological diseases [56]. There are several studies that have investigated the potential of heat-killed or inactivated microorganisms for the treatment of AD, which have used both human and animal models. The findings of seven of these studies are reported herein. One employed animal models, five involved clinical trials, of which two were in children, and finally one was conducted within an in vitro reconstructed human epidermis (RHE) (see Table S1B in the Supplementary Electronic Material for details).

Topical application of a formulation containing heat-treated *Lactobacillus johnsonii* NCC 533 (HT La1) was able to modulate endogenous antimicrobial peptides (AMP) expression and to inhibit the binding of *S. aureus* in an in vitro reconstructed human epidermis model (RHE). These results highlight the role of innate skin immunity in reducing *S. aureus* colonization in atopic skin [57].

An open-label clinical study in AD patients showed that the application of a lotion containing a heat-treated *Lactobacillus johnsonii* NCC 533 (HT La1) led to a decrease in the SCORAD score. This clinical improvement was associated with a reduction in the *S. aureus* viable count. In addition, the authors were able to establish a directly proportional correlation between the *S. aureus* skin concentrations and the lotion response [58].

In a double-blind clinical trial conducted on 60 patients suffering from moderate AD, topical application of an emollient containing biomass from the non-pathogenic bacteria *Vitreoscilla filiformis* lysate one month after the end of the treatment ameliorated the evolution of the average SCORAD score, which was significantly lower than that of the control patients treated with a generic emollient.

During one month of treatment, the level of *Staphylococcus* spp. decreased in treated subjects with the formulation enriched by *V. filiformis* biomass, demonstrating the normalization of the skin microbiota and the significant reduction in the number and severity of flare-ups compared to another formulation without bacterial biomass [59].

In a clinical trial on 179 children, oral administration of the bacterial lysate OM-85 of 21 strains from eight common respiratory pathogenic microorganisms (i.e., *Haemophilus influenzae, Streptococcus pneumoniae, Klebsiella ozaenae and pneumoniae, S. aureus, Streptococcus viridans* and *pyogenes* and *Neisseria catarrhalis*) showed an adjuvant therapeutic effect which led to significantly fewer new flares and delayed their onset. Indeed, these results showed an adjuvant therapeutic effect of a well-standardised bacterial lysate OM-85 on established AD [60].

An in vivo study on NC/Nga mice demonstrated that the oral administration of *Lactobacillus plantarum* lysates was able to restore the skin homeostasis of the treated animals. Indeed, after two months of treatment, there was a reduction in the formation of the horny layer and a decrease in skin thickening compared to untreated mice [61].

Kim et al. stressed the importance of clinical research in the study of AD. In their study, the authors tested *L. plantarum* K8 lysates formulation, both with in vitro/in vivo experiments, and in a clinical trial with the healthy volunteer. Preliminary data obtained in vitro with HaCaT cells and after 2 months of in vivo treatment with on DNCB-treated SKH-1 hairless mice demonstrated an attenuation of the stratum corneum formation and epidermal thickening of AD mice skin. These data were supported by the clinical study, where an improvement in the barrier function of the epidermis was observed in subjects who ate candies containing L. plantarum K8 lysate [62].

In a clinical trial, 606 infants at risk of atopy were treated with an oral application of bacterial lysate containing heat-killed *Escherichia coli* and *Enterococcus faecalis*. The results showed a reduced possibility of developing AD, suggesting that bacterial lysates prevent the development of this skin condition in children [63].

2.1.3. Microorganism-Derived Substances

The capacity of microorganism-derived compounds to inhibit allergic inflammation make them candidates for novel therapies for allergic diseases [64]. Among these compounds are bacteriocins, proteins and enzymes [65]. Several studies have highlighted the beneficial role of skin commensals due to the production of bacteriocins. Indeed, many members of the cutaneous microbiome can metabolise glycerol into antimicrobial compounds, such as bacteriocin, that inhibit *S. aureus* growth. Skin commensal coagulase-negative staphylococci (CoNS) are the primary producers, but there are also other microorganisms able to produce these compounds [66,67]. There are several studies, both in human and animal models. In this section, seven studies concerning the use of microorganism-derived substances for AD treatment are presented. Of these, four employed animal models, two were conducted in vitro, and one were conducted both in vitro and in vivo (see Table S1C in the Supplementary Electronic Material for details).

An in vitro study showed that cytoplasmic bacteriocins isolated from *S. epidermidis* selectively exhibited antimicrobial activity against *S. aureus* and methicillin-resistant *S. aureus* (MRSA). These findings suggest that these cytoplasmic bacteriocin compounds could potentially inhibit the growth of *S. aureus* and be used as a topical AD treatment [68].

In an in vivo model of AD on BALB/cAJcl mice, the oral administration of an exopolysaccharide (EPS) produced by *Lactobacillus paracasei* reduced ear swelling, produced a repression of ear interleukin-4 (T helper (Th) 2 cytokine) mRNA and decreased serum immunoglobulin E levels. These results suggest that *Lactobacillus paracasei*-derived EPS inhibits the catalytic activity of hyaluronidase promoting inflammatory reactions and is useful for improving type I and type IV allergies, including AD [69].

The commensal yeast *Malassezia globosa*, secretes a protease called '*Malassezia globosa* secreted Aspartyl Protease 1 (MgSAP1)', which, in vitro, can disrupt *S. aureus* biofilms by hydrolysing protein A. This study defined a role for the skin fungus Malassezia in inter-kingdom interactions and suggested that this fungus enzyme may be beneficial for skin health [70].

In a mouse model the topical application of p40, a particulate fraction from *Corynebacterium granulosum*, used in a formula with hyaluronic acid produced a significant reduction in ear thickness, weight, oedema, and leukocyte recruitment. These results suggest that p40-conjugated with hyaluronic acid may constitute an outstanding innovative dermatitis treatment [71].

In addition, other bacteria not belonging to the skin microbiota are able to produce antibiotics with properties useful for treating AD. An example would be the topical application of josamycin, a macrolide antibiotic derived from *Streptomyces narbonensis* subsp. *josamyceticus* which was applied to NC/Nga mice. In this case, the topical application of this antibiotic reduced the expression of proinflammatory cytokines demonstrating antimicrobial activity against *S. aureus* present on the skin of AD mice [72].

Another molecule with antibacterial activity, produced by *S. lugdunensis*, lugdunine, was tested in an in vivo experiment with shaved black-6 (C57BL/6) mice and it was able to reduce or completely eradicate *S. aureus* viable count both on the surface and in the deeper layers of the skin. The isolation and study of other lugdunin- or lugdunin-like molecules isolated from s commensal bacteria could represent a new therapeutic approach in the prevention and management of staphylococcal infections [73].

Similarly, an AD-like in vivo NC/Nga mice model demonstrated that the protein P14, isolated from *Lactobacillus casei*, can be used as an active immunomodulatory agent for treating patients with AD [74].

2.1.4. Skin Bacterial Transplantation

Although there are still few studies on the transplantation of skin bacteria (SBT), this particular type of bacteriotherapy that involves transplanting several skin microbiota from one individual to another has already provided promising results in both human clinical trials and in animal models [45]. Indeed this intriguing therapeutic potential has earned it the definition of the "future of eczema therapy" [75]. Herein, three human studies are reported focusing on skin microbiota transplantation for the treatment AD. Of these studies, one involved a clinical trial conducted on healthy volunteers to develop the technique for transferring the entire skin microbiota, another was carried out on adults, and the last one involved both adults and paediatric patients (see Table S1D in the Supplementary Electronic Material for details).

In a recent prospective pilot study, researchers attempted to perform a complete skin microbiota transplant that shifted the entire bacterial skin community of healthy volunteers from the forearm to the back in a unidirectional manner. Evidence of the transfer of a partial DNA signature was seen by comparing the bacterial species present in the arm with the mixed communities ('transplantation') that were absent in the back. This technique aimed to move viable skin organisms from one site to another and is worthy of further investigation [76].

The successful transplantation of *Roseomonas mucosa* was conducted in an open-label phase I/II safety and activity trial with adults and pediatric patients. The results demonstrated a significant decrease in disease severity, a reduction in steroid administration, and a viable *S. aureus* count [77]. All these finds were supported by a previous study in mice conducted by the same authors [20].

Najatsuji et al. conducted a clinical study by autologous CoNS transplantation isolated from AD patients *S. aureus* culture positive. After isolation, CoNS strains (*S. epidermidis* and *S. hominis*) were formulated in a cream base vehicle and applied to the forearm of the same subjects for 24 h. The results showed a significant decrease in *S. aureus* colonization at the microbial transplant site compared to the contralateral forearm treated with the bacteria-free vehicle alone. These observations were also confirmed by in vivo experiments on the back of C57BL6 mice. These findings show, once again, the role of commensal skin bacteria in protecting against colonisation by pathogens and how dysbiosis of the skin microbiome can contribute to the onset of the disease [19].

2.2. *Drug Delivery Systems*

It is often preferable to use non-invasive delivery to provide relief for AD [78]. Topical treatment is preferential to parenteral or oral administration because of better compliance and the reduction in drug concentrations and side effects [79]. Topically, DDSs deliver therapeutic agents or natural active compounds directly to the target site to maximise the benefits and minimise the risks associated with drugs. In this regard, in the last two decades, an interest in nano-based DDSs has developed. The latter have already been applied in the treatment of various diseases ranging from cancer to Alzheimer's [80].

The most common nano-based DDS carriers addressed in this manuscript, include polymeric nanoparticles (NPs), solid lipid nanoparticles (SLNs), lLiposomes, ethosomes, and elastic vesicles due to their small size (range from 1 to 1000 nm). They can penetrate through the *stratum corneum* and accumulate in the target site, improving the delivery of transported bioactive compounds and favouring higher drug retention, demonstrated by drug diffusion and permeation study profiles [79–82]. Although the dimensions are variable, desired therapeutic benefits, avoidance of off-target effects, and optimal localised delivery of drugs are achieved using nanocarriers <200 nm in size. Nanocarrier-mediated interventions have been well-reported for topical and transdermal applications [83]. Together, these approaches offer novel solutions, allowing: (i) the management of severe forms of AD, especially those not responsive to steroid therapy; (ii) improved performance of pharmacokinetic parameters such as permeation and controlled release; (iii) significant improvements in the patient's state of health; iv) a reduction in the dosage of the active ingredient with a consequent reduction in toxicity and an improved safety profile [84,85].

2.2.1. Nanoparticles

Nanoparticles (NP) are a broad class of DSS in the order of 100 nanometres with optimal rheological properties, antimicrobial effects and the ability to restore skin conditions [16,86]. For instance, NPs loaded with a lipid drug and/or made by lipophilic compounds (i.e., lipid NPs) ensure skin hydration and the occlusion effect in a size-dependent manner and can form a thin film on the skin surface, which allows for rehydration [87]. The complete biodegradation of lipid NPs and their biocompatible chemical features have secured them the title of nano-safe carriers [84]. Twelve studies concerning the use of NPs in AD treatment were identified for review. Only in vivo studies using animals were selected. Of the seventeen studies, one employed only in vivo animal models, four were conducted in vitro and ex vivo, six were conducted in vitro and in vivo, and one was conducted in vitro, ex vivo and in vivo. In vitro tests provided a characterisation and evaluation of the formulation (see Table S2A in the Supplementary Electronic Material for details).

An in vitro and ex vivo drug test performed using a jacketed Franz diffusion cell showed that nanoencapsulation of betamethasone valerate (BMV) into the chitosan nanoparticles (CS-NPs) displayed a Fickian diffusion type mechanism of release in the simulated skin surface. Drug permeation efficiency and the amount of BMV retained in the epidermis and the dermis was higher when compared to BMV solution alone. These results suggest that this formulation of betamethasone improved the therapeutic efficacy of the treatment of AD [88].

Tacrolimus-loaded thermosensitive solid lipid nanoparticles (TCR-SLN) in the dorsal skin of Sprague Dawley rats penetrated to a deeper layer than the control formula. The penetration test in vivo of the skin of white rabbits demonstrated that TCR-SLNs delivered more drug into deeper skin layers than the control, suggesting that thermosensitive SLNs could be employed for the delivery of difficult-to-permeate, poorly water-soluble drugs into deep skin layers [89].

In an in vitro test with a Franz static diffusion cell system and ex vivo on skin from Wistar albino rats, the application of 'hyaluronic acid-modified betamethasone encapsulated polymeric nanoparticles' (HA-BMV-CS-NPs) revealed that drug permeation efficiency of betamethasone was higher in the case of BMV-CS-NPs and that there was a greater amount of drug retained in the epidermis and the dermis. This complex could be a promising nano delivery system for efficient dermal targeting of BMV and improved anti-AD efficacy [90].

In a clinical trial that enrolled healthy volunteers treated with hydrocortisone hydroxytyrosol anti-oxidant-loaded chitosan nanoparticles (HA-HT-CSNPs) to evaluate systemic toxicity, the results of blood haematology, blood biochemistry, and adrenal cortico-thyroid hormone levels were not significant. This indicated non-systemic toxicity and supports the view that this formula could be used for AD treatment [91].

In vitro and in vivo permeation studies on Sprague Dawley rats with tacrolimus nanoparticles based on chitosan and combined with nicotinamide (FK506-NIC-CS-NPs), demonstrated that these nanoparticles significantly enhance tacrolimus permeation through and into the skin, and deposited more tacrolimus into the skin. Moreover, this system enhances the permeability of tacrolimus and plays an adjuvant role in anti-AD, reducing the dose of tacrolimus in treating AD, and is, therefore, a promising nanoscale system of tacrolimus for the effective treatment of AD [92].

Betamethasone Valerate incorporates in a lipidic carrier revealed an enhancement of the Betamethasone Valerate ratio in comparison with the control group and had an anti-inflammatory effect. The outcome of complete characterisation suggests that the developed formulation is efficient in a single daily dosage in the therapy of AD [93].

An in vitro/ex vivo test on NC/Nga mice skin demonstrated the anti-AD efficacy of tacrolimus-hyaluronic acid-charged nanoparticles. According to the author's findings, this formulation can be used as a promising therapeutic approach for patients who cannot be treated with steroid therapy, such as children and adults with steroid intolerance [94].

In an in vivo test with SKH-1 mice, the topical application of dendritic nano-multi-shell dendritic nanocarriers was evaluated as a deposit formulation for anti-inflammatory drugs in the skin. Both in vitro release and toxicological studies have confirmed the biocompatibility of the formulation, providing evidence of prolonged release of the active substance especially for anti-inflammatory drugs like those used in AD. Furthermore, no evidence of local or systemic toxic/adverse effects was observed [95].

An in vivo test on Wistar albino rats evaluated the penetration into the deep skin layers of cationic polymeric chitosan nanoparticles loaded with anti-inflammatory (hydrocortisone) and antimicrobial (hydroxytyrosol,) anti-inflammatory agents compared to a similar commercial formulation. The results proved a better performance in the local release of the active ingredients without involving the underlying tissues. In addition, no toxicity was found compared to the commercial formulation, providing substantial safety benefits [96].

In an in vivo test with NC/Nga mice, transcutaneous co-delivery based on nanocarrier hydrocortisone and hydroxytyrosol was studied as a possible therapy for the management of the immunological and histological issues of AD. The results of immunological and histological experiments conducted on the sera and biopsies of the tested mice confirmed this hypothesis [97].

Furthermore, a Silver-nano lipid complex incorporated into an o/w cream and a lotion showed a high adhesivity to the skin and bacterial surfaces, leading to a locally high concentration of silver ion killing bacteria, restoring the distorted skin barrier, and being much more useful than silver alone. Data were generated either by in vitro tests determining the colony-forming unit (CFU) count over time of *S. aureus* ATCC25923, or in vivo on BALB/c mice. This formula makes the drug more effective in terms of enhanced penetration and exploits the skin normalisation ability of the skincare sNLC formulation [16].

Another in vivo study in NC/Nga mice aimed to assess whether the transcutaneous administration of hydrocortisone nanoparticle could be considered a valid therapeutic approach in the management of dermatitis suggested a substantial reduction in inflammatory cascade mediators, accompanied by positive histological results on fibroblast infiltration and elastic fiber fragmentation, demonstrating how these formulations can promote and maintain the integrity of connective tissues especially in an injured skin like AD [98].

2.2.2. Liposomes, Ethosomes, and Elastic Vesicles

Liposomes and ethosomes can be defined as vesicular DDSs. Liposomes are spherical vesicles with particle sizes ranging from 30 nm to several micrometres consisting of single or multiple concentric lipid bilayers encapsulating an aqueous compartment. These formulations have been successfully applied for the management of AD due to their moisturising effect on the *stratum corneum* and their ability to act as bioactive compound carriers [85]. Rigid liposomes remain confined to the *stratum corneum*, resulting in the formation of a drug reservoir in the upper skin layers, and do not allow percutaneous absorption. More recently, efforts have been made to investigate vesicular lipid systems capable of facilitating drug penetration to the underlying skin layers, allowing transdermal absorption [99].

In contrast, ethosomes are made mainly of phospholipids with a high concentration of ethanol (20–50%) and water. Due to this composition, they have demonstrated remarkably high deformability features [100]. Moreover, ethosomes guarantee a more efficient transfer of the active principle through the skin (epidermis and dermis) than liposomes [15].

Finally, a further advance in the field of DDS is represented by the elastic vesicles used as a new topical and transdermal delivery system. Although the manufacturing method of these vehicles is very similar to that of liposomes, the presence of an 'activating' agent in the phospholipid bilayer gives it a high degree of elasticity. It has been demonstrated that the topical administration of elastic vesicles does not occlude the skin and easily permeates through the *stratum corneum* lipid lamellar regions due to skin hydration or by osmotic force. Furthermore, this DDS can be loaded with a wide range of small molecules, peptides and proteins [101].

Six applications of liposomes, ethosomes, and elastic vesicles in AD treatment are herein reported. Of them, one was conducted using only in vitro methods, one enrolled patients with AD, one were conducted by in vitro and ex vivo studies, one by in vitro and in vivo and in the last two an in vitro, ex vivo and in vivo methodology was adopted. In vitro tests have provided a characterisation and evaluation of the formulation (see Table S2B in the Supplementary Electronic Material for details).

In an in vitro test with a static Franz diffusion cell setup on the heat-separated human epidermis, the use of ultra-flexible lipid vesicles effectively delivered cyclosporin A into the skin. This study introduces a promising approach to the topical treatment of skin pathologies with an immune component [102].

In an in vitro test with a dialysis membrane and ex vivo with Wistar rat skin, the application of cyclo-ethosomes with fluocinolone acetonide (FA) showed maximum permeability as compared with an optimised reference ethosomal gel and control gel. These results suggest that β-cyclo-ethosomes could be a promising carrier for improvised penetration of fluocinolone acetonide via topical gel [103].

In an open-label pilot study of 20 patients with AD, the application of liposomal polyvinylpyrrolidone-iodine hydrogel showed that this strategy was well tolerated and led to an improvement in pain, quality of life, eczema area and severity. This formula has potential utility as an effective treatment for inflammatory skin conditions associated with bacterial colonisation [104].

An in vitro test with a dialysis membrane and ex vivo with Wistar rat skin revealed that nano ethosomal glycolic vesicles of triamcinolone acetonide have excellent permeation. Besides the histological analysis, the study confirmed the non-irritant potential. These results suggest that nano-ethosomal glycolic vesicles can be active non-irritant carriers for the improvised penetration of triamcinolone acetonide for potential topical therapeutics [105].

The pharmaco-dynamic evaluation of the ethosome-based topical delivery system of the antihistaminic drug cetirizine (measured by in vivo and ex vivo tests on BALB/c mice) showed a reduction in the scratching score, the erythema score, skin hyperplasia and the dermal eosinophil count. The data suggest that this formula could be an effective carrier for the dermal delivery of the antihistaminic drug, cetirizine, for the treatment of AD [106].

An in vivo and ex vivo tests on BALB/c mice, a topical formulation of levocetirizine based on flexible vesicles (FVs) showed a reduction in the scratching score and the erythema score in addition to the dermal eosinophil count [107].

3. Discussion

AD is a pathological skin condition that is becoming increasingly common in clinical dermatological practice. The pathogenesis is exacerbated by its complex aetiological mechanisms that are not yet fully understood, providing many opportunities for misinterpretation [108]. Among the different hypotheses, numerous studies have demonstrated that dysbiosis and skin barrier dysfunction contribute to the pathobiology of AD [109]. Immune dysregulation is another factor involved in the pathogenesis of AD and is closely related to the previous ones. Indeed skin colonisation of Staphylococcus aureus damages the skin barrier and induces inflammatory responses, on the other hand, local Th2 immune responses diminish barrier function, promoting bacterial dysbiosis [9].

Although it is common to associate skin dysbiosis with an increase in *S. aureus* abundance, more recent studies are converging on the opinion that AD skin microbiota is characterised by low bacterial diversity. The relative abundance of both *S. aureus* and *S. epidermidis* are elevated and the presence of *Propionibacterium* spp. is reduced, along with other genera (*Streptococcus*, *Acinetobacter*, *Corynebacterium* and *Prevotella*). Moreover, the absence of early colonisation with commensal staphylococci might precede AD presentation [31]. Skin dysbiosis contributes to skin barrier defects [12]. The latter promote easy penetration of numerous insults relevant to the development of the disease i.e., pathogens, toxins, allergens, irritants and pollutants. Accordingly, all the treatments (pharmacological and adjuvants) aim to minimise the number of exacerbations, the so-called 'flares', and reduce their duration and intensity [110]. To date, there is not a resolutive therapy that can take into account the complex pathogenic interplay between a patient's susceptible genes, their skin barrier abnormalities and their immune dysregulation [15].

The majority of AD patients are paediatric and when moderate-to-severe symptoms occur, current therapies have proven to be of limited efficacy and have several side effects [111–113]. For all these reasons, there has been a surge of interest from clinicians and the lay public in exploring targeted bacteriotherapy to treat this pathological skin condition [76]. Microbic therapies with microorganisms that are commensal of the healthy skin microbiota, or probiotics in conjunction with transplantation, could represent a new diagnostic and therapeutic target for AD [114–116]. Several studies have demonstrated that probiotic use has led to increased skin ceramides and has improved erythema, scaling and pruritus, suggesting that probiotics may be useful for the treatment of AD, especially for moderate to severe AD in children and adults [48,51,53,116]. Furthermore, specific probiotic strains have shown active immunomodulatory properties [59,117].

Restoring the skin microbiota homeostasis could also represent a new era in AD treatment [118]. The reconstitution of healthy microbial diversity can boost the right immune response and normal barrier function [7,32,119,120]. Similarly, other studies have demonstrated that commensal microorganisms can reduce *S. aureus* by bacteriocin production or competition mechanisms, improving AD symptoms. In this context, the development of antibiotic resistance by the *S. aureus* methicillin-resistant (MRSA) strain has considerable importance, not only from the point of view of infectious disease but also as it can influence the course of the disease. Bacteriocins from CoNS also exhibit antimicrobial activity against MRSA [72,121]. The clinical promise of transplanting commensal skin organisms from healthy individuals onto diseased skin, together with faecal microbiota transplantation to selectively target pathogenic *S. aureus,* thus modifying the diseased skin microbiome to attenuate the course of the disease, have been investigated, with promising results [16,76].

Furthermore, the therapeutic potential of DDSs based on nano-products has provided a new avenue for the prevention and treatment of inflammation and sequelae of skin diseases. Several studies have shown the effectiveness of nanoparticles, liposomes, ethosomes and vesicles in AD. This was particularly valid in recalcitrant form treatments, due to their unique characteristics, such as the improvement in pharmacokinetic parameters (targeted transdermal release of the active ingredient, permeation, retention, and diffusion) and physicochemical properties. These advances in pharmaceutical technology have led to improvements in both clinical symptoms and immune responses, along with better inhibition of inflammatory cascades mediators that positively impact patients' quality of life, with fewer adverse events reported and increased patient compliance [85,110,122,123].

4. Materials and Methods

The interest of the scientific community in research into novel targets for the development of effective therapeutic strategies in AD management has dramatically increased. For this reason, the bibliographic research for scientific papers specialised in the field of interest was conducted from 2014 to March 2020 on PubMed (the MEDLINE database), using the following keywords: 'atopic dermatitis', 'bacterial therapy', 'drug delivery system' and 'alternative therapy' alone and/or in combination. As a preliminary result, more than 300 documents were found. Of these, 24 papers on microbial therapy and 15 on nano-based DDSs were selected for review due to their relevance.

5. Conclusions

All the studies reviewed show enormous potential for AD treatment, so we can state that research into novel targets is key to the development of effective therapeutic strategies. Nevertheless, some limitations still need to be overcome. An aspect of primary importance in the advancement of scientific and technological innovation is the possibility of marketing the new formulations. To this end, there are different international regulations regarding bacterial formulations for medical use. The European Medical Device Directive (MD) (DDM 93/42) and subsequent amendments include MDs containing live microorganisms (especially those containing probiotics) for the management of AD [124]. On the other hand, the US Food and Drug Administration (FDA) has not approved any oral or topical microbial-based formulations for the treatment of dermatological condition [125].

Although the potential of bacteriotherapy for the treatment of AD seems to be clear, further studies will need to be conducted with the goals of recruiting more patients with different clinical characteristics and standardising the process to produce reliable data. Put differently, even if the topically used DDSs offer promising opportunities in dermal delivery, many questions arise, which remain to be explored and addressed, concerning, for example, their toxicological characteristics and the long-term safety of these technologies.

In vivo and in vitro assays are useful to identify the toxicity of dds because they help to establish the dose–response relationship [126]

However despite in vitro tests are useful for bypassing cell interactions that exist in vivo, in vivo toxicity testing is needed due to the difference between in vitro dosimetry and real topical exposure and additional innovative research is needed to address the cost-effectiveness and long-term safety of these nanoparticles [127].

Supplementary Materials: The following are available online at http://www.mdpi.com/1424-8247/13/11/411/s1, Table S1: Restoration of healthy skin microbiota, Table S2: Drug Delivery System (DDS) for AD treatment.

Author Contributions: Conceptualization: I.M., G.P.P., R.D.M.; Methodology: I.M. and G.P.P.; Investigation: I.M., G.P.P., M.A.C., N.V., L.P.; Resources: I.M., G.P.P., F.V., K.M.; Writing—original draft preparation: I.M. and G.P.P.; Writing—review and editing, I.M., G.P.P., K.M., D.Z., R.D.M.; Supervision: D.Z. and R.D.M.; Project administration: R.D.M. All authors have read and agreed to the published version of the manuscript.

Funding: This research received no external funding except donation covering journal APC.

Acknowledgments: The authors thank Aileens Pharma s.r.l for founding the journal APC and Professor Amy Muschamp for the language revision.

Conflicts of Interest: The authors declare no conflict of interest.

References

1. Weidinger, S.; Beck, L.; Bieber, T.; Kabashima, K.; Irvine, A. Atopic dermatitis. *Nat. Rev. Dis. Primers* **2018**, *4*, 1. [CrossRef]
2. Abuabara, K.; Yu, A.; Okhovat, J.P.; Allen, I.; Langan, S.M. The prevalence of atopic dermatitis beyond childhood: A systematic review and meta-analysis of longitudinal studies. *Allergy* **2018**, *73*, 696–704. [CrossRef]
3. Patel, N.; Feldman, S.R. Adherence in atopic dermatitis. In *Management of Atopic Dermatitis*; Springer: Berlin, Germany, 2017; pp. 139–159.
4. Wollenberg, A.; Schnopp, C. Evolution of conventional therapy in atopic dermatitis. *Immunol. Allergy Clin.* **2010**, *30*, 351–368. [CrossRef]
5. Guttman-Yassky, E.; Waldman, A.; Ahluwalia, J.; Ong, P.Y.; Eichenfield, L.F. Atopic dermatitis: Pathogenesis. *Semin. Cutan. Med. Surg.* **2017**, *36*, 100–103. [CrossRef] [PubMed]
6. Spergel, J.M. From atopic dermatitis to asthma: The atopic march. *Ann. Allergy Asthma Immunol.* **2010**, *105*, 99–106. [CrossRef] [PubMed]
7. Seite, S.; Bieber, T. Barrier function and microbiotic dysbiosis in atopic dermatitis. *Clin. Cosmet. Investig. Dermatol.* **2015**, *8*, 479. [CrossRef] [PubMed]
8. Patrick, G.J.; Archer, N.K.; Miller, L.S. Which Way Do We Go? Complex Interactions in Atopic Dermatitis Pathogenesis. *J. Investig. Dermatol.* **2020**, *396*, P345–P360.
9. Langan, S.M.; Irvine, A.D.; Weidinger, S. Atopic dermatitis. *Lancet* **2020**, *396*, 345–360. [CrossRef]
10. Nakahara, T.; Kido-Nakahara, M.; Tsuji, G.; Furue, M. Basics and recent advances in the pathophysiology of atopic dermatitis. *J. Dermatol.* **2020**. [CrossRef]
11. Proksch, E.; Brandner, J.M.; Jensen, J.M. The skin: An indispensable barrier. *Exp. Dermatol.* **2008**, *17*, 1063–1072. [CrossRef]
12. Kim, B.E.; Leung, D.Y. Significance of skin barrier dysfunction in atopic dermatitis. *Allergy Asthma Immunol. Res.* **2018**, *10*, 207–215. [CrossRef] [PubMed]
13. Drislane, C.; Irvine, A.D. The role of filaggrin in atopic dermatitis and allergic disease. *Ann. Allergy Asthma Immunol.* **2020**, *124*, 36–43. [CrossRef] [PubMed]
14. Hamid, Q.; Boguniewicz, M.; Leung, D. Differential in situ cytokine gene expression in acute versus chronic atopic dermatitis. *J. Clin. Investig.* **1994**, *94*, 870–876. [CrossRef] [PubMed]
15. Shao, M.; Hussain, Z.; Thu, H.E.; Khan, S.; Katas, H.; Ahmed, T.A.; Tripathy, M.; Leng, J.; Qin, H.-L.; Bukhari, S.N.A. Drug nanocarrier, the future of atopic diseases: Advanced drug delivery systems and smart management of disease. *Colloids Surf. B Biointerfaces* **2016**, *147*, 475–491. [CrossRef] [PubMed]
16. Keck, C.; Anantaworasakul, P.; Patel, M.; Okonogi, S.; Singh, K.; Roessner, D.; Scherrers, R.; Schwabe, K.; Rimpler, C.; Müller, R. A new concept for the treatment of atopic dermatitis: Silver–nanolipid complex (sNLC). *Int. J. Pharm.* **2014**, *462*, 44–51. [CrossRef]
17. Tham, E.H.; Koh, E.; Common, J.E.; Hwang, I.Y. Biotherapeutic Approaches in Atopic Dermatitis. *Biotechnol. J.* **2020**, e1900322. [CrossRef]

18. Chen, Y.E.; Fischbach, M.A.; Belkaid, Y. Skin microbiota–host interactions. *Nature* **2018**, *553*, 427–436. [CrossRef]
19. Nakatsuji, T.; Chen, T.H.; Narala, S.; Chun, K.A.; Two, A.M.; Yun, T.; Shafiq, F.; Kotol, P.F.; Bouslimani, A.; Melnik, A.V. Antimicrobials from human skin commensal bacteria protect against Staphylococcus aureus and are deficient in atopic dermatitis. *Sci. Transl. Med.* **2017**, *9*, 4680. [CrossRef]
20. Myles, I.A.; Williams, K.W.; Reckhow, J.D.; Jammeh, M.L.; Pincus, N.B.; Sastalla, I.; Saleem, D.; Stone, K.D.; Datta, S.K. Transplantation of human skin microbiota in models of atopic dermatitis. *JCI Insight* **2016**, *1*, e86955. [CrossRef]
21. Chau, T.A.; McCully, M.L.; Brintnell, W.; An, G.; Kasper, K.J.; Vinés, E.D.; Kubes, P.; Haeryfar, S.M.; McCormick, J.K.; Cairns, E. Toll-like receptor 2 ligands on the staphylococcal cell wall downregulate superantigen-induced T cell activation and prevent toxic shock syndrome. *Nat. Med.* **2009**, *15*, 641. [CrossRef]
22. Lai, Y.; Di Nardo, A.; Nakatsuji, T.; Leichtle, A.; Yang, Y.; Cogen, A.L.; Wu, Z.-R.; Hooper, L.V.; Schmidt, R.R.; Von Aulock, S. Commensal bacteria regulate Toll-like receptor 3–dependent inflammation after skin injury. *Nat. Med.* **2009**, *15*, 1377. [CrossRef] [PubMed]
23. Gallo, R.L.; Nakatsuji, T. Microbial symbiosis with the innate immune defense system of the skin. *J. Investig. Dermatol.* **2011**, *131*, 1974–1980. [CrossRef] [PubMed]
24. Williams, M.R.; Gallo, R.L. Evidence that human skin microbiome dysbiosis promotes atopic dermatitis. *J. Investig. Dermatol.* **2017**, *137*, 2460–2461. [CrossRef] [PubMed]
25. Totté, J.; Van Der Feltz, W.; Hennekam, M.; van Belkum, A.; Van Zuuren, E.; Pasmans, S. Prevalence and odds of Staphylococcus aureus carriage in atopic dermatitis: A systematic review and meta-analysis. *Br. J. Dermatol.* **2016**, *175*, 687–695. [CrossRef] [PubMed]
26. Geoghegan, J.A.; Irvine, A.D.; Foster, T.J. Staphylococcus aureus and atopic dermatitis: A complex and evolving relationship. *Trends Microbiol.* **2018**, *26*, 484–497. [CrossRef]
27. Kong, H.H.; Oh, J.; Deming, C.; Conlan, S.; Grice, E.A.; Beatson, M.A.; Nomicos, E.; Polley, E.C.; Komarow, H.D.; Murray, P.R. Temporal shifts in the skin microbiome associated with disease flares and treatment in children with atopic dermatitis. *Genome Res.* **2012**, *22*, 850–859. [CrossRef]
28. Dekio, I.; Sakamoto, M.; Hayashi, H.; Amagai, M.; Suematsu, M.; Benno, Y. Characterization of skin microbiota in patients with atopic dermatitis and in normal subjects using 16S rRNA gene-based comprehensive analysis. *J. Med. Microbiol.* **2007**, *56*, 1675–1683. [CrossRef]
29. Oh, J.; Freeman, A.F.; Park, M.; Sokolic, R.; Candotti, F.; Holland, S.M.; Segre, J.A.; Kong, H.H. The altered landscape of the human skin microbiome in patients with primary immunodeficiencies. *Genome Res.* **2013**, *23*, 2103–2114. [CrossRef]
30. Chng, K.R.; Tay, A.S.L.; Li, C.; Ng, A.H.Q.; Wang, J.; Suri, B.K.; Matta, S.A.; McGovern, N.; Janela, B.; Wong, X.F.C.C. Whole metagenome profiling reveals skin microbiome-dependent susceptibility to atopic dermatitis flare. *Nat. Microbiol.* **2016**, *1*, 16106. [CrossRef]
31. Bjerre, R.; Bandier, J.; Skov, L.; Engstrand, L.; Johansen, J. The role of the skin microbiome in atopic dermatitis: A systematic review. *Br. J. Dermatol.* **2017**, *177*, 1272–1278. [CrossRef]
32. Wollina, U. Microbiome in atopic dermatitis. *Clin. Cosmet. Investig. Dermatol.* **2017**, *10*, 51. [CrossRef] [PubMed]
33. Gelmetti, C.; Colonna, C. The value of SCORAD and beyond. Towards a standardized evaluation of severity? *Allergy* **2004**, *59* (Suppl. 78), 61–65. [CrossRef]
34. Wollenberg, A.; Barbarot, S.; Bieber, T.; Christen-Zaech, S.; Deleuran, M.; Fink-Wagner, A.; Gieler, U.; Girolomoni, G.; Lau, S.; Muraro, A. Consensus-based European guidelines for treatment of atopic eczema (atopic dermatitis) in adults and children: Part II. *J. Eur. Acad. Dermatol. Venereol.* **2018**, *32*, 850–878. [CrossRef] [PubMed]
35. Chun, P.I.F.; Lehman, H. Current and Future Monoclonal Antibodies in the Treatment of Atopic Dermatitis. *Clin. Rev. Allergy Immunol.* **2020**, *59*, 208–219. [CrossRef] [PubMed]
36. Boguniewicz, M. Biologics for Atopic Dermatitis. *Immunol. Allergy Clin.* **2020**, *40*, 593–607. [CrossRef] [PubMed]
37. Katoh, N.; Kataoka, Y.; Saeki, H.; Hide, M.; Kabashima, K.; Etoh, T.; Igarashi, A.; Imafuku, S.; Kawashima, M.; Ohtsuki, M. Efficacy and safety of dupilumab in Japanese adults with moderate-to-severe atopic dermatitis: A subanalysis of three clinical trials. *Br. J. Dermatol.* **2020**, *183*, 39–51. [CrossRef] [PubMed]

38. Newsom, M.; Bashyam, A.M.; Balogh, E.A.; Feldman, S.R.; Strowd, L.C. New and Emerging Systemic Treatments for Atopic Dermatitis. *Drugs* **2020**, *1*, 1–12.
39. Chatterjee, S.; Hui, P.C.-L.; Wat, E.; Kan, C.-W.; Leung, P.-C.; Wang, W. Drug delivery system of dual-responsive PF127 hydrogel with polysaccharide-based nano-conjugate for textile-based transdermal therapy. *Carbohydr. Polym.* **2020**, *236*, 116074. [CrossRef]
40. Eichenfield, L.F.; Tom, W.L.; Berger, T.G.; Krol, A.; Paller, A.S.; Schwarzenberger, K.; Bergman, J.N.; Chamlin, S.L.; Cohen, D.E.; Cooper, K.D. Guidelines of care for the management of atopic dermatitis: Section 2. Management and treatment of atopic dermatitis with topical therapies. *J. Am. Acad. Dermatol.* **2014**, *71*, 116–132. [CrossRef]
41. Shi, K.; Lio, P.A. Alternative treatments for atopic dermatitis: An update. *Am. J. Clin. Dermatol.* **2019**, *20*, 251–266. [CrossRef]
42. Olle, B. Medicines from microbiota. *Nat. Biotechnol.* **2013**, *31*, 309–315. [CrossRef] [PubMed]
43. Cinque, B.; La Torre, C.; Melchiorre, E.; Marchesani, G.; Zoccali, G.; Palumbo, P.; Di Marzio, L.; Masci, A.; Mosca, L.; Mastromarino, P. Use of probiotics for dermal applications. In *Probiotics*; Springer: Berlin, Germany, 2011; pp. 221–241.
44. Lew, L.; Liong, M. Bioactives from probiotics for dermal health: Functions and benefits. *J. Appl. Microbiol.* **2013**, *114*, 1241–1253. [CrossRef] [PubMed]
45. Hendricks, A.J.; Mills, B.W.; Shi, V.Y. Skin bacterial transplant in atopic dermatitis: Knowns, unknowns and emerging trends. *J. Dermatol. Sci.* **2019**, *95*, 56–61. [CrossRef] [PubMed]
46. Verrucci, M.; Iacobino, A.; Fattorini, L.; Marcoaldi, R.; Maggio, A.; Piccaro, G. Use of probiotics in medical devices applied to some common pathologies. *Ann. dell'Ist. Super. Sanità* **2019**, *55*, 380–385.
47. Blandino, G.; Fazio, D.; Di Marco, R. Probiotics: Overview of microbiological and immunological characteristics. *Expert Rev. Anti-Infect. Ther.* **2008**, *6*, 497–508. [CrossRef] [PubMed]
48. Knackstedt, R.; Knackstedt, T.; Gatherwright, J. The role of topical probiotics on skin conditions: A systematic review of animal and human studies and implications for future therapies. *Exp. Dermatol.* **2019**, *29*, 15–21. [CrossRef]
49. Kim, I.S.; Lee, S.H.; Kwon, Y.M.; Adhikari, B.; Kim, J.A.; Yu, D.Y.; Kim, G.I.; Lim, J.M.; Kim, S.H.; Lee, S.S. Oral Administration of β-Glucan and Lactobacillus plantarum Alleviates Atopic Dermatitis-Like Symptoms. *J. Microbiol. Biotechnol.* **2019**, *29*, 1693–1706. [CrossRef]
50. Navarro-López, V.; Ramírez-Boscá, A.; Ramón-Vidal, D.; Ruzafa-Costas, B.; Genovés-Martínez, S.; Chenoll-Cuadros, E.; Carrión-Gutiérrez, M.; de la Parte, J.H.; Prieto-Merino, D.; Codoñer-Cortés, F.M. Effect of oral administration of a mixture of probiotic strains on SCORAD index and use of topical steroids in young patients with moderate atopic dermatitis: A randomized clinical trial. *JAMA Dermatol.* **2018**, *154*, 37–43. [CrossRef]
51. Holowacz, S.; Guinobert, I.; Guilbot, A.; Hidalgo, S.; Bisson, J. A Mixture of Five Bacterial Strains Attenuates Skin Inflammation in Mice. *Anti-Inflamm. Anti-Allergy Agents Med. Chem.* **2018**, *17*, 125–137. [CrossRef]
52. Lim, S.K.; Kwon, M.-S.; Lee, J.; Oh, Y.J.; Jang, J.-Y.; Lee, J.-H.; Park, H.W.; Nam, Y.-D.; Seo, M.-J.; Roh, S.W. Weissella cibaria WIKIM28 ameliorates atopic dermatitis-like skin lesions by inducing tolerogenic dendritic cells and regulatory T cells in BALB/c mice. *Sci. Rep.* **2017**, *7*, 1–9. [CrossRef]
53. Wang, I.J.; Wang, J.Y. Children with atopic dermatitis show clinical improvement after Lactobacillus exposure. *Clin. Exp. Allergy* **2015**, *45*, 779–787. [CrossRef] [PubMed]
54. Niccoli, A.A.; Artesi, A.L.; Candio, F.; Ceccarelli, S.; Cozzali, R.; Ferraro, L.; Fiumana, D.; Mencacci, M.; Morlupo, M.; Pazzelli, P. Preliminary results on clinical effects of probiotic Lactobacillus salivarius LS01 in children affected by atopic dermatitis. *J. Clin. Gastroenterol.* **2014**, *48*, S34–S36. [CrossRef] [PubMed]
55. Drago, L.; De Vecchi, E.; Toscano, M.; Vassena, C.; Altomare, G.; Pigatto, P. Treatment of atopic dermatitis eczema with a high concentration of Lactobacillus salivarius LS01 associated with an innovative gelling complex: A pilot study on adults. *J. Clin. Gastroenterol.* **2014**, *48*, S47–S51. [CrossRef] [PubMed]
56. Piqué, N.; Berlanga, M.; Miñana-Galbis, D. Health benefits of heat-killed (Tyndallized) probiotics: An overview. *Int. J. Mol. Sci.* **2019**, *20*, 2534. [CrossRef] [PubMed]
57. Rosignoli, C.; Thibaut de Ménonville, S.; Orfila, D.; Béal, M.; Bertino, B.; Aubert, J.; Mercenier, A.; Piwnica, D. A topical treatment containing heat-treated Lactobacillus johnsonii NCC 533 reduces Staphylococcus aureus adhesion and induces antimicrobial peptide expression in an in vitro reconstructed human epidermis model. *Exp. Dermatol.* **2018**, *27*, 358–365. [CrossRef] [PubMed]

58. Blanchet-Réthoré, S.; Bourdès, V.; Mercenier, A.; Haddar, C.H.; Verhoeven, P.O.; Andres, P. Effect of a lotion containing the heat-treated probiotic strain Lactobacillus johnsonii NCC 533 on Staphylococcus aureus colonization in atopic dermatitis. *Clin. Cosmet. Investig. Dermatol.* **2017**, *10*, 249. [CrossRef] [PubMed]
59. Seité, S.; Zelenkova, H.; Martin, R. Clinical efficacy of emollients in atopic dermatitis patients–relationship with the skin microbiota modification. *Clin. Cosmet. Investig. Dermatol.* **2017**, *10*, 25. [CrossRef]
60. Bodemer, C.; Guillet, G.; Cambazard, F.; Boralevi, F.; Ballarini, S.; Milliet, C.; Bertuccio, P.; La Vecchia, C.; Bach, J.-F.; de Prost, Y. Adjuvant treatment with the bacterial lysate (OM-85) improves management of atopic dermatitis: A randomized study. *PLoS ONE* **2017**, *12*, e0161555. [CrossRef]
61. Kim, H.; Kim, H.R.; Kim, N.-R.; Jeong, B.J.; Lee, J.S.; Jang, S.; Chung, D.K. Oral administration of Lactobacillus plantarum lysates attenuates the development of atopic dermatitis lesions in mouse models. *J. Microbiol.* **2015**, *53*, 47–52. [CrossRef]
62. Kim, H.; Kim, H.R.; Jeong, B.J.; Lee, S.S.; Kim, T.-R.; Jeong, J.H.; Lee, M.; Lee, S.; Lee, J.S.; Chung, D.K. Effects of oral intake of kimchi-derived Lactobacillus plantarum K8 lysates on skin moisturizing. *J. Microbiol. Biotechnol.* **2015**, *25*, 74–80. [CrossRef]
63. Lau, S. Oral application of bacterial lysate in infancy diminishes the prevalence of atopic dermatitis in children at risk for atopy. *Benef. Microbes* **2014**, *5*, 147–149. [CrossRef] [PubMed]
64. Dunstan, J.; Brothers, S.; Bauer, J.; Hodder, M.; Jaksic, M.; Asher, M.; Prescott, S. The effects of Mycobacteria vaccae derivative on allergen-specific responses in children with atopic dermatitis. *Clin. Exp. Immunol.* **2011**, *164*, 321–329. [CrossRef] [PubMed]
65. Gupta, C.; Prakash, D.; Gupta, S. Natural useful therapeutic products from microbes. *Microbiol. Exp.* **2014**, *1*, 00006. [CrossRef]
66. Woo, T.E.; Sibley, C.D. The emerging utility of the cutaneous microbiome in the treatment of acne and atopic dermatitis. *J. Am. Acad. Dermatol.* **2019**. [CrossRef] [PubMed]
67. O'Sullivan, J.N.; Rea, M.C.; O'Connor, P.M.; Hill, C.; Ross, R.P. Human skin microbiota is a rich source of bacteriocin-producing staphylococci that kill human pathogens. *FEMS Microbiol. Ecol.* **2019**, *95*, fiy241. [CrossRef] [PubMed]
68. Jang, I.-T.; Yang, M.; Kim, H.-J.; Park, J.-K. Novel Cytoplasmic Bacteriocin Compounds Derived from Staphylococcus epidermidis Selectively Kill Staphylococcus aureus, Including Methicillin-Resistant Staphylococcus aureus (MRSA). *Pathogens* **2020**, *9*, 87. [CrossRef]
69. Noda, M.; Sultana, N.; Hayashi, I.; Fukamachi, M.; Sugiyama, M. Exopolysaccharide Produced by Lactobacillus paracasei IJH-SONE68 Prevents and Improves the Picryl Chloride-Induced Contact Dermatitis. *Molecules* **2019**, *24*, 2970. [CrossRef]
70. Li, H.; Goh, B.N.; Teh, W.K.; Jiang, Z.; Goh, J.P.Z.; Goh, A.; Wu, G.; Hoon, S.S.; Raida, M.; Camattari, A. Skin commensal Malassezia globosa secreted protease attenuates Staphylococcus aureus biofilm formation. *J. Investig. Dermatol.* **2018**, *138*, 1137–1145. [CrossRef]
71. Mangano, K.; Vergalito, F.; Mammana, S.; Mariano, A.; De Pasquale, R.; Meloscia, A.; Bartollino, S.; Guerra, G.; Nicoletti, F.; Di Marco, R. Evaluation of hyaluronic acid-P40 conjugated cream in a mouse model of dermatitis induced by oxazolone. *Exp. Ther. Med.* **2017**, *14*, 2439–2444. [CrossRef]
72. Matsui, K.; Tachioka, K.; Onodera, K.; Ikeda, R. Topical application of josamycin inhibits development of atopic dermatitis-like skin lesions in NC/Nga mice. *J. Pharm. Pharm. Sci.* **2017**, *20*, 38–47. [CrossRef]
73. Zipperer, A.; Konnerth, M.C.; Laux, C.; Berscheid, A.; Janek, D.; Weidenmaier, C.; Burian, M.; Schilling, N.A.; Slavetinsky, C.; Marschal, M. Human commensals producing a novel antibiotic impair pathogen colonization. *Nature* **2016**, *535*, 511–516. [CrossRef] [PubMed]
74. Kim, M.-S.; Kim, J.-E.; Yoon, Y.-S.; Kim, T.H.; Seo, J.-G.; Chung, M.-J.; Yum, D.-Y. Improvement of atopic dermatitis-like skin lesions by IL-4 inhibition of P14 protein isolated from Lactobacillus casei in NC/Nga mice. *Appl. Microbiol. Biotechnol.* **2015**, *99*, 7089–7099. [CrossRef] [PubMed]
75. Abbasi, J. Are bacteria transplants the future of eczema therapy? *JAMA* **2018**, *320*, 1094–1095. [CrossRef] [PubMed]
76. Perin, B.; Addetia, A.; Qin, X. Transfer of skin microbiota between two dissimilar autologous microenvironments: A pilot study. *PLoS ONE* **2019**, *14*, e0226857. [CrossRef] [PubMed]
77. Myles, I.A.; Earland, N.J.; Anderson, E.D.; Moore, I.N.; Kieh, M.D.; Williams, K.W.; Saleem, A.; Fontecilla, N.M.; Welch, P.A.; Darnell, D.A. First-in-human topical microbiome transplantation with Roseomonas mucosa for atopic dermatitis. *JCI Insight* **2018**, *3*, e120608. [CrossRef]

78. Wang, J.; Hui, P.; Kan, C.-W. Functionalized Textile Based Therapy for the Treatment of Atopic Dermatitis. *Coatings* **2017**, *7*, 82. [CrossRef]
79. Kakkar, V.; Saini, K. Scope of nano delivery for atopic dermatitis. *Ann. Pharmacol. Pharm.* **2017**, *2*, 1038.
80. Patra, J.K.; Das, G.; Fraceto, L.F.; Campos, E.V.R.; del Pilar Rodriguez-Torres, M.; Acosta-Torres, L.S.; Diaz-Torres, L.A.; Grillo, R.; Swamy, M.K.; Sharma, S.J. Nano based drug delivery systems: Recent developments and future prospects. *J. Nanobiotechnol.* **2018**, *16*, 71. [CrossRef]
81. Souto, E.B.; Dias-Ferreira, J.; Oliveira, J.; Sanchez-Lopez, E.; Lopez-Machado, A.; Espina, M.; Garcia, M.L.; Souto, S.B.; Martins-Gomes, C.; Silva, A.M. Trends in Atopic Dermatitis—From Standard Pharmacotherapy to Novel Drug Delivery Systems. *Int. J. Mol. Sci.* **2019**, *20*, 5659. [CrossRef]
82. Gupta, M.; Agrawal, U.; Vyas, S.P. Nanocarrier-based topical drug delivery for the treatment of skin diseases. *Expert Opin. Drug Deliv.* **2012**, *9*, 783–804. [CrossRef]
83. Dubey, V.; Mishra, D.; Dutta, T.; Nahar, M.; Saraf, D.; Jain, N. Dermal and transdermal delivery of an anti-psoriatic agent via ethanolic liposomes. *J. Control. Release* **2007**, *123*, 148–154. [CrossRef] [PubMed]
84. Puglia, C.; Bonina, F. Lipid nanoparticles as novel delivery systems for cosmetics and dermal pharmaceuticals. *Expert Opin. Drug Deliv.* **2012**, *9*, 429–441. [CrossRef] [PubMed]
85. Damiani, G.; Eggenhöffner, R.; Pigatto, P.D.M.; Bragazzi, N.L. Nanotechnology meets atopic dermatitis: Current solutions, challenges and future prospects. Insights and implications from a systematic review of the literature. *Bioact. Mater.* **2019**, *4*, 380–386. [CrossRef] [PubMed]
86. Khan, I.; Saeed, K.; Khan, I. Nanoparticles: Properties, applications and toxicities. *Arab. J. Chem.* **2019**, *12*, 908–931. [CrossRef]
87. Schäfer-Korting, M.; Mehnert, W.; Korting, H.-C. Lipid nanoparticles for improved topical application of drugs for skin diseases. *Adv. Drug Deliv. Rev.* **2007**, *59*, 427–443. [CrossRef]
88. Md, S.; Kuldeep Singh, J.K.A.P.; Waqas, M.; Pandey, M.; Choudhury, H.; Habib, H.; Hussain, F.; Hussain, Z. Nanoencapsulation of betamethasone valerate using high pressure homogenization–solvent evaporation technique: Optimization of formulation and process parameters for efficient dermal targeting. *Drug Dev. Ind. Pharm.* **2019**, *45*, 323–332. [CrossRef]
89. Kang, J.-H.; Chon, J.; Kim, Y.-I.; Lee, H.-J.; Oh, D.-W.; Lee, H.-G.; Han, C.-S.; Kim, D.-W.; Park, C.-W. Preparation and evaluation of tacrolimus-loaded thermosensitive solid lipid nanoparticles for improved dermal distribution. *Int. J. Nanomed.* **2019**, *14*, 5381. [CrossRef]
90. Pandey, M.; Choudhury, H.; Gunasegaran, T.A.; Nathan, S.S.; Md, S.; Gorain, B.; Tripathy, M.; Hussain, Z. Hyaluronic acid-modified betamethasone encapsulated polymeric nanoparticles: Fabrication, characterisation, in vitro release kinetics, and dermal targeting. *Drug Deliv. Transl. Res.* **2019**, *9*, 520–533. [CrossRef]
91. Siddique, M.I.; Katas, H.; Jamil, A.; Amin, M.C.I.M.; Ng, S.-F.; Zulfakar, M.H.; Nadeem, S.M. Potential treatment of atopic dermatitis: Tolerability and safety of cream containing nanoparticles loaded with hydrocortisone and hydroxytyrosol in human subjects. *Drug Deliv. Transl. Res.* **2019**, *9*, 469–481. [CrossRef]
92. Yu, K.; Wang, Y.; Wan, T.; Zhai, Y.; Cao, S.; Ruan, W.; Wu, C.; Xu, Y. Tacrolimus nanoparticles based on chitosan combined with nicotinamide: Enhancing percutaneous delivery and treatment efficacy for atopic dermatitis and reducing dose. *Int. J. Nanomed.* **2018**, *13*, 129. [CrossRef]
93. Nagaich, U.; Gulati, N. Preclinical assessment of steroidal nanostructured lipid carriers based gels for atopic dermatitis: Optimization and product development. *Curr. Drug Deliv.* **2018**, *15*, 641–651. [CrossRef] [PubMed]
94. Zhuo, F.; Abourehab, M.A.; Hussain, Z.J.C.P. Hyaluronic acid decorated tacrolimus-loaded nanoparticles: Efficient approach to maximize dermal targeting and anti-dermatitis efficacy. *Carbohydr. Polym.* **2018**, *197*, 478–489. [CrossRef] [PubMed]
95. Radbruch, M.; Pischon, H.; Ostrowski, A.; Volz, P.; Brodwolf, R.; Neumann, F.; Unbehauen, M.; Kleuser, B.; Haag, R.; Ma, N. Dendritic core-multishell nanocarriers in murine models of healthy and atopic skin. *Nanoscale Res. Lett.* **2017**, *12*, 1–12. [CrossRef] [PubMed]
96. Siddique, M.I.; Katas, H.; Amin, M.C.I.M.; Ng, S.-F.; Zulfakar, M.H.; Jamil, A. In-vivo dermal pharmacokinetics, efficacy, and safety of skin targeting nanoparticles for corticosteroid treatment of atopic dermatitis. *Int. J. Pharm.* **2016**, *507*, 72–82. [CrossRef]

97. Hussain, Z.; Katas, H.; Amin, M.C.I.M.; Kumolosasi, E. Efficient immuno-modulation of TH1/TH2 biomarkers in 2, 4-dinitrofluorobenzene-induced atopic dermatitis: Nanocarrier-mediated transcutaneous co-delivery of anti-inflammatory and antioxidant drugs. *PLoS ONE* **2014**, *9*, e113143. [CrossRef]
98. Hussain, Z.; Katas, H.; Amin, M.C.I.M.; Kumolosasi, E.; Sahudin, S. Downregulation of immunological mediators in 2, 4-dinitrofluorobenzene-induced atopic dermatitis-like skin lesions by hydrocortisone-loaded chitosan nanoparticles. *Int. J. Nanomed.* **2014**, *9*, 5143.
99. Peralta, M.F.; Guzmán, M.L.; Pérez, A.; Apezteguia, G.A.; Fórmica, M.L.; Romero, E.L.; Olivera, M.E.; Carrer, D.C. Liposomes can both enhance or reduce drugs penetration through the skin. *Sci. Rep.* **2018**, *8*, 1–11. [CrossRef]
100. Godin, B.; Touitou, E. Ethosomes: New prospects in transdermal delivery. *Crit. Rev. Ther. Drug Carr. Syst.* **2003**, *20*, 63–102. [CrossRef]
101. Benson, H.A. Vesicles for transdermal delivery of peptides and proteins. In *Percutaneous Penetration Enhancers Chemical Methods in Penetration Enhancement*; Springer: Berlin, Germany, 2016; pp. 297–307.
102. Carreras, J.J.; Tapia-Ramirez, W.E.; Sala, A.; Guillot, A.J.; Garrigues, T.M.; Melero, A. Ultraflexible lipid vesicles allow topical absorption of cyclosporin A. *Drug Deliv. Transl. Res.* **2019**, *24*, 1–12. [CrossRef]
103. Akhtar, N.; Verma, A.; Pathak, K. Investigating the penetrating potential of nanocomposite β-cycloethosomes: Development using central composite design, in vitro and ex vivo characterization. *J. Liposome Res.* **2018**, *28*, 35–48. [CrossRef]
104. Augustin, M.; Goepel, L.; Jacobi, A.; Bosse, B.; Mueller, S.; Hopp, M. Efficacy and tolerability of liposomal polyvinylpyrrolidone-iodine hydrogel for the localized treatment of chronic infective, inflammatory, dermatoses: An uncontrolled pilot study. *Clin. Cosmet. Investig. Dermatol.* **2017**, *10*, 373. [CrossRef] [PubMed]
105. Akhtar, N.; Verma, A.; Pathak, K. Feasibility of binary composition in development of nanoethosomal glycolic vesicles of triamcinolone acetonide using Box-behnken design: In vitro and ex vivo characterization. *Artif. Cells Nanomed. Biotechnol.* **2017**, *45*, 1123–1131. [CrossRef] [PubMed]
106. Goindi, S.; Dhatt, B.; Kaur, A. Ethosomes-based topical delivery system of antihistaminic drug for treatment of skin allergies. *J. Microencapsul.* **2014**, *31*, 716–724. [CrossRef] [PubMed]
107. Goindi, S.; Kumar, G.; Kaur, A. Novel flexible vesicles based topical formulation of levocetirizine: In vivo evaluation using oxazolone-induced atopic dermatitis in murine model. *J. Liposome Res.* **2014**, *24*, 249–257. [CrossRef] [PubMed]
108. Goddard, A.L.; Lio, P.A. Alternative, complementary, and forgotten remedies for atopic dermatitis. *Evid. Based Complement. Altern. Med.* **2015**, *2015*, 676897. [CrossRef] [PubMed]
109. Kim, J.; Kim, B.E.; Leung, D.Y. Pathophysiology of atopic dermatitis: Clinical implications. *Proc. Allergy Asthma Proc.* **2019**, *40*, 84–92. [CrossRef]
110. Kakkar, V.; Kumar, M.; Saini, K. An Overview of Atopic Dermatitis with a Focus on Nano-Interventions. *Innovations* **2019**, *1*, 2019.
111. Schneider, L.; Tilles, S.; Lio, P.; Boguniewicz, M.; Beck, L.; LeBovidge, J.; Novak, N.; Bernstein, D.; Blessing-Moore, J.; Khan, D. Atopic dermatitis: A practice parameter update 2012. *J. Allergy Clin. Immunol.* **2013**, *131*, 295–299.e227. [CrossRef]
112. Ring, J.; Alomar, A.; Bieber, T.; Deleuran, M.; Fink-Wagner, A.; Gelmetti, C.; Gieler, U.; Lipozencic, J.; Luger, T.; Oranje, A. Guidelines for treatment of atopic eczema (atopic dermatitis) part I. *J. Eur. Acad. Dermatol. Venereol.* **2012**, *26*, 1045–1060. [CrossRef]
113. Silverberg, J.I. Public health burden and epidemiology of atopic dermatitis. *Dermatol. Clin.* **2017**, *35*, 283–289. [CrossRef]
114. Balato, A.; Cacciapuoti, S.; Caprio, R.; Marasca, C.; Masarà, A.; Raimondo, A.; Fabbrocini, G. Human Microbiome: Composition and Role in Inflammatory Skin Diseases. *Arch. Immunol. Ther. Exp.* **2018**. [CrossRef] [PubMed]
115. Lacour, J.-P. Skin microbiota and atopic dermatitis: Toward new therapeutic options? In Proceedings of Annales de dermatologie et de venereologie. *Ann. Dermatol. Venereol.* **2015**, *142*, S18–S22. [CrossRef]
116. Kim, S.-O.; Ah, Y.-M.; Yu, Y.M.; Choi, K.H.; Shin, W.-G.; Lee, J.-Y. Effects of probiotics for the treatment of atopic dermatitis: A meta-analysis of randomized controlled trials. *Ann. Allergy Asthma Immunol.* **2014**, *113*, 217–226. [CrossRef] [PubMed]

117. Kano, H.; Kita, J.; Makino, S.; Ikegami, S.; Itoh, H. Oral administration of Lactobacillus delbrueckii subspecies bulgaricus OLL1073R-1 suppresses inflammation by decreasing interleukin-6 responses in a murine model of atopic dermatitis. *J. Dairy Sci.* **2013**, *96*, 3525–3534. [CrossRef]
118. Brandwein, M.; Fuks, G.; Israel, A.; Sabbah, F.; Hodak, E.; Szitenberg, A.; Harari, M.; Steinberg, D.; Bentwich, Z.; Shental, N. Skin Microbiome Compositional Changes in Atopic Dermatitis Accompany Dead Sea Climatotherapy. *Photochem. Photobiol.* **2019**, *95*, 1446–1453. [CrossRef]
119. Baviera, G.; Leoni, M.C.; Capra, L.; Cipriani, F.; Longo, G.; Maiello, N.; Ricci, G.; Galli, E. Microbiota in healthy skin and in atopic eczema. *BioMed Res. Int.* **2014**, *2014*, 436921. [CrossRef]
120. Paller, A.S.; Kong, H.H.; Seed, P.; Naik, S.; Scharschmidt, T.C.; Gallo, R.L.; Luger, T.; Irvine, A.D. The microbiome in patients with atopic dermatitis. *J. Allergy Clin. Immunol.* **2019**, *143*, 26–35. [CrossRef]
121. Okuda, K.-I.; Zendo, T.; Sugimoto, S.; Iwase, T.; Tajima, A.; Yamada, S.; Sonomoto, K.; Mizunoe, Y. Effects of bacteriocins on methicillin-resistant Staphylococcus aureus biofilm. *Antimicrob. Agents Chemother.* **2013**, *57*, 5572–5579. [CrossRef]
122. Sun, L.; Liu, Z.; Cun, D.; HY Tong, H.; Zheng, Y. Application of nano-and micro-particles on the topical therapy of skin-related immune disorders. *Curr. Pharm. Des.* **2015**, *21*, 2643–2667. [CrossRef]
123. Okada, H. Drug discovery by formulation design and innovative drug delivery systems (DDS). *Yakugaku Zasshi J. Pharm. Soc. JPN* **2011**, *131*, 1271–1287. [CrossRef]
124. Directive, C. 93/42/EEC of 14 June 1993 Concerning Medical Devices. *Official Journal of the European Communities*, 12 July 1993; OJ L 169.
125. Gottlieb, S. Statement from FDA Commissioner Scott Gottlieb, MD, on the Agency's Scientific Evidence on the Presence of Opioid Compounds in Kratom, Underscoring Its Potential for Abuse. Silver Spring MD Food Drug Adm. 2018. Available online: https://www.fda.gov/news-events/press-announcements/statement-fda-commissioner-scott-gottlieb-md-agencys-scientific-evidence-presence-opioid-compounds (accessed on 22 November 2020).
126. Dickinson, A.M.; Godden, J.M.; Lanovyk, K.; Ahmed, S.S. Assessing the safety of nanomedicines: A mini review. *Appl. In Vitro Toxicol.* **2019**, *5*, 114–122. [CrossRef]
127. Palmer, B.C.; DeLouise, L.A. Nanoparticle-enabled transdermal drug delivery systems for enhanced dose control and tissue targeting. *Molecules* **2016**, *21*, 1719. [CrossRef] [PubMed]

Article

3-Pentylcatechol, a Non-Allergenic Urushiol Derivative, Displays Anti-*Helicobacter pylori* Activity In Vivo

Hang Yeon Jeong [1], Tae Ho Lee [1], Ju Gyeong Kim [1], Sueun Lee [2], Changjong Moon [2], Xuan Trong Truong [3], Tae-Il Jeon [3] and Jae-Hak Moon [1,*]

1 Department of Food Science and Technology, Chonnam National University, 77 Yongbongro, Gwangju 61186, Korea; wjdgkddus@naver.com (H.Y.J.); xogh9954@naver.com (T.H.L.); wnrud0610@naver.com (J.G.K.)

2 Department of Veterinary Anatomy, College of Veterinary Medicine and BK21 FOUR Program, Chonnam National University, Gwangju 61186, Korea; leese@kiom.re.kr (S.L.); moonc@chonnam.ac.kr (C.M.)

3 Department of Animal Science, Chonnam National University, Gwangju 61186, Korea; trongxuan.vp@gmail.com (X.T.T.); tjeon@jnu.ac.kr (T.-I.J.)

* Correspondence: nutrmoon@jnu.ac.kr; Tel.: +82-62-530-2141

Received: 26 October 2020; Accepted: 11 November 2020; Published: 13 November 2020

Abstract: We previously reported that 3-pentylcatechol (PC), a synthetic non-allergenic urushiol derivative, inhibited the growth of *Helicobacter pylori* in an in vitro assay using nutrient agar and broth. In this study, we aimed to investigate the in vivo antimicrobial activity of PC against *H. pylori* growing in the stomach mucous membrane. Four-week-old male C57BL/6 mice (n = 4) were orally inoculated with *H. pylori* Sydney Strain-1 (SS-1) for 8 weeks. Thereafter, the mice received PC (1, 5, and 15 mg/kg) and triple therapy (omeprazole, 0.7 mg/kg; metronidazole, 16.7 mg/kg; clarithromycin, 16.7 mg/kg, reference groups) once daily for 10 days. Infiltration of inflammatory cells in gastric tissue was greater in the *H. pylori*-infected group compared with the control group and lower in both the triple therapy- and PC-treated groups. In addition, upregulation of cytokine mRNA was reversed after infection, upon administration of triple therapy and PC. Interestingly, PC was more effective than triple therapy at all doses, even at 1/15th the dose of triple therapy. In addition, PC demonstrated synergism with triple therapy, even at low concentrations. The results suggest that PC may be more effective against *H. pylori* than established antibiotics.

Keywords: urushiol; 3-pentylcatechol; 3-pentadecylcatechol; *Helicobacter pylori*; antimicrobial; triple therapy

1. Introduction

Helicobacter pylori infection is a major public health concern worldwide. This infection occurs in the gastric mucosa of more than 50% of the world's population [1] and it is directly associated with gastrointestinal disorders, including chronic gastritis, peptic ulcer disease, mucosa-associated lymphoid tissue (MALT) lymphoma, and gastric cancer [2–5]. Gastric cancer is the second leading cause of cancer-related mortality worldwide, following only lung cancer [6]. Furthermore, *H. pylori* infection is also associated with numerous extra-gastric disorders, such as cardiovascular, neurologic, hematologic, dermatologic, head and neck, and urogynecologic diseases, as well as diabetes mellitus and metabolic syndrome [7,8].

The international gold-standard treatment for *H. pylori* infection is triple therapy, comprising two antibiotics (usually selected from clarithromycin, metronidazole, amoxicillin, and tetracycline) and a proton-pump inhibitor, for 7–14 days [9–11]. However, the success rates of these *H. pylori* eradication therapies are less than 80%, and the failure rate of *H. pylori* eradication therapy has increased, primarily due to increased antibiotic resistance [12–14]. Another reason for treatment failure is patient non-adherence, owing to the complexity of the treatment: it involves the repeat administration of at least three drugs over a long period [15]. In addition, these drugs are associated with several side effects, including abdominal pain, nausea, and diarrhea [16]. The high cost of *H. pylori* treatment may also be a disadvantage [15]. Therefore, there is an urgent need for the development of safe and effective therapeutic agents for *H. pylori* infections.

Lacquer tree (*Toxicodendron vernicifluum* (Stokes) F.A. Barkley, Anacardiaceae) has been used for thousands of years as a protective surface-coating material and in traditional medicine in China, Japan, and Korea [17]. It is particularly effective for treating gastrointestinal disorders, such as gastritis and gastric cancer [17]. Urushiols are a group of compounds with alkyl side chains comprising 15 or 17 carbon atoms at the C-3 position of catechol. They are the major constituents of lacquer tree sap, accounting for 60–70% of the total content [18]. In addition to their various biological activities [19–23], urushiols display antimicrobial activity against *H. pylori* [24]. However, urushiols can also cause serious contact dermatitis [25–27], which is a limitation associated with their use.

Previously, we chemically synthesized catechol-type urushiol derivatives with different alkyl side chain lengths of $–C_5H_{11}$, $–C_{10}H_{21}$, $–C_{15}H_{31}$, and $–C_{20}H_{41}$ at the C-3 position (Figure 1) [28]. Among these compounds, 3-decylcatechol ($-C_{10}H_{21}$) and 3-pentadecylcatechol (PDC, natural type, $–C_{15}H_{31}$) induced contact dermatitis; however, 3-pentylcatechol (PC, $–C_5H_{11}$) and 3-eicosylcatechol (EC, $–C_{20}H_{41}$) did not [28]. In addition, PC and EC exhibited strong antioxidative activity and high affinity for phospholipid membranes [28]. Notably, PC demonstrated enhanced antimicrobial effects in agar and broth cultures against various microorganisms involved in food spoilage and pathogenicity [29]. In addition, PC inhibited *H. pylori* to a greater extent than nalidixic acid, erythromycin, tetracycline, and ampicillin, which have been used in *H. pylori* eradication therapy [29]. Moreover, unlike PDC (Part I), PC was absorbed in the blood after oral administration [30,31]. Therefore, PC is expected to effectively eradicate *H. pylori* in gastric tissue. In this study, the in vivo antimicrobial activity of PC against *H. pylori* was evaluated and compared to that of triple therapy.

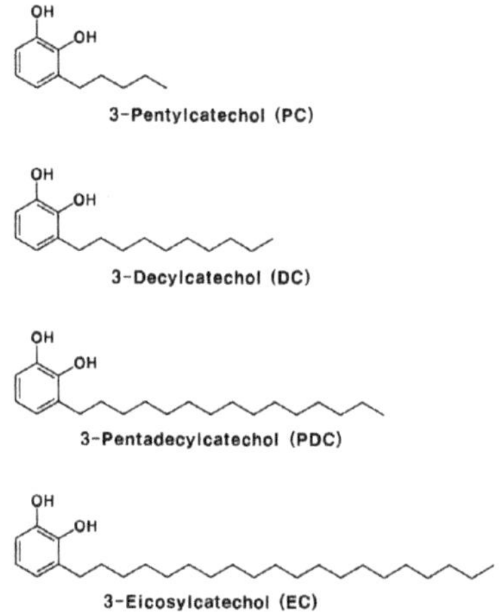

Figure 1. Structures of synthesized urushiol derivatives.

2. Results

2.1. Confirmation of Infection and Associated Gastric Disorders after H. pylori Inoculation

After 30 days of *H. pylori* Sydney Strain-1 (SS-1) inoculation, the colonization of *H. pylori* and associated gastric disorders in mouse gastric tissue were confirmed via quantitative polymerase chain reaction (qPCR) and histological analysis. The relative mRNA expression of the inflammatory cytokines, tumor necrosis factor alpha (Tnfα) and interleukin-1 beta (Il-1β) was upregulated to a greater extent in mice in the infected group than in mice in the control group (Figure 2). In addition, the expression of *H. pylori*-related genes, urease subunit alpha (ureA) and cytotoxin-associated gene A (cagA), was detected in mice in the infected group, but not in mice in the control group (Figure 2). Histological analysis revealed characteristics of gastritis, including inflammatory cell infiltration, erosion, and catarrhal inflammation in the gastric tissue of the infected group (Figure 3). These results indicate that *H. pylori* successfully colonized the stomachs of mice after inoculation and induced gastric disorders. Therefore, this animal model was used to investigate the in vivo antimicrobial activity of PC against *H. pylori*.

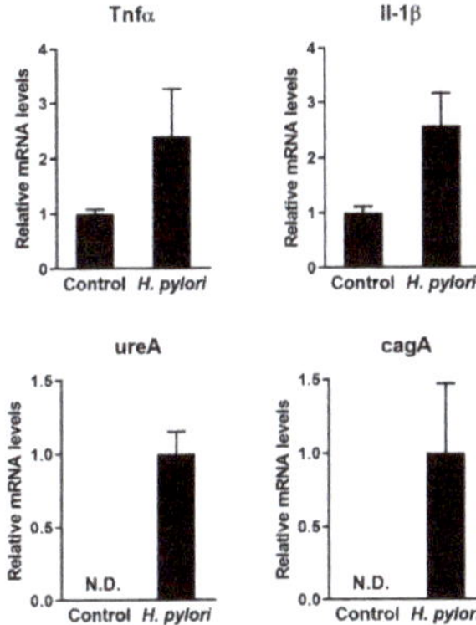

Figure 2. Expression levels of tumor necrosis factor alpha (Tnfα), interleukin-1 beta (Il-1β), urease subunit alpha (ureA), and cytotoxin-associated gene A (cagA) mRNA in mouse gastric tissue following *Helicobacter pylori* Sydney Strain-1 (SS-1) inoculation. *H. pylori* SS-1 was administered to C57BL/6 mice for 30 days. N.D., not detected.

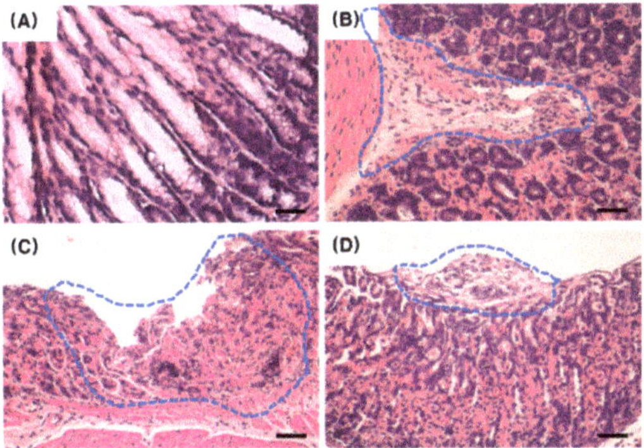

Figure 3. Histological analysis of hematoxylin and eosin-stained mouse gastric tissue after *H. pylori* SS-1 inoculation. *H. pylori* SS-1 was administered to C57BL/6 mice for 30 days. (**A**) uninfected control; (**B**) inflammatory cell infiltration (dotted line) in an *H. pylori*-infected mouse; (**C**) erosion (dotted line) in an *H. pylori*-infected mouse; (**D**) catarrhal inflammation (dotted line) in an *H. pylori*-infected mouse. Scale bar = 20 μm.

2.2. *Effect of PC on the Gastric Tissue Histology of H. pylori-Infected Mice*

We evaluated and graded the level of inflammatory cell infiltration in the gastric mucosa of *H. pylori*-infected mice via hematoxylin and eosin (H&E) staining (Figure 4). Grades of 0 to 3 were assigned, as follows: 0, normal; 1, mild; 2, moderate; 3, marked. All mice in the uninfected control group displayed a score of 0 (no infiltration of inflammatory cells), whereas those in the infected group displayed a score of 2 (moderate infiltration of inflammatory cells) and 3 (marked infiltration of inflammatory cells) in two mice each. The inflammation scores in all treatment groups were lower than those in the infected group. Interestingly, the scores were lower in mice treated with low doses of PC (1 and 5 mg/kg) compared with those treated with triple therapy.

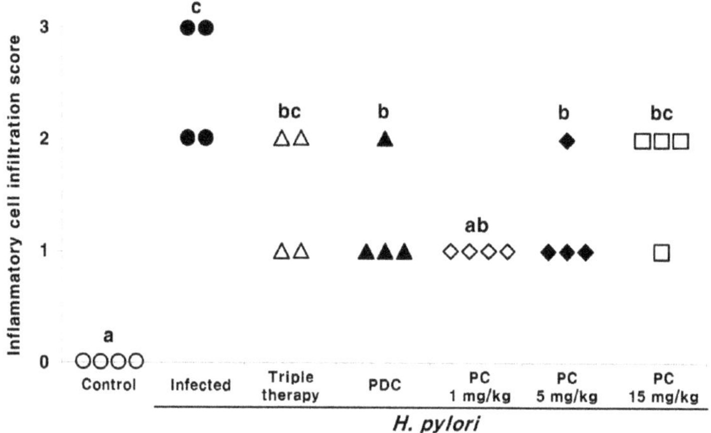

Figure 4. Effect of 3-pentylcatechol (PC) treatment on the histology of gastric tissue from *H. pylori*-infected mice (hematoxylin and eosin staining). Inflammatory cell infiltration was graded from 0 to 3: 0, normal; 1, mild; 2, moderate; 3, marked. ○, inflammatory cell infiltration score of control group; •, inflammatory cell infiltration score of *H. pylori*-infected group; Δ, inflammatory cell infiltration score of *H. pylori* + triple therapy-treated group; ▲, inflammatory cell infiltration score of *H. pylori* + 3-pentadecylcatechol (PDC)-treated group; ◊, inflammatory cell infiltration score of *H. pylori* + 1 mg/kg of PC-treated group; ♦, inflammatory cell infiltration score of *H. pylori* + 5 mg/kg of PC-treated group; □, inflammatory cell infiltration score of *H. pylori* + 15 mg/kg of PC-treated group. Different letters (a, b, and c) indicate a significant difference ($p < 0.05$), ascertained via the Tukey–Kramer test.

2.3. Effect of PC on H. pylori Eradication and Cytokine Expression

To assess the effect of PC therapy on *H. pylori* eradication, the mRNA expression of the *H. pylori* markers cagA, ureA, and neutrophil-activating protein A (napA) was assessed in pyloric antrum tissue via qPCR. As shown in Figure 5, all three *H. pylori*-related transcripts were detected in the infected mice, but not in the uninfected control mice. This suggests that PC can effectively eradicate *H. pylori*, even at a dose at 1/15th of the antibiotics used in triple therapy. This response was also observed when analyzing the mRNA expression of inflammatory cytokines Tnfα and Il-1β in the pyloric antrum tissue (Figure 6). The expression of both Tnfα and Il-1β was markedly upregulated in the infected group compared with the uninfected control group; however, these genes were significantly downregulated upon PC treatment and in the reference groups. Moreover, PC treatment reduced the levels of two inflammatory cytokines more efficiently than triple therapy. Notably, in mice treated with 1 and 5 mg/kg of PC, the mRNA expression of Tnfα and Il-1β was downregulated, similar to the observation in the uninfected control mice. These results suggest that PC effectively eradicates *H. pylori* in the gastric mucosa and also helps alleviate gastrointestinal disorders at much lower concentrations than conventional antibiotics.

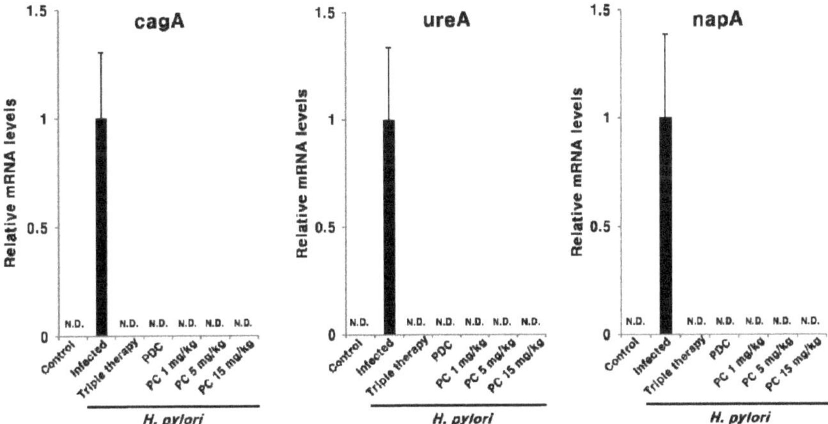

Figure 5. Effect of PC treatment on the expression of *H. pylori* cagA, ureA, and napA in the gastric tissue of the *H. pylori*-infected mice. N.D., not detected.

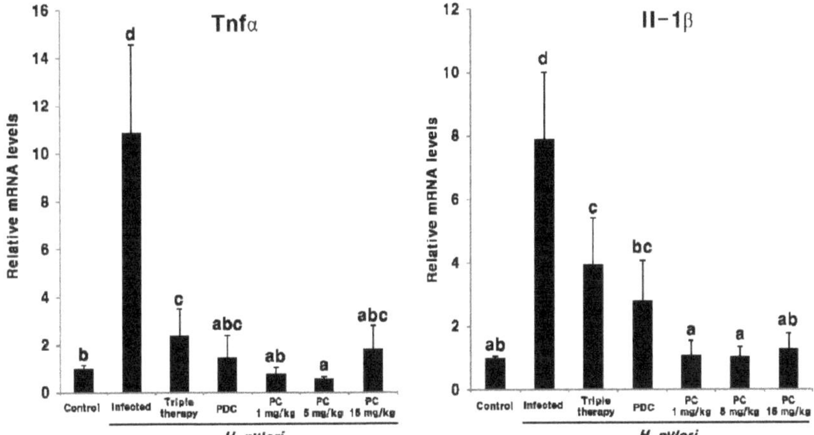

Figure 6. Effect of PC treatment on the expression of Tnfα and Il-1β mRNA in the gastric tissue of the *H. pylori*-infected mice. Different letters (a, b, c, and d) indicate a significant difference ($p < 0.05$), ascertained via the Tukey–Kramer test. PC, 3-pentylcatechol; PDC, 3-pentadecylcatechol.

2.4. Synergistic Effect of PC in Combination with Triple Therapy

Next, we evaluated the in vivo efficacy of PC in combination with triple therapy. The expression of *H. pylori*-related genes (cagA, ureA, and napA) was not completely suppressed in the triple therapy group when the antibiotic concentration was decreased (Figure 7). In contrast, when PC was administered with triple therapy, the expression of the *H. pylori*-related genes was not observed with all concentrations (Figure 7).

Next, we evaluated the synergistic effect of PC and triple therapy on the inflammatory response (Figure 8). When mice were treated with triple therapy alone, inflammation was not completely suppressed. However, when mice were treated with PC and triple therapy, cytokine expression decreased to a level similar to that observed in the uninfected control group. These results indicate that PC demonstrated synergism with conventional antibiotic therapy, suggesting that the use of antibiotics can be reduced in the treatment of *H. pylori*.

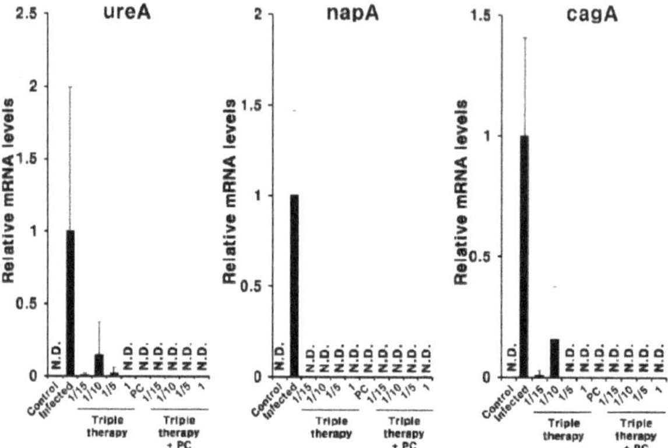

Figure 7. Expression of *H. pylori* ureA, napA, and cagA mRNA in the gastric tissue of the *H. pylori*-infected mice following combination treatment with PC and triple therapy. Different letters indicate a significant difference ($p < 0.05$), ascertained via the Tukey–Kramer test. Triple therapy was administered at four concentrations. 1, Existing concentration of triple therapy (metronidazole and clarithromycin: 16.7 mg/kg; omeprazole: 700 μg/kg); 1/5, one-fifth of the existing concentration of triple therapy; 1/10, one-tenth of the existing concentration of triple therapy; 1/15, one-fifteenth of the existing concentration of triple therapy. PC was administered at 1 mg/kg. N.D., not detected.

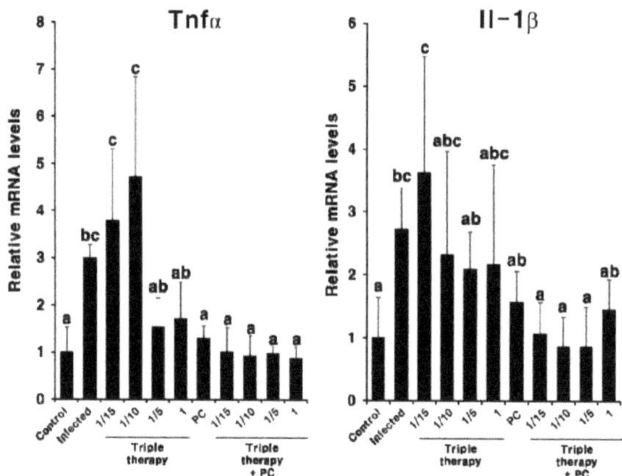

Figure 8. Expression levels of Tnfα and Il-1β mRNA in the gastric tissue of the *H. pylori*-infected mice after combination treatment with PC and triple therapy. Different letters (a, b, and c) indicate a significant difference ($p < 0.05$), ascertained via the Tukey–Kramer test. Triple therapy was administered at four concentrations. 1, Existing concentration of triple therapy (metronidazole and clarithromycin: 16.7 mg/kg; omeprazole: 700 μg/kg); 1/5, one-fifth of the existing concentration of triple therapy; 1/10, one-tenth of the existing concentration of triple therapy; 1/15, one-fifteenth of the existing concentration of triple therapy. PC was administered at 1 mg/kg.

2.5. Hepatotoxicity of PC

Plasma glutamate pyruvate transaminase (GPT) and glutamate oxaloacetate transaminase (GOT) levels were determined using commercial ELISA kits to evaluate the in vivo toxicity of PC after oral administration (Figure 9). No significant differences were observed between the PC-treated groups and the uninfected control group. These results indicate that PC does not cause liver toxicity.

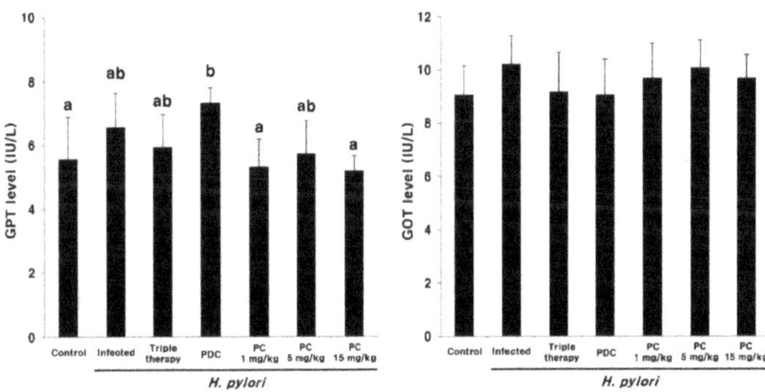

Figure 9. Plasma glutamate pyruvate transaminase (GPT) and glutamate oxaloacetate transaminase (GOT) levels after 3-pentylcatechol treatment. Different letters (a and b) indicate a significant difference ($p < 0.05$), ascertained via the Tukey–Kramer test.

3. Discussion

Urushiols are major constituents present in high concentrations in lacquer tree sap [18], with antimicrobial activity against *H. pylori* [24]. However, urushiols induce contact dermatitis [25–27], thereby limiting their application.

Previously, PC, a non-allergenic urushiol derivative (Figure 1), was chemically synthesized [28], and its antimicrobial activity against various food spoilage and pathogenic microorganisms was determined [29]. PC displayed marked antimicrobial effects in both agar and broth cultures [29]. In addition, PC demonstrated greater anti-*H. pylori* activity than nalidixic acid, erythromycin, tetracycline, and ampicillin, which have been widely used to eradicate *H. pylori* [29]. In the present study, we investigated the in vivo antimicrobial activity of PC against *H. pylori* and compared it with triple therapy, which is considered the international gold-standard treatment for *H. pylori* infections.

C57BL/6 mice were inoculated with *H. pylori* SS-1 to generate a model of *H. pylori* infection. In the pyloric antrum tissue, the increased expression of Tnfα and Il-1β mRNA, which are involved in *H. pylori*-induced inflammation [32], was more prominent in the infected group than in the control group (Figure 2). In addition, characteristics of gastritis were detected in the gastric tissue of the infected mice upon H&E staining (Figure 3).

To determine the in vivo antimicrobial activity of PC, mice were inoculated with *H. pylori* SS-1 for 60 days. Subsequently, three doses (1, 5, and 15 mg/kg) of PC were administered to the *H. pylori*-infected mice once daily for 10 days. Anti-*H. pylori* activity was compared between the PC-treated groups, the positive control group, the triple therapy group, and the group receiving PDC, a natural urushiol derivative. Mortality and inflammation upon *H. pylori* infection were assessed via qPCR and histological analysis of the pyloric antrum tissue, the major habitat of *H. pylori* [33].

Histological analysis of the gastric tissues following H&E staining (Figure 4) revealed that all uninfected mice appeared normal; in contrast, inflammatory cell infiltration increased in the infected group. Inflammation scores were reduced upon PC treatment, which was more effective than triple therapy (Figure 4).

The CagA toxin, encoded by cagA, is one of the most widely studied *H. pylori* virulence factors. The CagA effector protein is injected into host target cells via a type IV secretion system and is highly associated with inflammation and the development of gastric cancer [1]. The napA encodes the NapA protein, which activates neutrophils, prevents oxidative DNA damage [34], and regulates the adhesion of *H. pylori* to stomach mucin and host epithelial cells [35]. The ureA contributes to acid resistance in *H. pylori* via the production of ammonia through the enzymatic degradation of urea in the gastric environment [1]. *H. pylori*-related genes, cagA, ureA, and napA, were analyzed via qPCR to evaluate the extent of *H. pylori* eradication (Figure 5). All three genes were detected in the infected group only and not in the uninfected groups and those receiving treatment (Figure 5).

Therefore, these data indicate that *H. pylori* can be completely eradicated by PC at a much lower concentration than antibiotics. In addition, the expression of *H. pylori*-induced Tnfα and Il-1β mRNA was markedly downregulated following PC treatment (Figure 6). Moreover, the levels of these two inflammatory cytokines were effectively reduced in all the PC-treated groups compared with the triple therapy group (Figure 6).

A recent study showed that epidermal growth factor receptor signaling, implicated in gastric inflammation and carcinogenesis, remains activated following the eradication of *H. pylori* by antibiotics [36]. In addition, clarithromycin does not affect the expression of inflammatory markers in patients with atherosclerosis [37]. Knoop et al. (2016) reported that antibiotic therapy accelerates inflammation via the translocation of native intestinal bacteria [38]. Our results indicate that PC not only eradicates *H. pylori* but also improves *H. pylori*-induced gastritis. Although further studies are required to investigate the underlying mechanism of action, these results reflect the strong antioxidant activity and amphipathic structure of PC [28,39]. In addition, despite using a low concentration of PC, synergistic effects were observed with triple therapy (Figures 7 and 8). Thus, PC can markedly reduce the concentration of antibiotics used and can overcome issues associated with the misuse of antibiotics [16]. In addition, the poor treatment compliance of patients owing to the need to take large amounts of antibiotics, which is a major obstacle in the antibiotic treatment of *H. pylori* infections [15], can be improved.

The plasma levels of liver transaminases, GOT and GPT, are useful biomarkers of liver injury. These enzymes are released in the blood upon hepatocyte necrosis due to acute hepatitis, ischemic injury, or toxic injury [40]. In the present study, plasma GPT and GOT levels were determined following the oral administration of PC. No evidence of liver toxicity was observed following treatment with PC (Figure 9).

4. Materials and Methods

4.1. Chemicals

3-Pentylcatechol (PC) and 3-pentadecylcatechol (PDC) were chemically synthesized in accordance with our previous method [28]. Clarithromycin, metronidazole, and omeprazole were purchased from TCI Chemical Industry (Tokyo, Japan). All other chemicals and solvents were of analytical grade, unless specified otherwise.

4.2. H. pylori Strain and Culture Conditions

Mouse-adapted *H. pylori* Sydney Strain-1 (SS-1) was obtained from the Korean Culture Center of Microorganisms (KCCM, Seoul, Korea) and cultured on Columbia agar or in broth medium (MB cell, Seoul, Korea), containing 5% horse serum (Gibco, Gaithersburg, MD, USA). The culture was incubated at 37 °C in a 10% CO_2 incubator (MCO175, Sanyo, Osaka, Japan), and the bacteria were sub-cultured every 72 h [29]. Culture purity was assessed regularly.

4.3. Animals and Infection

All experimental procedures were approved by the Institutional Animal Care and Use Committee of Chonnam National University (no. CNU IACUC-YB-2012-26). Four-week-old C57BL/6 male mice were purchased from Samtako Bio Korea (Osan, Korea). Mice were reared in an environmentally controlled animal facility, operating on a 12:12 h dark/light cycle at 20 ± 1 °C and 55 ± 5% humidity, with ad libitum access to water and standard laboratory chow (Harlan Rodent diet, 2018S, by Samtako Bio Korea) [24].

Four mice per group were inoculated with *H. pylori* SS-1, which can effectively colonize the mouse gastric mucosa [41]. A total of 100 μL aliquots (10^8 CFU) of Columbia broth were administered to the mice for 60 days, three times every 2 days, using a zonde needle. After 30 days of inoculation, three mice were sacrificed to confirm infection. Blood was withdrawn from the abdominal aorta of the mice under light anesthesia (isoflurane) and collected in heparinized tubes. Plasma was obtained via centrifugation (2767× *g*, 4 °C, 15 min). The pyloric antrum of the stomach was harvested for quantitative polymerase chain reaction (qPCR) and histological analysis. Uninfected mice were administered the same volume of fresh Columbia broth; this group was considered the negative control. All samples were stored at −80 °C until use.

4.4. PC Treatment after H. pylori Infection

Following 60-day *H. pylori* inoculation, PC was administered to the infected mice with 100 μL of water once daily for 10 days [24]. Triple therapy and PDC, a natural form of urushiol, were used as reference groups. Infected mice were divided into seven experimental groups ($n = 4$): control group (uninfected, negative control); *H. pylori*-infected group; *H. pylori* infection + triple therapy treatment group; *H. pylori* infection + PDC 26.7 mg (83.3 μmol)/kg treatment group; *H. pylori* infection + PC 1 mg (5.6 μmol)/kg treatment group; *H. pylori* infection + PC 5 mg (27.8 μmol)/kg treatment group; and *H. pylori* infection + PC 15 mg (83.3 μmol)/kg treatment group. Triple therapy comprised omeprazole (700 μg/kg), metronidazole (16.7 mg/kg), and clarithromycin (16.7 mg/kg). After 10 days of treatment, the mice were euthanized and samples were harvested as described above.

To confirm the synergistic effect of PC with triple therapy, triple therapy was administered at four concentrations, as follows: existing concentration (metronidazole and clarithromycin: 16.7 mg/kg; omeprazole: 700 μg/kg), one-fifth, one-tenth, and one-fifteenth of the existing concentration. In contrast, PC was administered at the same concentration (1 mg/kg). In accordance with the above conditions, triple therapy and PC were administrated orally to the *H. pylori*-infected mice once daily for 5 days. The control and infection groups received distilled water under the same conditions. After 5 days of treatment, the mice were euthanized and samples were harvested, as described above.

4.5. Histological Examination

Gastric tissue was fixed in 4% (*w*/*v*) paraformaldehyde (PFA) in phosphate-buffered saline (PBS, pH 7.4) for 24 h, dehydrated in a graded ethanol series (70%, 80%, 90%, 95%, and 100%), cleared in xylene, embedded in paraffin, and sectioned into 5-μm-thick slices. Serial sections were stained with hematoxylin and eosin (H&E) and examined microscopically to determine whether the gastric mucosa contained any pathological lesions [42].

4.6. RNA Analysis

Total RNA was isolated from mouse gastric tissue using the TRI Reagent® (Molecular Research Center, Cincinnati, OH, USA). cDNA was synthesized using the ReverTra Ace® qPCR RT kit (Toyobo, Osaka, Japan), and qPCR amplification was accomplished using a Mx3000P qPCR System (Agilent Technologies, Santa Clara, CA, USA). Primer sequences are listed in Table 1. mRNA expression levels were normalized to those of the mouse ribosomal protein, Large, P0 (Rplp0), as the internal control, determined via the comparative threshold cycle method [43].

Table 1. Primers used in this study.

Gene	Sequence	
	Forward	**Reverse**
Rplp0	GTGCTGATGGGCAAGAAC	AGGTCCTCCTTGGTGAAC
Tnfα	CGAGTGACAAGCCTGTAGCC	AGCTGCTCCTCCACTTGGT
Il-1β	ATGAGAGCATCCAGCTTCAA	TGAAGGAAAAGAAGGTGCTC
cagA	CCGATCGATCCGAAATTTTA	CGTTCGGATTTGATTCCCTA
ureA	TGTTGGCGACAGACCGGTTCAAATC	GCTGTCCCGCTCGCAATGTCTAAGC
napA	CCATGTGCATAAAGCCACTG	GAGTTTGAGCGCTTCGGATA

Ribosomal protein (Rplp0), Large, P0; tumor necrosis factor alpha (Tnfα); interleukin-1 beta (Il-1β); cytotoxin-associated gene A (cagA); urease subunit alpha (ureA); neutrophil-activating protein A (napA).

4.7. Determination of Plasma Glutamate Oxaloacetate Transaminase (GOT) and Glutamate Pyruvate Transaminase (GPT) Levels

Plasma GPT and GOT levels were determined using GPT and GOT enzyme-linked immunosorbent assay (ELISA) kits (Asan Pharmaceutical, Seoul, Korea) in accordance with the manufacturer's instructions.

4.8. Statistical Analysis

Data are presented as the mean ± standard deviation and were determined using Statistical Package for Social Sciences (SPSS, IBM, Armonk, NY, USA) version 19.0. Statistically significant differences were ascertained using one-way analysis of variance, followed by the Tukey–Kramer and Student's *t*-tests. $p < 0.05$ was considered significant.

5. Conclusions

In summary, we compared the in vivo antimicrobial effects of PC and conventional triple therapy against *H. pylori* using a mouse model of *H. pylori* infection. PC completely eradicated *H. pylori*, even when administered at a dose 1/15th that of conventional antibiotics used for triple therapy. In addition, gastritis was rapidly alleviated upon PC treatment. Thus, PC may be a potential viable alternative to triple therapy for *H. pylori* and gastrointestinal disorders.

Author Contributions: Conceptualization, H.Y.J. and J.-H.M.; methodology, H.Y.J., T.H.L., J.G.K., S.L., C.M., X.T.T., T.-I.J., and J.-H.M.; validation, H.Y.J., T.H.L., J.G.K., S.L., C.M., X.T.T., T.-I.J., and J.-H.M.; formal analysis, H.Y.J., T.H.L., J.G.K., S.L., and X.T.T.; investigation, H.Y.J., T.H.L., and S.L.; resources, T.-I.J., C.M., and J.-H.M.; data curation, H.Y.J., T.-I.J., C.M., and J.-H.M.; writing—original draft preparation, H.Y.J. and T.H.L.; writing—review and editing, T.-I.J., C.M., and J.-H.M.; supervision, J.-H.M.; project administration, J.-H.M.; funding acquisition, J.-H.M. All authors have read and approved the final version of the manuscript.

Funding: This research was supported by the Basic Science Research Program through the National Research Foundation of Korea (NRF), funded by the Ministry of Education, Science, and Technology (no. NRF-2020R1A2C2013711).

Acknowledgments: We thank the members of Moon and Jeon's laboratory for their discussions and technical support.

Conflicts of Interest: The authors declare no conflict of interest.

References

1. Kusters, J.G.; van Vliet, A.H.M.; Kuipers, E.J. Pathogenesis of *Helicobacter Pylori* Infection. *Clin. Microbiol. Rev.* **2006**, *19*, 449–490. [CrossRef] [PubMed]
2. Marshall, B.J.; Warren, J.R. Unidentified Curved Bacilli in the Stomach of Patients with Gastritis and Peptic Ulceration. *Lancet* **1984**, *1*, 1311–1315. [CrossRef]
3. Uemura, N.; Okamoto, S.; Yamamoto, S.; Matsumura, N.; Yamaguchi, S.; Yamakido, M.; Taniyama, K.; Sasaki, N.; Schlemper, R.J. *Helicobacter Pylori* Infection and the Development of Gastric Cancer. *N. Engl. J. Med.* **2001**, *345*, 784–789. [CrossRef] [PubMed]
4. Goodwin, C.S. *Helicobacter Pylori* Gastritis, Peptic Ulcer, and Gastric Cancer: Clinical and Molecular Aspects. *Clin. Infect. Dis.* **1997**, *25*, 1017–1019. [CrossRef] [PubMed]
5. Parsonnet, J.; Hansen, S.; Rodriguez, L.; Gelb, A.B.; Warnke, R.A.; Jellum, E.; Orentreich, N.; Vogelman, J.H.; Friedman, G.D. *Helicobacter Pylori* Infection and Gastric Lymphoma. *N. Engl. J. Med.* **1994**, *330*, 1267–1271. [CrossRef] [PubMed]
6. Ferlay, J.; Shin, H.-R.; Bray, F.; Forman, D.; Mathers, C.; Parkin, D.M. Estimates of Worldwide Burden of Cancer in 2008: GLOBOCAN 2008. *Int. J. Cancer* **2010**, *127*, 2893–2917. [CrossRef] [PubMed]
7. Leontiadis, G.I.; Sharma, V.K.; Howden, C.W. Non-Gastrointestinal Tract Associations of *Helicobacter Pylori* Infection. *Arch. Intern. Med.* **1999**, *159*, 925–940. [CrossRef]
8. Franceschi, F.; Tortora, A.; Gasbarrini, G.; Gasbarrini, A. *Helicobacter Pylori* and Extragastric Diseases. *Helicobacter* **2014**, *19* (Suppl. 1), 52–58. [CrossRef]

9. Chey, W.D.; Wong, B.C.Y. American College of Gastroenterology Guideline on the Management of *Helicobacter Pylori* Infection. *Am. J. Gastroenterol.* **2007**, *102*, 1808–1825. [CrossRef]
10. Delchier, J.C.; Malfertheiner, P.; Thieroff-Ekerdt, R. Use of a Combination Formulation of Bismuth, Metronidazole and Tetracycline with Omeprazole as a Rescue Therapy for Eradication of *Helicobacter Pylori*. *Aliment. Pharmacol. Ther.* **2014**, *40*, 171–177. [CrossRef]
11. Chuah, S.-K.; Tsay, F.-W.; Hsu, P.-I.; Wu, D.-C. A New Look at Anti-*Helicobacter Pylori* Therapy. *World J. Gastroenterol.* **2011**, *17*, 3971–3975. [CrossRef] [PubMed]
12. Vakil, N.; Vaira, D. Treatment for *H. Pylori* Infection: New Challenges with Antimicrobial Resistance. *J. Clin. Gastroenterol.* **2013**, *47*, 383–388. [CrossRef] [PubMed]
13. Gong, E.J.; Yun, S.-C.; Jung, H.-Y.; Lim, H.; Choi, K.-S.; Ahn, J.Y.; Lee, J.H.; Kim, D.H.; Choi, K.D.; Song, H.J.; et al. Meta-Analysis of First-Line Triple Therapy for *Helicobacter Pylori* Eradication in Korea: Is It Time to Change? *J. Korean Med. Sci.* **2014**, *29*, 704–713. [CrossRef] [PubMed]
14. Kawai, T.; Takahashi, S.; Suzuki, H.; Sasaki, H.; Nagahara, A.; Asaoka, D.; Matsuhisa, T.; Masaoaka, T.; Nishizawa, T.; Suzuki, M.; et al. Changes in the First Line Helicobacter Pylori Eradication Rates Using the Triple Therapy-a Multicenter Study in the Tokyo Metropolitan Area (Tokyo Helicobacter Pylori Study Group). *J. Gastroenterol. Hepatol.* **2014**, *29* (Suppl. 4), 29–32. [CrossRef]
15. Ayala, G.; Escobedo-Hinojosa, W.I.; de la Cruz-Herrera, C.F.; Romero, I. Exploring Alternative Treatments for *Helicobacter Pylori* Infection. *World J. Gastroenterol.* **2014**, *20*, 1450–1469. [CrossRef]
16. Myllyluoma, E.; Veijola, L.; Ahlroos, T.; Tynkkynen, S.; Kankuri, E.; Vapaatalo, H.; Rautelin, H.; Korpela, R. Probiotic Supplementation Improves Tolerance to *Helicobacter Pylori* Eradication Therapy—A Placebo-Controlled, Double-Blind Randomized Pilot Study. *Aliment. Pharmacol. Ther.* **2005**, *21*, 1263–1272. [CrossRef]
17. Yang, Y.X.; Lohakare, J.; Chae, B. Effects of Lacquer (*Rhus Verniciflua*) Meal Supplementation on Layer Performance. *Asian-Australas. J. Anim. Sci.* **2007**, *20*. [CrossRef]
18. Hatada, K.; Kitayama, T.; Nishiura, T.; Nishimoto, A.; Simonsick, W.J.; Vogl, O. Structural Analysis of the Components of Chinese Lacquer "Kuro-Urushi". *Macromol. Chem. Phys.* **1994**, *195*, 1865–1870. [CrossRef]
19. Hong, S.H.; Suk, K.T.; Choi, S.H.; Lee, J.W.; Sung, H.T.; Kim, C.H.; Kim, E.J.; Kim, M.J.; Han, S.H.; Kim, M.Y.; et al. Anti-Oxidant and Natural Killer Cell Activity of Korean Red Ginseng (*Panax Ginseng*) and Urushiol (*Rhus Vernicifera* Stokes) on Non-Alcoholic Fatty Liver Disease of Rat. *Food Chem. Toxicol.* **2013**, *55*, 586–591. [CrossRef]
20. Kim, M.; Choi, Y.; Kim, W.; Kwak, S. Antioxidative Activity of Urushiol Derivatives from the Sap of Lacquer Tree (*Rhus Vernicifera* Stokes). *Korean J. Plant Res.* **1997**, *10*, 227–230.
21. Kim, D.; Jeon, S.; Seo, J. The Preparation and Characterization of Urushiol Powders (YPUOH) Based on Urushiol. *Prog. Org. Coat.* **2013**, *76*, 1465–1470. [CrossRef]
22. Kim, H.S.; Yeum, J.H.; Choi, S.W.; Lee, J.Y.; Cheong, I.W. Urushiol/Polyurethane-Urea Dispersions and Their Film Properties. *Prog. Org. Coat.* **2009**, *65*, 341–347. [CrossRef]
23. Choi, J.Y.; Park, C.S.; Choi, J.O.; Rhim, H.S.; Chun, H.J. Cytotoxic Effect of Urushiol on Human Ovarian Cancer Cells. *J. Microbiol. Biotechnol.* **2001**, *11*, 399–405.
24. Suk, K.T.; Baik, S.K.; Kim, H.S.; Park, S.M.; Paeng, K.J.; Uh, Y.; Jang, I.H.; Cho, M.Y.; Choi, E.H.; Kim, M.J.; et al. Antibacterial Effects of the Urushiol Component in the Sap of the Lacquer Tree (*Rhus Verniciflua* Stokes) on *Helicobacter Pylori*. *Helicobacter* **2011**, *16*, 434–443. [CrossRef]
25. Ma, X.M.; Lu, R.; Miyakoshi, T. Recent Advances in Research on Lacquer Allergy. *Allergol. Int.* **2012**, *61*, 45–50. [CrossRef]
26. Zepter, K.; Häffner, A.; Soohoo, L.F.; De Luca, D.; Tang, H.P.; Fisher, P.; Chavinson, J.; Elmets, C.A. Induction of Biologically Active IL-1 Beta-Converting Enzyme and Mature IL-1 Beta in Human Keratinocytes by Inflammatory and Immunologic Stimuli. *J. Immunol.* **1997**, *159*, 6203–6208.
27. Wakabayashi, T.; Hu, D.-L.; Tagawa, Y.-I.; Sekikawa, K.; Iwakura, Y.; Hanada, K.; Nakane, A. IFN-Gamma and TNF-Alpha Are Involved in Urushiol-Induced Contact Hypersensitivity in Mice. *Immunol. Cell Biol.* **2005**, *83*, 18–24. [CrossRef]
28. Kim, J.Y.; Cho, J.Y.; Ma, Y.K.; Lee, Y.G.; Moon, J.H. Nonallergenic Urushiol Derivatives Inhibit the Oxidation of Unilamellar Vesicles and of Rat Plasma Induced by Various Radical Generators. *Free Radic. Biol. Med.* **2014**, *71*, 379–389. [CrossRef]

29. Cho, J.-Y.; Park, K.Y.; Kim, S.-J.; Oh, S.; Moon, J.-H. Antimicrobial Activity of the Synthesized Non-Allergenic Urushiol Derivatives. *Biosci. Biotechnol. Biochem.* **2015**, *79*, 1915–1918. [CrossRef]
30. Lee, Y.G. Absorption and Metabolism of an Urushiol Derivative, 3-Pentylcathechol, in Rat Plasma. Master's Thesis, Chonnam National University, Gwangju, Korea, 2013.
31. Jeong, H.Y. Comparison of Absorption, Metabolism, and Bioactivity among Chemically Synthesized 3-Pentylcatechol and Its Glucosides. Master's Thesis, Chonnam National University, Gwangju, Korea, 2014.
32. Noach, L.A.; Bosma, N.B.; Jansen, J.; Hoek, F.J.; van Deventer, S.J.; Tytgat, G.N. Mucosal Tumor Necrosis Factor-Alpha, Interleukin-1 Beta, and Interleukin-8 Production in Patients with *Helicobacter Pylori* Infection. *Scand. J. Gastroenterol.* **1994**, *29*, 425–429. [CrossRef]
33. Olbe, L.; Hamlet, A.; Dalenbäck, J.; Fändriks, L. A Mechanism by Which *Helicobacter Pylori* Infection of the Antrum Contributes to the Development of Duodenal Ulcer. *Gastroenterology* **1996**, *110*, 1386–1394. [CrossRef] [PubMed]
34. Wang, G.; Hong, Y.; Olczak, A.; Maier, S.E.; Maier, R.J. Dual Roles of *Helicobacter Pylori* NapA in Inducing and Combating Oxidative Stress. *Infect. Immun.* **2006**, *74*, 6839–6846. [CrossRef] [PubMed]
35. Tzouvelekis, L.S.; Mentis, A.F.; Makris, A.M.; Spiliadis, C.; Blackwell, C.; Weir, D.M. *In Vitro* Binding of *Helicobacter Pylori* to Human Gastric Mucin. *Infect. Immun.* **1991**, *59*, 4252–4254. [CrossRef] [PubMed]
36. Sierra, J.C.; Asim, M.; Verriere, T.G.; Piazuelo, M.B.; Suarez, G.; Romero-Gallo, J.; Delgado, A.G.; Wroblewski, L.E.; Barry, D.P.; Peek, R.M.J.; et al. Epidermal Growth Factor Receptor Inhibition Downregulates *Helicobacter Pylori*-Induced Epithelial Inflammatory Responses, DNA Damage and Gastric Carcinogenesis. *Gut* **2018**, *67*, 1247–1260. [CrossRef] [PubMed]
37. Berg, H.F.; Maraha, B.; Scheffer, G.-J.; Peeters, M.F.; Kluytmans, J.A.J.W. Effect of Clarithromycin on Inflammatory Markers in Patients with Atherosclerosis. *Clin. Diagn. Lab. Immunol.* **2003**, *10*, 525–528. [CrossRef]
38. Knoop, K.A.; McDonald, K.G.; Kulkarni, D.H.; Newberry, R.D. Antibiotics Promote Inflammation through the Translocation of Native Commensal Colonic Bacteria. *Gut* **2016**, *65*, 1100–1109. [CrossRef] [PubMed]
39. Farooqui, T.F.; Farooqui, A.A. Molecular Mechanism Underlying the Therapeutic Activities of Propolis: A Critical Review. *Curr. Nutr. Food Sci.* **2010**, *6*, 186–199. [CrossRef]
40. Johnston, D.E. Special Considerations in Interpreting Liver Function Tests. *Am. Fam. Physician* **1999**, *59*, 2223–2230.
41. Lee, A.; O'Rourke, J.; De Ungria, M.C.; Robertson, B.; Daskalopoulos, G.; Dixon, M.F. A Standardized Mouse Model of *Helicobacter Pylori* Infection: Introducing the Sydney Strain. *Gastroenterology* **1997**, *112*, 1386–1397. [CrossRef]
42. Lee, S.; Yang, M.; Kim, J.; Son, Y.; Kim, J.; Kang, S.; Ahn, W.; Kim, S.-H.; Kim, J.-C.; Shin, T.; et al. Involvement of BDNF/ERK Signaling in Spontaneous Recovery from Trimethyltin-Induced Hippocampal Neurotoxicity in Mice. *Brain Res. Bull.* **2016**, *121*, 48–58. [CrossRef]
43. Go, D.-H.; Lee, Y.G.; Lee, D.-H.; Kim, J.-A.; Jo, I.-H.; Han, Y.S.; Jo, Y.H.; Kim, K.-Y.; Seo, Y.-K.; Moon, J.-H.; et al. 3-Decylcatechol Induces Autophagy-Mediated Cell Death through the IRE1α/JNK/P62 in Hepatocellular Carcinoma Cells. *Oncotarget* **2017**, *8*, 58790–58800. [CrossRef] [PubMed]

Article

Biofilm Inhibition and Eradication Properties of Medicinal Plant Essential Oils against Methicillin-Resistant *Staphylococcus aureus* Clinical Isolates

Fethi Ben Abdallah [1,2,*], Rihab Lagha [1,2] and Ahmed Gaber [1,3]

1. Department of Biology, College of Sciences, Taif University, P.O. Box 11099, Taif 21944, Saudi Arabia; rihab.k@tu.edu.sa (R.L.); a.gaber@tu.edu.sa (A.G.)
2. Unité de Recherche, Virologie & Stratégies Antivirales, UR17ES30, Institut Supérieur de Biotechnologie, University of Monastir, Monastir 5000, Tunisia
3. Department of Genetics, Faculty of Agriculture, Cairo University, Giza 12613, Egypt

* Correspondence: fetyben@tu.edu.sa

Received: 16 October 2020; Accepted: 4 November 2020; Published: 6 November 2020

Abstract: Methicillin-resistant *Staphylococcus aureus* is a major human pathogen that poses a high risk to patients due to the development of biofilm. Biofilms, are complex biological systems difficult to treat by conventional antibiotic therapy, which contributes to >80% of humans infections. In this report, we examined the antibacterial activity of *Origanum majorana*, *Rosmarinus officinalis*, and *Thymus zygis* medicinal plant essential oils against MRSA clinical isolates using disc diffusion and MIC methods. Moreover, biofilm inhibition and eradication activities of oils were evaluated by crystal violet. Gas chromatography–mass spectrometry analysis revealed variations between oils in terms of component numbers in addition to their percentages. Antibacterial activity testing showed a strong effect of these oils against MRSA isolates, and *T. zygis* had the highest activity succeeded by *O. majorana* and *R. officinalis*. Investigated oils demonstrated high biofilm inhibition and eradication actions, with the percentage of inhibition ranging from 10.20 to 95.91%, and the percentage of eradication ranging from 12.65 to 98.01%. *O. majorana* oil had the highest biofilm inhibition and eradication activities. Accordingly, oils revealed powerful antibacterial and antibiofilm activities against MRSA isolates and could be a good alternative for antibiotics substitution.

Keywords: methicillin-resistant *Staphylococcus aureus*; essential oils; *Origanum majorana*; *Rosmarinus officinalis*; *Thymus zygis*; antibacterial; biofilm inhibition and eradication

---

1. Introduction

Methicillin-resistant *Staphylococcus aureus* (MRSA) is considered a principal human pathogen and the most common cause of nosocomial infections. MRSA causes several diseases ranging from skin and soft tissue infections to serious invasive infections such as pneumonia, bacteremia, endocarditis and osteomyelitis [1]. The number of MRSA infections, which are more frequently associated with mortality than other bacterial infections, has increased considerably over recent years. *S. aureus* carries 20–40% mortality at 30 days despite appropriate treatment [2].

MRSA poses a high risk to patients due to the development of biofilm [3]. Biofilm is considered as major virulence factor and is an organized structure built by almost all bacteria that is composed of nucleic acids, lipids, proteins, and polysaccharides [4]. Biofilms contribute to >80% of human infections and *S. aureus* is considered as the leading species in biofilm-associated infections [5]. In Biofilm, MRSA like other bacteria, become more persistent in the host organism, environment, and medical surfaces, and show an increased resistance to antibiotics and host immune factors [6–8], which is an important medical problem. Therefore, the development of novel compounds to treat biofilm is urgently required; plant essentials oils (EOs) that act against bacterial biofilm are of great interest.

EOs are volatile compounds that have been used to combat a variety of infections during hundreds of years as a natural medicine. It has been shown that EOs possess several significant antimicrobial activities such as antibacterial, antiviral, antifungal, and anti-parasitic activities in addition to their antioxidant, antiseptic, and insecticidal properties [9,10].

Rosmarinus officinalis L., *Thymus zygis* L., and *Origanum majorana* L. belong to the Lamiaceae family. EOs obtained from aerial parts of the flowering stage of these plants, have been reported for their antibacterial activities against *S. aureus* [11,12] and their antibiofilm activities against uropathogenic *E. coli* [13]. Several reports have shown that tea tree, thyme, and peppermint EOs, are effective against planktonic [14] and biofilm [15,16] MRSA. In addition, Cáceres et al. [17] demonstrated high anti-biofilm activity of thymol-carvacrol-chemotype (II) oil from *Lippia origanoides* against *E. coli* and *S. epidermidis*. However, these oils did not alter the growth rate of planktonic bacteria. The antibacterial effect of EOs, which is manifested by alterations of the bacterial cell wall and cell membrane, depends of their chemical composition [18]. The cell membrane compositions play an important role in the high resistance of Gram-negative bacteria to EOs compared to Gram-positive [19]. The hydrophobic molecules penetrate easily into the cells due to cell wall structure in Gram-positive bacteria and act on the cell wall and within the cytoplasm [20].

This study aimed to investigate the antibacterial, biofilm inhibition, and eradication properties of *O. majorana*, *R. officinalis*, and *T. zygis* medicinal plants' EOs against clinical methicillin-resistant *Staphylococcus aureus*.

2. Results

2.1. Distribution of the MRSA Isolates

Thirty clinical MRSA isolates were collected from King Abdulaziz Specialist Hospital, Taif, Saudi Arabia. The isolates were obtained from infection sites: surgical site infection (SSI, n = 4), skin and soft tissue (SST, n = 12), blood (n = 1), nasal (n = 8) and burn (n = 5). The distribution of isolates based on the type of specimen is presented in Figure 1.

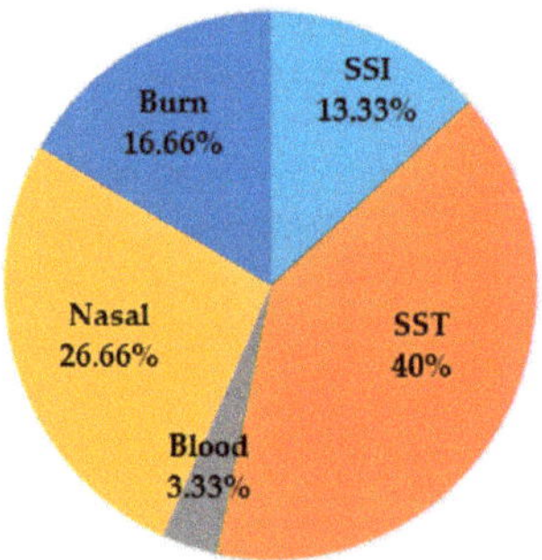

Figure 1. Distribution of MRSA isolates.

2.2. Chemical Composition of the Essential Oils

O. majorana, *T. zygis*, and *R. officinalis*' EOs chemical compositions are summarized in Table 1. In total 37 components were detected in these oils: 10 compounds in *R. officinalis* and 31 compounds in each of *O. majorana* and *T. zygis*.

GC-MS results showed variations between these oils regarding the compound numbers and their percentages. The major constituents of *O. majorana* were terpinen-4-ol (25.9%), γ-terpinene (16.9%), linalool (10.9%), sabinene (8%), and α-terpinene (7.7%); those of *R. officinalis* were α-pinene

(37.7%), bornyl acetate (9.1%), camphene (7.3%), borneol (5.5%), verbenone (5.4%), camphor (5.2%), and 1,8-cineole (4.7%). However, the main components of *T. zygis* were linalool (39.7%), terpinen-4-ol (11.7%), β-myrcene (8.6%), and γ-terpinene (7.6%).

Table 1. Chemical composition of the essential oils.

Components	*O. majorana* (%) [13]	*T. zygis* (%) [13]	*R. officinalis* (%)
α-Pinene	0.46	3.6	37.7
Sabinene	8	0.84	
β-Pinene	1.4	0.33	1.1
β-Myrcene	1.1	8.6	
α-phellandrene	0.30	0.48	
Limonene	3.5	2.6	4.1
Terpinen-4-ol	25.9	11.7	
Bornyl acetate		0.07	9.1
β-Caryophyllene	2.3	1.6	
α-Thujene	0.33	0.21	
Camphene	0.03	0.74	7.3
α-Terpinene	7.7	4.2	
p-Cymene	3.4	2.2	
1,8-Cineole	0.15		4.7
γ-Terpinene	16.9	7.6	
Terpinolene	1.7	2	
Linalool	10.9	39.7	1.8
Borneol		1.9	5.5
α-Terpineol	2.5	1.7	
Camphor		0.22	5.2
α-Humulene	0.05		
cis and *trans*-thujan-4-ol	2.2–2.3	0.88–2.2	
cis and *trans* piperitol	0.13–0.18	0.13–0.08	
Linalyl acetate	7	0.5	
Carvacrol	0.03	0.08	
Thymol	0.05	0.52	
Bicyclogermacrene	0.41	0.16	
Cis and *trans*-p-menth-2-en-1-ol	0.59–0.32	0.37–0.25	
Caryophyllene oxide	0.04		
Ocimene	0.07		
Spathulenol	0.01		
cis-Dihydrocarvone		0.17	
trans-Dihydrocarvone		0.2	
Verbenone			5.4

2.3. Antibacterial Activity of Essential Oils against MRSA

2.3.1. Disc Diffusion

The antibacterial activity of EOs against MRSA isolates was assessed by the disc diffusion method (Table 2). *T. zygis* EO has shown strong inhibitory activity on 80% of the strains, succeeded by *O. majorana* and *R. officinalis* that demonstrated a strong inhibitory action on 53.33% and 16.66% of the isolates, respectively. Moreover, according to the high percentage of anti-MRSA activity, *T. zygis* and *O. majorana'* EOs have a strong inhibitory action on 80% and 53.33% of the strains, respectively. However, *R. officinalis* had a slight inhibitory action on 46.66% of the strains. Thereby, *T. zygis'* EO appeared as the EO with the highest antibacterial activity, succeeded by *O. majorana* and *R. officinalis*. According to the type of specimen, globally, the same result was found as detected in cases of total isolates. *T. zygis* was regarded as an EO with strong inhibitory activity, succeeded by *O. majorana* and *R. officinalis*.

Table 2. Antimicrobial effect of EOs against MRSA isolates using disc diffusion.

Essential Oils	Inhibitory Action				
	Strong *n* (%)	Complete *n* (%)	Partial *n* (%)	Slight *n* (%)	No Action *n* (%)
O. majorana	16 (53.33%)	11 (36.66%)	3 (10%)		
T. zygis	24 (80%)	6 (20%)			
R. officinalis	5 (16.66%)	5 (16.66%)	6 (20%)	14 (46.66%)	

n: number of isolates.

2.3.2. Determination of MIC and MBC

Antibacterial activity of EOs was assessed by measuring MICs and MBCs for the 30 MRSA isolates and the reference strain. The values of MIC extended from 0.78 mg/mL to 1.56 mg/mL, while the MBC varied from 3.125 mg/mL to 12.5 mg/mL for *O. majorana* EO.

Concerning *T. zygis* EO, the MIC values ranged from 0.39 mg/mL to 0.78 mg/mL, while the MBC value was 3.125 mg/mL. However, The MIC values of *R. officinalis* varied from 0.78 mg/mL to 1.56 mg/mL, but the MBC value was 12.5 mg/mL. Compared to *O. majorana* and *R. officinalis*, *T. zygis* EO showed the greatest antibacterial activity against MRSA isolates. Person correlation (r) showed no significant correlation between the type of specimen and MICs of oils ($p > 0.05$).

2.4. Biofilm Formation

MRSA strains were tested for their potentialities to form biofilm on a polystyrene surface. Table 3 indicated that 96.66% of the isolates were able to form biofilm and were distributed as follow: 40% were highly positive biofilm producers with OD570 values ranged from 1.175 to 3.635, and 56.66% were low-grade positive with OD570 values extended from 0.113 to 0.87. However, out of the 30 isolates only one strain was isolated from the nasal sample was biofilm negative. Reference *S. aureus* ATCC 25923 was considered as a highly positive biofilm producer. Analysis of variance (ANOVA) indicated that there is no significant effect of the specimen on biofilm formation ($p > 0.05$).

Table 3. Biofilm formation ability of MRSA isolates on polystyrene surface.

Isolates	Specimen	OD570 ± SD	Biofilm Formation
1	Burn	0.24 ± 0.026	low-grade positive
2	Blood	2.609 ± 0.088	highly positive
3	SST	3.635 ± 0.052	highly positive
4	SST	1.437 ± 0.074	highly positive
5	Nasal	1.175 ± 0.03	highly positive
6	SSI	0.147 ± 0.028	low-grade positive
7	SST	0.135 ± 0.031	low-grade positive
8	Burn	1.971 ± 0.049	highly positive
9	Nasal	0.19 ± 0.079	low-grade positive
10	Nasal	1.378 ± 0.06	highly positive
11	SST	0.194 ± 0.075	low-grade positive
12	SST	1.554 ± 0.086	highly positive
13	SST	0.305 ± 0.021	low-grade positive
14	Nasal	2.157 ± 0.071	highly positive
15	Nasal	0.045 ± 0.007	Negative
16	SST	0.198 ± 0.078	low-grade positive
17	SSI	0.87 ± 0.023	low-grade positive
18	Nasal	0.221 ± 0.048	low-grade positive
19	Burn	0.428 ± 0.068	low-grade positive
20	Nasal	0.745 ± 0.018	low-grade positive
21	Nasal	0.319 ± 0.012	low-grade positive
22	SST	0.233 ± 0.087	low-grade positive
23	SST	0.788 ± 0.027	low-grade positive
24	SSI	0.642 ± 0.028	low-grade positive
25	SST	1.836 ± 0.038	highly positive
26	Burn	2.696 ± 0.054	highly positive
27	SST	0.418 ± 0.056	low-grade positive
28	Burn	0.438 ± 0.067	low-grade positive
29	SSI	0.113 ± 0.045	low-grade positive
30	SST	2.308 ± 0.039	highly positive
ATCC 25923		3.36 ± 0.098	highly positive

SST: skin and soft tissue; SSI: surgical site infection.

2.5. Biofilm Inhibition Activity of Essentials Oils

Biofilm inhibitory activities of *O. majorana*, *T. zygis*, and *R. officinalis'* EOs are summarized in Table 4. MRSA isolates that showed a biofilm formation potential were selected for this investigation. In all, 29 isolates considered as highly positive biofilm and low-grade positive biofilm in addition to the reference strain were used.

Table 4. Biofilm inhibition activity of EOs against MRSA isolates.

Isolates	Control OD570 ± SD	*O. majorana* OD570 ± SD	Inhibition (%)	*T. zygis* OD570 ± SD	Inhibition (%)	*R. officinalis* OD570 ± SD	Inhibition (%)
1	0.24 ± 0.026	0.061 ± 0.004 *	74.58	0.112 ± 0.015	53.33	0.238 ± 0.028	0
2	2.609 ± 0.088	2.603 ± 0.093	0	2.61 ± 0.019	0	2.595 ± 0.098	0
3	3.635 ± 0.052	2.701 ± 0.082	25.69	3.658 ± 0.01	0	2.744 ± 0.066	24.51
4	1.437 ± 0.074	0.131 ± 0.038 **	90.88	0.965 ± 0.022 **	32.84	0.728 ± 0.084 **	49.33
5	1.175 ± 0.03	0.048 ± 0.006 ***	95.91	0.1 ± 0.038 **	91.48	0.051 ± 0.004 ***	95.65
6	0.147 ± 0.028	0.03 ± 0.008 *	79.59	0.124 ± 0.043	15.64	0.132 ± 0.023	10.20
7	0.135 ± 0.031	0.114 ± 0.023	15.55	0.136 ± 0.073	0	0.134 ± 0.011	0
8	1.971 ± 0.049	0.105 ± 0.018 **	94.67	0.239 ± 0.088 **	87.87	0.346 ± 0.018 **	82.44
9	0.19 ± 0.079	0.049 ± 0.009 *	74.21	0.138 ± 0.043	27.36	0.146 ± 0.093	23.15
10	1.378 ± 0.06	1.387 ± 0.038	0	1.369 ± 0.054	0	0.828 ± 0.082 **	39.91
11	0.194 ± 0.075	0.02 ± 0.005 *	89.69	0.142 ± 0.058	26.80	0.111 ± 0.044	42.78
12	1.554 ± 0.086	0.162 ± 0.077 **	89.57	0.22 ± 0.077 **	85.84	0.17 ± 0.023 **	89.06
13	0.305 ± 0.021	0.027 ± 0.006 *	91.14	0.303 ± 0.032	0	0.301 ± 0.069	0
14	2.157 ± 0.071	1.935 ± 0.014	10.29	2.154 ± 0.04	0	1.724 ± 0.092	20.07
16	0.198 ± 0.078	0.038 ± 0.009 *	80.80	0.072 ± 0.008 *	63.63	0.053 ± 0.006 *	73.23
17	0.87 ± 0.023	0.043 ± 0.017 *	95.05	0.124 ± 0.089	85.74	0.16 ± 0.026	81.60
18	0.221 ± 0.048	0.105 ± 0.028	52.48	0.142 ± 0.037	35.74	0.122 ± 0.038	44.79
19	0.428 ± 0.068	0.053 ± 0.039	87.61	0.095 ± 0.002 *	77.80	0.226 ± 0.077	47.19
20	0.745 ± 0.018	0.055 ± 0.033 *	92.61	0.11 ± 0.082	85.23	0.562 ± 0.065	24.56
21	0.319 ± 0.012	0.072 ± 0.013 *	77.42	0.317 ± 0.075	0	0.085 ± 0.004 *	73.35
22	0.233 ± 0.087	0.231 ± 0.032	0	0.131 ± 0.012	43.77	0.089 ± 0.003 *	61.80
23	0.788 ± 0.027	0.554 ± 0.065	29.69	0.696 ± 0.028	11.67	0.465 ± 0.073	40.98
24	0.642 ± 0.028	0.192 ± 0.013	70.09	0.639 ± 0.087	0	0.241 ± 0.053	62.461
25	1.836 ± 0.038	1.325 ± 0.078	27.83	1.84 ± 0.042	0	1.821 ± 0.096	0
26	2.696 ± 0.054	2.196 ± 0.04	18.54	2.206 ± 0.07	18.17	2.012 ± 0.014	25.37
27	0.418 ± 0.056	0.254 ± 0.068	39.23	0.415 ± 0.069	0	0.065 ± 0.008 *	84.44
28	0.438 ± 0.067	0.034 ± 0.005 *	92.23	0.217 ± 0.016	50.45	0.069 ± 0.002 *	84.24
29	0.113 ± 0.045	0.11 ± 0.022	0	0.016 ± 0.006 *	85.84	0.114 ± 0.032	0
30	2.308 ± 0.039	1.533 ± 0.055	33.57	2.306 ± 0.086	0	0.907 ± 0.048 **	60.70
ATCC 25922	3.36 ± 0.098	1.838 ± 0.066	45.29	3.352 ± 0.014	0	3.345 ± 0.029	0

* Isolates changed from low-grade positive to biofilm negative after treatment with EOs. ** Isolates changed from highly positive to low-grade positive after treatment with EOs. *** Isolates changed from highly positive to biofilm negative after treatment with EOs.

O. majorana EO demonstrated an antibiofilm activity on 89.66% of the isolates (26 strains) in addition to the reference strain. Among them, 11 isolates (39.93%) were passed from low-grade positive to biofilm negative, three highly positive biofilm isolates (10.34%) become low-grade positive, and one strain isolated from nasal samples was changed from highly positive to biofilm negative after treatment with *O. majorana* EO. The percentage of inhibition ranged from 10.29 to 95.91%.

Concerning *R. officinalis* EO, we detected an antibiofilm effect on 79.31% of the isolates (23 strains). Furthermore, two groups of five isolates (17.24%) were changed. The first group was varied from low-grade positive to biofilm negative, and the second group was passed from highly positive to low-grade positive. In addition, the same isolate that changed from highly positive to biofilm negative under *O. majorana* EO, also became biofilm negative under *R. officinalis* EO. The percentage of inhibition extended from 10.20 to 95.65%.

Antibiofilm activity of *T. zygis* was observed on 62.06% of the isolates (18 strains). The percentage of biofilm inhibition ranged from 11.67 to 91.48%. Moreover, three low-grade positive isolates (10.34%) were changed to biofilm negative, and four highly positive isolates (13.79%) become low-grade positive.

The outcomes of this study indicated that *O. majorana* EO had the greatest antibiofilm activity against MRSA isolates succeeded by *R. officinalis* and *T. zygis*.

Person correlation (r) indicated a non-significant correlation between MIC and antibiofilm of EOs ($p > 0.05$). ANOVA test showed a non-significant effect of the specimen on biofilm inhibition ($p > 0.05$).

2.6. Biofilm Eradication Activity of Essentials Oils

The results of biofilm eradication activities of EOs are shown in Table 5. The same isolates selected for biofilm inhibitory investigation were used. *O. majorana*, *T. zygis*, and *R. officinalis* EOs showed eradication activities on 41.37% (12 strains) of the MRSA isolates independently of the specimen, including the reference strain. The highest percentage of eradication was recorded with *O. majorana*. The percentage of eradication ranged from 18.31 to 98.01%, from 12.65 to 94.39%, and from 13.45 to 92.69%, respectively, for *O. majorana*, *T. zygis*, and *R. officinalis* EOs.

Table 5. Biofilm eradication activity of EOs against MRSA isolates.

Isolates	Control OD570 ± SD	*O. majorana* OD570 ± SD	Eradication (%)	*T. zygis* OD570 ± SD	Eradication (%)	*R. officinalis* OD570 ± SD	Eradication (%)
1	0.418 ± 0.024	0.417 ± 0.015	0	0.415 ± 0.019	0	0.413 ± 0.021	0
2	3.13 ± 0.096	1.754 ± 0.055	43.96	1.626 ± 0.06	48.05	2.709 ± 0.084	13.45
3	4.362 ± 0.086	2.92 ± 0.071	33.05	3.468 ± 0.091	20.49	2.15 ± 0.079	50.71
4	1.939 ± 0.078	1.941 ± 0.069	0	1.931 ± 0.074	0	1.115 ± 0.063	42.49
5	1.41 ± 0.088	0.237 ± 0.025 **	83.19	0.079 ± 0.008 ***	94.39	0.103 ± 0.011 **	92.69
6	0.176 ± 0.016	0.078 ± 0.005 *	55.68	0.061 ± 0.004 *	65.34	0.131 ± 0.018	25.56
7	0.189 ± 0.012	0.185 ± 0.013	0	0.191 ± 0.015	0	0.186 ± 0.015	0
8	2.465 ± 0.083	0.37 ± 0.045 **	84.98	2.153 ± 0.079	12.65	1.143 ± 0.059	53.63
9	0.304 ± 0.056	0.308 ± 0.021	0	0.309 ± 0.022	0	0.302 ± 0.024	0
10	1.722 ± 0.066	0.386 ± 0.027 **	77.58	1.074 ± 0.055	37.63	0.174 ± 0.014 **	89.89
11	0.269 ± 0.013	0.262 ± 0.016	0	0.265 ± 0.025	0	0.271 ± 0.02	0
12	2.334 ± 0.073	0.687 ± 0.032 **	70.56	1.356 ± 0.058	41.90	2.328 ± 0.026	0
13	0.433 ± 0.026	0.431 ± 0.026	0	0.436 ± 0.031	0	0.43 ± 0.032	0
14	2.617 ± 0.078	0.052 ± 0.003 ***	98.01	2.614 ± 0.082	0	2.609 ± 0.079	0
16	0.286 ± 0.034	0.287 ± 0.02	0	0.281 ± 0.015	0	0.288 ± 0.019	0
17	1.249 ± 0.028	1.247 ± 0.063	0	1.241 ± 0.068	0	1.244 ± 0.064	0
18	0.298 ± 0.011	0.292 ± 0.018	0	0.294 ± 0.016	0	0.296 ± 0.023	0
19	0.676 ± 0.026	0.679 ± 0.039	0	0.671 ± 0.029	0	0.674 ± 0.038	0
20	0.894 ± 0.045	0.448 ± 0.028	49.88	0.613 ± 0.036	31.43	0.584 ± 0.034	34.67
21	0.516 ± 0.068	0.514 ± 0.04	0	0.513 ± 0.031	0	0.517 ± 0.041	0
22	0.319 ± 0.022	0.313 ± 0.024	0	0.317 ± 0.024	0	0.314 ± 0.021	0
23	1.194 ± 0.039	1.19 ± 0.052	0	1.195 ± 0.057	0	1.192 ± 0.051	0
24	0.808 ± 0.027	0.801 ± 0.048	0	0.805 ± 0.044	0	0.81 ± 0.046	0
25	2.249 ± 0.069	2.245 ± 0.069	0	1.824 ± 0.072	18.89	1.806 ± 0.058	19.69
26	3.396 ± 0.094	1.674 ± 0.049	50.70	1.785 ± 0.058	47.43	2.025 ± 0.06	40.37
27	0.568 ± 0.021	0.561 ± 0.034	0	0.567 ± 0.039	0	0.564 ± 0.039	0
28	0.535 ± 0.019	0.437 ± 0.025	18.31	0.337 ± 0.03	37	0.081 ± 0.006 *	84.85
29	0.133 ± 0.011	0.134 ± 0.019	0	0.136 ± 0.026	0	0.132 ± 0.011	0
30	2.989 ± 0.083	0.348 ± 0.023 **	88.35	1.342 ± 0.056	55.10	1.756 ± 0.053	41.25
ATCC 25922	4.32 ± 0.098	2.796 ± 0.085	35.27	1.972 ± 0.068	54.35	2.577 ± 0.087	40.34

* Isolates changed from low-grade positive to biofilm negative after treatment with EOs. ** Isolates changed from highly positive to low-grade positive after treatment with EOs. *** Isolates changed from highly positive to biofilm negative after treatment with EOs.

Based on biofilm categories and under *O. majorana* EO, five isolates (17.24%) were changed from highly positive to low-grade positive. Furthermore, a low-grade positive (isolate number 6) and a highly positive (isolate number 14) strains became biofilm negative. For *T. zygis*, only isolates numbers 5 and 6 were changed from highly positive and low-grade positive respectively to biofilm negative after treatment, while, *R. officinalis* caused modifications on three isolates. Among them, a low-grade positive (isolate number 28) was changed to biofilm negative, and two highly positive (isolates number 5 and 10) became low-grade positive.

Person correlation (r) indicated a non-significant correlation between MIC and biofilm eradication of EOs ($p > 0.05$). ANOVA test showed a non-significant effect of the specimen on biofilm eradication ($p > 0.05$).

3. Discussion

Infection caused by MRSA is considered a major public health threat in many countries and MRSA remains the principal cause of hospital and community-acquired infections [21]. This bacterium is accountable for numerous infections related to remarkable morbidity and mortality [22], such as bacteremia, pneumonia, and skin, soft tissue, surgical site, and urinary tract infections [23,24]. This study was conducted on 30 clinical MRSA isolates and results showed variability in the prevalence of the isolates. Indeed, most of the strains (40%) were recovered from SST followed by nasal, burn, SSI, and blood. Akanbi et al. [25] showed that the majority of MRSA strains were isolated from blood, wound, and urine specimens. In addition, Ghebremedhin et al. [26] demonstrated that MRSA was most found in surgical wound infections, succeeded by eye swabs, skin and soft tissue infections.

The ability of MRSA to develop resistance to every antibiotic to which it is exposed makes it a problem to human health [27]. Thereby the development of novel compounds is of great importance. Medicinal plant EOs have been largely used as a natural medicine to combat bacteria, fungi, viruses, and other pathogens [28]. Until now, about 3000 EOs are known, among them 300 are important for industries such as pharmaceutical, food, agronomic, cosmetics, and fragrance. In this work, we investigated the potential antibacterial activities of *O. majorana*, *T. zygis*, and *R. officinalis* medicinal plant EOs against MRSA clinical isolates by disc diffusion, MIC, and MBC techniques. The highest antibacterial activity was observed with *T. zygis*, followed by *O. majorana*, and *R. officinalis* EOs. This result is in agreement with the study of Lagha et al. [13], who showed that *T. zygis* possessed the strongest antimicrobial effect against uropathogenic *E. coli* in contrast to *O. majorana*, and *R. officinalis* EOs. According to biochemical composition, the greater effect of *T. zygis* is owing to linalool (39.7%), which has a strong effect against bacteria and fungi [29]. Regarding *O. majorana* EO, the antibacterial activity can be attributed to the monoterpene alcohol, terpinen-4-ol, as a major compound (25.9%) [30], which was found to be effective against MRSA [31]. According to Cordeiro et al. [32], terpinen-4-ol has a powerful antibacterial effect against *S. aureus*. This compound functions as a bactericidal by obstructing the synthesis of the cell wall. Moreover, other main components such as terpinen-4-ol (11.7%), β-myrcene (8.6%) and γ-terpinene (7.6%) are present in *T. zygis* in addition to linalool (10.9%), γ-terpinene (16.9%) and α-terpinene (7.7%) present in *O. majorana* may enhance the antibacterial effect of these oils. Concerning *R. officinalis* EO, which showed the lowest activity against MRSA isolates, its antibacterial activity may be related to α-pinene (37.7%) as a major constituent. The study of Leite et al. [33] showed antibacterial activity of α-pinene against *S. aureus* and *S. epidermidis*. Other reports [34,35] revealed the antibacterial activity of some EOs against Gram-negative and Gram-positive bacteria when α-Pinene is the major constituent. However, Utegenova et al. [36] demonstrated that α-pinene had low activity against MRSA, indicating that other components were probably responsible. Among them, in this study, 1,8-cineole (4.7%) altered the structure of *E. coli*, *S. enteritidis*, and *S. aureus* [37]. The antibacterial activity of *R. officinalis* could be attributed to the synergistic effect of camphor (5.2%), verbenone (5.4%), and borneol (5.5%) in addition to α-pinene and 1,8-cineole [38]. In general, whole essential oils have an important antimicrobial effect compared to the major compounds individually or collectively. This suggests that minor constituents are essential and may have a synergistic antibacterial effect [10].

MRSA were tested for their capacities to produce biofilm on polystyrene microplates and results indicated that 40% of the strains were highly positive biofilm and 56.66% were low-grade positive. Out of the 30 isolates, only one strain was biofilm negative, which indicates the high potentiality of the isolates to form biofilm. Biofilm, as a virulence factor that favors the chronicity of *S. aureus* infections, is accountable for more 65% of nosocomial infections and 80% of microbial infections [5]. Biofilm is related to various staphylococcal diseases, such as skin and soft tissue infections, nasal colonization, endocarditis and urinary tract infections [39]. Further, biofilm becomes a serious threat in the urology field due to its responsibility for the long persistence of bacteria in the genitourinary tract [40]. The high ability of the investigated isolates to form biofilm confirms the fact that *S. aureus* is the leading species in biofilm-associated infection.

Biofilm has been associated with medical devices and its treatment is becoming increasingly difficult due to the resistance to antibiotics and the immune system in addition to the spread of infection [39]. Thereby, the development of new therapeutic strategies, such as EOs, to inhibit or eradicate biofilm is great of interest. In this work, biofilm inhibitory activity of EOs showed that *O. majorana* had the highest antibiofilm activity (antibiofilm effect on 89.66% of the isolates) against MRSA isolates followed by *R. officinalis* and *T. zygis* that demonstrated activity on 79.31 and 62.06% of the isolates, respectively. EOs also showed a strong potential to inhibit biofilm, with percentage of inhibition ranging from 10.29 to 95.91%, from 10.20 to 95.65%, and from 11.67 to 91.48%, respectively for *O. majorana*, *R. officinalis*, and *T. zygis*. Based on our results, the oil with the highest growth inhibition activity was different from the oil with the highest biofilm inhibition effect, which indicates that the components involved in growth inhibition were different from those associated with biofilm inhibition. According to biochemical specificity, terpinen-4-ol present in *O. majorana* as major compound, has more inhibition of the biofilm formation process by MRSA isolates compared to α-pinene and linalool present, as the main components, in *R. officinalis* and *T. zygis* EOs, respectively.

This finding corroborates the recent data of Cordeiro et al. [32] showing that the strongest antibiofilm activity of terpinen-4-ol was against *S. aureus*. Other studies have also demonstrated that this compound possesses an excellent potential against biofilm formed by some pathogenic bacteria like *Pseudomonas aeruginosa* [41], *Streptococcus mutans*, *Lactobacillus acidophilus* [42], *Porphyromonas gingivalis*, and Fusobacterium nucleatum [43]. Biofilm inhibition properties of *O. majorana*, *R. officinalis*, and *T. zygis* EOs against MRSA suggest that the addition of these oils before biofilm formation eliminates planktonic cells and may reduce the polystyrene surface adherence, which becomes less susceptible to cell adhesion. Additionally, the modification of MRSA surface proteins caused by their interactions with oils inhibits the adhesion of this bacterium to the polystyrene surface, which is the initial attachment phase [44].

Preformed biofilms are difficult to eradicate by conventional antibiotic therapy. However, in the present study, *O. majorana*, *T. zygis*, and *R. officinalis* EOs showed high biofilm eradication activities on 41.37% of the MRSA isolates. *O. majorana* EO had the strongest effect, with the percentage of eradication going up to 98.01%, and seven isolates were changed their biofilm phenotype. It seems that the monoterpenoid terpinen-4-ol has an excellent potential to eradicate mature biofilm than α-pinene and linalool. The activity of these oils on mature biofilms was lower than their capacity to inhibit the formation of biofilms. This can be explained by the fact that the major constituents in these oils have an effect on the biofilm formation process more than on mature biofilm. This is in agreement with the report of Cordeiro et al. [32], showing that terpinen-4-ol is more efficient in inhibiting the formation of *S. aureus* biofilms than in breaking or eliminating mature biofilms. Many EOs, such as tea tree [45], eucalyptus [46], and cinnamon oil [47] have shown their effective ability to remove biofilm. Moreover, *R. officinalis* EO has reduced the quantity of *S. aureus* biofilm to 60.76% [48]. In general, EOs diffuse through polysaccharide matrix of the preformed biofilm and destabilize it because of higher intrinsic antimicrobial activities [44].

4. Materials and Methods

4.1. Bacterial Strains

Thirty clinical MRSA isolates were collected from King Abdulaziz Specialist Hospital, Taif, Saudi Arabia. The isolates were identified as *S. aureus*, as described previously [49]. The methicillin resistance phenotype was performed by the Vitek 2 system (bioMérieux, Durham, North CA, USA) in accordance with the British Society for Antimicrobial Chemotherapy (BSAC). Each isolate was considered as methicillin-resistant when the minimum inhibitory concentration (MIC) breakpoint of oxacillin was >2 mg/L and cefoxitin >4 mg/L. [50]. *S. aureus* ATCC 25922 was used as control.

4.2. Medicinal Plants Essential Oils

Three commercial EOs extracted from medicinal plants were investigated. These EOs were bought from Laboratoires OMEGA Pharma (Groupe Perrigo)—Phytosun Arôms (Châtillon, France) and kept at 4 °C in dark glass bottles till used. These oils were extracted from twigs of *R. officinalis* L. (M14302), and from the aerial parts of flowering stage of *T. zygis* L. subsp. *zygis* (M13184) and *O. majorana* L. (74K100C6). These EOs were carefully chosen for their antibacterial and/or antibiofilm actions, as stated previously [11–13] and their usage in traditional medicine.

4.3. Gas Chromatography—Mass Spectrometry Analysis

The GC-MS analysis was performed as described previously [51].

4.4. Antibacterial Activity of Essential Oils

4.4.1. Disc Diffusion

The agar disc diffusion method was used to evaluate the antibacterial activities of the EOs [52]. Briefly, an overnight cultures of MRSA cells grown at 37 °C were diluted to a density of 0.5 McFarland standards turbidity (DENSIMAT, Bio-merieux, Marcy l'Etoile, France) and were streaked onto Mueller–Hinton agar (MHA) plates using a sterile swab. A sterile filter disc (diameter 6 mm) was placed and then was impregnated by *R. officinalis*, *T. zygis*, and *O. majorana* EOs (10 μL/disc). The plates were maintained at 4 °C for 2 h and then incubated at 37 °C for 24 h. After incubation, the antibacterial activity was evaluated by determining the zone of growth inhibition throughout the discs.

Inhibitory action was categorized according to the zone of inhibition (ZI) as described previously [13–53]. The experiment was performed in triplicate, and the mean diameter of the inhibition zone was documented.

4.4.2. Minimum Inhibitory Concentration and Minimum Bactericidal Concentration

The minimal inhibition concentration (MIC) and the minimal bactericidal concentration (MBC) were assessed in triplicates on 96-well microtiter plates (Nunc, Roskilde, Denmark) as described previously [54]. A bacterial suspension at a density of 0.5 McFarland standards turbidity was prepared from an overnight culture. Then, a serial two-fold dilution for each EOs (50 mg/mL stock solution) was made in 5 mL of nutrient broth with concentration ranged from 0.012–50 mg/mL.

Each well of the 96-well plates contains 95 μL of nutrient broth and 5 μL of the bacterial inoculum. A 100 μL aliquot from the stock solutions of each EO was added into the first well. Then, 100 μL from the serial dilutions were transferred into the consecutive well. The negative control well contains 195 μL of nutrient broth without EO and 5 μL of the bacterial inoculum. The final volume in each well was 200 μL, and the plates were incubated at 37 °C for 18–24 h.

The MIC was defined as the lowest concentration of the EO at which the MRSA cells growth is inhibited. The MBC was determined by subculturing 20 μL from clear wells of the MIC test on MHA. MBC was defined as the lowest concentration of EOs, required to kill ≥99.9% of the initial bacterial inoculum [55].

4.5. Biofilm Formation

Biofilm formation by MRSA isolates was determined using crystal violet assay on U-bottomed 96-well microtiter plates, as detailed previously [56]. Each MRSA strain was tested three times. Wells with sterile TSB only were worked as controls. The optical density of each well was measured at 570 nm (OD570) using an automated Multiskan reader (GIO. DE VITA E C, Rome, Italy). Biofilm formation was interpreted as highly positive (OD570 ≥ 1), low-grade positive (0.1 ≤ OD570 < 1), or negative (OD570 < 0.1).

4.6. Biofilm Inhibition

EOs were tested for their potential to prevent biofilm formation by MRSA isolates. For the experiment, 100 µL of the EOs emulsified in TSB supplement with 2% glucose were put in the U-bottomed 96-well microtiter plate, including 100 µL of bacterial suspensions (10^8 CFU/mL) in each well. The final concentrations of the EOs were equal to MIC, and the final volume was 200 µL per well. The analyzes were performed three times. After incubation of microplates at 37 °C for 24 h, the formed biofilm was measured by crystal violet as described previously [56]. For the Controls wells, the inoculums volume and EOs were replaced by TSB and sterile water, respectively. Inhibition of biofilm was determined from the formula described by Jadhav et al. [57].

$$\%\ \text{Inhibition} = 100 - \left(\frac{\text{OD570 sample}}{\text{OD570 control}} \times 100\right)$$

4.7. Biofilm Eradication

In order to eradicate the preformed biofilm at the maturation stage (48 h biofilms), the plates were incubated for 48 h, the medium was changed after 24 h, and EOs were added at the same concentrations and at the last 24 h. Biofilms formed by bacteria that did not undergo any treatment were used as controls. Experimentally, the plates were incubated for 24 h at 37 °C to allow for biofilm attachment and growth. The following day, the non-adhered cells were removed from each well, and the adhered biofilm was rinsed two times with PBS. Then, 200 µL of TSB (2% glucose) with final concentrations of the EOs equivalent to MIC was added, and the plates were incubated for 24 h [44]. The biofilm was assessed by crystal violet, and eradication of biofilm was calculated as described in Section 4.6. The experiment was carried out in triplicate.

4.8. Statistical Analysis

Statistical analysis was performed using analysis of variance (ANOVA). Pearson's simple linear correlation coefficient (r) and their significance (p) were assessed using IBM SPSS (v20).

5. Conclusions

The outcomes of this study support the medical application of *O. majorana*, *T. zygis*, and *R. officinalis* EOs for the prevention and/or treatment of MRSA infections and diseases as an alternative to or combined with antibiotics. These EOs, provided from Laboratoires OMEGA Pharma–Phytosun Arôms (Châtillon, France), are used orally and in high concentrations (doses), corroborating their non-toxic effect. Generally, the therapeutic application of EOs is limited by their solubility, skin-sensitization synonymous allergic contact dermatitis, and their physicochemical stability due to the volatile components and the conversion of components by cyclization, isomerization, oxidation, or dehydrogenation reactions. Further adequate in vitro testing or in vivo preclinical experiments are warranted to establish safety, efficacy, potential adverse effects, and interaction with other drugs of *O. majorana*, *T. zygis*, and *R. officinalis* EOs before including them in clinical practice.

Author Contributions: F.B.A. conceived and designed the experiments; F.B.A. and R.L. performed the experiments; A.G. funding acquisition; F.B.A. writing—review and editing. All authors have read and agreed to the published version of the manuscript.

Funding: The authors appreciated Taif University Researchers Supporting Project number (TURSP-2020/39), Taif University, Taif, Saudi Arabia.

Acknowledgments: The authors are thankful to Taif University for supplying essential facilities and acknowledge the support of Taif University Researchers Supporting Project number (TURSP-2020/39), Taif University, Taif, Saudi Arabia.

Conflicts of Interest: The authors declare no conflict of interest.

References

1. Gordon, R.J.; Lowy, F.D. Pathogenesis of methicillin-resistant *Staphylococcus aureus* infection. *Clin. Infect. Dis.* **2008**, *46*, S350–S359. [PubMed]
2. Melzer, M.; Welch, C. Tirty-day mortality in UK patients with community-onset and hospital-acquired methicillin susceptible *Staphylococcus aureus* bacteraemia. *J. Hosp. Infect.* **2013**, *84*, 143–150. [PubMed]
3. Doebbeling, B. The epidemiology of methicillin-resistant *Staphylococcus aureus* colonisation and infection. *J. Chemother.* **1995**, *7*, 99–103. [PubMed]
4. Hall-Stoodley, L.; Costerton, J.W.; Stoodley, P. Bacterial biofilms: From the natural environment to infectious diseases. *Nat. Rev. Microbiol.* **2004**, *2*, 95–108. [PubMed]
5. Römling, U.; Balsalobre, C. Biofilm infections, their resilience to therapy and innovative treatment strategies. *J. Intern. Med.* **2012**, *272*, 541–561. [PubMed]
6. Simoes, M.; Bennett, R.N.; Rosa, E.A.S. Understanding antimicrobial activities of phytochemicals against multidrug resistant bacteria and biofilms. *Nat. Prod. Rep.* **2009**, *26*, 746–757. [PubMed]
7. Costerton, J.W.; Stewart, P.S.; Greenberg, E.P. Bacterial biofilms: A common cause of persistent infections. *Science* **1999**, *284*, 1318–1322.
8. Stewart, P.S. Mechanisms of antibiotic resistance in bacterial biofilms. *Int. J. Med. Microbiol.* **2002**, *292*, 107–113.
9. Ebadollahi, A.; Ziaee, M.; Palla, F. Essential oils extracted from different species of the *Lamiaceae* plant family as prospective bioagents against several detrimental pests. *Molecules* **2020**, *25*, 1556.
10. Burt, S. Essential oils: Their antibacterial properties and potential applications in foods—A review. *Int. J. Food Microbiol.* **2004**, *94*, 223–253.
11. Rota, M.C.; Herrera, A.; Martinez, R.M.; Sotomayor, J.A.; Jordan, M.J. Antimicrobial activity and chemical composition of *Thymus vulgaris*, *Thymus zygis*, and *Thymus hyemalis* essential oils. *Food Control* **2008**, *19*, 681–687.
12. Moumni, S.; Elaissi, A.; Trabelsi, A.; Merghni, A.; Chraief, I.; Jelassi, B.; Chemli, R.; Ferchichi, S. Correlation between chemical composition and antibacterial activity of some *Lamiaceae* species essential oils from Tunisia. *BMC Complement Med. Ther.* **2020**, *20*, 103.
13. Lagha, R.; Ben Abdallah, F.; AL-Sarhan, B.O.; Al-Sodany, Y. Antibacterial and Biofilm Inhibitory Activity of Medicinal Plant Essential Oils against *Escherichia coli* Isolated from UTI Patients. *Molecules* **2019**, *23*, 1161.
14. Nostro, A.; Blanco, A.R.; Cannatelli, M.A.; Enea, V.; Flamini, G.; Morelli, I.; Roccaro, A.S.; Alonzo, V. Susceptibility of methicillin-resistant staphylococci to oregano essential oil, carvacrol and thymol. *FEMS Microbiol. Lett.* **2004**, *230*, 191–195.
15. Brady, A.; Loughlin, R.; Gilpin, D.; Kearney, P.; Tunney, M. In vitro activity of tea-tree oil against clinical skin isolates of methicillin-resistant and -sensitive *Staphylococcus aureus* and coagulase-negative staphylococci. *J. Med. Microbiol.* **2006**, *55*, 1375–1380. [PubMed]
16. Jia, P.; Xue, Y.J.; Duan, X.J.; Shao, S.H. Effect of cinnamaldehyde on biofilm formation and sarA expression by methicillin-resistant *Staphylococcus aureus*. *Lett. Appl. Microbiol.* **2011**, *53*, 409–416. [PubMed]
17. Cáceres, M.; Hidalgo, W.; Stashenko, E.; Torres, R.; Ortiz, C. Essential Oils of Aromatic Plants with Antibacterial, Anti-Biofilm and Anti-Quorum Sensing Activities against Pathogenic Bacteria. *Antibiotics* **2020**, *9*, 147.

18. Huma, J.; Fohad, M.H.; Iqbal, A. Antibacterial and antibiofilm activity of some Essential oils and compounds against clinical strains of *Staphylococcus aureus*. *J. Biomed.* **2014**, *1*, 65–71.
19. Nazzaro, F.; Fratianni, F.; de Martino, L.; Coppola, R.; de Feo, V. Effect of essential oils on pathogenic bacteria. *Pharmaceuticals* **2013**, *6*, 1451–1474.
20. Tiwari, B.K.; Valdramidis, V.P.; O'Donnel, C.P.; Muthukumarappan, K.; Bourke, P.; Cullen, P.J. Application of natural antimicrobials for food preservation. *J. Food Chem.* **2009**, *57*, 5987–6000.
21. Chen, C.J.; Huang, Y.C. New epidemiology of *Staphylococcus aureus* infection in Asia. *Clin. Microbiol. Infect.* **2014**, *20*, 605–623. [CrossRef] [PubMed]
22. Ippolito, G.; Leone, S.; Lauria, F.N.; Nicastri, E.; Wenzel, R.P. Methicillin-resistant *Staphylococcus aureus*: The superbug. *Int. J. Infect. Dis.* **2010**, *14*, 7–11. [CrossRef]
23. Sit, P.S.; The, C.S.; Idris, N.; Sam, I.C.; Syed Omar, S.F.; Sulaiman, H.; Thong, K.L.; Kamarulzaman, A.; Ponnampalavana, S. Prevalence of methicillin-resistant *Staphylococcus aureus* (MRSA) infection and the molecular characteristics of MRSA bacteraemia over a two-year period in a tertiary teaching hospital in Malaysia. *BMC Infect. Dis.* **2017**, *17*, 1–14. [CrossRef]
24. Sganga, G.; Tascini, C.; Sozio, E.; Carlini, M.; Chirletti, P.; Cortese, F.; Gattuso, R.; Granone, P.; Pempinello, C.; Sartelli, M.; et al. Focus on the prophylaxis, epidemiology and therapy of methicillin-resistant *Staphylococcus aureus* surgical site infections and a position paper on associated risk factors: The perspective of an Italian group of surgeons. *World J. Emerg. Surg.* **2016**, *11*, 1–13. [CrossRef]
25. Akanbi, B.O.; Mbe, J.U. Occurrence of methicillin and vancomycin resistant *Staphylococcus aureus* in University of Abuja Teaching Hospital, Abuja, Nigeria. *Afr. J. Clin. Exp. Microbiol.* **2013**, *14*, 10–13. [CrossRef]
26. Ghebremedhin, B.; Olugbosi, M.O.; Raji, A.M.; Layer, F.; Bakare, R.A.; König, B.; König, W. Emergence of a community-associated methicillin-resistant *Staphylococcus aureus* strain with a unique resistance profile in Southwest Nigeria. *J. Clin. Microbiol.* **2009**, *47*, 2975–2980. [CrossRef]
27. Watkins, R.R.; Holubar, M.; David, M.Z. Antimicrobial resistance in methicillin-resistant *Staphylococcus aureus* to newer antimicrobial agents. *Antimicrob. Agents. Chemother.* **2019**, *63*, e01216-19. [CrossRef]
28. Hammer, K.; Carson, C.; Riley, T. Antimicrobial activity of essential oils and other plant extracts. *J. Appl. Microbiol.* **1999**, *86*, 985–990. [CrossRef]
29. Pattnaik, S.; Subramanyam, V.R.; Bapaji, M.; Kole, C.R. Antibacterial and antifungal activity of aromatic constituents of essential oils. *Microbios* **1997**, *89*, 39–46. [PubMed]
30. Ouedrhiria, W.; Mounyr, B.; Harkib, H.; Mojac, S.; Grechea, H. Synergistic antimicrobial activity of two binary combinations of marjoram, lavender, and wild thyme essential oils. *Int. J. Food Prop.* **2017**, *12*, 3149–3158. [CrossRef]
31. Thomsen, N.A.; Hammer, K.A.; Riley, T.V.; Van Belkum, A.; Carson, C.F. Effect of habituation to tea tree (*Melaleuca alternifolia*) oil on the subsequent susceptibility of *Staphylococcus* spp. to antimicrobials, triclosan, tea tree oil, terpinen-4-ol and carvacrol. *Int. J. Antimicrob. Agents* **2013**, *41*, 343–351. [CrossRef] [PubMed]
32. Cordeiro, L.; Figueiredo, P.; Souza, H.; Sousa, A.; Andrade-Júnior, F.; Medeiros, D.; Nóbrega, J.; Silva, D.; Martins, E.; Barbosa-Filho, J.; et al. Terpinen-4-ol as an Antibacterial and Antibiofilm Agent against *Staphylococcus aureus*. *Int. J. Mol. Sci.* **2020**, *21*, 4531. [CrossRef]
33. Leite, A.M.; Lima, E.O.; Souza, E.L.; Diniz, M.F.F.M.; Trajano, V.N.; Medeiros, I.A. Inhibitory effect of β-pinene, α-pinene and eugenol on the growth of potential infectious endocarditis causing gram-positive bacteria. *Braz. J. Pharm. Sci.* **2007**, *43*, 121–126. [CrossRef]
34. Nissen, L.; Zatta, A.; Stefanini, I.; Grandi, S.; Sgorbati, B.; Biavati, B.; Monti, A. Characterization and antimicrobial activity of essential oils of industrial hemp varieties (*Cannabis sativa* L.). *Fitoterapia.* **2010**, *81*, 413–419. [CrossRef]
35. Gachkar, L.; Yadegari, D.; Rezaei, M.B.; Taghizadeh, M.; Astaneh, S.A.; Rasooli, I. Chemical and biological characteristics of *Cuminum cyminum* and *Rosmarinus officinalis* essential oils. *Food Chem.* **2007**, *102*, 898–904. [CrossRef]
36. Utegenova, G.A.; Pallister, K.B.; Kushnarenko, S.V.; Özek, G.; Özek, T.; Abidkulova, K.T.; Kirpotina, L.N.; Schepetkin, I.A.; Quinn, M.T.; Voyich, J.M. Chemical Composition and Antibacterial Activity of Essential Oils from *Ferula* L. Species against Methicillin-Resistant *Staphylococcus aureus*. *Molecules* **2018**, *23*, 1679. [CrossRef]

37. Li, L.; Li, Z.W.; Yin, Z.Q.; Wei, Q.; Jia, R.Y.; Zhou, L.J.; Xu, J.; Song, X.; Zhou, Y.; Du, Y.H.; et al. Antibacterial activity of leaf essential oil and its constituents from *Cinnamomum longepaniculatum*. *Int. J. Clin. Exp. Med.* **2014**, *7*, 1721–1727. [PubMed]
38. Santoyo, S.; Cavero, S.; Jaime, L.; Ibanez, E.; Senorans, F.J.; Reglero, G. Chemical composition and antimicrobial activity of *Rosmarinus officinalis* L. essential oil obtained via supercritical fluidextraction. *J. Food Protect.* **2005**, *68*, 790–795. [CrossRef] [PubMed]
39. Paharik, A.E.; Horswill, A.R. The Staphylococcal Bioflm: Adhesins, Regulation, and Host Response. *Microbiol. Spectr.* **2016**, *4*. [CrossRef]
40. Tenke, P.; Kovacs, B.; Jackel, M.; Nagy, E. The role of biofilm infection in urology. *World J. Urol.* **2006**, *24*, 13–20. [CrossRef]
41. Bose, S.K.; Chauhan, M.; Dhingra, N.; Chhibber, S.; Harjai, K. Terpinen-4-ol attenuates quorum sensing regulated virulence factors and biofilm formation in *Pseudomonas aeruginosa*. *Future Microbiol.* **2020**, *15*, 127–142. [CrossRef]
42. Bordini, E.A.F.; Toron, C.C.; Francisconi, R.S.; Magalhães, F.A.C.; Huacho, P.M.M.; Bedran, T.L.; Pratavieira, S.; Spolidorio, L.C.; Spolidorio, D.P. Antimicrobial effects of terpinen-4-ol against oral pathogens and its capacity for the modulation of gene expression. *Biofouling* **2018**, *34*, 815–825. [CrossRef] [PubMed]
43. Maquera-Huacho, P.M.; Tonon, C.C.; Correia, M.F.; Francisconi, R.S.; Bordini, E.A.F.; Marcantonio, É.; Spolidorio, D.M.P. In vitro antibacterial and cytotoxic activities of carvacrol and terpinen-4-ol against biofilm formation on titanium implant surfaces. *Biofouling* **2018**, *34*, 699–709. [CrossRef]
44. Nostro, A.; Roccaro, A.S.; Bisignano, G.; Marino, A.; Cannatelli, M.A.; Pizzimenti, F.C.; Cioni, P.L.; Procopio, F.; Blanco, A.R. Effects of oregano, carvacrol and thymol on *Staphylococcus aureus* and *Staphylococcus epidermidis* biofilms. *J. Med. Microbiol.* **2007**, *56*, 519–523. [CrossRef]
45. Karpanen, T.J.; Worthington, T.; Hendry, E.R.; Conway, B.R.; Lambert, P.A. Antimicrobial efficacy of chlorhexidine digluconate alone and in combination with eucalyptus oil, tea tree oil and thymol against planktonic and biofilm cultures of *Staphylococcus epidermidis*. *J. Antimicrob. Chemother.* **2008**, *62*, 1031–1036. [CrossRef]
46. Hendry, E.R.; Worthington, T.; Conway, B.R.; Lambert, P.A. Antimicrobial efficacy of eucalyptus oil and 1,8-cineole alone and in combination with chlorhexidine digluconate against microorganisms grown in planktonic and biofilm cultures. *J. Antimicrob. Chemother.* **2009**, *64*, 1219–1225. [CrossRef]
47. Nuryastuti, T.; van der Mei, H.C.; Busscher, H.J.; Iravati, S.; Aman, A.T.; Krom, B.P. Effect of cinnamon oil on *icaA* expression and biofilm formation by *Staphylococcus epidermidis*. *Appl. Environ. Microbiol.* **2009**, *75*, 6850–6855. [CrossRef]
48. Ceylan, O.; Uğur, A.; Saraç, N.; Ozcan, F.; Baygar, T. The in vitro antibiofilm activity of *Rosmarinus officinalis* L. essential oil against multiple antibiotic resistant *Pseudomonas* sp. and *Staphylococcus* sp. *J. Food Agric. Environ.* **2014**, *12*, 82–86.
49. Murray, P.; Jorgenson, J.H.; Pfaller, M.A.M.; Yolken, R.H. *Manual of Clinical Microbiology*, 8th ed.; ASM Press: Washington, DC, USA, 2003.
50. Andrews, J.M.; Howe, R.A. BSAC standardized disc susceptibility testing method (version 10). *J. Antimicrob. Chemother.* **2011**, *66*, 2726–2757. [CrossRef]
51. Gooréa, S.G.; Ouattara, Z.A.; Yapi, A.T.; Békro, Y.A.; Bighelli, A.; Paoli, M.; Tomi, F. Chemical composition of the leaf oil of *Artabotrys jollyanus* from Côte d'Ivoire. *Rev. Bras. Farmacogn.* **2017**, *27*, 414–418. [CrossRef]
52. Bagamboula, C.; Uyttendaele, M.; Debevere, J. Inhibitory effect of thyme and basil essential oils, carvacrol, thymol, estragol, linalool and p-cymene towards *Shigella sonnei* and *S. flexneri*. *J. Food microbiol.* **2004**, *21*, 33–42. [CrossRef]
53. El-Deeb, B.; Elhariry, H.; Mostafa, N.Y. Antimicrobial Activity of Silver and Gold Nanoparticles Biosynthesized Using Ginger Extract. *Res. J. Pharm. Biol. Chem. Sci.* **2016**, *7*, 1085.
54. Gulluce, M.; Sahin, F.; Sokmen, M. Antimicrobial and antioxidant properties of the essential oils and methanol extract from *Mentha longifolia* L. *Food Chem.* **2007**, *103*, 1449–1456. [CrossRef]
55. Oulkheir, S.; Aghrouch, M.; El Mourabit, F.; Dalha, F.; Graich, H.; Amouch, F.; Ouzaid, K.; Moukale, A.; Chadli, S. Antibacterial Activity of Essential Oils Extracts from Cinnamon, Thyme, Clove and Geranium against a Gram Negative and Gram-Positive Pathogenic Bacteria. *J. Dis. Med. Plants* **2017**, *3*, 1–5.

56. Ben Abdallah, F.; Chaieb, K.; Zmantar, T.; Kallel, H.; Bakhrouf, A. Adherence assays and Slime production of *Vibrio alginolyticus* and *Vibrio parahaemolyticus*. *Braz. J. Microbiol.* **2009**, *40*, 394–398. [CrossRef] [PubMed]
57. Jadhav, S.; Shah, R.; Bhave, M.; Palombo, E.A. Inhibitory activity of yarrow essential oil on *Listeria* planktonic cells and biofilms. *J. Food Control* **2013**, *29*, 125–130. [CrossRef]

Publisher's Note: MDPI stays neutral with regard to jurisdictional claims in published maps and institutional affiliations.

Communication

Bromo-Cyclobutenaminones as New Covalent UDP-*N*-Acetylglucosamine Enolpyruvyl Transferase (MurA) Inhibitors

David J. Hamilton [1,2], Péter Ábrányi-Balogh [2], Aaron Keeley [2], László Petri [2], Martina Hrast [3], Tímea Imre [4], Maikel Wijtmans [1], Stanislav Gobec [3], Iwan J. P. de Esch [1] and György Miklós Keserű [2,*]

[1] Division of Medicinal Chemistry, Amsterdam Institute of Molecular and Life Sciences (AIMMS), Faculty of Science, Vrije Universiteit Amsterdam, De Boelelaan 1108, 1081 HZ Amsterdam, The Netherlands; d.j.hamilton@vu.nl (D.J.H.); m.wijtmans@vu.nl (M.W.); i.de.esch@vu.nl (I.J.P.d.E.)
[2] Medicinal Chemistry Research Group, Research Centre for Natural Sciences, Magyar tudósok krt 2, H-1117 Budapest, Hungary; abranyi-balogh.peter@ttk.hu (P.Á.-B.); aaron.keeley@ttk.hu (A.K.); petri.laszlo@ttk.hu (L.P.)
[3] Faculty of Pharmacy, University of Ljubljana, Aškerčeva 7, SI-1000 Ljubljana, Slovenia; martina.hrast@ffa.uni-lj.si (M.H.); Stanislav.Gobec@ffa.uni-lj.si (S.G.)
[4] MS Metabolomics Research Group, Research Centre for Natural Sciences, Magyar tudósok krt 2, H-1117 Budapest, Hungary; imre.timea@ttk.hu
* Correspondence: keseru.gyorgy@ttk.hu; Tel.: +36-1-382-6821

Received: 8 October 2020; Accepted: 30 October 2020; Published: 3 November 2020

Abstract: Drug discovery programs against the antibacterial target UDP-*N*-acetylglucosamine enolpyruvyl transferase (MurA) have already resulted in covalent inhibitors having small three- and five-membered heterocyclic rings. In the current study, the reactivity of four-membered rings was carefully modulated to obtain a novel family of covalent MurA inhibitors. Screening a small library of cyclobutenone derivatives led to the identification of bromo-cyclobutenaminones as new electrophilic warheads. The electrophilic reactivity and cysteine specificity have been determined in a glutathione (GSH) and an oligopeptide assay, respectively. Investigating the structure-activity relationship for MurA suggests a crucial role for the bromine atom in the ligand. In addition, MS/MS experiments have proven the covalent labelling of MurA at Cys115 and the observed loss of the bromine atom suggests a net nucleophilic substitution as the covalent reaction. This new set of compounds might be considered as a viable chemical starting point for the discovery of new MurA inhibitors.

Keywords: covalent inhibitor; MurA; cyclobutenaminone; antibacterial; irreversible

1. Introduction

MurA (UDP-*N*-acetylglucosamine enolpyruvyl transferase) is a key enzyme in the peptidoglycan biosynthesis that catalyzes the transfer of phosphoenolpyruvate (PEP) to UDP-*N*-acetylglucosamine (UNAG) [1]. Targeting the catalytic site of MurA leads to the inactivation of the enzyme that increases the osmotic vulnerability of bacteria [2]. MurA is a preferred antibacterial target, as there is no human orthologue for the enzyme. Known MurA inhibitors contain a three- (**1**,**3**), five-(**2**,**4**–**9**) or occasionally six-membered (**10**) heterocycle equipped with a halogen atom leaving group (**6**,**10**), or an epoxide ring (**1**,**3**) that are prone to nucleophilic substitution. Other inhibitors contain a double bond (**2**,**4**,**5**,**7**–**9**) that is available for Michael addition (Figure 1) [3–6].

Our attention was drawn to the cyclobutenone scaffold, as it also harbors a ring with electrophilic character. Cyclobutenones have received relatively little attention in the literature [7–12], and cyclobutyl compounds, in general, are underrepresented in most (fragment) screening libraries [13]. The ring strain of the cyclobutenone unit suggests a substantial reactivity as an electrophile [10,12,14]. Given the foreseen use as covalent fragments, the reactivity and stability in biological assays need to be balanced. Therefore, the electrophilic reactivity was carefully modulated by incorporating an amine functionality in the electrophilic core, giving cyclobutenaminones. As an additional advantage, appending the amine group to the core provides further chemical handles for growing any hit fragments. Last, cyclobutenaminones contain a double bond enabling the incorporation of, e.g., halogen atoms, that can target nucleophilic amino acid side chains, especially that of cysteines.

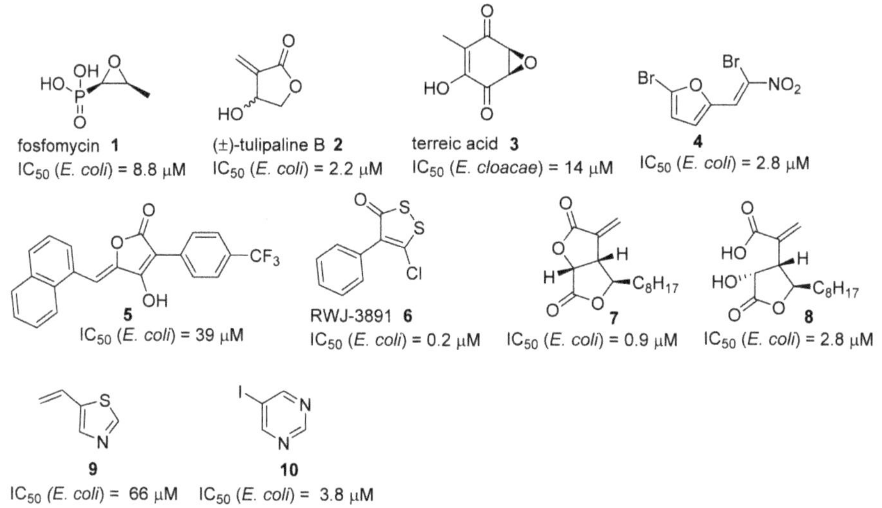

Figure 1. Known MurA inhibitors with small heterocyclic scaffolds.

In continuation of our interests in finding new MurA inhibitors and in the use of the cyclobutyl motif in drug discovery [6,15–17], here we describe cyclobutenaminone derivatives with carefully-modulated electrophilic character as new warheads for covalently targeting the Cys115 residue in the active site of MurA.

2. Results and Discussion

The synthesis of a small set of cyclobutenaminones was accomplished using a strategy based on that from Brand et al. (Scheme 1) [18]. The sequence began with ethoxyacetylene (**11**), which underwent a [2+2] cycloaddition [18–21] with the in situ generated ketene formed via the base-mediated HCl elimination from isobutyryl chloride. The two methyl groups were incorporated so as to restrict the nucleophilic character of enaminones (**14a–e**) to but one position in, e.g., the electrophilic bromination. The sequence furnished ethoxyenone (**12**) in moderate yield, which was transformed by acidic hydrolysis to 2,2-dimethylcyclobutane-1,3-dione (**13**) in high yield [18–21]. Next, dione **13** was condensed with various amines in the presence of AcOH as catalyst at 65 °C to generate the desired enaminones (**14a–e**) in moderate yields [18,22]. Bromination of all enaminones was achieved via electrophilic substitution using Br_2 and base at 0 °C to produce bromoenaminones **15a–e** [18,22]. Selected compounds were subjected to *N*-acylation via deprotonation of the secondary enaminone in tetrahydrofurane (THF) at −78 °C by sodium bis(trimethylsilyl)amide (NaHDMS), followed by subsequent trapping by the relevant acid chloride. Conceivably, the acylation of **14a** to **16a** could also take place at the nucleophilic vinylic position. However, the correct regiochemistry was proven by 2D NMR experiments. The acylations resulted in the corresponding *N*-acyl-cyclobutenaminones (**16a** and **17a–b**) in good yields.

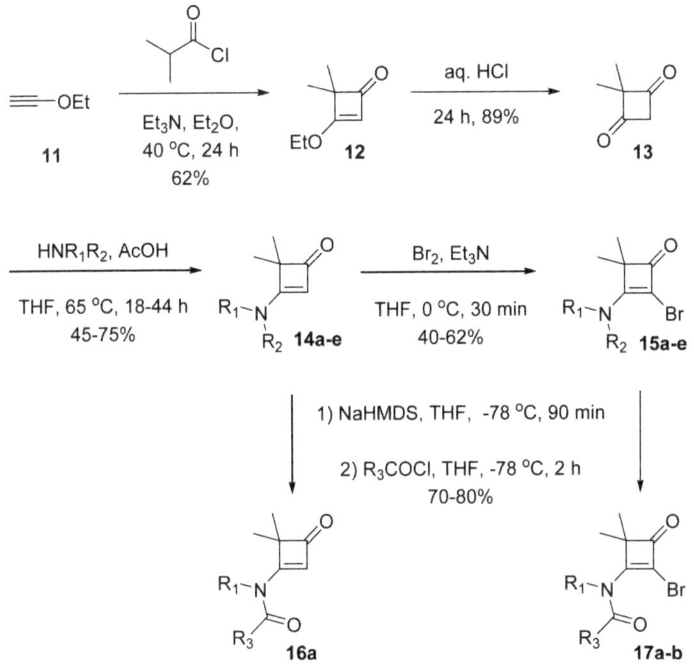

Scheme 1. General synthesis route to various (bromo)enaminones and subsequent *N*-acylation of secondary enaminones. Table 1 shows the different substituents introduced (R_1, R_2), while R_3 = Me or Ph.

Table 1. Structures and biological activity of synthesized compounds.

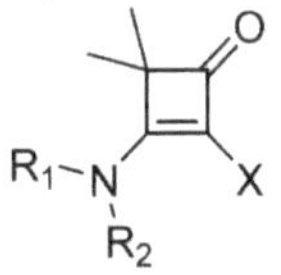

Compound	R_1	R_2	X	RA[a] [%] at 500 μM and IC_{50} [μM]
14a	Me	H	H	93 ± 4
14b	Me	Me	H	99 ± 1
14c	Et	Et	H	96 ± 2
14d	Me	PMB [b]	H	99 ± 3
14e	Me	F_3CCH_2	H	98 ± 7
16a	Me	PhCO	H	99 ± 8
12	-	-	-	98 ± 2
15a	Me	H	Br	100 ± 7
15b	Me	Me	Br	91 ± 5
15c	Et	Et	Br	88 ± 6
15d	Me	PMB[b]	Br	13 ± 1, IC_{50} = 363 ± 11
15e	Me	F_3CCH_2	Br	84 ± 4
17a	Me	PhCO	Br	12 ± 3, IC_{50} = 138 ± 9
17b	Me	MeCO	Br	8 ± 2, IC_{50} = 128 ± 10
fosfomycin	-	-	-	8.8 [3]

[a] RA% refers to the remaining activity in the MurA (*E. coli*) biochemical assay with a fragment concentration of 500 μM with 30 min preincubation at 37 °C; [b] PMB: 4-methoxy-benzyl.

A library of thirteen fragments was prepared, containing nonbrominated (six) and brominated (seven) cyclobutenaminones, all of which are novel to the best of our knowledge (Table 1). This library was then screened against MurA from *E. coli* in order to identify possible starting points for the future development of covalent MurA inhibitors. The screening showed that the cyclobutenaminones with a vinylic proton (**14a–d**) do not give any substantial inhibition. Indeed, the amine substituent selected for balancing reactivity and stability (vide supra) will likely deactivate the double bond by its electron-donating character. We postulated that *N*-trifluoroethylation or *N*-acylation of the nitrogen atom might reactivate the system towards nucleophiles by withdrawing electron density from the conjugated system, but the results on **14e** and **16a** did not support this postulate. As an intermediate in the synthetic route, ethoxycyclobutenone (**12**) was also tested as the ethoxy unit could serve as an improved leaving group, but to no avail. Next, bromination of the vinylic position was explored for activation, bearing in mind that this modification has been successfully applied already in Diels-Alder reactions for improving reactivity [23]. Gratifyingly, several brominated cyclobutenaminones (**15d**, **17a–b**) inhibit MurA from *E. coli* at the 500 μM screening concentration (RA < 15%). Compounds containing amines alkylated with small substituents (**15a–c**) do not show any affinity to the protein, but the incorporation of the 4-methoxybenzyl group (**15d**) increases the affinity to IC_{50} = 363 μM. Turning attention to electron withdrawing groups once again, it was found that *N*-trifluoroethylation has no effect (**15e**), but the *N*-acetyl- and *N*-benzoyl-methylamino derivatives (**17a** and **17b**) substantially inhibit MurA activity. The time-dependent IC_{50} values of these compounds after 30 min are 138 μM and 128 μM, respectively—a substantial effect for such small fragments, with the latter possessing only thirteen heavy atoms. The time dependency of the IC_{50} values (see Supplementary Table S1) and the enhanced electrophilicity caused by the electron-withdrawing substituents suggest a covalent mechanism of action, which was confirmed by proving the labeling on Cys115 by MS/MS measurements for both compounds (Scheme 2E,F, Supplementary Figure S1). The MIC (Minimal Inhibitory Concentrations) values for the antimicrobial action of all compounds were determined against *S. aureus* (ATCC 29213) and *E. coli* (ATCC 25922) bacterial strains. These values were > 625 μM, implying that although MurA inhibition is clearly related to antibiotic action, more finetuning on these structures is needed on the path to a potential new class of antibiotics.

In order to characterize this new electrophilic chemotype, the cysteine reactivity of compound **17a** was evaluated in a GSH (glutathione)-based cysteine surrogate assay (Scheme 2A,B) [15]. The reaction of **17a** with GSH gives an adduct in the HPLC-MS-based assay ($M + H^+$ = 535 Da], suggesting loss of the Br atom, and the conjugation reaction could be characterized with a rate constant of $k_{GSH} = 0.0128\ (M\ min)^{-1}$. Next, the selectivity of **17a** was explored using a nonapeptide assay (Scheme 2C,D) [15]. The KGDYHFPIC nonapeptide contains a cysteine but also other nucleophilic residues i.e., lysine, tyrosine, aspartate, proline, and histidine. As such, the nonapeptide can help to assess the selectivity between different biologically-relevant nucleophiles. In the case of **17a**, only the thiol group of the oligopeptide reacts with the warhead, indicating a high degree of cysteine specificity. To evaluate if the warhead is not too reactive for standard assay conditions, the aqueous stability of the compound (**17a**) was also investigated in PBS buffer (pH 7.4) [15]. The stability proves to be appropriate for biological investigations, as the $t_{1/2}$ value for the aqueous degradation was determined to be 36.5 h at room temperature. The stability and bioavailability of these structures is also supported by the fact that interestingly, the rather unique bromocyclobutenaminone core has been incorporated both in the α4β1/α4β7 integrin antagonist prodrug, Zaurategrast [24,25], which progressed to phase II clinical trials [26], as well as in related compounds [22].

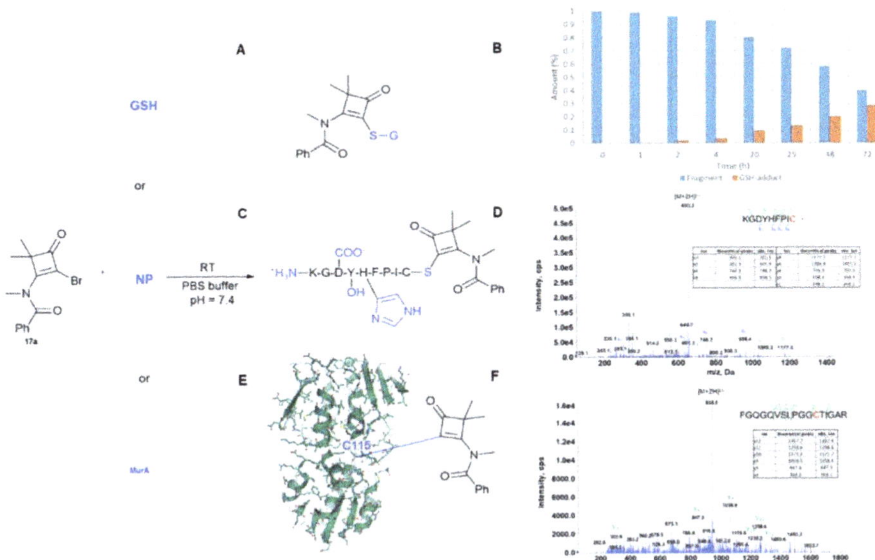

Scheme 2. Labelling of (**A**) glutathione (GSH), (**C**) KGDHFPIC nonapeptide and (**E**) MurA with fragment **17a** together with (**B**) the measured consumption of the fragment (blue columns) and the increasing amount of the GSH-adduct (orange columns) in the GSH-assay, (**D**) the MS/MS spectrum of the Cys-labelled nonapeptide indicating the Cys-labelling together with the theoretical and observed ion peaks and (**F**) the MS/MS spectrum of the digested MurA fragment (amino acids 104–120) labelled on Cys115 together with the theoretical and observed ion peaks. For **E** the 1UAE X-ray structure has been used [27].

3. Materials and Methods

3.1. Synthesis and Characterisation of Compounds

All starting materials were obtained from commercial suppliers (primarily being Sigma-Aldrich (Swijndrecht, The Netherlands), Fluorochem (Hadfield, Derbyshire, UK) and CombiBlocks (San Diego, CA, USA)) and used without purification. Anhydrous Et_2O, dichloromethane (DCM), acetonitrile (MeCN) and tetrahydrofurane (THF) were obtained by passing through an activated alumina column prior to use. All other solvents used were used as received unless otherwise stated. All reactions were carried out under a nitrogen atmosphere unless mentioned otherwise. TLC analyses were performed using Merck F_{254} (Merck KGaA, Darmstadt, Germany or VWR International B.V., Amsterdam, The Netherlands) aluminum-backed silica gel plates and visualized with 254 nm UV light or a potassium permanganate stain. Flash column chromatography was executed using Silicycle Siliaflash F_{60} silica gel (SiliCycle Inc., Quebec City, QC, Canada or Screening Devices, Amersfoort, The Netherlands) or by means of a Teledyne Isco CombiFlash (Teledyne Isco Inc., Lincoln, NE, USA or Beun de Ronde, Abcoude, The Netherlands) or a Biotage Isolera equipment using Biotage SNAP columns (Biotage AB, Uppsala, Sweden). All HRMS spectra were recorded on a Bruker micrOTOF mass spectrometer (Bruker Corp., Billerica, MA, USA) using ESI in positive-ion mode. All NMR spectra were recorded on either a Bruker Avance 300, Bruker Avance 500, or Bruker Avance 600 spectrometer (Bruker Corp., Billerica, MA, USA or Fällanden, Switzerland). The peak multiplicities are defined as follows: s, singlet; bs, broad singlet; d, doublet; t, triplet; q, quartet; p, pentet; dd, doublet of doublets; dt, doublet of triplets; td, a triplet of doublets; m, multiplet; app, apparent. The spectra were referenced to the internal solvent peak as follows: $CDCl_3$ (^1H = 7.26 ppm, ^{13}C = 77.16 ppm), DMSO-d6 (^1H = 2.50 ppm, ^{13}C = 39.52 ppm). IUPAC names were adapted from ChemDraw Professional 16.0 (PerkinElmer). Purities were measured with the aid of analytical LC–MS using a Shimadzu LC-20AD liquid chromatography pump system (Shimadzu Corp., Kyoto, Japan or 's Hertogenbosch, The Netherlands) with a Shimadzu SPDM20A diode array detector (Shimadzu Corp., Kyoto, Japan) with the MS detection performed with a Shimadzu LC-MS-2010EV mass spectrometer (Shimadzu Corp., Kyoto, Japan) operating in both positive and negative ionization mode. The column used was an XBridge (C18) 5 μm column (50 mm × 4.6 mm) (Waters Corp., Milford, MA, USA or Phenomenex, Utrecht, The Netherlands). The following solutions are used for the eluents. Solvent A: H_2O (+0.1% HCOOH) and solvent B: MeCN (+0.1% HCOOH). The eluent program used is as follows: flow rate: 1.0 mL/min, start 95% A in a linear gradient to 10% A over 4.5 min, hold 1.5 min at 10% A, in 0.5 min in a linear gradient to 95% A, hold 1.5 min at 95% A, total run time: 8.0 min. Compound purities were calculated as the percentage peak area of the analysed compound by UV detection at 254 nm.

3.2. *GSH Reactivity and Aqueous Stability Assay*

The assay was adapted from our former publication [15].

For the glutathione assay, 500 μM solution of the fragment (PBS buffer pH 7.4, 10% MeCN, 250 μL) with 200 μM solution of indoprofen (Merck KGaA, Darmstadt, Germany) as internal standard was added to 10 mM glutathione (Merck KGaA, Darmstadt, Germany) solution (dissolved in PBS buffer, 250 μL) in a 1:1 ratio. The final concentration was 250 μM fragment, 100 μM indoprofen, 5 mM glutathione and 5% MeCN (500 μL). The final mixture was analyzed by HPLC-MS (Shimadzu LCMS-2020) after 0, 1, 2, 4, 20, 25, 48, 72 h time intervals. Degradation kinetics were also investigated respectively using the previously described method, applying pure PBS buffer instead of the glutathione solution. In this experiment, the final concentration of the mixture was 250 μM fragment, 100 μM indoprofen and 5% MeCN. The AUC (area under the curve) values were determined via integration of HPLC spectra then corrected using the internal standard. The fragment AUC values were applied for ordinary least squares (OLS) linear regression and for computing the important parameters (kinetic rate constant, half-life time) a programmed excel (Visual Basic for Applications) was utilized. The data are expressed as means of duplicate determinations, and the standard deviations were within 10% of the given values.

The calculation of the kinetic rate constant for the degradation and corrected GSH-reactivity is as follows:

The reaction half-life for pseudo-first-order reactions is $t_{1/2} = \ln2/k$, where k is the reaction rate. In the case of competing reactions (reaction with GSH and degradation), the effective rate for the consumption of the starting compound is $k_{eff} = k_{deg} + k_{GSH}$. When measuring half-lives experimentally, the $t_{1/2(eff)} = \ln2/(k_{eff}) = \ln2/(k_{deg} + k_{GSH})$. In our case, the corrected k_{deg} and k_{eff} (regarding blank and GSH-containing samples, respectively) can be calculated by linear regression of the data points of the kinetic measurements. The corrected k_{GSH} is calculated by $k_{eff} - k_{deg}$, and finally, the half-life time is determined using the equation $t_{1/2(GSH)} = \ln2/k_{GSH}$.

3.3. *Oligopeptide Selectivity Assay*

The assay was adapted from our former publication [15].

For the nonapeptide assay, a 2 mM solution of the fragment (PBS buffer pH 7.4 with 20% MeCN) was added to 200 μM nonapeptide solution (PBS buffer pH 7.4) in a 1:1 ratio. The final assay mixture contained 1 mM fragment, 100 μM peptide and 10% MeCN. Based on the GSH reactivity, the applied incubation time was 24 h.

3.4. *LC-MS/MS Measurement and Data Analysis of the Nonapeptide Reactivity Assay*

A Sciex 6500 QTRAP triple quadrupole—linear ion trap mass spectrometer, equipped with a Turbo V Source in electrospray mode (AB Sciex Pte. Ltd., Framingham, MA, USA) and an Agilent 1100 Binary Pump HPLC system (Agilent Technologies, Waldbronn, Germany) equipped with an autosampler was used for LC-MS/MS analysis. Data acquisition and processing were performed using Analyst software version 1.6.2 (AB Sciex Pte. Ltd., Framingham, MA, USA). Chromatographic separation was achieved by Purospher STAR RP-18 endcapped (50 mm × 2.1 mm, 3μm) LiChocart® 55-2 HPLC Cartridge (Merck KGaA, Darmstadt, Germany). The sample was eluted with gradient elution using solvent A (0.1% HCOOH in water) and solvent B (0.1% HCOOH in MeCN). Flow rate was set to 0.5 mL/min. The initial condition was 5% B for 2 min, followed by a linear gradient to 95% B for 6 min, followed by holding at 95% B 6–8 min; and from 8 to 8.5 min back to the initial condition with 5% eluent B and held for 14.5 min. The column temperature was kept at room temperature and the injection volume was 10 μL. Nitrogen was used as the nebulizer gas (GS1), heater gas (GS2), and curtain gas with the optimum values set at 35, 45 and 45 (arbitrary units). The source temperature was 450 °C and the ion spray voltage set at 5000 V. The declustering potential value was set to 150 V. Information Dependent Acquisition (IDA) LC-MS/MS experiment was used to determine if the fragment binding was specific to thiol residues or not. An enhanced MS scan was applied as the survey scan and enhanced product ion (EPI) was the dependent scan. The collision energy in EPI experiments was set to 30 eV with a collision energy spread (CES) of 10 V. The identification of the binding position of the fragments to the nonapeptide was performed using GPMAW 4.2. software.

3.5. *Tryptic Digestion of MurA*

The tryptic digestion method was adapted from our former publication [15].

Briefly, 50 μL of MurA (42 μM) and 10 μL 0.2% (*w*/*v*) RapiGest SF (Waters Corp., Milford, MA, USA) solution buffered with 50 mM ammonium bicarbonate (NH_4HCO_3) were mixed (pH = 7.8). 4.5 μL of 45 mM DTT (~200 nmol) in 100 mM NH_4HCO_3 was added and the mixture kept at 37.5 °C for 30 min. After cooling the sample to room temperature, 7.5 μL of 100 mM iodoacetamide (750 nmol) in 100 mM NH_4HCO_3 was added and the mixture placed in the dark at room temperature for 30 min. The reduced and alkylated protein was then digested by 10 μL (1 mg mL^{-1}) trypsin (the enzyme-to-protein ratio was 1:10) (Sigma, St Louis, MO, USA). The sample was incubated at 37 °C overnight. To degrade the surfactant, 7 μL of HCOOH (500 mM) solution was added to the digested HDAC8 sample to obtain the final 40 mM (pH ≈ 2) solution which was incubated at 37 °C for 45 min. For LC-MS analysis, the acid-treated sample was centrifuged for 5 min at 13,000 rpm.

3.6. *LC-MS/MS Measurements on Digested MurA*

A QTRAP 6500 triple quadruple—linear ion trap mass spectrometer, equipped with a Turbo V source in electrospray mode (AB Sciex Pte. Ltd., Framingham, MA, USA) and an Agilent 1100 Binary Pump HPLC system (Agilent Technologies, Waldbronn, Germany) equipped with an autosampler was used for LC-MS/MS analysis. Data acquisition and processing were performed using Analyst software version 1.6.2 (AB Sciex Pte. Ltd., Framingham, MA, USA). Chromatographic separation was achieved by using the Discovery® BIO Wide Pore C-18-5 (250 mm × 2.1 mm, 5 μm). The sample was eluted with a gradient of solvent A (0.1% HCOOH in water) and solvent B (0.1% HCOOH in MeCN). The flow rate was set to 0.2 mL min^{-1}. The initial conditions for separation were 5% B for 7 min, followed by a linear gradient to 90% B for 53 min, followed by 90% B for 3 min; over 2 min back to the initial conditions with 5% eluent B retained for 10 min. The injection volume was 10 μL (300 pmol on the column).

An Information-Dependent Acquisiton (IDA) LC-MS/MS experiment was used to identify the modified tryptic MurA peptide fragments. Enhanced MS scan (EMS) was applied as the survey scan and an enhanced product ion (EPI) was the dependent scan. The collision energy in EPI experiments was set to rolling collision energy mode, where the actual value was set on the basis of the mass and charge state of the selected ion. Further IDA criteria: ions greater than: 400.00 *m*/*z*, which exceeds 106 counts, exclude former target ions for 30 s after 2 occurrence(s). In EMS and in EPI mode, the scan rate was 1000 Da/s as well. Nitrogen was used as the nebulizer gas (GS1), heater gas (GS2), and curtain gas with the optimum values set at 50, 40 and 40 (arbitrary units). The source temperature was 350 °C and the ion spray voltage was set at 5000 V. The declustering potential value was set to 150 V.

GPMAW 4.2. software was used to analyse a large number of MS-MS spectra and identify the modified tryptic MurA peptides.

3.7. MurA Assay

$MurA_{EC}$ protein was recombinant, expressed in *E. coli.* [28] The inhibition of MurA was monitored with the colorimetric malachite green method in which orthophosphate generated during the reaction is measured. MurA enzyme (*E. coli*) was pre-incubated with the substrate UNAG and compound for 30 min at 37 °C. The reaction was started by the addition of the second substrate PEP, resulting in a mixture with a final volume of 50 μL. The mixtures contained: 50 mM Hepes, pH 7.8, 0.005% Triton X-114, 200 μM UNAG, 100 μM PEP, purified MurA (diluted in 50 mM Hepes, pH 7.8) and 500 μM of each tested compound dissolved in DMSO. All compounds were soluble in the assay mixtures containing 5% DMSO (*v*/*v*). After incubation for 15 min at 37 °C, the enzyme reaction was terminated by adding Biomol® reagent (100μL) and the absorbance was measured at 650 nm after 5 min. All of the experiments were run in duplicate. Remaining activities (RAs) were calculated with respect to similar assays without the tested compounds and with 5% DMSO. The IC_{50} values, the concentration of the compound at which the remaining activity was 50%, were determined by measuring the remaining activities at seven different compound concentrations. The data are expressed as means of duplicate determinations, and the standard deviations were within 10% of the given values. A time-dependent inhibition assay was also performed. The IC_{50} values were determined at 0, 15 and 30 min of pre-incubation.

3.8. Antimicrobial Testing (MIC Determination)

Antimicrobial testing was carried out by the broth microdilution method in 96-well plate format following the CLSI guidelines and European Committee for Antimicrobial Susceptibility Testing recommendations. Bacterial suspension of specific bacterial strain, equivalent to 0.5 McFarland turbidity standard, was diluted with cation-adjusted Mueller Hinton broth to obtain a final inoculum of 10^5 CFU/mL. Compounds dissolved in DMSO and inoculum were mixed together and incubated for 20–24 h at 37 °C. After incubation the minimal inhibitory concentration (MIC) values were determined by visual inspection as the lowest dilution of compounds showing no turbidity. The MICs were determined against *S. aureus* (ATCC 29213) and *E. coli* (ATCC 25922) bacterial strains. Tetracycline was used as a positive control on every assay plate, showing a MIC of 0.5 μg/mL and 1 μg/mL for *S. aureus* and *E. coli*, respectively.

3.9. Chemical Syntheses

3-Ethoxy-4,4-dimethylcyclobut-2-en-1-one (**12**)

To a stirred solution of isobutyryl chloride (10.7 mL, 102 mmol) and ethoxyacetylene **11** (50.0 mL, 205 mmol, 40% wt in hexanes) in Et_2O (128 mL), Et_3N (21.4 mL, 154 mmol) was added slowly over 5 min. The mixture was stirred at rt for 30 min before being heated at 40 °C for 24 h. The mixture was then allowed to cool. The precipitate was filtered and the filtrate was concentrated in vacuo. The crude product was purified over silica gel with a gradient of 10–40% EtOAc/cHex to afford the title compound **2** (8.90 g, 62% yield) as a yellow oil.

^1H NMR (500 MHz, Chloroform-*d*) δ 4.78 (s, 1H), 4.21 (q, *J* = 7.1 Hz, 2H), 1.45 (t, *J* = 7.1 Hz, 3H), 1.24 (s, 6H). ^{13}C NMR (126 MHz, Chloroform-*d*) δ 194.1, 190.4, 102.4, 69.4, 60.1, 19.7, 14.3. LC-MS: RT = 3.33 min, 99+% (254 nm), *m*/*z* $[M + H]^+$ = 141. HRMS calculated for $C_8H_{13}O_2^+$ $[M + H]^+$ = 141.0910, found 141.0923.

2,2-Dimethylcyclobutane-1,3-dione (**13**)

To a flask containing enol ether **12** (8.40 g, 59.9 mmol) was added HCl (2.0 M in H_2O, 45.0 mL, 90.0 mmol) in one portion and the mixture was stirred vigorously at rt for 24 h. The product was extracted with DCM (3×). The organic layers were combined, dried over $MgSO_4$, and concentrated in vacuo to afford the title compound **3** (6.20 g, 92% yield) as a flaky brown solid.

^{1}H NMR (600 MHz, Chloroform-*d*) δ 3.92 (s, 2H), 1.28 (s, 6H). ^{13}C NMR (151 MHz, Chloroform-*d*) δ 207.0, 73.0, 60.4, 17.6. LC-MS: RT = 1.78 min, 98% (254 nm), *m/z* $[M + H]^+$ = 113. HRMS calculated for $C_6H_9O_2^+$ $[M + H]^+$ = 113.0597, found 113.0603.

3.10. *General Procedure A: Enaminone Formation*

To a solution of dione **13** (1.0 eq) in THF (0.50 M) was added amine (1.1 eq), AcOH (1.1 eq) and a spatula of Na_2SO_4. The reaction mixture was stirred at 65 °C for the indicated time. The reaction mixture was allowed to cool to rt. The solids were filtered and the filtrate concentrated in vacuo. The residue was taken up in EtOAc and washed with satd. aq. Na_2CO_3 and brine. The organic layer was dried over Na_2SO_4, filtered and concentrated *in vacuo*. The crude product was purified over silica gel using the indicated gradient of MeOH/DCM to afford the product enaminone.

3-(Methylamino)-4,4-dimethylcyclobut-2-en-1-one (**14a**)

This compound was prepared according to General Procedure **A** using dione **13** (388 mg, 3.46 mmol), $MeNH_2$ (2.0 M in THF, 1.90 mL, 3.81 mmol) and a reaction time of 40 h. Purification over silica gel using a gradient of 0–10% MeOH/DCM afforded the title compound **14a** (350 mg, 81%) as a pale brown solid.

Rotamers are observed in ratio ca. 1.0:0.1 in Chloroform-*d*. Only peaks corresponding to the major rotamer are reported. ^{1}H NMR (500 MHz, Chloroform-*d*) δ 5.63 (br s, 1H), 4.59 (s, 1H), 2.99 (d, *J* = 5.0 Hz, 3H), 1.24 (s, 6H). ^{13}C NMR (126 MHz, Chloroform-*d*) δ 192.2, 178.8, 95.7, 58.5, 31.8, 20.3. LC-MS: RT = 3.20 min, 99+% (254 nm), *m/z* $[M + H]^+$ = 126. HRMS calculated for $C_7H_{12}NO^+$ $[M + H]^+$ = 126.0913, found 126.0912.

3-(Dimethylamino)-4,4-dimethylcyclobut-2-en-1-one (**14b**)

This compound was prepared according to General Procedure **A** using dione **13** (200 mg, 1.78 mmol), Me_2NH (2.0 M in THF, 0.20 mL, 1.96 mmol) and a reaction time of 40 h. Purification over silica gel using a gradient of 0–10% MeOH/DCM afforded the title compound **14b** (191 mg, 77% yield) as a brown crystalline solid.

Rotamers are observed in ratio 1.0:1.0 in Chloroform-*d*. All peaks for both rotamers are reported. ^{1}H NMR (500 MHz, Chloroform-*d*) δ 4.53 (s, 1H), 3.07 (s, 3H), 2.99 (s, 3H), 1.32 (s, 6H). ^{13}C NMR (126 MHz, Chloroform-*d*) δ 190.9, 178.4, 96.0, 58.2, 40.1, 39.4, 21.2. LC-MS: RT = 2.34 min, 99 + % (254 nm), *m/z* $[M + H]^+$ = 140. HRMS calculated for $C_8H_{14}NO^+$ $[M + H]^+$ = 140.1064, found 140.1066.

3-(Diethylamino)-4,4-dimethylcyclobut-2-en-1-one (**14c**)

This compound was prepared according to General Procedure **A** using dione **13** (200 mg, 1.78 mmol), Et_2NH (0.20 mL, 1.96 mmol) and a reaction time of 40 h, followed by an additional portion of Et_2NH (0.10 mL, 0.89 mmol) and stirring for a further 5 h. Purification over silica gel using a gradient of 0–10% MeOH/DCM afforded the title compound **14c** (175 mg, 59% yield) as a brown oil.

Rotamers are observed in ratio 1.0:1.0 in Chloroform-*d*. All peaks for both rotamers are reported. ^{1}H NMR (500 MHz, Chloroform-*d*) δ 4.54 (s, 1H), 3.34 (q, *J* = 7.2 Hz, 2H), 3.27 (q, *J* = 7.2 Hz, 2H), 1.33 (s, 6H), 1.25 (t, *J* = 7.2 Hz, 3H), 1.22 (t, *J* = 7.2 Hz, 3H). ^{13}C NMR (126 MHz, Chloroform-*d*) δ 190.9, 177.5, 95.6, 58.4, 44.6, 44.1, 21.4, 14.2, 12.3. LC-MS: RT = 3.06 min, 99+% (254 nm), *m/z* $[M+H]^+$ = 168. HRMS calculated for $C_{10}H_{18}NO^+$ $[M + H]^+$ = 168.1377, found 168.1382.

3-((4-Methoxybenzyl)(methyl)amino)-4,4-dimethylcyclobut-2-en-1-one (**14d**)

This compound was prepared according to General Procedure **A** using dione 13 (200 mg, 1.78 mmol), (4-methoxybenzyl)-*N*-methylamine (0.29 mL, 1.96 mmol) and a reaction time of 40 h. Purification over silica gel using a gradient of 0–10% MeOH/DCM afforded the title compound **14d** (330 mg, 75% yield) as a viscous brown oil.

Rotamers are observed in ratio 1.0:1.0 in Chloroform-*d*. All peaks for both rotamers are reported. ^{1}H NMR (600 MHz, Chloroform-*d*) δ 7.18–7.12 (m, 2H), 6.93–6.88 (m, 2H), 4.70 (s, 1H), 4.59 (s, 1H), 4.42 (s, 2H), 4.29 (s, 2H), 3.82 (s, 3H), 3.81 (s, 3H), 2.95 (s, 3H), 2.80 (s, 3H), 1.38 (s, 6H), 1.34 (s, 6H). ^{13}C NMR (151 MHz, Chloroform-*d*) δ 190.9, 190.8, 178.4, 178.3, 159.6, 129.1, 128.7, 126.9, 126.7, 114.5, 114.4, 96.2, 95.9, 58.4, 58.3, 56.2, 55.4, 55.4, 55.3, 36.6, 36.3, 21.4, 21.1. LC-MS: RT = 3.62 min, 99+% (254 nm), *m*/*z* $[M + H]^+$ = 246. HRMS calculated for $C_{15}H_{20}NO_2^+$ $[M + H]^+$ = 246.1489, found 246.1489.

4,4-Dimethyl-3-(methyl(2,2,2-trifluoroethyl)amino)cyclobut-2-en-1-one (**14e**)

This compound was prepared according to General Procedure **A** using dione **13** (140 mg, 1.25 mmol), (2,2,2-trifluoroethyl)-methylamine (0.14 mL, 1.37 mmol) and a reaction time of 16 h. Purification over silica gel using a gradient of 0–10% MeOH/DCM afforded the title compound **14e** (197 mg, 76% yield) as a brown oil.

Rotamers are observed in ratio 1.0:0.8 in Chloroform-*d*. All peaks for both rotamers are reported. ^{1}H NMR (500 MHz, Chloroform-*d*) δ 4.67 (s, 1H), δ 4.70 (s, 1H), 3.80 (q, *J* = 8.6 Hz, 2H), 3.75 (q, *J* = 9.0 Hz, 2H), 3.20 (s, 3H), 3.09–3.10 (m, 3H), 1.35 (s, 6H), 1.32 (s, 6H). ^{13}C NMR (151 MHz, Chloroform-*d*) δ 190.63, 179.51, 179.47, 124.47 (q, *J* = 281.9 Hz), 123.70 (q, *J* = 281.9 Hz), 99.72, 99.03, 54.29 (q, *J* = 34.1 Hz), 53.51 (q, *J* = 34.1 Hz), 39.06, 21.19, 21.08. LC-MS: RT = 3.30 min, 99+% (254 nm), *m*/*z* $[M + H]^+$ = 208. HRMS calculated for $C_9H_{13}F_3NO^+$ $[M + H]^+$ = 208.0944, found 208.0947.

3.11. *General Procedure B: Bromination*

A solution of enaminone (1.0 eq) and Et_3N (2.0 eq) in THF (0.10 M) at 0 °C was treated dropwise with a solution of Br_2 (1.1 eq) in THF (2.0 mL). The reaction mixture was stirred at 0 °C for the indicated time. The mixture was diluted with EtOAc and washed with satd. aq. $NaHCO_3$ and brine. The organic layer was dried over $MgSO_4$, filtered and concentrated *in vacuo*. The crude product was purified over silica gel using the indicated gradient of MeOH/EtOAc followed by reversed-phase chromatography on C18 silica gel using the indicated gradient of MeCN/H_2O to afford the product bromoenaminone.

2-Bromo-4,4-dimethyl-3-(methylamino)cyclobut-2-en-1-one (**15a**)

This compound was prepared according to General Procedure B using enaminone **14a** (80 mg, 0.64 mmol) and a reaction time of 1 h. Purification over silica gel using a gradient of 0–10% MeOH/EtOAc and over reversed-phase C18 silica gel using a gradient of 0–100% MeCN/H_2O (+0.1% HCOOH) afforded the title compound **15a** (81 mg, 62% yield) as a pale yellow solid.

Rotamers are observed in ratio 1.0:0.6 in Chloroform-*d*. All peaks for both rotamers are reported. ^{1}H NMR (600 MHz, Chloroform-*d*) δ 5.97 (br s, 1H), 5.55 (br s, 1H), 3.30 (d, *J* = 5.2 Hz, 3H), 3.12 (d, *J* = 5.2 Hz, 3H), 1.38 (s, 6H), 1.24 (s, 6H). ^{13}C NMR (151 MHz, Chloroform-*d*) δ 187.8, 185.9, 177.6, 175.5, 72.8, 70.5, 59.1, 58.7, 31.6, 31.4, 20.8, 19.9. LC-MS: RT = 2.80 min, 99+% (254 nm), *m*/*z* $[M + H]^+$ = 204 (light isotope). HRMS calculated for $C_7H_{11}NOBr^+$ $[M + H]^+$ = 205.9998 (heavy isotope), found 205.9997.

2-Bromo-4,4-dimethyl-3-(methylamino)cyclobut-2-en-1-one (**15b**)

This compound was prepared according to General Procedure B using enaminone **14b** (80 mg, 0.58 mmol) and a reaction time of 1 h. Purification over silica gel using a gradient of 0–10% MeOH/EtOAc and over reversed-phase C18 silica gel using a gradient of 0–100% MeCN/H_2O (+0.1% HCOOH) afforded the title compound **15b** (59 mg, 47% yield) as a pale yellow solid.

Rotamers are observed in ratio 1.0:1.0 in Chloroform-*d*. All peaks for both rotamers are reported. ^{1}H NMR (600 MHz, Chloroform-*d*) δ 3.35 (s, 3H), 3.07 (s, 3H), 1.32 (s, 6H). ^{13}C NMR (151 MHz, Chloroform-*d*) δ 186.5, 174.5, 70.3, 58.7, 40.7, 39.6, 20.9. LC-MS: RT = 3.06 min, 99+% (254 nm), *m*/*z* $[M + H]^+$ = 218 (light isotope). HRMS calculated for $C_8H_{13}NOBr^+$ $[M + H]^+$ = 218.0175 (light isotope), found 218.0182.

2-Bromo-3-(diethylamino)-4,4-dimethylcyclobut-2-en-1-one (**15c**)

This compound was prepared according to General Procedure B using enaminone **14c** (80 mg, 0.48 mmol) and a reaction time of 1 h. Purification over silica gel using a gradient of 0–10% MeOH/EtOAc and over reversed-phase C18 silica gel using a gradient of 0–100% MeCN/H_2O (+0.1% HCOOH) afforded the title compound **15c** (49 mg, 42% yield) as a colourless oil.

Rotamers are observed in ratio 1.0:1.0 in Chloroform-*d*. All peaks for both rotamers are reported. ^{1}H NMR (600 MHz, Chloroform-*d*) δ 3.64 (q, *J* = 7.2 Hz, 1H), 3.35 (q, *J* = 7.2 Hz, 1H), 1.30 (t, *J* = 7.2 Hz, 3H), 1.28 (t, *J* = 7.2 Hz, 3H). ^{13}C NMR (151 MHz, Chloroform-*d*) δ 186.6, 173.9, 69.7, 58.8, 45.7, 43.0, 21.1, 14.4, 14.2. LC-MS: RT = 3.78 min, 97% (254 nm), *m*/*z* $[M + H]^+$ = 248 (heavy isotope). HRMS calculated for $C_{10}H_{17}NOBr^+$ $[M + H]^+$ = 246.0488 (light isotope), found 246.0488.

2-Bromo-3-((4-methoxybenzyl)(methyl)amino)-4,4-dimethylcyclobut-2-en-1-one (**15d**)

This compound was prepared according to General Procedure B using enaminone **14d** (100 mg, 0.41 mmol) and a reaction time of 2 h. Purification over silica gel using a gradient of 0–10% MeOH/EtOAc and over reversed-phase C18 silica gel using a gradient of 0–100% MeCN/H_2O (+0.1% HCOOH) afforded the title compound **15d** (51 mg, 39% yield) as a pale yellow oil.

Rotamers are observed in ratio 1:0.6 in Chloroform-*d*. All peaks for both rotamers are reported. ^{1}H NMR (600 MHz, Chloroform-*d*) δ 7.25–7.22 (m, 2H), 7.17–7.14 (m, 2H), 6.95–6.91 (m, 4H), 4.75 (s, 2H), 4.41 (s, 2H), 3.82 (s, 3H), 3.82 (s, 3H), 3.16 (s, 3H), 2.95 (s, 3H), 1.39 (s, 6H), 1.35 (s, 6H). ^{13}C NMR (151 MHz, Chloroform-*d*) δ 186.71, 186.68, 174.72, 174.28, 159.90, 159.82, 129.52, 128.88, 127.15, 125.90, 114.70, 114.60, 70.51, 70.45, 58.96, 58.93, 56.64, 55.50, 55.47, 54.65, 37.43, 36.27, 21.23, 20.94. LC-MS: RT = 4.24 min, 99+% (254 nm), *m*/*z* $[M + H]^+$ = 324 (light isotope). HRMS calculated for $C_{15}H_{19}NO_2Br^+$ $[M + H]^+$ = 326.0573 (heavy isotope), found 326.0573.

2-Bromo-4,4-dimethyl-3-(methyl(2,2,2-trifluoroethyl)amino)cyclobut-2-en-1-one (**15e**)

This compound was prepared according to General Procedure B using enaminone **14e** (80 mg, 0.39 mmol) and a reaction time of 1 h. Purification over silica gel using a gradient of 0–10% MeOH/EtOAc and over reversed-phase C18 silica gel using a gradient of 0–100% MeCN/H_2O (+0.1% HCOOH) afforded the title compound **15e** (32% yield) as a pale yellow solid.

Rotamers are observed in ratio 1:0.4 in Chloroform-*d*. All peaks for both rotamers are reported. ^{1}H NMR (600 MHz, Chloroform-*d*) δ 4.27 (q, *J* = 8.3 Hz, 2H), 3.79 (q, *J* = 8.3 Hz, 2H), 3.46 (s, 3H), 3.22 (s, 3H), 1.37 (s, 6H), 1.34 (s, 6H). ^{13}C NMR (151 MHz, Chloroform-*d*) δ 186.51, 175.87, 175.66, 123.95 (q, *J* = 282.1 Hz), 123.51 (q, *J* = 282.1 Hz), 74.86, 74.20, 59.70, 59.61, 54.57 (q, *J* = 34.0 Hz), 52.41 (q, *J* = 34.0 Hz), 40.05, 39.13, 20.93, 20.84. LC-MS: RT = 3.86 min, 99+% (254 nm), *m*/*z* $[M + H]^+$ = 286 (light isotope). HRMS calculated for $C_9H_{12}NOF_2Br^+$ $[M + H]^+$ = 286.0049 (light isotope), found 286.0044.

3.12. General Procedure C: Enaminone N-Acylation

To a solution of enaminone (1.0 eq) in THF (0.025 M) at −78 °C was added $NaN(SiMe_3)_2$ (1.0 M in THF, 1.5 eq) dropwise. The mixture was stirred at this temperature for 90 min before dropwise addition of the acid chloride (1.2 eq). The reaction mixture was stirred for a further 2 h at −78 °C. Brine was added slowly at this temperature whilst stirring vigorously and the mixture was allowed to warm to rt. The volatiles were removed in vacuo and the residue was partitioned between EtOAc and water. The organic layer was washed with brine, dried over Na_2SO_4, and concentrated *in vacuo*. The crude product was purified over silica gel to afford the product *N*-acyl-enaminone.

N-(4,4-dimethyl-3-oxocyclobut-1-en-1-yl)-*N*-methylbenzamide (**16a**)

This compound was prepared according to General Procedure C using enaminone **14a** (80 mg, 0.64 mmol) and BzCl (0.09 mL, 0.77 mmol). Purification over silica gel using a gradient of 20–80% EtOAc/cHex afforded the title compound **16a** (72% yield) as a white solid.

No significant rotamers are observed in NMR spectra. ^{1}H NMR (500 MHz, Chloroform-*d*) δ 7.56–7.52 (m, 3H), 7.49–7.45 (m, 2H), 4.83 (s, 1H), 3.35 (s, 3H), 1.41 (s, 6H). ^{13}C NMR (151 MHz, Chloroform-*d*) δ 194.2, 175.9, 171.1, 134.1, 132.0, 129.0, 127.9, 111.8, 62.7, 36.7, 21.4. LC-MS: RT = 3.85 min, 99% (254 nm), *m/z* $[M + H]^+$ = 230. HRMS calculated for $C_{14}H_{16}NO_2^+$ $[M + H]^+$ = 230.1176 found 230.1171.

The regiochemistry of the acylation (i.e., acylation of N atom and not of vinyl position) was proven by 2D NMR (HSQC + HMBC)—the singlet signal counting for 1H has an associated ^{13}C signal in HSQC and thus the vinyl position remains unsubstituted in the product.

N-(2-bromo-4,4-dimethyl-3-oxocyclobut-1-en-1-yl)-*N*-methylbenzamide (**17a**)

This compound was prepared according to General Procedure C using bromoenaminone **15a** (80 mg, 0.39 mmol) and BzCl (0.06 mL, 0.47 mmol). Purification over silica gel using a gradient of 20–70% EtOAc/cHex afforded the title compound **17a** (86 mg, 71% yield) as a white solid.

No significant rotamers are observed in NMR spectra. ^{1}H NMR (600 MHz, Chloroform-*d*) δ 7.62–7.54 (m, 3H), 7.53–7.46 (m, 2H), 3.61 (s, 3H), 1.43 (s, 6H). ^{13}C NMR (151 MHz, Chloroform-*d*) δ 191.4, 174.6, 170.1, 133.5, 132.5, 129.1, 128.5, 87.8, 63.5, 39.0, 21.2. LC-MS: RT = 4.55 min, 99+% (254 nm), *m/z* $[M + H]^+$ = 308 (light isotope). HRMS calculated for $C_{14}H_{15}BrNO_2^+$ $[M + H]^+$ = 310.0260 (heavy isotope), found 310.0254.

N-(2-bromo-4,4-dimethyl-3-oxocyclobut-1-en-1-yl)-*N*-methylacetamide (**17b**)

This compound was prepared according to General Procedure C using bromoenaminone **15a** (6 mg, 0.03 mmol) and AcCl (0.002 mL, 0.03 mmol). Extraction with DCM (5 mL) and water (2 × 1 mL) afforded the title compound **17b** (5.9 mg, 80% yield) as a white solid.

No significant rotamers are observed in NMR spectra. ^{1}H NMR (500 MHz, DMSO-*d6*) δ 3.19 (s, 3H), 1.90 (s, 3H), 1.11 (s, 6H). ^{13}C NMR (125 MHz, DMSO-*d6*) δ 186.3, 184.9, 175.3, 68.7, 40.9, 30.5, 20.8, 20.1. LC-MS: RT 1.53 min, 97% (254 nm), *m/z* $[M + H]^+$ = 246 (heavy isotope). HRMS calculated for deacetylated $C_7H_{11}BrNO^+$ $[M\text{-acetyl} + H]^+$ = 204.0018 (light isotope), found 204.0018.

4. Conclusions

A new electrophilic warhead chemotype, the bromocyclobutenaminone scaffold, was designed as a thiol-labelling agent. It was shown that the incorporation of the bromine atom in the cyclobutenaminone core, sometimes in conjunction with an electron withdrawing group on the nitrogen atom, turns the inactive fragments into novel and useful covalent probes. The investigation of a set of compounds against MurA protein from *E. coli* led to the identification of fragments with moderate inhibitory activity. These compounds represent promising starting points for hit optimization studies and fragment growing might lead to new and potent MurA inhibitors.

Supplementary Materials: The following are available online at http://www.mdpi.com/1424-8247/13/11/362/s1, Table S1. Results of time-dependent IC_{50} measurements of **17b**; Figure S1. The MS/MS spectrum of **17b** modified MurA enzyme peptide [amino acids 104–120] together with the annotation of the peaks; NMR, LC-MS and HRMS data of key compound **17a**.

Author Contributions: Conceptualization, S.G., I.J.P.d.E. and G.M.K.; methodology, P.Á.-B., M.W.; investigation, D.J.H., A.K., L.P., M.H. and T.I.; writing—original draft preparation, P.Á.-B., D.J.H. and M.W., writing—review and editing, P.Á.-B., D.J.H., M.W., S.G., I.J.P.d.E. and G.M.K. All authors have read and agreed to the published version of the manuscript.

Funding: The research was funded by H2020 MSCA FragNet (project 675899), SNN 125496, OTKA PD124598 and 2018-2.1.11-TÉT-SI-2018-00005, and the Slovenian Research Agency core funding P1-0208.

Acknowledgments: We thank Hélène Barreteau for *E. coli* MurA plasmid and Hans Custers for HRMS measurements. The authors are grateful for Krisztina Németh and Pál Szabó for contributing to the analytical experiments.

Conflicts of Interest: The authors declare no conflict of interest.

References

1. Bugg, T.D.H.; Walsh, C.T. Intracellular steps of bacterial cell wall peptidoglycan biosynthesis: Enzymology, antibiotics, and antibiotic resistance. *Nat. Prod. Rep.* **1992**, *9*, 199–215. [CrossRef] [PubMed]
2. Blake, K.L.; O'Neill, A.J.; Mengin-Lecreulx, D.; Henderson, P.J.F.; Bostock, J.M.; Dunsmore, C.J.; Simmons, K.J.; Fishwick, C.W.G.; Leeds, J.A.; Chopra, I. The nature of Staphylococcus aureus MurA and MurZ and approaches for detection of peptidoglycan biosynthesis inhibitors. *Mol. Microbiol.* **2009**, *72*, 335–343. [CrossRef] [PubMed]
3. Hrast, M.; Sosič, I.; Šink, R.; Gobec, S. Inhibitors of the peptidoglycan biosynthesis enzymes MurA-F. *Bioorg. Chem.* **2014**, *55*, 2–15. [CrossRef] [PubMed]
4. Chang, C.-M.; Chern, J.; Chen, M.-Y.; Huang, K.-F.; Chen, H.-H.; Yang, Y.-L.; Wu, S.-H. Avenaciolides: Potential MurA-Targeted Inhibitors against Peptidoglycan Biosynthesis in Methicillin-Resistant Staphylococcus aureus (MRSA). *J. Am. Chem. Soc.* **2015**, *137*, 267–275. [CrossRef]
5. Baum, E.Z.; Montenegro, D.A.; Licata, L.; Turchi, I.; Webb, G.C.; Foleno, B.D.; Bush, K. Identification and Characterization of New Inhibitors of the Escherichia coli MurA Enzyme. *Antimicrob. Agents Chemother.* **2001**, *45*, 3182–3188. [CrossRef] [PubMed]
6. Keeley, A.; Ábrányi-Balogh, P.; Hrast, M.; Imre, T.; Ilas, J.; Gobec, S.; Keserű, G.M. Heterocyclic electrophiles as new MurA inhibitors. *Arch. Pharm. Chem. Life. Sci.* **2018**, *351*, e1800184. [CrossRef]
7. Li, X.; Danishefsky, S.J. Cyclobutenone as a Highly Reactive Dienophile: Expanding Upon Diels–Alder Paradigms. *J. Am. Chem. Soc.* **2010**, *132*, 11004–11005. [CrossRef]
8. Paton, R.S.; Kim, S.; Ross, A.G.; Danishefsky, S.J.; Houk, K.N. Experimental Diels-Alder reactivities of cycloalkenones and cyclic dienes explained through transition-state distortion energies. *Angew. Chem. Int. Ed.* **2011**, *50*, 10366–10368. [CrossRef]
9. Ammann, A.A.; Rey, M.; Dreiding, A.S. Cyclobut-2-enones from Alkynes via Dichlorocyclobut-2-enones. *Helv. Chim. Acta* **1987**, *70*, 321–328. [CrossRef]
10. Huisgen, R.; Mayr, H. Reactions of cyclobutenones with nucleophilic reagents via vinylketen intermediates. *J. Chem. Soc. Chem. Commun.* **1976**, 55–56. [CrossRef]
11. Danheiser, R.L.; Savariar, S. A general method for the reductive dechlorination of 4,4-dichlorocyclobutenones. *Tetrahedron Lett.* **1987**, *28*, 3299–3302. [CrossRef]
12. Lumbroso, A.; Catak, S.; Sulzer-Mossé, S.; De Mesmaeker, A. Efficient access to functionalized cyclobutanone derivatives using cyclobuteniminium salts as highly reactive Michael acceptors. *Tetrahedron Lett.* **2015**, *56*, 2397–2401. [CrossRef]
13. Graaf, C.; Vischer, H.F.; de Kloe, G.E.; Kooistra, A.J.; Nijmeijer, S.; Kuijer, M.; Verheij, M.H.P.; England, P.J.; van Muijlwijk-Koezen, J.E.; Leurs, R.; et al. Small and colorful stones make beautiful mosaics: Fragment-based chemogenomics. *Drug Discov. Today* **2013**, *18*, 323–330. [CrossRef]
14. Khatik, G.L.; Kumar, R.; Chakraborti, A.K. Catalyst-Free Conjugated Addition of Thiols to α,β-Unsaturated Carbonyl Compounds in Water. *Org. Lett.* **2006**, *8*, 2433–2436. [CrossRef]
15. Ábrányi-Balogh, P.; Petri, L.; Imre, T.; Szijj, P.; Scarpino, A.; Hrast, M.; Mitrović, A.; Fonovič, U.P.; Németh, K.; Barreteau, H.; et al. A road map for prioritizing warheads for cysteine targeting covalent inhibitors. *Eur. J. Med. Chem.* **2018**, *160*, 94–107. [CrossRef]
16. Mihalovits, L.M.; Ferenczy, G.G.; Keserű, G.M. Catalytic Mechanism and Covalent Inhibition of UDP-N-Acetylglucosamine Enolpyruvyl Transferase (MurA): Implications to the Design of Novel Antibacterials. *J. Chem. Inf. Model.* **2019**, *59*, 5161–5173. [CrossRef]
17. Wijtmans, M.; Denonne, F.; Célanire, S.; Gillard, M.; Hulscher, S.; Delaunoy, C.; Vanhoutvin, N.; Bakker, R.A.; Defays, S.; Gérard, J.; et al. Histamine H3receptor ligands with a 3-cyclobutoxy motif: A novel and versatile constraint of the classical 3-propoxy linker. *Med. Chem. Commun.* **2010**, *1*, 39–44. [CrossRef]
18. Brand, S.; de Candole, B.C.; Brown, J.A. Efficient Synthesis of 3-Aminocyclobut-2-en-1-ones: Squaramide Surrogates as Potent VLA-4 Antagonists. *Org. Lett.* **2003**, *5*, 2343–2346. [CrossRef]
19. Wasserman, H.H.; Piper, J.U.; Dehmlow, E.V. Cyclobutenone derivatives from ethoxyacetylene. *J. Org. Chem.* **1973**, *38*, 1451–1455. [CrossRef]
20. Hasek, R.H.; Gott, P.G.; Martin, J.C. Ketenes III. Cycloaddition of ketenes to acetylenic ethers. *J. Org. Chem.* **1964**, *29*, 2510–2513. [CrossRef]

21. McCarney, C.C.; Ward, R.S.; Roberts, D.W. Reactions of ketenes with ethoxyalkynes: Synthesis of 2,2,4-trialkylcyclobutane-1,3-diones. *Tetrahedron* **1976**, *32*, 1189–1192. [CrossRef]
22. Philips, D.J.; Davenport, R.J.; Demaude, T.A.; Galleway, F.P.; Jones, M.W.; Knerr, L.; Perry, B.G.; Ratcliffe, A.J. Imidazopyriridines as VLA-4 integrin antagonists. *Bioorg. Med. Chem. Lett.* **2008**, *18*, 4146–4149. [CrossRef]
23. Ross, A.G.; Townsend, S.D.; Danishefsky, S.J. Halocycloalkenones as Diels–Alder Dienophiles. Applications to Generating Useful Structural Patterns. *J. Org. Chem.* **2013**, *78*, 204–210. [CrossRef] [PubMed]
24. Placebo Controlled Study in Subjects With Relapsing Forms of MS to Evaluate the Safety, Tolerability and Effects of CDP323. Available online: https://clinicaltrials.gov/ct2/show/NCT00484536 (accessed on 1 October 2020).
25. Chanteux, H.; Rosa, M.; Delatour, C.; Prakash, C.; Smith, S.; Nicolas, J.-M. In Vitro Hydrolysis and Transesterification of CDP323, an α4ß1/α4ß7 Integrin Antagonist Ester Prodrug. *Drug Metab. Dispos.* **2014**, *42*, 153–161. [CrossRef] [PubMed]
26. Schule, A.; Ates, C.; Palacio, M.; Stofferis, J.; Delatinne, J.-P.; Martin, B.; Lloyd, S. Monitoring and Control of Genotoxic Impurity Acetamide in the Synthesis of Zaurategrast Sulfate. *Org. Proc. Res. Dev.* **2010**, *14*, 1008–1014. [CrossRef]
27. Skarzynski, T.; Mistry, A.; Wonacott, A.; Hutchinson, S.E.; Kelly, V.A.; Duncan, K. Structure of UDP-N-acetylglucosamine enolpyruvyl transferase, an enzyme essential for the synthesis of bacterial peptidoglycan, complexed with substrate UDP-N-acetylglucosamine and the drug fosfomycin. *Structure* **1996**, *4*, 1465–1474. [CrossRef]
28. Rožman, K.; Lešnik, S.; Brus, B.; Hrast, M.; Sova, M.; Patin, D.; Barreteau, H.; Konc, J.; Janežič, D.; Gobec, S. Discovery of new MurA inhibitors using induced-fit simulation and docking. *Bioorg. Med. Chem. Lett.* **2017**, *27*, 944–949. [CrossRef] [PubMed]

Publisher's Note: MDPI stays neutral with regard to jurisdictional claims in published maps and institutional affiliations.

Article

Quinolizidine-Derived Lucanthone and Amitriptyline Analogues Endowed with Potent Antileishmanial Activity

Michele Tonelli [1,*], Anna Sparatore [2,*], Nicoletta Basilico [3,*], Loredana Cavicchini [3], Silvia Parapini [4], Bruno Tasso [1], Erik Laurini [5], Sabrina Pricl [5,6], Vito Boido [1] and Fabio Sparatore [1]

1 Dipartimento di Farmacia, Università degli Studi di Genova, Viale Benedetto XV, 3, 16132 Genova, Italy; tasso@difar.unige.it (B.T.); vito.boido@libero.it (V.B.); fabio.sparatore@alice.it (F.S.)
2 Dipartimento di Scienze Farmaceutiche, Università degli Studi di Milano, Via Mangiagalli 25, 20133 Milano, Italy
3 Dipartimento di Scienze Biomediche Chirurgiche e Odontoiatriche, Università degli Studi di Milano, Via Pascal 36, 20133 Milano, Italy; loredana.cavicchini@unimi.it
4 Dipartimento di Scienze Biomediche per la Salute, Università degli Studi di Milano, Via Mangiagalli 31, 20133 Milano, Italy; silvia.parapini@unimi.it
5 Molecular Biology and Nanotechnology Laboratory (MolBNL@UniTS), DEA, Piazzale Europa 1, 34127 Trieste, Italy; erik.laurini@dia.units.it (E.L.); sabrina.pricl@dia.units.it (S.P.)
6 Department of General Biophysics, Faculty of Biology and Environmental Protection, University of Lodz, 90-236 Lodz, Poland
* Correspondence: tonelli@difar.unige.it (M.T.); anna.sparatore@unimi.it (A.S.); nicoletta.basilico@unimi.it (N.B.)

Received: 4 October 2020; Accepted: 21 October 2020; Published: 25 October 2020

Abstract: Leishmaniases are neglected diseases that are endemic in many tropical and sub-tropical Countries. Therapy is based on different classes of drugs which are burdened by severe side effects, occurrence of resistance and high costs, thereby creating the need for more efficacious, safer and inexpensive drugs. Herein, sixteen 9-thioxanthenone derivatives (lucanthone analogues) and four compounds embodying the diarylethene substructure of amitriptyline (amitriptyline analogues) were tested in vitro for activity against *Leishmania tropica* and *L. infantum* promastigotes. All compounds were characterized by the presence of a bulky quinolizidinylalkyl moiety. All compounds displayed activity against both species of *Leishmania* with IC_{50} values in the low micromolar range, resulting in several fold more potency than miltefosine, comparable to that of lucanthone, and endowed with substantially lower cytotoxicity to Vero-76 cells, for the best of them. Thus, 4-amino-1-(quinolizidinylethyl)aminothioxanthen-9-one (**14**) and 9-(quinolizidinylmethylidene)fluorene (**17**), with selectivity index (SI) in the range 16–24, represent promising leads for the development of improved antileishmanial agents. These two compounds also exhibited comparable activity against intramacrophagic amastigotes of *L. infantum*. Docking studies have suggested that the inhibition of trypanothione reductase (TryR) may be at the basis (eventually besides other mechanisms) of the observed antileishmanial activity. Therefore, these investigated derivatives may deserve further structural improvements and more in-depth biological studies of their mechanisms of action in order to develop more efficient antiparasitic agents.

Keywords: *Leishmania tropica* and *infantum*; antileishmanial agents; lucanthone analogues; amitriptyline analogues; quinolizidine-derived compounds; molecular modelling studies

1. Introduction

Leishmaniases are neglected diseases that are endemic in many tropical and sub-tropical countries, leading annually to an estimated 700,000–1,000,000 new cases and 20,000–30,000 deaths [1]. Leishmaniases are caused by more than 20 species of protozoan parasites belonging to the genus *Leishmania*, which together with the genus *Trypanosoma*, belongs to the order *Trypanosomatidae*. Recent advances in the taxonomy, genetics, molecular and cellular biology and biochemistry of these organisms are well illustrated in two reviews [2,3]. The pathogen parasites (promastigotes) are transmitted to human and other mammalian hosts by the bites of infected female phlebotomine sandflies. In the mammalian host, the parasites differentiate into amastigote forms and affect skin, mucosa or internal tissues and organs to produce cutaneous (CL), muco-cutaneous (MC) and visceral (VL) leishmaniasis. The last form is fatal in absence of treatment. Current therapy is based on pentavalent antimonials; pentamidine; amphotericin B and its liposomal formulations used by parenteral route; and miltefosine, as the only oral agent. These drugs are burdened by heavy side effects, the occurrence of resistance and high costs, so that the need for novel, more efficacious, safe and inexpensive drugs is very stringent. Indeed, to meet this need, a number of studies are presently ongoing, exploring a very wide chemical space and also the possibility of the repositioning of known drugs. The structures of many interesting examples of investigational antileishmanial agents, able to hit different cellular targets, are illustrated in several reviews [4–9]. Interestingly, highly potent antileishmanial chemical classes, as the nitroimidazoles, the benzooxaboroles and the aminopyrazoles/pyrazolopyrimidines, have been identified, mainly within the Drugs for Neglected Diseases initiative (DNDi), Geneva, Switzerland. Orally active compounds from these series (such as DNDi-0690, DNDi-6148 and GSK-3186899) [10–12] are presently in clinical development (Figure 1).

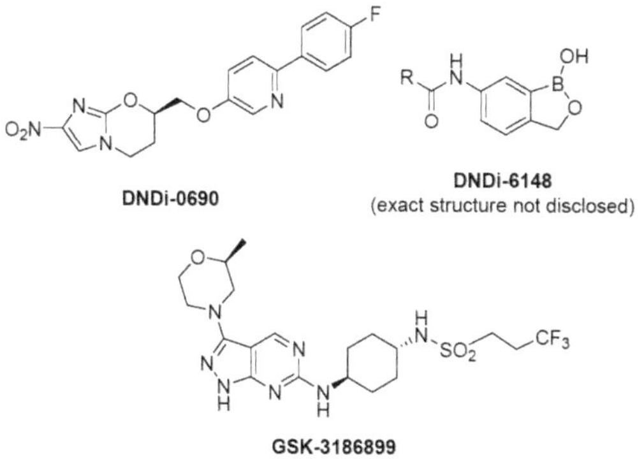

DNDi-0690

DNDi-6148
(exact structure not disclosed)

GSK-3186899

Figure 1. Antileishmanial drugs in clinical development.

Additionally, many tricyclic antidepressant and antipsychotic drugs (and structurally related compounds) have been shown to display various degrees of activity against different species of *Leishmania*, and/or to inhibit enzymes (such as trypanothione reductase) playing essential roles in parasite development and virulence [13–16]. Among these, clomipramine and cyclobenzaprine have been recently repurposed for the treatment of visceral leishmaniasis [15,16]. All these compounds are characterized by different kinds of linear tricyclic systems, such as phenothiazine, iminodibenzyl, thioxanthene, dibenzocycloheptane and sulphur isosteric analogues, to which an aliphatic basic side chain is attached to the central ring (Figure 2A).

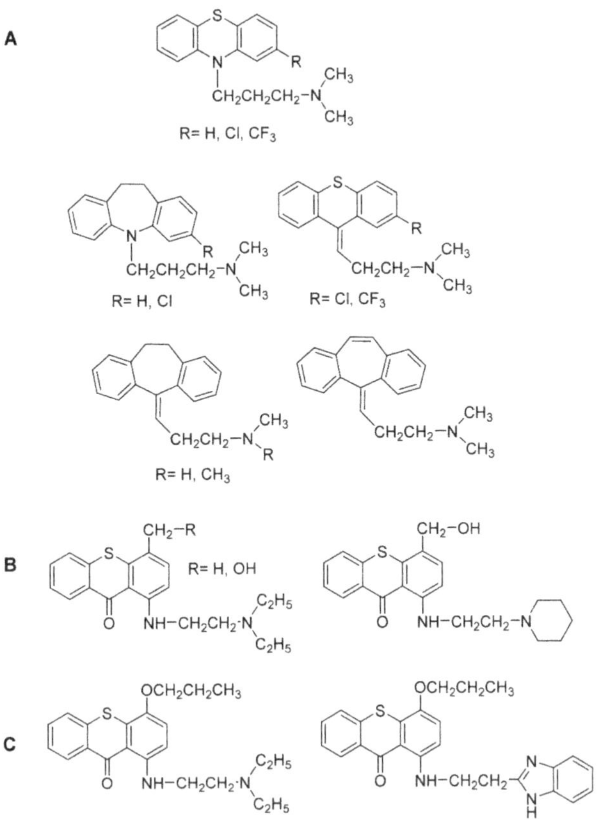

Figure 2. (A) Examples of basic derivatives of tricyclic systems displaying antileishmanial activity and inhibition of trypanothione reductase. (B) Basic derivatives of thioxanthene-9-one (lucanthone analogues) displaying antileishmanial activity. (C) Lucanthone analogues displaying dual inhibition of P-glycoprotein and cancer cell growth.

Thioxanthen-9-one is another tricyclic system whose derivatives exhibited antileishmanial activity, even featuring the basic side chain linked to a lateral ring (Figure 2B). Indeed, lucanthone, a drug largely used in the past for the treatment of schistosomiasis and presently (together with various analogues) under investigation as an antitumor agent, was shown to display activity against *L. major amazonensis* in tissue cultures a long time ago (1975) [17]. Although it was less powerful, this activity was also confirmed in vivo [18]. A few years later, this compound was also found to be active against intramacrophagic *L. tropica* amastigotes with IC_{50} = 0.93 µg/mL (2.47 µM) [19]. More recently (2011), through the screening of more than 4000 compounds, using a novel ex vivo splenic explant model system, an 1-(piperidinoethylamino)-analog of hycanthone (the active metabolite of lucanthone) was identified as a lead compound against *L. donovani*, with IC_{50} = 9.1 µM [20]. Hycanthone and its prodrug lucanthone are burdened with hepatic toxicity and mutagenicity [21–23], which were related to the presence or formation of a 4-hydroxymethyl group capable of alkylating the deoxyguanosine residue of DNA passing through the formation of a strongly electrophilic carbocation. Modification of the basic side chain and/or the introduction of substituents in position 6 of the thioxanthenone ring were shown to alter the mutagenicity, while retaining appreciable anti-schistosomal activity [24,25]. Even more, the replacement of the 4-methyl and 4-hydroxymethyl groups with other functional moieties limited the toxicological issues, as observed for the 4-propoxy analog of hycanthone (TXAI), investigated as an antitumor agent (Figure 2C) [26]. These data suggest that genetic and anti-parasitic activities of thioxanthenone derivatives may be dissociated from each other.

2. Results and Discussion

Thus, when pursuing our investigation on antiplasmodial [27–29] and antileishmanial [30,31] agents we deemed worthwhile the study of the antileishmanial activity of a set of lucanthone analogues with a modified substitution pattern, and of a few compounds embodying the diarylethene substructure of, e.g., chlorprothixene, amitriptyline and cyclobenzaprine. The studied compounds were characterized by the presence of the bulky quinolizidinylalkyl moieties, which were shown to improve the antiplasmodial and/or antileishmanial activity in the corresponding chloroquine and clofazimine analogs [30] and also in the set of 1-basic substituted 2-phenyl/benzyl benzimidazoles [31]. In position 4 of the thioxanthen-9-one nucleus, besides the methyl, a nitro or amino group was introduced, which were shown to improve the inhibitory activity in another field (against the lymphocytic leukemia P388 [32]). The considered compounds were obtained from our in-house library, having been synthesized and studied in the past as antimicrobial and anti-leukemia P388 agents [33,34]; as modulators of uptake and release of neurotransmitters [35–37]; and more recently, as dual inhibitors of cholinesterases and Aβ aggregation [38,39].

The structures of the presently investigated compounds are depicted in Figures 3 and 4.

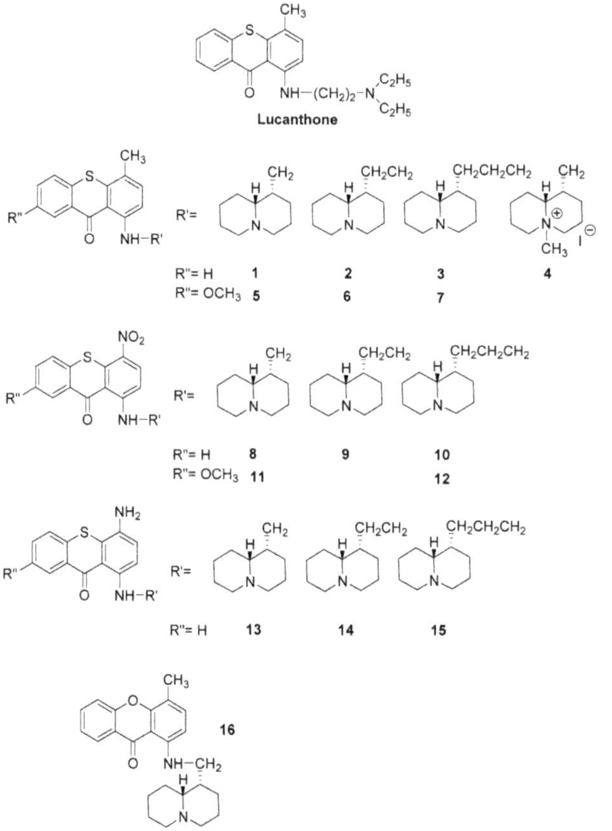

Figure 3. Investigated thioxanthen-9-one derivatives (lucanthone analogues) bearing quinolizidine-alkyl side chains.

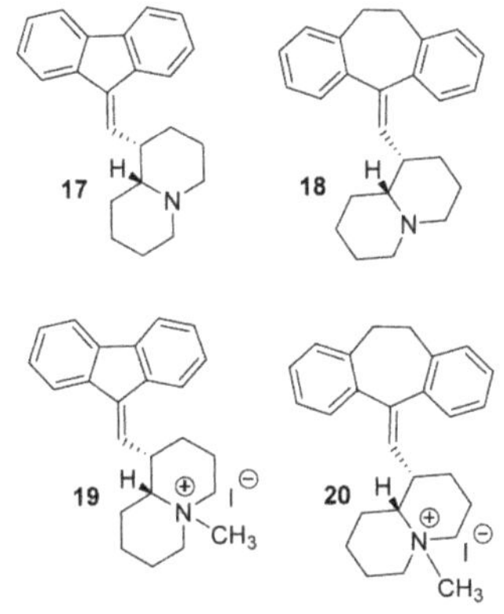

Figure 4. Investigated lupinylidene (quinolizidinyl-methylidene) derivatives of planar and corrugated tricyclic systems.

2.1. Chemistry

With the exception of **4** and **20**, all the compounds of Figures 3 and 4 were previously described according to the following references: **1–3**, **5–10** and **13–15** [34]; **11**, **12** and **16** [38]; **17** and **18** [37]; **19** [33]. The novel compounds **4** and **20** were prepared by treating compounds **1** and **18** with methyl iodide (Scheme 1).

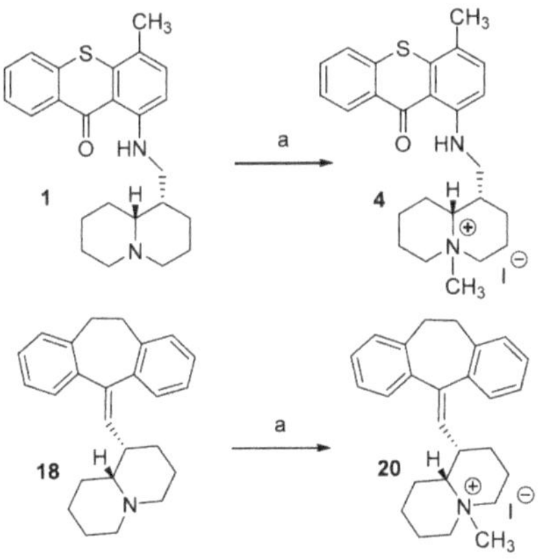

Scheme 1. Reagents and conditions: (a) CH_3I (excess), 24 h, r.t.

2.2. Biological Studies and SAR

2.2.1. Antileishmanial Activity Against *L. tropica* and *L. infantum* Promastigotes

Compounds in Figures 3 and 4 were tested in vitro against *Leishmania tropica* and *L. infantum* promastigotes using the MTT assay. Results are expressed as IC_{50} ± SD (μM) and reported in Table 1, together with the ratios between the IC_{50} of the reference drug (miltefosine) and that of each tested compound. All tested compounds displayed activity against both species of *Leishmania* and many of them exhibited IC_{50} less than 10 μM (65% and 50% versus *L. tropica* and *L. infantum*, respectively). Commonly, activity was higher for *L. tropica* than *L. infantum*. In comparison to miltefosine, all compounds were several–fold more potent, up to 17-fold against *L. tropica* and up to 9-fold against *L. infantum*. These results indicate that the introduction of a quinolizidinyl alkyl moiety on all the considered tricyclic systems is consistent with the expression of valuable antileishmanial activity, provided that, in the case of thioxanthenone, suitable substituents are present on position 4 and 7.

Table 1. In vitro data on antileishmanial activity against *Leishmania tropica* and *L. infantum* promastigotes of lucanthone and compounds **1–20**.

Compd.	IC_{50} (μM) [a] *L. tropica*	Ratio [b] IC_{50} Miltef./IC_{50} Compd.	IC_{50} (μM) [a] *L. infantum*	Ratio [b] IC_{50} Miltef./IC_{50} Compd.
Lucanthone	2.57 ± 1.03	16.8	3.50 ± 1.17	8.9
1	3.3 6 ± 1.53	12.9	3.87 ± 1.53	8.1
2	6.27 ± 2.75	6.9	10.60 ± 3.78	2.95
3	8.89 ± 3.02	4.87	17.19 ± 1.31	1.8
4	17.83 ± 5.61	2.4	13.00 ± 2.80	2.4
5	8.16 ± 5.23	5.3	8.99 ± 3.86	3.5
6	28.35 ± 14.34	1.5	>46	<0.7
7	12.00 ± 5.24	3.6	19.66 ± 5.79	1.6
8	19.24 ± 9.33	2.25	38.72 ± 7.67	0.8
9	4.64 ± 1.12	9.3	9.39 ± 4.34	3.3
10	11.05 ± 3.30	3.9	22.27 ± 3.30	1.4
11	6.35 ± 1.46	6.8	13.01 ± 0.97	2.4
12	7.39 ± 2.31	5.85	7.89 ± 2.41	3.95
13	2.87 ± 0.43	15.1	5.23 ± 1.83	6.0
14	3.80 ± 1.82	11.4	3.63 ± 0.78	8.6
15	3.72 ± 1.66	11.6	4.22 ± 1.95	7.4
16	6.56 ± 3.08	6.6	8.23 ± 1.51	3.8
17	3.49 ± 1.02	12.4	4.35 ± 0.37	7.2
18	24.25 ± 3.58	1.8	25.21 ± 1.92	1.24
19	10.52 ± 2.69	4.1	16.42 ± 0.59	1.90
20	7.21 ± 3.52	6.0	8.71 ± 0.68	3.58
Miltefosine	43.26 ± 11.36	1.0	31.26 ± 10.45	1.0

[a] The results are expressed as IC_{50} ± SD of at least three different experiments performed in duplicate. [b] Ratios between the IC_{50} of miltefosine and that of each compound against *L. tropica* or *L. infantum*.

Among the thioxanthenone derivatives, compounds **9** and **13–15** exhibited potency (IC_{50} in the range 2.87–9.39 μM) comparable to that of lucanthone and 1-(piperidinoethyl)-4-hydroxymethylthioxanthen-9-one (the novel lead compound of Osorio et al. [20]), but differently from two of the others, they should not be associated with mutagenic activity because they lack groups potentially leading to alkylating species.

The elongation of the polymethylene linker exerted different effects, depending on the nature of the substituents, in position 4. Among the 4-methyl derivatives, the activity decreased with the increasing number of methylene groups in the side chain (IC_{50}: from 3.36 to 8.89 μM and from 3.87 to 17.19 μM for the two *Leishmania* species), while among the 4-amino derivatives the activity remained practically unchanged (IC_{50} in the range 3–5 μM). In the set of 4-nitro derivatives, the activity varied for both species in an unaccountable way, with the worst IC_{50} values being for $n = 1$ (19.24 for *L. tropica* and 38.72 μM for *L. infantum*) and the best one being for $n = 2$ (IC_{50}= 4.64 and 9.39 μM for *L. tropica* and *L. infantum*, respectively, Table 1). In other words, for a given number of CH_2 in the side chain, the substitution of the 4-methyl with the 4-amino group, besides eliminating the risk of mutagenicity, always improved the activity of the relevant derivatives, while the presence of a 4-nitro moiety produced either a slight increase ($n = 2$) or a decrease ($n = 1$ and 3) of activity. In particular, the decrement of activity was very pronounced for $n = 1$, with IC_{50}= 19.24 and 38.72 μM against *L. tropica* and *L. infantum*, respectively. Additionally, the introduction of a methoxy group in position 7 had variable effects on the activity, which was enhanced in the case of 4-nitro derivatives and decreased for the 4-methyl derivatives. The decreasing effects was striking when $n = 2$ (compound **6**), with IC_{50} = 28.35 μM and >46 μM for *L. tropica* and *L. infantum*, respectively. Thus compound **6** displayed the lowest activity among all tested compounds, although it was by far the least toxic (CC_{50} = 137 μM) on rat skeletal myoblast cells (L6), with a still valuable selectivity index (SI = 4.8 and ≤3) for the two *Leishmania* species (Reto Brun, personal communication to A.S.).

When comparing compounds **1** and **16**, it was observed that the antileishmanial activity was maintained with only modest loss of potency, indicating that even the exchange of the sulfur bridge with an oxygen atom did not modify significantly the thioxanthenone physico-chemical interactions with the *Leishmania* target(s), while it was able to abolish the antileukemic activity of lucanthone, as observed by Blanz and French [40]. On the contrary, the quaternarization of compounds **1** to give **4** produced a strong reduction of the antileishmanial activity, suggesting that the presence of a fixed charge, while hampering the usual target interactions, might shift the molecule towards a different cellular target. Supposing that the antileishmanial activity of the thioxanthenone derivatives could be related to one or some of the mechanisms previously described for the anti-schistosomal and anti-tumor activities of lucanthone (DNA intercalation, inhibition of nucleic acid biosynthesis, inhibition of topoisomerase II and apurinic endonuclease-1 [41], induction of autophagy and apoptosis [42], alteration of cholesterol biosynthesis and localization [43]), the quaternization of the quinolizidine nitrogen of **1** might produce the shift from some of these mechanisms to the inhibition of choline uptake, as it is known for common quaternary ammonium surfactants [44,45] and for the peculiar ammonium salts bearing bulky moieties on the charged head or on the lipophilic tail [46,47].

It is worth noting that compound **1** was shown in the past [33] to exhibit a large spectrum of antibacterial activity with MIC in the same micromolar range of IC_{50} against *Leishmania* species; f.i. MIC: 2.5 μg/mL = 6.63 μM against *Staphylococcus aureus* and *Bacillus subtilis*, and 1.25 μg/mL = 3.32 μM against *Mycobacterium tuberculosis*. Lucanthone also displayed antibacterial activity but with MIC 10-fold higher (30–60 μM) [48,49].

The four *lupinylidene derivatives* **17–20** embody the 1,1-diarylethene substructure that characterizes amitriptyline, cyclobenzaprime, chlorprothixene and other antidepressant and antipsychotic drugs. Like most of these drugs, compounds **17–20** exhibited antileishmanial activity in the low micromolar range; in particular, the 9-lupinylidene fluorene **17** (with IC_{50} = 3.5 and 4.35 μM against *L. tropica* and *L. infantum*, respectively) was 6/7-fold more potent than the lupinylidene dibenzocycloheptadiene **18** (IC_{50} = 24.25 and 25.21 μM), possibly for being endowed with a more appropriate lipophilicity. The quaternization of these compounds produced a levelling effect on activity, decreasing the potency of the former and increasing that of the latter, so that the corresponding methyl iodides **19** and **20** were almost equipotent.

The antileishmanial activity of the abovementioned drugs has been shown to be related to the inhibition of trypanothione reductase (TryR) in the parasite [50–52], but also to modulation of the immune response in the host. It is reasonable to suppose that also the tertiary compounds **17** and **18** act through inhibition of TryR; however, in the quaternized compounds **19** and **20**, other mechanisms might modulate or replace the former, as discussed above for quaternized thioxanthenone **4**. Interestingly, the tertiary amitriptyline analog **18** (compound **17** was not tested) was shown in the past [35] to inhibit the choline uptake into rat brain synaptosomes at 1 μM concentration, at which amitriptyline was still completely ineffective. The inhibition of choline transport across the synaptosome membrane might be expected in some measure also at the level of the parasite cell membrane in an interplay with the intracellular TryR inhibition. With the quaternization of **18** to **20**, cell penetration and consequent TryR inhibition could be strongly reduced, but the basal activity on choline transport could be improved.

Lupinylidene fluorene **17** was not investigated as an inhibitor of choline uptake, but it was shown to display a large spectrum of anti-microbial activity with outstanding potency against *Mycobacterium tuberculosis H37Ra* (MIC = 0.49 μM) very close to that of isoniazide (MIC: 0.14–0.28 μM) [33]. The quaternization of **17** to **19** strongly affected the antimycobacterial activity (MIC = 83.9 μM) quite probably by reducing the cell wall crossing capability.

Even if the definition of the mechanism of the antileishmanial activity of the tested compounds is beyond the scope of this exploratory work, one cannot overlook the previous observation that lucanthone [53], the quinolizidinylalkylamino thioxanthenones [38] and the lupinylidene dibenzocycloheptadiene [39] inhibit AChE, and particularly BChE, with IC_{50} in the low micromolar and sub-micromolar range. The availability of choline for building up the phosphatidylcholine, the main component of *Leishmania* promastigote membranes [54,55], might be compromised by the inhibition of cholinesterases. Cholinesterases are known to be present even in non-motile unicellular organisms, where besides or instead of the hydrolytic function, they may play non-classical roles that are fundamental for cell survival [56–58]. The inhibition of cholinesterases has been recently claimed as another mechanism of action for some antileishmanial agents extracted from several plants [59–61], and the present results add further support to this hypothesis.

2.2.2. Cytotoxicity

To identify new potential candidates for the development of safe and effective antileishmanial drugs, the cytotoxicities of representative (most effective or structurally peculiar) compounds (**1**, **2**, **14**, **17**, **19** and **20**) were evaluated in a Vero-76 cell line. The cytotoxicity of lucanthone was also tested for comparison. The results in Table 2 show that all compounds exhibited CC_{50} higher than their corresponding IC_{50} values. The relevant selectivity index (SI)—a parameter that quantifies the preferential antileishmanial activity of a compound in relation to mammalian cell toxicity (CC_{50}/IC_{50})—is tabulated in Table 2. As can be seen from this Table, the 4-methyl-thioxanthen-9-ones (**1**, **2**, and lucanthone) exhibited, as expected, the highest cytotoxicity among the tested compounds. It is worth noting that compounds **1** and **2** were somewhat less toxic than lucanthone, suggesting a possible toxicity-lowering effect of the cumbersome quinolizidinylalkyl side chain in comparison to an open chain substituent. The replacement of the 4-methyl substituent with an amino group in compound **14**, while preserving good antileishmanial activity, led to a further substantial decrease of cytotoxicity, with a resultantly safer profile (SI = 16.2 and 16.9 for the two *Leishmania* species).

Table 2. In vitro cytotoxicity data against Vero-76 cells and selectivity index (SI) values for selected antileishmanial compounds of Table 1.

Compd.	CC_{50} [a] (μM) Vero-76	IC_{50} [b] (μM)		S.I. [c]	
		L. tropica	*L. infantum*	*L. tropica*	*L. infantum*
Lucanthone	12.4 ± 2.1	2.57 ± 1.03	3.50 ± 1.17	4.8	3.5
1	23.5 ±1.8	3.36 ± 1.53	3.87 ± 1.53	7.0	6.6
2	22.9 ± 2.3	6.27 ± 2.75	10.60 ± 3.78	3.7	2.2
14	61.4 ± 2.6	3.80 ± 1.82	3.63 ± 0.78	16.2	16.9
17	83.4 ± 3.0	3.48 ± 1.02	4.35 ± 0.37	23.9	19.2
19	82.7 ± 4.7	10.52 ± 2.69	16.42 ± 0.59	7.9	5.0
20	89.9 ± 3.7	7.21 ± 3.52	8.71 ± 0.68	12.5	10.3

[a] Compound concentration (μM) required to reduce the viability of mock-infected VERO-76 (monkey normal kidney) monolayers by 50%. The results are expressed as CC_{50} ± SD of three different experiments performed in duplicate. [b] See Table 1. [c] The selectivity index (SI) is expressed as the ratio between the CC_{50} value of each compound against Vero-76 cell line and the IC_{50} of each compound against *L. tropica* or *L. infantum* promastigotes.

On the other hand, the lupinylidene tricyclic derivatives **17**, **19** and **20** (the amitriptyline analogues, broadly speaking) displayed the lowest cytotoxic effects, with CC_{50} values in the range 80–90 μM, in the presence of either tertiary or quaternized nitrogen. However, taking into account the effects of quaternization on the activity, only the tertiary 9-lupinylidene fluorene **17** displayed a quite valuable SI value for both *Leishmania* species (23.9 and 16.2, respectively), resulting the most promising antileishmanial candidate among the whole set of studied compounds.

The cytotoxic concentration (CC_{50}) against THP-1 differentiated into macrophages (Table 3), used for testing the activity against the amastigote stage, is also reported. Interestingly, compounds **1**, **14**, **17** and lucanthone showed a SI trend against THP-1 cells (Table 3) comparable to that against Vero-76 cells.

Table 3. In vitro antileishmanial activity against intramacrophagic amastigotes of *L. infantum* and cytotoxicity against PMA-differentiated THP-1 (human acute monocytic leukemia cell line).

Compd.	*L. infantum* Amastigotes IC_{50} [a] (μM)	THP-1 CC_{50} [a,b] (μM)	SI [c]
1	3.49 ± 0.18	27.97 ± 8.00	8.0
14	2.13 ± 1.35	21.66 ± 8.83	10.2
17	2.70 ± 0.68	53.90 ± 3.98	20.0
Lucanthone	3.43 ± 0.93	15.17 ± 0.27	4.4

[a] Data are expressed as mean ± SD of three experiments in triplicate. [b] Compound concentration (μM) required to reduce the viability of PMA-differentiated THP-1 by 50%. [c] The selectivity index (SI) is expressed as the ratio between the CC_{50} value of each compound against THP-1 cell line and the IC_{50} against *L. infantum* amastigotes.

It is additionally observed that, based on the results of their previous testing for antileukemic activity [34], only a low to moderate level of in vivo toxicity should be expected. Indeed, no mortality was observed even when compounds **1**, **5**, **6**, **9** and particularly the amino derivatives **13–15** were injected i.p. at doses up to 200–350 mg/kg, once a day for five consecutive days, in a group of six mice previously inoculated with leukemia P388 cells. Particularly, for compound **14** no mortality was observed at a dose of 260 mg/kg (638 μmol/kg), and therefore this compound represents an interesting lead for improved lucanthone analogues.

2.2.3. Antileishmanial Activity against L. Infantum Amastigotes

Finally, for a better idea of their real value as antileishmanial agents, compounds **14** and **17**, displaying the most promising activity against promastigote stage and the highest SI values, together with lucanthone and the corresponding quinolizidine analog **1**, were tested against the intramacrophagic amastigote stage of *L. infantum*. The results indicate that the activity observed against *L. infantum* promastigotes is conserved (lucanthone and compound **1**) or even improved (**14** and **17**) against the corresponding amastigote stage (Table 3).

2.3. *Molecular Modelling Studies*

Amitriptyline and other related compounds have been identified as plausible inhibitors of the trypanothione reductase (TryR), an essential enzyme belonging to the antioxidant machinery of parasitic *Leishmania* [62,63]. TryR is a homodimer and it is active only in this aggregated form. Given some structural properties in common, we reasoned that the most potent compounds of this series, **14** and **17**, can exert their antileishmanial properties (mainly or besides other mechanisms) by efficiently binding TryR and consequently blocking its activity. In order to test our hypothesis, molecular dynamics (MD) simulations were performed on the corresponding complexes with TryR to shed light on the binding mechanisms of these compounds against their putative parasitic target (Figure 5). To validate our procedure and for comparison purposes, we applied the same computational procedure also to TryR in complex with lucanthone (Figure 5) and amitriptyline (Figure S1).

A putative binding site for these compounds was initially recognized on TryR following a consolidated protocol [64–66]. Through the MM/PBSA (molecular mechanics/Poisson–Boltzmann surface area) approach [67], we calculated each inhibitor/enzyme free energy of binding (ΔG_{bind}) and its enthalpic and entropic components (ΔH_{bind} and $-T\Delta S_{bind}$, respectively). The obtained values are in good agreement with their antileishmanial activity (Figure 5D–F and Table S1) yielding the following TryR affinity ranking: **14** (ΔG_{bind} = −8.73 kcal/mol) < **17** (ΔG_{bind} = −8.54 kcal/mol) <= lucanthone (ΔG_{bind} = −8.45 kcal/mol) << amitriptyline (ΔG_{bind} = −7.46 kcal/mol). Interestingly, all compounds share a common thermodynamics pattern; actually, their binding is robustly enthalpy driven characterized by favorable electrostatic and van der Waals interactions. On the other hand, the entropic components penalize the binding, as often detected in cases of small molecule/protein complexes. The precise binding mechanism and the specific ligand/protein interactions were elucidated through the per-residue binding free energy deconvolution (PRBFED) of the enthalpic terms (ΔH_{res}). The PRBFED analysis allowed us to identify the main aminoacid residues of TryR involved in the putative binding pocket (Figure 5D–F, Figure S1 and Table S2).

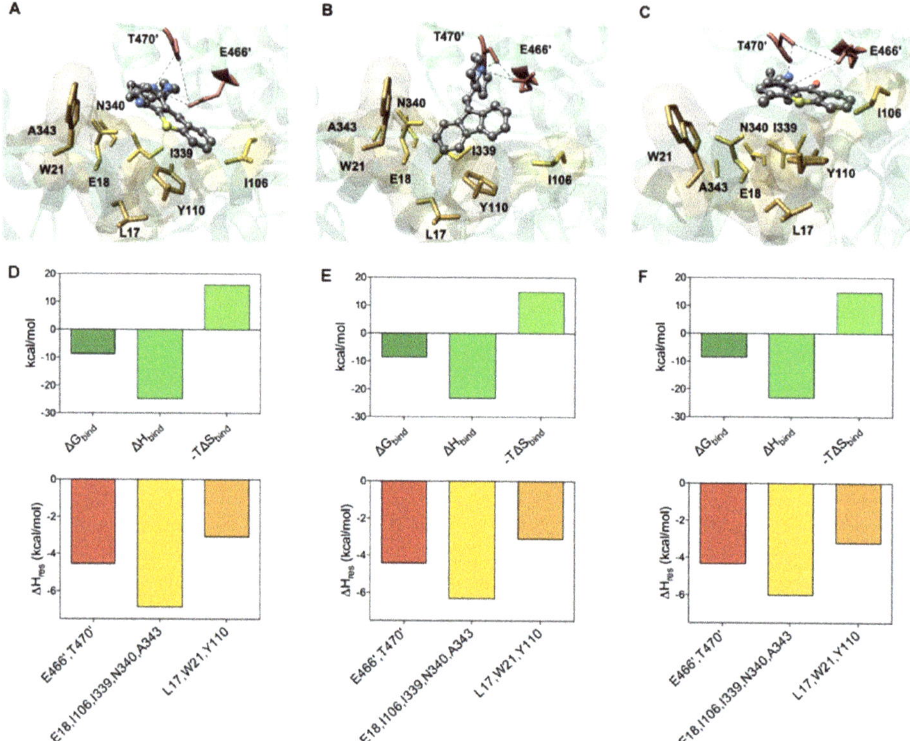

Figure 5. (top panel) Details of compounds **14** (**A**), **17** (**B**) and lucanthone (**C**) in the binding pocket of TryR. Compounds are shown as atom-colored sticks-and-balls (C, grey, N, blue, O, red). The side chains of the mainly interacting TryR residues are depicted as colored sticks and labeled as following: E466′ and T470′, firebrick; E18, W21, I339, N340 and A343, gold; L17, I106 and Y110, goldenrod. The hydrophobic pockets are also highlighted by their transparent van der Waals surface. Hydrogen atoms, water molecules, ions and counterions are omitted for clarity. (bottom panel) Calculated free energy of binding (ΔG_{bind}, forest green), and enthalpic (ΔH_{bind}, lime green) and entropic ($-T\Delta S_{bind}$, chartreuse) components (upper row) and PRBFED of the main involved amino acids (bottom row) of TryR in complex with **14** (**D**), **17** (**E**) and lucanthone (**F**).

The most peculiar interaction is definitively performed by the charged nitrogen atom present in all four of the compounds. Indeed, this protonated tertiary amine group is involved in a virtuous, interactive triangle with the side chain of the second monomer of TryR residues E466′ and T470′ through stable hydrogen bonds and a salt bridge. The length and the rigidity of the spacer between the nitrogen atom and the tricyclic moiety of the inhibitor can affect the efficiency of these interactions. In effect, the ylidenepropyl-amino spacer of amitriptyline is not able to provide the optimal addressing of the charged group towards the side chain of E466′ and T470′ with respect to the bulkiest quinolizidine moiety of **14** and **17** or the most flexible amino-ethyl spacer of lucanthone; accordingly, its ΔH_{res} values resulted the less favorable of the series (Figure 5D–F, Figure S1, and Table S2). On the other hand, it is already established that the dibenzocycloheptene ring of the amitriptyline can be aptly accommodated in the so-called TryR hydrophobic wall formed by residues L17, W21 and Y110 [62,63]. Our MD simulation confirmed this data (Figure S1) and our computational analysis allowed us to find other important TryR residues to improve the van der Waals interactions in this specific hydrophobic region (Figure S1). As listed in Table S2, the side chains of residues E18, I106, I339, N343 and A343 also had contributions to stabilizing the binding with the enzyme. It is worth to mention here that also the tricyclic scaffold of the other compounds can be encased in the same protein region with even better performance (Figure 5D–F, Table S2). In the case of compound **17**, this could be expected since the fluorene ring is very similar to the dibenzocycloheptene moiety, but for the thioxanthenone derivatives **14** and lucanthone this might not have been so obvious. Instead, our MD approach showed that both compounds can share a very similar TryR binding mode with the lupinylidene and amitriptyline derivatives. Finally, compound **14** was the best TryR binder and the amino substitution (-NH_2) in position 4 of the thioxanthenone ring of 14 played an important role. Actually, this amino group can establish a further polar interaction with the side chain of E18 leading to a slight yet significative improvement of its binding capability, as shown in Figure 5D and Table S2.

3. Materials and Methods

3.1. Chemistry

3.1.1. General Information

Chemicals, solvents and reagents used for the syntheses were purchased from Sigma-Aldrich or Alfa Aesar (Milan, Italy), and were used without any further purification. Melting points (uncorrected) were determined with a Büchi apparatus (Milan, Italy). ^{1}H NMR and ^{13}C NMR spectra were recorded with a Varian Gemini-200 spectrometer in $CDCl_3$; the chemical shifts were expressed in ppm (δ), coupling constants (J) in Hertz (Hz). Elemental analyses were performed on a Flash 2000 CHNS (Thermo Scientific, Milan Italy) instrument in the Microanalysis Laboratory of the Department of Pharmacy, University of Genova. Q = quinolizidine ring; Ar= aromatic.

3.1.2. General Procedure for the Synthesis of Quaternary Ammonium Iodides (**4** and **20**)

The compounds **1** [34] and **18** [37] (0.146 mmol) were reacted with iodomethane (0.5 mL, 8 mmol) at r.t. for 24 h with stirring. The reaction mixture was added with dry Et_2O, and the collected compound was washed with dry Et_2O affording the title quaternary ammonium salt.

(1*S*,9a*R*)-5-methyl-1-{[(4-methyl-9-oxo-9*H*-thioxanthen-1-yl)amino]methyl}-decahydroquinolizin-5 -ium iodide (**4**): orange crystals; yield: 92%. m.p. 147–150 °C (Et_2O an.); ^{1}H NMR (200MHz, $CDCl_3$): δ = 9.36 (s, NH–Ar, collapses with D_2O); 8.40 (d, *J* = 9.2, 1 ArH), 7.43–7.20 (m, 4 ArH), 6.72 (d, *J* = 9.2, 1 ArH), 4.21–3.00 (m, 8H of Q and 3.51, s, CH_3-N of Q), 2.95–2.83 (m, 1H, of Q), 2.40–1.40 (m, 9H of Q and 2.25, s, CH_3-Ar superimposed); ^{13}C NMR (50 MHz, $CDCl_3$): 182.4, 149.4, 137.1, 135.1, 134.7, 130.9, 128.6, 128.0, 125.1, 124.4, 119.7, 112.3, 107.3, 67.9, 65.0, 51.5, 49.8, 44.6, 31.7, 20.4, 19.7, 19.0, 18.7, 18.0. Anal. calcd for $C_{25}H_{31}IN_2OS$: C 56.18, H 5.85, N 5.24, S 6.00, found: C 56.09, H 6.15, N 5.15, S 6.39.

(1*S*,9a*R*)-1-[(10,11-dihydro-5*H*-dibenzo[a,d][7]annulen-5-ylidene)methyl]-5-methyl-decahydroquinolizin -5-ium iodide (**20**): pale yellow crystals; yield: 98%. m.p. 283–286 °C (Et_2O an.); ^{1}H NMR (200MHz, $CDCl_3$): δ = 7.50–6.97 (m, 8 ArH), 6.58–6.40 (m, 1H, HC=), 3.58–3.20 (m, 2Hα near N of Q and 3.36, s, CH_3-N of Q), 3.17–2.63 (m, 4H, 2CH_2-Ar), 2.36–1.00 (m, 14H of Q); ^{13}C NMR (50 MHz, $CDCl_3$): 139.8, 138.7, 138.6, 135.6, 135.4, 129.1, 128.9, 128.5, 128.1, 126.7, 126.6, 124.8, 124.6, 64.7, 64.4, 56.1, 38.0, 37.0, 32.9, 30.8, 29.7, 24.4, 23.7, 23.5, 20.5, 20.1. Anal. calcd for $C_{26}H_{32}IN$: C, 64.34; H, 6.64; N 2.89. Found: C, 64.27; H, 7.00; N 2.70.

3.2. Biological Tests

3.2.1. Antileishmanial Activity

Promastigote stage of *L. infantum* strain MHOM/TN/80/IPT1 (kindly provided by Dr. M. Gramiccia, ISS, Roma) and *L. tropica* (MHOM/SY/2012/ISS3130) were cultured in RPMI 1640 medium (EuroClone) supplemented with 10% heat-inactivated fetal calf serum (EuroClone, Milan Italy), 20 mM Hepes and 2 mM *L*-glutamine at 24 °C. To estimate the 50% inhibitory concentration (IC_{50}), the MTT (3-[4.5-dimethylthiazol-2-yl]-2.5-diphenyltetrazolium bromide) method was used [68,69]. Compounds were dissolved in DMSO and then diluted with medium to achieve the required concentrations. Drugs were placed in 96 wells round-bottom microplates and seven serial dilutions made. Miltefosine was used as reference anti-*Leishmania* drug. Parasites were diluted in complete medium to 5×10^6 parasites/mL and 100 μL of the suspension was seeded into the plates, incubated at 24 °C for 72 h and then 20 μL of MTT solution (5 mg/mL) was added into each well for 3 h. The plates were then centrifuged, the supernatants were discarded and the resulting pellets were dissolved in 100 μL of lysing buffer consisting of 20% (*w/v*) of a solution of SDS (Sigma), 40% DMF (Merck, Milan Italy) in H_2O. The absorbance was measured spectrophotometrically at a test wavelength of 550 nm and a reference wavelength of 650 nm. The results are all expressed as IC_{50}, which is the dose of compound necessary to inhibit parasite growth by 50%; each IC_{50} value is the mean of separate experiments performed in duplicate.

3.2.2. In Vitro Intracellular Amastigote Susceptibility Assays

THP-1 cells (human acute monocytic leukemia) were maintained in RPMI supplemented with 10% FBS (EuroClone), 50 μM 2-mercaptoethanol, 20 mM Hepes and 2 mM glutamine, at 37 °C in 5% CO_2. For *Leishmania* infections, THP-1 cells were plated at 5×10^5 cells/mL in 16-chamber Lab-Tek culture slides (Nunc, Milan, Italy) and treated with 0.1 μM phorbol myristate acetate (PMA, Sigma) for 48 h to achieve differentiation into macrophages. Cells were washed and infected with metacyclic *L. infantum* promastigotes at a macrophage/promastigote ratio of 1/10 for 24 h. Cell monolayers were then washed and incubated in the presence of test compounds for 72 h. Slides were fixed with methanol and stained with Giemsa. The percentages of infected macrophages among treated and non-treated cells were determined by light microscopy [70].

3.2.3. Cell Cytotoxicity Assays

THP-1 cells were plated at 5×10^5 cells/mL in 96 wells flat bottom microplates and treated with 0.1 μM PMA for 48 h to achieve differentiation into macrophages. Cells were then treated with serial dilutions of test compounds and cell proliferation evaluated using the MTT assay described for promastigotes. The results are expressed as CC_{50}, which is the dose of compound necessary to inhibit cell growth by 50%.

Vero-76 cells (ATCC CRL 1587 *Cercopithecus Aethiops*) were seeded at an initial density of 4×10^5 cells/mL in 24-well plates, in culture medium (Dulbecco's modified eagle's medium (D-MEM) with L-glutamine, supplemented with fetal bovine serum (FBS), 0.025 g/L kanamycin). Cell cultures were then incubated at 37 °C in a humidified, 5% CO_2 atmosphere in the absence or presence of serial dilutions of test compounds. Cell viability was determined after 48–96 h at 37 °C by the Crystal violet staining method. The results are expressed as CC_{50}, which is the concentration of compound necessary to inhibit cell growth by 50%. Each CC_{50} value is the mean and standard deviation of at least three separate experiments performed in duplicate.

3.3. Computational Methods

The 3D structure of the TryR from *Leishmania infantum* was obtained starting from the available Protein Data Bank file (pdb code: 2JK6 [71]) and optimized following a procedure previously described [64–66]. The optimized structures of the new tested compounds were docked into the putative binding pocket using Autodock 4.2.6/Autodock Tools1.4.61 [72]. The resulting complex was further energy minimized to convergence. The intermolecular complex was then solvated by a cubic box of TIP3P water molecules [73] and energy was minimized using a combination of molecular dynamics (MD) techniques [64–66]. Ten nanosecond molecular dynamics (MD) simulations at 298 K were then employed for system equilibration, and further, 50 ns MD simulations were run for data production. Following the MM/PBSA approach [67] each binding free energy value (ΔG_{bind}) was calculated as the sum of the electrostatic, van der Waals, polar solvation, nonpolar solvation, (ΔH_{bind}) and entropic contributions ($T\Delta S_{bind}$). The PRBFED analysis was carried out using the molecular mechanics/generalized Boltzmann surface area (MM/GBSA) approach [74] and was based on the same snapshots used in the binding free energy calculation. All simulations were carried out using the pmemd and pmemd.CUDA modules of Amber 18 [75], running on our own CPU/GPU calculation cluster. Molecular graphics images were produced using the UCSF Chimera package (v.1.14) [76]. All other graphs were obtained using GraphPad Prism (v. 6.0, GraphPad, La Jolla, CA, USA).

4. Conclusions

Sixteen 9-thioxanthenone derivatives (lucanthone analogues) and four compounds embodying the diarylethene substructure of amitriptyline (amitriptyline analogues) were tested in vitro for activity against *Leishmania tropica* and *L. infantum* promastigotes, and in a few cases also against intramacrophagic amastigotes of *L. infantum*. All compounds were characterized by the presence of a bulky quinolizidinylalkyl moiety, while differing for the tricyclic system to which the basic chain was connected. All compounds displayed activity against both species of *Leishmania* and most of them exhibited IC_{50} values lower than 10 μM, and were many-fold more potent than miltefosine. The six best compounds (**1**, **9**, **13**–**15** and **17**) displayed potency comparable to that of lucanthone (IC_{50} = 2.5 and 3.5 μM for the two *Leishmania* species), but their cytotoxicity versus the Vero 76 cells was always lower, with significant improvement of SI from 4.8–3.5 (lucanthone) to 16.2–16.9 and 23.6–19.2 for compounds **14** and **17**, respectively. These compounds exhibited comparable activity (and selectivity against THP-1 cells) against intramacrophagic amastigotes of *L. infantum*, and thus represent promising, structurally distinct leads for the development of improved antileishmanial agents. Docking studies suggest that the antileishmanial activity of compounds **14** and **17** may be related to the inhibition of trypanothione reductase, as is the case for other tricyclic compounds. However, lucanthone and the tested compounds were previously shown to potently inhibit the AChE, and particularly the BChE; thus, it is now put forward that this inhibitory property may have a notable additional role in the antiparasitic mechanism, either by reducing the availability of choline to build up the main component of promastigote membrane, or by inhibiting other non-classical functions of cholinesterases in the unicellular organisms.

Supplementary Materials: The following are available online at http://www.mdpi.com/1424-8247/13/11/339/s1. Table S1: In silico binding thermodynamics of compounds **14**, **17**, lucanthone and amitriptyline towards TryR. Table S2: Per-residue binding enthalpy decomposition (ΔH_{res}) for compounds **14**, **17**, lucanthone and amitriptyline towards TryR, Figure S1: Amitriptyline in the binding pocket of TryR.

Author Contributions: Conceptualization, F.S.; methodology and validation, M.T., A.S. and N.B.; software, E.L., S.P. (Sabrina Pricl); S.P. (Silvia Parapini), L.C., B.T., E.L. and V.B. participated to methodology; investigation, F.S., M.T., A.S., N.B. and S.P. (Sabrina Pricl); resources, M.T., A.S., N.B. and S.P. (Sabrina Pricl); writing—original draft preparation, F.S. and M.T.; writing—review and editing, F.S., M.T., A.S., N.B. and S.P. (Sabrina Pricl); visualization, M.T.; supervision, M.T., A.S. and N.B.; project administration, F.S., M.T., A.S. and N.B. All authors have read and agreed to the published version of the manuscript.

Funding: This research received no external funding.

Acknowledgments: This work was financially supported by the University of Genoa. The Authors thank O. Gagliardo for performing elemental analyses.

Conflicts of Interest: The authors declare no conflict of interest.

References

1. World Health Organization (WHO). Factsheet. Available online: https://www.afro.who.int/health-topics/Leishmaniasis (accessed on 2 March 2020).
2. Maslov, D.A.; Opperdoes, F.R.; Kostygov, A.Y.; Hashimi, H.; Lukeš, J.; Yurchenko, V. Recent advances in trypanosomatid research: Genome organization, expression, metabolism, taxonomy and evolution. *Parasitology* **2019**, *146*, 1–27. [CrossRef] [PubMed]

3. D'Avila-Levy, C.M.; Boucinha, C.; Kostygov, A.; Santos, H.L.; Morelli, K.A.; Grybchuk-Ieremenko, A.; Duval, L.; Votýpka, J.; Yurchenko, V.; Grellier, P.; et al. Exploring the environmental diversity of kinetoplastid flagellates in the high-throughput DNA sequencing era. *Mem. Inst. Oswaldo Cruz* **2015**, *110*, 956–965. [CrossRef] [PubMed]
4. Sangshetti, J.N.; Kalam Khan, F.A.; Kulkarni, A.A.; Arote, R.; Patilc, R.H. Antileishmanial drug discovery: Comprehensive review of the last 10 years. *RSC Adv.* **2015**, *5*, 32376–32415. [CrossRef]
5. Lee, S.M.; Kim, M.S.; Hayat, F.; Shin, D. Advances in the discovery of novel antiprotozoal agents. *Molecules* **2019**, *24*, 3886. [CrossRef]
6. Razzaghi-Asl, N.; Sepehri, S.; Ebadi, A.; Karami, P.; Nejatkhah, N.; Johari-Ahar, M. Insights into the current status of privileged N-heterocycles as antileishmanial agents. *Mol. Divers.* **2020**, *24*, 525–569. [CrossRef]
7. Andrade-Neto, V.V.; Ferreira Cunha-Junior, E.; Dos Santos Faioes, V.; Pereira, T.M.; Silva, R.L.; Leon, L.L.; Torres-Santos, E.C. Leishmaniasis treatment: Update of possibilities for drug repurposing. *Front. Biosci.* **2018**, *23*, 967–999.
8. Bekhit, A.A.; El-Agroudy, E.; Helmy, A.; Ibrahim, T.M.; Shavandi, A.; Bekhit, A.E.A. *Leishmania* treatment and prevention: Natural and synthesized drugs. *Eur. J. Med. Chem.* **2018**, *160*, 229–244. [CrossRef]
9. Mowbray, C.E. *Drug Discovery for Leishmaniasis*; Rivas, L., Gil, C., Eds.; Royal Society of Chemistry: London, UK, 2018; pp. 26–36.
10. Wijnant, G.-J.; Croft, S.L.; de la Flor, R.; Alavijeh, M.; Yardley, V.; Braillard, S.; Mowbray, C.; van Bocxlaer, K. Pharmacokinetics and pharmacodynamics of the nitroimidazole DNDI-0690 in mouse models of cutaneous Leishmaniasis. *Antimicrob. Agents Chemother.* **2019**, *63*, e00829-19. [CrossRef]
11. van den Kerkhof, M.; Mabille, D.; Chatelain, E.; Mowbray, C.E.; Braillard, S.; Hendrickx, S.; Maes, L.; Caljon, G. In vitro and in vivo pharmacodynamics of three novel antileishmanial lead series. *Int. J. Parasitol. Drugs Drug Resist.* **2018**, *8*, 81–86. [CrossRef]
12. Thomas, M.G.; De Rycker, M.; Ajakane, M.; Albrecht, S.; Álvarez-Pedraglio, A.I.; Boesche, M.; Brand, S.; Campbell, L.; Cantizani-Perez, J.; Cleghorn, L.A.T.; et al. Identification of GSK3186899/DDD853651 as a preclinical development candidate for the treatment of visceral Leishmaniasis. *J. Med. Chem.* **2019**, *62*, 1180–1202. [CrossRef]
13. Zilberstein, D.; Dwyer, D.M. Antidepressants cause lethal disruption of membrane function in the human protozoan parasite *Leishmania*. *Science* **1984**, *226*, 977–979. [CrossRef] [PubMed]
14. Evans, A.T.; Croft, S.L. Antileishmanial actions of tricyclic neuroleptics appear to lack structural specificity. *Biochem. Pharmacol.* **1994**, *48*, 613–616. [CrossRef]
15. da Silva Rodrigues, J.H.; Miranda, N.; Volpato, H.; Ueda-Nakamura, T.; Nakamura, C.V. The antidepressant clomipramine induces programmed cell death in *Leishmania amazonensis* through a mitochondrial pathway. *Parasitol. Res.* **2019**, *118*, 977–989. [CrossRef] [PubMed]
16. Ferreira Cunha-Júnior, E.; Andrade-Neto, V.V.; Lima, M.L.; da Costa-Silva, T.A.; Galisteo Junior, A.J.; Abengózar, M.A.; Barbas, C.; Rivas, L.; Almeida-Amaral, E.E.; Tempone, A.G.; et al. Cyclobenzaprine raises ROS levels in *Leishmania infantum* and reduces parasite burden in infected mice. *PLoS Negl. Trop. Dis.* **2017**, *11*, e0005281. [CrossRef]
17. Mattock, N.M.; Peters, W. The experimental chemotherapy of leishmaniasis, III Detection of antileishmanial activity in some new synthetic compounds in a tissue culture model. *Ann. Trop. Med. Parasitol.* **1975**, *69*, 449–462. [CrossRef]
18. Peters, W.; Trotter, E.R.; Robinson, B.L. The experimental chemotherapy of leishmaniasis, VII. *Ann. Trop. Med. Parasitol.* **1980**, *74*, 321–335. [CrossRef]
19. Berman, J.D.; Lee, L.S. Activity of oral drugs against *Leishmania tropica* in human macrophages in vitro. *Am. J. Trop. Med. Hyg.* **1983**, *32*, 947–951. [CrossRef]
20. Osorio, Y.; Travi, B.L.; Renslo, A.R.; Peniche, A.G.; Melby, P.C. Identification of small molecule lead compounds for visceral leishmaniasis using a novel ex vivo splenic explant model system. *PLoS Negl. Trop. Dis.* **2011**, *5*, e962. [CrossRef]
21. Cioli, D.; Pica-Mattoccia, L.; Archer, S. Antischistosomal drugs: Past, present... and future? *Pharmacol. Ther.* **1995**, *68*, 35–85. [CrossRef]
22. Russell, W.L. Results of tests for possible transmitted genetic effects of hycanthone in mammals. *J. Toxicol. Environ. Health* **1975**, *1*, 301–304. [CrossRef]

23. Hartman, P.E.; Hulbert, P.B.; Bueding, E.; Taylor, D.D. Microsomal activation to mutagens of antischistosomal methyl thioxanthenones and initial tests on a possibly non-mutagenic analogue. *Mutat. Res.* **1975**, *31*, 87–95. [CrossRef]
24. Hartman, P.E.; Hulbert, P.B. Genetic activity spectra of some antischistosomal compounds, with particular emphasis on thioxanthenones and benzothiopyranoindazoles. *J. Toxicol. Environ. Health* **1975**, *1*, 243–270. [CrossRef] [PubMed]
25. Palmeira, A.; Vasconcelos, M.H.; Paiva, A.; Fernandes, M.X.; Pinto, M.; Sousa, E. Dual inhibitors of P-glycoprotein and tumor cell growth: (Re)discovering thioxanthones. *Biochem. Pharmacol.* **2012**, *83*, 57–68. [CrossRef]
26. Barbosa, J.; Lima, R.T.; Sousa, D.; Gomes, A.S.; Palmeira, A.; Seca, H.; Choosang, K.; Pakkong, P.; Bousbaa, H.; Pinto, M.M.; et al. Screening a small library of xanthones for antitumor activity and identification of a hit compound which induces apoptosis. *Molecules* **2016**, *21*, 81. [CrossRef] [PubMed]
27. Tasso, B.; Novelli, F.; Tonelli, M.; Barteselli, A.; Basilico, N.; Parapini, S.; Taramelli, D.; Sparatore, A.; Sparatore, F. Synthesis and antiplasmodial activity of novel chloroquine analogues with bulky basic side chains. *ChemMedChem* **2015**, *10*, 1570–1583. [CrossRef] [PubMed]
28. Sparatore, A.; Basilico, N.; Casagrande, M.; Parapini, S.; Taramelli, D.; Brun, R.; Wittlin, S.; Sparatore, F. Antimalarial activity of novel pyrrolizidinyl derivatives of 4-aminoquinoline. *Bioorg. Med. Chem. Lett.* **2008**, *18*, 3737–3740. [CrossRef] [PubMed]
29. Sparatore, A.; Basilico, N.; Parapini, S.; Romeo, S.; Novelli, F.; Sparatore, F.; Taramelli, D. 4-Aminoquinoline quinolizidinyl- and quinolizidinylalkyl-derivatives with antimalarial activity. *Bioorg. Med. Chem.* **2005**, *13*, 5338–5345. [CrossRef] [PubMed]
30. Barteselli, A.; Casagrande, M.; Basilico, N.; Parapini, S.; Rusconi, C.M.; Tonelli, M.; Boido, V.; Taramelli, D.; Sparatore, F.; Sparatore, A. Clofazimine analogs with antileishmanial and antiplasmodial activity. *Bioorg. Med. Chem.* **2015**, *23*, 55–65. [CrossRef]
31. Tonelli, M.; Gabriele, E.; Piazza, F.; Basilico, N.; Parapini, S.; Tasso, B.; Loddo, R.; Sparatore, F.; Sparatore, A. Benzimidazole derivatives endowed with potent antileishmanial activity. *J. Enzyme Inhib. Med. Chem.* **2018**, *33*, 210–226. [CrossRef]
32. Archer, S.; Rej, R. Nitro and amino derivatives of lucanthone as antitumor agents. *J. Med. Chem.* **1982**, *25*, 328–331. [CrossRef]
33. Sparatore, A.; Veronese, M.; Sparatore, F. Quinolizidine derivatives with antimicrobial activity. *Farmaco Ed. Sci.* **1987**, *42*, 159–174.
34. Boido Canu, C.; Iusco, G.; Boido, V.; Sparatore, F.; Sparatore, A. Synthesis and antileukemic activity of 1-[(quinolizidinylalkyl)amino]4/7-R-thioxanthen-9-ones. *Farmaco* **1989**, *44*, 1069–1082. [PubMed]
35. Sparatore, A.; Marchi, M.; Maura, G.; Paudice, P.; Raiteri, M. Effects of some rigid analogues of imipramine and amitriptyline on the uptake of noradrenaline, serotonin and choline in rat brain synaptosomes. *Pharmacol. Res. Commun.* **1982**, *14*, 257–265. [CrossRef]
36. Sparatore, A.; Marchi, M.; Maura, G.; Raiteri, M. Effect of some quinolizidine derivatives on the release of serotonin, noradrenaline, dopamine and acetylcholine from rat brain synaptosomes. *Farmaco* **1989**, *44*, 1205–1216. [CrossRef] [PubMed]
37. Boido, V.; Sparatore, F. Derivatives of natural amino alcohols and diamines of pharmacologic interest. IV. Action of lupinylmagnesium chloride on aromatic and heterocyclic ketones. *Ann. Chim.* **1966**, *56*, 1603–1613.
38. Tonelli, M.; Catto, M.; Tasso, B.; Novelli, F.; Canu, C.; Iusco, G.; Pisani, L.; De Stradis, A.; Denora, N.; Sparatore, A.; et al. Multitarget therapeutic leads for Alzheimer's disease: Quinolizidinyl derivatives of bi- and tricyclic systems as dual inhibitors of cholinesterases and β-amyloid (Aβ) aggregation. *ChemMedChem* **2015**, *10*, 1040–1053. [CrossRef] [PubMed]
39. Tasso, B.; Catto, M.; Nicolotti, O.; Novelli, F.; Tonelli, M.; Giangreco, I.; Pisani, L.; Sparatore, A.; Boido, V.; Carotti, A.; et al. Quinolizidinyl derivatives of bi- and tricyclic systems as potent inhibitors of acetyl- and butyrylcholinesterase with potential in Alzheimer's disease. *Eur. J. Med. Chem.* **2011**, *46*, 2170–2184. [CrossRef]
40. Blanz, E.J., Jr.; French, F.A. A systematic investigation of thioxanthen-9-ones and analogs as potential antitumor agents. *J. Med. Chem.* **1963**, *6*, 185–191. [CrossRef]

41. Naidu, M.D.; Agarwal, R.; Pena, L.A.; Cunha, L.; Mezei, M.; Shen, M.; Wilson, D.M.; Liu, Y.; Sanchez, Z.; Chaudhary, P.; et al. Lucanthone and its derivative hycanthone inhibit apurinic endonuclease-1 (APE1) by direct protein binding. *PLoS ONE* **2011**, *6*, e23679. [CrossRef]
42. Carew, J.S.; Espitia, C.M.; Esquivel, J.A., 2nd; Mahalingam, D.; Kelly, K.R.; Reddy, G.; Giles, F.J.; Nawrocki, S.T. Lucanthone is a novel inhibitor of autophagy that induces cathepsin D-mediated apoptosis. *J. Biol. Chem.* **2011**, *286*, 6602–6613. [CrossRef]
43. Lima, R.; Sousa, D.; Gomes, A.; Mendes, N.; Matthiesen, R.; Pedro, M.; Marques, F.; Pinto, M.M.; Sousa, E.; Vasconcelo, M.H. The antitumor activity of a lead thioxanthone is associated with alterations in cholesterol localization. *Molecules* **2018**, *23*, 3301. [CrossRef]
44. Tischer, M.; Pradel, G.; Ohlsen, K.; Holzgrabe, U. Quaternary ammonium salts and their antimicrobial potential: Targets or nonspecific interactions? *ChemMedChem* **2012**, *7*, 22–31. [CrossRef] [PubMed]
45. Zufferey, R.; Mamoun, C.B. Choline transport in *Leishmania major* promastigotes and its inhibition by choline and phosphocholine analogs. *Mol. Biochem. Parasitol.* **2002**, *125*, 127–134. [CrossRef]
46. Duque-Benítez, S.M.; Ríos-Vásquez, L.A.; Ocampo-Cardona, R.; Cedeño, D.L.; Jones, M.A.; Vélez, I.D.; Robledo, S.M. Synthesis of novel quaternary ammonium salts and their in vitro antileishmanial activity and U-937 cell cytotoxicity. *Molecules* **2016**, *21*, 381. [CrossRef]
47. Papanastasiou, I.; Prousis, K.C.; Georgikopoulou, K.; Pavlidis, T.; Scoulica, E.; Kolocouris, N.; Calogeropoulou, T. Design and synthesis of new adamantyl-substituted antileishmanial ether phospholipids. *Bioorg. Med. Chem. Lett.* **2010**, *20*, 5484–5487. [CrossRef] [PubMed]
48. Weinstein, I.B.; Chernoff, R.; Finkelstein, I.; Hirschberg, E. Miracil D: An inhibitor of ribonucleic acid synthesis in Bacillus subtilis. *Mol. Pharmacol.* **1965**, *1*, 297–305. [PubMed]
49. Archer, S.; Yarinsky, A. Recent developments in the chemotherapy of schistosomiasis. *Prog. Drug Res.* **1972**, *16*, 11–66. [PubMed]
50. Krauth-Siegel, R.L.; Meiering, S.K.; Schmidt, H. The parasite-specific trypanothione metabolism of *Trypanosoma* and *Leishmania*. *Biol. Chem.* **2003**, *384*, 539–549. [CrossRef]
51. Kumar, S.; Ali, M.R.; Bawa, S. Mini review on tricyclic compounds as an inhibitor of trypanothione reductase. *J. Pharm. Bioallied Sci.* **2014**, *6*, 222–228.
52. Khan, M.O. Trypanothione reductase: A viable chemotherapeutic target for antitrypanosomal and antileishmanial drug design. *Drug Targ. Insights* **2007**, *2*, 129–146. [CrossRef]
53. Verdier, J.S.; Wolfe, A.D. Butyrylcholinesterase inhibition by miracil D and other compounds. *Biochem. Pharmacol.* **1986**, *35*, 1605–1608. [CrossRef]
54. Pulido, S.A.; Nguyen, V.H.; Alzate, J.F.; Cedeño, D.L.; Makurath, M.A.; Ríos-Vásquez, A.; Duque-Benítez, S.M.; Jones, M.A.; Robledo, S.M.; Friesen, J.A. Insights into the phosphatidylcholine and phosphatidylethanolamine biosynthetic pathways in *Leishmania* parasites and characterization of a choline kinase from *Leishmania infantum*. *Comp. Biochem. Physiol. B. Biochem. Mol. Biol.* **2017**, *213*, 45–54. [CrossRef] [PubMed]
55. Wassef, M.K.; Fioretti, T.B.; Dwyer, D.M. Lipid analyses of isolated surface membranes of *Leishmania donovani* promastigotes. *Lipids* **1985**, *20*, 108–115. [CrossRef] [PubMed]
56. Pezzementi, L.; Chatonnet, A. Evolution of cholinesterases in the animal kingdom. *Chem. Biol. Interact.* **2010**, *187*, 27–33. [CrossRef]
57. Karczmar, A.G. Cholinesterases (ChEs) and the cholinergic system in ontogenesis and phylogenesis, and non-classical roles of cholinesterases-A review. *Chem. Biol. Interact.* **2010**, *187*, 34–43. [CrossRef]
58. Halliday, A.C.; Greenfield, S.A. From protein to peptides: A spectrum of non-hydrolytic functions of acetylcholinesterase. *Protein Pept. Lett.* **2012**, *19*, 165–172. [CrossRef]
59. Lenta, B.N.; Vonthron-Sénécheau, C.; Weniger, B.; Devkota, K.P.; Ngoupayo, J.; Kaiser, M.; Naz, Q.; Choudhary, M.I.; Tsamo, E.; Sewald, N. Leishmanicidal and cholinesterase inhibiting activities of phenolic compounds from *Allanblackia monticola* and *Symphonia globulifera*. *Molecules* **2007**, *12*, 1548–1557. [CrossRef]
60. Vila-Nova, N.S.; Morais, S.M.; Falcão, M.J.C.; Bevilaqua, C.M.L.; Rondon, F.C.M.; Wilson, M.E.; Vieira, I.G.P.; Andrade, H.F. Leishmanicidal and cholinesterase inhibiting activities of phenolic compounds of *Dimorphandra gardneriana* and *Platymiscium floribundum*, native plants from Caatinga biome. *Pesq. Vet. Bras.* **2012**, *32*, 1164–1168. [CrossRef]
61. Mogana, R.; Adhikari, A.; Debnath, S.; Hazra, S.; Hazra, B.; Teng-Jin, K.; Wiart, C. The antiacetylcholinesterase and antileishmanial activities of *Canarium patentinervium* Miq. *Biomed Res. Int.* **2014**, *2014*, 903529. [CrossRef]

62. Benson, T.J.; McKie, J.H.; Garforth, J.; Borges, A.; Fairlamb, A.H.; Douglas, K.T. Rationally designed selective inhibitors of trypanothione reductase. Phenothiazines and related tricyclics as lead structures. *Biochem. J.* **1992**, *286*, 9–11. [CrossRef]
63. Venkatesan, S.K.; Shukla, A.K.; Dubey, V.K. Molecular docking studies of selected tricyclic and quinone derivatives on trypanothione reductase of *Leishmania infantum*. *J. Comput. Chem.* **2010**, *31*, 2463–2475. [PubMed]
64. Briguglio, I.; Loddo, R.; Laurini, E.; Fermeglia, M.; Piras, S.; Corona, P.; Giunchedi, P.; Gavini, E.; Sanna, G.; Giliberti, G.; et al. Synthesis, cytotoxicity and antiviral evaluation of new series of imidazo[4,5-g]quinoline and pyrido[2,3-g]quinoxalinone derivatives. *Eur. J. Med. Chem.* **2015**, *105*, 63–79. [CrossRef]
65. Carta, A.; Sanna, G.; Briguglio, I.; Madeddu, S.; Vitale, G.; Piras, S.; Corona, P.; Peana, A.T.; Laurini, E.; Fermeglia, M.; et al. Quinoxaline derivatives as new inhibitors of coxsackievirus B5. *Eur. J. Med. Chem.* **2018**, *145*, 559–569. [CrossRef] [PubMed]
66. Loddo, R.; Francesconi, V.; Laurini, E.; Boccardo, S.; Aulic, S.; Fermeglia, M.; Pricl, S.; Tonelli, M. 9-Aminoacridine-based agents impair the bovine viral diarrhea virus (BVDV) replication targeting the RNA-dependent RNA polymerase (RdRp). *Bioorg. Med. Chem.* **2018**, *26*, 855–868. [CrossRef] [PubMed]
67. Massova, I.; Kollman, P.A. Combined molecular mechanical and continuum solvent approach (MM-PBSA/GBSA) to predict ligand binding. *Perspect. Drug Discov. Des.* **2000**, *18*, 113–135. [CrossRef]
68. Mosmann, T. Rapid colorimetric assay for cellular growth and survival: Application to proliferation and cytotoxicity assays. *J. Immunol. Methods* **1983**, *65*, 55–63. [CrossRef]
69. Baiocco, P.; Ilari, A.; Ceci, P.; Orsini, S.; Gramiccia, M.; Di Muccio, T.; Colotti, G. Inhibitory effect of silver nanoparticles on trypanothione reductase activity and *Leishmania infantum* proliferation. *ACS Med. Chem. Lett.* **2010**, *2*, 230–233. [CrossRef] [PubMed]
70. Konstantinović, J.; Videnović, M.; Orsini, S.; Bogojević, K.; D'Alessandro, S.; Scaccabarozzi, D.; Terzić Jovanović, N.; Gradoni, L.; Basilico, N.; Šolaja, B.A. Novel aminoquinoline derivatives significantly reduce parasite load in *Leishmania infantum* infected mice. *ACS Med. Chem. Lett.* **2018**, *9*, 629–634. [CrossRef]
71. Baiocco, P.; Colotti, G.; Franceschini, S.; Ilari, A. Molecular basis of antimony treatment in leishmaniasis. *J. Med. Chem.* **2009**, *52*, 2603–2612. [CrossRef]
72. Morris, G.M.; Huey, R.; Lindstrom, W.; Sanner, M.F.; Belew, R.K.; Goodsell, D.S.; Olson, A.J. AutoDock4 and AutoDockTools4: Automated docking with selective receptor flexibility. *J. Comput. Chem.* **2009**, *30*, 2785–2791. [CrossRef]
73. Jorgensen, W.L.; Chandrasekhar, J.; Madura, J.D.; Impey, R.W.; Klein, M.L. The origin of layer structure artifacts in simulations of liquid water. *J. Chem. Phys.* **1983**, *79*, 926–935. [CrossRef]
74. Onufriev, A.; Bashford, D.; Case, D.A. Modification of the generalized born model suitable for macromolecules. *J. Phys. Chem. B* **2000**, *104*, 3712–3720. [CrossRef]
75. Case, D.A.; Ben-Shalom, I.Y.; Brozell, S.R.; Cerutti, D.S.; Cheatham, T.E.; Cruzeiro III, V.W.D.; Darden, T.A.; Duke, R.E.; Ghoreishi, D.; Gilson, M.K.; et al. *AMBER 2018*; University of California: San Francisco, CA, USA, 2018.
76. Pettersen, E.F.; Goddard, T.D.; Huang, C.C.; Couch, G.S.; Greenblatt, D.M.; Meng, E.C.; Ferrin, T.F. UCSF Chimera-A visualization system for exploratory research and analysis. *J. Comput. Chem.* **2004**, *25*, 1605–1612. [CrossRef]

Publisher's Note: MDPI stays neutral with regard to jurisdictional claims in published maps and institutional affiliations.

Article

Antibacterial Drug-Release Polydimethylsiloxane Coating for 3D-Printing Dental Polymer: Surface Alterations and Antimicrobial Effects

Hang-Nga Mai [1,†], Dong Choon Hyun [2,†], Ju Hayng Park [2], Do-Yeon Kim [3], Sang Min Lee [3] and Du-Hyeong Lee [1,4,*]

1 Institute for Translational Research in Dentistry, Kyungpook National University, Daegu 41940, Korea; maihangnga1403@knu.ac.kr

2 Department of Polymer Science and Engineering, Kyungpook National University, Daegu 41566, Korea; dong.hyun@knu.ac.kr (D.C.H.); pjh99279@naver.com (J.H.P.)

3 Department of Pharmacology, School of Dentistry, Kyungpook National University, Daegu 41940, Korea; dykim82@knu.ac.kr (D.-Y.K.); leeyang2324@naver.com (S.M.L.)

4 Department of Prosthodontics, School of Dentistry, Kyungpook National University, Daegu 41940, Korea

* Correspondence: deweylee@knu.ac.kr; Tel.: +82-53-600-7676

† These authors contributed equally to this work.

Received: 4 September 2020; Accepted: 10 October 2020; Published: 12 October 2020

Abstract: Polymers are the most commonly used material for three-dimensional (3D) printing in dentistry; however, the high porosity and water absorptiveness of the material adversely influence biofilm formation on the surface of the 3D-printed dental prostheses. This study evaluated the effects of a newly developed chlorhexidine (CHX)-loaded polydimethylsiloxane (PDMS)-based coating material on the surface microstructure, surface wettability and antibacterial activity of 3D-printing dental polymer. First, mesoporous silica nanoparticles (MSN) were used to encapsulate CHX, and the combination was added to PDMS to synthesize the antibacterial agent-releasing coating substance. Then, a thin coating film was formed on the 3D-printing polymer specimens using oxygen plasma and thermal treatment. The results show that using the coating substance significantly reduced the surface irregularity and increased the hydrophobicity of the specimens. Remarkably, the culture media containing coated specimens had a significantly lower number of bacterial colony formation units than the noncoated specimens, thereby indicating the effective antibacterial activity of the coating.

Keywords: 3D-printing; dental polymer; antibacterial agent; coating; mesoporous silica nanoparticles; polydimethylsiloxane

1. Introduction

Three-dimensional (3D) printing technology is an additive manufacturing method in which a 3D object is formed by adding successive layers of material [1,2]. Dentistry has benefited from the rapid expansion of 3D-printing methods, especially in the field of prosthesis manufacturing [3–5]. The 3D-printing technology used in dentistry is classified into four main categories: extrusion printing, inkjet printing, laser melting/sintering, and stereolithography printing [3]. Those printing methods are based on the principle of layered manufacturing, which is more suitable than conventional casting and milling methods for producing individualized complex structures [6]. Moreover, using a machining process with computer-aided design/computer-aided manufacturing (CAD/CAM) reduces manual labor and material waste [1]. Recent studies have shown that 3D-printed dental prostheses had a clinically acceptable degree of precision [3,6,7].

Along with the increasing accuracy of 3D-printing technology, various materials, including polymers, ceramics, and metallic powders, have been developed [4]. Regarding dental materials used in 3D printers, polymers are the most commonly used in dentistry for interim and definitive prostheses due to their suitable mechanical strength, highly biocompatible properties, and ease of manipulation [3,4]. However, a major clinical complication associated with dental polymer prostheses is dental plaque surface deposition, which comprises numerous oral microorganisms due to the porosity and water absorptiveness of the polymer material [8]. Specifically, as the 3D-printed prosthesis is generated layer-by-layer, micropores are created when air is trapped between the layers during the printing process or when the individual layers are incompletely fused [9]. During clinical occlusal adjustment and clinical use, these micropores are exposed on the surface because of the abrasion of the restorations, potentially becoming a host for bacterial growth [6]. The risk of material contamination by microorganisms is a critical limitation for the longevity of 3D-printed polymer prostheses [3].

Several strategies have been reported for the fabrication of antimicrobial polymers for 3D-printed polymers [10]. Direct incorporation of an antibacterial agent into dental polymers has been used to reduce plaque accumulation [11]. Chlorhexidine (CHX) is an antiseptic agent widely used in dentistry for its broad-spectrum antibacterial effects and nontoxicity toward mammalian cells [12]. CHX is effective in managing infected oral mucositis; thus, it has commonly been used to prevent dental plaques and to control infections as a topical agent in daily mouth rinse and as an irrigation solution in endodontic treatment [13]. When CHX was directly mixed with the polymer and cured, CHX release at the therapeutic dose was maintained for 28 days [14–16]. However, direct mixing may negatively affect the mechanical and surface properties of polymers in terms of polymerization degree, surface porosity, and water absorption [11,17]. Moreover, commercial polymers used for the subtractive manufacturing method are provided in a completely polymerized state, making it difficult to directly add CHX inside the substance without jeopardizing material integrity [18].

An alternative technique for inhibiting bacterial adherence is to change the surface property of the object by creating a coating film [19–21]. Coating layer formation does not sacrifice the mechanical properties and integrity of the material. The major mechanisms underlying the antibacterial activity of the coating layers include antiadhesion/bacterial-repelling and contact-killing [22]. Antiadhesion coating reduces the adhesion force between bacteria and the substrate to enable the easy removal of bacteria before the biofilm layer is formed [21]. Alternatively, in the contact-killing approach, antibacterial compounds are attached to the surface of the material by flexible, hydrophobic polymeric chains, which can kill bacteria upon contact [21,22]. Remarkably, using a polymer coating significantly reduced plaque biofilm formation on polymeric restorations with the coating layer exhibiting acceptable mechanical and chemical durability [20]. Azuma et al. [19] reported that silica coating with silica nanoparticles of various sizes was effective in decreasing bacterial adherence to polymeric restorations. While these previous coating materials showed antiadhesion potential, they passively hindered the adherence of early colonizers by increasing polymer surface hydrophobicity [19,20]. This strategy may physically suppress dental plaque formation and maturation; however, it provides no active antimicrobial agent to inhibit the vitality and growth of pathogenic microorganisms.

With the development of new drug-carrier materials, active antibacterial agent-releasing coating is now a significant focus in biomedical research [22]. In its most advanced form, the coating mediates antibacterial activity by releasing loaded antibacterial compounds over time to kill the bacteria in the surrounding [21,22]. In drug delivery systems, materials with a porous structure are recommended for incorporating drug-loaded particles [23,24]. Among the porous polymeric materials, polydimethylsiloxane (PDMS) has been used in the medical field due to its flexibility, biocompatibility, transparency, low cost, and ease of fabrication [23,25]. The incompatibility between the coated substrate and PDMS could cause a dewetting that may induce a greater surface roughness; however, the subsequent treatments of oxygen plasma could enhance the wettability of precured PDMS, inducing its uniform coating [26,27]. Moreover, the drug release rate of PDMS can be controlled by surface modifications using plasma treatment, which effectively influence the water penetration rate and functionalization of the polymer that contains drug [28]. The highly functionalized PDMS can be polymerized using thermal treatment or ultraviolet (UV) light activation for curing the coating layer, inducing a more chemically and physically stable coating layer on the substrate [27]. Thus, it would be an effective approach for tissue engineering or drug delivery systems in biological and biomedical applications [29]. In dentistry, PDMS coating has been used for tooth enamel and metallic dental implants as a hydrophobic layer to improve the antibacterial features [30]. However, no studies have examined the PDMS coating application for active antibacterial effects in dental 3D-printable polymers. Therefore, this study evaluated the effect of a newly developed CHX-loaded PDMS-based coating on the surface microstructure, surface wettability, and antibacterial activity of the 3D-printing dental polymer.

2. Results

2.1. Surface Characteristics

Figure 1 shows the microscopic images of specimens in different groups. Deep grooves and small defects were found on the surface of noncoated specimens, whereas no scratches were observed on the surface of coated specimens. Several different-sized silica particles were observed on the coating surface.

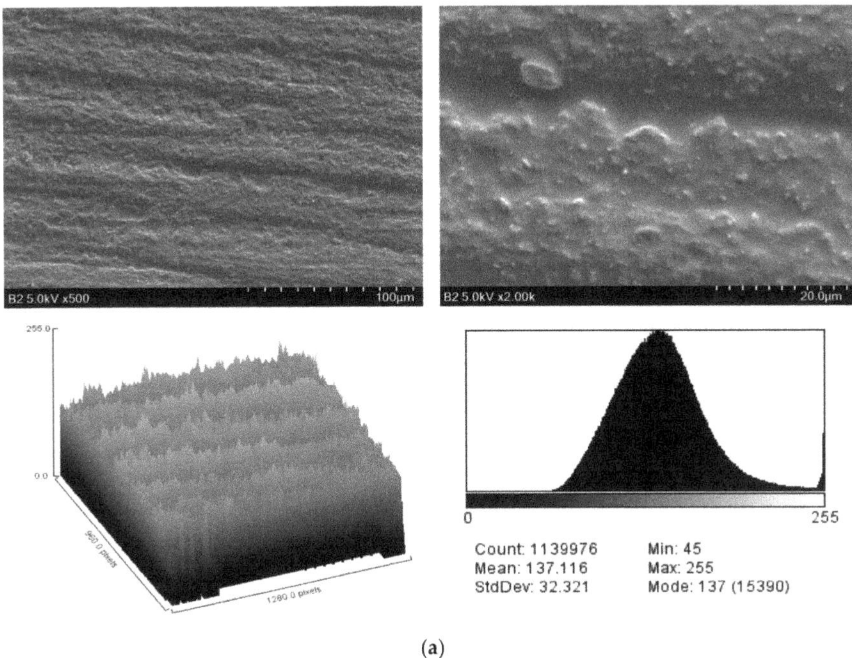

(a)

Figure 1. *Cont.*

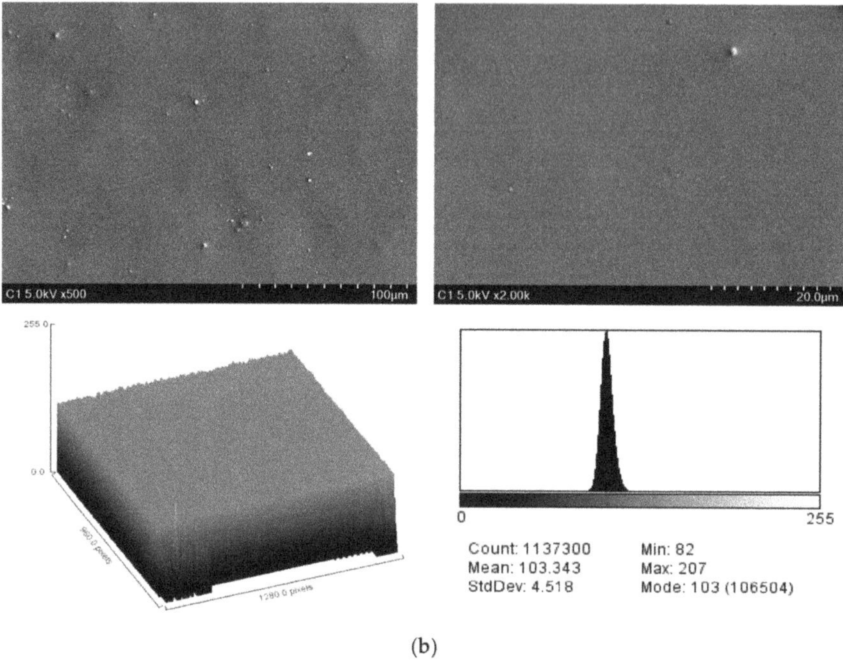

Figure 1. Scanning electron microscope (SEM) images at 500 × (left) and 2000 × (right) magnification and the resulting histogram of pixel brightness levels (0 = black, 255 = white) of: (**a**) noncoated specimen; and (**b**) coated specimens.

Table 1 presents the SEM image analysis for surface roughness obtained from the surface plot histograms. The SEM roughness index (SRI) values were significantly lower in the coated group than in the noncoated group ($p < 0.001$). Regarding surface wettability, the mean contact angles were significantly higher in the coated group than in the noncoated group ($p < 0.001$) (Figure 2 and Table 1).

Table 1. Effect of coating layer on SEM roughness index (SRI) and contact angle degree (CA) measurements.

Group	Mean (SD)		
	SRI	CA	*p*-Value
Noncoated	39.25 (3.69) ^{a}	120.22 (4.46) ^{a}	<0.001
Coated	15.41 (8.07) ^{b}	91.88 (8.19) ^{b}	

$^{a,\,b}$ Different superscript lowercase letters indicate a statistically significant difference within a column; SRI, SEM roughness index; CA, contact angle degree.

Figure 2. Contact angle of the specimens: (**a**) noncoated specimen; and (**b**) coated specimen.

2.2. Antimicrobial Activity

Because the specimen itself could affect bacterial growth, the incubation of *S. mutans* with noncoated specimen was a control in this experiment. Statistical analysis was performed using three technical replicates. Biological relevance of the results was confirmed using three independent experiments with similar results. Figure 3 shows the results of the *S. mutans* bacterial growth inhibition assay for the noncoated and coated groups. The numbers of bacterial colonies ($\times 10^4$ CFU/mL) of the noncoated and coated groups were 210.92 ± 8.02 and 70.76 ± 9.16, respectively. Independent *t*-test showed that the culture media containing the coated specimens had significantly lower CFU values than the culture media containing the noncoated specimens ($p < 0.001$).

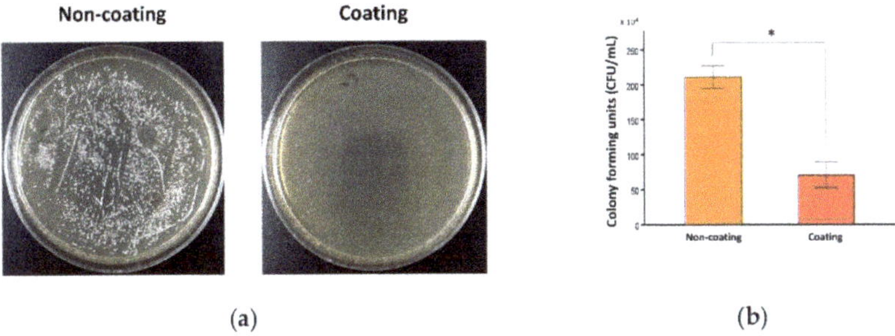

Figure 3. Bacterial growth inhibition assay of noncoated and coated groups: (**a**) bacterial colonization on agar plates; and (**b**) colony-forming units. Statistical analysis was performed using three independent technical replicates, inducing similar biological results. Independent *t*-test showed that the culture media containing the coated specimens had significantly lower CFU values than the culture media containing the noncoated specimens (* $p < 0.001$).

3. Discussions

This study evaluated the effect of a newly developed CHX-loaded PDMS-based coating substance on the surface properties and antibacterial ability of coated 3D-printing dental material. From the results of this study, applying the coating layer on the polymer specimens significantly reduced their surface irregularity while increasing the hydrophobicity and antibacterial activity.

Once a dental restoration is placed in the oral cavity, proteins from saliva cover the surface of the restoration as a film, followed by the attachment of free-floating bacteria to the film by microfilaments in the cell walls to form a biofilm on the restoration [31]. Restorations with greater surface roughness show higher biofilm formation because the irregular surface geometry offsets shear forces to the surface, thereby providing a favorable context for bacterial growth [32]. In the clinical context, surface smoothing of dental restoration by using polishing tools is necessary before restoration is placed in the mouth [33]. The microscopic images of this study showed sharp grooves and defects on the surface of specimens in the noncoated group due to the grinding motion during the polishing process. The result agrees with previous studies where the polishing tools caused some microdefects to the restoration surface as they removed material by abrasion [34]. In specimens given the coating substance, scratches and defects were unobserved, and the roughness value was significantly lower than that of the noncoated specimens. The results suggest that the coating decreases surface irregularity by filling the scratches resulting from polishing and the inherent micropores created by the 3D-printing process. Previous studies that evaluated the effect of different surface treatments on dental ceramic restoration also indicated a significant decrease in roughness when a thin layer of glazing material was coated on the restoration surface [35]. In addition, various micrometer-sized particles were observed on the surface of the coated specimens. Considering that the mesoporous silica nanoparticles (MSNs) dispersed in the solvent were nanosized, this result implies that some of the silica particles on the coated layer became aggregated and clustered. A silica coating layer needs to contain particles of an optimum size to reduce microorganism adherence [19,36]. Thus, further technical optimization is needed to improve the homogeneous distribution of the silica nanoparticles and to minimize their uncontrolled aggregation in the coating layer.

Surface free energy greatly influences the initial step of biofilm formation [37]. Lower surface energy of the material has weakened bacterial adhesion; thus, the bacteria that adhere to a material with low surface energy are more easily removed by an external force [37,38]. The work of adhesion can be calculated by measuring the contact angle of the liquid to the solid surface [39]. For a low wetting surface, the surface energy of the solid is weaker than the surface tension of the liquid, allowing the liquid to easily retain its droplet shape. Therefore, a higher contact angle is related to lower surface energy and low interfacial tension of the solid surface. In this study, the coated group showed a significantly higher contact angle than the noncoated group, thereby indicating a lower surface energy of the coated specimens than that of the noncoated specimens. This is due to the hydrophobicity and low surface energy of PDMS, which is a beneficial feature of the material that contributes to the low bacterial adhesion [40].

S. mutans is an important etiologic agent for initiating dental caries [41]. The acid produced from this bacteria decays tooth structure and induces restorative treatment failure [41]. Accordingly, dental restorations require antibacterial qualities to ensure long-term success. For this purpose, the MSN was used to encapsulate CHX in this study, and the CHX@MSN was then combined with PDMS to unprecedentedly synthesize an antibacterial coating substance for polymeric restorative 3D-printing dental material. The bacterial growth inhibition assay results showed that the coated specimens had antimicrobial activity against *S. mutans.* The antimicrobial activity may be due to CHX release from the coating substance. Adding CHX to dental restorative material could increase its antibacterial activity [14,15]. Yan et al. [12] incorporated CHX@MSN into a glass ionomer cement powder and showed that CHX was continuously released, and the antibiofilm effect was maintained up to 30 days. Remarkably, dental resin composites with CHX@MSN showed controlled release of CHX over a prolonged time, providing strong inhibition against *S. mutans* adherence [11]. However, the antibacterial activity of previous coatings has been limited to the surface of the coated objects. In our study, an elastic porous material, PDMS, was used to store and release the CHX@MSN particles. The encapsulation and drug-loading efficiency of the CHX@MSN were recorded at rates of 25.22% and 63.04%, with a stable CHX releasing rate of approximately 1.56 μg/mL within the first 24 h in a pilot study [32]. Because the synthesized coating layer could release the loaded CHX over time [42], the coating exhibited antibacterial effects in the surrounding areas not directly in contact with the surface of the restoration. Moreover, the CHX@MSN and PDMS materials used in this study have been reported to be relatively noncytotoxic [32]. Therefore, this active antibacterial protective film formation is expected to be a novel method for actively inhibiting bacterial inhabitation around the coated surface of restorations.

Note that the human oral environment is more complicated than the in vitro experimental context because the oral cavity temperature, food intake, and the pH and composition of saliva vary between subjects and even within a subject [43]. In addition, the mechanical properties and material stability of the coating layer were not investigated in this study. However, the mechanical properties of PDMS strongly depend on the mixing ratio of a base to a curing agent in PDMS mixture [26,27,44]. In this study, the PDMS mixture comprises a base and a curing agent at a weight ratio of 5:1 with the relative elastic modulus number of 3.59 MPa [44]. However, the mechanical properties of the coating can be significantly improved by incorporating MSN into the PDMS as the elastic modulus of MSN is at least four orders of magnitude larger than that of PDMS [45]. Moreover, the hydrogen bonds between MSNs and covalent bonds between MSN and PDMS may induce a strong resistance to mechanical deformation [46]. The findings in those previous studies indicates that the mechanical strength of the PDMS coating layer can be tuned by controlling the input parameters such as the curing agent and MSN. In this study, the surface microstructure, surface wettability, and antibacterial activity were immediately evaluated after coating and within 24 h of drug releasing; thus, the material was expected to be stable during this stage. A further study that focuses on investigating the mechanical properties of the coating should be conducted in clinically relevant conditions to extend the understanding of this new material.

4. Materials and Methods

4.1. Synthesis of the Coating Material (CHX@MSN-Loaded PDMS)

Figure 4 illustrates the fabrication process for the coating substance. Fifty milligrams of CHX (Sigma-Aldrich Co., St. Louis, MO, USA) were dissolved in 5 mL of absolute ethanol, and then dried MCM-41 mesoporous silica nanoparticles (MSN) (Sigma-Aldrich Co., St. Louis, MO, USA) with a pore volume of 0.98 cm^3/g, pore size of approximately 2.5 nm, and Brunauer–Emmett–Teller (BET) surface area below 1000 m^2/g were dispersed into the CHX solution. The mixture was sonicated for 10 min and incubated for 24 h at room temperature using a magnetic stirrer (Corning PC-420D, Fisher Scientific, Lowell, MA, USA) at a speed of 300 rpm. Next, to collect the CHX@MSN particles, the mixture was filtered (Labogene Scan Speed Mini, Lynge, Denmark) and then vacuum-dried (OV-11, JEIO Tech, Seoul, Korea).

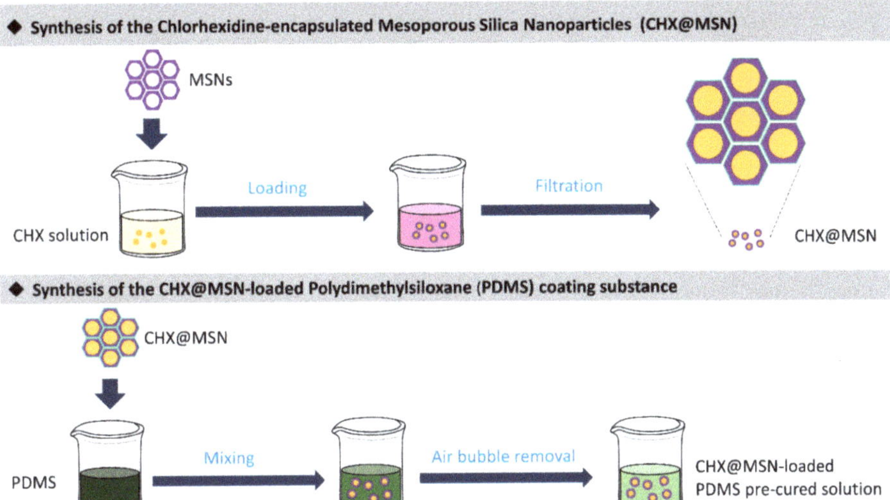

Figure 4. Synthesis of the CHX@MSN-loaded PDMS coating substance. CHX, Chlorhexidine; MSN, Mesoporous silica nanoparticles; CHX@MSN, Chlorhexidine encapsulated in mesoporous silica nanoparticles; PDMS, Polydimethylsiloxane.

The CHX@MSN particles were then mixed with PDMS (Sylgard 184, Dow Corning, Midland, MI, USA) solution at 0.4 wt% relative to the total PMDS mass. The PDMS mixture comprised a base and a curing agent at a weight ratio of 5:1. All components were blended in a SpeedMixerTM (FlackTek Inc., Landrum, SC, USA) for 3 × 30 s to form a precured coating paste, which was stored in the dark at room temperature and vacuumed for 20 min to remove bubbles before use.

4.2. Coating Procedure

Sixty disk-shape specimens ($N = 60$) with a thickness of 1.0 mm and diameter of 13.0 mm were designed using CAD software (Geomagic Design X, 3D Systems, Inc., Rock Hill, SC, USA) and printed with the photopolymer (RAYDent C&B, Ray Co., Hwaseong, Korea) using a digital light processing 3D printer (RAM500, Ray Co., Hwaseong, Korea). All specimens were postcured with a curing unit (RPC500, Ray Co., Hwaseong, Korea) for 20 min and polished with a 1000-grit silicon carbide abrasive paper (Buehler GmbH, Dusseldorf, Germany) for 60 s. Table 2 provides the composition of the photopolymer.

Table 2. Composition of the 3D-printing photopolymer.

Component *	Content (%)
α,α′-[(1-Methylethylidene)di-4,1-phenylene] bis [ω-[(2-methyl-1-oxo-2-propenyl) oxy] poly (oxy-1,2-ethanediyl)	20–35
7,7,9(or 7,9,9)-Trimethyl-4,13-dioxo-3,14-dioxa-5,12-diazahexadecane1,16-diyl 2-methyl-2-propenoate	20–28
2-Methyl-2-propenoic acid 1,2-ethanediylbis(oxy-2,1-ethanediyl) ester	20–25
Phenylbis (2,4,6-trimethylbenzoyl) phosphine oxide	1–10
Rutile (TiO2)	0.1–5

* As provided by the manufacturer.

Surface functionalization and coating process were performed in the coated group specimens (n = 30). The polished specimens were cleaned with isopropyl alcohol and treated with oxygen plasma (CUTE, Femto Science Co., Seoul, Korea) for 5 min (Figure 5a). The specimen surfaces were functionalized by immersion into 5% (v/v) 3-aminopropyltriethoxysilane (APTES) (Sigma-Aldrich Co., St. Louis, MO, USA) solution at 85 °C for 10 min (Figure 5b). The specimens were then coated by dipping in the precured coating solution using a dip coating equipment (KD Scientific, Holliston, MA, USA) at a lowering speed of 6000 μm/s and lifting speed of 1000 μm/s (Figure 5c). Subsequently, the specimens underwent thermal treatment in an oven (OV-11, JEIO Tech, Seoul, Korea) at 80 °C for 2 h (Figure 5d). The noncoated group specimens were defined as control (n = 30).

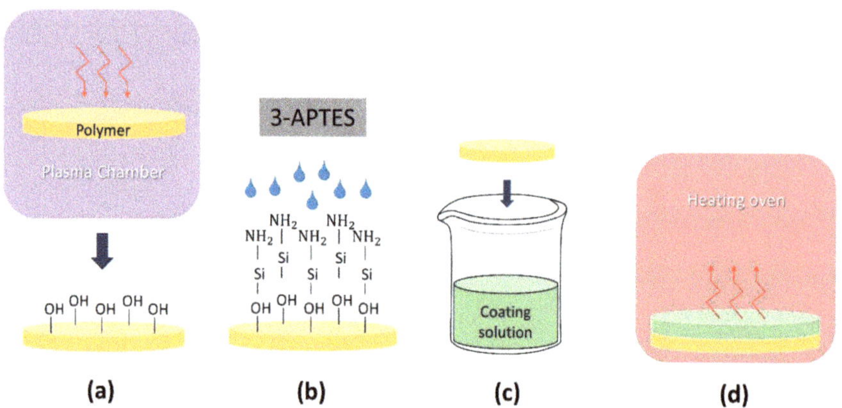

Figure 5. Surface functionalization and coating treatment: (**a**) oxygen plasma treatment; (**b**) APTES treatment; (**c**) dip coating; and (**d**) Thermal curing. APTES, 3-aminpropyltriethoxysilane; PMMA, polymethyl methacrylate; CHX@MSN, Chlorhexidine encapsulated in mesoporous silica nanoparticles; PDMS, Polydimethylsiloxane.

4.3. Evaluation of Surface Microstructure

Surface microstructure was evaluated using a scanning electron microscope (SEM) (S-4500, Hitachi Co., Tokyo, Japan). Specimens were coated with spotted platinum using a sputter coater (E-1030, Hitachi Co., Tokyo, Japan), and scanned at 5 kV at 500× and 2000× magnification. Ten locations in each specimen were randomly selected for image acquisition. The images were converted to 8-bit grayscale (black = 0, white = 225) using ImageJ software (version 1.52k, National Institute of Health, Bethesda, MD, USA). The pixel brightness data of each image were plotted as 256-level histograms to build a surface plot image, where the y-axis represents the 0–255 grayscale levels and the x-axis represents the pixel frequency. The histograms from all 10 images were combined, and the standard deviation of the pixel brightness was calculated as the SRI [47].

4.4. Measurement of Surface Wettability

Surface wettability was evaluated by measuring the contact angle (CA) in an air environment. A liquid droplet (5 μL) of distilled water was dispensed onto the specimen surface at a room temperature of 20 °C. The image of the droplet was immediately captured by a digital camera (Canon EOS 500D with Canon EF 100 mm f2.8 Macro USM Lens, Canon Inc., Tokyo, Japan) (Figure 6a), and the contact angles were determined from the corresponding pictures using the contact angle plugin of the ImageJ software. The CA degree was determined by measuring the tangent angle to the surface of the liquid droplet (Figure 6b). The mean value was calculated by averaging three individual measurements.

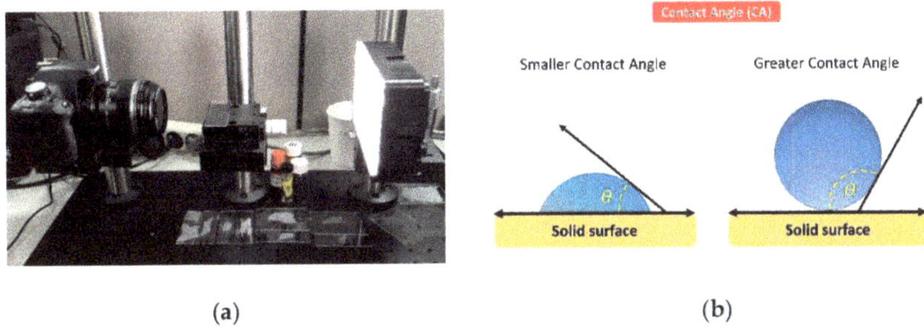

Figure 6. Measurement of surface wettability: (**a**) contact angle measurement system setting; and (**b**) contact angle (θ) measurement by the tangent angle of the liquid droplet to the surface.

4.5. Assessment of Antimicrobial Activities

To evaluate the antibacterial activity, a bacterial growth inhibition assay was performed. *Streptococcus mutans* ATCC 25175 (*S. mutans*) was inoculated into a 15-mL tube containing 10 mL of brain heart infusion (BHI) media (BHI, Mast Group, Bootle, UK) and incubated on a rotary shaker (150 rpm) at 37 °C. To determine the concentration of bacteria in culture media, the optical density at 600 nm (OD_{600}) of the cell suspensions was recorded using a DS-11+ apparatus (DeNovix, Wilmington, DE, USA). After 12 h, OD_{600} was checked to measure bacterial density in liquid media.

When the OD_{600} of the *S. mutans* culture reached approximately 0.4, aliquots of 500 μL were added to each well in a 24-well plate and incubated with noncoated or coated specimens at 37 °C on a shaker at 15 rpm for 24 h. Cultures without any test specimens were used as the control condition. Following incubation, appropriate dilutions of the cultures were plated on BHI agar plates using an ethanol-flamed bacterial spreader, and the plate was incubated for 24 h at 37 °C. Then, the colony-forming units per milliliter (CFU/mL) were calculated using the number of colonies observed and the dilution factor ($1{:}10^4$) for each well.

4.6. Statistical Analysis

Statistical calculations were performed using SPSS software (SPSS version 25.0, IBM Inc., Chicago, IL, USA). The measured values were expressed as the mean ± standard deviation. The independent *t*-test was conducted to compare differences between the groups. The statistical significance level was set at 0.05.

5. Conclusions

This study evaluated the effect of a newly developed CHX@MSN-loaded PDMS-based coating substance on the surface properties and antibacterial ability of coated 3D-printing dental material. From our results, it can be concluded that applying the coating layer on the polymer specimens significantly reduced their surface irregularity, while increasing the hydrophobicity and antibacterial activity.

Author Contributions: Conceptualization, H.-N.M. and D.-H.L.; methodology, D.C.H., J.H.P., and D.-Y.K.; formal analysis, D.C.H. and S.M.L.; writing—original draft preparation, H.-N.M.; writing—review and editing, D.-H.L.; and supervision, D.-H.L. All authors have read and agreed to the published version of the manuscript.

Funding: This work was supported by the National Research Foundation of Korea (NRF) grant funded by the Korea government (MSIT) (2020R1A2C4002518). The funders had no role in the study design, data collection and interpretation, or the decision to submit the work for publication.

Conflicts of Interest: The authors declare no conflict of interest.

References

1. Stansbury, J.W.; Idacavage, M.J. 3D printing with polymers: Challenges among expanding options and opportunities. *Dent. Mater.* **2016**, *32*, 54–64. [CrossRef] [PubMed]
2. Eleftheriadis, G.; Monou, P.K.; Andriotis, E.; Mitsouli, E.; Moutafidou, N.; Markopoulou, C.; Bouropoulos, N.; Fatouros, D. Development and Characterization of Inkjet Printed Edible Films for Buccal Delivery of B-Complex Vitamins. *Pharmaceuticals* **2020**, *13*, 203. [CrossRef] [PubMed]
3. Tahayeri, A.; Morgan, M.; Fugolin, A.P.; Bompolaki, D.; Athirasala, A.; Pfeifer, C.S.; Ferracane, J.L.; Bertassoni, L.E. 3D printed versus conventionally cured provisional crown and bridge dental materials. *Dent. Mater.* **2018**, *34*, 192–200. [CrossRef] [PubMed]
4. Tappa, K.; Jammalamadaka, U. Novel Biomaterials Used in Medical 3D Printing Techniques. *J. Funct. Biomater.* **2018**, *9*, 17. [CrossRef] [PubMed]
5. Saratti, C.M.; Rocca, G.T.; Krejci, I. The potential of three-dimensional printing technologies to unlock the development of new 'bio-inspired' dental materials: An overview and research roadmap. *J. Prosthodont. Res.* **2019**, *63*, 131–139. [CrossRef]
6. Park, J.-M.; Ahn, J.-S.; Cha, H.-S.; Lee, J.-H. Wear Resistance of 3D Printing Resin Material Opposing Zirconia and Metal Antagonists. *Materials* **2018**, *11*, 1043. [CrossRef]
7. Mai, H.N.; Lee, K.B.; Lee, D.H. Fit of interim crowns fabricated using photopolymer-jetting 3D printing. *J. Prosthet. Dent.* **2017**, *118*, 208–215. [CrossRef]
8. Boaro, L.C.C.; Campos, L.M.; Varca, G.H.C.; dos Santos, T.M.R.; Marques, P.A.; Sugii, M.M.; Saldanha, N.R.; Cogo-Müller, K.; Brandt, W.C.; Braga, R.R. Antibacterial resin-based composite containing chlorhexidine for dental applications. *Dent. Mater.* **2019**, *35*, 909–918. [CrossRef]
9. Belter, J.T.; Dollar, A.M. Strengthening of 3D printed fused deposition manufactured parts using the fill compositing technique. *PLoS ONE* **2015**, *10*, e0122915. [CrossRef]
10. Gonzalez-Henriquez, C.M.; Sarabia-Vallejos, M.A.; Hernandez, J.R. Antimicrobial Polymers for Additive Manufacturing. *Int. J. Mol. Sci.* **2019**, *20*, 1210. [CrossRef]
11. Zhang, J.; Wu, R.; Fan, Y.; Liao, S.; Wang, Y.; Wen, Z.; Xu, X. Antibacterial dental composites with chlorhexidine and mesoporous silica. *J. Dent. Res.* **2014**, *93*, 1283–1289. [CrossRef]
12. Yan, H.; Yang, H.; Li, K.; Yu, J.; Huang, C. Effects of Chlorhexidine-Encapsulated Mesoporous Silica Nanoparticles on the Anti-Biofilm and Mechanical Properties of Glass Ionomer Cement. *Molecules* **2017**, *22*, 1225. [CrossRef] [PubMed]
13. Van Strydonck, D.A.; Slot, D.E.; Van der Velden, U.; Van der Weijden, F. Effect of a chlorhexidine mouthrinse on plaque, gingival inflammation and staining in gingivitis patients: A systematic review. *J. Clin. Periodontol.* **2012**, *39*, 1042–1055. [CrossRef] [PubMed]
14. Anusavice, K.J.; Zhang, N.Z.; Shen, C. Controlled release of chlorhexidine from UDMA-TEGDMA resin. *J. Dent. Res.* **2006**, *85*, 950–954. [CrossRef] [PubMed]
15. Priyadarshini, B.M.; Selvan, S.T.; Lu, T.B.; Xie, H.; Neo, J.; Fawzy, A.S. Chlorhexidine Nanocapsule Drug Delivery Approach to the Resin-Dentin Interface. *J. Dent. Res.* **2016**, *95*, 1065–1072. [CrossRef] [PubMed]
16. Salim, N.; Moore, C.; Silikas, N.; Satterthwaite, J.D.; Rautemaa, R. Fungicidal amounts of antifungals are released from impregnated denture lining material for up to 28 days. *J. Dent.* **2012**, *40*, 506–512. [CrossRef] [PubMed]

17. Al-Haddad, A.; Roudsari, R.V.; Satterthwaite, J.D. Fracture toughness of heat cured denture base acrylic resin modified with Chlorhexidine and Fluconazole as bioactive compounds. *J. Dent.* **2014**, *42*, 180–184. [CrossRef] [PubMed]
18. Alghazzawi, T.F. Advancements in CAD/CAM technology: Options for practical implementation. *J. Prosthodont. Res.* **2016**, *60*, 72–84. [CrossRef]
19. Azuma, A.; Akiba, N.; Minakuchi, S. Hydrophilic surface modification of acrylic denture base material by silica coating and its influence on Candida albicans adherence. *J. Med. Dent. Sci.* **2012**, *59*, 1–7.
20. Fukunishi, M.; Inoue, Y.; Morisaki, H.; Kuwata, H.; Ishihara, K.; Baba, K. A Polymethyl Methacrylate-Based Acrylic Dental Resin Surface Bound with a Photoreactive Polymer Inhibits Accumulation of Bacterial Plaque. *Int. J. Prosthodont.* **2017**, *30*, 533–540. [CrossRef]
21. Rokaya, D.; Srimaneepong, V.; Sapkota, J.; Qin, J.; Siraleartmukul, K.; Siriwongrungson, V. Polymeric materials and films in dentistry: An overview. *J. Adv. Res.* **2018**, *14*, 25–34. [CrossRef] [PubMed]
22. Cloutier, M.; Mantovani, D.; Rosei, F. Antibacterial Coatings: Challenges, Perspectives, and Opportunities. *Trends Biotechnol.* **2015**, *33*, 637–652. [CrossRef]
23. Zhu, D.; Handschuh-Wang, S.; Zhou, X. Recent progress in fabrication and application of polydimethylsiloxane sponges. *J. Mater. Chem. A* **2017**, *5*, 16467–16497. [CrossRef]
24. Raghu, P.K.; Bansal, K.K.; Thakor, P.; Bhavana, V.; Madan, J.; Rosenholm, J.M.; Mehra, N.K. Evolution of Nanotechnology in Delivering Drugs to Eyes, Skin and Wounds via Topical Route. *Pharmaceuticals* **2020**, *13*, 167. [CrossRef]
25. Si, J.; Cui, Z.; Xie, P.; Song, L.; Wang, Q.; Liu, Q.; Liu, C. Characterization of 3D elastic porous polydimethylsiloxane (PDMS) cell scaffolds fabricated by VARTM and particle leaching. *J. Appl. Polym.* **2016**, *133*. [CrossRef]
26. Palchesko, R.N.; Zhang, L.; Sun, Y.; Feinberg, A.W. Development of polydimethylsiloxane substrates with tunable elastic modulus to study cell mechanobiology in muscle and nerve. *PLoS ONE* **2012**, *7*, e51499. [CrossRef] [PubMed]
27. Hyun, D.C.; Jeong, U. Substrate thickness: An effective control parameter for polymer thin film buckling on PDMS substrates. *J. Appl. Polym. Sci.* **2009**, *112*, 2683–2690. [CrossRef]
28. Yoshida, S.; Hagiwara, K.; Hasebe, T.; Hotta, A. Surface modification of polymers by plasma treatments for the enhancement of biocompatibility and controlled drug release. *Surf. Coat. Technol.* **2013**, *233*, 99–107. [CrossRef]
29. Zhang, H.; Chiao, M. Anti-fouling coatings of poly (dimethylsiloxane) devices for biological and biomedical applications. *J. Med. Biol. Eng.* **2015**, *35*, 143–155. [CrossRef]
30. Fornell, A.-C.; Sköld-Larsson, K.; Hallgren, A.; Bergstrand, F.; Twetman, S. Effect of a hydrophobic tooth coating on gingival health, mutans streptococci, and enamel demineralization in adolescents with fixed orthodontic appliances. *Acta Odontol. Scand.* **2002**, *60*, 37–41. [CrossRef]
31. Aguayo, S.; Marshall, H.; Pratten, J.; Bradshaw, D.; Brown, J.S.; Porter, S.R.; Spratt, D.; Bozec, L. Early Adhesion of Candida albicans onto Dental Acrylic Surfaces. *J. Dent. Res.* **2017**, *96*, 917–923. [CrossRef] [PubMed]
32. Mai, H.-N.; Hong, S.-H.; Kim, S.-H.; Lee, D.-H. Effects of different finishing/polishing protocols and systems for monolithic zirconia on surface topography, phase transformation, and biofilm formation. *J. Adv. Prosthodont.* **2019**, *11*, 81–87. [CrossRef] [PubMed]
33. Erdemir, U.; Sancakli, H.S.; Yildiz, E. The effect of one-step and multi-step polishing systems on the surface roughness and microhardness of novel resin composites. *Eur. J. Dent.* **2012**, *6*, 198. [CrossRef] [PubMed]
34. Jefferies, S.R. Abrasive finishing and polishing in restorative dentistry: A state-of-the-art review. *Dent. Clin. N. Am.* **2007**, *51*, 379–397. [CrossRef]
35. Rashid, H. The effect of surface roughness on ceramics used in dentistry: A review of literature. *Eur. J. Dent.* **2014**, *8*, 571. [CrossRef] [PubMed]
36. Salamanca, C.H.; Yarce, C.J.; Roman, Y.; Davalos, A.F.; Rivera, G.R. Application of nanoparticle technology to reduce the anti-microbial resistance through β-lactam antibiotic-polymer inclusion nano-complex. *Pharmaceuticals* **2018**, *11*, 19. [CrossRef]
37. Pereni, C.I.; Zhao, Q.; Liu, Y.; Abel, E. Surface free energy effect on bacterial retention. *Colloid Surf. B Biointerfaces* **2006**, *48*, 143–147. [CrossRef]

38. Busscher, H.J.; Weerkamp, A.H.; van der Mei, H.C.; Van Pelt, A.; de Jong, H.P.; Arends, J. Measurement of the surface free energy of bacterial cell surfaces and its relevance for adhesion. *Appl. Environ. Microbiol.* **1984**, *48*, 980–983. [CrossRef]
39. Schrader, M.E. Young-dupre revisited. *Langmuir* **1995**, *11*, 3585–3589. [CrossRef]
40. Ye, X.; Cai, D.; Ruan, X.; Cai, A. Research on the selective adhesion characteristics of polydimethylsiloxane layer. *AIP Adv.* **2018**, *8*, 095004. [CrossRef]
41. Colombo, A.P.V.; Tanner, A.C.R. The Role of Bacterial Biofilms in Dental Caries and Periodontal and Peri-implant Diseases: A Historical Perspective. *J. Dent. Res.* **2019**, *98*, 373–385. [CrossRef]
42. Mai, H.N.; Kim, D.Y.; Hyun, D.C.; Park, J.H.; Lee, S.M.; Lee, D.H. A New Antibacterial Agent-Releasing Polydimethylsiloxane Coating for Polymethyl Methacrylate Dental Restorations. *J. Clin. Med.* **2019**, *8*, 1831. [CrossRef] [PubMed]
43. Laske, M.; Opdam, N.J.M.; Bronkhorst, E.M.; Braspenning, J.C.C.; Huysmans, M. Risk Factors for Dental Restoration Survival: A Practice-Based Study. *J. Dent. Res.* **2019**, *98*, 414–422. [CrossRef] [PubMed]
44. Wang, Z.; Volinsky, A.A.; Gallant, N.D. Crosslinking effect on polydimethylsiloxane elastic modulus measured by custom-built compression instrument. *J. Appl. Polym. Sci.* **2014**, *131*. [CrossRef]
45. Liu, J.; Zong, G.; He, L.; Zhang, Y.; Liu, C.; Wang, L. Effects of fumed and mesoporous silica nanoparticles on the properties of sylgard 184 polydimethylsiloxane. *Micromachines* **2015**, *6*, 855–864. [CrossRef]
46. Shim, S.E.; Isayev, A.I. Rheology and structure of precipitated silica and poly (dimethyl siloxane) system. *Rheol. Acta* **2004**, *43*, 127–136. [CrossRef]
47. Banerjee, S.; Yang, R.; Courchene, C.E.; Conners, T.E. Scanning electron microscopy measurements of the surface roughness of paper. *Ind. Eng. Chem. Res.* **2009**, *48*, 4322–4325. [CrossRef]

Article

Cytotoxicity Effects of Water-Soluble Multi-Walled Carbon Nanotubes Decorated with Quaternized Hyperbranched Poly(ethyleneimine) Derivatives on Autotrophic and Heterotrophic Gram-Negative Bacteria

Nikolaos S. Heliopoulos [1,2], Georgia Kythreoti [1,3], Kyriaki Marina Lyra [1], Katerina N. Panagiotaki [1], Aggeliki Papavasiliou [1], Elias Sakellis [1], Sergios Papageorgiou [1], Antonios Kouloumpis [4], Dimitrios Gournis [4], Fotios K. Katsaros [1], Kostas Stamatakis [3] and Zili Sideratou [1,*]

1 Institute of Nanoscience and Nanotechnology, National Centre of Scientific Research "Demokritos", 15310 Aghia Paraskevi, Greece; nikosheliopoulos@gmail.com (N.S.H.); geokyth@bio.demokritos.gr (G.K.); kymarin@gmail.com (K.M.L.); knpanagiotaki@gmail.com (K.N.P.); a.papavasiliou@inn.demokritos.gr (A.P.); e.sakellis@inn.demokritos.gr (E.S.); s.papageorgiou@inn.demokritos.gr (S.P.); f.katsaros@inn.demokritos.gr (F.K.K.)

2 Department of Industrial Design & Production Engineering, University of West Attica, 12241 Egaleo, Attiki, Greece

3 Institute of Biosciences and Applications, National Centre of Scientific Research "Demokritos", 15310 Aghia Paraskevi, Greece; kstam@bio.demokritos.gr

4 Department of Material Science & Engineering, University of Ioannina, 45110 Ioannina, Greece; antoniokoul@gmail.com (A.K.); dgourni@uoi.gr (D.G.)

* Correspondence: z.sideratou@inn.demokritos.gr; Tel.: +30-210-6503616

Received: 14 July 2020; Accepted: 1 October 2020; Published: 6 October 2020

Abstract: Oxidized multi-walled carbon nanotubes (oxCNTs) were functionalized by a simple non-covalent modification procedure using quaternized hyperbranched poly(ethyleneimine) derivatives (QPEIs), with various quaternization degrees. Structural characterization of these hybrids using a variety of techniques, revealed the successful and homogenous anchoring of QPEIs on the oxCNTs' surface. Moreover, these hybrids efficiently dispersed in aqueous media, forming dispersions with excellent aqueous stability for over 12 months. Their cytotoxicity effect was investigated on two types of gram(−) bacteria, an autotrophic (cyanobacterium *Synechococcus* sp. PCC 7942) and a heterotrophic (bacterium *Escherichia coli*). An enhanced, dose-dependent antibacterial and anti-cyanobacterial activity against both tested organisms was observed, increasing with the quaternization degree. Remarkably, in the photosynthetic bacteria it was shown that the hybrid materials affect their photosynthetic apparatus by selective inhibition of the Photosystem-I electron transport activity. Cytotoxicity studies on a human prostate carcinoma DU145 cell line and 3T3 mouse fibroblasts revealed that all hybrids exhibit high cytocompatibility in the concentration range, in which they also exhibit both high antibacterial and anti-cyanobacterial activity. Thus, QPEI-functionalized oxCNTs can be very attractive candidates as antibacterial and anti-cyanobacterial agents that can be used for potential applications in the disinfection industry, as well as for the control of harmful cyanobacterial blooms.

Keywords: carbon nanotubes; quaternary ammonium groups; hyperbranched dendritic polymers; antibacterial properties; anti-cyanobacterial properties

1. Introduction

Carbon nanotubes (CNTs) have attracted significant scientific and technological interest due to their unique structural characteristics and their excellent electronic, mechanical, and thermal properties [1,2]. Based on these properties, they have been used in a wide range of applications, including fillers in composite materials, sensors, drug delivery systems, antibacterial agents, and others [3,4]. However, their poor dispersibility in solvents, especially in water, has prevented their widespread industrial use, and reduced their great potential. Attempts to overcome this problem have focused on the functionalization of their surface, using a variety of covalent and non-covalent modification strategies [5]. On one hand, various organic molecules, such as dendritic and linear polymers, have been covalently conjugated onto the CNT's convex surfaces and tips by chemical reactions [6–8] in order to reduce aggregates and size polydispersity. However, covalent functionalization causes damage to the conjugated π-electrons, leading to degradation of their properties. On the other hand, non-covalent functionalization, based on π–π stacking and ionic interactions between various molecules and the CNTs graphitic surface [9–11] does not affect their electronic structure, and has been achieved using a multitude of surfactants [6,7] and polymers [8,10], resulting in modified CNTs, compatible with specific solvents or targeted applications. In this context, their functionalization with dendritic polymers such as dendrons, dendrimers, and hyperbranched polymers, is expected to be a very promising strategy, when aiming to achieve increased water solubility. This strategy has already been applied to single-walled carbon nanotubes (SWCNTs) that were functionalized using dendritic polymers through non-covalent interactions, achieving enhanced water solubility [8,12,13]. However, only a few studies have addressed the non-covalent functionalization of multi-walled carbon nanotubes (MWCNTs) with dendritic polymers for increased water solubility [14].

Dendritic polymers are highly branched macromolecules of nanosized dimensions, consisting of repeating units and surface end groups [15]. Their properties depend on both the structural characteristics of the branches in their interior, and the large number of surface end groups. These polymers can incorporate a variety of organic compounds as well as inorganic ions in their interior, while the surface end groups are primarily susceptible to functionalization or even multi-functionalization, to yield a variety of novel materials with diversified, tailor-made properties such as drug delivery systems, antibacterial agents, etc. [16–18]. Additionally, these terminal groups exhibit the so-called polyvalency effect [19], which enhances their binding with various substrates, due to their close proximity to the dendritic polymers' scaffold.

Recently, several studies have focused on the potential applications of carbon nanomaterials (CNMs), taking advantage of their antibacterial properties [20–22]. Specifically, functionalized single-walled carbon nanotubes (SWCNTs) and multi-walled carbon nanotubes (MWCNTs) were found to exhibit significant antibacterial activity towards both gram-positive and gram-negative bacteria [23,24]. CNTs' properties, such as carbon nanotube diameter [25], length [26], aggregation [27], concentration [28], surface functionalization [29–31], etc. have been reported to influence their antibacterial activity. To that respect, aqueous dispersibility can critically influence antibacterial efficiency, as highly dispersed CNTs enhance their interaction with cells, leading to increased antibacterial properties. Indeed, it was found [32] that individually dispersed CNTs were more toxic to bacteria than CNTs aggregates, due to increased contact with bacterial cells.

Nowadays, there is a growing concern about the possible effects of nanomaterials, as end of life products, on organisms and ecosystems [33,34]. Apart from studies investigating the mechanisms of interaction between CNTs and various biomacromolecules (DNA, RNA, etc.), to identify possible causes of undesired effects, [35] research efforts have also focused on the possible impact of MWCNTs on photosynthetic pathways. Remarkably, in photosynthetic organisms, unlike bacteria, a favorable effect of MWCNTs was observed, e.g., in the development of cereals, and the production of vegetative biomass [36]. Studies on algae revealed that MWCNTs did not influence photosynthesis, and any negative effects were due to turbidity and the resulting reduction of the available light [37]. However, oxidized MWCNTs were found to be toxic to the marine green alga *Dunaliella tertiolecta*, as at concentrations between 1–10 mg/L they reduced algal growth, and affected the Photosystem (PS) II photochemical process and the cellular glutathione redox status [38]. Moreover, although cyanobacterial blooms have become a serious environmental problem, only recently, a novel nanomaterial based on MWCNTs, called Taunit, loaded with antibiotic chloramphenicol or herbicide diuron, was investigated as an effective anti-cyanobacterial agent [39]. It was found that the Taunit-diuron complex exhibited high biocide action against cyanobacterium *Synechocystis* sp. PCC 6803, higher than that of a Taunit-chloramphenicol complex.

In this study, aiming to develop water soluble MWCNTs with enhanced antibacterial/anti-cyanobacteria properties, oxidized multi-walled carbon nanotubes (oxCNTs) were non-covalently functionalized using a series of partially quaternized hyperbranched poly(ethyleneimine) derivatives, yielding novel water-soluble hybrid materials. Specifically, three positively charged derivatives of hyperbranched poly(ethyleneimine) (PEI) with a different degree of quaternization at the primary amino groups of PEI were prepared and interacted with the negatively charged oxidized CNTs through electrostatic interactions and van der Waals attraction forces. The obtained hybrid materials were physicochemically characterized by various techniques (FTIR, Raman, SEM, TEM, AFM, etc.). Their excellent aqueous stability was demonstrated using ζ-potential measurements, dynamic light scattering, and UV-vis spectroscopy. Additionally, their antibacterial and anti-cyanobacterial activity was investigated against two types of gram negative bacteria, an autotrophic (cyanobacterium *Synechococcus* sp. PCC 7942) and a heterotrophic (bacterium *Escherichia coli*), while their cytocompatibility was investigated on eukaryotic cell lines.

2. Results and Discussion

2.1. Synthesis and Characterization of QPEI-Functionalized oxCNTs

Positively charged stable aqueous suspensions of carbon nanotubes were prepared, applying quaternized hyperbranched poly(ethyleneimine) derivatives (QPEIs). A series of partially quaternized hyperbranched poly(ethyleneimine) derivatives with 30%, 50%, and 80% substitution degree of primary amino groups was prepared, following a method analogous to one previously described [40,41]. Initially, PEI was characterized by inverse-gate decoupling ^{13}C NMR. According to the literature [42] and comparing the integration of carbons of α-CH_2 relative to primary, secondary, and tertiary amine groups, the ratio of primary to secondary and to tertiary amines of PEI was found to be 1.00:1.18:1.01. The branching degree was found to be 0.68, and the average number of primary amine groups was determined to be 183. Based on the above, the introduction of α-hydroxyamine moieties, together with the trimethylammonium groups, to PEI was achieved by the reaction of the PEI primary amines with appropriate amounts of glycidyltrimethylammonium chloride, yielding three PEI derivatives, i.e., 30-QPEI, 50-QPEI, and 80-QPEI (Scheme 1). Their structures were confirmed by ^{1}H and ^{13}C NMR spectroscopy. Specifically, the new quintet appearing at 4.25 ppm was attributed to the proton of α-CH group relative to hydroxyl group, which was formed after an oxiran ring opening. Additionally, the α-CH_2 protons relative to the quaternary group appeared as a triplet at 3.45 ppm, and the protons of the quaternary methyl groups at 3.25 ppm, while a multiplet in the region 2.50–2.70 ppm was attributed to the PEI scaffold protons. Comparing the integration of peaks at 3.45 ppm and 2.70–2.50 ppm, the degree of quaternization at the primary amino group of PEI was calculated, and was found to be 30%, 50%, and 80% for 30-QPEI, 50-QPEI, and 80-QPEI, respectively.

Furthermore, ^{13}C NMR spectroscopy provided insights into the structure of QPEIs. The attachment of the α-hydroxyamine moiety at the PEI scaffold was confirmed by the peaks at 51.0 and 48.0 ppm, attributed to the α carbon of PEI, and relative to the newly formed amino group (α-CH_2NH-Q), close to secondary (known as $C_{1,2}$) and tertiary (known as $C_{1,3}$) amines, respectively. Additionally, the α-methylene and methyl groups, attached at the quaternary center, were observed at 71.5 and 57.0 ppm, respectively, while a peak at 67.5 ppm was attributed to the α carbon relative to the newly formed hydroxyl group (CH–OH).

PEI

QPEI

Scheme 1. Schematic representation of poly(ethyleneimine) (PEI) and the reaction scheme of quaternization.

Subsequently, these dendritic derivatives were interacted with oxidized CNTs in aqueous media (Scheme 2). The final hybrid nanomaterials, oxCNTs@30-QPEI, oxCNTs@50-QPEI, and oxCNTs@80-QPEI, were obtained after ultracentrifugation to remove the excess QPEIs, and physicochemically characterized using a variety of techniques, such as FTIR, RAMAN, TGA, SEM, TEM, AFM, etc.

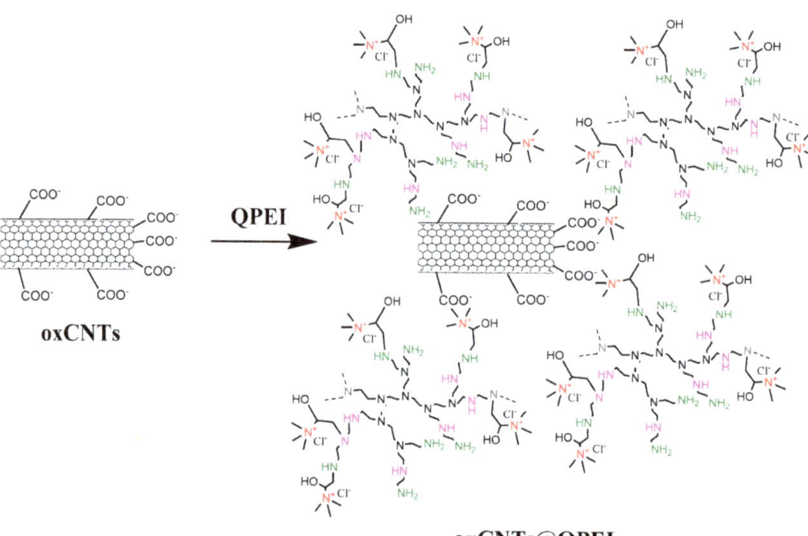

Scheme 2. Schematic representation of oxidized multi-walled carbon nanotubes (oxCNTs) decorated with quaternized hyperbranched poly(ethyleneimine) derivative (QPEI).

To investigate the successful attachment of QPEIs on the oxCNTs, initially, FTIR spectroscopy was employed. In Figures 1A and S1, the FTIR spectra of QPEIs, oxCNTs, and QPEI-functionalized oxCNTs are shown. The spectrum of oxCNTs shows a C=C stretching band at 1650 cm^{-1}, attributed to the CNTs graphite structure. Additionally, the presence of oxygen containing groups (carboxylates, carbonyl, hydroxyl, and epoxy groups) on the oxCNTs was confirmed by the appearance of a C=O stretching band at 1740 cm^{-1}, a broad OH stretching band centered at 3370 cm^{-1}, and a strong C–OH stretching band at 1100 cm^{-1}, as well as two peaks at 1565 and 1380 cm^{-1}, which are associated with the carboxylate anion stretch mode (asymmetrical and symmetrical vibrations of COO^-, respectively). Furthermore, the peaks at 2750 and 1255 cm^{-1}, attributed to the stretching vibrations of OC–H and C–O–C, respectively, revealed the presence of aldehydes and epoxy groups on oxCNTs [43]. On the other hand, as expected, the FTIR spectra of all QPEIs exhibited the sample characteristic bands. Those included the characteristic vibrations of PEI, i.e., at 3360 and 3270 cm^{-1}, attributed to the stretching vibration of primary and secondary amino groups, at 2940 and 2835 cm^{-1}, assigned to the asymmetrical and symmetrical vibrations of CH_2, at 1600 and 1560 cm^{-1}, attributed to the NH deformation mode of the primary and secondary amino groups, respectively, at 1455 and 760 cm^{-1}, corresponding to the bending and rocking mode of CH_2, respectively, and at 1115 and 1050 cm^{-1}, assigned to the asymmetrical and symmetrical vibrations of C–N [44]. Additionally, the most important stretching and deformation vibrations of CH_3 in the quaternary ammonium group $(CH_3)_3N^+$ at 3030 cm^{-1} and 1485 cm^{-1}, the asymmetrical stretching vibration of the whole $(CH_3)_3N^+$ group at 970 cm^{-1}, and the stretching vibrations of C–OH groups at 1100 cm^{-1} were observed [45]. The FTIR spectra of the QPEI-functionalized oxCNTs (Figures 1A and S1) revealed the existence of both oxCNTs and QPEIs, confirming their successful interaction.

The successful functionalization of oxCNTs was also confirmed by Raman measurements. The spectra of oxCNTs and QPEI-functionalized oxCNTs presented in Figure 1B, display the two main typical graphite bands at 1585 cm^{-1} (G band) and 1345 cm^{-1} (D band), attributed to the in-plane vibration of the sp^2-bonded carbon atoms in graphite layers, and to the defects presented in carbon nanotubes due to the conversion of carbon atoms from an sp^2 to an sp^3 hybridization state, respectively. Additionally, a band at ~2700 cm^{-1} (G′ band) is shown in Figure 1B, attributed to the D band overtone. As observed in Figure 1B, the Raman shifts of QPEI-functionalized oxCNTs compared to those of oxCNTs did not change, revealing that the graphitic structure of oxCNTs does not significantly alter after functionalization. Only the value of the intensity ratio of D- to G- bands (I_D/I_G), a measure of the defects present in a graphene structure during functionalization, slightly increased from 1.03 to 1.16, revealing successful polymer wrapping all over the graphite layer of CNTs [46]. In analogous studies involving covalent functionalization of multi-walled carbon nanotubes via third-generation dendritic poly(amidoamine) or amphiphilic poly(propyleneimine) dendrimers, a larger increase of the intensity ratio (I_D/I_G), was reported indicating that functionalization caused a larger defect of the graphitic network [47,48]. In contrast, in the present study oxCNTs were non-covalently functionalized with QPEI derivatives, retaining their surface almost intact.

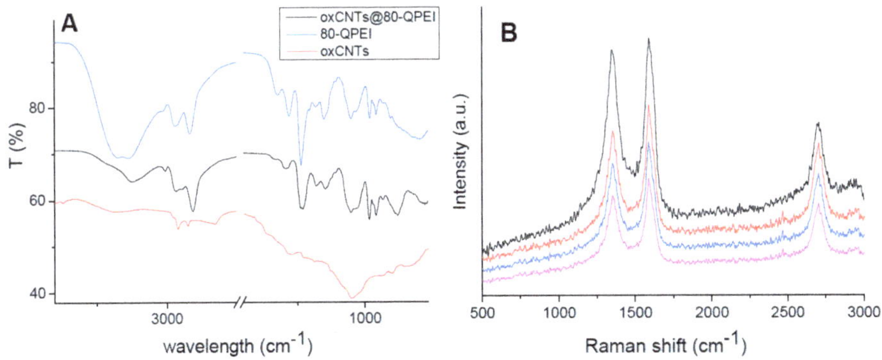

Figure 1. (**A**) FTIR spectra of oxCNTs, oxCNTs@80-QPEI, and 80-QPEI. (**B**) Raman spectra of oxCNTs (black), oxCNTs@30-QPEI (red), oxCNTs@50-QPEI (blue), and oxCNTs@80-QPEI (magenta).

The results of thermogravimetric analysis (TGA) provided further information on the QPEI content on the surfaces of oxCNTs (Figure 2). In the TGA curve of oxCNTs, two distinct decomposition regions were observed. Specifically, a weight loss corresponding to ~4% of the initial weight was recorded in the temperature range of 180–400 °C, due to the removal of oxygen-containing functional groups present on the graphitic framework. A second significant loss was observed at higher temperatures (>500 °C), and was attributed to the thermal degradation of the graphitic framework. In contrast, the QPEI-functionalized oxCNTs exhibited a significant weight loss (10–20%) up to 250 °C, due to both the removal of oxCNTs' oxygen-containing groups, and the partial PEI degradation. The weight loss for QPEI-functionalized oxCNTs in the temperature range 250–400 °C was significantly higher in comparison to oxCNTs (up to 60%), since together with the decomposition of graphitic lattice, QPEI molecules were removed from the graphitic framework. Therefore, TGA measurements provided further qualitative experimental evidence that functionalization had occurred, however, due to the oxCNT contribution to the thermal phenomena, the polymer content could not be quantified. Again, above ~500 °C, the sharp weight loss indicated the total thermal destruction of the graphitic network.

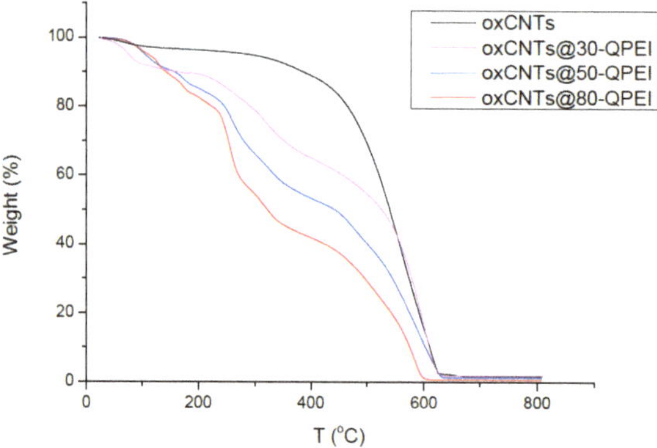

Figure 2. Thermogravimetric curves of oxCNTs and QPEI-functionalized oxCNTs.

Elemental analysis measurements were additionally performed to confirm and quantify the polymer content in the QPEI-functionalized oxCNTs. Given that the nitrogen signal in the final hybrid originated mainly from QPEI, the difference compared to the starting oxCNTs represented the amount of polymer attached to the CNTs. Therefore, the QPEI content in hybrids was calculated from the following formula:

$$\text{QPEI } (\% \, w/w) = (N_s - N_{CNTs})/(N_{QPEI} - N_{CNTs}) \times 100$$

where N_s, N_{QPEI}, and N_{CNTs}, were the nitrogen elemental mass fraction in QPEI-functionalized oxCNTs, QPEI, and oxCNTs, respectively [49]. The results are summarized in Table S1. According to the elemental analysis results, the actual value of QPEI weight fraction in oxCNTs@30-QPEI, oxCNTs@50-QPEI, and oxCNTs@80-QPEI was found to be 16.05%, 19.92%, and 23.23%, respectively.

The morphology of the QPEI-functionalized oxCNTs was studied by scanning electron (SEM), transmission electron (TEM), and atomic force (AFM) microscopies. Representative SEM micrographs are shown in Figure 3. It is clear that after functionalization with QPEIs, the morphology of the oxCNTs did not change significantly. Additionally, oxCNTs@QPEIs are shown to be well-dispersed and no aggregation of nanotubes was observed, as in the case of oxCNTs (Figure S2). Especially, functionalization of oxCNTs with 80-QPEI rendered them fully isolated, as shown in Figure 3 (images C and D).

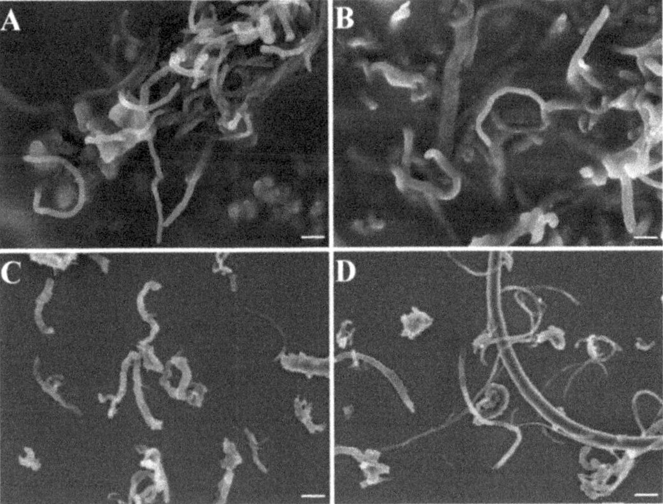

Figure 3. SEM images of oxCNTs@30-QPEI (**A**), oxCNTs@50-QPEI (**B**), and oxCNTs@80-QPEI (part **C**,**D**). The scale bar is 100 nm.

The morphology of QPEI-functionalized oxCNTs, as well as the presence of QPEIs on their surface, was studied by combining TEM bright-field imaging, EFTEM elemental mapping, and EELS spectroscopy. In Figure 4A functionalized carbon nanotubes are observed isolated, without any aggregation, as in the SEM images. These observations suggest that the QPEIs covered the surface of the nanotubes, improving their aqueous dispersibility and debundling. In the HRTEM images the structured graphite walls of oxCNTs can be observed, covered with an amorphous layer of QPEI polymer (Figure 4D,E). For this reason, electron energy loss spectroscopy (EELS) was employed to investigate the spatial distribution of nitrogen, observed in the bright field images of oxCNTs. An energy-filtered TEM (EFTEM) image (utilizing the three-window method), using the nitrogen K-edge at 401 eV electron energy loss, can be seen in Figure 4C, while Figure 4B is the bright field image of the same area. It is obvious that since the intensity of the maps corresponds to the concentration of N (red) that exclusively originated from QPEI, it can be concluded that oxCNTs were uniformly covered by QPEI. Additionally, in a typical background subtracted EELS spectrum the nitrogen K-edges recorded for oxCNTs@80-QPEI (Figure 4) are evidence for the presence of nitrogen atoms and the successful attachment of QPEI on the surface of oxCNTs.

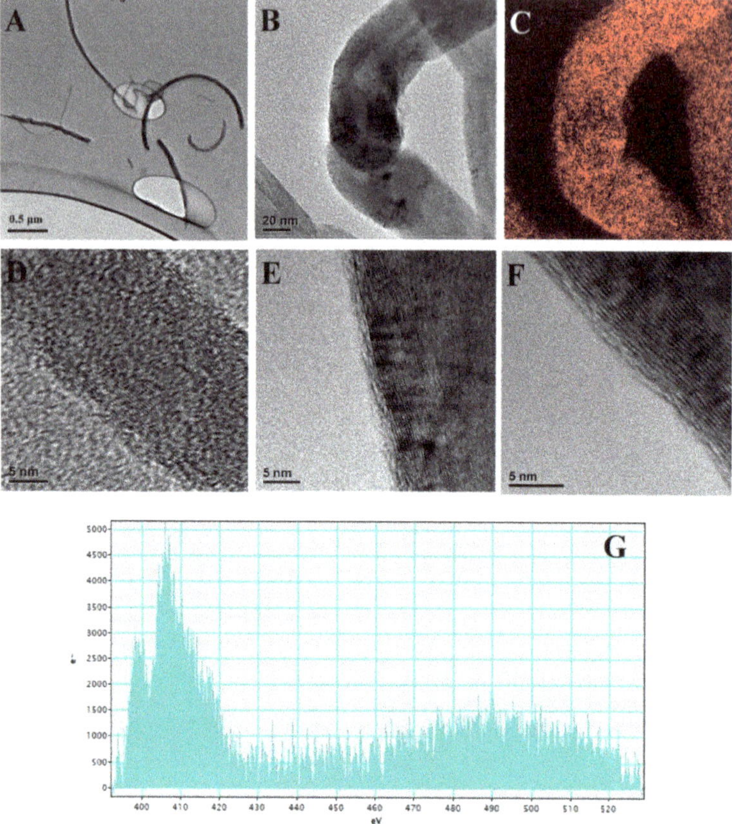

Figure 4. TEM bright field image of oxCNTs@50-QPEI (**A**). Bright field image (**B**) and the corresponding EFTEM compositional nitrogen N map (red, **C**) of oxCNTs@50-QPEI, HRTEM images: images of oxCNTs@80-QPEI (**D**–**F**) and a typical background subtracted EELS spectrum of nitrogen K- edges, recorded for oxCNTs@80-QPEI (**G**).

AFM images of oxCNTs@50-QPEI and oxCNTs@80-QPEI, deposited on Si-wafer (Figure 5) show the morphological features of oxCNTs at the nanoscale after the interaction with the polymers. The AFM images (height and 3D) of nanocomposites reveal the successful attachment (wrapping) of polymer on the oxCNTs sidewalls. As derived from topographical section analysis, an overlay of 10–25 nm is observed from each side of nanotube in the case of oxCNTs@50-QPEI, while the size of the polymeric coating is much higher, and easily distinguishable in the case of oxCNTs@80-QPEI, corresponding to an average of 25–40 nm (Figure 5). Moreover, the average diameter of oxCNTs@80-QPEI, as derived from height analysis, is about 40–50 nm, a value much higher than that of oxCNTs in the absence of 80-QPEI, which ranges between 15 and 25 nm (Figure S2).

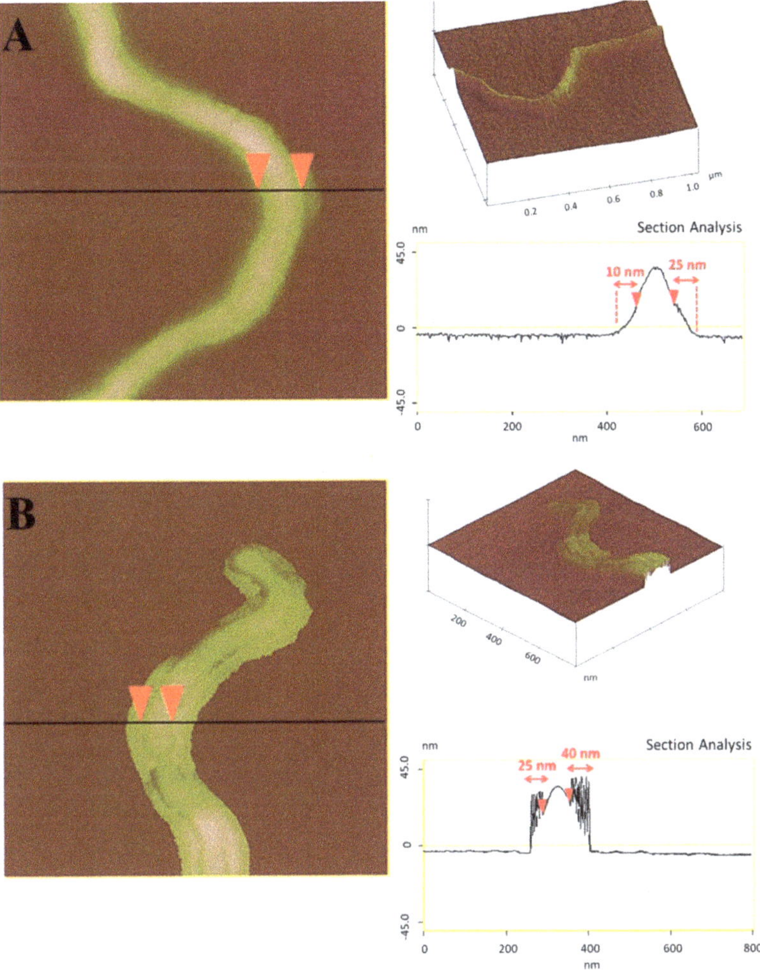

Figure 5. AFM images (height, profile section analysis, and 3D) of oxCNTs@50-QPEI (**A**) and oxCNTs@80-QPEI (**B**), showing the coverage of QPEI derivatives in the sidewalls of oxCNTs.

2.2. Colloidal Stability of the CNTs Dispersions

CNTs have an extremely strong tendency to aggregate in water due to their high surface energy, making them difficult to disperse in aqueous media resulting in the formation of large bundles [50]. Although the dispersibility of CNTs in aqueous media has been shown to increase following (i) various oxidation processes [51], and (ii) using high concentrations of various surfactants [6,7] or polymers [8,9], the resulting dispersions are only stable for short time. In this study, the negatively charged oxidized CNTs, modified with positively charged QPEIs through electrostatic interactions and van der Waals attraction forces, resulted in functionalized oxCNTs with high positive charge contents, and able to form stable aqueous dispersions. All QPEIs derivatives enhance the aqueous dispersibility of oxCNTs, as revealed by visual observation over time (Figure 6, upper part). It is obvious that stable dispersions of oxCNTs were obtained after functionalization with QPEIs for at least twelve months (Figure 6, upper part), while oxCNTs had precipitated within one month after the sonication process. This was achieved thanks to the presence of quaternary ammonium groups on the surface of the oxCNTs that provide a high compatibility with aqueous media due to their strong hydrophilicity, while preventing the CNTs' aggregation due to electrostatic repulsion. In an analogous study, involving SWCNTs, Grunlan, J.C. et al. observed that after non-covalent functionalization with PEI, SWCNTs exhibited poor aqueous stability, attributed to PEIs hyperbranched structure that sterically reduced electrostatic interactions [52]. The same behavior was observed in this study for PEI functionalized oxCNTs. In contrast, quaternized PEI derivatives behave differently, probably due to their higher positively charged moieties content.

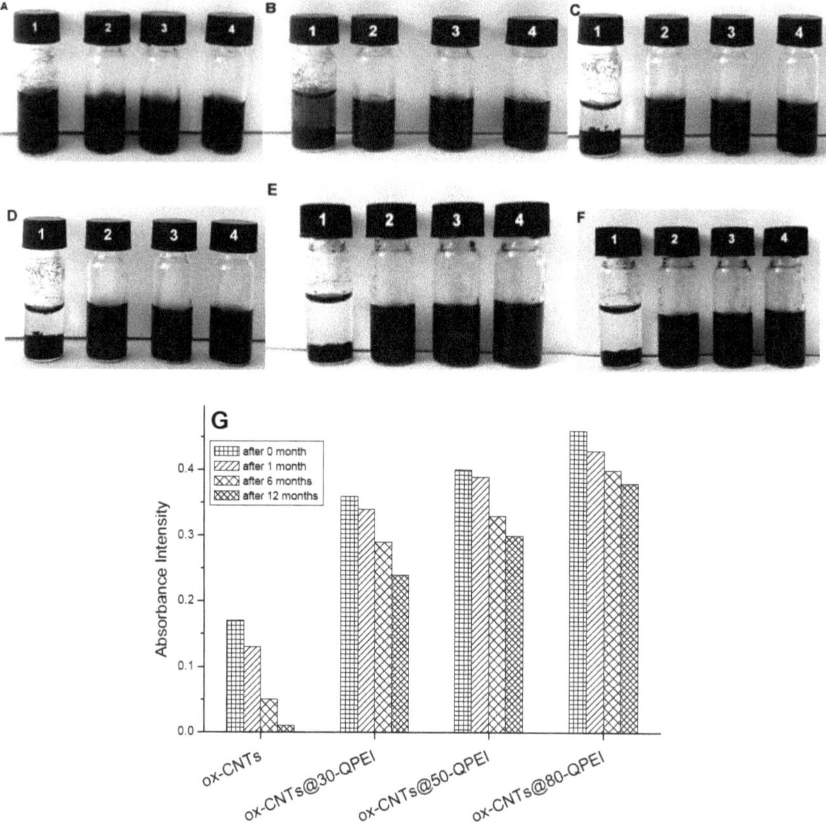

Figure 6. Dispersion state of (1) oxCNTs, (2) oxCNTs@30-QPEI, (3) oxCNTs@50-QPEI, and (4) oxCNTs@80-QPEI in water (5 mg/mL), (**A**) immediately after sonication, and after quiescent settling for (**B**) 2 weeks, (**C**) 1 month, (**D**) 3 months, (**E**) 6 months, and (**F**) 12 months (upper part). (**G**) Sedimentation behavior of oxCNTs and QPEI-functionalized oxCNTs at different aging times (lower part).

It is known that bundled CNTs, unlike individual ones, are not active in the UV–vis region [53] allowing the investigation of their dispersibility using UV–vis absorption spectroscopy. Figure S3 shows the UV–vis spectra of the oxCNTs@30-QPEI, oxCNTs@50-QPEI, and oxCNTs@80-QPEI aqueous dispersions. The dispersions have a characteristic absorption peak at 263 nm, affected by the p-plasmon absorption of carbon nanomaterials. The higher absorption, caused by the p-plasmon from the oxCNTs@80-QPEI, demonstrates more efficient debundling of oxCNTs by 80-QPEI, compared to other carbon nanomaterials [54]. Furthermore, the evaluation of colloidal stability of the CNTs was attained by UV–vis spectroscopy, again after investigation of the characteristic absorption of CNTs at 263 nm. Figure 6G presents the optical density (O.D.) of the oxCNTs and QPEI-functionalized oxCNTs dispersed in water within the storage periods. It was obvious that the dispersions of all QPEI-functionalized oxCNTs were stable for at least twelve months, since their optical densities were reduced only by 10–20% compared to the initial O.D. On the other hand, the O.D. of the oxCNTs dispersion was reduced by 90% after 12 months storage. The O.D. reduction of oxCNTs dispersion was attributed to the gradual formation of CNTs agglomerates, some of which subsequently aggregated and finally precipitated. These findings are in line with the visual inspection of the CNT dispersions presented in Figure 6A–F. Moreover, the dispersion of oxCNTs@80-QPEI is the most stable since only a 10% reduction of O.D. was observed after 12 months storage. To the best of our knowledge the stability achieved is one of the highest reported in the literature, and its importance lies in the fact that the aqueous stability of such hybrid nanomaterials is of paramount importance in several industrial applications.

2.3. Characterization of the QPEI-Functionalized oxCNTs Dispersions

Dynamic light scattering (DLS) and ζ-potential measurements can provide information on nanomaterials regarding their size distribution, and also their surface charge. Although, DLS measurement is appropriate for determination of the spherical particle diameter, it can also be used to determine the hydrodynamic diameter of nanotubes, based on the assumption that an equivalent hydrodynamic diameter (D_h) of a sphere has the same diffusion properties as the CNT. [55] Even though one cannot determine absolute values, a relative size comparison can be obtained for similarly shaped materials [56]. Thus, comparing the aggregate size of oxCNTs to those of QPEI-functionalized oxCNTs, it is obvious that debundling of the oxCNTs took place after their interaction with the QPEIs, since the value of hydrodynamic diameter of the oxCNTs decreased from 1300 nm to 150 nm when 80-QPEI was used (Figure 7).

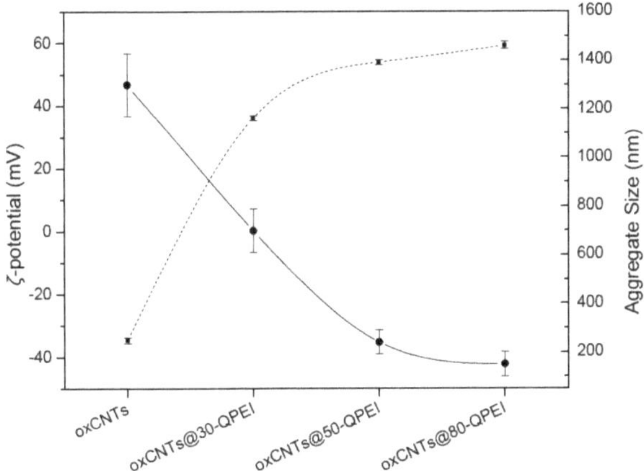

Figure 7. Mean hydrodynamic diameter (solid line) and ζ-potential values (dot line) of oxCNTs and QPEI-functionalized oxCNTs dispersions (0.05 mg/mL, pH = 7.0).

Zeta potential values of the oxCNTs dispersions at pH = 7.0 are given in Figure 7. As expected, the aqueous dispersion of oxCNTs has a ζ-potential value around −34 mV, due to their negative surface charges. After modification with QPEI, the ζ-potential values of the QPEI-functionalized oxCNTs dispersions increased to positive, reaching the value of +60 mV for oxCNTs@80-QPEI, offering further evidence that the positively charged QPEIs were successfully attached onto the oxCNTs surface. It should be noted that all ζ-potential values were higher than +30 mV, indicating stable aqueous colloidal suspensions, in which the CNTs' aggregation was prevented due to electrostatic repulsion, in line with the results of UV–vis and DLS measurements (see above) [57].

2.4. Evaluation of Antibacterial and Anti-Cyanobacterial Activity

The cytotoxicity of QPEI-functionalized oxCNTs was assessed against two types of Gram-negative bacteria, i.e., the heterotrophic bacterial strain *Escherichia coli* XL1-blue and the autotrophic cyanobacterium *Synechococcus* sp. PCC 7942.

2.4.1. Cytotoxicity Effects of oxCNTs@QPEIs on *Escherichia coli* XL1-Blue Bacteria

Escherichia coli growth was investigated by monitoring the fluorescence intensity of bacterial cells suspensions at 37 °C that express red fluorescent protein (RFP). The excellent dispersibility of oxCNTs@QPEIs renders the commonly used turbidity measurement inapplicable, as functionalized CNTs, especially at high concentrations, contribute to the final measurement. Thus, by employing the inherent fluorescence of RFP, the antibacterial activity can be precisely assessed, even in the presence of nanoparticle dispersions, as in the case of oxCNTs@QPEIs. Comparing the fluorescence intensity of each bacterial suspension containing oxCNTs@QPEIs at a certain time, with the initial fluorescence intensity corresponding to the initial *Escherichia coli* population (at OD_{600} = 0.4), bacterial growth could be directly determined. Figure 8 depicts the *Escherichia coli* growth after 6 h incubation in the presence of oxCNTs and oxCNTs@QPEIs at different concentrations, ranging from 5 to 400 μg/mL, as a function of the fluorescence intensity of RFP at 590 nm (excitation: 545 nm), and normalized with the initial fluorescence intensity of control (100% fluorescence intensity). Contrary to the increase of the fluorescence intensity upon untreated bacteria growth (Figure S4), a decrease in the intensity was observed, for all samples, revealing bacterial growth inhibition. However, as shown in Figure 8, the oxCNTs exhibited low antibacterial activity, which is in accordance with the literature [58]. On the other hand, all QPEI-functionalized oxCNTs inhibited bacterial growth in a dose-dependent manner, displaying significantly higher antibacterial activity than the oxCNTs, and which increased upon substitution of PEI from 30% to 80%. In a further attempt to quantify the antibacterial activity of oxCNTs@QPEIs, the 50% inhibitory concentrations (IC-50) were calculated (Table 1). It was found that the lowest IC-50 was observed in the case of oxCNTs@80-QPEI (28.4 μg/mL), which showed that oxCNTs@80-QPEI exhibited the highest antibacterial activity amongst the other hybrid materials, due to both the increased aqueous dispersibility and the higher positive quaternary ammonium group content.

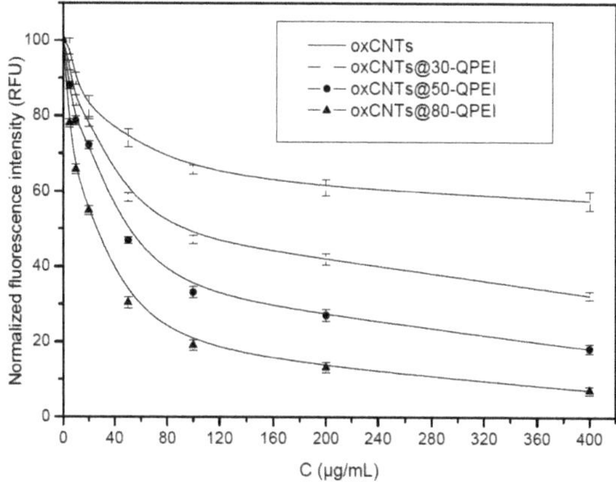

Figure 8. Cytotoxicity of oxCNTs and QPEI-functionalized oxCNTs against gram-negative *Escherichia coli* XL1-blue bacteria. Fluorescence intensity of RFP at 590 nm (excitation: 545 nm) after bacteria incubation for 6 h with oxCNTs@QPEIs at concentrations ranging from 5 to 400 μg/mL, normalized with the initial fluorescence intensity, corresponding to initial *E.coli* population (at OD_{600} = 0.4). Error bars represent mean ± SD for at least three independent experiments.

Table 1. IC-50 values of QPEI-functionalized oxCNTs on *Escherichia coli* XL1-blue bacteria.

Samples	**IC-50 (μg/mL)**
oxCNTs@30-QPEI	93.2
oxCNTs@50-QPEI	50.1
oxCNTs@80-QPEI	28.4

The morphology of *Escherichia coli* after 6-h incubation at 37 °C with oxCNTs@QPEIs was investigated by scanning electron microscopy (SEM). In Figure 9, SEM images of control (untreated cells) and cells treated with oxCNTs@QPEIs at 50% inhibitory concentrations are presented, showing significant changes in cell morphology. Specifically, the treated cells lost their cellular integrity, and are shown more clustered, while their cell walls seem rougher and damaged in all cases (Figure 9B–D) compared to the untreated cells (Figure 9A), which appear to be intact, with a smooth surface. Moreover, Figure 9D shows the most severe effect of oxCNTs@80-QPEI on the bacterial cell wall and membrane, in which the cell walls seem to be ruptured and bacterial cell lysis is clearly observed probably due to membrane damage.

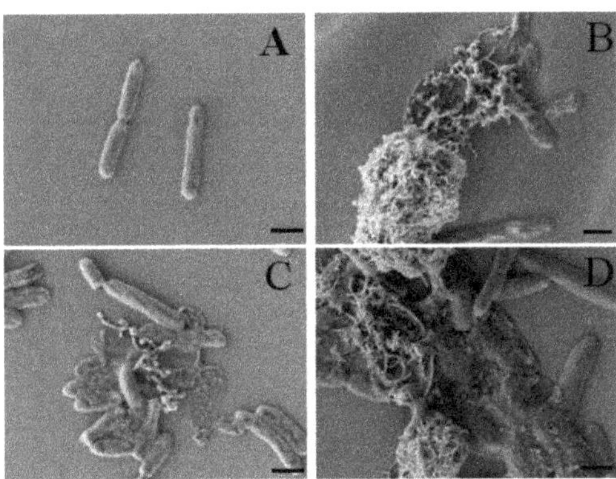

Figure 9. SEM images of *Escherichia coli* bacteria: untreated cells (**A**) and cells after 6-h incubation time at 37 °C with oxCNTs@30-QPEI (**B**), oxCNTs@50-QPEI (**C**), and oxCNTs@80-QPEI (**D**) at 50% inhibitory concentrations. The scale bar is 1 μm.

It is known that highly dispersed carbon nanotubes, mainly single wall carbon nanotubes, are able to interact strongly with bacteria through van der Waals forces, forming bacteria-CNTs aggregations [20,25]. This fact results in bacterial death due to either inhibition of transmembrane electron transfer, or to penetration leading to rupture or deformation of cell walls and membranes, which alter the bacterial metabolic processes [59]. Moreover, SWCNTs and MWCNTs containing various types of surface groups were investigated [29] regarding their antibacterial activity towards gram-negative and gram-positive bacteria. It was found that SWCNTs functionalized with hydroxyl and carboxyl surface groups exhibited improved antimicrobial activity against both gram-positive and gram-negative bacteria. However, MWCNTs containing the same functional surface groups did not exhibit any significant antibacterial effect [29]. On the contrary, covalently functionalized MWCNTs with positive moieties such as amines, arginines, and lysines, [60,61] or MWCNTs combined with surfactant molecules, such as dioctyl sodium sulfosuccinate [32], hexadecyltrimethylammonium bromide, triton X-100, and sodium dodecyl sulfate [7], exhibited enhanced antibacterial properties, due to enhanced interactions with bacterial membranes and the improved aqueous dispersibility and stability. In this study, similar antibacterial behavior of QPEI-functionalized oxCNTs was observed due to the high positive quaternary ammonium group content. These positive groups, as in the case of surfactant molecules or other positive functional groups, not only improved the debundling of

MWCNTs, which favors the strong interaction between the bacteria and MWCNTs, but also enhanced the penetration of MWCNTs though cell membranes, resulting in cell lysis and death.

Moreover, it is known that the quaternary ammonium moieties efficiently interact with the negatively charged groups of bacterial walls or cytoplasmic membranes, mainly through electrostatic as well as with secondary hydrophobic interactions, leading to dysfunction in cellular processes and probably to cell death [62]. Highly functionalized polymers with quaternary ammonium groups have been found to be more effective antibacterial agents than their low molecular weight analogues, as their higher charge densities lead to stronger interactions with the negatively charged bacteria walls [63]. For example, poly(propylene imine) dendrimers bearing 16 quaternary ammonium groups per molecule were found to exhibit two orders of magnitude greater antimicrobial activity than their mono-functional counterparts [64]. In agreement with this, the 80-QPEI derivative containing the highest content of quaternary ammonium groups provided the highest polycationic character to oxCNTs, in regards to the other QPEI derivatives. This effect probably induces the strongest interaction with bacteria, and the highest permeability of the cell membrane, and thus oxCNTs@80-QPEI exhibited the best antibacterial activity against the *Escherichia coli* bacteria compared to the other hybrid materials.

2.4.2. Cytotoxicity Effects of oxCNTs@QPEIs on *Synechococcus* sp. PCC 7942 Cyanobacteria

The antibacterial activity of oxCNTs@QPEIs was further assessed against the cyanobacterium *Synechococcus* sp. PCC 7942, a very widespread bacterial strain in the aquatic environment. In general, cyanobacteria (gram negative bacteria) are prokaryotic organisms that perform oxygenic photosynthesis similar to higher plants. They are the oldest and one of the largest and most important groups of bacteria on earth. Cyanobacteria (except prochlorophytes) contain only Chl α, the molecule which makes photosynthesis possible, by passing its energized electrons on to molecules during sugar synthesis [65]. However, several cyanobacterial strains are known to produce a wide range of toxic secondary metabolites (hepatotoxins, neurotoxins, cytotoxins, dermatotoxins, and irritant toxins), which could be harmful to animals and potentially dangerous to humans [66].

In this study, cyanobacteria *Synechococcus* sp. PCC 7942 cell proliferation was monitored by measurement of the Chl α concentration every 24 h, for seven days. Figure 10 shows the cell proliferation of the unicellular cyanobacterium *Synechococcus* sp. PCC 7942 cells in the presence of increasing concentrations of oxCNTs@30-QPEI, oxCNTs@50-QPEI, and oxCNTs@80-QPEI. For comparison reasons, analogous experiments were performed using oxCNTs, and their effect on the cyanobacteria cell proliferation is shown in Figure S5. It is clear that oxCNTs did not inhibit the cyanobacterial cell proliferation in contrast to all oxCNTs@QPEIs. This can be attributed again to the strong positive character of oxCNTs@QPEIs that intensified the interaction with the cyanobacteria membrane, and resulted in higher cell penetration compare to oxCNTs. In order to quantify these results, the effective concentration for 50% inhibition (IC-50), which shows the ability of cells to proliferate under the toxic effect of the oxCNTs@QPEIs, was calculated from the cyanobacterial cell proliferation curves in the presence of each hybrid material (Figure S6 and Table 2). Based on the interpretation of the experimental data using a non-linear regression of the four-parameter logistic function (Figure S6), it is obvious that cell proliferation is concentration dependent, while oxCNTs@80-QPEI exhibited the most promising anti-cyanobacterial activity, compared to the other two hybrid materials. This implies that upon increasing the degree of quaternization, the proliferation rate of cyanobacteria cells decreases. Therefore the anti-cyanobacterial properties, as in case of *Escherichia coli*, depend, not only on the concentration, but also on the degree of quaternization.

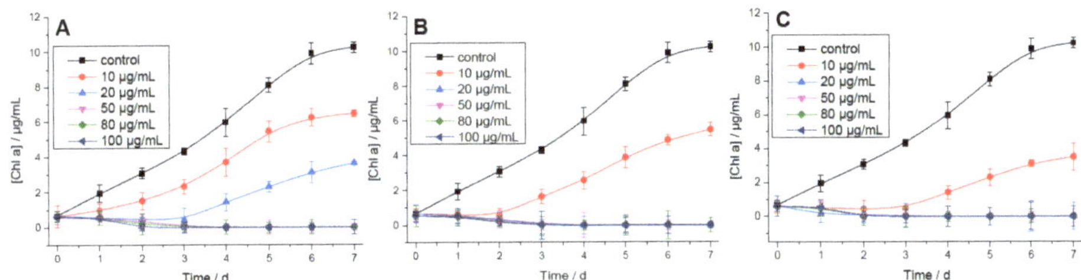

Figure 10. Effect of oxCNTs@QPEIs on cell proliferation of cyanobacteria *Synechococcus* sp. PCC 7942. Growth curves of cyanobacteria in the presence of different concentrations of: oxCNTs@30-QPEI (**A**), oxCNTs@50-QPEI (**B**), and oxCNTs@80-QPEI (**C**). Error bars represent mean ± SD for at least three independent experiments.

Table 2. IC-50 values of oxCNTs@QPEIs on cyanobacterium *Synechococcus* sp. PCC 7942.

Samples	IC-50 (μg/mL)
oxCNTs@30-QPEI	12.4
oxCNTs@50-QPEI	≤10
oxCNTs@80-QPEI	<10

Triggered by these results, it was interesting to evaluate the effect of oxCNTs@QPEIs on the photosynthetic apparatus of cyanobacteria. Therefore, the activity of Photosystem (PS) I and II was assessed in the presence of oxCNTs@80-QPEI, the material exhibiting the best antibacterial performance, to investigate the consequences of oxCNTs@QPEIs on the integrity of the photosynthetic apparatus in terms of photoinduced electron transport.

Specifically, selective detection of the PSII [67] and PSI electron transporting activities [68] was performed on *Synechococcus* sp. PCC 7942 bacteria treated with lysozyme (permeaplasts) at room temperature [69]. Using *Synechococcus* permeaplasts, oxymetrically photoinduced electron transport activities, across both PSII (electron donor: water; post-PSII electron acceptor: p-benzoquinone) and PSI (post- PSII inhibitor: DCMU; post-PSII electron donor: diaminodurene and ascorbate; post-PSI electron acceptor: methyl viologen) were measured. It was found that upon increase in concentration of oxCNTs@80-QPEI, the rate of oxygen evolution decreases (Table S2), indicating that PSII and PSI electron transport activities depend on the QPEI-functionalized oxCNT concentration. The inhibition of PSI and PSII by oxCNTs and oxCNT@80-QPEI is shown in Figure 11. Similarly to results previously reported in the literature, [70], at high concentrations (250 μg/mL) the oxCNTs used in this study inhibited the PSII and PSI by 37.6% and 95.7%, respectively, while at lower concentrations (20 μg/mL) the PSII is practically unaffected and the PSI activity is reduced by almost 44.8%. However, the impact of oxCNT@80-QPEI was much higher. The novel hybrid with an 80% quaternization degree inhibited the PSII by 16.7% at concentration 20 μg/mL, and by around 53.9% at 250 μg/mL. In the case of the PSI, the effect was even more significant, exhibiting almost complete inhibition (more than 97%), even at low concentrations (20 μg/mL).

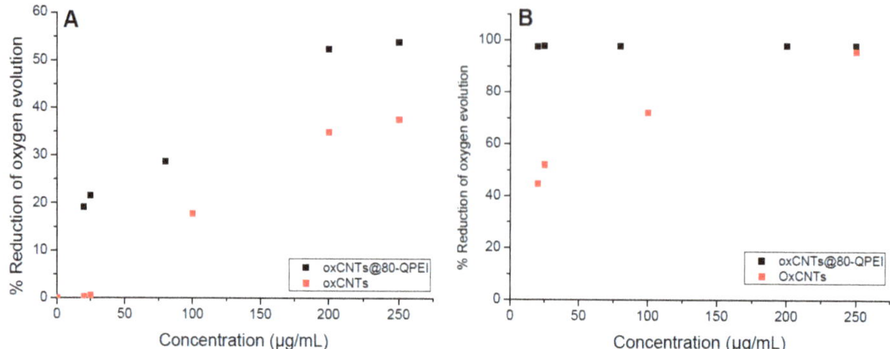

Figure 11. Effect of oxCNTs and oxCNTs@80-QPEI on the photosynthetic electron transport activities of PSII (**A**) and PSI (**B**) in *Synechococcus* sp. PCC 7942 permeaplasts.

The remarkable decrease in the IC-50 values of oxCNTs@QPEIs on cyanobacterial cell proliferation (Table 2) indicates that photosynthetic electron transport of both PSII and PSI is functionally impaired in cyanobacterial cells. Although, oxCNTs inhibit the photosynthetic redox reactions, and in the case of PSI almost fully prevent its activity, the effect of QPEI is significant. This may be attributed to the polycationic character of QPEI, as such compounds are known to completely inhibit PSI reactions, while leaving PSII relatively unaffected [71]. However, under the test conditions oxCNTs did not inhibit cyanobacterial cell proliferation (Figure S5), implying that the oxCNTs could not penetrate the cyanobacteria membranes. On the contrary, as mentioned above, the oxCNTs@QPEIs, exhibited high toxicity against cyanobacteria as a result of their efficient cell penetration (Figure 10).

Furthermore, to elucidate the potential alternative patterns for electron flow to and from PSI, the $P700^+$ transients in the presence of oxCNTs@80-QPEI were investigated. Table S3 shows the amounts of functional PSI complexes, estimated as the photooxidizable form of the PSI ($P700^+$) reaction center, measured as ΔA820/A820 [72], where an 80% inhibition by oxCNTs@80-QPEI at high concentrations (250 μg/mL) can be observed. In a previous study, MWCNTs were successfully applied for direct transfer of electrons in isolated spinach thylakoids and cyanobacteria Nostoc sp. [73], pointing out that the results obtained in this study may also be associated with an interruption of the electron transport, due to the presence of oxCNTs. Concomitantly, the lower steady state photooxidation of P700 by far red light, might also be considered as an indication that oxCNTs@QPEIs quench the $P700^+$. Based on the above, the observed enhanced anti-cyanobacterial effect of QPEI-functionalized oxCNTs on cyanobacterium *Synechococcus* sp. PCC 7942 is ascribed to the selective inhibition of PSI.

2.5. Cell Viability Assay

To investigate the cytotoxicity of QPEI-functionalized oxCNTs, the human prostate carcinoma DU145 cell line and the 3T3 mouse fibroblasts were employed. For this purpose, these cells were incubated for 24 h with oxCNTs and oxCNTs@QPEIs at concentrations below and above their IC-50 values and cell viability was assessed, employing the standard MTT assay. The percent cell viability caused by all derivatives is presented in Figure 12. It is obvious that all oxCNTs@QPEIs were not toxic at their IC-50 values, and significantly less lethal than the parent oxCNTs, while at higher concentrations (100–200 μg/mL) only oxCNTs@30-QPEI exhibited slight toxicity (~70% cell survival). It should be noted that at 200 μg/mL, a concentration much higher than their IC-50 values, oxCNTs@50-QPEI and oxCNTs@80-QPEI did not display any cytotoxicity. These results suggest that all oxCNTs@QPEIs simultaneously exhibited both low cytotoxicity and enhanced antibacterial/anti-cyanobacterial properties.

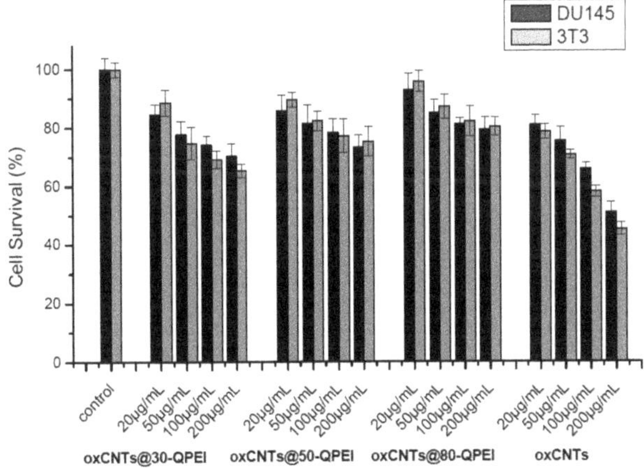

Figure 12. Cytotoxicity of oxCNTs and QPEI-functionalized oxCNTs on DU145 and 3T3 cells following incubation at various concentrations for 24 h, as determined by MTT assays. Data are expressed as mean ± SD of eight independent values obtained from at least three independent experiments.

3. Materials and Methods

3.1. Chemicals and Reagents

Hyperbranched poly(ethyleneimine) (PEI) with molecular weight 25 KDa (Lupasol® WF, water-free, 99%) and oxidized multi-walled carbon nanotubes were kindly donated by BASF (Ludwigshafen, Germany) and Glonatech S.A. (Athens Greece), respectively. Glycidyltrimethylammonium chloride, dialysis tubes (molecular weight cut-off: 1200) and triethylamine were obtained from Sigma-Aldrich (St. Louis, MA, USA). D-MEM low glucose with phenol red, L-glutamine, phosphate buffer saline (PBS), fetal bovine serum (FBS), penicillin/streptomycin, and trypsin/EDTA were purchased from BIOCHROM (Berlin, Germany). Thiazolyl blue tetrazolium bromide (MTT) and isopropanol were purchased from Merck KGaA (Calbiochem®, Darmstadt, Germany).

3.2. Synthesis of Quaternized Hyperbranched Poly(ethyleneimine) Derivatives

Quaternized derivatives of hyperbranched poly(ethyleneimine), with different substitution degrees of primary amino groups were prepared by a method previously described [40,41]. In brief, to an aqueous solution (20 mL) of PEI (5 mM), an aqueous mixture (10 mL) containing, 8, 12, or 18 mmol glycidyltrimethylammonium chloride and 16, 24, or 36 mmol triethylamine, respectively, was added. The reaction was completed after two days at room temperature and the final quaternized derivatives with 30% (30-QPEI), 50% (50-QPEI), and 80% (80-QPEI) degree substitution of primary amino groups were received after dialysis against deionized water and lyophilization. The introduction of the quaternary moieties at the external surface of the parent PEI was confirmed by ^{1}H and ^{13}C NMR spectroscopy. Additionally, the degree of quaternization at the primary amino groups of PEI was calculated by the integration of peaks at 3.45 ppm and 2.70–2.50 ppm in the ^{1}H NMR spectra.

^{1}H NMR: (500 MHz, D_2O) δ (ppm) = 4.25 (broad s, CH–OH), 3.45 (m, $CH_2N^+(CH_3)_3$), 3.25 (s, CH_3), 2.70–2.50 (m, CH_2 of PEI scaffold).

^{13}C NMR (125.1 MHz, D_2O): δ (ppm) = 71.5 ($CH_2N^+CH_3)_3$), 67.5 (CH–OH), 57.0 (CH_3), 55–51.0 (CH_2 of PEI scaffold), 51.0 and 48.0 (CH_2NH-Q primary and secondary, respectively), 42.0 and 40.0 (CH_2NH_2 close to secondary ($C_{1,2}$) and tertiary ($C_{1,3}$) amine, respectively).

3.3. Preparation of QPEI-Functionalized oxCNTs

The functionalization of oxCNTs was achieved by a method previously described [41]. Specifically, 50 mg of oxCNT powder was dispersed in 50 mL of an aqueous solution, containing an excess quantity (150 mg) of each quaternized derivative, and the resulting dispersions were ultrasonicated for 15 min (Hielscher UP200S high intensity ultrasonic processor equipped with a standard sonotrode (3 mm tip-diameter) at 50% amplitude and 0.5 cycles) and stirred for a further 12 h, at room temperature. The final hybrid materials, oxCNTs@30-QPEI, oxCNTs@50-QPEI, and oxCNTs@80-QPEI, were received after ultracentrifugation at 45,000 rpm, followed by thorough washing with water to remove the unreacted QPEI derivatives and lyophilization.

3.4. Characterization of QPEI-Functionalized oxCNTs

FTIR spectra were recorded using a Nicolet 6700 spectrometer (Thermo Scientific, Waltham, MA, USA) equipped with an attenuated total reflectance accessory with a diamond crystal (Smart Orbit, Thermo Electron Corporation, Madison, WI, USA). Raman spectra were obtained using a micro-Raman system RM 1000 Renishaw (laser excitation line at 532 nm, Nd-YAG) in the range of 400–2000 cm^{-1}. AFM images were obtained in tapping mode, with a 3D Multimode Nanoscope, using Tap-300G silicon cantilevels with a <10 nm tip radius and a ≈20–75 N/m force constant. Samples were deposited onto silicon wafers (P/Bor, single side polished) by drop casting from ethanol solutions. Scanning electron microscopy (SEM) images were recorded using a Jeol JSM 7401F field emission scanning electron microscope equipped with a gentle beam mode. Transmission electron micrographs were taken using a Philips C20 TEM instrument equipped with a Gatan GIF 200 energy filter for electron energy loss elemental mapping. For the sample preparation, a drop of oxCNTs@QPEIs aqueous solution (0.1 mg/mL) was casted on a PELCO® Formvar grid and was left to evaporate. Thermogravimetric analyses (TGA) were carried out on a Setaram SETSYS Evolution 17 system at a 5 °C/min heating rate under oxygen atmosphere. Elemental analysis (EA) was measured by a Perkin Elmer 240 CHN elemental analyzer.

3.5. Preparation and Characterization of QPEI-Functionalized oxCNTs Aqueous Dispersions

The suspensions of polymer-functionalized oxCNTs were prepared by adding 10 mg of oxCNTs@30-QPEI, oxCNTs@50-QPEI, or oxCNTs@80-QPEI into 2 mL pure water. The suspensions were then ultra-sonicated for 5 min using a Hielscher UP200S high intensity ultrasonic processor at 40% amplitude and 0.5 cycles. Each sample was centrifuged at 1500 rpm for 15 min, and then the supernatant was diluted with pure water before measurement.

ζ-potential measurements were performed using a ZetaPlus -Brookhaven Instruments Corp. In a typical experiment, an aqueous 0.05 mg/mL dispersion of QPEI-functionalized oxCNTs was used, ten ζ-potential measurements were collected, and the results were averaged. Dynamic light scattering studies were carried out on an AXIOS-150/EX (Triton Hellas) system equipped with a 30 mW laser source, and an avalanche photodiode detector at 90° angle. In a typical experiment, an aqueous 0.05 mg/mL dispersion of QPEI-functionalized oxCNTs was used, at least ten measurements were collected, and the data were analyzed using the CONTIN algorithm to obtain the hydrodynamic radii distribution.

UV-vis spectra of the aqueous dispersions (1 mg/mL) were obtained by a Cary 100 Conc UV-visible spectrophotometer (Varian Inc., Mulgrave, Australia) in the range of 200–600 nm. Additionally, the colloid stability of the functionalized carbon nanotubes was evaluated at static conditions for 1, 6, and 12 months. Specifically, the dispersions obtained as described previously, were placed in vertically standing tubes and stored at room temperature. At each time point, 100 μL of the stock solutions from the very upper part was taken diluted in 1 mL water and the optical density (O.D.) of these dispersions was measured using UV-vis spectroscopy.

3.6. *Escherichia coli* Growth Inhibition Assay

The antibacterial activity of QPEI-functionalized oxCNTs was obtained by a bacteria growth inhibition assay. *Escherichia coli* XL1-blue bacteria expressing red fluorescent protein (RFP), from a plasmid-encoded gene, were grown in Luria–Bertani (LB) broth at 37 °C overnight, in a Stuart SI500 orbital shaker at approximately 200 rpm shaking speed in aerobic conditions. The culture was subsequently diluted to an optical density (O.D.) of 0.4 at 600 nm. The QPEI-functionalized oxCNTs were homogeneously dispersed in distilled water by sonication and added to the bacterial culture at concentrations ranging from 5 to 400 μg/mL. The assay was performed in a 96-well plate format in a 200 μL final volume. Fluorescence intensity over growth of untreated bacteria revealed that the optimum incubation time was 6 h, as the intensity reached a plateau (Figure S4). Thus, plates were incubated at 37 °C, shaking at 100 rpm in aerobic conditions for 6 h, and bacterial growth was monitored using the fluorescence intensity of red fluorescent protein, which was recorded at an emission wavelength of 590 nm by an Infinite M200 plate reader (Tecan group Ltd., Männedorf, Switzerland) at an excitation wavelength of 545 nm. In order to eliminate the effect of CNTs in the measured intensities, the initial values (at 0 h), although minor compared to those obtained after 6 h, were subtracted from the final measurements. For each treatment eight replicates were used and three independent experiments were performed. Untreated bacteria were used as control, representing 100% fluorescence intensity in Figure 8.

3.7. SEM Analysis of the Cellular Morphology

The morphology of the *Escherichia coli* bacteria after treatment with QPEI-functionalized oxCNTs was characterized by scanning electron microscopy (Jeol JSM 7401F Field Emission SEM). Specifically, cells were incubated with QPEI-functionalized oxCNTs at their 50% inhibitory concentration (IC-50), fixed with 3% glutaraldehyde in sodium cacodylate buffer (100 mM, pH = 7.1) for 6 h, transferred to a poly(L-lysine) coated glass cover slip, dehydrated using ethanol gradient (twice of 50%, 70%, 95%, and 100% ethanol for 10 min each), drying, and coated with gold in a sputter coater [74].

3.8. *Synechococcus* sp. PCC7942 Cyanobacteria Growth Inhibition Assay

In vitro anti-cyanobacterial activity of QPEI-functionalized oxCNTs was screened against *Synechococcus* sp. PCC 7942 bacteria. The unicellular cyanobacterium *Synechococcus* sp. PCC7942 was purchased from the Collection Nationale de Cultures de Microorganismes (CNCM), Institut Pasteur, Paris, France. The cyanobacterial cells were cultured in BG11 that additionally contained 20 mM HEPES-NaOH (pH = 7.5). The cultures were illuminated with white light from fluorescent lamps, providing a photosynthetic active radiation (PAR) of 100 μmol photons m^{-2} s^{-1}, and were aerated with air containing 5% (*v*/*v*) CO_2 in an orbital incubator (Galenkamp INR-400) at 31 °C [75]. QPEI-functionalized oxCNTs dispersed in distilled water using ultrasonication were added to the bacterial culture at concentrations ranged from 10 to 100 μg/mL. Cyanobacterial cell proliferation was monitored in terms of concentration of Chl α, determined in *N*,*N*-dimethylformamide (DMF) extracts [76]. To extract Chl α, the cell suspensions were centrifuged, DMF was added to the residue, and the resulting clear supernatant DMF extract was obtained after a second centrifugation.

Toxicity tests were performed in three replicate experiments using at least five geometrically scaled dilutions for each compound concentration. The cyanobacteria culture was inoculated in each test solution in the exponential growth phase at concentrations of approximately 1 μg Chl α/mL. The toxicity of the oxCNTs@QPEIs was evaluated as the effective concentrations (μg/mL) of the test substance inhibiting cell proliferation by 50% (IC-50) relative to the control cultures; in this test, the IC-50 values were calculated by the area under the growth curves (biomass) for each concentration of the hybrid materials, using non-linear regression of a 4-parameters logistic function. The related data are presented in Figure S6.

3.9. Measurements of Photosystem I and II Electron Transport Activities

Photo-induced electron transport rates were determined in *Synechococcus* permeaplasts [68] at room temperature oxymetrically (for each Photosystem) with a Clark-type oxygen electrode (DW1; Oxygraph, Hansatech, King's Lynn, UK). *Synechococcus* sp. PCC 7942 bacteria were treated with lysozyme to obtain permeaplasts before the measurement of the photosynthetic electron transport activities [69]. The instrument was fitted with a slide projector to provide actinic illumination of samples. PSI activity was determined by measuring the rate of oxygen uptake, in the presence of the post-PSII electron transfer inhibitor 3-(3,4-dichlorophenyl)-1,10-dimethylurea (DCMU), using Na ascorbate/diaminodurene as an electron donor to PSI and methyl viologen as a post-PSI electron acceptor and mediator of oxygen uptake [77]. The reaction mixture (1 mL, in buffered BG11) contained permeaplasts (5 μg Chl α), diaminodurene (1 mM), Na-ascorbate (2 mM), methyl viologen (0.15 mM), and DCMU (0.01 mM). PSII activity was determined by measuring the rate of oxygen evolution, with water as electron donor and p-benzoquinone as post-PSII electron acceptor. The reaction mixture (1 mL in buffered BG11) contained permeaplasts (5 μg Chl α/mL) and p-benzoquinone (1 mM).

The redox state of P700 was determined *in vivo* using a PAM-101-modulated fluorometer (Heinz Walz GmbH, Effeltrich, Germany), equipped with an ED-800T emitter-detector, and PAM-102 units, following the procedure of Schreiber at al. [64]. The redox state of P700 was evaluated as the absorbance change around 820 nm (ΔA820/A820).

3.10. Cell Cytotoxicity

In this study, human prostate carcinoma cell line DU145 and 3T3 mouse fibroblasts, obtained from the American Type Culture Collection (ATCC, Manassas, VA, USA), were used. Cells were grown in low glucose supplemented D-MEM, containing 10% FBS, penicillin/streptomycin solution (100 U/mL + 100 μg/mL), and 2 mM L-Glutamine. Cells were incubated at 37 °C in a humidified atmosphere, containing 5% CO_2 and sub-cultured, twice a week after detaching with a solution containing 0.05% (*w/v*) trypsin and 0.02% (*w/v*) EDTA. The cytotoxicity of oxCNTs and the QPEI functionalized oxCNTs was assessed employing MTT assay. DU145 cancer cells and 3T3 mouse fibroblasts were inoculated (10^4 cells/well) into 96-well plates and incubated in complete media for 24 h. Cells were then treated with various concentrations of oxCNTs@QPEIs for 24 h. The mitochondrial redox function (translated as cell viability) of all cell groups was measured by the MTT assay. In brief, cell media was replaced with complete media containing MTT (10 μg/mL) and incubated at 37 °C in a 5% CO_2 humidified atmosphere for 3 h. Then, the supernatant containing MTT was discarded and the produced formazan crystals were dissolved in isopropyl alcohol (100 μL per well) under shaking for 10 min at 100 rpm in a Stuart SI500 orbital shaker. Finally, the endpoint absorbance measurements at 540 nm were carried out, employing an Infinite M200 plate reader (Tecan group Ltd., Männedorf, Switzerland). Eight replicates were performed for each concentration, and the experiment was repeated in triplicate. The relative cell viability was calculated as cell survival percentage compared to cells that were treated only with complete media (control). Blank values measured in wells with isopropyl alcohol and no cells, were in all cases subtracted.

4. Conclusions

In this study, negatively charged oxidized multi-walled carbon nanotubes (oxCNTs) were modified with positively charged quaternized hyperbranched poly(ethyleneimine) derivatives (QPEIs), through non-covalent functionalization. Specifically, three derivatives of hyperbranched poly(ethyleneimine), with a 30, 50, and 80% substitution degree of primary amino groups, were prepared and, subsequently, physically interacted with oxCNTs, yielding three novel QPEI functionalized oxCNTs, with QPEI loading ranged between 16–23%, approximately. Structural characterization of these hybrid materials using a variety of techniques, such as FTIR, RAMAN, SEM, TEM, AFM, etc., revealed the successful and homogenous anchoring of QPEIs on the oxCNTs surface. Furthermore, the microscopic techniques revealed the effective wrapping of the QPEI over the ox-CNTs. Contrary to previous studies on non-covalent functionalization of CNTs with PEI, the obtained hybrids efficiently dispersed in aqueous media, forming dispersions with excellent aqueous stability for over 12 months. To evaluate the antibacterial and anti-cyanobacterial properties of these hybrids, two types of gram(−) bacteria, an autotrophic (cyanobacterium *Synechococcus* sp. PCC 7942) and a heterotrophic (bacterium *Escherichia coli*), were used. It was found that all materials exhibited an enhanced, dose-dependent antibacterial and anti-cyanobacterial activity against both test organisms. The obtained IC-50 values were much lower compared to oxidized MWCNTs, revealing that the non-covalent attachment of QPEIs strongly induces the antibacterial/anti-cyanobacterial properties of the hybrid materials. These improved properties were attributed to the polycationic character of the oxCNTs@QPEIs, which enables the effective interaction of the hybrids with the bacteria membranes, facilitating their internalization into the cells. Moreover, the excellent aqueous dispersibility and stability of the hybrids, upon increasing the quaternization degree, further enhanced their activity. Indeed, the QPEI derivative containing the highest content of quaternary ammonium groups (80-QPEI) exhibited the highest performance, compared to the other QPEI derivatives. In the case of the photosynthetic bacteria, it was shown that the hybrid materials affect their photosynthetic apparatus by selective inhibition of the Photosystem (PS) I electron transport activity, while also reducing the photosynthetic electron transport in PSII. To the best of our knowledge, the QPEI functionalized hybrids are the first materials exhibiting strong anti-cyanobacterial properties, without the use of any antibiotic/herbicide. Furthermore, cytotoxicity studies on human prostate carcinoma DU145 cell line and the 3T3 mouse fibroblasts were performed, revealing that all hybrids exhibit high cytocompatibility in the concentration range in which they also exhibit high antibacterial and anti-cyanobacterial properties. These results suggest that QPEI-functionalized oxCNTs can be very attractive candidates as antibacterial and anti-cyanobacterial agents that can be used for potential applications in the disinfection industry, as well as for control of harmful cyanobacterial blooms.

Supplementary Materials: The following are available online at http://www.mdpi.com/1424-8247/13/10/293/s1, Figure S1: FTIR spectra of oxCNTs, 30-QPEI, oxCNTs@30-QPEI, 50-QPEI and oxCNTs@50-QPEI, Figure S2: SEM images (upper part), AFM image and profile section (lower part) of oxCNTs, Figure S3: UV–vis absorption spectra of oxCNTs (a), oxCNTs@30-QPEI (b), oxCNTs@50-QPEI (c) and oxCNTs@80-QPEI (d) in aqueous solution (1 mg/mL), Figure S4: Fluorescence intensity change of RFP at 590 nm (excitation: 545 nm) upon *Escherichia coli* XL1-blue bacteria growth. Figure S5: Effect of oxCNTs on cell proliferation of cyanobacteria *Synechococcus* sp. PCC 7942 in the presence of different concentrations. Error bars represent mean ± SD for at least three independent experiments. Figure S6: Plot of the area under the growth curves of *Synechococcus* sp. PCC 7942 cells for each concentration of oxCNTs@PEIs versus the corresponding concentration as well as the relevant IC-50 calculations. Table S1: Elemental analysis results of ox-CNTs, QPEI and QPEI-functionalized ox-CNTs. Table S2: Photosystem II and I electron transport activities measured on *Synechococcus* sp. PCC 7942 permeaplasts in the presence of oxCNTs@80-QPEI. Table S3: Effects of oxCNTs@80-QPEI on the steady state oxidation of P700 (ΔA820/A820) by FR light in *Synechococcus* sp. PCC 7942 cells.

Author Contributions: Conceptualization, Z.S.; Data curation, N.S.H., G.K., K.M.L., K.N.P., A.P., E.S. and A.K.; Formal analysis, N.S.H., G.K., K.M.L., K.N.P., A.P., E.S., S.P. and A.K.; Investigation, D.G. and F.K.K.; Methodology, N.S.H., G.K., K.M.L., K.N.P., A.P., E.S., S.P., A.K., F.K.K. and K.S.; Project administration, D.G. and Z.S.; Resources, K.S.; Supervision, F.K.K. and Z.S.; Validation, K.N.P.; Visualization, S.P.; Writing—original draft, F.K.K., K.S. and Z.S.; Writing—review & editing, S.P., D.G., F.K.K. and Z.S. All authors have read and agreed to the published version of the manuscript.

Funding: This work was partially financed by the Greek General Secretariat for Research and Technology, under the frame of EuroNanoMed III, ANNAFIB project (MIS 5053890) and by the internal project entitled: "Synthesis and characterization of nanostructured materials for environmental applications" (EE11968). K.M.L. also thanks the Greek State Scholarships Foundation (MIS 5000432, contract number: 2018-050-0502-13820).

Acknowledgments: This research was supported by the Greek General Secretariat for Research and Technology, under the frame of EuroNanoMed III, ANNAFIB project (MIS 5053890) and by the internal project entitled: "Synthesis and characterization of nanostructured materials for environmental applications" (EE11968). K.M.L. acknowledges financial support from the Greek State Scholarships Foundation, program "Enhancement of human scientific resources through implementation of PhD research" with resources of the European program "Development of human resources, Education and lifelong learning", 2014–2020, co-funded by the European Social Fund and Greek State (MIS 5000432, contract number: 2018-050-0502-13820).

Conflicts of Interest: The authors declare no conflict of interest.

References

1. Dresselhaus, M.S.; Dresselhaus, G.; Avouris, P. *Carbon Nanotubes: Synthesis, Structure, Properties and Applications*; Springer: Berlin, Germany, 2001.
2. Novoselov, K.S.; Fal, V.; Colombo, L.; Gellert, P.; Schwab, M.; Kim, K. A roadmap for graphene. *Nature* **2012**, *490*, 192–200. [CrossRef] [PubMed]
3. Ji, H.; Sun, H.; Qu, X. Antibacterial applications of graphene-based nanomaterials: Recent achievements and challenges. *Adv. Drug Deliv. Rev.* **2016**, *105*, 176–189. [CrossRef] [PubMed]
4. Goenka, S.; Sant, V.; Sant, S. Graphene-based nanomaterials for drug delivery and tissue engineering. *J. Control. Release* **2014**, *173*, 75–88. [CrossRef] [PubMed]
5. Breuer, O.; Uttandaraman, S. Big returns from small fibers: A review of polymer/carbon nanotube composites. *Polym. Compos.* **2004**, *25*, 630–645. [CrossRef]
6. Soleyman, R.; Hirbod, S.; Adeli, M. Advances in the biomedical application of polymer-functionalized carbon nanotubes. *Biomater. Sci.* **2015**, *3*, 695–711. [CrossRef]
7. Baia, Y.; Park, I.S.; Lee, S.J.; Bae, T.S.; Watari, F.; Uo, M.; Lee, M.H. Aqueous dispersion of surfactant-modified multiwalled carbon nanotubes and their application as an antibacterial agent. *Carbon* **2011**, *49*, 3663–3671. [CrossRef]
8. Sun, J.-T.; Hong, C.-Y.; Pan, C.-Y. Surface modification of carbon nanotubes with dendrimers or hyperbranched polymers. *Polym. Chem.* **2011**, *2*, 998–1007. [CrossRef]
9. Tuncel, D. Non-covalent interactions between carbon nanotubes and conjugated polymers. *Nanoscale* **2011**, *3*, 3545–3554. [CrossRef]
10. Bilalis, P.; Katsigiannopoulos, D.; Avgeropoulos, A.; Sakellariou, G. Non-covalent functionalization of carbon nanotubes with polymers. *RSC Adv.* **2014**, *4*, 2911–2934. [CrossRef]
11. Ata, M.S.; Poon, R.; Syed, A.M.; Milne, J.; Zhitomirsky, I. New developments in non-covalent surface modification, dispersion and electrophoretic deposition of carbon nanotubes. *Carbon* **2018**, *130*, 584–598. [CrossRef]
12. Star, A.; Stoddart, J.F. Dispersion and Solubilization of Single-Walled Carbon Nanotubes with a Hyperbranched Polymer. *Macromolecules* **2002**, *35*, 7516–7520. [CrossRef]
13. Caminade, A.-M.; Majoral, J.-P. Dendrimers and nanotubes: A fruitful association. *Chem. Soc. Rev.* **2010**, *39*, 2034–2047. [CrossRef] [PubMed]
14. Chen, M.-L.; Chen, M.-L.; Chen, X.-W.; Wang, J.-H. Functionalization of MWNTs with Hyperbranched PEI for Highly Selective Isolation of BSA. *Macromol. Biosci.* **2010**, *10*, 906–915. [CrossRef]
15. Fréchet, J.M.J.; Tomalia, D.A. *Dendrimers and Other Dendritic Polymers*; J Wiley & Sons: Chichester, UK, 2001.
16. Pedziwiatr-Werbicka, E.; Milowska, K.; Dzmitruk, V.; Ionov, M.; Shcharbin, D.; Bryszewska, M. Dendrimers and hyperbranched structures for biomedical applications. *Eur. Polym. J.* **2019**, *119*, 61–73. [CrossRef]
17. Paleos, C.M.; Tsiourvas, D.; Sideratou, Z. Triphenylphosphonium decorated liposomes and dendritic polymers: Prospective second generation drug delivery systems for targeting mitochondria. *Mol. Pharm.* **2016**, *13*, 2233–2241. [CrossRef] [PubMed]
18. Yudovin-Farber, I.; Golenser, J.; Beyth, N.; Weiss, E.I.; Domb, A.J. Quaternary ammonium polyethyleneimine: Antibacterial activity. *J. Nanomater.* **2010**, *2010*. [CrossRef]
19. Mammen, M.; Choi, S.-K.; Whitesides, G.M. Polyvalent interactions in biological systems: Implications for design and use of multivalent ligands and inhibitors. *Angew. Chem. Int. Ed.* **1998**, *37*, 2755–2794. [CrossRef]
20. Maleki Dizaj, S.; Mennati, A.; Jafari, S.; Khezri, K.; Adibkia, K. Antimicrobial activity of carbon-based nanoparticles. *Adv. Pharm. Bull.* **2015**, *5*, 19–23.
21. Al-Jumaili, A.; Alancherry, S.; Bazaka, K.; Jacob, M.V. Review on the antimicrobial properties of carbon nanostructures. *Materials* **2017**, *10*, 1066. [CrossRef]
22. Maas, M. Carbon Nanomaterials as Antibacterial Colloids. *Materials* **2016**, *9*, 617. [CrossRef]
23. Kang, S.; Pinault, M.; Pfefferle, L.D.; Elimelech, M. Single-walled carbon nanotubes exhibit strong antimicrobial activity. *Langmuir* **2007**, *23*, 8670–8673. [CrossRef] [PubMed]

24. Mocan, T.; Matea, C.T.; Pop, T.; Mosteanu, O.; Buzoianu, A.D.; Suciu, S.; Puia, C.; Zdrehus, C.; Iancu, C.; Mocan, L. Carbon nanotubes as anti-bacterial agents. *Cell. Mol. Life Sci.* **2017**, *74*, 3467–3479. [CrossRef] [PubMed]
25. Kang, S.; Herzberg, M.; Rodrigues, D.F.; Elimelech, M. Antibacterial effects of carbon nanotubes: Size does matter! *Langmuir* **2008**, *24*, 6409–6413. [CrossRef] [PubMed]
26. Yang, C.; Mamouni, J.; Tang, Y.; Yang, L. Antimicrobial activity of single-walled carbon nanotubes: Length effect. *Langmuir* **2010**, *26*, 16013–16019. [CrossRef]
27. Kang, S.; Mauter, M.S.; Elimelech, M. Physicochemical determinants of multiwalled carbon nanotube bacterial cytotoxicity. *Environ. Sci. Technol.* **2008**, *42*, 7528–7534. [CrossRef]
28. Arias, L.R.; Yang, L.J. Inactivation of bacterial pathogens by carbon nanotubes in suspensions. *Langmuir* **2009**, *25*, 3003–3012. [CrossRef]
29. Baek, S.; Joo, S.H.; Su, C.; Toborek, M. Antibacterial effects of graphene- and carbon-nanotube-based nanohybrids on *Escherichia coli*: Implications for treating multidrug-resistant bacteria. *J. Environ. Manag.* **2019**, *247*, 214–223. [CrossRef]
30. Xia, L.; Xu, M.; Cheng, G.; Yang, L.; Guo, Y.; Li, D.; Fang, D.; Zhang, Q.; Liu, H. Facile construction of Ag nanoparticles encapsulated into carbon nanotubes with robust antibacterial activity. *Carbon* **2018**, *130*, 775–781. [CrossRef]
31. Atiyah, A.A.; Haider, A.J.; Dhahi, R.M. Cytotoxicity properties of functionalised carbon nanotubes on pathogenic bacteria. *IET Nanobiotechnol.* **2019**, *13*, 597–601. [CrossRef]
32. Baia, Y.; Park, I.S.; Lee, S.J.; Wen, P.S.; Bae, T.S.; Lee, M.H. Effect of AOT-assisted multi-walled carbon nanotubes on antibacterial activity. *Colloids Surf. B* **2012**, *89*, 101–107. [CrossRef]
33. Deng, R.; Zhu, Y.; Hou, J.; White, J.C.; Gardea-Torresdey, J.L.; Lin, D. Antagonistic toxicity of carbon nanotubes and pentachlorophenol to *Escherichia coli*: Physiological and transcriptional responses. *Carbon* **2019**, *145*, 658–667. [CrossRef]
34. Trompeta, A.-F.A.; Preiss, I.; Ben-Ami, F.; Benayahu, Y.; Charitidis, C.A. Toxicity testing of MWCNTs to aquatic organisms. *RSC Adv.* **2019**, *9*, 36707–36716. [CrossRef]
35. Ganguly, P.; Breen, A.; Pillai, S.C. Toxicity of nanomaterials: Exposure, pathways, assessment, and recent advances. *ACS Biomater. Sci. Eng.* **2018**, *4*, 2237–2275. [CrossRef]
36. Wang, X.P.; Han, H.Y.; Liu, X.Q.; Gu, X.X.; Chen, K.; Lu, D.L. Multi-walled carbon nanotubes can enhance root elongation of wheat (*Triticum aestivum*) plants. *J. Nanopart. Res.* **2012**, *14*, 841–850. [CrossRef]
37. Schwab, F.; Bucheli, T.D.; Lukhele, L.P.; Magrez, A.; Nowack, B.; Sigg, L.; Knauer, K. Are carbon nanotube effects on green algae caused by shading and agglomeration? *Environ. Sci. Technol.* **2011**, *45*, 6136–6144. [CrossRef]
38. Wei, L.; Thakkar, M.; Chen, Y.; Ntim, S.A.; Mitra, S.; Zhang, X. Cytotoxicity effects of water dispersible oxidized multiwalled carbon nanotubes on marine alga *Dunaliella Tertiolecta*. *Aquat. Toxicol.* **2010**, *100*, 194–201. [CrossRef]
39. Timofeeva, A.V.; Tashlitsky, V.N.; Tkachev, A.G.; Baratova, L.A.; Koksharova, O.A. Nanocomplexes on the basis of taunit associated with biocides as effective anti-cyanobacterial agents. *Russ. J. Plant Physiol.* **2017**, *64*, 833–838. [CrossRef]
40. Sideratou, Z.; Tsiourvas, D.; Paleos, C.M. Quaternized poly(propylene imine) dendrimers as novel pH-sensitive controlled-release systems. *Langmuir* **2000**, *16*, 1766–1769. [CrossRef]
41. Sapalidis, A.; Sideratou, Z.; Panagiotaki, K.N.; Sakellis, E.; Kouvelos, E.P.; Papageorgiou, S.; Katsaros, F. Fabrication of antibacterial PVA nanocomposite films containing dendritic polymer functionalized multi-walled carbon nanotubes. *Front. Mater.* **2018**, *5*. [CrossRef]
42. Cao, X.; Li, Z.; Song, X.; Cui, X.; Cao, P.; Liu, H.; Cheng, F.; Chen, Y. Core-shell type multiarm star poly(ε-caprolactone) with high molecular weight hyperbranched polyethylenimine as core: Synthesis, characterization and encapsulation properties. *Eur. Polym. J.* **2008**, *44*, 1060–1070. [CrossRef]
43. Bellamy, L. *The Infra-Red Spectra of Complex Molecules*; Springer: Amsterdam, The Netherlands, 1975. [CrossRef]
44. Arkas, M.; Tsiourvas, D. Organic/inorganic hybrid nanospheres based on hyperbranched poly(ethyleneimine) encapsulated into silica for the sorption of toxic metal ions and polycyclic aromatic hydrocarbons from water. *J. Hazard. Mater.* **2009**, *170*, 35–42. [CrossRef] [PubMed]

45. Pigorsch, E. Spectroscopic characterisation of cationic quaternary ammonium starches. *Starke* **2009**, *61*, 129–138. [CrossRef]
46. Wepasnick, K.A.; Smith, B.A.; Bitter, J.L.; Fairbrother, D.H. Chemical and structural characterization of carbon nanotube surfaces. *Anal. Bioanal. Chem.* **2010**. [CrossRef] [PubMed]
47. Yuan, W.; Jiang, G.; Che, J.; Qi, X.; Xu, R.; Chan-Park, M.B. Deposition of Silver Nanoparticles on Multiwalled Carbon Nanotubes Grafted with Hyperbranched Poly(amidoamine) and Their Antimicrobial Effects. *J. Phys. Chem. C* **2008**, *112*, 18754–18759. [CrossRef]
48. Murugan, E.; Vimala, G. Effective functionalization of multiwalled carbon nanotube with amphiphilic poly(propyleneimine) dendrimer carrying silver nanoparticles for better dispersability and antimicrobial activity. *J. Colloid Interface Sci.* **2011**, *357*, 354–365. [CrossRef] [PubMed]
49. Zhou, X.; Chen, Z.; Yan, D.; Lu, H. Deposition of Fe–Ni nanoparticles on polyethyleneimine-decorated graphene oxide and application in catalytic dehydrogenation of ammonia borane. *J. Mater. Chem.* **2012**, *22*, 13506–13516. [CrossRef]
50. Zhang, N.; Xie, J.; Guers, M.; Varadan, V.K. Chemical bonding of multiwalled carbon nanotubes to SU-8 via ultrasonic irradiation. *Smart Mater. Struct.* **2003**, *12*, 260–263. [CrossRef]
51. Schierz, A.; Zänker, H. Aqueous suspensions of carbon nanotubes: Surface oxidation, colloidal stability and uranium sorption. *Environ. Pollut.* **2009**, *157*, 1088–1094. [CrossRef] [PubMed]
52. Etika, K.C.; Cox, M.A.; Grunlan, J.C. Tailored dispersion of carbon nanotubes in water with pH-responsive polymers. *Polymer* **2010**, *51*, 1761–1770. [CrossRef]
53. Yu, J.; Grossiord, N.; Koning, C.E.; Loos, J. Controlling the dispersion of multi-wall carbon nanotubes in aqueous surfactant solution. *Carbon* **2007**, *45*, 618–623. [CrossRef]
54. Zhang, W.; Chen, M.; Gong, X.; Diao, G. Universal water-soluble cyclodextrin polymer–carbon nanomaterials with supramolecular recognition. *Carbon* **2013**, *61*, 154–163. [CrossRef]
55. Moon, Y.K.; Lee, J.; Lee, J.K.; Kim, T.K.; Kim, S.H. Synthesis of length-controlled aerosol carbon nanotubes and their dispersion stability in aqueous solution. *Langmuir* **2009**, *25*, 1739–1743. [CrossRef] [PubMed]
56. Schwyzer, I.; Kaegi, R.; Sigg, L.; Nowack, B. Colloidal stability of suspended and agglomerate structures of settled carbon nanotubes in different aqueous matrices. *Water Res.* **2013**, *47*, 3910–3920. [CrossRef] [PubMed]
57. Bhattacharjee, S. DLS and zeta potential—What they are and what they are not? *J. Control. Release* **2016**, *235*, 337–351. [CrossRef] [PubMed]
58. Chen, H.; Wang, B.; Gao, D.; Guan, M.; Zheng, L.; Ouyang, H.; Chai, Z.; Zhao, Y.; Feng, W. Broad-Spectrum Antibacterial Activity of Carbon Nanotubes to Human Gut Bacteria. *Small* **2013**, *9*, 2735–2746. [CrossRef] [PubMed]
59. Liu, D.; Mao, Y.; Ding, L. Carbon nanotubes as antimicrobial agents for water disinfection and pathogen control. *J. Water Health* **2018**, *16*, 171–180. [CrossRef]
60. Zardini, H.Z.; Amiri, A.; Shanbedi, M.; Maghrebi, M.; Baniadam, M. Enhanced antibacterial activity of amino acids-functionalized multi walled carbon nanotubes by a simple method. *Colloids Surf. B* **2012**, *92*, 196–202. [CrossRef]
61. Zardini, H.Z.; Davarpanah, M.; Shanbedi, M.; Amiri, A.; Maghrebi, M.; Ebrahimi, L. Microbial toxicity of ethanolamines—Multiwalled carbon nanotubes. *J. Biomed. Mater. Res. Part A* **2014**, *102*, 1774–1781. [CrossRef]
62. Gottenbos, B.; van der Mei, H.C.; Klatter, F.; Nieuwenhuis, P.; Busscher, H.J. In Vitro and In Vivo antimicrobial activity of covalently coupled quaternary ammonium silane coatings on silicone rubber. *Biomaterials* **2002**, *23*, 1417–1423. [CrossRef]
63. Tamayo-Belda, M.; González-Pleiter, M.; Pulido-Reyes, G.; Martin-Betancor, K.; Leganés, F.; Rosal, R.; Fernández-Piñas, F. Mechanism of the toxic action of cationic G5 and G7 PAMAM dendrimers in the cyanobacterium Anabaena sp. PCC7120. *Environ. Sci. Nano* **2019**, *6*, 863–878. [CrossRef]
64. Chen, C.Z.S.; Beck-Tan, N.C.; Dhurjati, P.; van Dyk, T.K.; LaRossa, R.A.; Cooper, S.L. Quaternary ammonium functionalized poly(propylene imine) dendrimers as effective antimicrobials: Structure-activity studies. *Biomacromolecules* **2000**, *1*, 473–480. [CrossRef] [PubMed]
65. Herrero, A.; Flores, E. *The Cyanobacteria: Molecular Biology, Genomics and Evolution*; Caister Academic Press: Norfolk, UK, 2008.
66. Wieg, C.; Pflugmacher, S. Ecotoxicological effects of selected cyanobacterial secondary metabolites a short review. *Toxicol. Appl. Pharmacol.* **2005**, *203*, 201–218.

67. Vernon, L.P.; Shaw, E.R. Photoreduction of 2,6-dichlorophenolindophenol by diphenylcarbazide: A Photosystem 2 reaction catalyzed by tris-washed chloroplasts and subchloroplast fragments. *Plant Physiol.* **1969**, *44*, 1645–1649. [CrossRef] [PubMed]
68. Papageorgiou, G.C. Rapid permeabilization of anacystis nidulans to electrolytes. *Meth. Enzymol.* **1988**, *167*, 259–262.
69. Kumazawa, S.; Mitsui, A. Photosynthetic activities of a synchronously grown aerobic N_2-fixing unicellular cyanobacterium, *Synechococcus* sp. Miami BG 043511. *J. Gen. Microbiol.* **1992**, *138*, 467–472. [CrossRef]
70. Zheng, M.; Diner, B.A. Solution Redox Chemistry of Carbon Nanotubes. *J. Am. Chem. Soc.* **2004**, *126*, 15490–15494. [CrossRef]
71. Brand, J.; Baszynski, T.; Crane, F.L.; Krogmann, D.W. Selective inhibition of photosynthetic reactions by polycations. *J. Biol. Chem.* **1972**, *247*, 2814–2819.
72. Schreiber, U.; Klughammer, C.; Neubauer, C. Measuring P700 absorbance changes around 830 nm with a new type of pulse modulation system. *Z. Naturforsch. C* **1988**, *43*, 686–698. [CrossRef]
73. Sekar, N.; Umasankar, Y.; Ramasamy, R.P. Photocurrent generation by immobilized cyanobacteria via direct electron transport in photo-bioelectrochemical cells. *Phys. Chem. Chem. Phys.* **2014**, *16*, 7862–7871. [CrossRef]
74. Goldbeck, J.C.; Victoria, F.N.; Motta, A.; Savegnago, L.; Jacob, R.G.; Perin, G.; Lenardão, E.J.; da Silva, W.P. Bioactivity and morphological changes of bacterial cells after exposure to 3-(p-chlorophenyl)thio citronellal. *LWT* **2014**, *59*, 813–819. [CrossRef]
75. Stamatakis, K.; Papageorgiou, G.C. The osmolality of the cell suspension regulates phycobilisome-to-photosystem I excitation transfers in Cyanobacteria. *Biochim. Biophys. Acta* **2001**, *1506*, 172–181. [CrossRef]
76. Moran, P. Formulae for determination of chlorophyllous pigments extracted with N,N- Dimethylformamide. *Plant Physiol.* **1982**, *69*, 1376–1381. [CrossRef] [PubMed]
77. Trebst, A.; Pistorius, E. Photosynthetische reaktionen in UV-bestrahlten chloroplasten. *Z. Naturforsch.* **1965**, *20b*, 885–889. [CrossRef]

Article

Potential of Cell-Free Supernatant from *Lactobacillus plantarum* NIBR97, Including Novel Bacteriocins, as a Natural Alternative to Chemical Disinfectants

Sam Woong Kim [1], Song I. Kang [1], Da Hye Shin [1], Se Yun Oh [1], Chae Won Lee [2], Yoonyong Yang [2], Youn Kyoung Son [2], Hee-Sun Yang [2], Byoung-Hee Lee [2], Hee-Jung An [3], In Sil Jeong [4,*] and Woo Young Bang [2,*]

1. Gene Analysis Center, Gyeongnam National University of Science & Technology, Jinju 52725, Korea; swkim@gntech.ac.kr (S.W.K.); mole160104@naver.com (S.I.K.); nini1114@naver.com (D.H.S.); ks-sy0809@naver.com (S.Y.O.)
2. National Institute of Biological Resources (NIBR), Environmental Research Complex, Incheon 22689, Korea; chaewon326@korea.kr (C.W.L.); tazemenia@korea.kr (Y.Y.); sophy004@korea.kr (Y.K.S.); moeicy@korea.kr (H.-S.Y.); dpt510@korea.kr (B.-H.L.)
3. Department of Pathology, CHA Bundang Medical Center, CHA University, Seongnam 13496, Korea; hjahn@cha.ac.kr
4. Center for Immune Cell Research, CHA Advanced Research Institute, Seongnam 13488, Korea

* Correspondence: insiljeong@gmail.com (I.S.J.); wybang@korea.kr (W.Y.B.); Tel.: +82-31-881-7374 (I.S.J.); +82-32-590-7206 (W.Y.B.)

Received: 6 August 2020; Accepted: 21 September 2020; Published: 23 September 2020

Abstract: The recent pandemic of coronavirus disease 2019 (COVID-19) has increased demand for chemical disinfectants, which can be potentially hazardous to users. Here, we suggest that the cell-free supernatant from *Lactobacillus plantarum* NIBR97, including novel bacteriocins, has potential as a natural alternative to chemical disinfectants. It exhibits significant antibacterial activities against a broad range of pathogens, and was observed by scanning electron microscopy (SEM) to cause cellular lysis through pore formation in bacterial membranes, implying that its antibacterial activity may be mediated by peptides or proteins and supported by proteinase K treatment. It also showed significant antiviral activities against HIV-based lentivirus and influenza A/H3N2, causing lentiviral lysis through envelope collapse. Furthermore, whole-genome sequencing revealed that NIBR97 has diverse antimicrobial peptides, and among them are five novel bacteriocins, designated as plantaricin 1 to 5. Plantaricin 3 and 5 in particular showed both antibacterial and antiviral activities. SEM revealed that plantaricin 3 causes direct damage to both bacterial membranes and viral envelopes, while plantaricin 5 damaged only bacterial membranes, implying different antiviral mechanisms. Our data suggest that the cell-free supernatant from *L. plantarum* NIBR97, including novel bacteriocins, is potentially useful as a natural alternative to chemical disinfectants.

Keywords: AMP; antimicrobial activity; antiviral activity; bacteriocin; COVID-19; disinfectant; *Lactobacillus plantarum*; plantaricin

1. Introduction

Severe acute respiratory syndrome coronavirus 2 (SARS-CoV-2), responsible for the global pandemic of coronavirus disease 2019 (COVID-19), is the foremost concern among recent global health issues [1]. For prevention of this infection, disinfectants have been widely used—mainly because SARS-CoV-2, like other coronaviruses and enveloped viruses, is surrounded by a fragile outer lipid envelope, which makes it more susceptible to disinfectants than non-enveloped viruses such as rotavirus, norovirus, and poliovirus [2]. Accordingly, the pandemic of COVID-19 has led to a large surge in demand for disinfectants, especially chemical disinfectants such as alcohol- or chlorine-based formulas for the disinfection of hands or environmental surfaces [3–5]. Although chemical disinfectants are considered very effective, they could be hazardous to users if they are not properly handled; for example, alcohol-based disinfectants are flammable and can be harmful to humans if they enter the body [3]. For this reason, there is increasing interest in disinfectants based on natural products.

Lactic acid bacteria, traditionally used in fermented foods, have been considered as interesting resources to contribute to developing a safe alternative to biocides, which are potentially hazardous to humans, because they produce diverse antimicrobial substances and are seldom hazardous to humans [6,7]; most are approved by the U.S. Food and Drug Administration as GRAS (Generally Recognized as Safe). As typical antimicrobial substances, they secrete lactic acid with bacteriocins and antimicrobial peptides (AMPs), which are produced by most microbes [6,7]. In particular, bacteriocins, such as nisin, sakacin, plantaricin, and leucocin from lactic acid bacteria have been reported to have antibacterial activity against foodborne bacteria, such as *Escherichia coli*, *Salmonella enterica*, and *Listeria monocytogenes*, and thus many studies have highlighted their application as natural alternatives to artificial preservatives and antibiotics [6,8–10]. In addition, several bacteriocins have shown antiviral activities against pathogenic viruses such as poliovirus, herpes simplex virus, and influenza viruses [10–12]. Accordingly, the cell-free supernatant, including the bacteriocins and lactic acid, has potential as a natural alternative to chemical disinfectants, although there have been no attempts to apply it as a disinfectant as of yet. To the best of our knowledge, this report is the first that addresses these issues.

In this study, we first suggest that the cell-free supernatant from *Lactobacillus plantarum* NIBR97, a lactic acid bacterium isolated from kimchi, a Korean fermented food, could potentially be useful for disinfection against both pathogenic bacteria and viruses, mediated by bacteriocins as well as lactic acid. Through the genomic analysis of the NIBR97 strain, we discovered novel bacteriocins functioning as antibacterial and antiviral peptides. Our study will provide important information that will guide new strategies to replace chemical disinfectants with natural substances.

2. Results

2.1. Antibacterial Activity of Cell-Free Supernatant from L. plantarum NIBR97

Lactobacillus plantarum NIBR97 were screened from kimchi as a strain with superior antibacterial activity, and its cell-free supernatant was further used for the examination of antibacterial activity, as shown in Figure 1. The minimum inhibitory concentrations (MIC50 and MIC90) were determined as 30.04 and 67.43 μg total proteins/mL against *Salmonella enterica* Serovar Enteritidis (*S.* Enteritidis), respectively, which indicates significantly higher antibacterial activity than the three *Lactobacillus plantarum* strains, KCTC33131, KCTC21004, and KCTC13093, with higher MIC50 and MIC90 values than NIBR97 (Figures 1A and S1A). The cell-free supernatant also showed MIC50s and MIC90s against *Salmonella* Gallinarum, *Edwardsiella tarda*, *Pasteurella multocida*, and *Streptococcus iniae* (Figures 1B and S1B), implying antibacterial activity against broad pathogenic bacteria. In addition, when *Escherichia coli* and *Staphylococcus aureus* were treated with the cell-free supernatant for 5 min, they showed a reduction of at least 99.9% ($\geq 3 \log_{10}$) of the total count in the original inoculum (Figure 1C), indicating bactericidal activity and potential as a disinfectant.

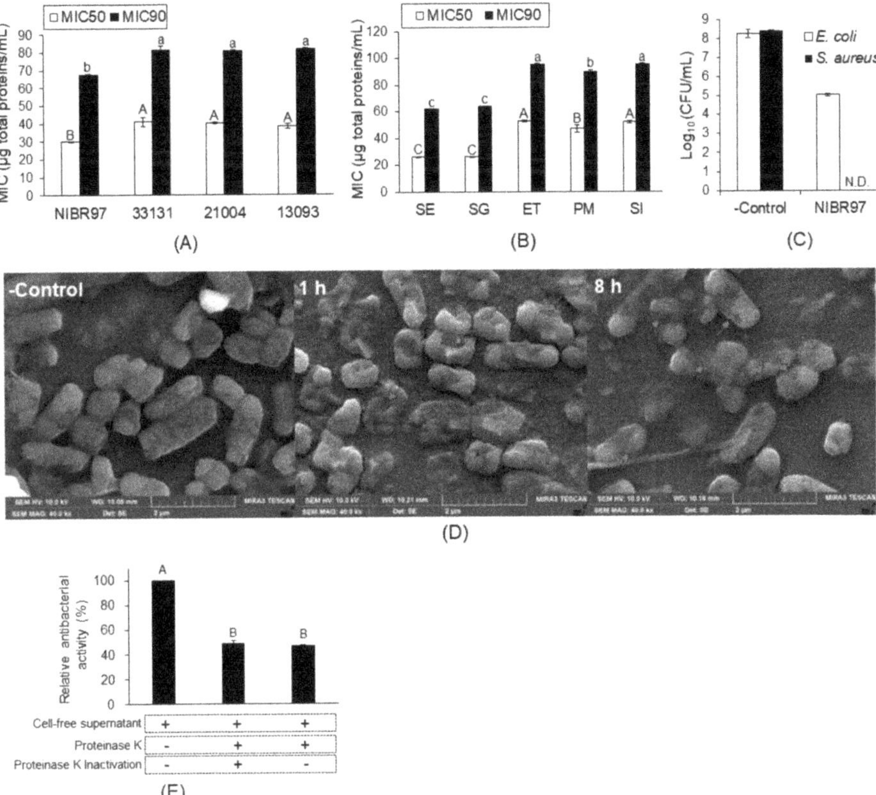

Figure 1. Antibacterial activity of cell-free supernatant from *Lactobacillus plantarum* NIBR97. Antibacterial activities of the cell-free supernatant from the *L. plantarum* strains NIBR97, KCTC33131, KCTC21004, and KCTC13093 were examined against *Salmonella* Enteritidis, whose MIC50s and MIC90s were determined (**A**). The MIC50s and MIC90s of the cell-free supernatant were determined against *Salmonella* Gallinarum (SG), *Edwardsiella tarda* (ET), *Pasteurella multocida* (PM), and *Streptococcus iniae* (SI), as well as *S.* Enteritidis (SE) (**B**). For bactericidal activity, *Escherichia coli* and *Staphylococcus aureus* were treated with the cell-free supernatant (126.6 μg total proteins/mL) for 5 min, and then were counted to determine the titer (Log_{10} (colony-forming unit (CFU)/mL) and reduction rate (%) (**C**). For scanning electron microscopy, *S.* Enteritidis was treated without (control) or with the cell-free supernatant (70.8 μg total proteins/mL, MIC against *S.* Enteritidis) for 1 h and 8 h (**D**). The red arrows indicate the pores forming in the *Salmonella* membrane. (**E**) To investigate the effect of protease on the antibacterial activity of the cell-free supernatant, we added the proteinase K (100 μg/mL) to the cell-free supernatant at 70.8 μg total proteins/mL and the treated sample was used to examine its antibacterial activity against *S.* Enteritidis. In (**E**), the plus (+) mark indicates the treatment of cell-free supernatant or proteinase K, whereas the minus (−) mark does no treatment, and the proteinase K was inactivated at 80 °C for 10 min (+) or not (−). The different letters (A, B, C, a, b and c) in the graphs ((**A**), (**B**), (**C**) and (**E**)) represent significant differences ($p < 0.05$) and in (**A**) and (**B**), the capital (A, B and C) and small letters (a, b c) indicate the significant differences in MIC50 and MIC90 data, respectively.

In order to prove the antibacterial activity against pathogenic bacteria with the cell-free supernatant from *L. plantarum* NIBR97, we observed the *S.* Enteritidis treated with the cell-free supernatant using scanning electron microscopy (SEM). As shown in Figure 1D, the SEM images revealed that the cell-free supernatant effectively caused the Salmonella death via pore formation by cellular penetrating peptides, as is the case for typical AMPs [13]. Furthermore, when the cell-free supernatant was treated with proteinase K, its antibacterial activity against S. Enteritidis decreased by about 50% compared with the control without the proteinase K treatment (Figure 1E). Therefore, we suggest that proteins or peptides play major roles for the antibacterial activities of cell-free supernatant from *L. plantarum* NIBR97.

2.2. *Antiviral Activity of Cell-Free Supernatant from L. plantarum NIBR97*

To assess its antiviral activity, the cell-free supernatant from *L. plantarum* NIBR97 was exposed to green fluorescent protein (GFP)-labeled lentiviruses, based on human immunodeficiency virus (HIV), which causes acquired immunodeficiency syndrome (AIDS), for 5 min and 24 h, as shown in Figure 2. When the GFP-labeled lentiviruses, treated with the cell-free supernatant, infected the HEK-293T cells (human host cells), they were observed by fluorescence microscopy to decrease dose- and time-dependently within the host cells (Figure 2A, the GFP images) without any cytotoxic effect on the human host cells (Figure 2A, the Bright images). SEM also confirmed its antiviral activity by showing that the cell-free supernatant effectively causes lentiviral lysis through the collapse of envelopes after 5 min (Figure 2B). In addition, when the human influenza A virus subtype H3N2 (A/H3N2) was treated with the cell-free supernatant, it showed a reduction of at least 99.5% of the total count of its original inoculums, which increased until 99.999% with treatment time (Table 1). These results indicate that the cell-free supernatant from *L. plantarum* NIBR97 has superior antiviral activity, as well as potential as a disinfectant.

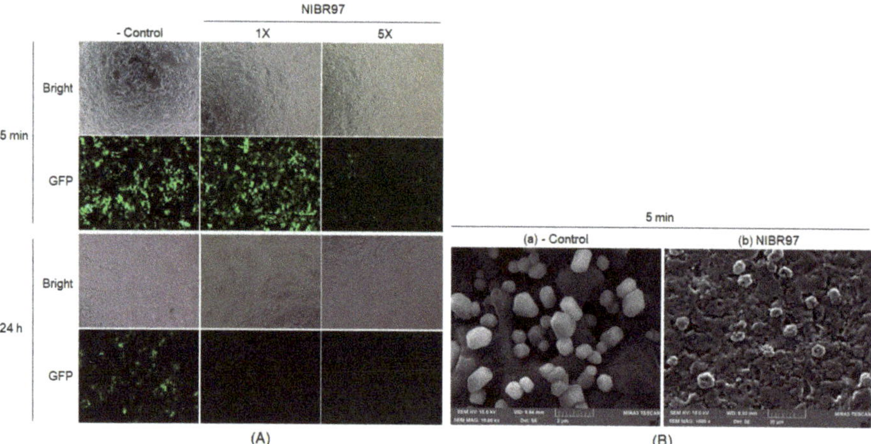

Figure 2. Antiviral activity of cell-free supernatant from *L. plantarum* NIBR97. (**A**) For fluorescence microscopy, we treated GFP-labeled lentiviruses with the cell-free supernatant for 5 min and 24 h, and then were infected in HEK-293T human host cells. The 1× and 5× correspond to the concentrations treated to the lentiviruses, 79.15 and 395.75 μg total proteins/mL, respectively. The bright-field images (Bright) indicate the HEK-293T cells, and the green signals in the fluorescent images (GFP) represent the GFP-labeled lentiviruses. (**B**) For scanning electron microscopy, the GFP-labeled lentiviruses were treated without (**a**) or with the cell-free supernatant (395.75 μg total proteins/mL) (**b**) for 5 min.

Table 1. Disinfection activity of the cell-free supernatant from *L. plantarum* NIBR97 against A/H3N2.

Treatments [1]	10 min [1]		30 min [1]		18 h [1]	
	Titer [2]	Reduction [3]	Titer [2]	Reduction [3]	Titer [2]	Reduction [3]
Water	5.66	0	5.45	0	5.34	0.21
NIBR97	3.27	99.594	<0.51	>99.999	<0.51	>99.999

[1] The A/H3N2 viruses were treated with water, a negative control, or the cell-free supernatant (NIBR97) for 10 min, 30 min, and 18 h; [2] and [3] indicate the viral titer (log_{10}CCID50) and reduction (%), respectively.

2.3. Discovery of Novel Bacteriocins by the Genomic Analysis of L. plantarum NIBR97

Analysis of the whole-genome sequence for the *L. plantarum* NIBR97 was carried out by the PacBio RS II (Pacific Biosciences, Menlo Park, CA, USA) sequencing platform to identify the AMPs from the NIBR97. The NIBR97 genome identified from de novo assembly was composed of a single circular bacterial chromosome and four plasmids, containing 2927 predicted open reading frames (ORFs), 68 tRNAs, and 16 rRNAs (Table 2 and Figure S2). Among the ORFs, 10 were identified to encode homologous proteins with known AMPs via an NCBI (National Center for Biotechnology Information) homology BLAST (Basic Local Alignment Search Tool) (Table S1). Furthermore, their expression in *L. plantarum* NIBR97 was confirmed by the transcriptomic data (Table S2). In detail, the five ORFs—orf02155, orf02163, orf02164, orf02421, and orf00645—were found to have 100% identities with plantaricin N, F, and E, as well as bacteriophage holing and lysozyme, known previously as AMPs from the *Lactobacillus* genus (Table S1). Five ORFs—orf00467, orf01336, orf01363, orf01599, and orf01790—which were previously uncharacterized until now, were discovered in this study to consist of amino acid sequences with high positives (>60%) with AMPs undiscovered in *L. plantarum* strains: grammistin Pp3, indolicidin, bactofencin A, hymenochirin-5B, and latarcin-2a, (Table S1). Thus, we herein designated the AMPs as novel bacteriocins called plantaricin (Pln) 1, 2, 3, 4, and 5 (Table S1). Interestingly, their structural models revealed that the three plantaricins—Pln 1, 4 and 5—form helix structures, and the two plantaricins—Pln 2 and 3—form random coil structures (Figure 3), similar to typical AMPs [14,15], implying that they may have antibacterial activities.

2.4. Antibacterial and Antiviral Activities of Plantaricins from L. plantarum NIBR97

To confirm whether the five Plns function as AMPs, we assessed their synthetic peptides for antibacterial activity against *Salmonella* Typhimurium (Figure S3). Among them, Pln 5 exhibited the highest antimicrobial activity, showing the lowest MIC50 compared with others, whereas Pln 4 showed the lowest antimicrobial activity (Figure S3). In addition, the Pln 3 and 5 were identified to inhibit the growth of *Salmonella* Enteritidis (Figure 4A) and were observed by SEM to effectively cause cellular lysis by damaging the membrane of *S.* Enteritidis via pore formation (Figure 4B), as did the cell-free supernatant from *L. plantarum* NIBR97 (Figure 1D).

Table 2. Summary of the de novo genome assembly of *L. plantarum* NIBR97.

Items	Contig 1	Contig 2	Contig 3	Contig 4	Contig 5
Form	A circular chromosome	A circular plasmid	A circular plasmid	A linear plasmid	A linear plasmid
Length [1]	3,022,780	61,378	32,520	7394	6876
GC [2]	44.74	39.22	39.59	34.33	35.67
ORF [3]	2816	60	32	10	9
rRNA [4]	16	0	0	0	0
tRNA [5]	68	0	0	0	0

[1] and [2] indicate the length (bp, base pair) and GC (guanine-cytosine) contents (%) of contig in the form, respectively; [3], [4], and [5] represent the number of predicted open reading frames (ORFs), rRNA, and tRNA, respectively.

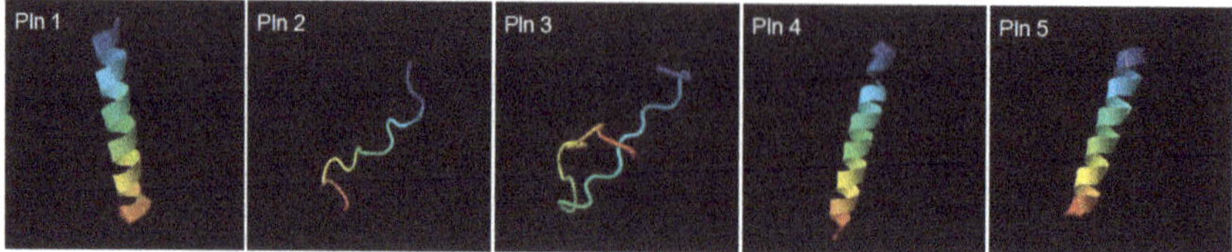

Figure 3. Structural models of plantaricins. Pln 1, 2, 3, 4, and 5 comprise the amino acid sequences VLGSLIGSVGIGVLSSLAARYK, IYPEKQPEEPVRR, KKSRRCQVYNNGMPTGMYTSC, PIVREPFKAMAVGIILAVMSGLLVT, and KAKKRFLRNRLSQQARKARTK, respectively. Pln 1, 4, and 5 form helix structures, and Pln 2 and 3 form random coil structures. The structures of Pln 1, 2, 3, 4, and 5 were predicted by the automated I-TASSER server (https://zhanglab.ccmb.med.umich.edu/I-TASSER/).

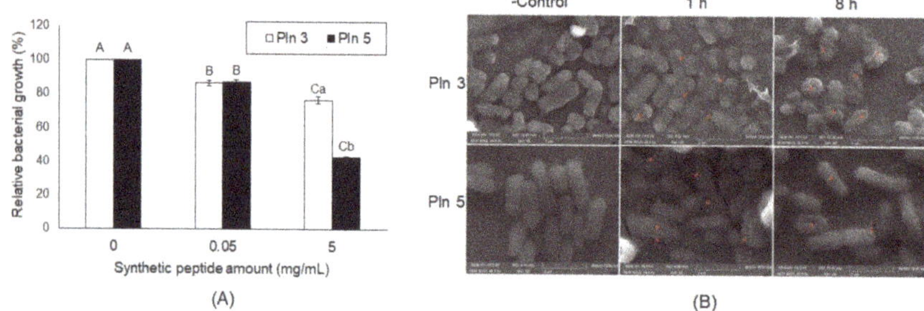

Figure 4. Antibacterial activity of plantaricin 3 and 5 against *S.* Enteritidis. Pln 3 and 5 were synthesized according to the amino acid sequences in Figure 3, and further examined for their antibacterial activity against *S.* Enteritidis (**A**). The *y*-axis and different letters (A, B, C, a and b) in the graphs represent the relative bacterial growth (%) and significant differences ($p < 0.05$), respectively. In (**A**), the capital (A, B and C) and small letters (a and b) indicate the significant differences between different concentrations (0-5 mg/mL) and the ones between Pln 3 and 5, respectively. (**B**) For scanning electron microscopy, *S.* Enteritidis was treated without or with synthetic Pln 3 or 5 (5 mg/mL) for 1 h and 8 h. The red arrows indicate the pores forming in the *Salmonella* membrane.

The synthetic Pln 3 and 5 were further examined for antiviral activity against GFP-labeled lentiviruses. The synthetic peptides exhibited a cytotoxicity on the human host cells when the lentiviruses were treated with 5 μg/μL of synthetic peptides, but not with ≈2.5 μg/μL of synthetic peptides (Figure 5, the Bright images). The fluorescence microscopy revealed that the lentiviruses decreased considerably within the host cells when they were treated with the Pln 3 or 5 for 24 h, but not for 5 min (Figure 5, the GFP images). This suggests that Pln 3 and 5 can considerably suppress viral infection in host cells. Interestingly, SEM revealed that Pln 3 effectively caused lentiviral lysis through the collapse of the envelopes (Figure 6), as the cell-free supernatant did (Figure 2B), whereas Pln 5 did not (Figure 6). This implies that Pln 3 and 5 may exert their antiviral role through different mechanisms.

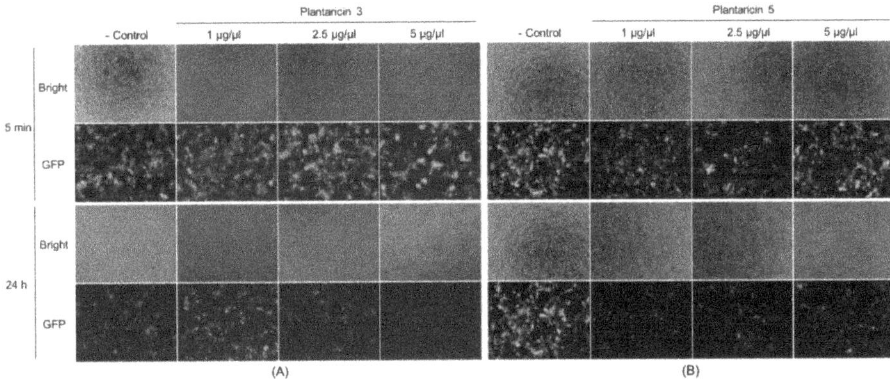

Figure 5. Fluorescence micrographs of HEK-293T cells infected with GFP-labeled lentiviruses treated with synthetic Pln 3 and 5. The lentiviruses were treated without (control) or with the synthetic peptides (1 to 5 μg/μL) Pln 3 (**A**) and 5 (**B**) for 5 min or 24 h, and then the HEK-293T human host cells were infected. The bright-field images (Bright) indicate the HEK-293T cells, and the green signals in the fluorescent images (GFP) represent the GFP-labeled lentiviruses.

Figure 6. Scanning electron micrographs of the GFP-labeled lentiviruses treated with synthetic Pln 3 and 5. The lentiviruses were treated without or with the synthetic peptides at 5 μg/μL for 24 h.

3. Discussion

Lactobacillus plantarum is one of the most widespread lactic acid bacteria species and is largely used for the production of fermented products of animal and plant origin [16]. Moreover, some strains are known to produce several natural antibacterial substances, such as bacteriocins, organic acids (mainly lactic and acetic acid), and hydrogen peroxide [17,18], and thus many studies have highlighted their application as preservatives and antibiotics [6,8–10]. Here, we investigated their potential as a natural alternative to chemical disinfectants.

In this study, the NIBR97 strain was screened from kimchi, a Korean fermented food, and its cell-free supernatant was identified to have higher antibacterial activity against *Salmonella* bacteria than other *L. plantarum* strains (Figure 1A), as well as possessing antibacterial activities against a broad range of pathogenic bacteria (Figure 1B). It exhibited significant disinfection activities against the human pathogens influenza A virus H3N2, *Escherichia coli*, and *Staphylococcus aureus*, reducing them by at least 99.9% of the total count of their original inoculums within 30 min (Figure 1C and Table 1). These results indicate that the cell-free supernatant from *L. plantarum* NIBR97 has potential as a natural disinfectant, and thus further investigations were performed to identify the antimicrobial substances, such as AMPs, in the NIBR97 strain.

AMPs are small peptides composed of 10 to 40 amino acids, which cause microbial membrane modification via either pore formation by cell-penetrating property through a barrel stave or a toroidal pore mechanism, or through a non-pore carpet-like mechanism [13,19]. Our scanning electron micrographs of *S.* Enteritidis showed clearly that the cell-free supernatant from the NIBR97 formed a pore on the *Salmonella* surface (Figure 1D), as do typical AMPs [13]. Proteinase K treatment of the cell-free supernatant led to a considerable decrease in its antibacterial activity against both *S.* Enteritidis and *S.* Gallinarum (Figure 1E). Thus, these results confirm that the antibacterial activities of the cell-free supernatant from the NIBR97 are mediated mainly by its proteins or peptides, functioning as AMPs. The scanning electron micrographs of GFP-labeled lentivirus showed that the cell-free supernatant causes lentiviral lysis through envelope collapse (Figure 2A), but it was unclear whether the AMPs were involved in the envelope collapse of the virus.

Finally, to identify AMPs from the NIBR97 strains, we performed whole-genome sequencing, which revealed that the 10 ORFs encoded AMPs, including known forms (plantaricin E, F, N; bacteriophage holin; lysozyme) and novel forms (Pln 1 to 5 (Table S1)). In the case of the known AMPs, plantaricin E, F, and N are bacteriocins produced in *Lactobacillus plantarum* C11 [20]; holin, produced by bacteriophages, triggers and controls the degradation of the cell wall of the host bacteria [21]; and lysozyme functions as 1,4-beta-N-acetylmuramidase, an antimicrobial enzyme, and has been found mainly in *Lactobacillus rhamnosus* strains [22]. Interestingly, five ORFs were discovered as novel bacteriocins in this study (Figure 3) and were designated as Pln 1, 2, 3, 4, and 5. They were further confirmed as being expressed in the NIBR97 strain through transcriptomic sequencing (Table S2), and even their synthetic peptides exhibited antibacterial activity against *Salmonella* Typhimurium (Figure S3). The synthetic Pln 3 and 5 also inhibited the growth of *S.* Enteritidis and effectively caused cellular lysis through damage to the *Salmonella* membrane via pore formation (Figure 4), suggesting that they function as AMPs. However, the synthetic Plns showed overall lower antibacterial activities than antibiotics such as octenidine when their MICs were compared with each other (Figure S3) [23]. This is presumably because the Plns were not synthesized on the basis of complete amino acid sequences for the optimal antibacterial activity but were done on the basis of minimal sequences for the activity; thus, it is further necessary to identify the mature peptide sequence responsible for the optimal antibacterial activity, following the signal peptide cleavage. Moreover, Pln 3 and 5 were identified to suppress lentiviral infection in human host cells (Figure 5). Collectively, these results suggest that the cell-free supernatant from *L. plantarum* NIBR97 may include AMPs, such as Pln 3 and 5, exhibiting antibacterial and antiviral activities. However, Pln 3 and 5 were observed by SEM to act differentially in the suppression of viral infection; Pln 3 had a significant effect on the viral shape through the collapse of the viral envelope, which suggests that it may cause direct damage to the envelope. In contrast, Pln 5 had little effect on it (Figure 6), which implies that it may interfere with the interaction between viruses and host cells [24,25].

Noticeably, Pln 3 and 5 suppressed viral infection when used against lentivirus for 24 h, but not for 5 min (Figure 5), which indicates that long exposure is required for their antiviral role. Although the Plns exhibited low antibacterial activities as mentioned above, during long expose (i.e., 24 h), they may also contribute significantly to the antibacterial activities of cell-free supernatant, together with other AMPs discovered by genomic analysis of NIBR97, which is strongly supported by the proteinase K treatment leading to a considerable decrease (>50%) in antibacterial activity of the cell-free supernatant (Figure 1E). Furthermore, this is confirmed by Figure S4—the cell-free supernatant from the *E. coli* Top10 strain (Invitrogen, Carlsbad, CA, USA), harboring each *Pln* gene cloned, showed significant antibacterial activities against both Gram-negative and Gram-positive bacteria, whereas very little antibacterial activity was detected in the negative control, that is, treatment with the cell-free supernatant from the strain without the *Pln* genes (Figure S4). Meanwhile, the disinfection activity of the cell-free supernatant during short exposures (i.e., within 30 min), as shown in Figure 1C and Table 1, was presumably because the lactic acid may have functioned as a disinfectant during the short exposure. This is supported by the data, showing that the cell-free supernatant contained considerable lactic acids (≈2%) when the NIBR97 strain was cultured in the de Man, Rogosa and Sharpe (MRS) medium, consisting of 5% solutes and 95% water, for 24 h (Figure S5), and by a previous report stating that they induce sudden severe acid stress, leading to a shock of oxidative stress and resulting in the destabilization of the bacterial membrane [26]. Therefore, the cell-free supernatant may exert its role as a disinfectant, mainly through lactic acid during short exposure (i.e., within 30 min), while it does so through an integrated effect between the lactic acid and the various AMPs during long exposure (i.e., 24 h).

4. Materials and Methods

4.1. Materials

As susceptible bacteria to AMPs, *S.* Enteritidis, *S.* Gallinarum, *Edwardsiella tarda, Pasteurella multocida,* and *Streptococcus iniae* were obtained from Dr. Jin Hur (Chonbuk National University, Iksan, Korea) and Dr. Tae Sung Jung (Gyeongsang National University, Jinju, Korea). The *Lactobacillus plantarum* strains KCTC33131, KCTC21004, and KCTC13093, as well as the susceptible bacteria *Escherichia coli* ATCC 10536 and *Staphylococcus aureus* ATCC 6538, were purchased from KCTC (Korean Collection for Type of Cultures, Daejeon, Korea). The human influenza A/H3N2 was provided by the Korea Centers for Disease Control and Prevention (KCDC, Chungcheongbuk-do, Korea). The plasmids for lentiviral packaging (two packaging vectors, pRSV-Rev and pCgpV, and an envelope vector, pCMV-VSV-G) and for a positive control of transduction (pSIH1-H1-siLUC-copGFP) were purchased from Cellbiolab (San Diego, CA, USA) and System Biosciences (Palo Alto, CA, USA), respectively. The five synthetic peptides—plantaricin 1 to 5—were purchased from Cosmogenetech Inc. (Seoul, Korea).

4.2. Analysis of the Minimal Inhibitory Concentration (MIC50 and MIC90)

L. plantarum NIBR97 was incubated at 37 °C for 24 h in an MRS liquid medium (10 g/L peptone, 8 g/L meat extract, 4 g/L yeast extract, 20 g/L d(+)-glucose, 2 g/L dipotassium hydrogen phosphate, 5 g/L sodium acetate trihydrate, 2 g/L triammonium citrate, 0.2 g/L magnesium sulfate heptahydrate, and 0.05 g/L manganous sulfate tetrahydrate). The cultural broth was centrifuged for 20 min at 2000× *g*, and the centrifugal supernatant was collected and then sterilized by a 0.22 μm filtration. The sterilized fluid was either applied directly for the examination of antimicrobial activity or fractionated and stored at −80 °C until use. The assessment of antimicrobial activity on a microtiter plate was performed by some modification of the dilution assay of Wiegand et al. [27]. The MIC50 and MIC90 were expressed as total proteins equivalent (μg) per volume (mL) of the sample, and the effect of proteinase K treatment was examined by a previously described procedure [28].

4.3. Measurement of Bactericidal Activity

The susceptible bacterial strains *Escherichia coli* ATCC 10536 and *Staphylococcus aureus* ATCC 6538 were adjusted into 1.5 to 5.0 × 10^8 CFU/mL after pre-culture, and 10% sucrose was used as an interfering agent, 0.25 M KH_2PO_4 (pH 7.2) was used as a neutralizing agent, and 20 mg/mL proteinase K was used to degrade the AMPs. For the bactericidal activity assay, we mixed 100 μL of prepared susceptible bacterial solution, 100 μL 10% sucrose, and 800 μL cell-free supernatant (126.6 μg total proteins/mL) from *L. plantarum* NIBR97 and reacted the mixture at 20 °C for 5 min. An aliquot (100 μL) of the reaction solution was mixed with 800 μL 0.25 M KH_2PO_4 (pH 7.2), 5 μL proteinase K, and 100 μL distilled water, and then reacted at 20 °C for 5 min. The surviving cells were counted by serial dilution of the treated solution and incubation on an Luria-Bertani (LB) plate.

4.4. Scanning Electron Microscopy (SEM)

The *S.* Enteritidis was treated with the cell-free supernatant (70.8 μg total proteins/mL, MIC against *S.* Enteritidis) from *L. plantarum* culture or the synthetic peptides, Pln 3 (1 μg/μL) or Pln 5 (1 μg/μL), for 0, 1, and 8 h, and the lentivirus was assessed with the cell-free supernatant (15.8 μg total proteins/mL) for 5 min and with the synthetic peptides Pln 3 (5 μg/μL) or Pln 5 (5 μg/μL) for 24 h. The treated bacteria and viruses were observed by a scanning electron microscope according to previously described procedures [28].

4.5. Antiviral Analysis Against Influenza A/H3N2

For the antiviral test, we co-incubated 0.1 mL of the A/H3N2 soup (2–4 × 10^5 viruses/μL) with 0.9 mL of the cell-free supernatant (142.5 μg total proteins/mL) for 10 min, 30 min, and 18 h at 25 °C. After the co-incubation, the cell-free supernatant-A/H3N2 mixture was 10-fold serially diluted to infect Madin–Darby canine kidney (MDCK) cells (3 × 10^4 cells per well) and, thereafter, the cell culture infectious dose (CCID50) and the viral reduction were determined by cytopathic effect (CPE) and plaque assays, as previously described [29].

4.6. Antiviral Analysis Against GFP-Labeled Lentivirus

For the production of GFP-labeled lentivirus, we transfected 5 μg of pRSV-Rev, 5 μg of pCMV-VSV-G, 5 μg of pCgpV, and 15 μg of pSIH1-H1-siLUC-copGFP plasmids into HEK-293T cells (6 × 10^6 cells per well) using lipofectamine 2000 (Invitrogen, Carlsbad, CA, USA). The lentiviral supernatants were harvested 72 h after transfection, filtered through Millex-GP 0.45 μm filters (Millipore, Schwalbach, Germany), and concentrated using Retro-Concentin Retroviral Concentration Reagent (System Biosciences, Palo Alto, CA, USA). The titer of lentiviruses was measured with a QuickTiter Lentivirus Titer Kit (Cellbiolabs, San Diego, CA, USA) and stored at −80 °C.

For the anti-viral test, we co-incubated 2 μL of lentivirus soup (2.8 × 10^6 lentiviruses/μL) with 2 μL of test sample for 5 min and 24 h at 25 °C. After the co-incubation, 2 μL from the total 4 μL of the test sample–lentivirus mixture was infected in HEK-293T cells (1 × 10^4 cells per well). Expression of the copGFP protein was observed at day 3 after infection with a Zeiss 510 fluorescence microscope (Carl Zeiss Co., Oberkochen, Germany).

4.7. Analysis of the Genome

Genomic analysis of *L. plantarum* NIBR97 was performed by previously described procedures. In detail, genomic DNA from the NIBR97 was extracted and sequenced by previously described procedures [28]. De novo assembly and putative gene coding sequences (CDSs) from the assembled contigs was performed by the hierarchical genome assembly process (HGAP, Version 3) workflow [30] and the bacterial genome was checked by MUMmer 3.5 [31], identifying that the genome comprises a single circular DNA chromosome of 3,022,780 bp with four plasmids by trimming one of the self-similar ends for manual genome closure (Table 2). Putative gene coding sequences (CDSs) from the assembled contigs were identified by Glimmer v3.02 [32], and the obtained ORFs were examined by Blastall alignment (http://www.ncbi.nlm.nih.gov/books/NBK1762). Gene ontology annotations of the ORFs were assigned by Blast2GO software [33]. In addition, ribosomal RNAs and transfer RNAs were separated by RNAmmer 1.2 and tRNAscan-SE 1.4 [34,35]. Finally, the whole-genome sequence data were deposited as Sequence Read Archive (SRA) data in GenBank (SRA no., SRR12344691; BioProject no., PRJNA647132).

4.8. Statistical Analysis

The mean values were separated by the probability difference option according to significant differences. The results are exhibited as least square means with standard deviations. Duncan's multiple range tests (MRT) were applied for verification of significant differences ($p < 0.05$) between sample types. All the analyses were performed by the SAS statistical software package (version 9.1, SAS Inst., Inc., Cary, NC, USA), for which differences were considered significant at $p < 0.05$.

5. Conclusions

Together, our data showed that the cell-free supernatant from *L. plantarum* NIBR97, producing novel bacteriocins, has superior antibacterial and antiviral activities during both short and long exposures, which suggests that it is potentially useful as a natural material to completely or partially replace chemical disinfectants.

Supplementary Materials: The following are available online at http://www.mdpi.com/1424-8247/13/10/0266/s1, Figure S1. Antibacterial activity of cell-free supernatant from *L. plantarum* NIBR97. Figure S2. Overall features of the *L. plantarum* NIBR97 genome (contig 1) and plasmids (contig 2 to 5). Figure S3. Antibacterial activity of synthetic plantaricins identified from the *L. plantarum* NIBR97 genome. Figure S4 Antibacterial activity of the cell-free supernatant from *E. coli*. Top10 strain, harboring each *Pln* gene. Figure S5. The content of lactic acid in the cell-free supernatant from *L. plantarum* NIBR97. Table S1. Identification of ORFs predicted as antimicrobial peptides (AMPs) from the genome assembly data of *L. plantarum* NIBR97. Table S2. Transcriptomic analysis results of AMPs from *L. plantarum* NIBR97.

Author Contributions: W.Y.B., I.S.J., and S.W.K. conceived and designed the experiments; S.I.K., D.H.S., S.Y.O., Y.Y., and S.W.K. performed the experiments; C.W.L., Y.K.S., H.-S.Y., and B.-H.L. analyzed the data; S.W.K., B.-H.L., H.-J.A., I.S.J., and W.Y.B contributed reagents/materials/analysis tools; I.S.J. and W.Y.B. wrote the paper. All authors have read and agreed to the published version of the manuscript.

Funding: This research was funded mainly by the National Institute of Biological Resources (NIBR), the Ministry of Environment (MOE) of the Republic of Korea, grant number NIBR202019103. I.S.J. was supported by the National Research Foundation of Korea (NRF) grant funded by the Korean government (Ministry of Science and ICT) (no. 2020R1C1C1007796).

Acknowledgments: The *S.* Gallinarum, pathogenic *E. coli*, and *S. iniae* that are susceptible to AMPs were obtained from Jin Hur (Chonbuk National University, Iksan, Republic of Korea) and Tae Sung Jung (Gyeongsang National University, Jinju, Republic of Korea).

Conflicts of Interest: The authors declare no conflict of interest.

References

1. World Health Organization. *Coronavirus Disease (COVID-19): Situation Report, 150*; World Health Organization: Geneva, Switzerland, 2020.
2. World Health Organization. *Cleaning and Disinfection of Environmental Surfaces in the Context of COVID-19: Interim Guidance*; World Health Organization: Geneva, Switzerland, 2020.
3. Atolani, O.; Baker, M.T.; Adeyemi, O.S.; Olanrewaju, I.R.; Hamid, A.A.; Ameen, O.M.; Oguntoye, S.O.; Usman, L.A. COVID-19: Critical discussion on the applications and implications of chemicals in sanitizers and disinfectants. *EXCLI J.* **2020**, *19*, 785. [PubMed]
4. Pradhan, D.; Biswasroy, P.; Ghosh, G.; Rath, G. A review of current interventions for COVID-19 prevention. *Arch. Med. Res.* **2020**, *51*, 363–374. [CrossRef]
5. Berardi, A.; Perinelli, D.R.; Merchant, H.A.; Bisharat, L.; Basheti, I.A.; Bonacucina, G.; Cespi, M.; Palmieri, G.F. Hand sanitisers amid CoViD-19: A critical review of alcohol-based products on the market and formulation approaches to respond to increasing demand. *Int. J. Pharm.* **2020**, *584*, 119431. [CrossRef] [PubMed]
6. Ibrahim, O.O. Classification of Antimicrobial Peptides Bacteriocins, and the Nature of Some Bacteriocins with Potential Applications in Food Safety and Bio-Pharmaceuticals. *EC Microbiol.* **2019**, *15*, 591–608.
7. Stanojević-Nikolić, S.; Dimić, G.; Mojović, L.; Pejin, J.; Djukić-Vuković, A.; Kocić-Tanackov, S. Antimicrobial activity of lactic acid against pathogen and spoilage microorganisms. *J. Food Process. Preserv.* **2016**, *40*, 990–998. [CrossRef]
8. Vieco-Saiz, N.; Belguesmia, Y.; Raspoet, R.; Auclair, E.; Gancel, F.; Kempf, I.; Drider, D. Benefits and inputs from lactic acid bacteria and their bacteriocins as alternatives to antibiotic growth promoters during food-animal production. *Front. Microbiol.* **2019**, *10*, 57. [CrossRef]
9. Ahmad, V.; Khan, M.S.; Jamal, Q.M.S.; Alzohairy, M.A.; Al Karaawi, M.A.; Siddiqui, M.U. Antimicrobial potential of bacteriocins: In therapy, agriculture and food preservation. *Int. J. Antimicrob. Agents* **2017**, *49*, 1–11. [CrossRef]
10. Hashim, H.; Sikandar, S.; Khan, M.A.; Qurashi, A.W. Bacteriocin: The avenues of innovation towards applied microbiology. *Pure Appl. Biol. (PAB)* **2019**, *8*, 460–478. [CrossRef]
11. Cerqueira, J.; Dimitrov, S.; Silva, A.; Augusto, L. Inhibition of Herpes simplex virus 1 and Poliovirus (PV-1) by bacteriocins from Lactococcus lactis subsp. lactis and Enterococcus durans strains isolated from goat milk. *Int. J. Antimicrob. Agents* **2018**, *51*, 33–37.
12. Ermolenko, E.; Desheva, Y.; Kolobov, A.; Kotyleva, M.; Sychev, I.; Suvorov, A. Anti–Influenza Activity of Enterocin B In vitro and Protective Effect of Bacteriocinogenic Enterococcal Probiotic Strain on Influenza Infection in Mouse Model. *Probiotics Antimicrob. Proteins* **2019**, *11*, 705–712. [CrossRef]

13. Park, S.-C.; Park, Y.; Hahm, K.-S. The role of antimicrobial peptides in preventing multidrug-resistant bacterial infections and biofilm formation. *Int. J. Mol. Sci.* **2011**, *12*, 5971–5992. [CrossRef] [PubMed]
14. Jenssen, H.; Hamill, P.; Hancock, R.E. Peptide antimicrobial agents. *Clin. Microbiol. Rev.* **2006**, *19*, 491–511. [CrossRef] [PubMed]
15. O'Connor, P.M.; O'Shea, E.F.; Cotter, P.D.; Hill, C.; Ross, R.P. The potency of the broad spectrum bacteriocin, bactofencin A, against staphylococci is highly dependent on primary structure, N-terminal charge and disulphide formation. *Sci. Rep.* **2018**, *8*, 1–8.
16. Vescovo, M.; Bottazzi, V.; Torriani, S.; Dellaglio, F. Basic characteristics, ecology and application of Lactobacillus plantarum [in the production of fermented foods of animal and plant origin]: A review. *Ann. Microbiol. Enzimol. (Italy)* **1993**, *43*, 261–284.
17. Tremonte, P.; Pannella, G.; Succi, M.; Tipaldi, L.; Sturchio, M.; Coppola, R.; Luongo, D.; Sorrentino, E. Antimicrobial activity of Lactobacillus plantarum strains isolated from different environments: A preliminary study. *Int. Food Res. J.* **2017**, *24*, 852–859.
18. Dinev, T.; Beev, G.; Tzanova, M.; Denev, S.; Dermendzhieva, D.; Stoyanova, A. Antimicrobial activity of Lactobacillus plantarum against pathogenic and food spoilage microorganisms: A review. *Bulg. J. Vet. Med.* **2018**, *21*, 253–268. [CrossRef]
19. Fjell, C.D.; Hiss, J.A.; Hancock, R.E.; Schneider, G. Designing antimicrobial peptides: Form follows function. *Nat. Rev. Drug Discov.* **2012**, *11*, 37–51. [CrossRef]
20. Diep, D.B.; Håvarstein, L.S.; Nes, I.F. Characterization of the locus responsible for the bacteriocin production in Lactobacillus plantarum C11. *J. Bacteriol.* **1996**, *178*, 4472–4483. [CrossRef]
21. Young, R. Bacteriophage holins: Deadly diversity. *J. Mol. Microbiol. Biotechnol.* **2002**, *4*, 21–36.
22. Nissilä, E.; Douillard, F.P.; Ritari, J.; Paulin, L.; Järvinen, H.M.; Rasinkangas, P.; Haapasalo, K.; Meri, S.; Jarva, H.; De Vos, W.M. Genotypic and phenotypic diversity of Lactobacillus rhamnosus clinical isolates, their comparison with strain GG and their recognition by complement system. *PLoS ONE* **2017**, *12*, e0176739.
23. Karpiński, T.M. Efficacy of octenidine against Pseudomonas aeruginosa strains. *Eur. J. Biolog. Res.* **2019**, *9*, 135–140.
24. Hsieh, I.-N.; Hartshorn, K.L. The role of antimicrobial peptides in influenza virus infection and their potential as antiviral and immunomodulatory therapy. *Pharmaceuticals* **2016**, *9*, 53. [CrossRef] [PubMed]
25. Ahmed, A.; Siman-Tov, G.; Hall, G.; Bhalla, N.; Narayanan, A. Human antimicrobial peptides as therapeutics for viral infections. *Viruses* **2019**, *11*, 704. [CrossRef] [PubMed]
26. Desriac, N.; Broussolle, V.; Postollec, F.; Mathot, A.-G.; Sohier, D.; Coroller, L.; Leguerinel, I. Bacillus cereus cell response upon exposure to acid environment: Toward the identification of potential biomarkers. *Front. Microbiol.* **2013**, *4*, 284. [CrossRef] [PubMed]
27. Wiegand, I.; Hilpert, K.; Hancock, R.E. Agar and broth dilution methods to determine the minimal inhibitory concentration (MIC) of antimicrobial substances. *Nat. Protoc.* **2008**, *3*, 163. [CrossRef] [PubMed]
28. Kim, S.W.; Ha, Y.J.; Bang, K.H.; Lee, S.; Yeo, J.-H.; Yang, H.-S.; Kim, T.-W.; Lee, K.P.; Bang, W.Y. Potential of Bacteriocins from Lactobacillus taiwanensis for Producing Bacterial Ghosts as a Next Generation Vaccine. *Toxins* **2020**, *12*, 432. [CrossRef]
29. Jang, Y.; Shin, J.S.; Lee, J.-Y.; Shin, H.; Kim, S.J.; Kim, M. In Vitro and In Vivo Antiviral Activity of Nylidrin by Targeting the Hemagglutinin 2-Mediated Membrane Fusion of Influenza A Virus. *Viruses* **2020**, *12*, 581. [CrossRef] [PubMed]
30. Chin, C.-S.; Alexander, D.H.; Marks, P.; Klammer, A.A.; Drake, J.; Heiner, C.; Clum, A.; Copeland, A.; Huddleston, J.; Eichler, E.E. Nonhybrid, finished microbial genome assemblies from long-read SMRT sequencing data. *Nat. Methods* **2013**, *10*, 563–569. [CrossRef] [PubMed]
31. Kurtz, S.; Phillippy, A.; Delcher, A.L.; Smoot, M.; Shumway, M.; Antonescu, C.; Salzberg, S.L. Versatile and open software for comparing large genomes. *Genome Biolog.* **2004**, *5*, R12. [CrossRef]
32. Delcher, A.L.; Bratke, K.A.; Powers, E.C.; Salzberg, S.L. Identifying bacterial genes and endosymbiont DNA with Glimmer. *Bioinformatics* **2007**, *23*, 673–679. [CrossRef]
33. Conesa, A.; Götz, S.; García-Gómez, J.M.; Terol, J.; Talón, M.; Robles, M. Blast2GO: A universal tool for annotation, visualization and analysis in functional genomics research. *Bioinformatics* **2005**, *21*, 3674–3676. [CrossRef] [PubMed]

34. Lagesen, K.; Hallin, P.; Rødland, E.; Stærfeldt, H.; Rognes, T.; Ussery, D. RNammer: Consistent annotation of rRNA genes in genomic sequences. *Nucleic Acids Res.* **2007**, *35*, 3100–3108. [CrossRef] [PubMed]
35. Lowe, T.M.; Eddy, S.R. tRNAscan-SE: A program for improved detection of transfer RNA genes in genomic sequence. *Nucleic Acids Res.* **1997**, *25*, 955–964. [CrossRef] [PubMed]

Article

Screening of Bacterial Quorum Sensing Inhibitors in a *Vibrio fischeri* LuxR-Based Synthetic Fluorescent *E. coli* Biosensor

Xiaofei Qin [1,2], Celina Vila-Sanjurjo [2,3], Ratna Singh [4], Bodo Philipp [5] and Francisco M. Goycoolea [2,6,*]

1 Department of Bioengineering, Zhuhai Campus of Zunyi Medical University, Zhuhai 519041, China; iamxfqin@njtech.edu.cn
2 Laboratory of Nanobiotechnology, Institute of Plant Biology and Biotechnology, University of Münster, Schlossplatz 8, D-48143 Münster, Germany; celina.vila@usc.es
3 Department of Pharmacology, Pharmacy and Pharmaceutical Technology, University of Santiago de Compostela, Campus Vida, s/n, 15782 Santiago de Compostela, Spain
4 Laboratory of Molecular Phytopathology and Renewable Resources, Institute of Plant Biology and Biotechnology, University of Münster, Schlossplatz 8, D-48143 Münster, Germany; singhr@uni-muenster.de
5 Institute of Molecular Microbiology and Biotechnology, University of Münster, Corrensstraße 3, D-48149 Münster, Germany; bodo.philipp@uni-muenster.de
6 School of Food Science and Nutrition, University of Leeds, Leeds LS2 9JT, UK
* Correspondence: F.M.Goycoolea@leeds.ac.uk; Tel.: +44-1133-431412

Received: 13 August 2020; Accepted: 18 September 2020; Published: 22 September 2020

Abstract: A library of 23 pure compounds of varying structural and chemical characteristics was screened for their quorum sensing (QS) inhibition activity using a synthetic fluorescent *Escherichia coli* biosensor that incorporates a modified version of lux regulon of *Vibrio fischeri*. Four such compounds exhibited QS inhibition activity without compromising bacterial growth, namely, phenazine carboxylic acid (PCA), 2-heptyl-3-hydroxy-4-quinolone (PQS), 1*H*-2-methyl-4-quinolone (MOQ) and genipin. When applied at 50 μM, these compounds reduced the QS response of the biosensor to 33.7% ± 2.6%, 43.1% ± 2.7%, 62.2% ± 6.3% and 43.3% ± 1.2%, respectively. A series of compounds only showed activity when tested at higher concentrations. This was the case of caffeine, which, when applied at 1 mM, reduced the QS to 47% ± 4.2%. In turn, capsaicin, caffeic acid phenethyl ester (CAPE), furanone and polygodial exhibited antibacterial activity when applied at 1mM, and reduced the bacterial growth by 12.8% ± 10.1%, 24.4% ± 7.0%, 91.4% ± 7.4% and 97.5% ± 3.8%, respectively. Similarly, we confirmed that *trans*-cinnamaldehyde and vanillin, when tested at 1 mM, reduced the QS response to 68.3% ± 4.9% and 27.1% ± 7.4%, respectively, though at the expense of concomitantly reducing cell growth by 18.6% ± 2.5% and 16% ± 2.2%, respectively. Two QS natural compounds of *Pseudomonas aeruginosa*, namely PQS and PCA, and the related, synthetic compounds MOQ, 1H-3-hydroxyl-4-quinolone (HOQ) and 1H-2-methyl-3-hydroxyl-4-quinolone (MHOQ) were used in molecular docking studies with the binding domain of the QS receptor TraR as a target. We offer here a general interpretation of structure-function relationships in this class of compounds that underpins their potential application as alternatives to antibiotics in controlling bacterial virulence.

Keywords: compounds screening; quorum sensing inhibition; antibacterial; molecular docking

1. Introduction

Bacteria communicate with secreted chemical signalling molecules that act as autoinducers in a phenomenon known as quorum sensing (QS) which allows them to fulfil a variety of functions, including bioluminescence, virulence and biofilm formation, among others [1]. In the case of Gram-negative bacteria, the autoinducer molecules belong mostly to the family of the acyl homoserine lactones (AHLs), which accumulate to particular threshold concentrations once the population of cells grow to sufficient [2,3].

AHL synthesis relies on the synthase LuxI family, while AHL reception depends on LuxR-type transcriptional regulators, which include the nominal LuxR protein from *Vibrio fischeri*, but also the related TraR and LasR from *Agrobacterium tumefaciens* and *P. aeruginosa*, among others. These act as transcriptional activators or repressors of the target QS genes [4–7]. The canonical LuxR protein from *V. fischeri* comprises two domains. The N-terminal domain is responsible of AHL binding and also can mediate protein dimerisation [8]. In contrast, the C-terminal domain contains a helix-turn-helix (HTH) motif, which is thought to make sequence-specific binding to DNA and to drive RNA polymerase binding to target promoters [9]. Much research has been conducted in recent years on mutant- LuxR-type proteins [10–16]. Among these, the TraR protein from *Agrobacterium tumefaciens* has been thoroughly studied, and its crystal structure has been solved, revealing the presence of the above mentioned two functional domains. In TraR, the N-terminal domain binds to N-3-oxooctanoyl-L-homoserine lactone (OOHL), and the C-terminal domain interacts with the DNA binding domain of the tra box [7,17]. TraR is a dimer in the presence of OOHL, and the TraR-OOHL-*tra* DNA ternary complex can be used as a prototype for the large family of AHL-induced transcription activators. The LasR protein of *Pseudomonas aeruginosa* shares 70% homology with TraR of *A. tumefaciens*, and the 3D model of the TraR active site closely resembles the X-ray structure of the LasR active site [18]. The signal binding sites in both apo-proteins are highly accessible, so TraR constitutes a useful model receptor which allows predicting the ability of putative QS inhibitors (QSIs) to block QS-based mechanisms in the human pathogen *P. aeruginosa* [19,20].

The development of strategies aimed to block or disrupt QS is gaining momentum. These efforts are directed to inhibit the production of virulence factors at the site of infection by dismantling the collective power of bacterial pathogens, an approach known as quorum quenching (QQ) or QS inhibition [21,22]. In the last years, QSIs have attracted significant attention from the scientific community and are considered as potential weapons and new generation antimicrobials in the therapeutic arsenal against infections caused by drug-resistant bacteria [23–25].

There are three major approaches to target bacterial QS using QSIs: (i) destruction of the signalling molecule, (ii) inhibition of signal production, and (iii) inhibition of the receptor [26]. To the date, most of the literature on QQ is centred in the investigation of AHL-degrading enzymes [27] and more abundantly, in QSIs that targeted to specific QS regulators of diverse species [26]. These QSIs may structurally resemble the natural ligand or may on the contrary, have an entirely different molecular structure [28,29]. QSIs include compounds of diverse sources, both natural and synthetic, and include fungal metabolites, plant substances, antibiotics, and synthetic derivatives of QS autoinducers or natural antagonists, to name a few [26–29]. Identification of QSIs is commonly performed through the screening of compound libraries and biosensor-based analysis of the QS response alone or combined with computer modelling analysis. These methods have allowed expanding the catalogue of available QSIs, which includes a variety of AHL analogues, brominated furanones, polyphenolic compounds and polypeptides [23,26,30–32].

Regarding computer modelling, docking-based screening of candidate QSIs is normally carried out by using a genetic algorithm on a library of 2344 compounds and calculates their binding free energy, hence identifying those candidates able to interact with target conserved residues in the binding site of a LuxR-type model receptor [33]. In-silico screening of ligand databases has thus become an important strategy towards the discovery of novel QSIs. GRID molecular interaction fields (GRID-MIFs) is accepted as an efficient method in virtual screening of candidate molecules which target protein binding sites. It is a computational procedure for detecting energetically favourable binding regions in proteins and small molecule drugs of known 3D structure. The energies are calculated using the electrostatic, hydrogen-bond, Lennard Jones, and entropic interactions of chemically selective probes with the chosen biological target. The program works by defining a three-dimensional grid of points that contains the chosen substrate binding site. The above mentioned calculations are repeated for each node in the three-dimensional grid and for each probe being considered. The results of these calculations are a collection of three-dimensional matrices, one for each probe-target interaction. A detailed description of the GRID program, the force field parameters, and details of calculations can be found elsewhere [34,35]. Briefly, a grid is projected inside the protein regions and cavities of interest. The probes are functional groups that can move stepwise from grid point to grid point. The calculated interaction energies of the probes are computed to create the MIFs, which represents the potential interaction of the protein with a particular chemical group. GRID-MIFs is considered as a high-throughput screening method to virtually analyse protein-ligand interactions [36,37].

In this study, we have screened 23 potential QS inhibition compounds, representing five groups according to their chemical structures. We found that seven of them were potent, dose-dependent inhibitors of the *V. fisheri* LuxR-based system expressed on recombinant *E. coli* biosensor cells, including two unprecedented ones. Another five compounds have shown antibacterial activity, while the remaining eleven compounds were inert at the tested doses. Moreover, we performed in silico GRID-MIFs-based computational molecular docking on the 3D crystal structure of TraR that allowed us to propose a hypothesis on the disparate effects observed experimentally for compounds of chemically related structure. We propose that our biosensor- and docking-based approaches have a high potential for drug designing purposes.

2. Results

2.1. Screening a Panel of Potential Quorum Sensing Inhibitors

We selected a panel of 23 pure compounds with the potential to act as inhibitors of LuxR-AHL-mediated QS. These comprised the following groups, namely: Group (1) lactone analogues, Group (2) aromatic ring structures, Group (3) heterocyclic compounds, Group (4) *Pseudomonas spp*- relevant compounds, and Group (5) structurally unrelated compounds. Figure 1 shows the chemical structures of the cognate AHL molecules of LuxR and TraR regulators, namely *N*-3-oxohexanoyl-L-homoserine lactone ($3OC_6HSL$) and OOHL, respectively.

All 23 compounds were tested using the *E. coli* Top10 pSB1A3-BBaT9002 biosensor, which expresses a synthetic version of the *lux* regulon of *V. fischeri* and produces a fluorescent as a response to external (3OC6HSL) [38]. The rate of the density-normalised fluorescence of the *E. coli* biosensor as a function of the 3OC6HSL concentration displays a Hill behaviour, with a k_{Hill} of $7.48 \times 10^{-10} \pm 9.03 \times 10^{-11}$ M. At 3OC6HSL concentrations higher than 1×10^{-9} M, the fluorescence response is saturated, while the fluorescence levels are undetectable at a 3OC6HSL concentration of 1×10^{-10} M [38]. We evaluated the QS inhibitory effect of our panel of compounds by their ability to reduce the fluorescence response of this *E. coli* biosensor without compromising cell growth.

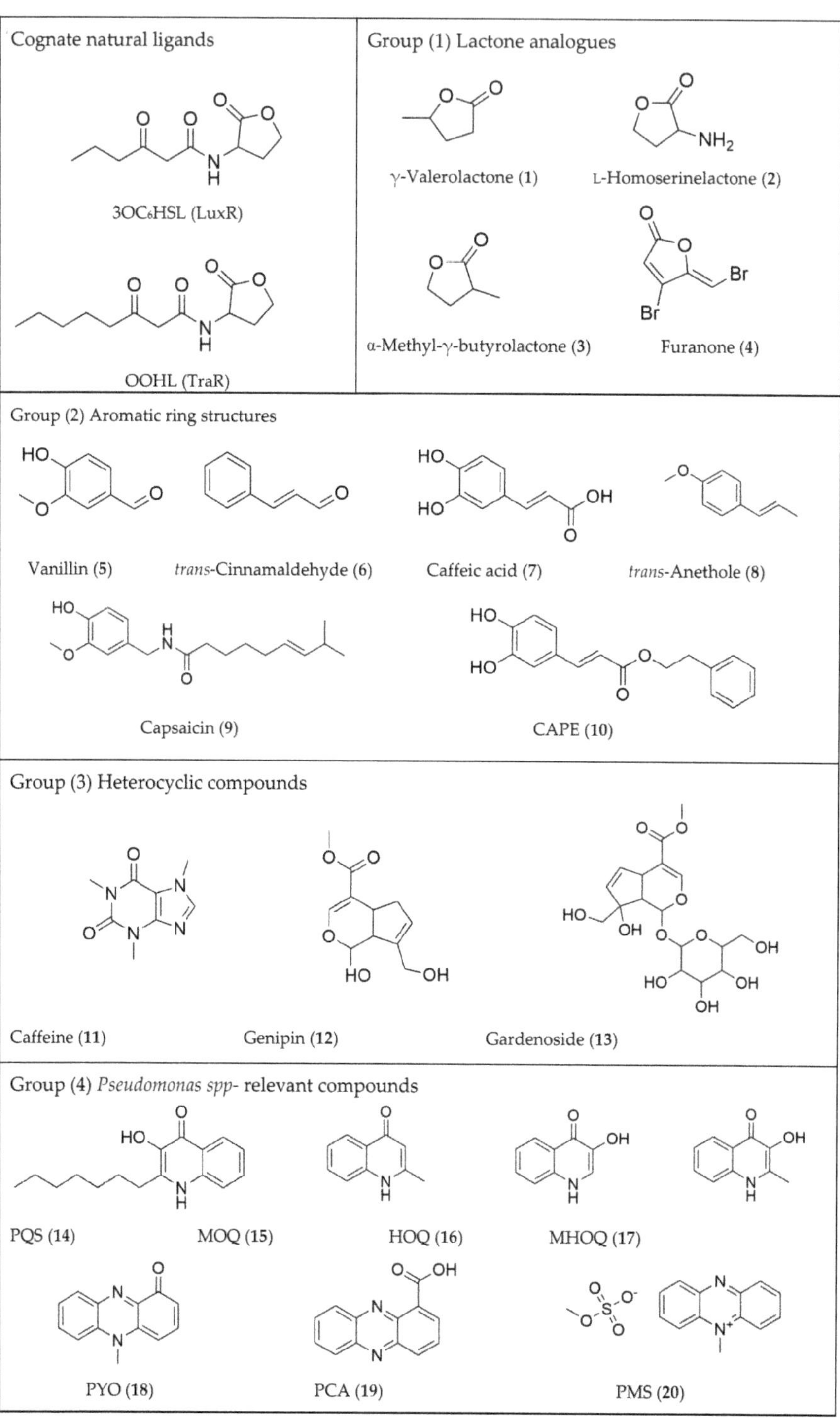

Figure 1. *Cont.*

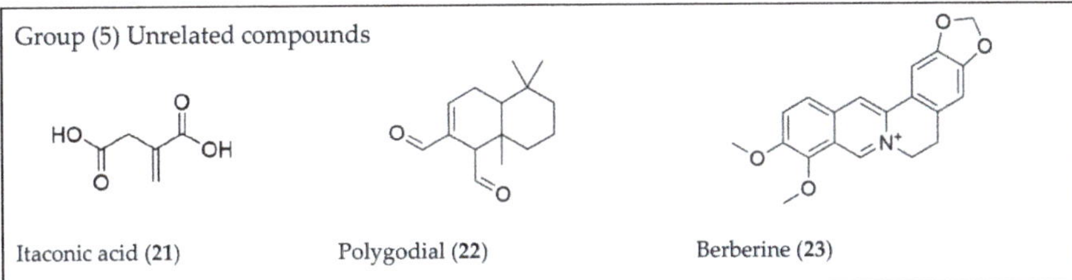

Figure 1. Chemical structures of studied compounds. Group (1) lactone analogues, Group (2) aromatic ring structures, Group (3) heterocyclic compounds, Group (4) *Pseudomonas* spp.-relevant compounds, and Group (5) structurally unrelated compounds. In the Figure are also shown the structure of natural LuxR and TraR ligands, namely 3OC6HSL and OOHL. Other details of the series of compounds are given in Materials and Methods section.

Screening of the 23 compounds revealed that they could be classified into three main categories based on their QS inhibition activity, relative to their effects on bacterial growth. The first category refers to compounds that have not shown QS inhibition nor antibacterial activities (i.e., no apparent effect on the fluorescent response and growth of the *E. coli* biosensor). The second includes compounds with the ability to reduce the QS response of the biosensor without compromising cell growth. While the third one comprises compounds with the ability to reduce the QS response at the expense of hampering cell growth. Figure 2 shows representative dose-response curves of compounds belonging to the classes 1, 2, 3, namely gardenoside (13) (Figure 2a,d,g), caffeine (11) (Figure 2b,e,h) and furanone (4) (Figure 2c,f,i) on the fluorescence (Figure 2a–c), growth (Figure 2d–f) and density-normalized fluorescence (Figure 2g–i) of the *E. coli* biosensor. A selection of dose-response effects of other compounds is available as Supporting Information, as discussed below.

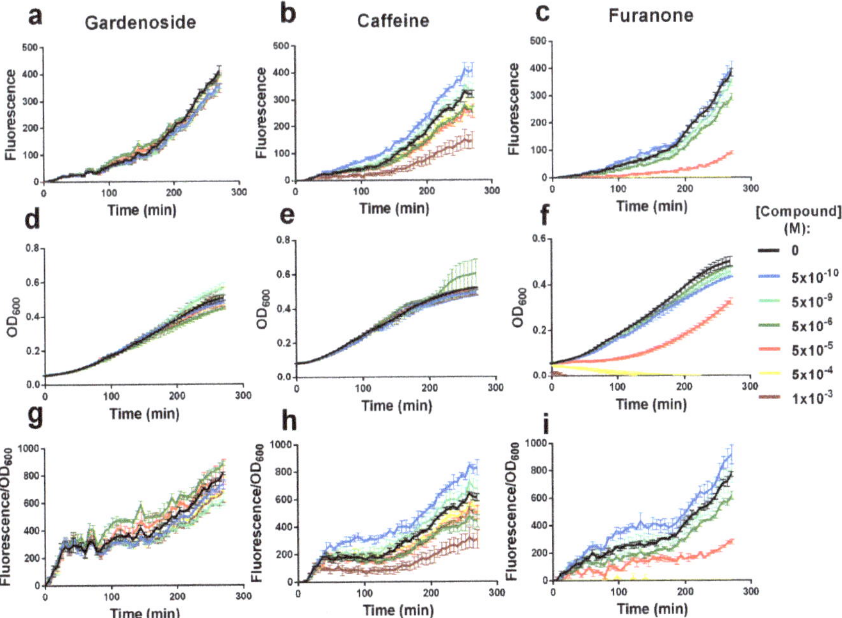

Figure 2. Effect of increasing concentrations of gardenoside, caffeine and furanone on the fluorescence (**a–c**), growth (**d–f**) and density-normalised fluorescence (**g–i**) of the *E. coli* biosensor over time.

The three compounds are chosen as representatives of the following categories: no inhibition (gardenoside; **a**,**d**,**g**), QS inhibition in the absence of growth reduction (caffeine; **b**,**e**,**h**) and QS and growth inhibition (furanone; **c**,**f**,**i**). Results from additional experiments on other compounds are available in Supporting Information. Data shows the mean and standard deviation of a representative experiment with triplicated treatments.

Next, we calculated the end-point effect of the 23 candidate compounds on the QS-based response and growth of the biosensor, relative to those of untreated cells. Figure 3 shows end-point, relative reduction of density-normalised fluorescence and cell density of compounds tested at fixed concentrations of 5×10^{-5} M (Figure 3a,b) and 1×10^{-3} M (Figure 3c,d). The rationale behind the chosen concentrations was based on the maximum water solubility of the compounds. Relative values close to 1.0 indicate none or negligible effect of a given compound on the density-normalised response and/or growth of the biosensor. In turn, we considered that relative fluorescence values significantly lower than 1.0 as were a diagnose of inhibition of the QS-based response, proven that the relative OD_{600} values stayed close to 1.0. This condition applied to vanillin (**5**), caffeine (**11**) and *trans*-cinnamaldehyde (**6**) applied at 1×10^{-3} M (Figure 3a,b); and to phenazine carboxylic acid (PCA, **19**), 2-heptyl-3-hydroxy-4-quinolone (PQS, **14**), genipin (**12**) and 1*H*-2-methyl-4-quinolone (MOQ, **15**) applied at 5×10^{-5} M. In all cases, the relative QS and OD_{600} data were compared statistically against itaconic acid, due to its negligible effect on both parameters.

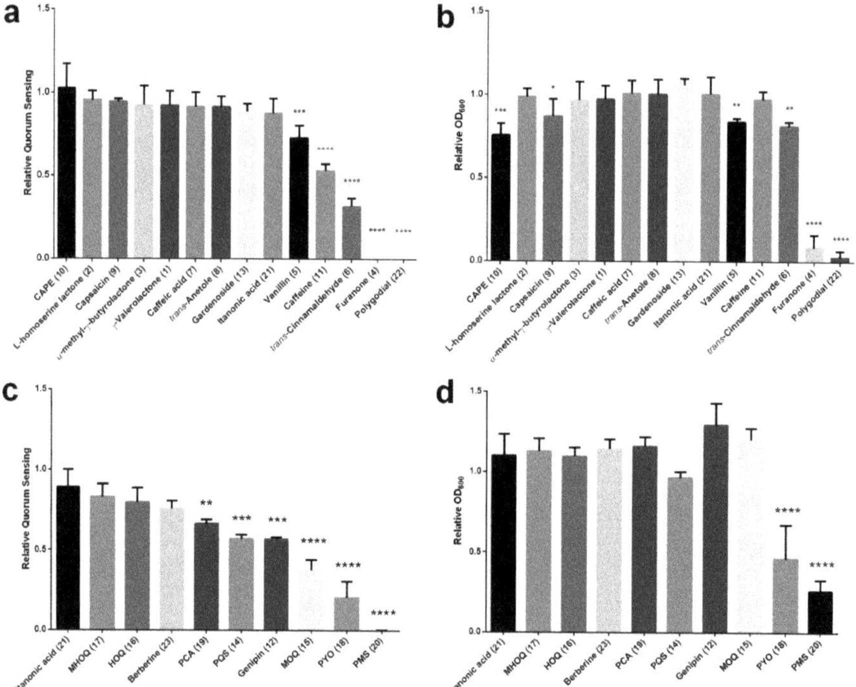

Figure 3. End-point effect of the 23 candidate compounds on the QS-based response and growth of the *E. coli* biosensor. (**a**) Effect of compounds **10, 2, 9, 3, 1, 7, 8, 13, 21, 5, 11, 6, 4, 22**; applied at 1×10^{-3} M on the density-normalised fluorescence of treated cells relative to control cells.

Relative fluorescence was calculated as follows: the mean of the last ten values of density-normalised fluorescence over time, corresponding to 246–300 min of incubation (see Figure 2) was divided by the corresponding values of untreated cells. (**b**) Effect of compounds **10, 2, 9, 3, 1, 7, 8, 13, 21, 5, 11, 6, 4, 22**; applied at 1×10^{-3} M on cell density of treated cells relative to control cells. Relative OD_{600} was calculated as follows: the mean of the last 10 OD_{600} values over time, corresponding to 246–300 min of incubation (see Figure 2) was divided by the corresponding values of untreated cells. (**c**). Effect of compounds **21, 17, 16, 23, 19, 14, 12, 15, 18, 20**; applied at 5×10^{-5} M on the density-normalised fluorescence of treated cells relative to control cells. Relative fluorescence was calculated as in (**a**,**d**) Effect of compounds **21, 17, 16, 23, 19, 14, 12, 15, 18, 20**; applied at 5×10^{-5} M on cell density of treated cells relative to control cells. Relative OD_{600} was calculated as in (**b**) *t*-Student statistical comparisons were made using itaconic acid as a reference treatment (* $p < 0.05$, ** $p < 0.01$, *** $p < 0.001$ and **** $p < 0.0001$). Data show the mean and standard deviation of three independent experiments with triplicated treatments.

Figure 3a shows that the compounds *trans*-cinnamaldehyde (**6**) ($p < 0.0001$), caffeine (**10**) ($p < 0.0001$), and, to a less extent, vanillin (**5**) ($p < 0.001$) can significantly reduce the QS-based fluorescence of the biosensor when applied at 1×10^{-3} M. Importantly, the QS inhibitory effect of caffeine (10) was not accompanied by any compromise of cell growth (Figure 3b). As for vanillin (**5**) ($p < 0.01$) and *trans*-cinnamaldehyde (**6**) ($p < 0.01$), they slightly reduced cell growth by 16.0% and 18.6%, respectively (Figure 3b). Furanone (**4**) ($p < 0.0001$), and polygodial (21) ($p < 0.0001$), which abolished the QS-based fluorescence of the biosensor at the tested concentration, they concomitantly exerted a dramatic antibacterial effect (cf. Figure 3a,b). Figure 3c shows that compounds PCA (**19**) ($p < 0.01$), PQS (**14**) ($p < 0.001$), genipin (**12**) ($p < 0.001$) and MOQ (**15**) ($p < 0.0001$) significantly reduced the density-normalized fluorescence of the biosensor when tested at 5×10^{-5} M (Figure 3a) and, importantly, they did not exert a significant effect on bacterial growth (Figure 3b). By contrast, compounds PYO (**18**) ($p < 0.0001$) and PMS (**20**) ($p < 0.0001$), showed a strong deleterious effect both on the QS response of the biosensor and its growth (cf. Figure 3c,d). We carried out a series of experiments to confirm that inhibition of the density-normalised fluorescence observed with PYO (**18**) and genipin (**12**) at sub-lethal doses was in fact due to interference with the QS response of the biosensor and not due to GFP fluorescence quenching. To this end, we used a control *E. coli* strain Top10 pBCA9445-jtk28282::sfGFP, which constitutively expresses a super folded version of GFP (sfGFP). We compared the effects of both compounds in the *E. coli* biosensor and control strains (Figures S2–S8). In our hands, neither of them at concentrations in the range 1×10^{-8} M to 1×10^{-4} M (Figures S3–S5) acted as quenchers of sfGFP fluorescence in the control strain *E. coli* Top10 pBCA9445-jtk28282::sfGFP. We observed a high degree of experimental variability and not apparent dose-response effects of PYO at sub-lethal concentrations ranging from 1×10^{-8} to 1×10^{-4} M (Figure S5). The implications of these disparate observations are further considered in the Discussion section.

2.2. Computation of GRID-MIFs

The three-dimensional molecular crystal structure of TraR receptor was obtained from the protein data bank (PDB No. 1L3L). Information about the binding region where the natural ligand OOHL interacts with TraR was described elsewhere [17]. Favorable interaction points at the binding site of the receptor was studied with GRID-MIFs. To define the GRID maps at TraR binding site, the Autogrid utility inbuilt in AutoDockTools 1.5.6 software was applied. Three different chemical probes (HD, HA, DRY) were used, representing the potentially significant functional group at the binding site. In Figure 4a the GRID-MIFs generated at TarR binding site is shown. Here, the green patch generated by DRY probe accounts for the favourable hydrophobic interaction ligand and receptor; blue contours were generated by HD (donor) probe responsible for favourable hydrogen bonds between receptor and ligand, and red contours were generated by HA (acceptor) probe, which informs about favourable hydrogen bonds between ligand and receptor amino acid residues. Comparing with the natural substrate OOHL (Figure 4b) the dry probe matches with the ring and all carbons in the substrate, the HD (blue) probe matches the NH group of substrate and HA (red) matches the O functional group present in the substrate.

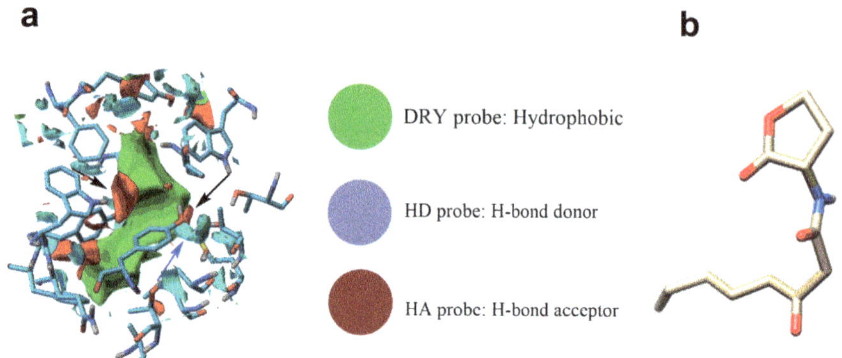

Figure 4. (**a**) GRID-MIFs for TraR protein with DRY probe (green) showing favorable hydrophobic interaction sites. GRID-MIFs for TraR protein with HD (blue) and HA (red) probes showing favorable hydrogen bond (blue for hydrogen bond donor and red for hydrogen bond acceptor) binding sites. (**b**) Natural substrate OOHL.

2.3. Docking and Interaction of Selected Compounds

We docked six selected compounds, based on their generated GRID-MIFs, over the binding domain of TraR. We included the cognate OOHL ligand as a reference. This enabled to gain Information on dock score, hydrogen bond and hydrophobic interactions. We selected ligand conformations using that of the natural ligand OOHL into the binding site as a reference (Figure 5a). Thus, we fixed docked ligand binding poses to that of the natural ligand. In other words, superimposition of the docked ligand on OOHL conformation in TraR's crystal structure implied that the aromatic ring of the ligand coincided with the lactone ring of the natural substrate. Table 1 lists the estimated binding energy values for the docked compounds. Strikingly, PCA (19) and PQS (14) showed a more negative docking score (i.e., predicted free energy of binding) than that estimated for the cognate ligand OOHL, which would in principle indicate a stronger affinity of these two compounds towards the binding site of TraR (Table 1). Figure 5b shows the docking-predicted binding pose of PCA into TraR's binding pocket. Two ligand conformations were identified for PQS (14), namely PQS-conf A (Figure 5c) and PQS-conf B (Figure 5d and Table 1). Both conformations were similar to OOHL binding pose, only differing in ring flip, which affects hydrogen bond interactions between ligand and receptor (cf. Figure 5a–c). The estimated docking score for PQS (14)-conf A was more negative than that of OOHL, whereas the predicted value for PQS-conf B was less negative, indicating a potential stronger affinity of conf B vs conf A towards the binding site of TraR (Table 1). Docking scores of smaller ligands, namely MOQ, HOQ and MHOQ were less negative than that of OOHL, predicting poorer affinities for these ligands towards TraR (Table 1).

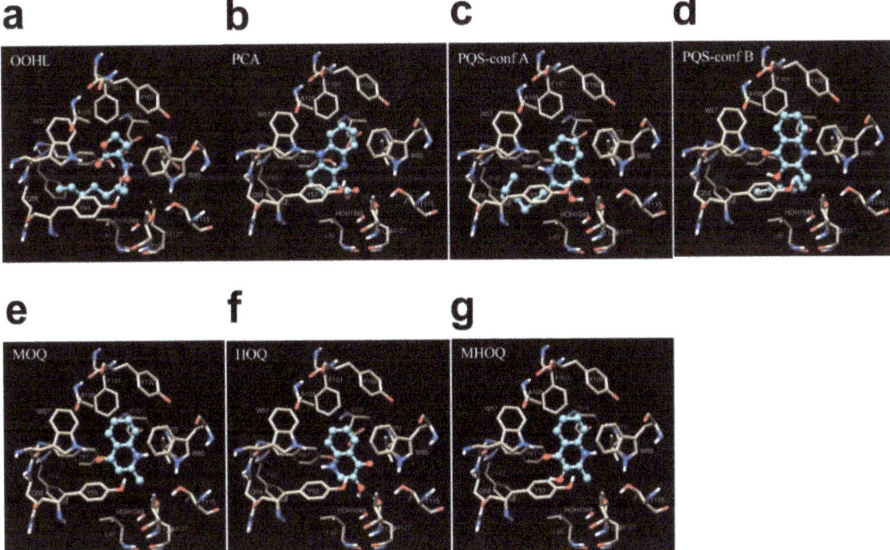

Figure 5. Interaction of TraR receptor with (**a**) OOHL; (**b**) PCA, (**c**) and (**d**) PQS; (**e**) MOQ; (**f**) HOQ and (**g**) MHOQ. (**c**) shows us the PQS conf A and (**d**) shows PQS conf B. Close-up view of all compounds binding site, oxygen carbon and nitrogen are colored in red yellow and blue, respectively, hydrogen bonds are shown as blue line.

Table 1. Cumulative results for the compounds docked against TraR protein.

Compound	Hydrogen Bonding Interactions	Dock Score (Kcal/mol)	Hydrophobic Interactions
OOHL	4 (including Trp 57)	−7.58	Leu 40, Tyr 53, Gln 58, Trp 85, Phe101, Ala 105, Ile 110.
MOQ (**15**)	1 (including Trp 57)	−6.16	Tyr 53, Val 72, Trp 85, Phe 101, Tyr 102, Ala 105, Met 127
HOQ (**16**)	0	−5.72	Trp 57, Tyr 61, Asp 70, Val 72, Trp 85, Phe 101, Met 127
MHOQ (**17**)	1 (including Trp 57)	−5.83	Tyr 53, Asp 70, Val 72, Trp 85, Phe 101, Tyr 102, Ala 105, Met 127, Thr 129
PCA (**19**)	3 (including Trp 57)	−8.01	Leu 40, Tyr 53, Tyr 61, Val 72, Trp 85, Phe101, Tyr 102, Ala 105, Met 127
PQS (**14**) (conf A)	2	−8.04	Leu 40, Tyr 53, Tyr 61, Val 72, Val 73, Trp 85, Phe 101, Tyr 102, Ala 105, Met127, Thr 129
PQS (**14**) (conf B)	1 (including Trp 57)	−6.59	Ala 38, Leu 40, Ala 49, Thr 51, Gln 58, Tyr 61, Phe 62, Val 72, Trp 85, Phe 101, Met127

We studied in detail the docking-predicted poses of selected candidates to elucidate the mechanisms of receptor-ligand possible interactions (Table 1). Figure 5 shows interaction maps of the six selected compounds (Table 1) with TraR's binding pocket. Clear from Figure 5a are the presence of four hydrogen bonds established between the cognate substrate OOHL and specific amino acids at the binding pocket of the receptors The H- bonds are located as follows: (1) between Trp 57 in TraR and keto group present in the lactone ring, (2) between Tyr 53 in TraR and the 1-keto group of the lactone, (3) between Asp 70 in TraR and the imino group in OOHL and (4) between Tyr 61 in TraR and the 3-keto group of OOHL (Figure 5a). Apart from these bonds, the cognate ligand also establishes strong hydrophobic interactions with nearby residues, as shown in Table 1. Altogether, it seems that both hydrophobic and hydrogen bond interactions are important for ligand selectivity. Our docking results confirm previous docking predictions of binding and pose conformations for OOHL in TraR's binding site [18]. Figure 5b shows the docking-predicted map for PCA (19)-TraR interaction (19). Importantly, our predictions show the presence of three hydrogen bonds with TraR's binding pocket, namely: (1) between Trp 57 in TraR and the deprotonated N present in the aromatic group of PCA, (2) and (3) between Thr 129 in TraR, an intermediate water molecule and the carboxylic acid functional group of PCA (Figure 5b). Again, our docking predicted several hydrophobic interactions between specific aminoacids of the binding site of TraR and the aromatic and heterocyclic rings of PCA (19) (Table 1). Figure 5c shows the predicted maps for PQS-confA and the binding site of TraR. In this case, two hydrogen bonds are predicted to form, namely: (1) between Thr129 and the keto group present in the aromatic ring of PQS (14), (2) between a hydroxyl group in PQS and a molecule of water (Figure 5c). In turn the predicted map for PQS-conf B and TraR interaction (Figure 5d shows one hydrogen bond formed between Trp 57 in TraR and the keto group present in the aromatic ring ofPQS (14). Both PQS-conf A and -confB are expected to establish strong hydrophobic interactions with the binding pocket of TraR, as listed in Table 1 and the long carbon chain of PQS in both poses cover most of the hydrophobic patch generated by GRID (cf. Figures 4 and 5c,d). Panels e, f and g of Figure 5, illustrate the prediction interaction maps for TraR and MOQ (15), HOQ (16) and MHOQ (17), respectively. Docking-predicted poses in the binding site of the receptor are very similar (cf. Figure 5e–g); hence, the predicted interaction sites for the three of them did not very much. However, in the case of both MOQ (15) and MOQH (17) an hydrogen bond was predicted between Trp 57 in TraR and the keto group in the aromatic ring of the ligands (cf. Figure 5e,g), while this interaction was not observed in HOQ (15) (Figure 5f). Strikingly, a relative inhibition of the QS-based response of the *E. coli* biosensor was found to be significantly higher for MOQ (15) than that of HOQ (16) and MOQH (17) (Figure S9). This, in principle would indicate that the specific position of H-bonding between these kind of molecules and the binding site of TraR would not necessarily explain *per se* towards LuxR-based proteins.

3. Discussion

QSIs are gaining momentum as potent alternatives to the use of classical antibiotics in the context of rising resistance spread and with enormous potential in many fields, from food science to agriculture and medicine [39,40]. QSIs are structurally diverse molecules that can be synthesised or extracted from natural sources. Here, we screened a total of 23 chemically diverse compounds with potential QS inhibition activity to identify their impact on LuxR-regulated QS models. To this end, we used a well-established *E. coli* Top10 biosensor assay and in silico modelling to recreate the structural interactions between the ligand candidates and the LuxR-like receptor.

The candidate compounds were classified into five broad groups according to with their chemical structure (Table 1). Group 1 comprised the alkyl-substituted lactones (Table 1): γ-valerolactone (**1**), L-homoserine lactone (**2**), α-methyl-γ-butyrolactone (**3**), and furanone **4**. In this group, only furanone (4) showed a significant inhibition (≥37.5%) of the fluorescent- QS-based response of the biosensor at concentrations $\geq 5 \times 10^{-6}$ M 37.5% (Figure 3a and Figure S10). At 1×10^{-3} M, furanone completely abolished the QS-based response of the biosensor, albeit a dramatic, deleterious effect on cell growth (Figure 3b and Figure S10).

γ-Butyrolactones were the first class of small signalling molecules identified from Gram-positive species *Streptomyces griseus*. They were found to induce streptomycin production and sporulation [41]. Even though the chemical structure of γ-butyrolactones is rather similar to that of AHLs, except for the carbon side-chain, it is known that the receptors of γ-butyrolactone do not bind to AHL receptors and that AHL do not bind to γ-butyrolactone receptors. Also, both receptors have low structural similarity [41]. The functions of the signalling molecules also differ as AHLs i.a. are known to regulate QS in Gram-negative bacteria, while γ-butyrolactones mainly regulate the production of antibiotics and differentiation.

From our results, we found that neither α-methyl-γ-butyrolactone, γ-valerolactone nor L-homo-serine lactone showed any QS inhibition activity. These results are consistent with the notion that the carbon side chain is important for binding to the LuxR receptor. Also, it is not unexpected that γ-valerolactone did not show activity, as the methyl substituent is in a different position of the lactone ring than that in 3-oxohexanoylhomoserine. As suggested in previous studies, our results support the proposal that the LuxR receptor does not recognise α-methyl-γ-butyrolactone nor other related lactones (e.g., γ-valerolactone).

Halogenated furanones which resemble AHLs structurally were first discovered in the marine red alga *Delisea pulchra* and proved to preventing swarming motility with the opportunistic human pathogen *Proteus mirabilis* and *S. liquefaciens* without affect growth rate [42,43]. These compounds were the first QSIs found to occur in nature. After that, Manefield et al. further obtained proof-of-principle of the QS inhibition activity of the same type of halogenated furanones by specifically interfering with AHL-mediated gene expression at the level of the LuxR protein [44]. Furthermore, Ren et al. found that (5Z)-4-bromo-5-(bromomethylene)-3-butyl-2(5*H*)-furanone is a non-specific intercellular signal antagonist [45]. Later, Defoirdt et al. proved that this furanone disrupts quorum sensing-regulated gene expression in *Vibrio harveyi* by decreasing the DNA-binding activity of the transcriptional regulator protein LuxR [46]. Besides halogenated furanones with side alkyl chains that have QS inhibition activity, some synthetic halogenated furanones lacking alkyl chain also have anti-biofilm formation activity through interference with QS, and have been found to protect surfaces from being colonised by *S. epidermidis* [47]. Compound (5Z)-4-bromo-5-(bromomethylene)-3-butyl-2(5*H*)-furanone included in our library, a synthetic derivate of natural furanone has been reported as a potent antagonist of bacterial QS that not only can increase bacterial susceptibility to tobramycin and SDS when it is applied for action with *P. aeruginosa* biofilms but also can inhibit virulence factor expression because of targeting QS systems [48]. Several previous studies have shown that halogenated furanones can interact with the LuxR protein and induce conformational changes due to rapid proteolytic degradation of the furanone-LuxR complex, which in turn destabilises the AHL-dependent transcriptional activator [49]. However, independent studies confirmed that halogenated furanones without alkyl chains were strongly toxic to the planktonic cell. While furanones with long alkyl chains were shown not to reduce the biofilm formation [50,51]. The present study confirmed that furanone without alkyl chains has high toxicity against the *E. coli* Top10 pSB1A3-BBaT9002 strain, as it reduced 50% the bacterial growth when dosed at 5×10^{-5} M. In fact, at 1×10^{-3} M, indeed, furanone can kill all bacteria, diagnostic that the lack of alkyl chain is at play in the induced toxicity. It has been suggested that the increase of water solubility may explain this effect [51]. Interestingly, when dosed at 5×10^{-6} M, the results showed that furanone could reduce the fluorescence production at no detrimental expense of the bacterial growth (Figure 2c). A possible explanation to these phenomena might be that some compounds when applied at sub-lethal concentrations will not kill bacterial but delay the QS activity because bacteria become weaker than at the normal condition. This phenomenon gave us the illusion that those compounds were QSIs, but when we increased the dose, we noticed that they had an antibacterial activity. The observed antimicrobial effect at furanone concentrations $\geq 5 \times 10^{-6}$ M (Figure S10) prompted us to search QSIs with lower associated toxicity. Closer attention should be paid to this phenomenon to discriminate between potential QSIs more carefully.

Group 2 candidates included compounds with at least one aromatic ring. Vanillin (**5**), proposed as a less toxic alternative to brominated furanones, revealed a relative reduction of the biosensor's fluorescent response of 27.1 % when applied at 1×10^{-3} M (Figure 3a and Figure S7), an effect that was accompanied by a discrete, 16.0% reduction of cell growth (Figure 3b and Figure S7). Belonging to the same group, the well-reported QSI, *trans*-cinnamaldehyde (**6**) revealed a potent capacity to reduce the QS-based response of the biosensor to up to 68.3% (Figure 3a and Figure S7) when applied at 1×10^{-3} M, with a slight reduction of cell growth (18.6%; Figure 3b and Figure S7). Surprisingly, related compounds caffeic acid (**7**) and *trans*-anethole (**8**) were apparently innocuous to the *E. coli* biosensor even at concentrations of 1×10^{-3} M (Figure 3a,b) while capsaicin (**9**) and CAPE (**10**) inhibited cell growth by 12.8 and 24.4%, respectively (Figure 3b).

Vanillin (**5**) is a well-known food flavouring agent and is a major constituent of vanilla pods. Its QS inhibition activity has recently been demonstrated in *Aeromonas hydrophila*, *Agrobacterium tumifaciens* NTL-4, *Chromobacterium violaceum* CV026 when applied at 250 μg/mL (1.64 mM), where it showed significant inhibition in short-chain AHLs (C4-HSL (69%) and 3OC8-HSL (59.8%)), followed by C6-HSL(32%), and C8-HSL (28%), but lower inhibition in long-chain AHLs (C14-HSL (13.5%) and C10-HSL (12%)). It also reduces the biofilm formation on the reverse osmosis membrane of *A hydrophila* [52]. Our own experiments confirmed that vanillin at 1×10^{-3} M can inhibit the QS activity with 3OC6HSL 27.1%. It has been speculated that the possible mechanism whereby vanillin inhibits QS activity is that it interferes with the binding of the short-chain AHLs to their cognate receptor [52]. Our study offers experimental evidence that vanillin may also interfere with binding of 3OC6HSL to its cognate LuxR receptor.

trans-Cinnamaldehyde (**6**) is a component of cinnamon and cassia essential oils, and it is commonly present in food as a flavouring agent and fungicide. In a previous study, it was shown that 200 μM cinnamaldehyde can reduce by 70% the fluorescence intensity due to the expression of GFP, induced by 3-oxo-C6-HSL in a bioreporter *E. coli* ATCC 33,456 pJBA89, and also effective at inhibiting AI-2 mediated QS [53]. Furthermore, Brackman et al. proved *trans*-cinnamaldehyde at concentrations < 1 mM shifts the SDS-PAGE mobility of LuxR-DNA. They concluded that *trans*-cinnamaldehyde and cinnamaldehyde derivatives interfere with AI-2 based QS in various *Vibrio* spp. by decreasing the DNA-binding ability of LuxR [54]. Recent molecular docking analysis studies suggested that *trans*-cinnamaldehyde interacts with LasI and EsaI substrate binding sites thus inducing the QS inhibition activity [55]. Our own studies also confirm that *trans*-cinnamaldehyde inhibits the LuxR-mediated GFP production in the *E. coli* Top 10 biosensor by 68.3% at 1×10^{-3} M concentration. This strain does not express the LuxI gene, thus it cannot produce LuxI synthase of 3-oxo-C6-HSLs but can constitutively overexpress LuxR protein. In our related study, we have shown that *trans*-cinnamaldehyde not only inhibits the expression of GFP, but it also retards its kinetics [38]. We suggest that there are maybe two mechanisms at play for QS inhibition involving LuxR. Firstly, the three-carbon aliphatic side chain of *trans*-cinnamaldehyde could interfere with the binding of 3OC6HSL to LuxR receptor; secondly, it can also decrease the DNA-binding ability of LuxR dimers, as previously suggested.

trans-Cinnamaldehyde (**6**)-related compounds, namely caffeic acid (**7**) and *trans*-anethole (**8**), were also tested on their QS inhibition activity. The results showed that neither of them had QS inhibition activity nor antibacterial effect at the tested concentration of 1×10^{-3} M. Caffeic acid is a phenolic acid that has various documented beneficial biological properties. Besides being a powerful antioxidant, it also has anticancer, anti-inflammatory and antiviral activities [56]. *trans*-Anethole, in turn, is also a natural component of anise seeds and fruits and it is used as a flavouring ingredient. Until now, its antibacterial and QS inhibitory activities have not been tested.

Comparing the chemical structures of caffeic acid and *trans*-anethole with that of *trans*-cinnamaldehyde, we can clearly see that neither the catechol phenolic nor the aromatic ester groups of caffeic acid and trans-anethole, respectively, have any effect on increasing the binding affinity with the LuxR protein. These results may reflect the importance of the unsubstituted aromatic ring in *trans*-cinnamaldehyde to make π-π interactions with the receptor residues at the binding site of LuxR. In this regard, the phenolic residues of caffeic acid can act as the H-bond donors, but not acceptors, at the hydrogen bond acceptor binding domain of the binding pocket of LuxR, hence, decreasing its overall binding efficiency. Similarly, in trans-anethole, the presence of the methyl ester substituent at the aromatic ring may be is enough to prevent its efficient binding with the binding site of LuxR. However, vanillin, with one phenolic, one methyl ester and an aldehyde substituent at the aromatic ring, does show QS inhibition activity as discussed above. Therefore, our study is consistent with the idea that the role of the presence of H-bond acceptor groups along with the the π-π interactions of the aromatic ring, is what determines the overall affinity of compounds to bind with LuxR receptor.

Capsaicin (**9**), structurally related to vanillin (**5**), is a natural alkaloid extracted from fruit of Capsicum family and it is responsible for the pungency of chili peppers. Its structure has an aromatic ring and a long hydrophobic aliphatic chain with a polar amide group. Most studies with capsaicin have addressed its pharmacology and clinical applications [57]. Only a few studies have focused on the inhibition activity of the growth of the gastric pathogen *Helicobacter pylori* [58,59]. One such study found that capsaicin did not inhibit the growth of a human fecal commensal *E. coli* strain [58]. So far, the QS inhibitory activity of capsaicin has not been documented. Our results show that even at the concentration of 1×10^{-3} M, capsaicin did not significantly reduce the QS activity of the *E. coli* Top10 pSB1A3-BBaT9002 biosensor, and it decreased the bacterial growth only by 12.8% ($p < 0.05$). If we compare the structures of capsaicin with that of vanillin, we can observe that they share identical substitution positions in their aromatic ring, except that vanillin has not the long aliphatic chain, but only a strong aldehyde H-bond acceptor group. Our results indicate that the increase in the length of the aliphatic chain does not lead to an increase in binding affinity. CAPE (**10**) is the main active component of propolis extract. Its chemical structure is described as the ester of caffeic acid and phenetyl alcohol; hence, it is a catechol ring with an ester chain bearing another aromatic ring. CAPE has known bioactivities such as antimicrobial, anti-inflammatory and cytotoxic activities. About the antimicrobial activity of CAPE, it has been reported that can inhibit the 60% *E. coli* growth when the concentration is over 60.6 μM. This has been explained as the result of the synthesis of reactive oxygen species that damage the outer membrane of the bacteria-induced by CAPE [60,61]. Our experiments contrast with this study, as we found that CAPE only inhibits the growth of the *E. coli* Top10 pSB1A3-BBaT9002 strain by 24.4% when applied at 1×10^{-3} M. As a possible explanation for the observed discrepancies between the results of our study and the previous ones may stem in the distinct protocols of the assays to quantify the bacterial growth rate. Overall, no QS inhibition activity was shown on the surviving bacteria after treatment with CAPE.

Group 3, comprised by heterocyclic compounds, revealed candidates with important QSI capacity, namely caffeine (**11**) and genipin (**12**), which have been shown to reduce the QS-based response of the biosensor by 47% and by 43.3% when applied at $1 \text{x} 10^{-3}$ M and $5 \text{x} 10^{-5}$ M, respectively (Figure 3a,c, Figures S7 and S8) and with negligible effects on cell growth (Figure 3b–d, Figures S7 and S8). Belonging to the same group, gardenoside (**13**) did not show any apparent effect on the *E. coli* biosensor at the doses tested (Figure 3a,b).

Caffeine (**11**) is yet another alkaloid which occurs in coffee cherry and tea. Documented uses of caffeine include as a pesticide to kill certain larvae, insects and beetles. Some studies have shown that caffeine, applied at 2.5 mg/mL (≈12.8 mM), can retard the growth of *E. coli*, *Enterobacter aerogenes*, *Proteus vulgaris*, and *P. aeruginosa* within a short time. The first time caffeine was tested as QS inhibition compound against *C. violaceum* CV026 and *P. aeruginosa* PA01 strains, it was found that when applied at 0.3–1.0 mg/mL (≈1.5–5.0 mM), it inhibits violacein production in *C violaceum* CV026, and short chain AHLs production and swarming motility in *P. aeruginosa* PA01 [62]. Our experiments also proved that caffeine, applied at 1×10^{-3} M to the *E. coli* Top10 pSB1A3-BBaT9002 biosensor strain, it can inhibit 47% GFP production without affecting bacterial growth. Our results, along with previous studies [62], suggest that caffeine has a broad spectrum of QS inhibition activities in different bacterial species; interestingly, all share in common to contain the AHL-regulated QS system. As earlier reported by Norizan et al. Caffeine did not degrade C6-HSL [62], so we hypothesise here that it can be a competitor that binds with AHL receptors because the keto groups from the aromatic ring can also form strong hydrogen bonds with type I QS LuxI/LuxR receptors.

Also in Group 3, genipin (**12**) is an iridoid compound isolated from *Gardenia jasminodies* Ellis fruits. It is the aglycone derivative of geniposide. It was initially identified as a protein cross-linking agent but can also inhibit the production of nitric oxide by downregulating the activity of nuclear factor-κB (NF-κB) [63]. It is also a cell-permeable inhibitor with anti-inflammatory and anti-angiogenic activity mediated by the induction of apoptosis in hepatoma and hepatocarcinoma cell lines [64]. A number of studies have also shown that genipin cross-links chitosan nano- and microparticles, thus allowing them to be used for QS inhibition and the delivery of antimicrobial drugs [65,66]. We found, for the first time, that genipin on its own can inhibit GFP production in the *E. coli* Top10 pSB1A3-BBaT9002 biosensor, without exhibiting significant toxicity, more effectively than a diverse spectrum of alternative compounds. Knowing that genipin possesses inherent fluorogenic properties, we wondered whether the observed effect on the fluorescent response of our biosensor could be due to an artefact associated with the direct fluorescence quenching [67]. We performed a series of extra experiments where we tested genipin on the control *E. coli* strain Top10 pBCA9445-jtk28282::sfGFP strain (Supporting Information). We found that genipin concentrations ranging from 9.95×10^{-10} to 1.19×10^{-4} M did not exert significant fluorescence quenching on the control strain (Figures S2 and S3). Nevertheless, the fact that the fluorescence over growth profiles of both bacterial strains widely differ (Figure S1) make these comparisons a difficult task. Moreover, both strains express different GFP species (namely GFPmut3b and sfGFP in the biosensor and control strain, respectively) and their expression vectors widely diverge (see Supporting Information) [68,69]. Future efforts should be focused on validating the effects of genipin by using a control *E. coli* Top10 strain expressing GFPmut3b constitutively from a modified version of the pSB1A3-BbaT9002 plasmid. On our hands, genipin and related compounds are promising candidates for the development of novel therapeutic approaches based on the inhibition of QS in bacteria. Our data showed a significant difference between the QS inhibition activities of genipin and the closely related glycosylated iridoid compound, gardenoside. These two iridoids share similar structures, differing only with respect to the glycosylation in the hydroxyl group of C1, and in the hydroxyl group at position C8 of the heterocyclic structure, in gardenoside. This result may have a biological significance in host-pathogen interactions via inhibition of QS, as glucose oxidase (GOD) is synthesised by many plants, fungi and bacteria. This hypothesis would need further experimental validation.

Quinolone- and phenazine- like compounds belonging to Group 4 (Table 1) are reported to play important roles in QS-regulated phenotypes of *Pseudomonas spp* [70]. These comprise the alkylquinolone PQS (**14**) signal of *P. aeruginosa* and three structurally-related synthetic quinolones, namely MOQ (**15**), HOQ (**16**) and MHOQ (**17**) (Table 1); bearing a bicyclic core structure with different substituents. Among these three structurally-related synthetic compounds, only MOQ demonstrated significant relative QS inhibition activity of 62.2%. The rest of tested compounds, namely HOQ and MHOQ, did not have any detectable activity, neither QS inhibition nor antibacterial (Figure 3c,d and Figure S9). Also, compound PQS comprised not only one six-carbon ring but also a long carbon chain at the six-member heterocyclic ring, which showed an important QSI activity, by inhibiting the QS-based response of the biosensor by 43.1%, with a negligible effect on cell growth (Figure 3c,d and Figure S8). Compounds PYO (**18**), PCA (**19**) and PMS (**20**) are similarly with three rings, but only PYO and PMS have high GFP inhibition and high toxicity, compound PCA also has QQ activity, inhibiting GFP production by 33.7%, but no toxicity (Figure 3c,d).

Group 4, comprised heterocyclic compounds, some of which can be produced naturally, such as the *Pseudomonas* spp. metabolites (PQS, PCA, PYO), while the rest are synthetic [70]. It is well known that phenazines have antimicrobial activity [71]. Moreover, some of these compounds are known to act as signalling molecules that regulate the QS systems in *P. aeruginosa* as discussed in detail below.

Compound PQS has been found as a QS signal that participates in the *P. aeruginosa* QS network and acts as a link between las and rhl quorum sensing, which either directly or indirectly, mediates the expression of 182 genes in *P. aeruginosa* [72,73]. Besides the intraspecific signalling role of PQS in *P. aeruginosa* involving interactions with its cognate LasR-like PqsR receptor and non-cognate-LuxR-like-RhlI, there is no evidence to the date on PQS interference with other LuxR-based QS regulation circuitries. Despite the fact that there is currently no crystal structure available for RhlI, it is known that it shows significant sequence divergence from TraR, our prototype for docking studies. Thus, it is unknown how PQS binds to RhlI in *P. aeruginosa*. A recent paper from Mukherjee et al. explored ligand-receptor binding of PqsR with C4HSL, by generating a homology model based on the *E. coli* SdiA structure, which is a LuxR-like closest homolog of PqsR [74]. In SdiA and other LuxR-type proteins, the highly conserved amino acids Trp68 and Asp81 interact with the amide group-oxygen and the amide group-nitrogen, respectively, of the cognate C4HSL. Other conserved residues, such as Tyr72 and Trp96, are required for hydrophobic and van der Waals interactions with the ligands [74]. These interactions seem to correlate with our docking predictions of OOHL establishing H-bonding with Trp57 and hydrophobic interactions with Tyr53 and Trp85 in the binding site of TraR (Figure 5a and Table 1). A similar scheme of PQS predicted H-bonding and hydrophobic interactions with Trp57 and Tyr61 (Figure 5c,d and Table 1) would in principle serve as a rationale for a strong affinity of PQS to LuxR homologs and the observed quorum quenching effect on the biosensor's LuxR-regulated response (Figure 3c and Figure S8).

Phenazines are a well-known family of pigments that are secreted from *P. aeruginosa*. Among phenazines, PYO is widely known due to its cytotoxic and redox activities [75]. Importantly, PYO is a terminal signalling factor in the QS network of *P. aeruginosa* [76]. PYO is also an intercellular signal that triggers specific responses in enteric *E. coli* and *Salmonella enterica*, via the SoxR regulon. Whether this kind of signal transduction is also involved in our *E. coli* biosensor in the presence of PYO is a question that we cannot elucidate at present [77]. Also, it is well known that PYO interacts with molecular oxygen to form ROS species that change the redox balance of the cells and that GFP fluorescence can be affected by the presence of ROS species [78,79]. To shed some light on extra effects of PYO on fluorescence quenching and/or the metabolism of *E. coli* Top 10 cells, we decided to perform extra experiments applying increasing PYO concentrations on both the *E. coli* Top10 pSB1A3-T9002 (Figure S4) biosensor and the *luxR*-deficient *E. coli* Top10 pBCA9445-jtk28282::sfGFP (Figures S5 and S6). Strikingly, we found a lack of dose-response effect of PYO on the fluorescence of both strains (Figures S4–S6), together with strong variability among experiments (cf. Figures S5 and S6). These disparate results could arise from some of the transductory and/or redox activity of PYO on our *E. coli* Top10 cells and need further investigation.

PCA (**19**) is yet another redox-active phenazine pigment that is produced from *P. aeruginosa* [80]. PCA is known to be a broad-spectrum activity compound that inhibits the growth of several plant pathogenic species (e.g., *Corynebacterium fascians, Agrobacterium tumefaciens, Erwinia aroideae, Diplodia zeae*) [81,82]. Further studies have found that PCA is precursor for more complex phenazine metabolites, such as 1-hydroxyphenazine, phenazine-1-carboxamide and PYO [76]. Yun Wang et al. found that PCA may shift the redox equilibrium between Fe(III) and Fe(II) in *P. aeruginosa* [83]. Even though PCA and PYO, both are redox-active phenazines, in our study, we found that, at 5×10^{-5} M, PYO has strong toxicity to *E. coli* Top 10 pSB1A3-BBaT9002 strain, while PCA only inhibits the fluorescence intensity but has no effect on bacterial growth. The different effect of PYO and PCA can be attributed to the fact that PCA may help the *E. coli* alleviate Fe(III) limitation by reducing it to ferrous iron [Fe(II)], thus promoting the bacterial growth [83], thus allowing to observe the QS inhibition activity. We will discuss further this aspect in the context of the in silico molecular docking results below.

PMS is a simple phenazonium salt and an electron acceptor and carrier in biochemical oxidation/reduction studies [84,85]. It is also a O_2^- generating agent that can increase intracellular H_2O_2 levels and lead to the formation of free radicals that can affect bacterial growth [85]. This explains why in our experiments PMS shows strong toxic effect against *E. coli* Top 10 pSB1A3-BBaT9002 strain. As explained above, this strain is lacking the SoxR regulon to confer resistance against redox stress.

Previous studies have proved that phenazines are antibiotic compounds that can inhibit microbial growth because of the redox-active effect. From our own studies, it can be argued that when phenazines (namely, PCA, PYO, PMS) are applied at the same concentration of 5×10^{-5} M, only PYO and PMS are toxic to the *E. coli* Top10 pSB1A3-BBaT9002 strain, as we mentioned above. Interestingly, PCA, does not inhibit bacterial growth but has QS inhibition activity. Several studies have also noticed this phenomenon. Morales et al. found that lower concentrations of PCA, PYO and PMS inhibited the fungal yeast-to-filament transition and affected the development of *C. albicans* wrinkled colony biofilms but allowed growth. However, those phenazines have anti-candida activity when the concentration is higher than 500 μM, which means that those compounds have different biological effects at different concentrations [86]. Furthermore, Skindersoe et al. used sub-inhibitory concentrations of antibiotics in *P. aeruginosa* and found that lower doses of antibiotics could modulate gene expression, so that they interfere with QS signalling [87]. To the best of our knowledge, this dual concentration-dependent activity of phenazines had not been reported before to operate in a mutant *E. coli* Top10 pSB1A3-BBaT9002 strain. However, recently, it has been hypothesised that it is a general mechanism of action of many compounds [88].

For the identification of functional group and their arrangement in the binding site required for binding ligand, GRID map was generated by using three different chemical probes i.e., H bond donor (HD), H bond acceptor (HA) and DRY probe. Grid-MIFs generated for TraR indicated (Figure 4) that acceptor interaction points and hydrophobic patches are dominant in comparison to donor interactions points at the binding. Comparing the functional group present in ligands with the GRID-MIFs (Figure 4) it is clear that because of big hydrophobic patch in the center of cavity, hydrophobic interaction from carbon either from aromatic ring or long chain carbon make very favorable interaction. Apart this one donor group at one side of aromatic ring makes favorable interaction with the Trp 57. Hydrogen bond interaction with Trp 57 is identified as important interaction in receptor substrate interaction. Apart from HOQ we found this interaction in all other five ligands. Whereas at the other side of ring one donor or one acceptor would also make favorable interaction with the Thr129 or water.

Consequently, we identified that in the same ringside two functional groups e.g. an OH group just next to the O (acceptor group), do not make a favourable interacting group. This can be observed in the docking score results for HOQ and MHOQ both with lower scores than MOQ. This might explain that in vitro HOQ and MHOQ, did not exhibit QS inhibition activity, while MOQ, that lacks the 3-hydroxyl group did exhibit QS inhibition activity experimentally. PQS showed better score, i.e., PQS-conf A (−8.04) and PQS-conf B (−6.59), in comparison to HOQ and MHOQ because of the long alkyl chain and an overall more favourable hydrophobic interaction (Table 1). In the PQS-conf A (−8.04), the H-bond between the Trp 57 and ligand is missing in this conformation, but because of PQS has a longer alkyl chain than OOHL, a H-bond with Thr 129 and putative H-bond with water, PQS-conf A has a better score than the other compounds. Whereas in PQS-conf B (−6.59) the H-bond is present between ligand O and Trp 57, but because of the OH just next to O is not a favourable interaction (according to GRID-MIFs), hence its lower score. This analysis clearly indicates that OH next to O is not a favourable functional group for interaction in ligand PQS, which affects the pose and docking score. Moreover, because of the H-bond interactions and more hydrophobic interactions in comparison to natural ligand OOHL, this compound is better over other ligands. Along with PQS, PCA has also shown a high dock score compared to other ligands. We argue that this is the result of the combined effect of the three aromatic rings making more hydrophobic interaction, the deprotonated N present in the aromatic group involved in H-bond interaction with Trp57, and the carboxylate function group involved in H-bond interaction with Thr129 and also probably with water. These favourable interactions might explain the high QS inhibition activity observed for PQS and PCA.

Finally, Group 5 contained a more diverse collection of organic molecules with complex structures. Experimentally, with the exception of compounds itaconic acid at 1×10^{-3} M and berberine at 5×10^{-5} M, none of these compounds showed QS inhibition activity. However, we found that polygodial exhibited strong bacterial growth inhibition. Polygodial is a bicyclic sesquiterpene dialdehyde, isolated from different traditional medicinal plants such as *Polygonum hydropiper* and *P. punctatum* [89]. Kubo et al. showed that polygodial has the antibacterial activity against various bacteria, not only as a surfactant to form the pyrrole with primary amine groups at the plasma membrane, thereby disturbing the balance of the membranes, but also may react with various intracellular components when it enters into the cells after the membrane damaged [89]. We also proved that polygodial has high antibacterial activity that suppressed almost completely the growth of *E. coli* Top10 pSB1A3-BBaT9002 strain when dosed at 1×10^{-3} M.

Berberine is an isoquinoline-type alkaloid isolated from *Coptidis rhizomaand* ("huang lian" in Chinese), a plant used in traditional Chinese medicine, and from other plants. It has been reported that when the concentration is at 30-45 μg/mL could exhibit an antibacterial effect and inhibit biofilm formation of *Staphylococcus epidermidis*. Whether the biofilm formation inhibition of berberine observed in Gram-positive bacteria is connected with the QS regulation is not confirmed [90]. However, recent studies have shown that berberine inhibits the QS in Gram-negative bacteria including antimicrobial-resistant *E. coli* strains, *P. aeruginosa PA01*, *C. violaceum* and *Salmonella enterica* [91,92]. Sun et al. investigated the QS inhibition activity of berberine in antimicrobial-resistant (AMR) *E. coli* strains and found that berberine inhibited biofilm formation and downregulated QS-related genes *luxS*, *pfS*, *hflX*, *ftsQ*, and *ftsE* of AMR *E. coli* strains at 1/2 (640 μg/mL) or 1/4 (320 μg/mL) minimal inhibitory concentration (MIC) [90]. Thus the tested berberine concentrations of berberine were tested by Sun et al. were ≥ 9.5 times higher than ours (cf. ≥ 160 μg/mL and 16.8 μg/mL in Sun et al.'s and our study, respectively). Moreover, the AMR *E. coli* QS system is a LuxS/AI-mediated system, unrelated to the LuxR-based circuitry present in our biosensor. Under our setting, we found no QS inhibition at a berberine concentration of 50μM (16.8 μg/mL). The lower concentration tested in our assays may explain the observed lack of QS inhibition activity. We decided to limit berberine concentration to 50 μM due to solubility problems at higher concentrations. Further efforts should be focused on testing the QS inhibition potential of berberine and other related compounds at concentrations comparable to those of Sun et al.'s and exploring whether the strong effect observed on LuxS-based QS systems can be extrapolated to LuxR-regulated circuitries.

4. Materials and Methods

4.1. Library of Tested Chemical Compounds

Compounds were selected according to their chemical structure and were divided into five groups. They were in all cases of high purity (≥90%) and were either commercially available or synthesised. They were shipped in glass vials as powders or in liquid form and were dissolved in water or organic solution (ethanol or methanol) before use. The details for each compound are given in Table 2. Each is assigned a reference number used throughout this manuscript. 3-Oxohexanoyl-homoserine lactone (3OC6HSL) and all other chemicals were of analytical grade and, unless otherwise stated, were purchased from Merck KGaA (Darmstadt, Germany).

Table 2. List of screened compounds.

Number	Compounds	Group [1]	Solvent/Method	Supplier
1	γ-Valerolactone	Group (1) Lactone analogues	Water	Sigma (St. Louis, MO, USA)
2	L-Homoserine lactone		Water	Santa Cruz Biotechnology
3	α-Methyl-γ-butyrolactone		Water	Sigma (St. Louis, MO, USA)
4	Furanone((Z-)-4-Bromo-5-(bromomethy-lene)-2 (5*H*)-furanone)		First dissolved in ethanol, then diluted with water	Sigma (St. Louis, MO, USA)
5	Vanillin	Group (2) Aromatic ring structures	Water	Sigma (St. Louis, MO, USA)
6	*trans*-Cinnamaldehyde		First dissolved in ethanol, then diluted with water	Sigma (St. Louis, MO, USA)
7	Caffeic acid		First dissolved with ethanol, then diluted with water	Sigma (St. Louis, MO, USA)

Table 2. *Cont.*

Number	Compounds	Group [1]	Solvent/Method	Supplier
8	*trans*-Anethole		Water	Sigma (St. Louis, MO, USA)
9	Capsaicin		First dissolved with ethanol, then diluted with water	Merck KGaA (Darmstadt, Germany)
10	CAPE (caffeic acid phenethyl ester)		First dissolved with ethanol, then diluted with water	Merck KGaA (Darmstadt, Germany)
11	Caffeine	Group (3) Heterocyclic compounds	water	Merck KGaA (Darmstadt, Germany)
12	Genipin		Water	Challenge Bioproducts Co., Ltd.
13	Gardenoside		water	Nanjing Zelang Medical Technology Co.,Ltd
14	PQS (2-heptyl-3-hydroxy-4-quinolone)	Group (4) Quinolone- and phenazine-based compounds relevant to QS systems of *Pseudomonas* spp	First dissolved with methanol, then diluted with water	Merck KGaA (Darmstadt, Germany)
15	MOQ (1*H*-2-methyl-4-quinolone)		First dissolved with methanol, then diluted with water	Prof. Fetzner's [2]
16	HOQ (1*H*-3-hydroxyl-4-quinolone)		First dissolved with methanol, then diluted with water	Prof. Fetzner's [3]
17	MHOQ (1*H*-2-methyl-3-hydroxyl-4-quinolone)		First dissolved with methanol, then diluted with water	Prof. Fetzner's [4]
18	PYO (pyocyanine)		First dissolved with methanol, then diluted with water	Merck KGaA (Darmstadt, Germany)
19	PCA (Phenazine carboxylic acid)		First dissolved with methanol, then diluted with water	Key Organics Ltd. (Camelford, UK)
20	PMS (Phenazine methosulfate)		First dissolved with methanol, then diluted with water	Sigma (St. Louis, MO, USA)
21	Itaconic acid	Group (5) Structurally unrelated compounds	First dissolved in ethanol, then diluted with water	Sigma (St. Louis, MO, USA)
22	Polygodial		Water	Santa Cruz Biotechnology
23	Berberine		First dissolved in ethanol, then diluted with water	Sigma (St. Louis, MO, USA)

[1] The Group column refers to the classification based on chemical structural features (Figure 1), as explained in the text; [2] Synthesised by Prof. Susane Fetzner according to the method of Eiden et al. [93]. HPLC and UV absorption analysis indicated a purity of over 90%; [3] Synthesised by Prof. Susane Fetzner according to the method of Evans and Eastwood [94]. HPLC and UV absorption analysis indicated a purity of over 90%; [4] Synthesised by Prof. Susane Fetzner according to the method of Cornforth and James [95]. HPLC and UV absorption analysis indicated a purity of over 90%.

4.2. Bacterial Strains

The QQ activity of the 23 selected compounds was determined using the *E. coli* Top 10 strains listed below. The BioBrick standard biological sequence BBa_T9002, ligated into vector psb1a3 (http://partsregistry.org/Part:BBa_T9002), was a gift from Prof. Anderson (UC Berkeley, CA, USA). The sequence BBa_T9002 was introduced by chemical transformation into *E. coli* Top 10 (Invitrogen, Life Technologies Co., Leicestershire, UK) and single-colony cultures from the transformed strain were stored as 30% glycerol stocks at −80 °C as described in Section 2.3 below. The sequence BBa_T9002 comprised the transcription factor (LuxR), which is constitutively expressed, but it is active only in the presence of the exogenous autoinducer signalling molecule 3OC6HSL. At an adequate concentration, two molecules of 3OC6HSL bind to two molecules of LuxR and activate the expression of GFP (output), under the control of the lux pR promoter from *Vibrio fischeri*. The fluorescence biosensor was calibrated for different 3OC6HSL concentrations, as described in our previous studies [4]. An *E. coli* strain Top10 (Invitrogen, Life Technologies Co., U.K.) was transformed with plasmid pBCA9445-jtk2828, carrying a superfolder version of the *gfp* gene (*sfgfp*), which was kindly donated by Prof. Anderson Lab (UC Berkeley, Berkeley, CA, USA). The transformed strain expresses sfGFP constitutively and was used as control culture to test possible fluorescence quenching artefacts of genipin and PYO that could account for the effects observed in the fluorescence *E. coli* Top10 pSB1A3-BBaT9002 biosensor (Supporting Information).

4.3. Growth Media and Glycerol Stocks Preparation

Bacterial strains were cultivated using on Luria-Bertani (LB) and M9 minimal medium purchased from BD GmbH (Heidelberg, Germany). We inoculated 10 mL of LB broth supplemented with 200 μg/mL ampicillin with a single colony from a freshly streaked plate of Top10 containing BBa_T9002 and incubated the culture for 18 h at 37 °C, shaking at 100 rpm. Glycerol stocks were prepared as described in our previous studies [38]. Briefly, a 500 μL aliquots of overnight bacterial culture were mixed with 500 μL 30% sterile glycerol in 1.5 mL plastic vials and stored at −80 °C. Prior to each experiment, an aliquot of a glycerol stock from the single culture was diluted 1:1000 into 20 mL M9 minimal medium supplemented with 0.2% casamino acids, 1 mM thiamine hydrochloride and 200 μg/mL ampicillin (AppliChem GmbH, city, Germany). The culture was maintained under the same conditions until the OD600 reached ~0.15 (~5 h).

4.4. E. coli Top10 Fluorescent Biosensor Assay

Each tested compound was dissolved in MilliQ water or 100% organic solution (ethanol or methanol) according to their solubility at a high concentration of 200 mM, then diluted with MilliQ water to produce samples at six concentrations: 2×10^{-2}, 1×10^{-2}, 1×10^{-3}, 1×10^{-4}, 1×10^{-7} and 1×10^{-8} M; however, some compounds can only be prepared at a maximal concentration of 50 μM given by their water solubility. The 3OC6HSL was dissolved in acetonitrile to a stock concentration of 100 mM and stored at −20 °C kept in a sealed glass vial. Prior to each experiment, serial dilutions from the AHL stock solution were prepared in water to produce solutions with a concentration ranging from 100 mM to 10 nM. We then mixed 10 μL 3OC6HSL solution with 10 μL of the diluted compounds at different concentrations in the wells of a flat-bottomed 96-well plate (cat. # M3061, Greiner Bio-One, city, state abbrev if USA, country) and each well was then filled with 180 μL aliquots of the bacterial culture to test for QS inhibition activity. The final inhibitor concentrations, therefore, ranged from 1×10^{-3} M to 5×10^{-10} M. Several controls were also set up. Blank 1 contained 180 μL of M9 medium and 20 μL of MilliQ water to measure the absorbance background. Blank 2 wells contained 180 μL of bacterial culture and 20 μL of MilliQ water to measure the absorbance background-corrected for the cells. Finally, positive control wells contained 10 μL of water plus 10 μL 3OC6HSL solution and 180 μL of the bacterial culture to measure the fluorescence background. The plates were incubated in a Safire Tecan-F129013 Microplate Reader (Tecan, Crailsheim, Germany) at 37 °C and fluorescence measurements were taken automatically using a repeating procedure (λ_{ex} = 480 nm and λ_{em} = 510 nm, 40 μs, 10 flashes, gain 100, top fluorescence), absorbance measurements (OD_{600}) (λ = 600 nm absorbance filter, 10 flashes) and shaking (5 s, orbital shaking, high speed). The interval between measurements was 6 min. For each experiment, the fluorescence intensity (FI) and OD_{600} data were corrected by subtracting the values of absorbance and fluorescence backgrounds and expressed as the average for each treatment. Data were presented as FI/OD_{600} versus incubation time. All measurements were taken in triplicate.

4.5. Protein Structure File, Ligand Database

The X-ray crystal structure of *Agrobacterium tumefaciens* TraR was downloaded from the Protein Data Bank (PDB ID 1L3L) and used for computer docking. All the water molecules were removed except one molecule in the binding pocket, which plays an important role in interaction and forms the hydrogen bond with the autoinducer OOHL of TraR protein. To define the grid box of TraR protein, OOHL was used as a ligand to select spheres and also followed with the Information from the previous study [18]. The 2D structures of six compounds (OOHL, MOQ, HOQ, MHOQ, PCA and PQS) were drawn manually using Marvin sketch v6.1.3 (ChemAxon Ltd., Budapest, Hungary) and saved as MDL mol files. The mol files were merged into a single mol file and likewise converted to 3D structures using Discovery Studio 3.5 client software. PyMol was used for visualisation and molecular modelling.

4.6. Molecular Docking Studies

For the generation of GRID-MIFs (molecular interaction fields) at the TarR binding site where a given chemical group can interact favourably, Autogrid program inbuilt in AutoDockTools 1.5.6. was used. For MIF generation, mainly three probed were applied i.e., DRY probe representing hydrophobic interaction, HA probe to representing H bond acceptor groups, and HD probe to representing H bond donor groups. Docking guided by the grid map was performed using Autodock tool. Fifty conformations were generated for each docked substrate. Binding scores between the ligand and protein was evaluated using the autodock utility autoscorer considering the hydrogen bond forces, electrostatic forces, van der Waals forces, solvation energy and entropy.

4.7. Statistical Analysis

All the experiments were performed in triplicates to validate reproducibility and the P values were calculated statistically by Student's *t*-test. Values were expressed as mean ±SD. A comparison analysis was performed between tests and control.

5. Conclusions

In this study, we have screened the QS inhibition activity of a library of 23 structurally different compounds against an *E. coli* Top10 pSB1A3-BBaT9002 reporter of AHL-regulated QS. This library included a selection of natural and synthetic compounds that occur naturally in plants and in bacteria species such as *P. aeruginosa*. We were able to establish cues of structure-function relationships for compounds with QS inhibitory activity (e.g., *trans*-cinnamaldehyde, vanillin, caffeine, PQS, PCA). We showed, for the first time, that genipin and MOQ have QS inhibition activity. We also conducted molecular simulations using GRID-MIFs on a selection of compounds (e.g., MOQ, HOQ, MHOQ). Our results aid in the future rational design of novel QS inhibition compounds. For example, the introduction of a 3-methyl group in MOQ may increase the binding affinity substantially to the TraR receptor and hence the QS inhibition activity. This hypothesis could be validated experimentally in future studies. The results of this study may pave the way to future works aimed to fully realise the potential of QS inhibition as an alternative strategy to overcome antimicrobial resistance and biofilms in clinical and other settings.

Supplementary Materials: The following are available online at http://www.mdpi.com/1424-8247/13/9/263/s1.

Author Contributions: Conceptualiazation, F.M.G., X.Q., C.V.-S. and R.S.; methodology, X.Q. and C.V.-S.; formal analysis, X.Q. and R.S.; resources, B.P. and C.V.-S.; writing—original draft preparation, X.Q.; writing—review and editing X.Q., B.P., R.S., C.V.-S. and F.M.G.; supervision, F.M.G. and B.P.; project administration, F.M.G.; funding acquisition, F.M.G. All authors have read and agreed to the published version of the manuscript.

Funding: This research was funded by the Deutsche Forschungsgemeinschaft, grant number: GRK 1549 and the APC was funded by Ideas: European Research Council: FP7 613931. The authors acknowledge financial support by the Open Access Publication Fund of the Westfälische Wilhelms-Universität Münster, Germany.

Acknowledgments: X.Q. was recipient of a fellowship from China Scholarship Council. CVS was supported by a pre-doctoral fellowship of the Xunta de Galicia and by a FPU fellowship of the "Ministerio de Educación y Ciencia" of Spain, by a research fellowship of the DAAD (Germany), and research fellowship of the Fundación Pedro Barrié de la Maza (Spain). We acknowledge support from D.F.G., Germany (Project GRK 1549 International Research Training Group 'Molecular and Cellular GlycoSciences'); the research leading to these results has also received funding from the European Union's Seventh Framework Programme for research, technological development and demonstration under grant agreement no. 613931. We are also indebted to Antje von Schaewen for generous access to the Safire Tecan-F129013 Microplate Reader.

Conflicts of Interest: The authors declare no conflict of interest.

References

1. Fuqua, W.C.; Winans, S.C.; Greenberg, E.P. Quorum Sensing in Bacteria: The LuxR-LuxI Family of Cell Density-Responsive Transcriptional Regulatorst. *J. Bacteriol.* **1994**, *176*, 269–275. [CrossRef] [PubMed]
2. Rutherford, S.T.; Bassler, B. Bacterial Quorum Sensing: Its Role in Virulence and Possibilities for Its Control. *Cold Spring Harb. Perspect. Med.* **2012**, *2*, a012427. [CrossRef] [PubMed]
3. Bassler, B.; Losick, R. Bacterially Speaking. *Cell* **2006**, *125*, 237–246. [CrossRef] [PubMed]
4. Passador, L.; Cook, J.; Gambello, M.; Rust, L.; Iglewski, B. Expression of Pseudomonas aeruginosa virulence genes requires cell-to-cell communication. *Science* **1993**, *260*, 1127–1130. [CrossRef]
5. Ochsner, U.A.; Reiser, J. Autoinducer-mediated regulation of rhamnolipid biosurfactant synthesis in Pseudomonas aeruginosa. *Proc. Natl. Acad. Sci. USA* **1995**, *92*, 6424–6428. [CrossRef]
6. Singh, P.K.; Schaefer, A.L.; Parsek, M.R.; Moninger, T.O.; Welsh, M.J.; Greenberg, E.P. Quorum-sensing signals indicate that cystic fibrosis lungs are infected with bacterial biofilms. *Nature* **2000**, *407*, 762–764. [CrossRef]
7. Vannini, A.; Volpari, C.; Gargioli, C.; Muraglia, E.; Cortese, R.; De Francesco, R.; Neddermann, P.; Di Marco, S. The crystal structure of the quorum sensing protein TraR bound to its autoinducer and target DNA. *EMBO J.* **2002**, *21*, 4393–4401. [CrossRef]
8. Hanzelka, B.L.; Greenberg, E.P. Evidence that the N-terminal region of the Vibrio fischeri LuxR protein constitutes an autoinducer-binding domain. *J. Bacteriol.* **1995**, *177*, 815–817. [CrossRef]
9. Stevens, A.M.; Dolan, K.M.; Greenberg, E.P. Synergistic binding of the Vibrio fischeri LuxR transcriptional activator domain and RNA polymerase to the lux promoter region. *Proc. Natl. Acad. Sci. USA* **1994**, *91*, 12619–12623. [CrossRef]
10. Chai, Y.; Winans, S.C. Site-directed mutagenesis of a LuxR-type quorum-sensing transcription factor: Alteration of autoinducer specificity. *Mol. Microbiol.* **2004**, *51*, 765–776. [CrossRef]
11. Choi, S.H.; Greenberg, E.P. The C-terminal region of the Vibrio fischeri LuxR protein contains an inducer-independent lux gene activating domain. *Proc. Natl. Acad. Sci. USA* **1991**, *88*, 11115–11119. [CrossRef] [PubMed]
12. Kiratisin, P.; Tucker, K.D.; Passador, L. LasR, a Transcriptional Activator of Pseudomonas aeruginosa Virulence Genes, Functions as a Multimer. *J. Bacteriol.* **2002**, *184*, 4912–4919. [CrossRef] [PubMed]
13. Lamb, J.R.; Patel, H.; Montminy, T.; Wagner, V.E.; Iglewski, B.H. FunctionalDomains of the RhlR Transcriptional Regulator ofPseudomonasaeruginosa. *J. Bacteriol.* **2003**, *185*, 7129–7139. [CrossRef] [PubMed]
14. Luo, Z.-Q.; Smyth, A.J.; Gao, P.; Qin, Y.; Farrand, S.K. Mutational Analysis of TraR. *J. Boil. Chem.* **2003**, *278*, 13173–13182. [CrossRef] [PubMed]
15. Shadel, G.S.; Young, R.F.; Baldwin, T.O. Use of regulated cell lysis in a lethal genetic selection in Escherichia coli: Identification of the autoinducer-binding region of the LuxR protein from Vibrio fischeri ATCC 7744. *J. Bacteriol.* **1990**, *172*, 3980–3987. [CrossRef] [PubMed]
16. Slock, J.; VanRiet, D.; Kolibachuk, D.; Greenberg, E.P. Critical regions of the Vibrio fischeri luxR protein defined by mutational analysis. *J. Bacteriol.* **1990**, *172*, 3974–3979. [CrossRef]
17. Zhang, R.-G.; Pappas, K.M.; Brace, J.L.; Miller, P.C.; Oulmassov, T.; Molyneaux, J.M.; Anderson, J.C.; Bashkin, J.K.; Winans, S.C.; Joachimiak, A. Structure of a bacterial quorum-sensing transcription factor complexed with pheromone and DNA. *Nature* **2002**, *417*, 971–974. [CrossRef]
18. Müh, U.; Hare, B.J.; Duerkop, B.A.; Schuster, M.; Hanzelka, B.L.; Heim, R.; Olson, E.R.; Greenberg, E.P. A structurally unrelated mimic of a Pseudomonas aeruginosa acyl-homoserine lactone quorum-sensing signal. *Proc. Natl. Acad. Sci. USA* **2006**, *103*, 16948–16952. [CrossRef]
19. Koch, B.; Liljefors, T.; Persson, T.; Nielsen, J.; Kjelleberg, S.; Givskov, M. The LuxR receptor: The sites of interaction with quorum-sensing signals and inhibitors. *Microbiology* **2005**, *151*, 3589–3602. [CrossRef]

20. Ding, X.; Yin, B.; Qian, L.; Zeng, Z.; Yang, Z.; Li, H.; Lu, Y.; Zhou, S. Screening for novel quorum-sensing inhibitors to interfere with the formation of Pseudomonas aeruginosa biofilm. *J. Med. Microbiol.* **2011**, *60*, 1827–1834. [CrossRef]
21. Grandclément, C.; Tannières, M.; Moréra, S.; Dessaux, Y.; Faure, D. Quorum quenching: Role in nature and applied developments. *FEMS Microbiol. Rev.* **2015**, *40*, 86–116. [CrossRef] [PubMed]
22. Martin, C.A.; Hoven, A.D.; Cook, A.M. Therapeutic frontiers: Preventing and treating infectious diseases by inhibiting bacterial quorum sensing. *Eur. J. Clin. Microbiol. Infect. Dis.* **2008**, *27*, 635–642. [CrossRef] [PubMed]
23. Kociolek, M. Quorum-Sensing Inhibitors and Biofilms. *Anti Infect. Agents Med. Chem.* **2009**, *8*, 315–326. [CrossRef]
24. Choudhary, S.; Schmidt-Dannert, C. Applications of quorum sensing in biotechnology. *Appl. Microbiol. Biotechnol.* **2010**, *86*, 1267–1279. [CrossRef]
25. Clatworthy, A.E.; Pierson, E.; Hung, D.T. Targeting virulence: A new paradigm for antimicrobial therapy. *Nat. Methods* **2007**, *3*, 541–548. [CrossRef]
26. Kalia, V.C. Quorum sensing inhibitors: An overview. *Biotechnol. Adv.* **2013**, *31*, 224–245. [CrossRef]
27. Chen, F.; Gao, Y.; Chen, X.; Yu, Z.; Li, X. Quorum Quenching Enzymes and Their Application in Degrading Signal Molecules to Block Quorum Sensing-Dependent Infection. *Int. J. Mol. Sci.* **2013**, *14*, 17477–17500. [CrossRef]
28. Wang, Z.; Yu, P.; Zhang, G.; Xu, L.; Wang, D.; Wang, L.; Zeng, X.; Wang, Y. Design, synthesis and antibacterial activity of novel andrographolide derivatives. *Bioorg. Med. Chem.* **2010**, *18*, 4269–4274. [CrossRef]
29. Qin, X.; Thota, G.K.; Singh, R.; Balamurugan, R.; Goycoolea, F.M. Synthetic homoserine lactone analogues as antagonists of bacterial quorum sensing. *Bioorg. Chem.* **2020**, *98*, 103698. [CrossRef]
30. Yang, L.; Rybtke, M.T.; Jakobsen, T.H.; Hentzer, M.; Bjarnsholt, T.; Givskov, M.; Tolker-Nielsen, T. Computer-Aided Identification of Recognised Drugs as Pseudomonas aeruginosa Quorum-Sensing Inhibitors. *Antimicrob. Agents Chemother.* **2009**, *53*, 2432–2443. [CrossRef]
31. Zeng, Z.; Qian, L.; Cao, L.; Tan, H.; Huang, Y.; Xue, X.; Shen, Y.; Zhou, S. Virtual screening for novel quorum sensing inhibitors to eradicate biofilm formation of Pseudomonas aeruginosa. *Appl. Microbiol. Biotechnol.* **2008**, *79*, 119–126. [CrossRef] [PubMed]
32. Annapoorani, A.; Umamageswaran, V.; Parameswari, R.; Pandian, S.K.; Ravi, A.V. Computational discovery of putative quorum sensing inhibitors against LasR and RhlR receptor proteins of Pseudomonas aeruginosa. *J. Comput. Mol. Des.* **2012**, *26*, 1067–1077. [CrossRef] [PubMed]
33. Tan, S.Y.-Y.; Chua, S.L.; Chen, Y.; Rice, S.A.; Kjelleberg, S.; Nielsen, T.E.; Yang, L.; Givskov, M. Identification of Five Structurally Unrelated Quorum-Sensing Inhibitors of Pseudomonas aeruginosa from a Natural-Derivative Database. *Antimicrob. Agents Chemother.* **2013**, *57*, 5629–5641. [CrossRef] [PubMed]
34. Goodford, P.J. A Computational procedure for determining energetically favorable binding sites on biologically important macromolecules. *J. Med. Chem.* **1985**, *28*, 849–857. [CrossRef] [PubMed]
35. Carosati, E.; Sciabola, S.; Cruciani, G. Hydrogen Bonding Interactions of Covalently Bonded Fluorine Atoms: From Crystallographic Data to a New Angular Function in the GRID Force Field. *J. Med. Chem.* **2004**, *47*, 5114–5125. [CrossRef] [PubMed]
36. Ahlström, M.M.; Ridderström, M.; Luthman, A.K.; Zamora, I. Virtual Screening and Scaffold Hopping Based on GRID Molecular Interaction Fields. *J. Chem. Inf. Model.* **2005**, *45*, 1313–1323. [CrossRef]
37. Sciabola, S.; Stanton, R.V.; Mills, J.E.; Flocco, M.M.; Baroni, M.; Cruciani, G.; Perruccio, F.; Mason, J.S. High-Throughput Virtual Screening of Proteins Using GRID Molecular Interaction Fields. *J. Chem. Inf. Model.* **2009**, *50*, 155–169. [CrossRef]
38. Sanjurjo, C.V.; Engwer, C.; Qin, X.; Hembach, L.; Verdía-Cotelo, T.; Remuñán-López, C.; Vila-Sanjurjo, A.; Goycoolea, F.M. A single intracellular protein governs the critical transition from an individual to a coordinated population response during quorum sensing: Origins of primordial language. *bioRxiv* **2016**, 074369. [CrossRef]
39. Skandamis, P.N.; Nychas, G.-J. Quorum Sensing in the Context of Food Microbiology. *Appl. Environ. Microbiol.* **2012**, *78*, 5473–5482. [CrossRef]

40. Galloway, W.R.; Hodgkinson, J.T.; Bowden, S.; Welch, M.; Spring, D.R. Applications of small molecule activators and inhibitors of quorum sensing in Gram-negative bacteria. *Trends Microbiol.* **2012**, *20*, 449–458. [CrossRef]
41. Takano, E. γ-Butyrolactones: Streptomyces signalling molecules regulating antibiotic production and differentiation. *Curr. Opin. Microbiol.* **2006**, *9*, 287–294. [CrossRef] [PubMed]
42. Gram, L.; De Nys, R.; Maximilien, R.; Givskov, M.; Steinberg, P.; Kjelleberg, S. Inhibitory Effects of Secondary Metabolites from the Red Alga Delisea pulchra on Swarming Motility of Proteus mirabilis. *Appl. Environ. Microbiol.* **1996**, *62*, 4284–4287. [CrossRef] [PubMed]
43. Givskov, M.; De Nys, R.; Manefield, M.; Gram, L.; Maximilien, R.; Eberl, L.; Molin, S.; Steinberg, P.; Kjelleberg, S. Eukaryotic interference with homoserine lactone-mediated prokaryotuc signalling. *J. Bacteriol.* **1996**, *178*, 6618–6622. [CrossRef] [PubMed]
44. Manefield, M.; De Nys, R.; Naresh, K.; Roger, R.; Givskov, M.; Peter, S.; Kjelleberg, S. Evidence that halogenated furanones from Delisea pulchra inhibit acylated homoserine lactone (AHL)-mediated gene expression by displacing the AHL signal from its receptor protein. *Microbiology* **1999**, *145*, 283–291. [CrossRef] [PubMed]
45. Ren, D.; Sims, J.J.; Wood, T.K. Inhibition of biofilm formation and swarming of Escherichia coli by (5Z)-4-bromo-5-(bromomethylene)-3-butyl-2(5H)-furanone. *Environ. Microbiol.* **2001**, *3*, 731–736. [CrossRef]
46. Defoirdt, T.; Miyamoto, C.M.; Wood, T.K.; Meighen, E.A.; Sorgeloos, P.; Verstraete, W.; Bossier, P. The natural furanone (5Z)-4-bromo-5-(bromomethylene)-3-butyl-2(5H)-furanone disrupts quorum sensing-regulated gene expression in Vibrio harveyi by decreasing the DNA-binding activity of the transcriptional regulator protein luxR. *Environ. Microbiol.* **2007**, *9*, 2486–2495. [CrossRef]
47. Lönn-Stensrud, J.; Landin, M.A.; Benneche, T.; Petersen, F.C.; Scheie, A.A. Furanones, potential agents for preventing Staphylococcus epidermidis biofilm infections? *J. Antimicrob. Chemother.* **2009**, *63*, 309–316. [CrossRef]
48. Hentzer, M.; Wu, H.; Andersen, J.B.; Riedel, K.; Rasmussen, T.B.; Bagge, N.; Kumar, N.; Schembri, M.A.; Song, Z.; Kristoffersen, P.; et al. Attenuation of Pseudomonas aeruginosa virulence by quorum sensing inhibitors. *EMBO J.* **2003**, *22*, 3803–3815. [CrossRef]
49. Manefield, M.; Rasmussen, T.B.; Henzter, M.; Andersen, J.B.; Steinberg, P.; Kjelleberg, S.; Givskov, M. Halogenated furanones inhibit quorum sensing through accelerated LuxR turnover. *Microbiology* **2002**, *148*, 1119–1127. [CrossRef]
50. Steenackers, H.P.; Levin, J.; Janssens, J.C.; De Weerdt, A.; Balzarini, J.; Vanderleyden, J.; De Vos, D.E.; De Keersmaecker, S.C.J. Structure–activity relationship of brominated 3-alkyl-5-methylene-2(5H)-furanones and alkylmaleic anhydrides as inhibitors of Salmonella biofilm formation and quorum sensing regulated bioluminescence in Vibrio harveyi. *Bioorganic Med. Chem.* **2010**, *18*, 5224–5233. [CrossRef]
51. Janssens, J.C.A.; Steenackers, H.; Robijns, S.; Gellens, E.; Levin, J.; Zhao, H.; Hermans, K.; De Coster, D.; Verhoeven, T.L.; Marchal, K.; et al. Brominated Furanones Inhibit Biofilm Formation by Salmonella enterica Serovar Typhimurium. *Appl. Environ. Microbiol.* **2008**, *74*, 6639–6648. [CrossRef] [PubMed]
52. Ponnusamy, K.; Paul, D.; Kweon, J.H. Inhibition of Quorum Sensing Mechanism andAeromonas hydrophilaBiofilm Formation by Vanillin. *Environ. Eng. Sci.* **2009**, *26*, 1359–1363. [CrossRef]
53. Niu, C.; Afre, S.; Gilbert, E.S. Subinhibitory concentrations of cinnamaldehyde interfere with quorum sensing. *Lett. Appl. Microbiol.* **2006**, *43*, 489–494. [CrossRef] [PubMed]
54. Brackman, G.; Defoirdt, T.; Miyamoto, C.; Bossier, P.; Van Calenbergh, S.; Nelis, H.J.; Coenye, T. Cinnamaldehyde and cinnamaldehyde derivatives reduce virulence in Vibrio spp. by decreasing the DNA-binding activity of the quorum sensing response regulator LuxR. *BMC Microbiol.* **2008**, *8*, 149. [CrossRef] [PubMed]
55. Chang, C.-Y.; Krishnan, T.; Wang, H.; Chen, Y.; Yin, W.-F.; Chong, Y.-M.; Tan, L.Y.; Chong, T.M.; Chan, K.-G. Non-antibiotic quorum sensing inhibitors acting against N-acyl homoserine lactone synthase as druggable target. *Sci. Rep.* **2014**, *4*, 7245. [CrossRef] [PubMed]
56. Furuya, T.; Arai, Y.; Kino, K. Biotechnological Production of Caffeic Acid by Bacterial Cytochrome P450 CYP199A2. *Appl. Environ. Microbiol.* **2012**, *78*, 6087–6094. [CrossRef]
57. Hayman, M.; Kam, P.C. Capsaicin: A review of its pharmacology and clinical applications. *Curr. Anaesth. Crit. Care* **2008**, *19*, 338–343. [CrossRef]

58. Jones, N.L.; Shabib, S.; Sherman, P.M. Capsaicin as an inhibitor of the growth of the gastric pathogen Helicobacter pylori. *FEMS Microbiol. Lett.* **1997**, *146*, 223–227. [CrossRef]
59. Zeyrek, F.Y.; Oguz, E. In vitro activity of capsaicin against Helicobacter pylori. *Ann. Microbiol.* **2005**, *55*, 125–127.
60. Lee, H.S.; Lee, S.Y.; Park, S.H.; Lee, J.H.; Ahn, S.K.; Choi, Y.M.; Choi, D.J.; Chang, J.-H. Antimicrobial medical sutures with caffeic acid phenethyl ester and their in vitro/in vivo biological assessment. *MedChemComm* **2013**, *4*, 777. [CrossRef]
61. Murtaza, G.; Karim, S.; Akram, M.R.; Khan, S.A.; Azhar, S.; Mumtaz, A.; Bin Asad, M.H.H. Caffeic Acid Phenethyl Ester and Therapeutic Potentials. *BioMed Res. Int.* **2014**, *2014*, 1–9. [CrossRef] [PubMed]
62. Norizan, S.N.M.; Yin, W.-F.; Chan, K.-G. Caffeine as a Potential Quorum Sensing Inhibitor. *Sensors* **2013**, *13*, 5117–5129. [CrossRef]
63. Koo, H.-J.; Song, Y.S.; Kim, H.-J.; Lee, Y.-H.; Hong, S.-M.; Lim, S.-J.; Kim, B.-C.; Jin, C.; Lim, C.-J.; Park, E.-H. Anti-inflammatory effects of genipin, an active principle of gardenia. *Eur. J. Pharmacol.* **2004**, *495*, 201–208. [CrossRef] [PubMed]
64. Kim, B.-C.; Kim, H.-G.; Lee, S.-A.; Lim, S.; Park, E.-H.; Kim, S.-J.; Lim, C.-J. Genipin-induced apoptosis in hepatoma cells is mediated by reactive oxygen species/c-Jun NH2-terminal kinase-dependent activation of mitochondrial pathway. *Biochem. Pharmacol.* **2005**, *70*, 1398–1407. [CrossRef] [PubMed]
65. Sanjurjo, C.V.; David, L.; Remuñán-López, C.; Goycoolea, F.M.; Vila-Sanjurjo, A. Effect of the ultrastructure of chitosan nanoparticles in colloidal stability, quorum quenching and antibacterial activities. *J. Colloid Interface Sci.* **2019**, *556*, 592–605. [CrossRef] [PubMed]
66. Lin, Y.-H.; Tsai, S.-C.; Lai, C.-H.; Lee, C.-H.; He, Z.S.; Tseng, G.-C. Genipin-cross-linked fucose–chitosan/heparin nanoparticles for the eradication of Helicobacter pylori. *Biomaterials* **2013**, *34*, 4466–4479. [CrossRef]
67. Hwang, Y.-J.; Larsen, J.; Krasieva, T.B.; Lyubovitsky, J.G. Effect of Genipin Crosslinking on the Optical Spectral Properties and Structures of Collagen Hydrogels. *ACS Appl. Mater. Interfaces* **2011**, *3*, 2579–2584. [CrossRef]
68. Canton, B.; Labno, A.; Endy, D. Refinement and standardisation of synthetic biological parts and devices. *Nat. Biotechnol.* **2008**, *26*, 787–793. [CrossRef]
69. Pe'delacq, J.-D.; Cabantous, S.; Tran, T.; Terwilliger, T.C.; Waldo, G.S. Corrigendum: Engineering and characterisation of a superfolder green fluorescent protein. *Nat. Biotechnol.* **2006**, *24*, 1170. [CrossRef]
70. Lin, J.-S.; Cheng, J.; Wang, Y.; Shen, X. The Pseudomonas Quinolone Signal (PQS): Not Just for Quorum Sensing Anymore. *Front. Microbiol.* **2018**, *8*. [CrossRef]
71. Price-Whelan, A.; Dietrich, L.E.; Newman, D.K. Rethinking 'secondary' metabolism: Physiological roles for phenazine antibiotics. *Nat. Methods* **2006**, *2*, 71–78. [CrossRef] [PubMed]
72. Pesci, E.C.; Milbank, J.B.J.; Pearson, J.P.; McKnight, S.; Kende, A.S.; Greenberg, E.P.; Iglewski, B.H. Quinolone signaling in the cell-to-cell communication system of Pseudomonas aeruginosa. *Proc. Natl. Acad. Sci. USA* **1999**, *96*, 11229–11234. [CrossRef] [PubMed]
73. McKnight, S.L.; Iglewski, B.H.; Pesci, E.C. The Pseudomonas Quinolone Signal Regulates rhl Quorum Sensing in Pseudomonas aeruginosa. *J. Bacteriol.* **2000**, *182*, 2702–2708. [CrossRef] [PubMed]
74. Mukherjee, S.; Moustafa, D.A.; Stergioula, V.; Smith, C.D.; Goldberg, J.B.; Bassler, B. The PqsE and RhlR proteins are an autoinducer synthase–receptor pair that control virulence and biofilm development in Pseudomonas aeruginosa. *Proc. Natl. Acad. Sci. USA* **2018**, *115*, E9411–E9418. [CrossRef] [PubMed]
75. Mahajan-Miklos, S.; Tan, M.; Rahme, L.G.; Ausubel, F.M. Molecular Mechanisms of Bacterial Virulence Elucidated Using a Pseudomonas aeruginosa– Caenorhabditis elegans Pathogenesis Model. *Cell* **1999**, *96*, 47–56. [CrossRef]
76. Dietrich, L.E.; Price-Whelan, A.; Petersen, A.; Whiteley, M.; Newman, D.K. The phenazine pyocyanin is a terminal signalling factor in the quorum sensing network of Pseudomonas aeruginosa. *Mol. Microbiol.* **2006**, *61*, 1308–1321. [CrossRef]
77. Seo, S.; Gao, Y.; Kim, N.; Szubin, R.; Yang, J.; Cho, B.-K.; Palsson, B.O. Revealing genome-scale transcriptional regulatory landscape of OmpR highlights its expanded regulatory roles under osmotic stress in Escherichia coli K-12 MG1655. *Sci. Rep.* **2017**, *7*, 2181. [CrossRef]
78. Jagmann, N.; Brachvogel, H.-P.; Philipp, B. Parastic growth of Pseudomonas aeruginosa in co-culture with the chitinolytic bacterium Aeromonas hydrophila. *Environ. Microbiol.* **2010**, *12*, 1787–1802. [CrossRef]

79. Bou-Abdallah, F.; Chasteen, N.D.; Lesser, M.P. Quenching of superoxide radicals by green fluorecent protein. *Biochim. Biophys. Acta* **2006**, *1760*, 1960–1965.
80. Mavrodi, D.V.; Bonsall, R.F.; Delaney, S.M.; Soule, M.J.; Phillips, G.; Thomashow, L.S. Functional Analysis of Genes for Biosynthesis of Pyocyanin and Phenazine-1-Carboxamide from Pseudomonas aeruginosa PAO1. *J. Bacteriol.* **2001**, *183*, 6454–6465. [CrossRef]
81. Haynes, W.C.; Stodola, F.H.; Locke, J.M.; Pridham, T.G.; Conway, H.F.; Sohns, V.E.; Jackson, R.W. PSEUDOMONAS AUREOFACIENS KLUYVER AND PHENAZINE α-CARBOXYLIC ACID, ITS CHARACTERISTIC PIGMENT. *J. Bacteriol.* **1956**, *72*, 412–417. [CrossRef] [PubMed]
82. Mazzola, M.; Cook, R.J.; Thomashow, L.S.; Weller, D.M.; Pierson, L.S. Contribution of phenazine antibiotic biosynthesis to the ecological competence of fluorescent pseudomonads in soil habitats. *Appl. Environ. Microbiol.* **1992**, *58*, 2616–2624. [CrossRef]
83. Wang, Y.; Wilks, J.C.; Danhorn, T.; Ramos, I.; Croal, L.; Newman, D.K. Phenazine-1-Carboxylic Acid Promotes Bacterial Biofilm Development via Ferrous Iron Acquisition. *J. Bacteriol.* **2011**, *193*, 3606–3617. [CrossRef]
84. Pierson, L.S.; Pierson, E.A. Metabolism and function of phenazines in bacteria: Impacts on the behavior of bacteria in the environment and biotechnological processes. *Appl. Microbiol. Biotechnol.* **2010**, *86*, 1659–1670. [CrossRef] [PubMed]
85. Hassett, D.J.; Ma, J.-F.; Elkins, J.G.; McDermott, T.R.; Ochsner, U.A.; West, S.E.H.; Huang, C.-T.; Fredericks, J.; Burnett, S.; Stewart, P.S.; et al. Quorum sensing in Pseudomonas aeruginosa controls expression of catalase and superoxide dismutase genes and mediates biofilm susceptibility to hydrogen peroxide. *Mol. Microbiol.* **1999**, *34*, 1082–1093. [CrossRef] [PubMed]
86. Morales, D.K.; Grahl, N.; Okegbe, C.; Dietrich, L.E.; Jacobs, N.J.; Hogan, D.A. Control of Candida albicans Metabolism and Biofilm Formation by Pseudomonas aeruginosa Phenazines. *mBio* **2013**, *4*, 00526-12. [CrossRef] [PubMed]
87. Skindersoe, M.E.; Alhede, M.; Phipps, R.; Yang, L.; Jensen, P.Ø.; Rasmussen, T.B.; Bjarnsholt, T.; Tolker-Nielsen, T.; Høiby, N.; Givskov, M. Effects of Antibiotics on Quorum Sensing in Pseudomonas aeruginosa. *Antimicrob. Agents Chemother.* **2008**, *52*, 3648–3663. [CrossRef] [PubMed]
88. Schertzer, J.W.; Boulette, M.L.; Whiteley, M. More than a signal: Non-signaling properties of quorum sensing molecules. *Trends Microbiol.* **2009**, *17*, 189–195. [CrossRef]
89. Kubo, I.; Fujita, K.-I.; Lee, S.H.; Ha, T.J. Antibacterial activity of polygodial. *Phytother. Res.* **2005**, *19*, 1013–1017. [CrossRef]
90. Wang, X.; Yao, X.; Zhu, Z.-A.; Tang, T.; Dai, K.; Sadovskaya, I.; Flahaut, S.; Jabbouri, S. Effect of berberine on Staphylococcus epidermidis biofilm formation. *Int. J. Antimicrob. Agents* **2009**, *34*, 60–66. [CrossRef]
91. Sun, T.; Li, X.-D.; Hong, J.; Liu, C.; Zhang, X.-L.; Zheng, J.-P.; Xu, Y.-J.; Ou, Z.-Y.; Zheng, J.-L.; Yu, D.-J. Inhibitory Effect of Two Traditional Chinese Medicine Monomers, Berberine and Matrine, on the Quorum Sensing System of Antimicrobial-Resistant Escherichia coli. *Front. Microbiol.* **2019**, *10*, 2584. [CrossRef] [PubMed]
92. Aswathanarayan, J.B.; Vittal, R.R. Inhibition of biofilm formation and quorum sensing mediated phenotypes by berberine in Pseudomonas aeruginosa and Salmonella typhimurium. *RSC Adv.* **2018**, *8*, 36133–36141. [CrossRef]
93. Eiden, F.; Wendt, R.; Fenner, H. ChemInform Abstract: PYRONES AND PYRIDONES, PART 74. QUINOLYLIDENE DERIVATIVES. *Chem. Informationsdienst* **1978**, *9*, 561–568. [CrossRef]
94. Evans, D.; Eastwood, F. Synthesis of an arylhydroxytetronimide and of 3-Hydroxy-4(1H)-quinolone derivatives. *Aust. J. Chem.* **1974**, *27*, 537. [CrossRef]
95. Cornforth, J.W.; James, A.T. Structure of a naturally occurring antagonist of dihydrostreptomycin. *Biochem. J.* **1956**, *63*, 124–130. [CrossRef] [PubMed]

Article

In Vitro Selective Growth-Inhibitory Activities of Phytochemicals, Synthetic Phytochemical Analogs, and Antibiotics against Diarrheagenic/Probiotic Bacteria and Cancer/Normal Intestinal Cells

Tomas Kudera [1], Ivo Doskocil [2], Hana Salmonova [2], Miloslav Petrtyl [3], Eva Skrivanova [2] and Ladislav Kokoska [1,*]

1 Department of Crop Sciences and Agroforestry, Faculty of Tropical AgriSciences, Czech University of Life Sciences Prague, Kamycka 129, 16500 Praha-Suchdol, Czech Republic; kuderat@ftz.czu.cz

2 Department of Microbiology, Nutrition and Dietetics, Faculty of Agrobiology, Food and Natural Resources, Czech University of Life Sciences Prague, Kamycka 129, 16500 Praha-Suchdol, Czech Republic; doskocil@af.czu.cz (I.D.); salmonova@af.czu.cz (H.S.); skrivanovae@af.czu.cz (E.S.)

3 Department of Zoology and Fisheries, Faculty of Agrobiology, Food and Natural Resources, Czech University of Life Sciences Prague, Kamycka 129, 16500 Praha-Suchdol, Czech Republic; petrtyl@af.czu.cz

* Correspondence: kokoska@ftz.czu.cz; Tel.: +420-224-382-180

Received: 21 July 2020; Accepted: 31 August 2020; Published: 3 September 2020

Abstract: A desirable attribute of novel antimicrobial agents for bacterial diarrhea is decreased toxicity toward host intestinal microbiota. In addition, gut dysbiosis is associated with an increased risk of developing intestinal cancer. In this study, the selective growth-inhibitory activities of ten phytochemicals and their synthetic analogs (berberine, bismuth subsalicylate, ferron, 8-hydroxyquinoline, chloroxine, nitroxoline, salicylic acid, sanguinarine, tannic acid, and zinc pyrithione), as well as those of six commercial antibiotics (ceftriaxone, ciprofloxacin, chloramphenicol, metronidazole, tetracycline, and vancomycin) against 21 intestinal pathogenic/probiotic (e.g., *Salmonella* spp. and bifidobacteria) bacterial strains and three intestinal cancer/normal (Caco-2, HT29, and FHs 74 Int) cell lines were examined in vitro using the broth microdilution method and thiazolyl blue tetrazolium bromide assay. Chloroxine, ciprofloxacin, nitroxoline, tetracycline, and zinc pyrithione exhibited the most potent selective growth-inhibitory activity against pathogens, whereas 8-hydroxyquinoline, chloroxine, nitroxoline, sanguinarine, and zinc pyrithione exhibited the highest cytotoxic activity against cancer cells. None of the tested antibiotics were cytotoxic to normal cells, whereas 8-hydroxyquinoline and sanguinarine exhibited selective antiproliferative activity against cancer cells. These findings indicate that 8-hydroxyquinoline alkaloids and metal-pyridine derivative complexes are chemical structures derived from plants with potential bioactive properties in terms of selective antibacterial and anticancer activities against diarrheagenic bacteria and intestinal cancer cells.

Keywords: plant compounds; diarrhea; antibacterial; anticancer; selectivity

1. Introduction

The lack of an effective and safe antimicrobial therapy for diarrheagenic bacterial infections is a global health concern, especially in developing countries for children under the age of five years [1]. Although mortality associated with bacterial diarrhea is low in developed countries, the increased incidence rates of inflammatory bowel disease and colorectal cancer have also been attributed to gut dysbiosis that can result from chronic intestinal infections [2]. Currently, infectious diarrhea is treated using conventional drugs belonging to various classes of antibiotics, such as ceftriaxone, chloramphenicol, ciprofloxacin, tetracycline, metronidazole, and vancomycin. However, the irrational use of antibiotics, including incorrect dose prescription, has led to the development of drug resistance in several pathogens. Additionally, the applications of conventional antibiotics are limited, especially among children in developing countries, owing to high cost and increased risk of side effects including gut dysbiosis. Therefore, there is a need to identify novel antimicrobial agents for infectious bacterial diarrhea to overcome the limitations of conventional antimicrobial drugs [3].

Over the last decades, plant-derived products have become a mainstay in providing novel chemical scaffolds for the development of anti-infective drugs, and therefore antidiarrheal medicinal plants and their bioactive components can be examined first [4]. However, the therapeutic effect of products derived from antidiarrheal medicinal plants is not necessarily based on their antimicrobial activity against the causative agents as other mechanisms can be considered important, such as antimotility and antisecretory effects, for their useful applications [5]. The chance of finding a new plant-derived compound with promising antibacterial activity could therefore be enhanced by the chemotaxonomic approach by examining such novel medicinal plants that are taxonomically related to the species known to bear specific types of phytochemicals with growth-inhibitory effect on diarrheagenic bacteria [6]. In a recent review paper, Kokoska et al. [7] described the in vitro antimicrobial properties and clinical efficacy of plant-derived compounds with their synthetic analogs that are present in products available in the international market as over-the-counter pharmaceuticals, dietary supplements, and herbal medicines for intestinal infections. Benzylisoquinoline alkaloid berberine (e.g., *Hydrastis canadensis*), simple phenol bismuth subsalicylate, the analog of salicylic acid derived from salicin (*Salix alba*), and polyphenol tannic acid (e.g., *Quercus* spp.) were mentioned as examples of the efficient agents with potent activity against some of the gut bacterial pathogens.

As it has recently been proposed, investigations of new antimicrobial agents should be focused on the identification of structures with lowered toxicity to indigenous intestinal microbiota including probiotic bacterial strains. Since there is a common association between dysbiosis and the use of antibiotics, the agents selectively acting against pathogenic microorganisms can prevent the risk of developing diseases, such as chronic bowel inflammation and intestinal carcinoma [2]. Although the composition of the whole gut microbiota comprises a large number of microorganisms forming a complex ecosystem, a preliminary screening of selective antibacterial activity of newly tested agents could be performed by testing the in vitro susceptibilities of particular representatives of each of the three dominant bacterial phyla that can be found in human intestines, and which probiotic function have been described. Among them, we can recognize the strains such as *Bifidobacterium* spp. (Actinobacteria), *Lactobacillus* spp. (Firmicutes) and *Bacteroides fragilis* (Bacteroidetes) [8,9]. Selective in vitro growth inhibitory effect of a plant-derived compound was for example described in the study of Novakova et al. [10], where anticlostridial effect of 8-hydroxyquinoline (*Microstachys corniculata*) was comparably higher than the activities revealed against different strains of bifidobacteria. Chloroxine, the synthetic 8-hydroxyquinoline derivative, is the antimicrobial agent that has also been used as an oral formulation for infectious diarrhea, and disorders of the intestinal microbiota [11]. In dysbiosis, the increased abundance of *Fusobacterium nucleatum* and *Faecalibacterium prausnitzii* is positively and negatively correlated with the risk of intestinal carcinogenesis, respectively [12]. The repression of gut microbiota enhances the susceptibility of host intestinal cells to diarrheagenic bacterial infection and chemical-induced cytotoxicity. Thus, novel antimicrobial agents should not exhibit cytotoxic activity against normal intestinal cells [13]. For example, quinolone antibiotics were reported to inhibit the growth of both bacterial and eukaryotic cells through the same mechanism and consequently enhance the risk of eliciting cytotoxic response [14,15]. On the other hand, the antiproliferative activity of these agents with selective cytotoxic effects against intestinal cancer cells would be a suitable feature in cases of dysbiosis associated with carcinogenesis. There are limited studies on the anticancer effects of conventional antibiotics. For example, Bourikas et al. [16] reported that ciprofloxacin is potent to inhibit the proliferation of intestinal cancer cell line HT29 in vitro. On the other hand, a strong in vitro antiproliferative activity of plant-derived compounds against various cancer cell lines has extensively been reported, and some are used as a scaffold for anticancer drugs. For example, the alkaloid camptothecin, extracted from the bark of *Camptotheca acuminata*, is currently used as a cytostatic agent for the treatment of colon cancer [17]. We, therefore, suggest that the phytochemicals with known in vitro growth-inhibitory activity against some of the diarrheagenic bacteria could potentially exhibit selective cytotoxic effects on cancer cells. Amongst them, the anticancer effect has generally been reported for quinoline alkaloids [18]. In addition to those earlier mentioned, benzylisoquinoline alkaloid sanguinarine (*Sanguinaria canadensis*) [7,19], and 8-hydroxyquinoline derivatives ferron [20] and nitroxoline [21] are examples of antimicrobial drugs with potent anticancer activity. Another type of synthetic phytochemical analog with reported antimicrobial and anticancer properties is metal-pyridine derivative complex zinc pyrithione (pyrithione found in *Polyalthia nemoralis*) [7,22].

In this study, we compared the selective antibacterial (diarrheagenic/probiotic strains) and cytotoxic (cancer/normal intestinal cells) activities of phytochemicals (alkaloids and phenolics) and their synthetic analogs with those of antibiotics in vitro. For each test compound, values of minimum inhibitory concentrations (MICs), half-maximal inhibitory concentration (IC_{50}), and 80% inhibitory concentration of proliferation (IC_{80}) were assessed. The means of these values ($\overline{x}$-MIC, $\overline{x}$-IC_{50}, and $\overline{x}$-IC_{80}), each defined for a particular type of strain/cell line, were used for calculation of selectivity index (SI) between activities against normal intestinal cells and diarrheagenic strains (SIa), probiotic and diarrheagenic strains (SIb), and normal and cancer intestinal cells (SIc). The aim was to obtain some of the missing data on particular in vitro activities of the test compound, to assist in identifying phytochemicals with scaffolds that possess a potent combination of bioactivities. They could subsequently be utilized in future chemotaxonomic investigation of antidiarrheal medicinal plants with their bioactive components considered for new chemotherapeutic agents against diarrheal infections and associated intestinal cancer diseases. Our results show that 8-hydroxyquinoline alkaloids and zinc pyrithione possess in vitro selective antibacterial properties against diarrheagenic bacteria comparable to ciprofloxacin and tetracycline, with additional in vitro antiproliferative activity against cancer intestinal cell lines. However, in contrary to antibiotics, these compounds generally possess increased cytotoxicity to normal intestinal cells.

2. Results

2.1. Antibacterial Activity

As far as the antibacterial activity of antibiotics against diarrheagenic strains is considered, ciprofloxacin and tetracycline exhibited strong growth-inhibitory effect ($\overline{x}$-MICs = 2 ± 4 and 4.8 ± 8 µg/mL, respectively), while chloramphenicol and ceftriaxone exhibited moderate growth-inhibitory activities ($\overline{x}$-MIC = 16.5 ± 34 and 61.8 ± 141 µg/mL, respectively). Regarding the significant degree of variation between MICs of these compounds, particular types of pathogenic strains were highly susceptible while some other species were distinctly more resistant. For example, all gram-negative diarrheagenic bacteria were highly susceptible to ciprofloxacin (MICs = 0.016–0.125 µg/mL) and ceftriaxone (MICs = 0.062–0.5 µg/mL). In contrast, their MICs produced against gram-positive pathogens were comparably higher; for the former ranging from 1 to 16 µg/mL and for the latter in the range of 4–512 µg/mL. At least half of the diarrheagenic bacteria were inhibited at the low MICs (1–4 µg/mL) by chloramphenicol and tetracycline, whereas the variations were particularly caused by the weak activities revealed against *Enterococcus faecalis* (MIC = 128 µg/mL) and *Clostridium perfringens* (32 µg/mL), respectively. At the low concentration (MICs = 0.5 µg/mL), tetracycline also inhibited *Clostridium difficile* and *Bacillus cereus*. Although metronidazole and vancomycin were generally inactive against diarrheagenic bacteria ($\overline{x}$-MIC = 651.4 ± 459 and 512.9 ± 404, respectively), they produced a strong inhibitory effect against both clostridial species tested (MICs = 0.5–8 µg/mL), whereas the former also exhibited strong activity against *Escherichia coli* (0.062 µg/mL). The synthetic analogs of phytochemicals, namely, zinc pyrithione, nitroxoline, and chloroxine, exhibited strong to moderate growth-inhibitory activity against all diarrheagenic bacteria ($\bar{x}$-MICs = 7.1 ± 4, 12 ± 10, and 24 ± 19 μg/mL, respectively). The activities of these compounds against particular pathogenic bacteria did not have significant differences, however, some of the strains were comparably more susceptible. Regarding that, *B. cereus*, *E. coli*, *Shigella flexneri*, and *Vibrio parahaemolyticus* were highly susceptible to zinc pyrithione (MICs = 1–4 μg/mL). The MIC of chloroxine against both *B. cereus* and *C. difficile* was 8 μg/mL. Nitroxoline exhibited strong growth-inhibitory activities (MICs = 2–4 μg/mL) against *B. cereus*, clostridial species, *E. coli*, and *S. flexneri*. Although 8-hydroxyquinoline did not produce significant antibacterial activity against the pathogens ($\bar{x}$-MIC = 224.4 ± 181), its growth-inhibitory activities were strong against *E. faecalis* (MIC = 4 μg/mL) and *Listeria monocytogenes* (MIC = 1 μg/mL).

Subsequently, the growth-inhibitory activities of different test compounds against probiotic strains were evaluated. Probiotic bacteria exhibited high susceptibility to chloramphenicol ($\bar{x}$-MIC = 6.2 ± 4 μg/mL) and medium susceptibility to tetracycline, nitroxoline, zinc pyrithione, ciprofloxacin, ceftriaxone, sanguinarine, and vancomycin ($\bar{x}$-MICs = 19.8 ± 19–47.8 ± 78 μg/mL). The MICs (2–4 μg/mL) of both chloramphenicol and vancomycin against bifidobacteria were similarly low, but the latter produced significantly lower activities against lactobacilli (MICs = 64–256 μg/mL) and *B. fragilis* (MIC = 32 μg/mL). With the exception of *Bifidobacterium breve* and *Bifidobacterium longum* ssp. *longum* (MICs = 8–32 μg/mL), the remaining bifidobacteria were also susceptible to ceftriaxone (MICs = 1–4 μg/mL). However, considering the MIC range (MICs = 0.5–32 μg/mL) shown against lactobacilli and the low activity against *B. fragilis* (MIC = 128 μg/mL), the variation of activities of this drug against probiotic bacteria is quite high. The exceptionally strong activities against *B. fragilis* were revealed by metronidazole and tetracycline (MICs = 0.5 μg/mL), whereas the MIC (4 μg/mL) of nitroxoline against this bacterium was the same as in the case of chloramphenicol. In general, berberine (MICs ≥ 32 μg/mL), ferron (MICs ≥ 64 μg/mL), and phenolic compounds (MICs ≥ 64 μg/mL) did not exhibit significant antibacterial activities against any of the 21 strains. The complete data on growth-inhibitory activities of test compounds against diarrheagenic and probiotic strains, including calculated mean values ($\bar{x}$-MIC) are presented in Table 1.

Table 1. In vitro selective inhibitory activities of phytochemicals, their synthetic analogs, and antibiotics against intestinal bacteria and cells.

Cultures Tested		Alkaloids							Phenolic Compounds			Antibiotics					
		BR	SG	8HQ	CLX	NXL	FRN	ZP	SA	BS	TA	CF	CP	MA	VM	CA	TC
Bacterial strain/MIC (μg/mL)	BC	- [a]	128	512	8	4	512	4	-	512	512	128	1	128	256	8	0.5
	CD	-	64	128	8	2	64	8	256	128	-	64	16	0.5	2	4	0.5
	CP	256	128	128	16	4	64	8	-	512	512	4	1	8	1	4	32
	EF	-	32	4	16	32	-	8	-	-	-	512	2	-	-	128	4
	EC	-	256	128	16	4	-	4	-	-	-	0.062	0.062	0.062	-	4	1
	ECS	-	128	256	64	32	-	8	-	-	512	0.5	0.016	-	512	8	4
	LM	-	16	1	32	16	512	8	-	256	-	32	4	512	8	8	2
	SF	-	64	128	16	2	-	1	-	-	-	0.5	0.016	-	-	4	2
	SE	-	256	256	64	16	-	8	-	-	-	0.25	0.031	-	512	4	4
	ST	512	512	512	16	16	-	8	-	512	-	0.25	0.031	-	512	8	4
	VP	512	32	128	16	8	256	4	-	-	128	0.125	0.062	-	256	2	1
	YE	-	256	512	16	8	-	16	-	-	512	0.25	0.125	-	-	16	2
	$\bar{x}$-DB ± SD	874.7 ± 256	156 ± 137	224.4 ± 181	24 ± 19	12 ± 10	714.7 ± 388	7.1 ± 4	960 ± 212	757.3 ± 332	778.7 ± 307	61.8 ± 141	2 ± 4	651.4 ± 459	512.9 ± 404	16.5 ± 34	4.8 ± 8
	BF	-	32	32	32	4	128	8	128	32	-	128	8	0.5	32	4	0.5
	BA	128	16	512	128	16	256	16	-	32	-	1	8	64	2	4	64
	BLC	32	32	-	512	32	512	16	128	64	-	2	32	32	2	4	32
	BBF	64	32	512	-	16	512	8	-	64	-	4	16	-	4	4	16
	BB	64	32	-	64	32	512	16	-	64	256	32	64	32	4	4	16
	BL	32	64	512	64	16	256	16	512	64	128	8	16	8	2	4	2
	LC	64	32	-	128	32	256	16	512	128	512	32	32	-	256	16	8
	LR	-	32	-	128	16	512	64	512	64	512	0.5	32	256	64	8	32
	LRM	64	64	-	128	16	512	32	512	128	-	32	4	-	64	8	8
	$\bar{x}$-PB ± SD	277.3 ± 400	37.3 ± 15	743.1 ± 343	245.3 ± 306	20 ± 9	384 ± 148	21.3 ± 16	597.3 ± 336	71.1 ± 33	725.3 ± 352	26.6 ± 38	23.6 ± 18	384.9 ± 458	47.8 ± 78	6.2 ± 4	19.8 ± 19

Table 1. *Cont.*

Cultures Tested			Alkaloids							Phenolic Compounds			Antibiotics					
			BR	SG	8HQ	CLX	NXL	FRN	ZP	SA	BS	TA	CF	CP	MA	VM	CA	TC
Cell line (μg/mL)	IC_{50} ± SD	HT29	5 ± 1	0.9 ± 0.2	1.3 ± 0.3	3.7 ± 0.3	2.6 ± 0.3	86.3 ± 12	0.6 ± 0.01	-	461.9 ± 17	35.9 ± 4.9	-	130.3 ± 13	-	-	271.1 ± 1	392.9± 20
		Caco-2	19.4 ± 2.9	0.8 ± 0.1	0.3 ± 0.1	1.3 ± 0.04	1.1 ± 0.03	55.2 ± 4.8	0.7 ± 0.2	-	45.3 ± 5	27.6 ± 1.5	-	69.9 ± 4.9	-	-	439.3 ± 4	70.4 ± 15
		$\bar{x}$-CC ± SD	12.2 ± 7	0.8 ± 0.05	0.8 ± 0.5	2.5 ± 1.2	1.8 ± 0.8	70.8 ± 16	0.6 ± 0.05	-	253.6 ± 208	31.7 ± 4	-	100.1 ± 30	-	-	355.2 ± 84	231.6± 161
		FHs 74 Int	1 ± 0.1	1 ± 0.1	10.7 ± 0.2	0.5 ± 0.02	0.4 ± 0.05	22.6 ± 3.3	0.3 ± 0.1	73.2 ± 4.6	8.7 ± 1.3	5.9 ± 1.2	-	51.8 ± 27	-	-	30.7 ± 5.6	14.7 ± 2.3
	IC_{80} ± SD	HT29	42.1 ± 7.3	1.8 ± 0.4	4.8 ± 2.8	5 ± 0.7	3.7 ± 0.4	149.4 ± 6	0.7 ± 0.01	-	-	43 ± 2.2	-	-	-	-	-	-
		Caco-2	78.9 ± 0.3	1.5 ± 0.1	0.9 ± 0.1	5.9 ± 0.7	5 ± 1.5	139.9 ± 26	0.7 ± 0.01	-	454.9 ± 43	-	-	-	-	-	-	-
		$\bar{x}$-CC ± SD	*60.5 ± 18.4*	*1.6 ± 0.2*	*2.9 ± 2*	*5.4 ± 0.5*	*4.4 ± 0.7*	*144.7 ± 4.8*	*0.7 ± 0*	-	*740 ± 285*	*534 ± 491*	-	-	-	-	-	-
		FHs 74 Int	26.4 ± 0.8	1.9 ± 0.2	20.3 ± 2.4	2 ± 0.1	0.8 ± 0.05	46 ± 2.2	0.5 ± 0.03	206.9 ± 69	31.4 ± 6.6	10.2 ± 2.4	-	129.5 ± 24	-	-	-	108.2 ± 5
SI		(a)	−1.5	−1.9	−1	−1.1	−1.2	−1.2	−1.1	−0.7	−1.4	−1.9	1.2	1.8	0.2	0.3	1.8	1.4
		(b)	−0.5	−0.6	0.5	1	0.2	−0.3	0.5	−0.2	−1	−0.03	−0.4	1.1	−0.2	−1	−0.4	0.6
		(c)	−1.1	0.1	1.1	−0.7	−0.6	−0.5	−0.4	−1.1	−1.5	−0.7	0	−0.3	0	0	−1.1	−1.2

MIC: minimum inhibitory concentration; IC_{50}: half maximal inhibitory concentration; IC_{80}: 80% inhibitory concentration of proliferation; SD. standard deviation. [a] Not active ($MIC/IC_{50/80}$ > 512 μg/mL, the value 1024 μg/mL was used for average calculation). BR: berberine, SG: sanguinarine, 8HQ: 8-hydroxyquinoline, CLX: chloroxine, NXL: nitroxoline, FRN: ferron, ZP: zinc pyrithione, SA: salicylic acid, BS: bismuth subsalicylate, TA: tannic acid, CF: ceftriaxone, CP: ciprofloxacin, MA: metronidazole, VM: vancomycin, CA: chloramphenicol, TC: tetracycline. BC: *Bacillus cereus*, CD: *Clostridium difficile*, CP: *Clostridium perfringens*, EF: *Enterococcus faecalis*, EC: *Escherichia coli*, ECS: *E. coli* 0175:H7, LM: *Listeria monocytogenes*, SF: *Shigella flexneri*, SE: *Salmonella* Enteritidis, ST: *Salmonella* Typhimurium, VP: *Vibrio parahaemolyticus*, YE: *Yersinia enterocolitica*, BF: *Bacteroides fragilis*, BA: *Bifidobacterium adolescentis*, BLC: *Bifidobacterium animalis* spp. *lactis*, BBF: *Bifidobacterium bifidum*, BB: *Bifidobacterium breve*, BL: *Bifidobacterium longum* ssp. *longum*, LC: *Lactobacillus casei*, LR: *Lactobacillus reuteri*, LRM: *Lactobacillus rhamnosus*. $\bar{x}$ -DB: mean MIC for diarrheagenic bacteria, $\bar{x}$ *-PB*: mean MIC for probiotic bacteria, $\bar{x}$ -CC: mean $IC_{50/80}$ for intestinal cancer cells, FHs 74 Int (intestinal normal cells), SD: standard deviation. SI (Selective Index): **(a)** normal cells/diarrheagenic bacteria, **(b)** probiotic bacteria/diarrheagenic bacteria, **(c)** normal cells/cancer cells.

2.2. Cytotoxic Effect

Amongst the test compounds, only alkaloids and related structures exhibited strong cytotoxic activity, while other agents, especially antibiotics, exhibited moderate or no cytotoxic activity. Considering the antiproliferative effect of antibiotics on normal intestinal cells (FHs 74 Int), ceftriaxone, metronidazole, and vancomycin were not cytotoxic at all tested concentrations (IC_{50} and IC_{80} > 512 μg/mL), whereas tetracycline (IC_{50} = 14.7 ± 2.3 μg/mL; IC_{80} = 108.2 ± 5 μg/mL), chloramphenicol (IC_{50} = 30.7 ± 5.6 μg/mL; IC_{80} > 512 μg/mL), and ciprofloxacin (IC_{50} = 51.8 ± 27 μg/mL; IC_{80} = 129.5 ± 24 μg/mL) were moderately cytotoxic. In case of phytochemicals and their synthetic analogs, salicylic acid (IC_{50} = 73.2 ± 4.6 μg/mL; IC_{80} = 206.9 ± 69 μg/mL), ferron (IC_{50} = 22.6 ± 3.3 μg/mL; IC_{80} = 46 ± 2.2 μg/mL), and 8-hydroxyquinoline (IC_{50} = 10.7 ± 0.2 μg/mL; IC_{80} = 20.3 ± 2.4 μg/mL), revealed moderately cytotoxic effects against FHs 74 Int, whereas the other compounds were cytotoxic (IC_{50} values = 0.3 ± 0.1–1 ± 0.1 μg/mL; IC_{80} values = 0.5 ± 0.03–26.4 ± 0.8 μg/mL). Considering the antiproliferative effect on cancer intestinal cells, zinc pyrithione, 8-hydroxyquinoline, and sanguinarine were cytotoxic to HT29 (IC_{50} values = 0.6, 1.3, and 0.9 μg/mL, respectively) and Caco-2 (IC_{50} values = 0.7, 0.3 and 0.8 μg/mL, respectively) cells. Nitroxoline (IC_{50} = 1.1 μg/mL) and chloroxine (IC_{50} = 1.3 μg/mL) exhibited comparable cytotoxic activity against Caco-2 cells. Zinc pyrithione had the lowest $\overline{x}$-IC_{50} value (0.6 ± 0.05 μg/mL) against cancer cells, followed by 8-hydroxyquinoline (0.8 ± 0.5 μg/mL), sanguinarine (0.8 ± 0.05 μg/mL), nitroxoline (1.8 ± 0.8 μg/mL), and chloroxine (2.5 ± 1.2 μg/mL). Berberine ($\overline{x}$-IC_{50} = 12.2 ± 7 μg/mL), tannic acid ($\overline{x}$-IC_{50} = 31.7 ± 4 μg/mL), and ferron ($\overline{x}$-IC_{50} = 70.8 ± 16 μg/mL), produced moderate cytotoxic activity, while salicylic acid and bismuth subsalicylate ($\overline{x}$-IC_{50} ≥ 253.6 ± 208 μg/mL) did not exhibit significant cytotoxic activity against cancer cells. At relatively high concentrations, some antibiotics exhibited antiproliferative activity against cancer cells, namely: ciprofloxacin, tetracycline, and chloramphenicol ($\overline{x}$-IC_{50} = 100.1 ± 30–355.2 ± 84 μg/mL). The complete data on the antiproliferative activities of test compounds against normal and cancer intestinal cells, including calculated mean values for the latter ($\overline{x}$-IC_{50} and $\overline{x}$-IC_{80}), are presented in Table 1.

2.3. Selective Toxicity

The selective antibacterial activities against the pathogens with relatively lower activity against probiotic strains (SIb values range from 0.2–1.1) was revealed by most of the agents exhibiting strong to moderate inhibitory effects on diarrheagenic bacteria, namely: nitroxoline, zinc pyrithione, tetracycline, chloroxine, and ciprofloxacin. In contrast, chloramphenicol and ceftriaxone were more toxic to probiotic strains (SIbs = −0.4 for both). Although the antibacterial activity of berberine, ferron, phenolic compounds, and sanguinarine was in cases of both diarrheagenic and probiotic strains generally insignificant, the results show that these agents were rather toxic to the latter (SIb values range from −1 to −0.03). Due to the minor cytotoxicity revealed against FHs 74 Int, none of the antibiotics exhibited an increased toxicity to normal intestinal cells at the inhibitory concentrations active against diarrheagenic bacteria (SIa values = 0.2–1.8), especially ciprofloxacin and chloramphenicol. In contrast, all of the phytochemicals and their synthetic analogs revealed cytotoxicity to normal intestinal cells at the concentrations they were generally inactive against diarrheagenic bacteria (SIa values range from −1.9 to −0.7). Only 8-hydroxyquinoline (SIc = 1.1) and sanguinarine (SIc = 0.1) exhibited selective antiproliferative activity against cancer cells with the decreased cytotoxic effect on normal intestinal cells. Except these two, other tested compounds were more toxic to normal than to cancer intestinal cells (SIcs = from −1.5 to −0.3), or in the case of ceftriaxone, metronidazole, and vancomycin, they did not show any selectivity (SIcs = 0), as they did not inhibit cell lines at any concentration tested. The data on selective toxicities, including all calculated SI values are presented in Table 1. The curves of in vitro selective concentration-dependent effect of ciprofloxacin, chloroxine, nitroxoline, tetracycline, and zinc pyrithione on the growth of diarrheagenic and probiotic bacteria and of 8-hydroxyquinoline on intestinal normal and cancer cells proliferation are shown in Figure 1.

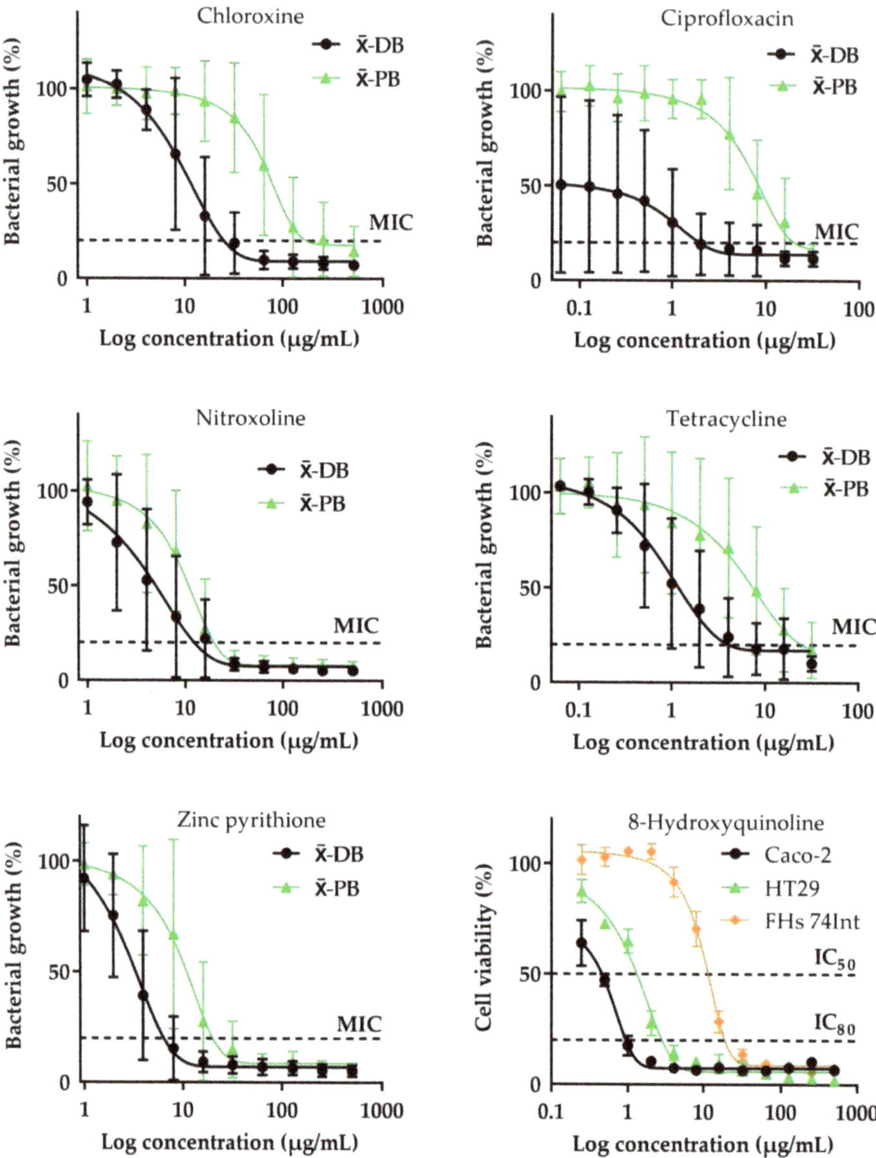

Figure 1. Selective concentration-dependent effect of chloroxine, ciprofloxacin, nitroxoline, tetracycline, and zinc pyrithione on the growth of diarrheagenic and probiotic bacteria and of 8-hydroxyquinoline on intestinal normal and cancer cells proliferation in vitro. MIC: minimum inhibitory concentration; IC_{50}: half maximal inhibitory concentration; IC_{80}: 80% inhibitory concentration of proliferation. $\bar{x}$ -DB: mean MIC for 12 diarrheagenic bacteria, $\bar{x}$ -PB: mean MIC for 9 probiotic bacteria; Caco-2 and HT29: intestinal cancer cells; Fhs74 Int: intestinal normal cells.

2.4. Principal Component Analysis (PCA)

The correlation between biological activities and chemical structures of the tested compounds and their groups (antibiotics, phenolic compounds, alkaloids, and related structures) was analyzed using PCA (Figure 2). Although the compounds were distributed equally in all quadrants, detailed analysis revealed specific patterns. The closest correlation was observed in the lower-left quadrant between four antibiotics (ceftriaxone, ciprofloxacin, chloramphenicol, and tetracycline), which indicated that these antibiotics exhibited strong growth-inhibitory activity against diarrheagenic and probiotic strains but did not exhibit distinct cytotoxic activity against normal and cancer intestinal cells. The second highest correlation observed in the lower right quadrant indicated that zinc pyrithione and 8-hydroxyquinolines (excluding ferron) exhibited selective antibacterial activity against diarrheagenic strains along with strong to moderate cytotoxic activity against both types of tested cell lines. The correlation observed in the upper right quadrant indicated that tannic acid, benzylisoquinoline alkaloids, and ferron exhibited moderate to no antipathogenic effect (with negative SIbs) and overall moderate to strong cytotoxic activity. The upper left quadrant contains the remaining antibiotics (metronidazole and vancomycin) and both simple phenols with minimal correlation. These agents exhibited moderate to no growth-inhibiting activity against diarrheagenic strains and practically no cytotoxic activity. Whereas all alkaloids are distributed in lower and upper right quadrants indicating their capability to reveal any type of the tested bioactivities, phenols are spread in the right and left upper quadrants which shows their lack of significant antibacterial activity but a certain degree of cytotoxicity. In contrast, antibiotics are concentrated in the lower-left quadrant slightly overlapping the upper left one, therefore they usually display a significant antibacterial effect which is rarely accompanied by cytotoxicity.

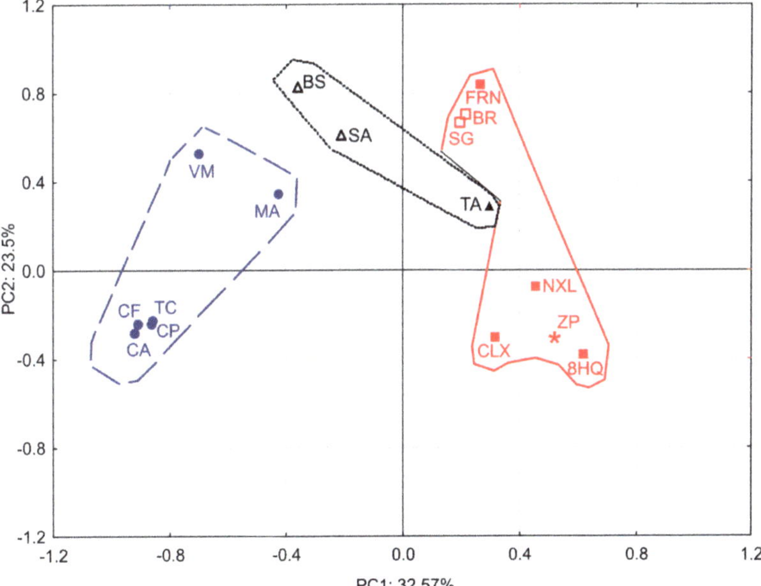

Figure 2. Principal component analysis of antibacterial and cytotoxic activities of phytochemicals, their synthetic analogues, and antibiotics against intestinal bacteria and cells in vitro. BR: berberine, SG: sanguinarine, 8HQ: 8-hydroxyquinoline, CLX: chloroxine, NXL: nitroxoline, FRN: ferron, ZP: zinc pyrithione, SA: salicylic acid, BS: bismuth subsalicylate, TA: tannic acid, CF: ceftriaxone, CP: ciprofloxacin, MA: metronidazole, VM: vancomycin, CA: chloramphenicol, TC: tetracycline. Antibiotics [——-, •]; Phenolic compounds [··· ··· ·, poly- (▲), simple (Δ)]; Alkaloids and related structures [——, benzylisoquinolines (□), 8-hydroxyquinolines (■), metal-pyridine derivative complex (*).

3. Discussion

The in vitro growth-inhibitory properties of the tested antibiotics have previously been reported for a number of diarrheagenic and probiotic bacteria. However, the bacterial strains tested as well as the methods with criteria used for antimicrobial activity assessment vary frequently among the previous studies. Moreover, the studies reporting the in vitro susceptibilities of probiotic bacteria deal more with clinical isolates [23], and less with standard strains [24]. The present study, therefore, provides the data on in vitro selective antibacterial activities of these antibiotics that can be fairly compared with the same data obtained for phytochemicals and their synthetic analogs. We suggest that the reason behind the increased resistance of probiotic strains differ for particular compounds that showed a selective antipathogenic effect. The growth-inhibitory activities of fluoroquinolones against Gram-positive bacteria are reported to be lower than those against Gram-negative bacteria [15]. Consistent with this finding, bifidobacteria and lactobacilli (Gram-positive) were generally less susceptible to ciprofloxacin than Gram-negative diarrheagenic bacteria that predominate over Gram-positive pathogens in this study. The decreased susceptibility of bifidobacteria to tetracycline might be caused by the presence of specific antibiotic resistance genes [25]. Similar to other third-generation cephalosporins [26], the growth-inhibitory activity of ceftriaxone against Gram-negative bacteria was higher than that against Gram-positive bacteria. However, as a result of significant resistance of the tested Gram-positive pathogens and susceptibility of bifidobacteria, ceftriaxone showed increased toxicity to probiotic strains. The growth-inhibitory activities of some alkaloids and related structures were comparable with those of antibiotics. The antibacterial activity of 8-hydroxyquinoline alkaloids is mediated through the chelation of metals that function as co-factors in various enzymes, which results in the inhibition of RNA synthesis. We suggest that probiotic strains (mainly bifidobacteria) are more resistant to 8-hydroxyquinolines as they are able to sequester iron from the environment [10]. The selective antibacterial activity of 8-hydroxyquinoline against diarrheagenic pathogens seems to be enhanced with chlorine halogenation or by the presence of a nitro group and decreased with iodine halogenation and the presence of a sulfo group, as respectively observed for chloroxine, nitroxoline, and ferron in our study. The in vitro selective anticlostridial effect of 8-hydroxyquinoline with increased resistance of bifidobacteria was previously described in studies of Novakova et al. [10,27,28], Skrivanova et al. [29], and Kim et al. [30]. However, data on its in vitro growth-inhibitory effects against a broader selection of diarrheagenic bacteria are limited. The present study also provides new data on in vitro antibacterial activities of chloroxine against diarrheagenic bacteria in addition to those previously published [31,32]. It has been reported that Endiaron, a chloroxine-containing antimicrobial product used for infectious diarrhea, exhibits antimicrobial activity against the pathogens and does not affect the host indigenous microbiota [11], which is in agreement with the increased resistance of probiotic bacteria described in the present study. Interestingly, the antipathogenic activity of nitroxoline, used to treat urinary tract infections, was higher than that of chloroxine. However, the antibacterial selectivity of nitroxoline against diarrheagenic strains was lower. Out of the intestinal bacteria tested herein, there are only data on in vitro inhibitory effects of nitroxoline against *E. coli* and *E. faecalis* that have been reported before [33]. In spite of zinc pyrithione being only used topically for dermatological infections [7], in relation to the plant compounds and their synthetic analogs in this study, it exhibited the highest growth-inhibitory activity against diarrheagenic bacteria with lowered toxicity to probiotic bacteria. According to our best knowledge, this is the first report on in vitro selective antibacterial activities of zinc pyrithione on intestinal diarrheagenic and probiotic bacteria. Although there is limited knowledge on the mechanism underlying the antibacterial activity of zinc pyrithione, the mechanism may be similar to that of 8-hydroxyquinolines [34]. The weak antimicrobial activities of phenols against diarrheagenic strains are consistent with those reported in previous studies. The effectiveness of phenols in infectious diarrhea may be based on other mechanisms, such as astringent, mucosa-protective, and anti-inflammatory properties, or inhibition of pathogenic enterotoxins [7]. Although clinical studies on extensively used phytochemical berberine have reported comparably higher efficiency than certain antibiotics (e.g., chloramphenicol) [35], our results did not show its significant in vitro antibacterial activity. A possible reason for this discordance is that berberine rather neutralizes diarrheagenic action of bacteria by inhibiting their enterotoxins, as described by Sack and Froelich [36]. The MICs of both benzylisoquinoline alkaloids against some diarrheagenic bacteria reported in this study were higher than those reported in previous studies. This may be because the inoculum density used in this study was higher than that used in previous studies [7,37].

The mechanism underlying the antiproliferative activity of some antibiotics, such as ciprofloxacin, tetracycline, and chloramphenicol, may be similar to that underlying antimicrobial activity [14,38,39]. Previous studies have evaluated the antiproliferative activity of chemicals derived from ciprofloxacin and tetracycline against cancer cells and suggested their applications in cancer therapy [14,40]. Consistent with the results of this study, previous studies have revealed that other antibiotics are not cytotoxic to eukaryotic cells [41–43]. The antitumor activities of some plant compounds and their synthetic analogs have been investigated previously. In the case of 8-hydroxyquinoline and its derivatives, the interaction with metal ions, namely copper and iron, and their transportation into cells has been reported as crucial for its antiproliferative activity [44]. Freitas et al. [45] reported that 8-hydroxyquinoline derivatives with potent anticancer potential often contain halogen substituents. However, in the present study, nitroxoline exhibited stronger antiproliferative activity against cancer cells than chloroxine and ferron. Previous studies have reported that the underlying mechanism of antitumor activity of zinc pyrithione and sanguinarine involves the inhibition of proteasomal deubiquitinases and microtubule depolymerization, respectively [19,22]. All of the above-mentioned alkaloids and related structures revealed increased toxicity to normal intestinal cells in comparison with their antipathogenic effect, which limits their applications in treating bacterial diarrhea. Although there are no studies reporting oral toxicity or toxicity to the digestive system from berberine, chloroxine, and nitroxoline, we suggest that their safety profile should be further examined and the potential protective role of indigenous gut microbiota against these cytotoxic chemicals should be more deeply studied. Zinc pyrithione, 8-hydroxyquinoline, and sanguinarine are not part of any product intended for internal use. Hence, their dose-dependent toxicological profile and oral safety must be carefully elucidated before any consideration for their application for treating infectious diarrhea associated with intestinal cancer [46].

4. Materials and Methods

4.1. Chemicals

Phytochemicals (berberine chloride, 8-hydroxyquinoline, salicylic acid, tannic acid, and sanguinarine chloride) and their synthetic analogs [chloroxine (5,7-dichloroquinolin-8-ol), nitroxoline (5-nitroquinolin-8-ol), ferron (7-iodo-8-hydroxyquinoline-5-sulfonic acid), bismuth subsalicylate, and zinc pyrithione], as well as antibiotics (ceftriaxone sodium, ciprofloxacin, chloramphenicol, metronidazole, tetracycline, and vancomycin hydrochloride), used in this study were purchased from Sigma-Aldrich (Prague, Czech Republic). Dimethyl sulfoxide (DMSO) (Sigma-Aldrich, Prague, Czech Republic) was used to prepare the stock solutions of all test compounds, except those of metronidazole, salicylic acid, vancomycin, and zinc pyrithione, which were prepared using distilled water. Stock solutions of chloramphenicol, tannic acid, and tetracycline were prepared using 96% ethanol (Sigma-Aldrich, Prague, Czech Republic). The chemical structures of individual compounds tested are shown in Figure 3.

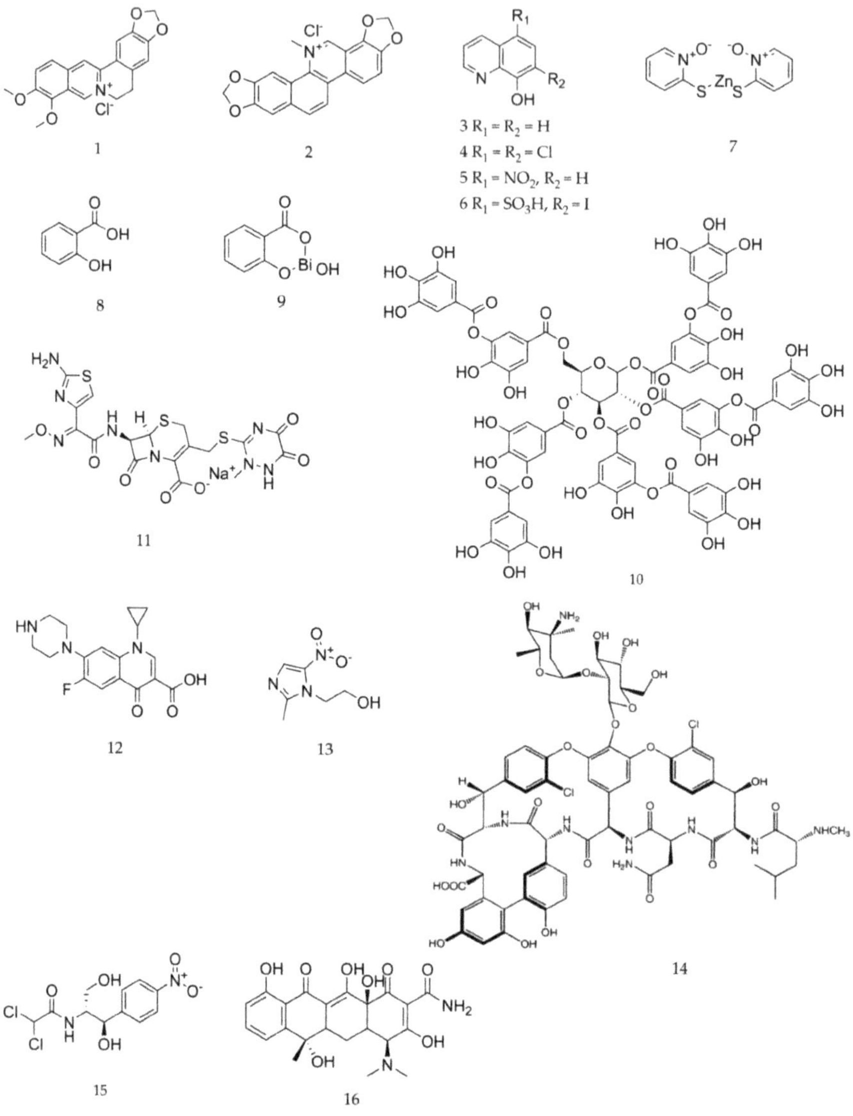

Figure 3. The chemical structures of the tested phytochemicals, their synthetic analogs, and antidiarrheal antibiotics. **1**. berberine chloride, **2**. sanguinarine chloride, **3**. 8-hydroxyquinoline, **4**. chloroxine, **5**. nitroxoline, **6**. ferron, **7**. zinc pyrithione, **8**. salicylic acid, **9**. bismuth subsalicylate, **10**. tannic acid, **11**. ceftriaxone sodium, 12. ciprofloxacin, **13**. metronidazole, **14**. vancomycin hydrochloride, **15**. chloramphenicol, **16**. tetracycline.

4.2. Bacterial Strains and Growth Media

The intestinal bacterial type strains were obtained from the American Type Culture Collection (ATCC, Rockville, MD, USA), Czech Collection of Microorganisms (CCM, Brno, Czech Republic), German Collection of Microorganisms and Cell Cultures (DSMZ, Braunschweig, Germany), and National Collection of Type Cultures (NCTC, London, UK). In accordance with the diversity of diarrheagenic Gram-positive and Gram-negative bacteria responsible for globally distributed foodborne, waterborne, and nosocomial infections [3,47], the following 12 strains were used in this study: *B. cereus* (ATCC 14579), *C. difficile* (DSMZ 12056), *C. perfringens* (DSMZ 11778), *E. faecalis* (ATCC 29212), *E. coli* (ATCC 25922), *E. coli* 0175:H7 (NCTC 12900), *L. monocytogenes* (ATCC 7644), *S. flexneri* (ATCC 12022), *Salmonella enterica* ssp. *enterica* serovar Enteritidis (ATCC 13076), *S. enterica* ssp. *enterica* serovar Typhimurium (ATCC 14028), *V. parahaemolyticus* (ATCC 17802), and *Yersinia enterocolitica* (ATCC 9610). The following nine bacterial strains, which belong to three predominant bacterial phyla in the human gut and exhibit probiotic functions [8,9], were used in this study: *Bacteroides fragilis* (ATCC 25285), *Bifidobacterium adolescentis* (DSMZ 20087), *Bifidobacterium animalis* spp. *lactis* (DSMZ 10140), *Bifidobacterium bifidum* (ATCC 29521), *B. breve* (ATCC 15700), *B. longum* ssp. *longum* (DSMZ 20219), *Lactobacillus casei* (DSMZ 20011), *Lactobacillus reuteri* (CCM 3625), and *Lactobacillus rhamnosus* (CCM 7091). As the maintenance and growth medium, Mueller–Hinton broth (Oxoid, Basingstoke, UK) was used for bacteria that grow aerobically (*E. faecalis* supp. 1% glucose, *V. parahaemolyticus* supp. 3% NaCl). The anaerobic bacteria (clostridia and bifidobacteria), including facultative species (lactobacilli), were cultured in Wilkins–Chalgren broth (Oxoid, Basingstoke, UK) supplemented with 5 g/L soya peptone and 0.5 g/L cysteine.

4.3. Cell Cultures

One representative normal intestinal cell line (FHs 74 Int (ATCC CCL 241)) and two cancer intestinal cell lines (Caco-2 (ATCC HTB 37) and HT29 (ATCC HTB 38)) were purchased from ATCC (Rockville, MD, USA). Normal cells were cultured in Hybri-Care medium supplemented with 10% fetal bovine serum, 1% sodium bicarbonate, 1% non-essential amino acids, 30 ng/mL of epidermal growth factor, and 1% penicillin-streptomycin solution (10,000 units/mL and 100 mg/mL, respectively). The cancer cells were cultured in Dulbecco's modified Eagle's medium (DMEM) supplemented with 1% sodium pyruvate, 10% fetal bovine serum, 1% sodium bicarbonate, 1% non-essential amino acids, and 1% penicillin-streptomycin solution (10,000 units/mL and 100 mg/mL, respectively) (all purchased from Sigma-Aldrich, Prague, Czech Republic). The cultures were incubated at 37 °C and 5% CO_2. The culture medium was replaced every 2–3 days and cells were passaged every 7 days.

4.4. Antibacterial Assay

The growth-inhibitory activities of the test compounds against aerobic and anaerobic bacterial strains were evaluated by the broth microdilution method using 96-well microtiter plates, following the protocols of CLSI guidelines [48] and Hecht et al. [49], respectively. Prior to testing, the strains were sub-cultured in the appropriate media at 37 °C for 24 h. Obligate anaerobes and lactobacilli were cultured for 48 h using Whitley A35 Anaerobic Workstation (Don Whitley Scientific, Bingley, UK). The anaerobic conditions were created by the supply of a standard anaerobic gas mixture of 10% H_2, 10% CO_2, and 80% N_2 (Linde Gas, Prague, Czech Republic). Test agents were diluted 2-fold in appropriate growth media using the Freedom EVO 100 automated pipetting platform (Tecan, Männedorf, Switzerland) or multichannel pipette (Eppendorf, Hamburg, Germany) (initial concentration of 512 µg/mL). After the bacterial cultures reached an inoculum density of 1.5×10^8 CFU/mL by 0.5 McFarland standard using Densi-La-Meter II (Lachema, Brno, Czech Republic), the 96-well plates were inoculated (5 µL/well). The plates containing the volatile compound, 8-hydroxyquinoline, were covered using EVA capmats (Micronic, Lelystad, Netherlands) after inoculation to prevent evaporation [50]. Bacterial cultures in microplates were incubated by employing the same protocols as used for their cultivation prior to the test. The optical density of the cultures was measured at 405 nm ($OD_{450\ nm}$) using a Cytation 3 Imaging Reader (BioTek, Winooski, VT, USA) before and after the growth. The lowest concentration (µg/mL) of test compounds at which the bacterial growth was inhibited by ≥80% was defined as MIC. All tests were performed as three independent experiments each carried out in triplicate. The data are presented as median/mode. As a result of experiments performed without dissolved test compounds, DMSO and 96% ethanol (both from Sigma-Aldrich, Prague, Czech Republic) did not inhibit bacterial growth of any strain at the tested concentrations (≤1%).

4.5. Cytotoxicity Assay

The antiproliferative activities of test compounds against normal and cancer intestinal lines were evaluated using the modified thiazolyl blue tetrazolium bromide (MTT) cytotoxicity assay developed by Mosmann et al. [51]. The cancer (2.5×10^3) and normal intestinal (2.5×10^5) cells were seeded in a 96-well microtiter plate for 24 h. Cells were incubated with two-fold serially diluted test compounds (0.25–512 µg/mL) for 72 h. Plates containing 8-hydroxyquinoline were covered using EVA capmats. Next, the cells were incubated with MTT reagent (1 mg/mL) (Sigma-Aldrich, Prague, Czech Republic) in DMEM or Hybri-Care medium for an additional 2 h at 37 °C and 5% CO_2. The medium with MTT was removed and the intracellular formazan product was dissolved in 100 µL of DMSO. The absorbance was measured at 555 nm using a Tecan Infinite M200 spectrometer (Tecan, Männedorf, Switzerland), and the percentage of viability was calculated when compared to an untreated control. Antiproliferative activity of the test compounds was represented as IC_{50} (µg/mL). Three independent experiments (two replicates each) were performed for every test. Data are presented as mean ± standard deviation. DMSO was used as a positive control at the highest concentration with and without EVA capmats. The solvents did not affect the viability of normal and cancer intestinal cell lines at the tested concentration (≤1%).

4.6. Calculations and Statistics

For comparison of microbiological and toxicological data, IC_{80} was calculated as equivalent to the MIC endpoint, defined as 80% bacterial growth inhibition [52]. Subsequently, $\overline{x}$-MIC, $\overline{x}$-IC_{50}, and $\overline{x}$-IC_{80} values (±standard deviations) were calculated to quantify the inhibitory activity of test compounds against diarrheagenic/probiotic bacteria and cancer cells, respectively. After that, SIa (normal intestinal cells/ diarrheagenic strains), SIb (probiotic/diarrheagenic strains), and SIc (normal/cancer intestinal cells) was calculated using the formulas below.

$$SIa = \log (X_1/Y_1),$$

$$SIb = \log (X_2/Y_1)$$

$$SIc = \log (X_3/Y_2)$$

where: X_1 = IC_{80} against FHs 74 Int; X_2 = $\overline{x}$-MIC against probiotic strains; X_3 = IC_{50} against FHs 74 Int; Y_1 = $\overline{x}$-MIC against diarrheagenic strains; Y_2 = $\overline{x}$-IC_{50} against cancer intestinal cells. The SI values > 0 and <0 indicate selective toxicity against diarrheagenic strains/cancer cell lines and probiotic strains/normal cell lines, respectively.

The correlation between the combination of activities revealed by test compounds and their chemical classes was analyzed using PCA with Statistica 13 software [53]. All data for particular activities were grouped into four types of targets (cancer cells, diarrheagenic strains, normal cells, and probiotic strains) using MIC and IC_{80} values. There was no adjustment of PCA parameters for this analysis. For the calculation of each $\overline{x}$-MIC, $\overline{x}$-IC_{50}, $\overline{x}$-IC_{80} and for PCA, values greater than the maximum tested concentration (512 µg/mL) were replaced with 1024 µg/mL.

5. Conclusions

In summary, ciprofloxacin, 8-hydroxyquinoline alkaloids (chloroxine and nitroxoline), tetracycline, and zinc pyrithione exhibited a significant selective growth-inhibitory activity against diarrheagenic bacteria with lowered toxicity to probiotic bacteria in vitro. 8-Hydroxyquinoline, chloroxine, nitroxoline, sanguinarine, and zinc pyrithione also exhibited a strong cytotoxic effect, whereas the antiproliferative action of 8-hydroxyquinoline and sanguinarine were selective to cancer intestinal cells. These findings indicate that 8-hydroxyquinoline alkaloids and metal-pyridine derivative complexes are chemical structures with promising bioactive properties in terms of in vitro selective antibacterial and anticancer activities which could be utilized in future chemotaxonomic investigation of antidiarrheal medicinal plants and their bioactive components. These could be further investigated as possible new chemotherapeutic agents against diarrheal infections and associated intestinal cancer diseases. However, in vivo studies on the toxicity of these compounds with more complex animal models will be needed before their consideration to be used for this purpose.

Author Contributions: Conceptualization, T.K. and L.K.; methodology, T.K., L.K., and I.D.; software, M.P.; validation, L.K., I.D., H.S., M.P., and E.S.; formal analysis, T.K., L.K., I.D., H.S., M.P., and E.S.; investigation, T.K., L.K., I.D., H.S., and E.S.; resources, L.K., H.S., and E.S.; data curation, T.K., L.K., I.D., and M.P.; writing—original draft preparation, T.K. and L.K.; writing—review and editing, I.D., H.S., M.P., and E.S.; visualization, T.K. and M.P.; supervision, L.K.; project administration, L.K.; funding acquisition, L.K. and E.S. All authors have read and agreed to the published version of the manuscript.

Funding: This research was funded by the Czech University of Life Sciences Prague, project IGA 20205001, and European Regional Development Fund, project CZ.02.1.01/0.0/0.0/16_019/0000845.

Conflicts of Interest: The authors declare no conflict of interest.

References

1. World Health Organization. Diarrhoeal Disease. Available online: https://www.who.int/en/news-room/fact-sheets/detail/diarrhoeal-disease (accessed on 19 March 2020).
2. Garrett, W.S. The gut microbiota and colon cancer. *Science* **2019**, *364*, 1133–1135. [CrossRef] [PubMed]
3. Diniz-Santos, D.R.; Silva, L.R.; Silva, N. Antibiotics for the empirical treatment of acute infectious diarrhea in children. *Braz. J. Infect. Dis.* **2006**, *10*, 217–227. [CrossRef] [PubMed]
4. White, N.J. Qinghaosu (Artemisinin): The Price of Success. *Science* **2008**, *320*, 330–334. [CrossRef] [PubMed]
5. Formiga, R.D.O.; Quirino, Z.G.M.; Diniz, M.D.F.F.M.; Marinho, A.F.; Tavares, J.F.; Batista, L. Maytenus Erythroxylon Reissek (Celastraceae) ethanol extract presents antidiarrheal activity via antimotility and antisecretory mechanisms. *World J. Gastroenterol.* **2017**, *23*, 4381–4389. [CrossRef] [PubMed]
6. Hao, D.C.; Xiao, P.G. Pharmaceutical resource discovery from traditional medicinal plants: Pharmacophylogeny and pharmacophylogenomics. *Chin. Herb. Med.* **2020**, *12*, 104–117. [CrossRef]
7. Kokoska, L.; Kloucek, P.; Leuner, O.; Novy, P. Plant-Derived Products as Antibacterial and Antifungal Agents in Human Health Care. *Curr. Med. Chem.* **2019**, *26*, 5501–5541. [CrossRef]
8. Sun, F.; Zhang, Q.; Zhao, J.; Zhang, H.; Zhai, Q.; Chen, W. A potential species of next-generation probiotics? The dark and light sides of *Bacteroides fragilis* in health. *Food Res. Int.* **2019**, *126*, 108590. [CrossRef]
9. Behnsen, J.; Deriu, E.; Sassone-Corsi, M.; Raffatellu, M. Probiotics: Properties, Examples, and Specific Applications. *Cold Spring Harb. Perspect. Med.* **2013**, *3*, a010074. [CrossRef]
10. Novakova, J.; Vlkova, E.; Bonusova, B.; Rada, V.; Kokoska, L. In vitro selective inhibitory effect of 8-hydroxyquinoline against bifidoba cteria and clostridia. *Anaerobe* **2013**, *22*, 134–136. [CrossRef]
11. WikiZero: Chloroxine. Available online: https://wikizero.com/en/Chloroxine (accessed on 10 August 2020).
12. Bruneau, A.; Baylatry, M.T.; Joly, A.C.; Sokol, H. Gut microbiota: What impact on colorectal carcinogenesis and treatment? *Bull. Cancer* **2018**, *105*, 70–80. [CrossRef]
13. Sambruy, Y.; Ferruzza, S.; Ranaldi, G.; De Angelis, I. Intestinal cell culture models: Applications in toxicology and pharmacology. *Cell Biol. Toxicol.* **2001**, *17*, 301–317. [CrossRef]
14. Elsea, S.H.; Osheroff, N.; Nitiss, J.L. Cytotoxicity of quinolones toward eukaryotic cells. Identification of topoisomerase-II as the primary cellular target for the quinolone CP-115,953 in yeast. *J. Biol. Chem.* **1992**, *267*, 13150–13153. [PubMed]
15. Oliphant, C.M.; Green, G.M. Quinolones: A comprehensive review. *Am. Fam. Physician* **2002**, *65*, 455–464. [PubMed]
16. Bourikas, L.A.; Kolios, G.; Valatas, V.; Notas, G.; Drygiannakis, I.; Pelagiadis, I.; Manousou, P.; Klironomos, S.; A Mouzas, I.; Kouroumalis, E. Ciprofloxacin decreases survival in HT-29 cells via the induction of TGF-β1 secretion and enhances the anti-proliferative effect of 5-fluorouracil. *Br. J. Pharmacol.* **2009**, *157*, 362–370. [CrossRef] [PubMed]
17. Zeng, X.H.; Li, Y.H.; Wu, S.S.; Hao, R.L.; Li, H.; Ni, H.; Han, H.B.; Lin, L. New and Highly Efficient Column Chromatographic Extraction and Simple Purification of Camptothecin from *Camptotheca acuminata* and *Nothapodytes pittosporoides*. *Phytochem. Anal.* **2013**, *24*, 623–630. [CrossRef]

18. Arafa, R.K.; Hegazy, G.H.; Piazza, G.; Abadi, A.H. Synthesis and in vitro antiproliferative effect of novel quinoline-based potential anticancer agents. *Eur. J. Med. Chem.* **2013**, *63*, 826–832. [CrossRef]
19. Slaninova, I.; Pencikova, K.; Urbanova, J.; Slanina, J.; Taborska, E. Antitumour activities of sanguinarine and related alkaloids. *Phytochem. Rev.* **2013**, *13*, 51–68. [CrossRef]
20. Khalifa, N.; Eweas, A.; Al-Omar, M.A.; Hozzein, W.N. Synthesis and antimicrobial activity of some novel 8-hydroxy-7-iodoquinoline-5-sulfonamide derivatives. *J. Pure Appl. Microbiol.* **2014**, *8*, 629–637.
21. Kos, J.; Mitrovic, A. Nitroxoline: Repurposing its antimicrobial to antitumor application. *Acta Biochim. Pol.* **2019**, 521–531. [CrossRef]
22. Zhao, C.; Chen, X.; Yang, C.; Zang, D.; Lan, X.; Liao, S.; Zhang, P.; Wu, J.; Li, X.; Liu, N.; et al. Repurposing an antidandruff agent to treating cancer: Zinc pyrithione inhibits tumor growth via targeting proteasome-associated deubiquitinases. *Oncotarget* **2017**, *8*, 13942–13956. [CrossRef]
23. Merriam, C.V.; Citron, D.M.; Tyrrell, K.L.; Warren, Y.A.; Goldstein, E.J. In vitro activity of azithromycin and nine comparator agents against 296 strains of oral anaerobes and 31 strains of *Eikenella corrodens*. *Int. J. Antimicrob. Agents* **2006**, *28*, 244–248. [CrossRef] [PubMed]
24. Kheadr, E.; Bernoussi, N.; Lacroix, C.; Fliss, I. Comparison of the sensitivity of commercial strains and infant isolates of bifidobacteria to antibiotics and bacteriocins. *Int. Dairy J.* **2004**, *14*, 1041–1053. [CrossRef]
25. Moubareck, C.A.; Gavini, F.; Vaugien, L.; Butel, M.J.; Doucet-Populaire, F. Antimicrobial susceptibility of bifidobacteria. *J. Antimicrob. Chemother.* **2005**, *55*, 38–44. [CrossRef] [PubMed]
26. Sharma, R.; Park, T.E.; Moy, S. Ceftazidime-Avibactam: A Novel Cephalosporin/β-Lactamase Inhibitor Combination for the Treatment of Resistant Gram-negative Organisms. *Clin. Ther.* **2016**, *38*, 431–444. [CrossRef]
27. Novakova, J.; Dzunkova, M.; Musilova, S.; Vlkova, E.; Kokoska, L.; Moya, A.; D'Auria, G.; Unkova, M.D. Selective growth-inhibitory effect of 8-hydroxyquinoline towards *Clostridium difficile* and *Bifidobacterium longum* subsp. *longum in co-culture analysed by flow cytometry. J. Med. Microbiol.* **2014**, *63*, 1663–1669. [CrossRef]
28. Novakova, J.; Vlkova, E.; Salmonova, H.; Pechar, R.; Rada, V.; Kokoska, L. Anticlostridial agent 8-hydroxyquinoline improves the isolation of faecal bifidobacteria on modified Wilkins-Chalgren agar with mupirocin. *Lett. Appl. Microbiol.* **2016**, *62*, 330–335. [CrossRef]
29. Skrivanova, E.; Van Immerseel, F.; Hovorkova, P.; Kokoska, L. In Vitro Selective Growth-Inhibitory Effect of 8-Hydroxyquinoline on *Clostridium perfringens* versus Bifidobacteria in a Medium Containing Chicken Ileal Digesta. *PLoS ONE* **2016**, *11*, e0167638. [CrossRef]
30. Kim, Y.M.; Jeong, E.Y.; Lim, J.H.; Lee, H.S. Antimicrobial effects of 8-quinolinol. *J. Food Sci. Biotechnol.* **2006**, *15*, 817–819.
31. Prapasarakul, N.; Tummaruk, P.; Niyomtum, W.; Tripipat, T.; Serichantalergs, O. Virulence Genes and Antimicrobial Susceptibilities of Hemolytic and Nonhemolytic *Escherichia coli* Isolated from Post-Weaning Piglets in Central Thailand. *J. Vet. Med. Sci.* **2010**, *72*, 1603–1608. [CrossRef]
32. Tranter, R.W. The in vitro activity of halquinol against *Vibrio cholerae*. *J. Trop. Med. Hyg.* **1968**, *71*, 146–149.
33. Sobke, A.; Makarewicz, O.; Baier, M.; Bar, C.; Pfister, W.; Gatermann, S.; Pletz, M.; Forstner, C. Empirical treatment of lower urinary tract infections in the face of spreading multidrug resistance: In vitro study on the effectiveness of nitroxoline. *Int. J. Antimicrob. Agents* **2018**, *51*, 213–220. [CrossRef] [PubMed]
34. Chandler, C.J.; Segel, I.H. Mechanism of the Antimicrobial Action of Pyrithione: Effects on Membrane Transport, ATP Levels, and Protein Synthesis. *Antimicrob. Agents Chemother.* **1978**, *14*, 60–68. [CrossRef] [PubMed]
35. Lahiri, S.C.; Dutta, N.K. Berberine and chloramphenicol in the treatment of cholera and severe diarrhoea. *J. Indian Med. Assoc.* **1967**, *48*, 1–11.
36. Sack, R.B.; Froehlich, J.L. Berberine inhibits intestinal secretory response of *Vibrio cholerae* and *Escherichia coli* enterotoxins. *Infect. Immun.* **1982**, *35*, 471–475. [CrossRef]
37. Hamoud, R.; Reichling, J.; Wink, M.R. Synergistic antibacterial activity of the combination of the alkaloid sanguinarine with EDTA and the antibiotic streptomycin against multidrug resistant bacteria. *J. Pharm. Pharmacol.* **2014**, *67*, 264–273. [CrossRef]
38. Chukwudi, C.U. rRNA Binding Sites and the Molecular Mechanism of Action of the Tetracyclines. *Antimicrob. Agents Chemother.* **2016**, *60*, 4433–4441. [CrossRef] [PubMed]

39. Hu, D.; Han, Z.; Li, C.; Lv, L.; Cheng, Z.; Liu, S. Florfenicol induces more severe hemotoxicity and immunotoxicity than equal doses of chloramphenicol and thiamphenicol in Kunming mice. *Immunopharmacol. Immunotoxicol.* **2016**, *38*, 472–485. [CrossRef] [PubMed]
40. Onoda, T.; Ono, T.; Dhar, D.K.; Yamanoi, A.; Nagasue, N. Tetracycline analogues (doxycycline and COL-3) induce caspase-dependent and -independent apoptosis in human colon cancer cells. *Int. J. Cancer* **2005**, *118*, 1309–1315. [CrossRef]
41. Dewdney, J.M. Effects of beta-lactam antibiotics on eukaryotic cells. *Cell Boil. Toxicol.* **1986**, *2*, 509–511. [CrossRef]
42. Eisenstein, B.I.; Schaechter, M. DNA and Chromosome Mechanics. In *Schaechter's Mechanisms of Microbial Disease*; Schaechter, M., Engleberg, N.C., DiRita, V.J., Eds.; Lippincott Williams & Wilkins: Philadelphia, PA, USA, 2007; p. 28.
43. Hanaki, H.; Kuwahara-Arai, K.; Boyle-Vavra, S.; Daum, R.S.; Labischinski, H.; Hiramatsu, K. Activated cell-wall synthesis is associated with vancomycin resistance in methicillin-resistant *Staphylococcus aureus* clinical strains Mu3 and Mu50. *J. Antimicrob. Chemother.* **1998**, *42*, 199–209. [CrossRef]
44. Oliveri, V.; Vecchio, G. 8-Hydroxyquinolines in medicinal chemistry: A structural perspective. *Eur. J. Med. Chem.* **2016**, *120*, 252–274. [CrossRef] [PubMed]
45. Freitas, L.B.D.O.; Borgati, T.F.; Gil, R.F.; Ruiz, A.; Marchetti, G.M.; De Carvalho, J.E.; Da Cunha, E.F.; Ramalho, T.D.C.; Alves, R.B. Synthesis and antiproliferative activity of 8-hydroxyquinoline derivatives containing a 1,2,3-triazole moiety. *Eur. J. Med. Chem.* **2014**, *84*, 595–604. [CrossRef] [PubMed]
46. Lazar, V.; Ditu, L.-M.; Pircalabioru, G.G.; Gheorghe, I.; Curutiu, C.; Holban, A.M.; Picu, A.; Petcu, L.; Chifiriuc, M.C. Aspects of Gut Microbiota and Immune System Interactions in Infectious Diseases, Immunopathology, and Cancer. *Front. Immunol.* **2018**, *9*, 1830. [CrossRef] [PubMed]
47. Rajkovic, A.; Jovanovic, J.; Monteiro, S.; Decleer, M.; Andjelkovic, M.; Foubert, A.; Beloglazova, N.; Tsilla, V.; Sas, B.; Madder, A.; et al. Detection of toxins involved in foodborne diseases caused by Gram-positive bacteria. *Compr. Rev. Food Sci. Food Saf.* **2020**, *19*, 1605–1657. [CrossRef]
48. Clinical and Laboratory Standards Institute. *Methods for Dilution Antimicrobial Susceptibility Tests for Bacteria that Grow Aerobically*, 3rd ed.; Approved Standard; Clinical and Laboratory Standards Institute: Wayne, PA, USA, 2015.
49. Hecht, D.W. Antimicrobial agents and susceptibility testing: Susceptibility testing of anaerobic bacteria. In *Manual of Clinical Microbiology*, 7th ed.; Murray, P.R., Baron, E.J., Pfaller, M.A., Tenover, F.C., Yolken, R.H., Eds.; ASM Press: Washington, DC, USA, 1999; pp. 1555–1562.
50. Houdkova, M.; Rondevaldova, J.; Doskocil, I.; Kokoska, L. Evaluation of antibacterial potential and toxicity of plant volatile compounds using new broth microdilution volatilization method and modified MTT assay. *Fitoterapia* **2017**, *118*, 56–62. [CrossRef]
51. Mosmann, T. Rapid colorimetric assay for cellular growth and survival: Application to proliferation and cytotoxicity assays. *J. Immunol. Methods* **1983**, *65*, 55–63. [CrossRef]
52. Houdkova, M.; Urbanova, K.; Doskocil, I.; Rondevaldova, J.; Novy, P.; Nguon, S.; Chrun, R.; Kokoska, L. In vitro growth-inhibitory effect of Cambodian essential oils against pneumonia causing bacteria in liquid and vapour phase and their toxicity to lung fibroblasts. *S. Afr. J. Bot.* **2018**, *118*, 85–97. [CrossRef]
53. TIBCO Software Inc. Statistica, Data Analysis Software System, Version 13.0. Available online: http://statistica.io (accessed on 30 September 2017).

Article

3-Amino-5-(indol-3-yl)methylene-4-oxo-2-thioxothiazolidine Derivatives as Antimicrobial Agents: Synthesis, Computational and Biological Evaluation

Volodymyr Horishny [1], Victor Kartsev [2], Vasyl Matiychuk [3], Athina Geronikaki [4,*], Petrou Anthi [4], Pavel Pogodin [5], Vladimir Poroikov [5], Marija Ivanov [6], Marina Kostic [6], Marina D. Soković [6] and Phaedra Eleftheriou [7]

1. Department of Chemistry, Danylo Halytsky Lviv National Medical University, Pekarska 69, 79010 Lviv, Ukraine; vgor58@ukr.net
2. InterBioScreen, 142432 Chernogolovka, Moscow Region, Russia; vkartsev@ibscreen.chg.ru
3. Department of Chemistry, Ivan Franko National University of Lviv, Kyryla i Mefodia 6, 79005 Lviv, Ukraine; v_matiychuk@ukr.net
4. School of Pharmacy, Aristotle University of Thessaloniki, 54124 Thessaloniki, Greece; anthi.petrou.thessaloniki1@gmail.com
5. Institute of Biomedical Chemistry, Pogodinskaya Street 10 Bldg.8, 119121 Moscow, Russia; pogodinpv@gmail.com (P.P.); vladimir.poroikov@ibmc.msk.ru (V.P.)
6. Mycological Laboratory, Department of Plant Physiology, Institute for Biological Research, Siniša, Stanković-National Institute of Republic of Serbia, University of Belgrade, Bulevar Despota Stefana 142, 11000 Belgrade, Serbia; marija.smiljkovic@ibiss.bg.ac.rs (M.I.); marina.kostic@ibiss.bg.ac.rs (M.K.); mris@ibiss.bg.ac.rs (M.D.S.)
7. Department of Biomedical Sciences, School of Health Sciences, International Hellenic University, Sindos, 57400 Thessaloniki, Greece; eleftheriouphaedra@gmail.com

* Correspondence: geronik@pharm.auth.gr; Tel.: +30-23-1099-7616

Received: 27 July 2020; Accepted: 28 August 2020; Published: 1 September 2020

Abstract: Herein we report the design, synthesis, computational, and experimental evaluation of the antimicrobial activity of fourteen new 3-amino-5-(indol-3-yl) methylene-4-oxo-2-thioxothiazolidine derivatives. The structures were designed, and their antimicrobial activity and toxicity were predicted in silico. All synthesized compounds exhibited antibacterial activity against eight Gram-positive and Gram-negative bacteria. Their activity exceeded those of ampicillin and (for the majority of compounds) streptomycin. The most sensitive bacterium was *S. aureus* (American Type Culture Collection ATCC 6538), while *L. monocytogenes* (NCTC 7973) was the most resistant. The best antibacterial activity was observed for compound **5d** (Z)-N-(5-((1H-indol-3-yl)methylene)-4-oxo-2-thioxothiazolidin-3-yl)-4-hydroxybenzamide (Minimal inhibitory concentration, MIC at 37.9–113.8 μM, and Minimal bactericidal concentration MBC at 57.8–118.3 μM). Three most active compounds **5d, 5g,** and **5k** being evaluated against three resistant strains, Methicillin resistant *Staphilococcus aureus* (MRSA), *P. aeruginosa*, and *E. coli*, were more potent against MRSA than ampicillin (MIC at 248–372 μM, MBC at 372–1240 μM). At the same time, streptomycin (MIC at 43–172 μM, MBC at 86–344 μM) did not show bactericidal activity at all. The compound **5d** was also more active than ampicillin towards resistant *P. aeruginosa* strain. Antifungal activity of all compounds exceeded those of the reference antifungal agents bifonazole (MIC at 480–640 μM, and MFC at 640–800 μM) and ketoconazole (MIC 285–475 μM and MFC 380–950 μM). The best activity was exhibited by compound **5g**. The most sensitive fungal was *T. viride* (IAM 5061), while *A. fumigatus* (human isolate) was the most resistant. Low cytotoxicity against HEK-293 human embryonic kidney cell line and reasonable selectivity indices were shown for the most active compounds **5d**, **5g**, **5k**, **7c** using thiazolyl blue tetrazolium bromide MTT assay.

The docking studies indicated a probable involvement of *E. coli* Mur B inhibition in the antibacterial action, while CYP51 inhibition is likely responsible for the antifungal activity of the tested compounds.

Keywords: indole; thioxothiazolidine; antibacterial activity; antifungal activity; computer-aided prediction; docking; Mur B; CYP 51

---

1. Introduction

Infectious diseases affect large populations and cause significant morbidity and mortality [1]. They represent a global indirect load on public health security and an impact on socio-economic stability worldwide. Bacterial, fungal, and viral infections have monopolized the dominant factors of death and disability of millions of humans for centuries. They are presently plaguing and even ravaging populations worldwide each year with performances far surpassing wars [2].

It should be mentioned that several dozen new infections have grown and affected the health of billions of people over the world, mainly in developing countries [3]. Unfortunately, there are no successful pharmaceuticals or vaccines for many of these new infections [3].

The treatment of infectious disease is still an imperative and demanding problem due to the growing number of multi-drug resistant pathogens, especially Gram-positive bacteria. Due to this, the lack of effective antimicrobial drugs, morbidity, and mortality notably increased [4].

Drug resistance causes vast human suffering, and now it is one of the most significant challenges of the twenty-first century. Species such as the methicillin-resistant *S. aureus* and vancomycin-resistant enterococci have emerged due to the irrational or overuse of antimicrobial agents [5].

The pathogens, including *Enterococcus faecium, Staphylococcus aureus, Klebsiella pneumoniae, Acinetobacter baumannii, Pseudomonas aeruginosa*, and *Enterobacter spp.* which also called ESCAPE pathogens, are of particular importance since they play a significant role affecting several human organs including the lung and urinary system. Besides, they exhibited increased resistance to clinically used antibiotics [6].

Numerous of these pathogens are Gram-negative bacteria, which are of specific concern due to their resistance of up to 50% against carbapenems that have been reported in some developing countries [6]. Despite the availability of some new antibiotics against Gram-positive pathogens, no treatment of these pathogens with a new class of compounds has been introduced in the last 40 years. Therefore, to overcome the resistance, the discovery of safer and more effective antimicrobial agents with a different mechanism of action is still an urgent need [7].

The interest in thiazolidine-based compounds attracted the attention of medicinal chemists, and a plethora of them have been studied to evaluate pharmacological properties [8–10]. Despite the appearance of some controversial opinions regarding the analysis of the molecular mechanism of their action, prominent representatives among the developed drug-like molecules are thiazolidinone derivatives [11,12] since they are a valuable source of building blocks for the development of novel molecules [13–15].

N-(4-oxo-2-thioxothiazolidin-3-yl)carboxamides exhibit antimicrobial [16–20] and antitumor [21–23] actions, are dual COX-1/2 and 5-LOX inhibitors [24,25], non-nucleoside inhibitors of Hepatitis C NS5b RNA polymerase [26,27] and HIV-1 reverse transcriptase inhibitors [28].

The combination of the thiazolidinone ring with other pharmacologically promising heterocycles has been a warranted approach for developing new "drug-like" molecules with the desired activity profile [29–31]. Our previous studies showed that thiazolidinone core with indole fragment in one molecule gave the compounds with high antimicrobial activity [19].

On the other hand, indole derivatives represent another scaffold widely spread in nature with a broad spectrum of biological activities. The indole ring was found not only in natural compounds but also in diverse semisynthetic and synthetic drug-like molecules [32,33].

They exhibit antimicrobial [34–39], anti-inflammatory [40,41], COX inhibitory [42,43] anticancer [44–46], antiviral [47,48], anti-HIV [49,50], and antidiabetic [51] activities. Among the natural compounds containing the indolene fragment, several imidazoline and imidazolidine alkaloids are known, which have a wide spectrum of biological activity, including antibacterial. Thus, indole-containing azahydantoins 1-6 from sponges and streptomycetes have a potent antibacterial and antiseptic action (Figure 1) [52–54]:

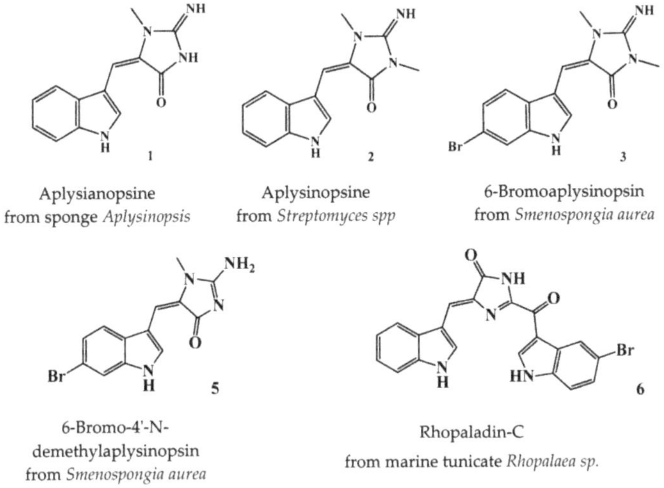

1 Aplysianopsine from sponge *Aplysinopsis*

2 Aplysinopsine from *Streptomyces spp*

3 6-Bromoaplysinopsin from *Smenospongia aurea*

5 6-Bromo-4'-N-demethylaplysinopsin from *Smenospongia aurea*

6 Rhopaladin-C from marine tunicate *Rhopalaea sp.*

Figure 1. Structure of indole-containing azahydantoins 1-5 from sponges and streptomycetes.

It is also known that synthetic thiohydantoin (rhodanine) analogs **7**, **8** (Figure 2), exhibit pronounced antibacterial properties [55].

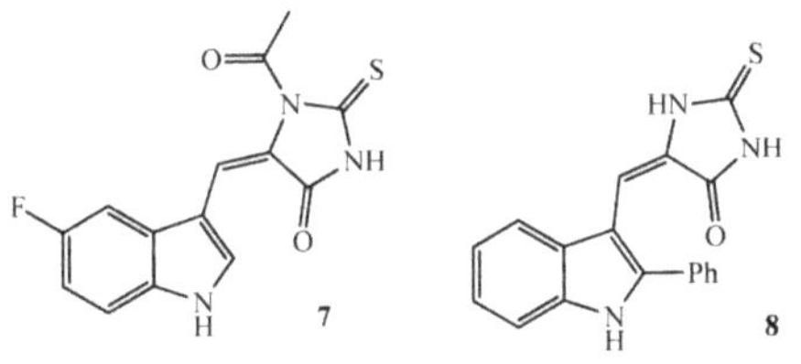

Figure 2. Synthetic thiohydantoin analogues.

Therefore, the design and development of hybrid molecules combining thiazolidinone and indole cores in the same structure is a promising approach. Taking into account all issues mentioned above and encouraging results obtained in our earlier studies [19], in this paper, we present the synthesis and biological evaluation of new (1H-indole-3-yl-methylene)-4-oxo-2-thioxothiazolidin derivatives with potent antimicrobial activity.

2. Results and Discussion

2.1. *In Silico Antimicrobial Activity Estimation*

2.1.1. Antibacterial Activity

Using AntiBac-Pred [56] one of the predictive web services of Way2Drug platform [57], activity against at least one strain of bacteria was predicted for each of the fourteen designed compounds with Pa-Pi values in the range from 0.001 to 0.309. According to the prediction results, the highest probability of antibacterial activity against the *Bacillus subtilis subsp. subtilis* str. 168 was estimated for derivatives **7a** and **5b** (Pa-Pi values are 0.309 and 0.305, respectively).

Similarly, we estimated in silico the probability of antibacterial activity for the reference drugs streptomycin and ampicillin. For both reference drugs, wide antibacterial action was predicted. For the top-10 predictions of streptomycin Pa-Pi values vary from 0.905 to 0.947; for ampicillin—from 0.712 to 0.989. Contrary, for relatively new antibacterial agent trifolirhizin, which structure was disclosed only on July 7, 2020 (Clarivate Analytics Integrity [58]), the top-10 predictions Pa-Pi values vary from 0.369 to 0.552.

2.1.2. Antifungal Activity

Using web service AntiFun-Pred [59], activity against at least one of the fungal species was predicted for six of the fourteen studied compounds with Pa-Pi values ranging from 0.001 to 0.112. The results show that among the studied compounds, derivatives 5a (Pa-Pi against *Trichophyton mentagrophytes* equals 0.112) and 7a (Pa-Pi against *Candida equals* 0.101) have better chances to be found active in biological evaluation of the antifungal activity.

The results of in silico antimicrobial activity assessment are given in the supplementary file PASSweb_results_13mols.xlsx. Small Pa-Pi values reflect the novelty of the analyzed compounds compared to those included in the PASS training set.

Similarly, for the reference drug ketoconazole wide antifungal action was predicted with Pa-Pi values in the range 0.622–0.812 (top-10 predictions), while for the new antifungal agent drimenin disclosed on 12 June 2020 (Clarivate Analytics Integrity [58]), only two antifungal activity were predicted with Pa-Pi values 0.007 and 0.030.

2.1.3. Acute Rat Toxicity

Using web service based on GUSAR software [60,61], acute rat toxicity with regards to different administration routes was estimated for the studied compounds. LD50 values and toxicity classes are given in Table 1. Most of the predictions indicate that the studied compounds belong to the fifth or fourth rodent toxicity classes.

Table 1. In silico assessments of acute rat toxicity.

Compound ID	LD_{50}, mg/kg				Toxicity Class			
	IP	IV	Oral	SC	IP	IV	Oral	SC
5a	809.8	402.5	1218	780.4 *	5	5	4	4 *
5b	680.4	309	1266	1434	5	5	4	5
5c	980.3	311.5	1325	619.9 *	5	5	4	4 *
5d								
5e	1263	466	843.9	440 *	NT	5	4	4 *
5f	1266	502.2	469.2	477.7 *	NT	5	4	4 *
5g	1010 *	371.9	192.4 *	397.6 *	5 *	5	3 *	4 *
5h	1282	448.5	1001	1588	NT	5	4	5
5i	1299	476.5	732.3	545.4 *	NT	5	4	4 *
5j	1258	381.6	196.2 *	422 *	NT	5	3 *	4 *
5k	1031 *	398.3	202.8 *	442 *	5 *	5	3 *	4 *
7a	1033	236.6	593.7	1644 *	5	4	4	5 *
7b	1061	287.2	720 *	862.1 *	5	4	4 *	4 *
7c	1180 *	210.7	765.1 *	682 *	5 *	4	4 *	4 *

Notes: *: Calculated for compounds that do not correspond to the model's applicability domain; thus, they are less reliable than unmarked ones. NT: Non-Toxic.

2.2. Chemistry

The starting N-(4-oxo-2-thioxothiazolidin-3-yl) -carbamides **3a-d** was prepared by reacting the acid hydrazides **1a-d** with trithiocarbonyl diglycolic acid (Scheme 1). The reaction was carried out in a medium of boiling aqueous alcohol. The yield of the products was 83–97%.

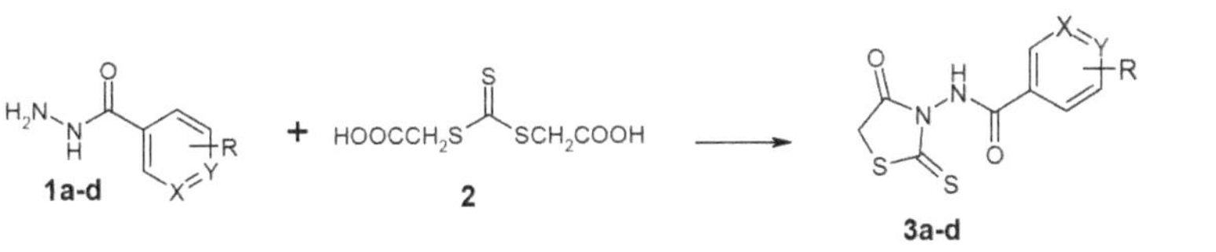

1a, 3a R = 2-OH, X = Y = CH; **1b, 3b** R = 4-OH, X = Y = CH; **1c, 3c** X = N, R = H, Y = CH;
1d, 3d Y = N, R = H, X = CH

Scheme 1. Synthesis of initial compounds.

The titled compounds were synthesized according to the process shown in Scheme 2.

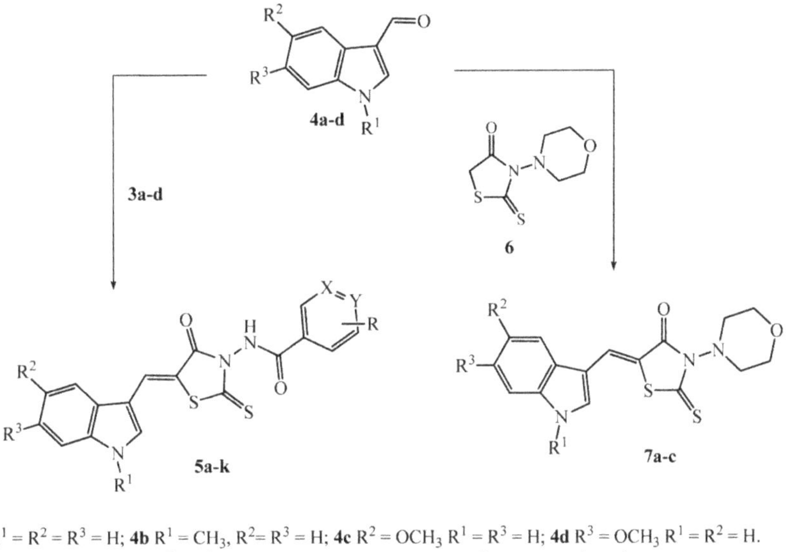

4a $R^1 = R^2 = R^3 = H$; **4b** $R^1 = CH_3$, $R^2 = R^3 = H$; **4c** $R^2 = OCH_3$ $R^1 = R^3 = H$; **4d** $R^3 = OCH_3$ $R^1 = R^2 = H$.
5a R = 2-OH, $R^1 = CH_3$, $R^2 = R^3 = H$, X = Y = CH; **5b** R = 2-OH, $R^2 = OCH_3$, $R^1 = R^3 = H$, X = Y = CH;
5c R = 2-OH,$R^3 = OCH_3$, $R^1 = R^2 = H$, X = Y = CH; **5d** R = 4-OH, $R^1 = R^2 = R^3 = H$, X = Y = CH;
5e X = N, $R = R^1 = R^2 = R^3 = H$, Y = CH; **5f** X = N, $R^1 = CH_3$, $R = R^2 = R^3 = H$, Y = CH;
5g X = N, $R^2 = OCH_3$, $R = R^1 = R^3 = H$, Y = CH; **5h** Y = N, $R = R^1 = R^2 = R^3 = H$, X = CH;
5i Y = N, $R^1 = CH_3$, $R = R^2 = R^3 = H$, X = CH; **5j** Y = N, $R^2 = OCH_3$, $R = R^1 = R^3 = H$, X = CH;
5k Y = N, $R^3 = OCH_3$, $R = R^1 = R^2 = H$, X = CH.
7a $R^1 = R^2 = R^3 = H$, **7b** $R^1 = CH_3$, $R^2 = R^3 = H$; **7c** $R^2 = OCH_3$, $R^1 = R^3 = H$.

Scheme 2. Synthesis of final compounds.

The reaction of N-(4-oxo-2-thioxothiazolidin-3-yl)carbamides **3a–d** with indole-3-carbaldehydes **4a-d** in acetic acid in the presence of an ammonium acetate catalyst afforded with high yield 5-[(R-1*H*-indol-3-yl)methylene]-4-oxo-2-thioxothiazolidin-3-ylcarbamides **5a–k**, while upon reaction of indole-3-carbaldehydes **4a–d** with 3-morpholino-2-thioxothiazolidin-4-one **6** in the same conditions 5-[(R-1*H*-indol-3-yl) methylene] -3-morpholin-4-yl-2-thioxothiazolidin-4-ones **7a–c** were obtained.

All compounds were characterized by IR, ^{1}H and ^{13}C NMR spectroscopy. In the IR spectra of compounds **3a–d, 5a–k,** and **7a, 7c,** the carbonyl group of the 4-thiazolidone ring absorbs at 1753.21–1690.53 cm^{-1}, and the thiocarbonyl group—at 1608.56–1556.48 cm^{-1}. The absorption band of the carbonyl group of the amide fragment of **3a–d** and **5a–k** is located at 1689.56–1654.84 cm^{-1}.

In the starting 3-substituted 2-thione-4-thiazolidones, the amide proton NH-CO of the compounds **3a–d** appears as a singlet in the range 11.95-10.91 ppm, and the cyclic methylene group resonates as a singlet or quartet at 4.55–4.48 ppm. etc. In the target products **5a–k**, the amide proton is in the range of 11.85–11.12 ppm. The 5-methylidene proton CH = of compounds **5a–k** and **7a, 7c** resonates in the form of a singlet at 8.20–7.94 ppm, which, according to the literature [9,62], is characteristic of the Z isomer. The singlet NH of the protons of the indole ring appeared in the range 12.31–12.06 ppm.

2.3. Biological Evaluation

2.3.1. Antibacterial Activity

Compounds **5a–k** and **7a–c** were evaluated for antibacterial activity, by microdilution method to determine the minimal bacteriostatic and bactericidal concentrations. As reference compounds, we used ampicillin and streptomycin, which are both broad-spectrum antibiotics commonly applied to treat different conditions. Antibacterial activity of tested compounds is shown in Table 2 with MIC values in the range of 36.5–211.5 μM and MBC at 73.3–282.0 μM. According to the order of activity which can be presented as: **5d** > **5g** > **5k** > **5j** > **5c** > **5h** > **5e** > **5f** > **5a** > **7c** > **7b** > **5b** > **7a** > **5i** the best activity is achieved for compound **5d** with MIC at 37.9–113.8 μM and MBC at 75.9–151.7 μM. The lowest antibacterial activity was observed for compound **5i** with MIC values in the range of 76.1–152.1 μM and MBC at 152.1–304.2 μM. The most sensitive bacterium appeared to be *S.aureus* (ATCC 6538), *En. cloacae* (ATCC 35,030) was the second most sensitive, while *S.Typhimirium* was the most resistant one. Another resistant strain was Gram-negative bacterium *S. Typhimurium* (ATCC 13,311).

Compound **7b** exhibited good activity against *B. cereus* with MIC and MBC at 41.7 and 83.4 μM respectively. Compound **5d** appeared to be potent against *S. aureus* (ATCC 6538), *P. aeruginosa* (ATCC 27,853), and *En. cloacaei* (ATCC 35,030) with MIC at 37.9 μM and MBC at 75.9 μM. It also showed good activity against *B. cereus* with MIC and MBC at 55.6 and 75.9 μM respectively. Compound **5h** appeared to be potent against *En. cloacae* and *P. aeruginosa* (ATCC 27,853) with MIC and MBC at 39.4 and 78.9 μM. Good activity against these two species and S. aureus (ATCC 6538) was also shown by compound 5j (MIC/MBC 58.6/73.1 μM). Good activity against *S. aureus* (ATCC 6538), also exhibited by compound **7b** with MIC at 41.7 μM and MBC at 83.4 μM. On the other hand, compound **5g** exhibited good activity against *En. cloacae* (ATCC 35030), *S. aureus* (ATCC 6538), and *S. typhimurium* (ATCC 13311) with MIC/MBC values 36.5/73.1 and 53.6/73.1 μM, respectively. It is worth to notice that all compounds appeared to be more potent than ampicillin against all bacteria used and more active than streptomycin against all bacteria except *B. cereus* and *S. typhimurium* (ATCC 13,311).

The structure-activity studies revealed that the most beneficial for antibacterial activity is the presence of hydroxybenzamide (**5d**) on the N-atom of (Z)-5-((5-methoxy-1H-indol-3-yl)methylene)-3-morpholino-2-thioxothiazolidin-4-one. Introduction of the 5-methoxy group to indole ring and replacement of hydroxybenzamide by nicotinamide (**5g**) decreased a little activity while shifting of methoxy group from position 5 to position 6 of indole ring and replacement of nicotinamide by isonicotinamide led to less active compound **5k** compared to compound **5g**.

On the other hand, the isonicotinamide derivative of (Z)-5-((1-methyl-indol-3-yl)methylene)-2-thioxothiazolidin-4-one **(5i)** appeared to be the less active compound. It was observed that for (Z)-5- [(1H-indol-3-yl)methylene]-2-thioxothiazolidin-4-one (**5h**) as well as for (Z)-N-5-[(1-methyl-1H-indol-3-yl)methylene]-4-oxo-2-thioxothiazolidin-4-one (**5g**) derivatives isonicotinamide substituent is endowed with better activity. The opposite was observed for 6-methoxy indole derivatives where more preferable is nicotinamide as a substituent (**5k**). Between methylindole derivatives (**5a, 5f, 5i**), more favorable for activity was nicotinamide substituent (**5f**), followed by benzamide, (**5a**) while isonicotinamide (**5i**) had a negative effect on antibacterial activity. For -2-hydroxybenzamides derivatives more preferable for antibacterial activity appeared to be 6-methoxy substitution of indole ring (**5c**) followed by methylidole (**5a**) while 5-methoxy substitution on indole ring was negative leading to one of the less active compounds (**5b**). In the case of 3-morpholino-2-thioxothiazolidin-4- one derivatives (**7a–c**), which were among the less active compound, it seems that 5-methoxy substitution on indole ring is preferable than methylindole or indole ring.

Thus, it can be concluded that the most favorable effect on the antibacterial activity of the target compounds is provided by the introduction into the molecule of an unsubstituted indolidene and 6-methoxyindolidene fragment. In addition, the nature of the substituent at position 3 of the thiazolidine ring has a direct influence on the enhancement of the antibacterial action. An increase in the antibacterial effect is observed from the use of 4-hydroxybenzamide and isonicotinamide substitutes.

From all mentioned above, it is evident that the antibacterial activity of these compounds depends not only on substituent and its position in the indole ring but also on substituent on the N-atom of 2-thioxothiazolidin-4-one ring.

Table 2. Antibacterial activity of compounds **5a–k** and **7a–c** (MIC/MBC in μM).

Com/d ID		*B.c*	*M.f*	*S.a*	*L.m*	*En.cl*	*P.a*	*S.T*	*E.coli*
5a	MIC	73.3 ± 0.4	109.9 ± 0.1	73.3 ± 0.3	146.5 ± 1.0	73.3 ± 0.8	73.3 ± 0.08	73.3 ± 0.08	109.9 ± 0.1
	MBC	146.5 ± 1.0	146.5 ± 2.0	146.5 ± 2.0	293.0 ± 4.0	146.5 ± 1.0	146.5 ± 1.0	146.5 ± 1.0	146.5 ± 2.0
5b	MIC	70.5 ± 0.4	105.8 ± 0.8	70.5 ± 0.8	105.8 ± 1.5	70.5 ± 0.8	70.5 ± 0.8	141.0 ± 1.2	211.5 ± 2.0
	MBC	141.0 ± 1.0	141.0 ± 2.0	141.0 ± 1.0	141.0 ± 2.0	141.0 ± 1.0	141.0 ± 0.1	282.0 ± 3.0	282.0 ± 2.0
5c	MIC	68.2 ± 0.8	102.6 ± 1.0	68.4 ± 0.5	68.4 ± 0.4	68.4 ± 0.4	68.4 ± 0.4	102.6 ± 1.0	102.6 ± 1.5
	MBC	136.8 ± 1.0	136.8 ± 1.5	136.8 ± .1.0	136.8 ± 1.0	136.8 ± 1.0	136.8 ± 1.0	136.8 ± 1.5	136.8 ± 2.0
5d	MIC	55.6 ± 0.2	113.8 ± 0.8	37.9 ± 0.2	113.8 ± 0.8	37.9 ± 0.4	37.9 ± 0.2	113.8 ± 1.0	75.6 ± 0.4
	MBC	75.9 ± 0.4	151.7 ± 2.0	75.9 ± 0.5	151.7 ± 2.0	75.9 ± 0.8	75.9 ± 0.6	151.7 ± 1.0	151.7 ± 1.0
5e	MIC	78.9 ± 0.2	118.3 ± 1.0	57.8 ± 0.4	78.9 ± 0.5	78.9 ± 0.1	78.9 ± 0.6	78.9 ± 0.8	118.3 ± 1.0
	MBC	157.7 ± 0.8	157.7 ± 2.0	78.9 ± 0.8	157.7 ± 1.5	157.7 ± 2.0	157.7 ± 1.5	157.7 ± 1.0	157.7 ± 2.0
5f	MIC	114.1 ± 1.0	114.1 ± 8.0	114.1 ± 1.0	76.1 ± 0.4	76.1 ± 0.8	76.1 ± 0.3	114.1 ± 1.5	114.1 ± 1.0
	MBC	152.1 ± 2.0	152.1 ± 1.0	152.1 ± 1.5	152.1 ± 1.0	152.1 ± 1.0	152.1 ± 1.0	152.1 ± 2.0	152.1 ± 1.0
5g	MIC	73.1 ± 1.0	73.1 ± 1.0	53.6 ± 0.4	73.1 ± 0.8	36.5 ± 0.5	109.6 ± 1.0	53.6 ± 0.6	109.6 ± 1.0
	MBC	146.2 ± 1.0	146.2 ± 1.0	73.1 ± 0.8	146.2 ± 1.6	73.1 ± 1.0	146.2 ± 1.2	73.1 ± 1.0	146.2 ± 2.0
5h	MIC	78.9 ± 0.5	118.3 ± 1.5	118.3 ± 1.0	78.9 ± 0.8	39.4 ± 0.5	39.4 ± 0.6	78.9 ± 0.6	118.3 ± 1.5
	MBC	157.7 ± 1.0	157.7 ± 2.0	157.7 ± 2.0	157.7 ± 1.0	78.9 ± 0.8	78.9 ± 0.8	157.7 ± 1.2	157.7 ± 1.0
5i	MIC	114.1 ± 1.0	114.1 ± 1.5	76.1 ± 0.8	152.1 ± 1.0	76.1 ± 0.8	76.1 ± 0.8	152.1 ± 1.0	114.1 ± 1.0
	MBC	152.1 ± 1.0	152.1 ± 2.0	152.1 ± 2.0	304.2 ± 4.0	152.1 ± 1.2	152.1 ± 1.2	304.2 ± 2.0	152.1 ± 1.0
5j	MIC	73.1 ± 0.5	109.6 ± 1.0	58.6 ± 0.4	146.2 ± 0.8	58.6 ± 0.6	58.6 ± 0.8	109.6 ± 1.0	109.6 ± 1.0
	MBC	146.2 ± 1.0	146.2 ± 2.0	73.1 ± 0.8	292.3 ± 0.2	73.1 ± 0.6	73.1 ± 0.06	146.2 ± 1.0	146.2 ± 2.0
5k	MIC	73.1 ± 0.5	109.6 ± 1.0	58.6 ± 0.4	109.6 ± 1.5	58.6 ± 0.6	58.6 ± 0.6	109.6 ± 1.5	109.6 ± 2.0
	MBC	146.2 ± 1.0	146.2 ± 2.0	73.1 ± 0.8	146.2 ± 1.0	73.1 ± 0.8	73.1 ± 0.8	146.2 ± 2.0	146.2 ± 2.0
7a	MIC	130.3 ± 1.0	130.3 ± 1.5	63.7 ± 0.4	86.9 ± 0.4	86.9 ± 1.0	86.9 ± 1.0	173.7 ± 2.0	130.3 ± 2.0
	MBC	173.7 ± 2.0	173.7 ± 2.0	86.9 ± 0.8	173.7 ± 1.5	173.7 ± 1.5	173.7 ± 1.5	347.4 ± 4.0	173.7 ± 1.5
7b	MIC	41.7 ± 0.2	125.2 ± 1.0	41.7 ± 0.2	166.9 ± 1.0	83.4 ± 0.9	61.2 ± 0.5	166.9 ± 2.0	125.2 ± .1.0
	MBC	83.4 ± 0.4	166.9 ± 2.0	83.4 ± 0.8	333.9 ± .2.0	166.9 ± 1.0	83.4 ± 1.0	333.9 ± 4.0	166.9 ± 2.0
7c	MIC	79.9 ± 0.4	119.8 ± 1.5	79.9 ± 1.0	159.8 ± 1.0	58.6 ± 0.4	58.6 ± 0.4	119.8 ± 1.0	119.8 ± 1.5
	MBC	159.8 ± 1.0	159.8 ± 2.0	159.8 ± 1.4	319.6 ± 2.0	79.9 ± 1.0	79.9 ± 1.0	159.8 ± 2.0	159.8 ± .2.0
Am.	MIC	248.0 ± 3.0	248.0 ± 2.0	248.0 ± 2.0	372.0 ± 4.0	248.0 ± 3.0	744.0 ± 9.0	248.0 ± 3.0	372.0 ± 4.0
	MBC	372.0 ± 4.0	372.0 ± 4.0	372.0 ± 2.0	744.0 ± 8.0	372.0 ± 3.0	1240 ± 2	492.0 ± 6.0	492.0 ± 8.0
Str.	MIC	43.0 ± 0.8	86.0 ± 1.0	172.0 ± 2.0	258.0 ± 4.0	43.0 ± 0.3	172.0 ± 3.0	172.0 ± 3.0	172.0 ± 2.0
	MBC	86.0 ± 1.0	172.0 ± 2.0	344.0 ± 4.0	516.0 ± 4.0	86.0 ± 0.6	344.0 ± 3.0	344.0 ± 3.0	344.0 ± 2.0

MIC–minimal inhibitory concentration, MBC–minimal bactericidal concentration, *B.c.-B.cereus* (clinical isolate), *M.f.-M.flavus* (ATCC 10,240), *S.a.-S.aurues* (ATCC 6538), *l.m.-L.monocytogenes* (NCTC 7973), *E.c.-E.coli* (ATTC 35210, *En.c.-En.cloaca* (ATCC 3503), *P.a.-P.aeruginosa* (ATCC 27,853), *S.T.-S.Typhimurium* (ATCC 13,311).

Three most active compounds were also evaluated against the resistant strains, including MRSA, *P. aeruginosa*, and *E. coli*, (Table 3). From the obtained results, it is evident that all three compounds were more active against MRSA than ampicillin, while streptomycin did not show any bactericidal activity. The compound **5d** was also more active than ampicillin towards resistant *P. aeruginosa* strain.

Table 3. Antibacterial activity against resistant strains (MIC/MBC in μM).

Compound ID		Resistant Strains		
		MRSA	*P.aeruginosa*	*E.coli*
5d	MIC	1260 ± 0.8	315 ± 9.0	1260 ± 21
	MBC	2520 ± 0.1	630 ± 8.0	2502 ± 22
5g	MIC	1220 ± 18	610 ± 5.0	1220 ± 19
	MBC	2440 ± 0.2	1202 ± 21	2440 ± 16
5k	MIC	1220 ± 0.6	610 ± 10	1220 ± 0.6
	MBC	2440 ± 22	1220 ± 21	2440 ± 22
Streptomycin	MIC	172.0 ± 21	86 ± 12	172 ± 21
	MBC	-	172 ± 14	344 ± 42
Ampicilline	MIC	-	572 ± 64	572 ± 78
	MBC	/	/	/

2.3.2. Antifungal Activity

All compounds also showed antifungal activity with MIC values ranging from 9.7 to 347.4 μM and MFC at 19.5–694.8 μM.The antifungal activity of compounds is shown in Table 4 and follows the order: **5g > 7c > 7b > 5d > 5b > 5e > 5k > 5f > 5j > 5c > 5i > 5a > 7a > 5h**. Compound **5g** displayed the best activity with MIC values in the range of 9.7–73.1 μM and MFC at 36.5–146.2 μM, while compound **5h** exhibited the lowest potential with MIC and MFC at 28.9–315.5 μM and 39.4–630.9 μM respectively. It was observed that similar to bacteria, fungi showed different sensitivity towards compounds tested. Thus, the most sensitive fungal strain appeared to be *T. viride* (IAM 5061), while the most resistant filamentous A. fumigatus. The behavior of compounds towards fungi species was different, too.

Several compounds showed very good activity against some species. For example, compound **5d** exhibited good activity against the most resistant *A. fumigatus* (MIC/MFC at .20.2/37.9 μM, while compound **7b** against *T. viride* (IAM 5061), *P. cyclpoium var verucosum* (food isolate) and all Aspergillus species except *A. fumigatus* (human isolate) with MIC at 22.3 μM and MFC at 41.4 μM. Compound **5g** exhibited excellent activity against *T. viride* (IAM 5061) (MIC/MFC at 0.97/1.95 μmol/mL × 10^{-2}). Additionally, good activity was achieved for compound **5g** against *A. versicolor* (ATCC 11730), *A. ochraceus* (ATCC 12066), *P. funiculosum* (ATCC 36839) with MIC and MFC at 19.5 μM and 36.5 μM respectively. Compound **5c** appeared to be potent against *A. ochraceus* (ATCC 12066) and *T. viride* (IAM 5061) (MIC/MFC at 18.8/35.3 μM whereas compound **7c** exhibited very good activity against *T. viride* (IAM 5061)with MIC at 10.7 μM and MFC at 21.3 μM and also good activity against *A. ochraceus* (ATCC 12066) and *P. funiculosum* (ATCC 36839 (MC/MFC at 23.1/39.9 μM. The potential of ketoconazole was at MIC 285-475 μM and MFC at 380–950 μM. Bifonazole displayed MIC at 480-640 μM and MFC at 640–800 μM. It should be mentioned that all compounds appeared to be more potent than ketoconazole and bifonazole. Only compound **7a** against *A. fumigatus (human isolate)* was less active than bifonazole.

According to the analysis of the structure-activity relationships, the most beneficial for antifungal activity is the presence of the 5-methoxy group in indole ring as well as nicotinamide as a substituent of the side chain (**5g**). In contrast, the presence of isonicotinamide in methylindole (**5i**) derivative appeared to be detrimental. Shifting of 5-OMe of compound **5g** to position 6 of indole and replacement of nicotinamide by 2-hydroxybenzamide resulted in compound **5c** with decreased activity. Removal of methoxy group and introduction of morpholino moiety to the N atom of thioxothiazolidinone **(7a)** decreased more activity.

In indole derivatives (**5d**, **5e**, **5h**), the presence of 4-hydroxybenzamide was favorable for antifungal activity, while isonicotinamide substituent had a negative effect. On the contrary, for methylindole derivatives (**5a**, **5f**, **5i**), the negative impact was observed with the presence of 2-hydroxybenzamide, while in the case of the 5-methoxy indole derivatives (**5b**, **5j**) it was the opposite. Finally, for the derivatives with morpholino moiety, the best activity was observed with the presence of the 5-methoxy group in the indole ring. The indole derivative was one of the less potent.

Thus, as in the case of antibacterial activity, antifungal activity depends not only on substitution in the indole ring but also on substituent on the N-atom of the 2-thioxothiazolidinone ring. In the series of (Z)-5-((5-methoxy-1H-indol-3-yl)methylene)-3-morpholino-2-thioxothiazolidin-4-one derivatives the most important structural features which enhanced the antifungal activity are again 4-hydroxybenzamide and 1H-indole moiety as well as nicotinamide and 5- and 6-methoxyindole moieties. On the other hand, in the series of indole 3-methylene morpholino-2-thioxothiazolidin-4-one derivatives, the presence of the 5-OCH3 group in the indole ring enhance the antifungal activity.

Table 4. Antifungal activity of compounds **5a–k** and **7a–c** (MIC/MFC in μM).

Com. ID		*A.f*	*A.v*	*A.o*	*A.n*	*T.v*	*P.o*	*P.f*	*Pvc*
5a	MIC	293.0 ± 2.2	36.6 ± 0.4	26.9 ± 0.1	53.7 ± 0.6	26.9 ± 0.1	36.6 ± 0.2	36.6 ± 0.2	109.9 ± 0.1
	MFC	586.1 ± 7.0	73.3 ± 0.8	36.6 ± 0.2	73.3 ± 0.8	36.6 ± 0.3	73.3 ± 0.5	73.3 ± 0.8	146.5 ± 0.2
5b	MIC	35.2 ± 0.6	35.2 ± 0.2	25.9 ± 0.2	35.2 ± 0.2	25.9 ± 0.2	35.2 ± 0.5	51.7 ± 0.5	35.2 ± 0.2
	MFC	70.5 ± 0.6	70.5 ± 0.4	35.2 ± 0.4	70.5 ± 0.8	35.2 ± 0.2	70.5 ± 0.5	70.5 ± 0.5	70.5 ± 0.5
5c	MIC	282.0 ± .2.0	35.3 ± 0.2	18.8 ± 0.2	25.9 ± 0.1	18.8 ± 0.2	35.3 ± 0.2	35.3 ± 0.2	35.3 ± 0.2
	MFC	564.1 ± 4.0	68.2 ± 0.4	35.3 ± 0.2	35.3 ± 0.2	35.3 ± 0.5	68.2 ± 0.5	68.2 ± 0.5	68.2 ± 0.5
5d	MIC	202 ± 0.1	37.9 ± 0.2	27.8 ± 0.1	37.9 ± 0.0	27.8 ± 0.2	37.9 ± 0.5	37.9 ± 0.2	37.9 ± 0.2
	MFC	37.9 ± 0.2	75.6 ± 0.4	37.9 ± 0.5	75.6 ± 0.5	37.9 ± 0.5	75.6 ± 0.5	75.6 ± 0.5	75.6 ± 0.5
5e	MIC	39.4 ± 0.2	39.4 ± 0.2	21.0 ± 0.1	39.4 ± 0.5	21.0 ± 0.1	39.4 ± 0.5	39.4 ± 0.5	39.4 ± 0.5
	MFC	78.9 ± 0.4	78.9 ± 0.4	39.4 ± 0.2	78.9 ± 1.0	39.4 ± 0.5	78.9 ± 0.5	78.9 ± 0.5	78.9 ± 0.5
5f	MIC	76.1 ± 0.4	38.0 ± 0.2	27.9 ± 0.1	38.0 ± 0.5	27.9 ± 0.2	55.8 ± 0.5	55.8 ± 0.5	55.8 ± 0.5
	MFC	152.1 ± 0.1	76.1 ± 0.4	38.0 ± 0.5	76.1 ± 0.8	38.0 ± 0.5	76.1 ± 1.0	76.1 ± 0.8	76.1 ± 1.0
5g	MIC	73.1 ± 0.4	19.5 ± 0.2	19.5 ± 0.1	14.6 ± 0.1	9.7 ± 0.01	36.5 ± 0.5	19.5 ± 0.1	26.8 ± 0.1
	MFC	146.2 ± 0.1	36.5 ± 0.4	36.5 ± 0.5	19.5 ± 0.1	19.5 ± 0.08	73.1 ± 0.8	36.5 ± 0.5	36.5 ± 0.2
5h	MIC	315.5 ± 2.5	78.9 ± 1.0	39.4 ± 0.2	78.9 ± 0.5	28.9 ± 0.1	39.4 ± 0.5	78.9 ± 0.5	118.3 ± 1.0
	MFC	630.9 ± 8.0	157.7 ± 1.0	78.9 ± 0.8	157.7 ± 1.0	39.4 ± 0.2	78.9 ± 0.5	157.7 ± 1.0	157.7 ± 2.0
5i	MIC	152.1 ± 1.0	38.0 ± 0.4	38.0 ± 0.0	38.0 ± 0.0	27.9 ± 0.2	76.1 ± 0.5	55.8 ± 0.5	76.1 ± 0.5
	MFC	304.2 ± .2.0	76.1 ± 0.0	76.1 ± 1.0	76.1 ± 0.8	38.0 ± 0.5	152.1 ± 1.0	76.1 ± 0.8	152.1 ± 1.0
5j	MIC	146.2 ± 1.0	35.5 ± 0.2	53.6 ± 0.4	26.8 ± 0.2	35.5 ± 0.5	35.5 ± 0.4	35.5 ± 0.4	35.5 ± 0.6
	MFC	292.3 ± .2.0	73.1 ± 0.8	73.1 ± 0.8	35.5 ± 0.5	73.1 ± 1.0	73.1 ± 1.0	73.1 ± 1.0	73.1 ± 1.0
5k	MIC	35.5 ± 0.4	35.5 ± 0.2	35.5 ± 0.2	35.5 ± 0.5	35.5 ± 0.2	35.5 ± 0.5	35.5 ± 0.4	35.5 ± 0.2
	MFC	73.1 ± 0.8	73.1 ± 0.8	73.1 ± 1.0	73.1 ± 0.5	73.1 ± 0.8	73.1 ± 0.8	73.1 ± 0.5	73.1 ± 0.8
7a	MIC	347.4 ± 2.0	43.4 ± 0.2	43.4 ± 0.2	43.4 ± 0.2	31.8 ± 0.2	43.4 ± 0.2	43.4 ± 0.4	86.9 ± 1.0
	MFC	694.8 ± 4.0	86.9 ± 1.0	86.9 ± 0.5	86.9 ± 1.0	43.4 ± 0.5	86.9 ± 1.0	86.9 ± 1.0	173.7 ± 2.0
7b	MIC	22.3 ± 0.1	22.3 ± 0.1	22.3 ± 0.1	41.7 ± 0.5	22.3 ± 0.2	41.7 ± 0.5	41.7 ± 1.0	22.3 ± 0.1
	MFC	41.7 ± 0.2	41.7 ± 0.2	41.7 ± 0.5	83.4 ± .1.0	41.7 ± 0.8	83.4 ± 1.0	83.4 ± 1.0	41.7 ± 0.4
7c	MIC	79.9 ± 0.4	21.3 ± 0.1	16.0 ± 0.1	21.3 ± 0.2	10.7 ± 0.2	21.3 ± 0.1	21.3 ± 1.0	58.6 ± 0.4
	MFC	159.8 ± 1.0	39.9 ± 0.5	21.3 ± 0.1	39.9 ± 0.2	21.3 ± 0.5	39.9 ± 0.2	39.9 ± 1.0	79.9 ± 1.0
Ket.	MIC	380 ± 12	2850 ± 68	380 ± 12	380 ± 8.0	475 ± 58	3800 ± 58	380 ± 16	380 ± 12
	MFC	950 ± 23	3800 ± 84	950 ± 12	950 ± 6.0	570 ± 86	3800 ± 48	950 ± 26	950 ± 23
Bif.	MIC	480 ± 22	480 ± .2	480 ± 28	480 ± 12	640 ± 28	480 ± 20	640 ± 12	480 ± 22
	MFC	640 ± 3.4	640 ± 0.8	800 ± 1.8	640 ± 2.3	800 ± 3.8	640 ± 1.6	800 ± 2.1	640 ± 3.4

MIC–minimal inhibitory concentration, MFC–minimal fungicidal concentration. *A.fum.-A.fumigatus* (human isolate), *A.v.-A.versicolor* (ATCC 11730), *A.o.-A.ochraceus* (ATCC 12066), *A.n.-A.niger* (ATCC 6275), *T.v.-T.viride* (IAM 5061), *P.f.-P.funiculosum* (ATCC 36839), *P.o.-P.ochrochloron* (ATCC 9112), *P.v.c.-P.cyclpoium var. verucosum* (food isolate).

2.3.3. Cytotoxicity Assessment

Low toxicity and selectivity of action of antimicrobial compounds is a crucial pre-requisite for further development. Thus, we studied the cytotoxicity of the most active compounds. MTT analysis was performed on the HEK-293 human embryonic kidney cell line. The cells were cultured in DMEM medium supplemented with 10% fetal bovine serum. The cells were inoculated into a 96-well plate at a concentration of $5 \cdot 10^4$/mL ($5 \cdot 10^3$ per well, 100 μL each). After one day of culture, compound preparations were added, and the results were obtained after a 72 h culture period. The compounds were added at four concentrations (25, 50, 100, and 250 μM). Since compound solutions contained DMSO, control cultures containing only DMSO at the final concentration obtained when the appropriate volume of compound solution was added were performed.

Although the compounds do not exhibit statistically significant concentration-dependent toxicity up to 100 μM (Figure 3), they show some toxicity at higher concentrations. The average CC_{50} values obtained from three different experiments are given in Tables 5 and 6. The SI index is also shown in Tables 5 and 6.

Compound **5g** and **7c** exhibited the best SI index for anti-fungal activity while compound **5d** exhibited the best SI index for anti-bacterial activity.

We compared the CC_{50} values of compounds **5d**, **5k**, **5g**, **7c** with cytotoxicity of the reference drugs obtained in the HEK-293 human embryonic kidney cell line. For antibacterials streptomycin, ampicillin and antifungal bifonazole CC_{50} exceeded 100 μM [63,64]; for antifungal ketoconazole CC_{50} = 60 μM [65]. Thus, cytotoxicity of the most active compounds in our study is comparable or lower than cytotoxicity of the reference antimicrobial drugs.

Table 5. Antibacterial activity (MIC), cytotoxicity (CC_{50}), and selectivity indices (SI) of compounds **5d, 5g, 5k, 7c.**

ID	CC_{50}		*B.c*	*M.f*	*S.a*	*L.m*	*En.cl*	*P.a*	*S.T*	*E.coli*
5d	252 ± 1.5	MIC	55.6 ± 0.2	113.8 ± 0.8	37.9 ± 0.2	113.8 ± 0.8	37.9 ± 0.4	37.9 ± 0.2	113.8 ± 1.0	75.6 ± 0.4
		SI	4.5	2.2	6.7	2.2	6.7	6.7	2.2	3.3
5g	256 ± 6.21	MIC	73.1 ± 0.1	73.1 ± 1.0	53.6 ± 0.4	73.1 ± 0.8	36.5 ± 0.5	109.6 ± 1.0	53.6 ± 0.6	109.6 ± 1.0
		SI	3.5	3.5	4.8	3.5	7.0	2.3	4.8	2.3
5k	252 ± 1.89	MIC	73.1 ± 0.5	109.6 ± 1.0	58.6 ± 0.4	109.6 ± 1.5	58.6 ± 0.6	58.6 ± 0.6	109.6 ± 1.5	109.6 ± 2.0
		SI	3.5	2.3	4.3	2.3	4.3	4.3	2.3	2.3
7c	225± 1.87	MIC	79.9 ± 0.4	119.8 ± 1.5	79.9 ± 1.0	159.8 ± 1.0	58.6 ± 0.4	58.6 ± 0.4	119.8 ± 1.0	119.8 ± 1.5
		SI	2.8	1.9	2.8	1.4	3.8	3.8	1.9	1.9

Table 6. Antifungal activity (MIC), cytotoxicity (CC_{50}), and selectivity indices (SI) of compounds **5d, 5g, 5k, 7c.**

Com.	CC_{50} (μM)		*A.f*	*A.v*	*A.o*	*A.n*	*T.v*	*P.o*	*P.f*	*Pvc*
5d	252 ± 1.5	MIC	202 ± 0.1	37.9 ± 0.2	27.8 ± 0.1	37.9 ± 0.2	27.8 ± 0.2	37.9 ± 0.5	37.9 ± 0.2	37.9 ± 0.2
		SI	1.3	6.7	9.1	6.7	9.1	6.7	6.7	6.7
5g	256 ± 6.21	MIC	73.1 ± 0.2	19.5 ± 0.2	19.5 ± 0.2	14.6 ± 0.1	9.7 ± 0.1	36.5 ± 0.5	19.5 ± 0.1	26.8 ± 0.1
		SI	3.5	13.1	13.1	17.5	26.4	7.0	13.1	9.6
5k	252 ± 1.89	MIC	35.5 ± 0.4	35.5 ± 0.2	35.5 ± 0.2	35.5 ± 0.5	35.5 ± 0.2	35.5 ± 0.5	35.5 ± 0.4	35.5 ± 0.2
		SI	7.1	7.1	7.1	7.1	7.1	7.1	7.1	7.1
7c	225 ± 1.87	MIC	79.9 ± 0.4	21.3 ± 0.1	16.0 ± 0.1	21.3 ± 0.2	10.7 ± 0.2	21.3 ± 0.1	21.3 ± 1.0	58.6 ± 0.4
		SI	2.8	10.6	14.1	10.6	21.0	10.6	10.6	3.8
Ket.	60 *	MIC	380 ± 12	285 ± 68	380 ± 12	380 ± 8.0	475 ± 58	380 ± 58	380 ± 16	380 ± 12
		SI	0.158	0.210	0.158	0.158	0.126	0.158	0.158	0.158

* 24 h.

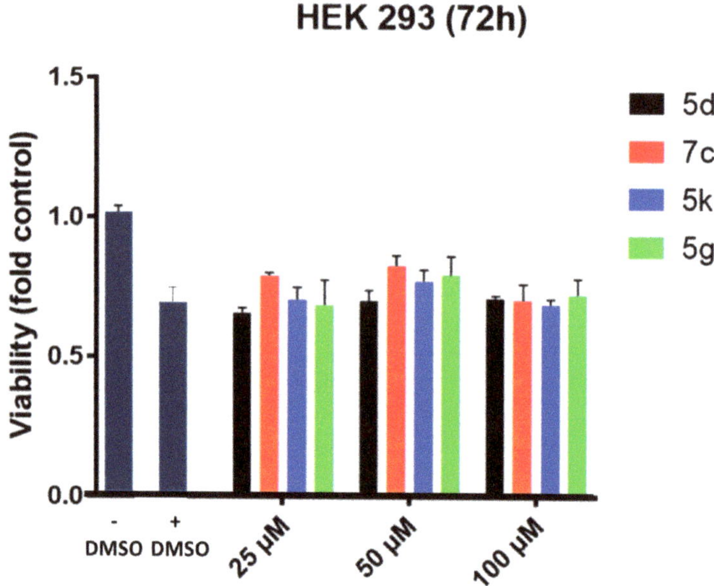

Figure 3. MTT assay results for compounds **5d, 5k, 5g, 7c.** According to the results, all compounds did not show statistically significant, concentration-dependent cytotoxicity at concentrations up to 100 μM. The stable decrease in viability observed can be attributed to dimethyl soulfoxide (DMSO,) present at stable concentration at all compound samples.

2.4. Docking Studies

Since the mechanism of antimicrobial action of our compounds is not known, to choose the proteins as potential targets, we based on the literature. It was found that benzothiazole derivatives are mentioned as Gyrase inhibitors [66–68]. On the other hand, according to the literature, thiazolidinones act as MurB inhibitors [69–72]. Furthermore, prediction of the mechanism of action by computer program PASS indicated Thymidylate kinase as the probable antibacterial target. On the other hand, several publications mentioned thiazolidinone and indole derivatives as 14^{α}-lanosterol demethylase inhibitors [73–75]. Thus, taking all these into account, we proposed *E. coli* DNA Gyrase, Thymidylate kinase, and *E. coli* MurB enzymes as antibacterial targets, with CYP51 as the antifungal target.

2.4.1. Docking to Antibacterial Targets

The docking studies revealed that estimated binding energy to *E. coli* DNA Gyrase (−2.59 to −6.54 kcal/mol) as well as to thymidylate kinase (−1.55 to −4.12 kcal/mol), were higher than that to *E. coli* MurB (−7.07 to −10.93 kcal/mol). Therefore, it may be resolved that *E. coli* MurB is the most suitable enzyme where binding scores were consistent with biological activity (Table 7).

Table 7. Molecular docking binding energies.

Comp.	Est. Binding Energy (kcal/mol)			I-H *E. coli MurB*	Residues *E. coli MurB*
	E.coli DNA Gyrase 1KZN	*Thymidylate Kinase 4QGG*	*E. coli MurB 2Q85*		
5a	−4.63	-	−8.22	2	Gly122, Ser228
5b	−3.12	-	−7.70	2	Arg213, Asn232
5c	−5.39	−2.13	-9.16	2	Gly122, Ser228
5d	−6.21	−4.12	−10.93	2	Ser228, Arg326
5e	−6.28	−2.39	−8.97	2	Arg213, Ser228
5f	−5.46	−1.55	−8.74	2	Gly122, Ser228
5g	-6.54	-3.26	−10.88	3	Gly122, Ser228, Asn232
5h	−6.11	−1.24	−9.12	2	Gly122, Ser228
5i	−3.69	−1.15	−7.07	2	Arg213, Arg326
5j	−5.52	−3.25	−9.21	2	Gly122, Ser228
5k	−5.63	−2.96	−9.83	2	Arg213, Ser228
7a	−2.59	-	−7.28	2	Gly122, Arg213
7b	−3.67	-	−7.75	2	Ser228, Asn232
7c	−4.28	-	−7.88	2	Gly122, Ser228

The docking pose of the most active compound **5d** in *E. coli* MurB enzyme showed two favorable hydrogen bond interactions. The first one is between the oxygen atom of the C=O group of the compound and the hydrogen of the side chain of Ser228. The second one between the oxygen atom of -OH group of the compound and the side chain of Arg326 (distances 2.17 Å and 1.99 Å, respectively). The fused rings interact hydrophobically with the residues Tyr189, Asn232, Leu289, Ala123, Leu217, and Arg213, while the benzene ring interacts hydrophobically with the residues Asn50, Ser115, Ile118, Ile121, Gln119 and Glu324 (Figure 4). These interactions stabilize the complex compound-enzyme and play a crucial role in the increased inhibitory activity of compound **5d** Moreover, the hydrogen bond formation with the residue Ser228 is essential for the inhibitory action of the compounds; thus, this residue takes part in the proton transfer at the second stage of peptidoglycan synthesis [76].

The second most active compound, **5g,** also forms the hydrogen bond interaction with the residue Ser228 that explains its high inhibitory action (Figure 2). Detailed analysis of the docking pose of the two most active compounds showed that they similarly bind MurB, and they insert deeper to the binding center of the enzyme than FAD, forming a hydrogen bond with the residue Ser228 (Figure 5).

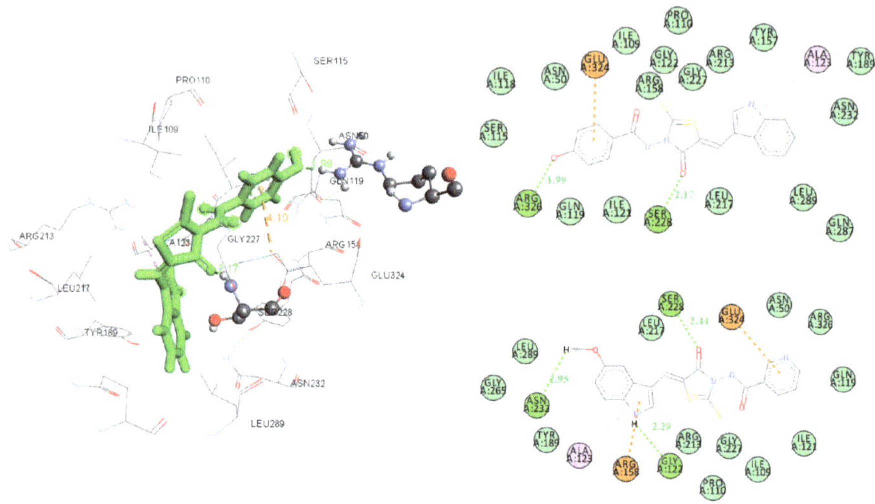

Figure 4. Docked conformation of the most active compound **5d** in *E.coli* MurB **(Left)**. 2D diagrams of the most active compounds **5d** (up) and **5g** (down) in *E.coli* MurB **(Right)**.

The same behavior was observed in the case of docking of the most active compound among 5-(1*H*-indol-3-ylmethylene)-4-oxo-2-thioxothiazolidin-3-yl)alkane carboxylic acids [19] and 5-adamantane thiadiazole-based thiazolidinones [70]. Again, the formation of the hydrogen bond between the C=O group and Ser228 was observed. Thus, the obtained results support previous data [69–72] that MurB maybe is the most appropriate target for the antibacterial activity for this chemical series.

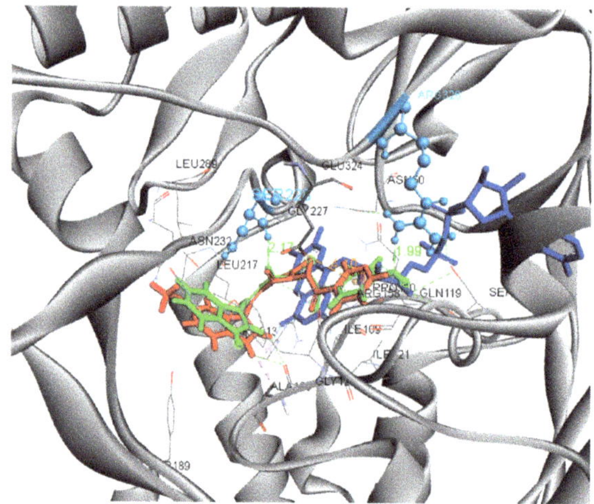

Figure 5. Docked conformation of compounds **5d** (green), **5g** (red) and FAD (blue) in *E.Coli* MurB.

2.4.2. Docking to Lanosterol 14α-demethylase of *C. albicans*

All the synthesized compounds and the reference drug ketoconazole were docked to lanosterol 14α-demethylase of *C. albicans* (Table 8).

Table 8. Molecular docking binding energies.

N/N	Est. Binding Energy (kcal/mol) CYP51 of *C. albicans* PDB ID: 5V5Z	I-H	Residues CYP51 of *C. albicans* PDB ID: 5V5Z	Interactions with HEM601
5a	−7.65	1	Tyr132	Hydrophobic
5b	−9.74	1	Tyr132	Ionizable, Hydrophobic
5c	−8.13	2	Tyr64, Tyr132	Hydrophobic
5d	−10.22	2	Tyr118, Tyr132	Ionizable, Hydrophobic
5e	−9.15	1	Tyr132	Ionizable, Hydrophobic
5f	−8.79	2	Tyr118, Tyr132	Hydrophobic
5g	−11.55	1	Tyr132	Fe binding, Ionizable, Hydrophobic
5h	−7.11	-	-	Ionizable, Hydrophobic
5i	−7.84	1	Tyr132	Ionizable, Hydrophobic
5j	−8.72	2	Tyr118, Met508	Hydrophobic
5k	−9.24	2	Tyr64, Tyr118	Hydrophobic
7a	−7.08	1	Tyr132	Hydrophobic
7b	−10.36	1	Tyr132	Ionizable, Hydrophobic
7c	−10.84	2	Tyr118, Tyr132	Ionizable, Hydrophobic
ketoconazole	−8.23	1	Tyr64	Ionizable, Hydrophobic

Docking results showed that all the synthesized compounds might bind to $CYP51_{Ca}$ close to those of the reference drug ketoconazole. Compound **5g** is located inside the enzyme alongside to heme group, interacting with the Fe of the heme group of $CYP51_{Ca}$ throughout its atom N of the pyridine ring. Moreover, compound **5g** forms a hydrogen bond between the oxygen of $-OCH_3$ substituent and the hydrogen of the side chain of Tyr132. Hydrophobic interactions were detected between residues Thr122, Phe126, Tyr132, and Ile131 and the fused rings of the compound **5g**, also between Leu376, Thr311 and the benzene ring of the compound. Furthermore, compound **5g** interacts hydrophobically throughout its benzene ring with the heme group of the enzyme, and also it forms a positive ionizable bond with it (Figure 6). Interaction with the heme group was also observed with the benzene ring of ketoconazole, which forms positive ionizable interactions (Figures 6 and 7). However, compound **5g** forms a more stable complex of the ligand with enzyme indicating its interaction with the Fe, which is probably why compound **5g** showed high antifungal activity.

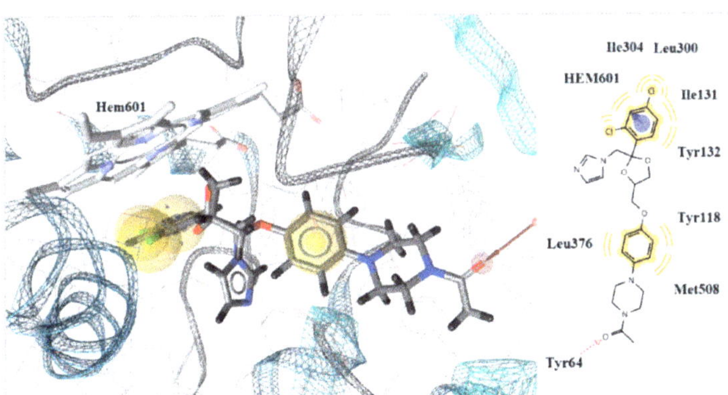

Figure 6. Docked conformation of ketoconazole in lanosterol 14alpha-demethylase of *C. albicans* ($CYP51_{ca}$).

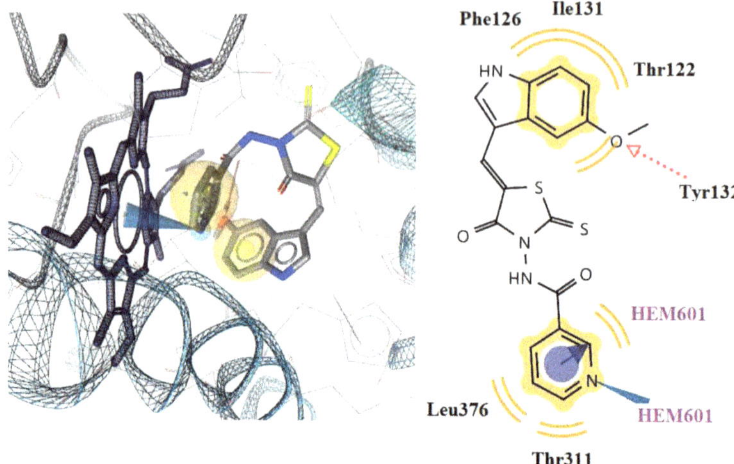

Figure 7. Docked conformation of compound **5g** in lanosterol 14alpha-demethylase of *C. albicans* ($CYP51_{ca}$).

It should be mentioned that the tested compounds interact more strongly with the heme group of the enzyme $CYP51_{Ca}$ because the heme's Fe is involved in this interaction. In the case of our previous work [19], the most active compound interacts with the heme but throughout its benzene ring and the $-NO_2$ group, forming pi and negative ionizable interactions with the heme group, respectively. In the case of 5-adamantane thiadiazole-based thiazolidinones **[72]**, again, the most active compound form positive interactions between the heme group and heterocyclic rings of the compound. Thus, it can be concluded that thiazolidinone derivatives, in general, can interact with the heme of $CYP51_{Ca}$ in the same way as ketoconazole interacts.

3. Materials and Methods

All starting materials were purchased from Merck and used without purification. NMR spectra were determined with Varian Mercury VX-400" (Varian Co., Palo Alto, CA, USA) and AM-300 Bruker 300 MHz. spectrometers in DMSO-d_6. MS (ESI) spectra were recorded on an LC-MS system-HPLC Agilent 1100 (Agilent Technologies Inc., Santa, Clara, CA USA) equipped with a diode array detector Agilent LC\MSD SL. Parameters of analysis: Zorbax SB-C18 column (1.8 μM, 4.6–15 mm, PN 821975-932), solvent water–acetonitrile mixture (95:5), 0.1% of aqueous trifluoroacetic acid; eluent flow 3 mL min^{-1}; injection volume 1 μL; IR spectra were recorded on a Vertex 70 Bruker" (Bruker, Karlsruhe, Germany) spectrometer in KBr pellets. Melting points were determined in open capillary tubes and are uncorrected.

3.1. In Silico Biological Activity Evaluation

Antimicrobial activity and toxicity of the designed compounds have been estimated in silico using web services available on the Way2Drug portal [56]. These services are based on the PASS (Prediction of Activity Spectra for Substances) and GUSAR (General Unrestricted Structure-Activity Relationships) software, which is described in detail elsewhere [60,61]. It is essential to mention that PASS-based services provide the assessments of the compound's activity as the difference between the probabilities for the chemical compound with a particular structure to display activity (Pa) and do not display this activity (Pi). By default, in PASS, all activities with Pa > Pi are considered as probable. High Pa-Pi values reflect the high structural similarity of the analyzed compound to the structures included in the training set with those activities. Since our goal was not finding close analogs of the earlier discovered antimicrobial agents, we considered compounds with small Pa-Pi values as the promising hits for experimental testing. If the experiment will confirm their activity, there is a chance to find a New Chemical Entity. GUSAR-based service [60,61] provides the quantitative assessment of acute rat toxicity expressed as LD_{50} values for four routes of administration: intraperitoneal (IP), intravenous (IV), oral, and subcutaneous (SC).

3.2. Chemistry

3.2.1. General Procedure for the Preparation of N-(4-oxo-2-thioxothiazolidin-3-yl) carbamides **3a–d**

In a round-bottom flask equipped with a reflux condenser, 0.05 mol of trithiocarbonyl diglycolic acid, 0.05 mol of the corresponding hydrazide and alcohol-water mixture (1:1) were placed and boiled for 3 h. The reaction mixture is cooled, the precipitate is filtered off and recrystallized.

2-Hydroxy-N-(4-oxo-2-thioxothiazolidin-3-yl)benzamide 3a. Yield 97%; m.p. 104–106 °C (CH_3COOH-H_2O 2:1). IR (cm^{-1}): 3342.48 (OH), 1751.28 (C=O), 1657.74 (C=O), 1608.56 (C=S).). ^1H NMR (400 MHz, DMSO-d_6, ppm) δ 11.62–10.91 (br.s, 2H, NH, OH), 7.88 (dd, J = 8.0, 1.6 Hz, 1H, H_6 benzene), 7.52–7.46 (m, 1H, H_3 benzene), 7.05–6.95 (m, 1H, 2H, H_4 +H_5, aromatic), 4.48 (q, J = 18.7 Hz, 2H, CH_2). ^{13}C NMR (101 MHz, DMSO, ppm) δ 199.90, 170.25, 164.91, 157.93, 134.62, 129.83, 119.44, 117.19, 115.07, 33.38. Anal. Calcd. for $C_{10}H_8N_2O_3S_2$ (%): C, 44.77; H, 3.01; N, 10.44; S, 23.90 Found (%):C, 44.88; H, 3.09; N, 10.37; S, 23.95.

4-Hydroxy-*N*-(4-oxo-2-thioxothiazolidin-3-yl)benzamide 3b. Yield 87%; m.p. 207–209 °C (CH_3COOH).). IR (cm^{-1}): 3259.54 (OH), 3166(NH), 1739.71 (C=O), 1667.38(C=O), 1583.48 (C=S). ^1H NMR (400 MHz, DMSO-d_6, ppm) δ 11.25 (s, 1H, NH), 10.29 (s, 1H, OH), 7.81 (dd, J = 9.1, 2.3 Hz, 2H, H_2 +H_6, benzene), 6.88 (dd, J = 9.1, 2.3 Hz, 2H, H_3 +H_5, benzene), 4.51 (s, 2H, CH_2). ^{13}C NMR (101 MHz, DMSO, ppm) δ 200.33, 170.54, 163.93, 161.44, 129.96, 121.44, 115.24, 33.32. Anal. Calcd. for $C_{10}H_8N_2O_3S_2$ (%): C, 44.77; H, 3.01; N, 10.44; S, 23.90 Found (%):C, 44.69; H, 2.95; N, 10.36; S, 23.81.

N-(4-Oxo-2-thioxothiazolidin-3-yl)nicotinamide 3c. Yield 83%; m.p. 190 °C decomp.(C_2H_5OH). IR (cm^{-1}): 1753.21 (C=O), 1687.63 (C=O), 1556.48 (C=S). ^1H NMR (400 MHz, DMSO-d_6, ppm) δ 11.95 (s, 1H, NH), 8.97–8.73 (m, 2H, H_2 +H_4, pyridine), 7.99–7.70 (m, 2H, H_5 +H_6, pyridine), 4.55 (s, 2H, CH_2).). ^{13}C NMR (101 MHz, DMSO, ppm) δ 199.81, 170.20, 163.25, 150.76, 137.93, 121.31, 119.56, 33.55. Anal. Calcd. for $C_9H_7N_3O_2S_2$ (%): C, 42.68; H, 2.79; N, 16.59; S, 25.32 Found (%):C, 42.79; H, 2.70; N, 16.48; S, 25.45.

N-(4-Oxo-2-thioxothiazolidin-3-yl)isonicotinamide 3d. Yield 85%; m.p. 193 °C decomp. (C_2H_5OH). IR (cm^{-1}): 1753.21(C=O), 1678.95 (C=O), 1556.48 (C=S).^1H NMR (400 MHz, DMSO-d_6, ppm) δ 11.95 (s, 1H, NH), 8.87–8.79 (m, 2H, H_3 +H_5, pyridine), 7.85–7.80 (m, 2H, H_2 +H_6, pyridine), 4.55 (s, 2H, CH_2). ^{13}C NMR (101 MHz, DMSO, ppm) δ 199.81, 170.19, 163.25, 150.76, 137.93, 121.39, 119.56, 33.58. Anal. Calcd. for $C_9H_7N_3O_2S_2$ (%): C, 42.68; H, 2.79; N, 16.59; S, 25.32 Found (%):C, 42.77; H, 2.85; N, 16.76; S, 25.26.

3.2.2. General Procedure 5-[(R-1H-indol-3-ylmethylene)-4-oxo-2-thioxothiazolidin-3-yl] carbamides **5a–k** and 5-(R-1H-indol-3-ylmethylene)-3-morpholin-4-yl-2-thioxothiazolidin-4-ones **7a–c**

In a round-bottom flask equipped with a reflux condenser, 2.5 mmol of 3-substituted 2-thioxo-4-oxothiazolidine **3a-d** or **6**, 3.3 mmol of the corresponding aldehyde **1a-d,** 2.5 mmol of ammonium acetate and 5 mL of acetic acid are placed. The reaction mixture is boiled for 2 h, cooled, the precipitate is filtered off, washed with acetic acid and water, dried and recrystallized.

2-Hydroxy-*N*-{(5*Z*)-5-[(1-methyl-1*H*-indol-3-yl)methylene]-4-oxo-2-thioxothiazolidin-3-yl}benzamide 5a. Yield 98%; m.p. 265–267 °C (DMFA-CH_3COOH). IR (cm^{-1}): 3272.08 (OH), 1700.17 (C=O), 1656.77 (C=O), 1588.3 (C=C), 1573.84 (C=S). ^1H-NMR (300 MHz, DMSO-d_6, ppm) δ 11.40 (s, 1H, NH-CO), 11.31 (s, 1H, OH), 8.12 (s, 1H, CH=), 8.01 (d, J = 7.9 Hz, 1H, H_6 benzene), 7.91 (d, J = 4.2 Hz, 2H, H_4 +H_7, indole), 7.54–7.41 (m, 2H, H_3 benzene + H_2 indole), 7.37–7.22 (m, 2H, H_5 +H_6, indole), 6.97 (dd, J = 17.3, 8.1 Hz, 2H, H_4 +H_5, benzene), 4.00 (s, 3H, CH_3N). ^{13}C NMR

(101 MHz, DMSO, ppm) δ 189.71, 164.99, 163.08, 158.00, 136.96, 134.65, 134.45, 129.90, 127.29, 127.01, 123.54, 122.01, 119.46, 118.78, 117.22, 115.13, 111.01, 109.93, 33.50. ESI-MS [m/z]: $[M + H]^+ = 411.0$; $[M - H]^- = 408.2$. Anal. Calcd. for $C_{20}H_{15}N_3O_3S_2$ (%): C, 58.66; H, 3.69; N, 10.26; S, 15.66 Found (%): 58.75 H, 3.62; N, 10.21; S, 15.52.

2-Hydroxy-*N*-{(5*Z*)-5-[(5-methoxy-1*H*-indol-3-yl)methylene]-4-oxo-2-thioxothiazolidin-3-yl}benzamide 5b. Yield 77%; m.p. 239–241 °C (DMFA-CH_3COOH).). IR (cm^{-1}): 3234.47 (OH), 1699.21 (C=O), 1654.84 (C=O), 1585.41 (C=C, C=S). ^{1}H-NMR (300 MHz, DMSO-d_6, ppm) δ 12.15 (s, 1H, NH), 11.45 (s, 1H, NH-CO), 11.35 (s, 1H, OH), 8.18 (s, 1H, CH=), 8.00 (d, *J* = 7.8 Hz, 1H, H_6 benzene), 7.74 (d, *J* = 2.2 Hz, 1H, H_4 indole), 7.50–7.33 (m, 3H, H_3 benzene + H_2 +H_7, indole), 7.04–6.90 (m, 2H, H_4 +H_5, benzene), 6.83 (d, *J* = 8.4 Hz, 1H, H_6 indole), 3.87 (s, 3H, CH_3O ^{13}C NMR (101 MHz, DMSO, ppm) δ 189.72, 165.10, 163.07, 158.10, 155.45, 134.64, 131.12, 131.09, 129.80, 128.22, 127.84, 119.43, 117.25, 115.07, 113.67, 113.38, 111.09, 110.44, 100.58, 55.85. ESI-MS [m/z]: $[M + H]^+ = 426.0$; $[M - H]^- = 424.0$. Anal. Calcd. for $C_{20}H_{15}N_3O_4S_2$ (%): C, 56.46; H, 3.55; N, 9.88; S, 15.07 Found (%):C, 56.57; H, 3.49; N, 9.96; S, 15.15.

2-Hydroxy-*N*-{(5*Z*)-5-[(6-methoxy-1*H*-indol-3-yl)methylene]-4-oxo-2-thioxothiazolidin-3-yl}benzamide 5c. Yield 80%; m.p. 268–270 °C (DMFA-CH_3COOH). IR (cm^{-1}): 3227.72 (OH), 1705.96 (C=O), 1670.27 (C=O), 1591.2 (C=C), 1576.73 (C=S). ^{1}H-NMR (300 MHz, DMSO-d_6, ppm) δ 12.06 (s, 1H, NH), 11.45 (s, 1H, NH-CO),), 11.34 (s, 1H, OH), 8.10 (s, 1H, CH=), 7.99 (d, *J* = 7.9 Hz, 1H, H_6 benzene), 7.75 (d, *J* = 8.7 Hz, 1H, H_4 indole), 7.69 (d, *J* = 1.9 Hz, 1H, H_3 benzene), 7.48-7,42 (m, 1H, H_2 indole), 7.05–6.91 (m, 3H, H_7 indole +H_4 + H_5, benzene), 6. 83 (d, *J* = 8.5 Hz, 1H, H_5 indole), 3.84 (s, 3H, CH_3O^{13}C NMR (101 MHz, DMSO, ppm) δ 189.71, 165.00, 163.10, 158.01, 156.93, 137.32, 134.64, 130.31, 129.87, 127.93, 120.68, 119.49, 119.44, 117.22, 115.11, 111.63, 111.13, 111.06, 95.41, 55.28. ESI-MS [m/z]: $[M + H]^+ = 426.0$; $[M - H]^- = 424.0$.. Anal. Calcd. for $C_{20}H_{15}N_3O_4S_2$ (%): C, 56.46; H, 3.55; N, 9.88; S, 15.07 Found (%):C, 56.39; H, 3.51; N, 9.80; S, 15.01.

4-Hydroxy-*N*-[(5*Z*)-5-(1*H*-indol-3-ylmethylene)-4-oxo-2-thioxothiazolidin-3-yl]benzamide 5d. Yield 90%; m.p. > 275 °C (DMFA-CH_3COOH). IR (cm^{-1}): 3369.48 (OH), 3225.79 (NH), 1694.38 (C=O), 1668.35 (C=O), 1591.2 (C=C), 1573.84 (C=S). ^{1}H-NMR (300 MHz, DMSO-d_6, ppm) δ 12.23 (s, 1H, NH), 11.12 (s, 1H, NH-CO), 9.89 (s, 1H, OH), 8.13 (s, 1H, CH=), 7.92–7.81 (m, 3H, H_2 + H_6, benzene + H_4 indole), 7.79 (d, *J* = 3.0 Hz, 1H, H_7 indole), 7.53–7.45 (m, 1H, H_2 indole), 7.28–7.15 (m, 2H, H_5 + H_6, indole), 6.85 (d, *J* = 8.7 Hz, 2H, H_3 +H_5, benzene). ^{13}C NMR (101 MHz, DMSO, ppm) δ 190.12, 164.06, 163.36, 161.49, 136.48, 131.20, 129.97, 127.80, 126.77, 123.49, 121.69, 121.50, 118.64, 115.30, 112.61, 111.16, 110.97. ESI-MS [m/z]: $[M + H]^+ = 396.0$; $[M - H]^- = 394.0$. Anal. Calcd. for $C_{19}H_{13}N_3O_3S_2$ (%): C, 57.71; H, 3.31; N, 10.63; S, 16.22 Found (%):C, 57.62; H, 3.37; N, 10.55; S, 16.16.

***N*-[(5*Z*)-5-(1*H*-Indol-3-ylmethylene)-4-oxo-2-thioxothiazolidin-3-yl]nicotinamide 5e.** Yield 90%; m.p. > 275 °C (DMFA-CH_3COOH). IR (cm^{-1}): 3485.2 (NH), 1696.31 (C=O), 1674.13 (C=O), 1596.98 (C=C), 1577.7 (C=S). ^{1}H-NMR (500 MHz, DMSO-d_6, ppm) δ 12.31 (s, 1H, NH), 11.75 (s, 1H, NH-CO), 9.15 (d, *J* = 1.7 Hz, 1H, H_4 pyridine), 8.77 (dd, *J* = 4.8, 1.4 Hz, 1H, H_2 pyridine), 8.33 (d, *J* = 8.0 Hz, 1H, H_6 pyridine), 8.17 (s, 1H, CH=), 7.90 (d, *J* = 7.6 Hz, 1H, H_5 pyridine), 7.84 (d, *J* = 3.0 Hz, 1H, H_4 indole), 7.57–7.48 (m, 2H, H_2 +H_7, indole), 7.27–7.18 (m, 2H, H_5 +H_6, indole). ^{13}C NMR (101 MHz, DMSO, ppm) δ 189.54, 163.32, 162.94, 150.78, 137.89, 136.41, 131.49, 128.38, 126.78, 123.56, 121.78, 121.38, 118.68, 112.65, 110.98, 110.72. ESI-MS [m/z]: $[M + H]^+ = 381.0$; Anal. Calcd. for $C_{18}H_{12}N_4O_2S_2$ (%): C, 56.83; H, 3.18; N, 14.73; S, 16.86 Found (%):C, 56.71; H, 3.24; N, 14.80; S, 16.79.

***N*-{(5*Z*)-5-[(1-Methyl-1*H*-indol-3-yl)methylene]-4-oxo-2-thioxothiazolidin-3-yl}nicotinamide 5f.** Yield 94%; m.p. 270–272 °C (DMFA-CH_3COOH). IR (cm^{-1}): 3241.22 (NH), 1706.92 (C=O), 1681.85 (C=O), 1589.27 (C=C), 1572.87 (C=S). ^{1}H-NMR (300 MHz, DMSO-d_6, ppm) δ 11.75 (s, 1H, NH-CO), 9.16 (d, *J* = 1.9 Hz, 1H, H_4 pyridine), 8.77 (dd, *J* = 4.8, 1.3 Hz, 1H, H_2 pyridine), 8.36–8.31 (m, 1H, H_6 pyridine), 8.12 (s, 1H, CH=), 7.96 (s, 1H, H_5 pyridine), 7.91 (d, 1H, *J* = 7.9 Hz, H_4 indole), 7.58–7.47 (m, 2H, H_2 + H_7, indole), 7.32 (t, *J* = 7.5 Hz, 1H, H_6 indole), 7.26 (t, *J* = 7.4 Hz, 1H, H_5 indole), 4.00 (s, 3H, CH_3N). ^{13}C NMR (101 MHz, DMSO, ppm) δ 189.64, 163.34, 163.01, 153.40, 148.60, 136.99, 135.59, 134.68, 127.57, 127.30, 126.77, 123.97, 123.60, 122.10, 118.80, 111.08, 110.55, 109.95, 33.54. ESI-MS [m/z]:

$[M + H]^+$ = 395.0; $[M - H]^-$ = 394.0. Anal. Calcd. for $C_{19}H_{14}N_4O_2S_2$ (%): C, 57.85; H, 3.58; N, 14.20; S, 16.26 Found (%):C, 57.94; H, 3.51; N, 14.15; S, 16.35.

***N*-{(5Z)-5-[(5-Methoxy-1*H*-indol-3-yl)methylene]-4-oxo-2-thioxothiazolidin-3-yl}nicotinamide 5g.** Yield 00%; m.p. 199–201 °C. IR (cm^{-1}): 3254.72 (NH), 1718.49 (C=O), 1681.85 (C=O), 1585.41 (C=C), 1576.73 (C=S). ^{1}H-NMR (300 MHz, DMSO-d_6, ppm) ^{1}H-NMR (300 MHz, DMSO-d_6, ppm) δ 12.17 (s, 1H, NH), 11.72 (s, 1H, NH-CO), 9.17 (s, 1H, H_4 pyridine), 8.77 (d, *J* = 3.0 Hz, 1H, H_2 pyridine), 8.35 (d, *J* = 7.5 Hz, 1H, H_6 pyridine), 8.19 (s, 1H, CH=), 7.74 (s, 1H, H_5 pyridine), 7.59–7.49 (m, 1H, H_2 indole), 7.46–7.30 (m, 2H, H_4 +H_7 indole), 6.83 (d, *J* = 8.5 Hz, 1H, H_6 indole), 3.86 (s, 3H, CH_3O). ^{13}C NMR (101 MHz, DMSO, ppm) δ 189.67, 163.30, 163.01, 155.51, 153.36, 148.61, 135.57, 131.33, 131.15, 128.76, 127.86, 126.82, 123.94, 113.69, 113.41, 111.12, 110.01, 100.64, 55.54. ESI-MS [m/z]: $[M + H]^+$ = 411.0; $[M - H]^-$ = 409.0. Anal. Calcd. for $C_{19}H_{14}N_4O_3S_2$ (%): C, 55.60; H, 3.44; N, 13.65; S, 15.62 Found (%): C, 55.49; H, 3.39; N, 13.58; S, 15.67.

***N*-[(5Z)-5-(1*H*-Indol-3-ylmethylene)-4-oxo-2-thioxothiazolidin-3-yl]isonicotinamide 5h.** Yield 86%; m.p. > 275 °C Yield 86%; m.p. > 275 °C (DMFA-CH_3COOH). IR (cm^{-1}): 3196.86 (NH), 1718.49 (C=O), 1672.2 (C=O), 1594.09 (C=C), 1576.73 (C=S). ^{1}H-NMR (300 MHz, DMSO-d_6, ppm) δ 12.27 (s, 1H, NH), 11.79 (s, 1H, NH-CO), 8.78 (d, *J* = 5.8 Hz, 2H, H_2 +H_6, pyridine), 8.17 (s, 1H, CH=), 7.94–7.86 (m, 3H, H_3 +H_5, pyridine +H_4 indole), 7.82 (s, 1H, H_7 indole), 7.51 (d, *J* = 7.1 Hz, 1H, H_2 indole), 7.30–7.15 (m, 2H, H_5 +H_6, indole). ^{13}C NMR (101 MHz, DMSO, ppm) δ 189.53, 163.32, 162.94, 150.76, 137.94, 136.42, 131.46, 128.34, 126.77, 123.54, 121.76, 121.38, 118.66, 112.65, 110.99, 110.78. ESI-MS [m/z]: $[M + H]^+$ = 381.0; $[M - H]^-$ = 379.0. Anal. Calcd. for $C_{18}H_{12}N_4O_2S_2$ (%): C, 56.83; H, 3.18; N, 14.73; S, 16.86 Found (%):C, 56.89; H, 3.26; N, 14.65; S, 16.88.

***N*-{(5Z)-5-[(1-Methyl-1*H*-indol-3-yl)methylene]-4-oxo-2-thioxothiazolidin-3-yl}isonicotinamide 5i.** Yield 95%; m.p. 269–271 °C (DMFA-CH_3COOH). IR (cm^{-1}): 3217.11 (NH), 1710.78 (C=O), 1674.13 (C=O), 1587.34 (C=C), 1570.95 (C=S).^{1}H-NMR (300 MHz, DMSO-d_6, ppm) δ 11.81 (s, 1H, NH-CO), 8.79 (d, *J* = 5.9 Hz, 2H, H_2 +H_6, pyridine), 8.13 (s, 1H, CH=), 7.99–7.86 (m, 4H, H_3 +H_5, pyridine + H_4 +H_7, indole), 7.51 (d, *J* = 7.9 Hz, 1H, H_2 indole), 7.37–7.20 (m, 2H, H_5 +H_6, indole), 4.00 (s, 3H, CH_3N). ^{13}C NMR (101 MHz, DMSO, ppm) δ 189.49, 163.32, 162.90, 150.78, 137.88, 137.00, 134.73, 127.68, 127.30, 123.61, 122.12, 121.38, 118.81, 111.09, 110.46, 109.95, 33.55. ESI-MS [m/z]: $[M + H]^+$ = 395.0; $[M - H]^-$ = 393.0. Anal. Calcd. for $C_{19}H_{14}N_4O_2S_2$ (%): C, 57.85; H, 3.58; N, 14.20; S, 16.26 Found (%):C, 57.78; H, 3.53; N, 14.28; S, 16.17.

***N*-{(5Z)-5-[(5-Methoxy-1*H*-indol-3-yl)methylene]-4-oxo-2-thioxothiazolidin-3-yl}isonicotinamide 5j.** Yield 89%; m.p. 261–263 °C (CH_3COOH).). IR (cm^{-1}): 3199.75 (NH), 1706.92 (C=O), 1676.06 (C=O), 1588.3 (C=C, C=S). ^{1}H-NMR (300 MHz, DMSO-d_6, ppm) δ 12.20 (s, 1H, NH), 11.83 (s, 1H, NH-CO), 8.78 (d, *J* = 5.4 Hz, 2H, H_2 +H_6, pyridine), 8.20 (s, 1H, CH=), 7.90 (d, *J* = 5.4 Hz, 2H, H_3 +H_5, pyridine), 7.76 (d, *J* = 3.0 Hz, 1H, H_4 indole), 7.44–7.33 (m, 2H, H_2 +H_7, indole), 6.83 (dd, *J* = 8.9, 1.7 Hz, 1H, H_6 indole), 3.86 (s, 3H, CH_3O). ^{13}C NMR (101 MHz, DMSO, ppm) δ 189.50, 163.29, 162.91, 155.49, 150.78, 137.91, 131.39, 131.12, 128.89, 127.88, 121.39, 113.72, 113.42, 111.12, 109.87, 100.58, 55.51. ESI-MS [m/z]: $[M + H]^+$ = 411.0; $[M - H]^-$ = 409.0. Anal. Calcd. for $C_{19}H_{14}N_4O_3S_2$ (%): C, 55.60; H, 3.44; N, 13.65; S, 15.62 Found (%): C, 55.52; H, 3.47; N, 13.73; S, 15.55.

***N*-{(5Z)-5-[(6-Methoxy-1*H*-indol-3-yl)methylene]-4-oxo-2-thioxothiazolidin-3-yl}isonicotinamide 5k.** Yield 89%; m.p. 275–277 °C (CH_3COOH). IR (cm^{-1}): 3550.78 (NH), 3346.34 (NH), 1725.24 (C=O), 1689.56 (C=O), 1596.98 (C=C), 1576.73 (C=S). ^{1}H-NMR (300 MHz, DMSO-d_6, ppm) δ 12.11 (s, 1H, NH), 11.85 (s, 1H, NH-CO), 8.78 (d, *J* = 5.3 Hz, 2H,H_2 +H_6, pyridine), 8.11 (s, 1H, CH=), 7.89 (d, *J* = 5.3 Hz, 2H, H_3 +H_5, pyridine), 7.74 (dd, *J* = 13.7, 5.6 Hz, 2H, H_2 +H_4, indole), 6.96 (s, 1H, H_7 indole), 6.83 (d, *J* = 8.6 Hz, 1H, H_5 indole), 3.84 (s, 3H, CH_3O). ^{13}C NMR (101 MHz, DMSO, ppm) δ 189.49, 163.30, 162.91, 156.97, 150.78, 137.88, 137.36, 130.66, 128.61, 121.38, 120.66, 119.52, 111.71, 111.17, 110.52, 95.46, 55.29. ESI-MS [m/z]: $[M + H]^+$ = 411.0; $[M - H]^-$ = 409.0. Anal. Calcd. for $C_{19}H_{14}N_4O_3S_2$ (%): C, 55.60; H, 3.44; N, 13.65; S, 15.62 Found (%): C, 55.73; H, 3.49; N, 13.57; S, 15.54.

(5Z)-5-(1*H*-Indol-3-ylmethylene)-3-morpholin-4-yl-2-thioxothiazolidin-4-one7a. Yield 86%; m.p. 273–275 °C (DMFA:CH_3COOH). IR (cm^{-1}): 3247.97 (NH), 1690.53 (C=O), 1594.09 (C=C), 1575.77 (C=S).^{1}H-NMR (300 MHz, DMSO-d_6, ppm) δ 12.10 (s, 1H, NH), 7.98 (s, 1H, CH=), 7.85 (d, *J* = 6.8 Hz, 1H, H_4 indole), 7.66 (d, *J* = 2.8 Hz, 1H, H_7 indole), 7.51–7.45 (m, 1H, H_2 indole), 7.27–7.13 (m, 2H, H_5 +H_6, indole), 3.81 (s, 6H, morpholine), 3.06 (s, 2H, morpholine). ^{13}C NMR (101 MHz, DMSO, ppm) δ 189.92, 165.21, 136.32, 130.51, 126.72, 126.03, 123.34, 121.50, 118.48, 112.53, 111.68, 110.95, 66.56, 50.14. ESI-MS [m/z]: $[M + H]^+$ = 346.2; $[M - H]^-$ = 344.2. Anal. Calcd. for $C_{16}H_{15}N_3O_2S_2$ (%): C, 55.63; H, 4.38; N, 12.16; S, 18.56 Found (%):C, 55.74; H, 4.32; N, 12.24; S, 18.49.

(5Z)-5-[(1-Methyl-1*H*-indol-3-yl)methylene]-3-morpholin-4-yl-2-thioxo-thiazolidin-4-one 7b was prepared according to [42].

(5Z)-5-[(5-Methoxy-1*H*-indol-3-yl)methylene]-3-morpholin-4-yl-2-thioxo-thiazolidin-4-one7c. Yield 82%; m.p. 250–252 °C (CH_3COOH). IR (cm^{-1}): 3163.11 (NH), 1690.53 (C=O), 1580.59 (C=C, C=S). ^{1}H-NMR (300 MHz, DMSO-d_6, ppm) δ 12.08 (s, 1H, NH), 7.99 (s, 1H, CH=), 7.61 (s, 1H, H_4 indole), 7.34 (d, *J* = 3.4 Hz, 2H, H_2 +H_7, indole), 6.81 (d, *J* = 8.7 Hz, 1H, H_6 indole), 4.05–3.57 (m, 9H, CH_3O, morpholine), 3.03 (s, 2H, morpholine). ^{13}C NMR (101 MHz, DMSO, ppm) δ 189.87, 165.20, 155.29, 131.04, 130.51, 127.75, 126.61, 113.52, 113.30, 111.04, 110.74, 100.34, 66.56, 55.46, 50.11. ESI-MS [m/z]: $[M + H]^+$ = 376.0; $[M - H]^-$ = 374.0. Anal. Calcd. for $C_{17}H_{17}N_3O_3S_2$ (%): C, 54.38; H, 4.56; N, 11.19; S, 17.08 Found (%):C, 54.31; H, 4.62; N, 11.04; S, 17.15.

3.3. *Antibacterial Activity Evaluation*

Bacterial strains utilized include Gram-negative: *Salmonella typhimurium* (ATCC 13311) *Pseudomonas aeruginosa* (ATCC 27853), *Escherichia coli* (ATCC 35210), *Enterobacter cloacae* (ATCC 35030) and Gram-positive bacteria: *Micrococcus flavus* (ATCC 10240), *Bacillus cereus* (isolated clinically), *Staphylococcus aureus* (ATCC 6538), and *Listeria monocytogenes* (NCTC 7973) bacteria. Pathogens were provided from the Mycological Laboratory, Institute for Biological Research "Siniša Stankovic" National institute of Republic of Serbia Belgrade. Resistant strains used were MRSA, *E. coli*, and *P. aeruginosa* [77,78].

For the determination of minimum inhibitory (MIC) and minimum bactericidal concentrations, the microdilution method, as previously described [77–79]. The minimum inhibitory (MIC) and minimum bactericidal (MBC) concentrations were determined by the modified microdilution method as previously reported [77–79]. Briefly, the fresh overnight culture of bacteria was adjusted to a concentration of 1×10^5 CFU/mL. The tested compounds were dissolved in 5% DMSO and serially diluted in tryptic soy broth (TSB) medium with bacterial inoculum (1.0×10^4 CFU per well). The microplates were incubated for 24 h at 37 °C. The MIC of the samples was detected following the addition of 40 μL of iodonitrotetrazolium chloride (INT) (0.2 mg/mL) and incubation at 37 °C for 30 min. The lowest concentration that produced a significant inhibition of the growth of the bacteria in comparison with the positive control was identified as the MIC. MBC was determined by serial sub-cultivation of 10 μL into microplates containing 100 μL of TSB. The lowest concentration that shows no growth after this sub-culturing was identified as the MBC, indicating 99.5% death of the original inoculum. Streptomycin and ampicillin were used as positive controls.

3.4. *Antifungal Evaluation*

The following fungi were used: Aspergillus niger (ATCC 6275), Aspergillus ochraceus (ATCC 12066), Aspergillus fumigatus (human isolate), Aspergillus versicolor (ATCC 11730), Penicillium funiculosum (ATCC 36839), Penicillium ochrochloron (ATCC 9112), Trichoderma viride (IAM 5061), Penicillium verrucosum var. cyclopium (food isolate). The organisms were obtained from the Mycological Laboratory, Department of Plant Physiology, Institute for Biological Research "Siniša Stankovic", National institute of Republic of Serbia, Belgrade, Serbia. All experiments were performed in duplicate and repeated three times, as previously described [80,81].

The fungal spores were washed from the surface of agar plates with sterile 0.85% saline containing 0.1% Tween 80 (*v/v*). The spore suspension was adjusted with sterile saline to a concentration of approximately 1.0×10^5 in a final volume of 100 μL per well. MIC determinations were performed by a serial dilution technique using 96-well microtiter plates. The examined compounds were serially diluted in broth Malt medium (MA), after which inoculum was added. The microplates were incubated for 72 h at 28 °C. The lowest concentrations without visible growth (at the binocular microscope) were defined as MICs. The fungicidal concentrations (MFCs) were determined by serial subcultivation of 2 μL of tested fractions dissolved in medium and inoculation into microtiter plates containing 100 μL of broth per well and further incubation 72 h at 28 °C. The lowest concentration with no visible growth was defined as MFC, indicating 99.5% killing of the original inoculum. The fungicides bifonazole and ketoconazole were used as positive controls.

3.5. Docking Studies

The program AutoDock 4.2® software was used for the docking simulation. The free energy of binding (ΔG) of *E. coli* DNA GyrB, Thymidylate kinase, *E. coli* MurA, *E. coli* primase, *E. coli* MurB, DNA topo IV, and CYP51 of *C. albicans*, in complex with the inhibitors were generated using this molecular docking program. The X-ray crystal structures data of all the enzymes used were obtained from the Protein Data Bank (PDB ID: 1KZN, AQGG, 1DDE, JV4T, 2Q85, 1S16, and 5V5Z respectively). All procedures were performed according to our previous paper [78].

3.6. Cytotoxicity

HEK 293 cells were cultured in DMEM medium, supplemented with 10% fetal calf serum (Sigma Chemical Co., St. Louis, MO, USA), 50 μg/mL streptomycin (Sigma Chemical Co.), and 50 units/mL penicillin (Sigma Chemical Co.) in 5% CO_2-containing humidified atmosphere at 37°C. Since compound solutions contained DMSO, control cultures containing only DMSO at the final concentration obtained when the appropriate volume of compound solution was added were performed.

MTT Assay for Determination of Cell Viability

MTT assay based on the colorimetric measurement of formazan formed after reducing MTT by cellular NAD(P)H-dependent oxidoreductases was used to examine the cytotoxic activity of the compounds. Briefly, the cells were seeded into 96-well plates in 100 μL of complete culture medium at a concentration of 5,000 substrate-dependent cells per well and left incubated overnight as described above. The formulations to be tested (100 μL aliquots) were added to the culture medium at different concentrations and left incubated for 72 h. The MTT assay was performed following the manufacturer's recommendations and assessed using an EL ×800 absorbance reader (BioTek Instruments; Winooski, VT, USA).

4. Conclusions

Eleven 5-[(R-1*H*-indol-3-yl)methylene]-4-oxo-2-thioxo-thiazolidin-3-ylcarbamides **5a-k** and three 5-[(R-1*H*-indol-3-yl) methylene] -3-morpholin-4-yl-2-thioxothiazolidin-4-ones **7a-c** were designed, synthesized and evaluated in silico and experimentally for their antimicrobial action against panel of Gram positive, Gram negative bacteria and fungi.

It should be mentioned that all compounds appeared to be more potent than ampicillin against all bacteria tested and then streptomycin against all bacteria except *B. cereus (isolated clinically M. flavus* (ATCC 10240), and *En. cloacae* (ATCC 35030). The most sensitive bacteria was found to be *S. aureus* (ATCC 6538), while *L. monocytogenes* (NCTC 7973) was the most resistant one. Compounds also appeared to be active against three resistant strains MRSA, *E. coli*, and *P. aeruginosa* showing better activity against MRSA than both reference drugs while against the other two resistant strains better than ampicillin.

Concerning antifungal action, the tested compounds exhibited very good activity against all the fungal species tested, being more active than ketoconazole and bifonazole. The most sensitive fungal strain appeared to be *T. viride* (IAM 5061), while the most resistant filamentous *A. fumigatus* (human isolate).

It can be observed that the growth of both Gram-negative and Gram-positive bacteria and fungi responded differently to the tested compounds, which indicates that different substituents may lead to different modes of action or that the metabolism of some bacteria/fungi was better able to overcome the effect of the compounds or adapt to it.

Docking analysis to DNA Gyrase, Thymidylate kinase and *E.coli* MurB indicated a probable involvement of MurB inhibition in the antibacterial mechanism of compounds tested while docking analysis to 14α-lanosterol demethylase (CYP51) and tetrahydrofolate reductase of *Candida albicans* indicated a likely implication of CYP51 reductase at the antifungal activity of the compounds and secondary involvement of dihydrofolate reductase inhibition at the mechanism of action of the most active compounds.

Since the most active compounds **5d**, **5g**, **5k**, **7c** demonstrated the low cytotoxicity against HEK-293 human embryonic kidney cell line and reasonable selectivity index, this chemical series looks promising for investigations as the antimicrobial agents.

Finally, compounds **5d** (Z)-N-(5-((1H-indol-3-yl)methylene)-4-oxo-2-thioxothiazolidin-3 -yl)-4-hydroxybenzamide and **5g** (Z)-N-(5-((5-methoxy-1H-indol-3-yl)methylene)-4-oxo-2-thioxothiazolidin-3-yl)nicotinamide as well as **7c** (Z)-5-((5-methoxy-1H-indol-3-yl)methylene)-3-morpholino-2-thioxothiazolidin-4-one can be considered as lead compounds for further development of more potent and safe antibacterial and antifungal agents.

Supplementary Materials: The following are available online at http://www.mdpi.com/1424-8247/13/9/229/s1, Supplementary file PASSweb_results_13mols.xlsx: Predictions of antimicrobial activity and acute rat toxicity.

Author Contributions: Conceptualization, A.G. and V.K.; methodology, V.H.; software, P.A. and P.P.; formal analysis, V.M.; investigation, M.I., M.K. and M.D.S; data curation, A.G., V.P., M.D.S. and P.E., original draft preparation, A.G. and P.P.; review & editing, A.G. and V.P.; supervision, A.G. and V.P. All authors have read and agreed to the published version of the manuscript.

Funding: This research was supported by the Serbian Ministry of Education, Science and Technological Development for financial support (project No. 451-03-68/2020-14/200007).

Acknowledgments: Computational predictions of biological activity by AntiBac-Pred, AntiFun-Pred and AcuTox web-services (P.P. and V.P.) were performed in the framework of the Russian State Academies of Sciences Fundamental Research Program for 2013–2020.

Conflicts of Interest: The authors declare no conflict of interest.

References

1. Michaud, C.M. Global Burden of Infectious Diseases. *Encycl. Microbiol.* **2009**, 444–454.
2. Nii-Trebi, N.I. Emerging and neglected infectious diseases: Insights, advances, and challenges. *Biomed. Res. Int.* **2017**, *2017*, 5245021. [CrossRef] [PubMed]
3. Mukherjee, S. Emerging Infectious Diseases: Epidemiological Perspective. *Indian J. Dermatol.* **2017**, *62*, 459–467.
4. Ventola, C.L. The antibiotic resistance crisis: Part 1: Causes and threats. *Pharm. Ther.* **2015**, *40*, 277–283.
5. Michael, C.A.; Dominey-Howes, D.; Labbate, M. The antimicrobial resistance crisis: Causes, consequences, and management. *Front. Public Health* **2014**, *16*, 145. [CrossRef]
6. Rice, L.B. Federal funding for the study of antimicrobial resistance in nosocomial pathogens: No ESKAPE. *J. Infect. Dis.* **2008**, *197*, 1079–1081. [CrossRef]
7. Holmes, A.H.; Moore, L.S.; Sundsfjord, A.; Steinbakk, M.; Regmi, S.; Karkey, A.; Guerin, P.J.; Piddock, L.J. Understanding the mechanisms and drivers of antimicrobial resistance. *Lancet* **2016**, *387*, 176–187. [CrossRef]
8. Tripathi, A.C.; Gupta, S.J.; Fatima, G.N.; Sonar, P.K.; Verma, A.; Saraf, S.K. 4- Thiazolidinones: The Advances Continue . . . . *Eur. J. Med. Chem.* **2014**, *7*, 52–57. [CrossRef]

9. Kaminskyy, D.; Kryshchyshyn, A.; Lesyk, R. 5-Ene-4-thiazolidinones–An efficient tool in medicinal chemistry. *Eur. J. Med. Chem.* **2017**, *140*, 542–594. [CrossRef]
10. Kaminskyy, D.; Kryshchyshyn, A.; Lesyk, R. Recent developments with rhodanine as a scaffold for drug discovery. *Expert Opin. Drug Discov.* **2017**, *12*, 1233–1252.
11. Baell, B. Observations on screening-based research and some concerning trends in the literature. *Future Med. Chem.* **2010**, *2*, 1529–1546.
12. Mendgen, T.; Steuer, C.; Klein, C.D. Privileged scaffolds or promiscuous binders: A comparative study on rhodanines and related heterocycles in medicinal chemistry. *J. Med. Chem.* **2012**, *55*, 743–753. [CrossRef] [PubMed]
13. Morphy, R.; Rankovic, Z. Designed multiple ligands. An emerging drug discovery paradigm. *J. Med. Chem.* **2005**, *48*, 6523–6543. [CrossRef] [PubMed]
14. Fortin, S.; Bérubé, G. Advances in the development of hybrid anticancer drugs. *Expert Opin. Drug Discov.* **2013**, *8*, 1029–1047.
15. Kryshchyshyn, A.; Roman, O.; Lozynskyi, A.; Lesyk, R. Thiopyrano[2,3-d]thiazoles as new efficient scaffolds in medicinal chemistry. *Sci. Pharm.* **2018**, *86*, 26. [CrossRef] [PubMed]
16. Cong, N.T.; Nhan, H.T.; Van Hung, L.; Thang, T.D.; Kuo, P.C. Synthesis and antibacterial activity of analogs of 5-arylidene-3-(4-methylcoumarin-7-yloxyacetylamino)-2-thioxo-1,3-thiazoli-din-4-one. *Molecules* **2014**, *19*, 13577–13586. [CrossRef]
17. Song, M.-X.; Deng, X.-Q.; Li, Y.-R.; Zheng, C.-J.; Hong, L.; Piao, H.-R. Synthesis and biological evaluation of (E)-1-(substituted)-3-phenylprop-2-en-1-ones bearing rhodanines as potent anti-microbial agents. *J. Enzyme Inhib. Med. Chem.* **2014**, *29*, 647–653. [CrossRef]
18. Krátký, M.; Vinšová, J.; Stolaříková, J. Antimicrobial activity of rhodanine-3-acetic acid derivatives. *Bioorg. Med. Chem.* **2017**, *25*, 1839–1845. [CrossRef]
19. Horishny, V.; Kartsev, V.; Geronikaki, A.; Matiychuk, V.; Petrou, A.; Glamoclija, J.; Ciric, A.; Sokovic, M. 5-(1H-Indol-3-ylmethylene)-4-oxo-2-thioxothiazolidin-3-yl)alkancarboxylic Acids as Antimicrobial Agents: Synthesis, Biological Evaluation, and Molecular Docking Studies. *Molecules* **2020**, *25*, 1964. [CrossRef]
20. Incerti, M.; Vicini, P.; Geronikaki, A.; Eleftheriou, P.; Tsagkadouras, A.; Zoumpoulakis, P.; Fotakis, C.; Ćirić, A.; Glamočlija, J.; Soković, M. New N-(2-phenyl-4-oxo-1,3-thiazolidin-3-yl)-1,2-benzothiazole -3-carboxamides and Acetamides as Antimicrobial Agents. *Med. Chem. Commun.* **2017**, *8*, 2142–2154. [CrossRef]
21. Ozen, C.; Ceylan-Unlusoy, M.; Aliary, N.; Ozturk, M.; Bozdag-Dundar, O. Thiazolidinedione or Rhodanine: A Study on Synthesis and Anticancer Activity Comparison of Novel Thiazole Derivatives. *Pharm. Pharm. Sci.* **2017**, *20*, 415–427. [CrossRef]
22. Fu, H.; Hou, X.; Wang, L.; Dun, Y.; Yang, X.; Fang, H. Design, synthesis and biological evaluation of 3-aryl-rhodanine benzoic acids as anti-apoptotic protein Bcl-2 inhibitors. *Bioorg. Med. Chem. Lett.* **2015**, *25*, 5265–5269. [CrossRef] [PubMed]
23. Havrylyuk, D.; Zimenkovsky, B.; Lesyk, R. Synthesis and Anticancer Activity of Novel Nonfused Bicyclic Thiazolidinone Derivatives. *Phosphorus Sulfur* **2009**, *184*, 638–650. [CrossRef]
24. El-Miligy, M.; Hazzaa, A.; El-Messmary, H.; Nassra, R.A. El-Hawash, Soad, New hybrid molecules combining benzothiophene or benzofuran with rhodanine as dual COX-1/2 and 5-LOX inhibitors: Synthesis, biological evaluation and docking study. *Bioorg. Chem.* **2017**, *72*, 102–115. [CrossRef] [PubMed]
25. R. Atta-Allah, S.; Nassar, I.F.; El-Sayed, W.A. Design, synthesis and anti-inflammatoryvel 5-(Indol-3-yl)-thiazolidinone derivatives as COX-2 inhibitors. *J. Pharmacol. Ther. Res.* **2020**, *5*, 1–16.
26. Powers, J.P.; Piper, D.E.; Li, Y.; Mayorga, V.; Anzola, J.; Chen, J.M.; Jaen, J.C.; Lee, G.; Liu, J.; Peterson, M.G.; et al. SAR and mode of action of novel non-nucleoside inhibitors of hepatitis C NS5b RNA polymerase. *J. Med. Chem.* **2006**, *49*, 1034–1046. [CrossRef] [PubMed]
27. Ramkumar, K.; Yarovenko, V.N.; Nikitina, A.S.; Zavarzin, I.V.; Krayushkin, M.M.; Kovalenko, L.V.; Esqueda, A.; Odde, S.; Neamati, N. Design, synthesis and structure-activity studies of rhodanine derivatives as HIV-1 integrase inhibitors. *Molecules* **2010**, *15*, 3958–3992. [CrossRef]
28. Petrou, A.; Eleftheriou, P.; Geronikaki, A.; Akrivou, M.G.; Vizirianakis, I. Novel thiazolidin-4-ones as potential non-nucleoside inhibitors of HIV-1 reverse transcriptase. *Molecules* **2019**, *24*, 3821. [CrossRef]
29. Kryshchyshyn, A.; Kaminskyy, D.; Roman, O.; Kralovics, R.; Karpenko, O.; Lesyk, R. Synthesis and anti-leukemic activity of pyrrolidinedione-thiazolidinone hybrids. *Ukr Biochem. J.* **2020**, *92*, 108–119.

30. Havrylyuk, D.; Zimenkovsky, B.; Vasylenko, O.; Gzella, A.; Lesyk, R. Synthesis of new 4- thiazolidinone-, pyrazoline-, and isatin-based conjugates with promising antitumor activity. *J. Med. Chem.* **2012**, *55*, 8630–8641. [CrossRef]
31. Havrylyuk, D.; Roman, O.; Lesyk, R. Synthetic approaches, structure activity relationship and biological applications for pharmacologically attractive pyrazole/pyrazoline–thiazolidine– based hybrids. *Eur. J. Med. Chem.* **2016**, *113*, 145–166. [CrossRef]
32. Saini, T.; Kumar, S.; Narasimhan, B. Central nervous system activities of indole derivatives: An overview. *Cent. Nerv. Syst. Agents Med. Chem.* **2016**, *16*, 19–28. [CrossRef] [PubMed]
33. Singh, P.; Singh, O.M. Recent progress in biological activities of indole and indole alkaloids. *Mini Rev. Med. Chem.* **2018**, *18*, 9–25.
34. Kaur, J.; Utreja, D.; Jain, N.; Sharma, S. Recent Developments in the Synthesis and Antimicrobial Activity of Indole and Its Derivatives. *Curr. Org. Synth.* **2019**, *16*, 17–37. [CrossRef]
35. Bathula, C.; Tripathi, S.; Srinivasan, R.; Jha, K.K.; Ganguli, A.; Chakrabarti, G.; Singh, S.; Munshi, P.; Sen, S. Synthesis of novel 5-arylidenethiazolidinones with apoptotic properties via a three component reaction using piperidine as a bifunctional reagent. *Org. Biomol. Chem.* **2016**, *14*, 8053–8063. [CrossRef]
36. Sayed, M.; El-Dean, A.; Ahmed, M.; Hassanien, R. Synthesis of some heterocyclic compounds derived from indole as antimicrobial agents. *Synth. Commun.* **2018**, *48*, 413–421. [CrossRef]
37. Jain, P.; Utreja, D.; Sharma, P. An efficacious synthesis of N-1–, C-3–substituted indole derivatives and their antimicrobial studies. *J. Hetrocyclic Chem.* **2020**, *57*, 428–435. [CrossRef]
38. Shaikh, T.M.A.; Debebe, H. Synthesis and Evaluation of Antimicrobial Activities of Novel N-Substituted Indole Derivatives. *J. Chem.* **2020**, 1–9, Article ID 4358453, 9 pages.
39. Kumar, P.; Singh, S.; Rizki, M.; Pratama, F. Synthesis of some novel 1H-indole derivatives with antibacterial activity and antifungal activity. *Lett. Appl. NanoBioScience* **2020**, *9*, 961–967.
40. Liu, Z.; Tang, L.; Zhu, H.; Xu, T.; Qiu, C.; Zheng, S.; Gu, Y.; Feng, J.; Zhang, Y.; Liang, G. Design, Synthesis, and Structure–Activity Relationship Study of Novel Indole-2-carboxamide Derivatives as Anti-inflammatoryAgents for the Treatment of Sepsis. *J. Med. Chem.* **2016**, *59*, 4637–4650. [CrossRef]
41. Li, S.; Wang, Z.; Xiao, H.; Bian, Z.; Wang, J. Enantioselective synthesis of indole derivatives byRh/Pd relay catalysis and their anti-inflammatory evaluation. *Chem. Commun.* **2020**, *56*, 7573–7576. [CrossRef]
42. Abdellatif, K.R.A.; Elsaady, M.Y.; Amin, N.H.; Hefny, A.A. Design, Synthesis and biological evaluation of some novel indole derivatives as selective COX-2 inhibitors. *J. Appl. Pharm. Sci.* **2017**, *7*, 69–77.
43. Bhat, M.A.; Al-Omar, M.A.; Raish, M. Indole Derivatives as Cyclooxygenase Inhibitors: Synthesis, Biological Evaluation and Docking Studies. *Molecules* **2018**, *23*, 1250. [CrossRef] [PubMed]
44. Sidhu, J.S.; Singla, R.; Jaitak, V. Indole Derivatives as Anticancer Agents for Breast Cancer Therapy: A Review. *Anticancer Agents Med. Chem.* **2015**, *16*, 160–173. [CrossRef] [PubMed]
45. El-Sharief, A.M.S.; Ammar, Y.A.; Belal, A.; El-Sharief, M.A.S.; Mohamed, Y.A.; Ahmed, B.M.; Mehany, A.B.M.; Elhag Ali, G.A.M.; Ragab, A. Design, synthesis, molecular docking and biological activity evaluation of some novel indole derivatives as potent anticancer active agents and apoptosis inducers. *Bioorg. Chem.* **2019**, *85*, 399–412. [CrossRef]
46. Cascioferro, S.; Li Petri, G.; Parrino, B.; El Hassouni, B.; Carbone, D.; Arizza, V.; Perricone, U.; Padova, A.; Funel, N.; Peters, G.J.; et al. 3-(6-Phenylimidazo [2,1-b][1,3,4]thiadiazol-2-yl)-1H-Indole Derivatives as New Anticancer Agents in the Treatment of Pancreatic Ductal Adenocarcinoma. *Molecules* **2020**, *25*, 329. [CrossRef] [PubMed]
47. Zhang, M.-Z.; Chen, Q.; Yang, G.F. A review on recent developments of indole-containing antiviral agents. *J. Med. Chem.* **2015**, *89*, 421–441. [CrossRef] [PubMed]
48. Bardiot, D. Discovery of Indole Derivatives as Novel and Potent Dengue Virus Inhibitors. *J. Med. Chem.* **2018**, *61*, 8390–8401. [CrossRef]
49. Che, Z.; Tian, Y.; Liu, S.; Hu, M.; Che, G. Discovery of N-arylsulfonyl-3-acylindole benzoyl hydrazone derivatives as anti-HIV-1 agents. *Braz. J. Pharm. Sci* **2019**, *54*, e17543. [CrossRef]
50. Sanna, G.; Madeddu, S.; Giliberti, G.; Piras, S.; Struga, M.; Wrzosek, M.; Kubiak-Tomaszewska, G.; Koziol, A.E.; Savchenko, O.; Lis, T.; et al. Synthesis and Biological Evaluation of Novel Indole-Derived Thioureas. *Molecules* **2018**, *23*, 2554. [CrossRef]

51. Ramya, V.; Vembu, S.; Ariharasivakumar, G.; Gopalakrishnan, M. Synthesis, Characterisation, Molecular Docking, Anti-microbial and Anti-diabetic Screening of Substituted 4-indolylphenyl-6-arylpyrimidine-2-imine Derivatives. *Drug Res. (Stuttg)* **2017**, *67*, 515–526. [CrossRef]
52. Tymiak, A.A.; Rinehart, K.L.; Bakus, G.J. Constituents of morphologically similar sponges: Aplysina and Smenospongia species. *Tetrahedron* **1985**, *41*, 1039–1047. [CrossRef]
53. Djura, P.; Stierle, D.B.; Sullivan, B. Some metabolites of the marine sponges Smenospongia aurea and Smenospongia (ident.Polyfibrospongia) echina. *J. Org.Chem.* **1980**, *45*, 1435–1441. [CrossRef]
54. Fattorusso, E.; Lanzotti, V.; Magno, S.; Novellino, E. Tryptophan derivatives from a Mediterranean anthozoan, Astroides calycularis. *J. Nat. Prod.* **1985**, *48*, 924–927. [CrossRef]
55. Buyukbingol, E.; Suzen, S.; Klopman, G. Studies on the synthesis and structure–activity relationships of 5-(3′-indolyl)-2-thiohydantoin derivatives as aldose reductase enzyme inhibitors. *Farmaco* **1994**, *49*, 443–447. [PubMed]
56. Pogodin, P.V.; Lagunin, A.A.; Rudik, A.V.; Druzhilovskiy, D.S.; Filimonov, D.A.; Poroikov, V.V. AntiBac-Pred: A Web Application for Predicting Antibacterial Activity of Chemical Compounds. *J. Chem. Inf. Model.* **2019**, *59*, 4513–4518. [CrossRef] [PubMed]
57. Poroikov, V.; Filimonov, D.; Gloriozova, T.; Lagunin, A.; Druzhilovskiy, D.; Rudik, A.; Stolbov, L.; Dmitriev, A.; Tarasova, O.; Ivanov, S.; et al. Computer-aided prediction of biological activity spectra for organic compounds: The possibilities and limitations. *Russ. Chem. Bull.* **2019**, *68*, 2143–2154. [CrossRef]
58. Clarivate Analytics Integrity. Available online: https://integrity.clarivate.com/integrity/ (accessed on 21 July 2020).
59. Antifungal Activity Predictor. Available online: http://www.way2drug.com/micf (accessed on 21 July 2020).
60. Filimonov, D.A.; Zakharov, A.V.; Lagunin, A.A.; Poroikov, V.V. QNA-based 'Star Track' QSAR approach. *SAR QSAR Env. Res.* **2009**, *20*, 679–709. [CrossRef]
61. Lagunin, A.; Zakharov, A.; Filimonov, D.; Poroikov, V. QSAR Modelling of Rat Acute Toxicity on the Basis of PASS Prediction. *Mol. Inform.* **2011**, *30*, 241–250. [CrossRef]
62. Bruno, G.; Costantino, L.; Curinga, C.; Maccari, R.; Monforte, F.; Nicolo, F.; Ottana, R.; Vigorita, M.G. Synthesis and Aldose Reductase Inhibitory Activity of5-Arylidene-2,4-thiazolidinediones. *Bioorg. Med. Chem.* **2002**, *10*, 1077–1084. [CrossRef]
63. Abdeen, S.; Kunkle, T.; Salim, N.; Ray, A.-M.; Mammadova, N.; Summers, C.; Stevens, M.; Ambrose, A.J.; Park, Y.; Schultz, P.G.; et al. Sulfonamido-2-arylbenzoxazole GroEL/ES Inhibitors as Potent Antibacterials against Methicillin-Resistant Staphylococcus aureus (MRSA). *J. Med. Chem.* **2018**, *61*, 7345–7357. [CrossRef]
64. Menozzi, G.; Merello, L.; Fossa, P.; Ranise, A.; Mosti, L.; Bondavalli, F.; Loddo, R.; Murgioni, C.; Mascia, V.; La Colla, P.; et al. Synthesis, antimicrobial activity and molecular modeling studies of halogenated 4-[1H-imidazol-1-yl(phenyl)methyl]-1,5-diphenyl-1H-pyrazoles. *Bioorg. Med. Chem.* **2004**, *12*, 5465–5483. [CrossRef] [PubMed]
65. Vieira, F.T.; de Lima, G.M.; Maia, J.R.; Speziali, N.L.; Ardisson, J.D.; Rodrigues, L.; Correa, A., Jr.; Romero, O.B. Synthesis, characterization and biocidal activity of new organotin complexes of 2-(3-oxocyclohex-1-enyl)benzoic acid. *Eur. J. Med. Chem.* **2010**, *45*, 883–889. [CrossRef] [PubMed]
66. Gjorgjieva, M.; Tomasicč, T.; Barancokova, M.; Katsamakas, S.; Ilas, J.; Tammela, P.; Masicč, L.P.; Kikelj, D. Discovery of Benzothiazole Scaffold-Based DNA Gyrase B Inhibitors. *J. Med. Chem.* **2016**, *59*, 8941–8954. [CrossRef] [PubMed]
67. Parvathy, N.G.; Manju, P.; Mukesh, M.; Thomas, L. Design, synthesis and molecular docking studies of benzothiazole derivatives as anti microbial agents. *Int. J. Pharm. Pharm. Sci.* **2013**, *5*, 101–106.
68. Ren, Y.; Zhang, L.; Zhou, C.H.; Geng, R.X. Recent Development of Benzotriazole-based Medicinal Drugs. *Med. Chem.* **2014**, *4*, 640–662. [CrossRef]
69. Andres, C.J.; Bronson, J.J.; D'Andrea, S.V.; Walsh, A.W. 4-thiazolidinones: Novel inhibitors of the bacterial enzyme MurB. *Bioorg. Med. Chem. Lett.* **2000**, *10*, 715–717. [CrossRef]
70. Ahmed, S.; Zayed, M.F.; El-Messery, S.M.; Al-Agamy, M.H.; Abdel-Rahman, H.M. Design, Synthesis, Antimicrobial Evaluation and Molecular Modeling Study of 1,2,4-Triazole-Based 4-Thiazolidinones. *Molecules* **2016**, *21*, 568. [CrossRef]
71. Pitta, E.; Tsolaki, E.; Geronikaki, A.; Petrovic, J.; Glamočlija, J.; Sokovic, M.; Crespan, E.; Maga, G.; Bhunia, S.S.; Saxena, A.K. 4-Thiazolidinone derivatives as potent antimicrobial agents: Microwave-assisted synthesis, biological evaluation and docking studies. *MedChemComm* **2015**, *6*, 319–326. [CrossRef]

72. Karanth, S.; Narayana, B.; Kodandoor, S.C.; Sarojini, B.K. 2-{[(4-Hydroxy-3,5-dimethoxyphenyl)methylidene]hydrazinylidene}-4-oxo-1,3-thiazolidin-5-yl Acetic Acid. *Molbank* **2018**, *2018*, 2-9 M1009. [CrossRef]
73. Stana, A.; Vodnar, D.C.; Tamaian, R.; Pîrnău, A.; Vlase, L.; Ionuț, I.; Oniga, O.; Tiperciuc, B. Design, Synthesis and Antifungal Activity Evaluation of New Thiazolin-4-ones as Potential Lanosterol 14α-Demethylase Inhibitors. *Int. J. Mol. Sci.* **2017**, *18*, 177. [CrossRef]
74. Incerti, M.; Vicini, P.; Geronikaki, A.; Eleftheriou, P.; Tsagkadouras, A.; Zoumpoulakis, P.; Fotakis, C.; Ćirić, A.; Glamočlija, J.; Soković, M. New N-(2-phenyl-4-oxo-1,3-thiazolidin-3-yl)-1,2-benzothiazole-3-carboxamides and acetamides asantimicrobial agents. *Med. Chem. Commun.* **2017**, *8*, 2142. [CrossRef] [PubMed]
75. Can, N.O.; Çevik, U.A.; Sağlık, B.N.; Levent, S.; Korkut, B.; Özkay, Y.; Kaplancıklı, Z.A.; Koparal, A.S. Synthesis, Molecular Docking Studies, and Antifungal Activity Evaluation of New Benzimidazole-Triazoles as Potential Lanosterol 14α-Demethylase Inhibitors. *J. Chem.* **2017**, 1–15, Article ID 9387102, 15 pages. [CrossRef]
76. Benson, T.E.; Walsh, C.T.; Massey, V. Kinetic characterization of wild-type and S229A mutant MurB: Evidence for the role of Ser 229 as a general acid. *Biochemistry* **1997**, *36*, 796–805. [CrossRef] [PubMed]
77. Kartsev, V.; Lichitsky, B.; Geronikaki, A.; Petrou, A.; Smiljkovic, M.; Kostic, M.; Radanovic, O.; Soković, M. Design, synthesis and antimicrobial activity of usnic acid derivatives. *MedChemComm* **2018**, *9*, 870–882. [CrossRef]
78. Fesatidou, M.; Zagaliotis, P.; Camoutsis, C.; Petrou, A.; Eleftheriou, P.; Tratrtat, C.; Haroun, M.; Geronikaki, A.; Ciric, A.; Sokovic, M. 5-Adamantan thiadiazole-based thiazolidinones as antimicrobial agents. Design, synthesis, molecular docking and evaluation. *Bioorg. Med. Chem.* **2018**, *26*, 4664–4676. [CrossRef]
79. Kostić, M.; Smiljković, M.; Petrović, J.; Glamočilija, J.; Barros, L.; Ferreira, I.C.F.R.; Ćirić, A.; Soković, M. Chemical, nutritive composition and a wide range of bioactive properties of honey mushroom Armillaria mellea (Vahl: Fr.) Kummer. *Food Function* **2017**, *8*, 3239–3249. [CrossRef]
80. Kritsi, E.; Matsoukas, M.T.; Potamitis, C.; Detsi, A.; Ivanov, M.; Sokovic, M.; Zoumpoulakis, P. Novel Hit Compounds as Putative Antifungals: The Case of Aspergillus fumigatus. *Molecules* **2019**, *24*, 3853. [CrossRef]
81. Aleksić, M.; Stanisavljević, D.; Smiljković, M.; Vasiljević, P.; Stevanović, M.; Soković, M.; Stojković, D. Pyrimethanil: Between efficient fungicide against Aspergillus rot on cherry tomato and cytotoxic agent on human cell lines. *Ann. App. Bio.* **2019**, *175*, 228–235. [CrossRef]

Communication

A Fluorinated Analogue of Marine Bisindole Alkaloid 2,2-Bis(6-bromo-1*H*-indol-3-yl)ethanamine as Potential Anti-Biofilm Agent and Antibiotic Adjuvant Against *Staphylococcus aureus*

Raffaella Campana [1], Gianmarco Mangiaterra [2], Mattia Tiboni [1], Emanuela Frangipani [1], Francesca Biavasco [2], Simone Lucarini [1,*] and Barbara Citterio [1,*]

[1] Department of Biomolecular Sciences, University of Urbino Carlo Bo, 61029 Urbino, Italy; raffaella.campana@uniurb.it (R.C.); mattia.tiboni@uniurb.it (M.T.); emanuela.frangipani@uniurb.it (E.F.)

[2] Department of Life and Environmental Sciences, Polytechnic University of Marche, 60131 Ancona, Italy; g.mangiaterra@pm.univpm.it (G.M.); f.biavasco@staff.univpm.it (F.B.)

* Correspondence: simone.lucarini@uniurb.it (S.L.); barbara.citterio@uniurb.it (B.C.); Tel.: +39-0722-303-333 (S.L.); +39-0722-304-962 (B.C.)

Received: 12 August 2020; Accepted: 25 August 2020; Published: 26 August 2020

Abstract: Methicillin resistant *Staphylococcus aureus* (MRSA) infections represent a major global healthcare problem. Therapeutic options are often limited by the ability of MRSA strains to grow as biofilms on medical devices, where antibiotic persistence and resistance is positively selected, leading to recurrent and chronic implant-associated infections. One strategy to circumvent these problems is the co-administration of adjuvants, which may prolong the efficacy of antibiotic treatments, by broadening their spectrum and lowering the required dosage. The marine bisindole alkaloid 2,2-bis(6-bromo-1*H*-indol-3-yl)ethanamine (**1**) and its fluorinated analogue (**2**) were tested for their potential use as antibiotic adjuvants and antibiofilm agents against *S. aureus* CH 10850 (MRSA) and *S. aureus* ATCC 29213 (MSSA). Both compounds showed antimicrobial activity and bisindole **2** enabled 256-fold reduction (ΣFICs = 0.5) in the minimum inhibitory concentration (MIC) of oxacillin for the clinical MRSA strain. In addition, these molecules inhibited biofilm formation of *S. aureus* strains, and compound **2** showed greater eradicating activity on preformed biofilm compared to **1**. None of the tested molecules exerted a viable but non-culturable cells (VBNC) inducing effect at their MIC values. Moreover, both compounds exhibited no hemolytic activity and a good stability in plasma, indicating a non-toxic profile, hence, in particular compound **2**, a potential for in vivo applications to restore antibiotic treatment against MRSA infections.

Keywords: MRSA; marine bisindole alkaloids; antibiofilm activity; adjuvants agents; VBNC cells

1. Introduction

The diffusion of methicillin resistant *Staphylococcus aureus* (MRSA) strains is considered among the most common causes of health care-associated infections (HAIs) in hospitalized people all around the world [1]. Treatment of MRSA infections is limited, mainly because these microorganisms can develop resistance to multiple antibiotics [2,3]. Moreover, MRSA infections can be worsened by the ability of MRSA strains to grow as matrix-enclosed communities called biofilms, which promote adhesion and favor long-term survival on both biotic and abiotic surfaces [4,5].

Another strategy that MRSA have developed to resist antibiotic treatment is the ability to enter a state of dormancy, either as viable but non-culturable cells (VBNC) or as antibiotic persisters (APs) [6,7]. The complex architecture of biofilms, presenting different stressful microenvironments, may lead to a spatial-physiologic heterogeneity of the embedded bacterial population [8,9] and, in this peculiar survival structure, both VBNC cells and AP can be present. Considering that VBNC cells have been shown to tolerate a wide variety of stressors, including starvation, growth inhibiting temperatures, suboptimal salinity, suboptimal pH and antibiotics [10], their role in biofilm biology is of great concern in view of developing new therapeutic approaches. One of these includes the use of antibiotic adjuvants (i.e., molecules that have the potential to improve the effectiveness of an antibiotic against which bacteria have developed resistance), by reducing the bacterial resistance to it, hence prolonging the lifespan of life-saving drugs [11].

In this context, marine sessile organisms can represent a suitable source of bioactive products that are normally used to contrast bacteria diffused in the surrounding water [12]. These molecules have shown activity and selectivity against a wide spectrum of pharmacological targets, and their structures are often used as leads in drug discovery and development [13]. Among the various structural classes, bisindole alkaloids have attracted the attention of many researchers for their biological activities [14–16], especially their antimicrobial and antibiofilm activities [17–19]. Most of them present bromine and/or chlorine substitutions but there are no known examples of natural marine alkaloids containing fluorine. One of the possible reasons of the absence of fluorometabolites could be due to the very low abundance of fluoride ion (1.3 ppm) in the oceans with respect to chloride (Cl^- = 20,000 ppm) and bromide (Br^- = 70 ppm). On the other hand, about 20–25% of drugs on the market contain at least one fluorine atom. The very special effects of fluorine are very difficult to fully rationalize and its high presence in anthropogenic bioactive molecules has generally arisen from intense structure–activity relationship studies [20]. However, some of the effects of fluorine substitution are relatively straightforward to interpret such as the ability of fluorinated molecules to suppress metabolism relative to their hydrocarbon analogues. Moreover, organo-fluorine compounds are biologically and chemically more stable than the corresponding chlorine and bromine containing compounds [21].

As a part of our ongoing investigations of the biological activities and possible applications of the marine bisindole alkaloid 2,2-bis(6-bromo-1*H*-indol-3-yl)ethanamine **1** [14–18] and for all the above described reasons, in the present work we focus our studies on natural compound **1** and its fluorinated analogue **2** [18] (Figure 1) to determine their potential as antibiotic adjuvants and antibiofilm agents against methicillin-susceptible (MSSA) and methicillin-resistant (MRSA) *S. aureus*.

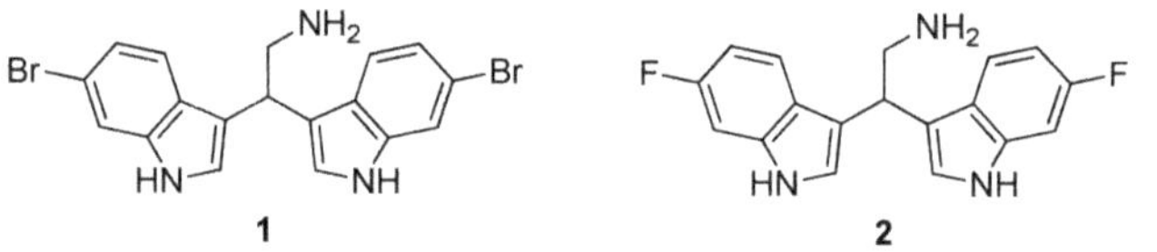

Figure 1. Marine bisindole alkaloid 2-bis(6-bromo-3-indolyl) ethylamine **1** and its fluorinated analogue **2**.

The experimental design included four different steps: (i) preliminary determination of the minimum inhibitory concentration (MIC) of each compound; (ii) assessment of their antimicrobial activity in combination with an antibiotic; (iii) determination of their antibiofilm properties in terms of both inhibition of biofilm formation and disruption of preformed biofilms and (iv) investigation of their activity against VBNC *S. aureus* forms.

2. Results

2.1. Chemistry

Marine bisindole alkaloid **1** and fluorinated analogue **2** were synthesized as previously reported [18] and described in Scheme 1.

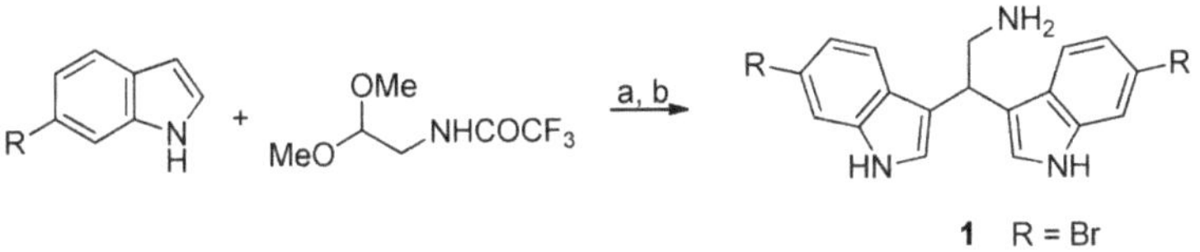

OMe
MeO NHCOCF_3
a, b
NH_2
R
HN NH
1 R = Br
2 R = F

Scheme 1. Reaction conditions: (**a**) diphenyl phosphate, acetonitrile, 80 °C, 24 h and (**b**) K_2CO_3, methanol, reflux, 2 h.

2.2. Antibacterial and Adjuvants Activities of Bisindoles **1** and **2**

The MIC of the compounds **1** and **2** against the MRSA (*S. aureus* CH 10850) and the MSSA (*S. aureus* ATCC 29213) strains is reported in Table 1, along with their hemolytic activity. In detail, the MIC of compound **1** was 2 µg/mL for both *S. aureus* strains, while compound **2** exhibited MIC values of 32 and 16 µg/mL against *S. aureus* CH 10850 and *S. aureus* ATCC 29213, respectively. The toxicity of both compounds toward mammalian cells, assessed by determining their ability to lyse human erythrocytes, resulted to be very low, with values of 2.96% ± 0.02% and 2.32% ± 0.06%, respectively.

Table 1. Antimicrobial activity (minimum inhibitory concentration (MIC), µg/mL) and hemolytic activity (%) of the examined compounds **1** and **2**.

Compound	MIC (µg/mL)		Hemolysis (%)
	S. aureus CH 10850 (MRSA)	*S. aureus* ATCC 29213	
1	2	2	2.96 ± 0.02
2	32	16	2.32 ± 0.06

When tested in association with oxacillin against *S. aureus* CH 10850, compound **1** caused a MIC reduction from 256 to 128 µg/mL, with a ΣFICs = 1.0, indicating additivity (Table 2). More promising results were obtained with the combination oxacillin-compound **2**, that showed a MIC decrease of oxacillin (from 256 to 1 µg/mL) with ΣFICs of 0.5, indicating a synergistic effect (Table 2).

Table 2. Reduction of oxacillin resistance in *S. aureus* CH 10850 (methicillin resistant *Staphylococcus aureus* (MRSA)) exerted by compound **1** and **2** and determination of the synergistic activity as assessed by checkerboard assay (ΣFICs).

Compound	Concentration (µg/mL)	Oxacillin MIC (µg/mL)	ΣFIC	
None	-	256		
1	1	128	1.0	Additive
2	16	1	0.5	Synergistic

2.3. Antibacterial Activity in Plasma

One of the main issues with indole and their derivatives is their instability in vivo due to oxidation [22,23]. Thus, the MIC of both compounds was evaluated after preincubation (0, 3 and 6 h) in 50% human plasma at 37 °C. MIC values of compound **2** were in line with what found in both *S. aureus* strains, while a 16-fold increase was observed for compound **1** (Figure 2). However, it is very remarkable that all antimicrobial activities remained constant even after 6 h of plasma preincubation, hence showing a good stability in the physiologically relevant time intervals (Figure 2).

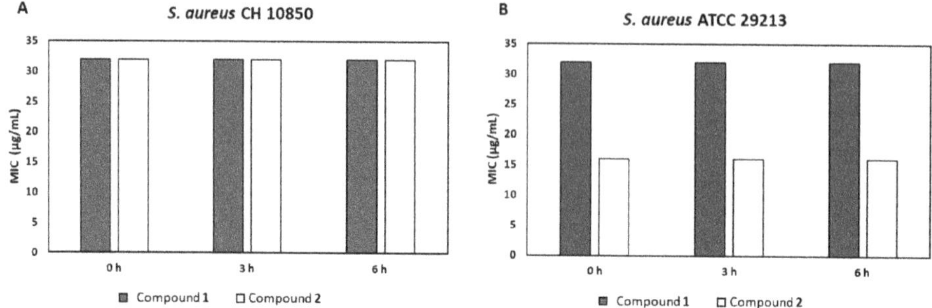

Figure 2. MIC of compounds **1** and **2** against *S. aureus* CH 10850 (**A**) and *S. aureus* ATCC 29213 (**B**), after preincubation in 50% blood plasma for 0, 3 and 6 h at 37 °C. Conservative estimates of three trials for each compound are shown.

2.4. Antibiofilm Activity

Compound **1** was able to inhibit the biofilm formation of *S. aureus* CH 10850 and *S. aureus* ATCC 29213 at their MIC values (2 µg/mL), causing a biofilm reduction of 52.6% and 49.6%, respectively (Table 3). When tested at 2× MIC (4 µg/mL), no biofilm formation was observed in the wells after 24 h of incubation (100% of biofilm formation inhibition). Compound **2** at its MIC (32 µg/mL) was able to inhibit (61.0%) the biofilm formation of *S. aureus* CH 10850, reaching a 76.5% inhibition when used at 64 µg/mL (2× MIC). *S. aureus* ATCC 29213 biofilm formation, although less susceptible (58.9% inhibition) than *S. aureus* CH 10850 to the MIC (16 µg/mL) of compound **2**, was completely (100%) inhibited by 32 µg/mL (2× MIC) of the same compound (Table 3). Concerning the biofilm-disrupting ability, compound **1** successfully removed 37.5% of *S. aureus* CH 10850 biofilm and the 28.0% of *S. aureus* ATCC 29213 biofilm, after a 30 min treatment. Under the same experimental conditions, compound **2** exerted a greater disaggregating activity, removing 56.3% and 53.9% of preformed biofilm of *S. aureus* CH 10850 and *S. aureus* ATCC 29213 biofilms, respectively (Table 3).

Table 3. Antibiofilm activities of compounds **1** and **2** against *S. aureus* CH 10850 and *S. aureus* ATCC 29213.

S. aureus Strain	Biofilm Formation Inhibition				Disaggregating Ability	
	Compound 1		Compound 2		Compound 1	Compound 2
	MIC	2× MIC	MIC	2× MIC	MIC	MIC
CH 10850 (MRSA)	52.6%	100%	61.0%	76.5%	37.5%	56.3%
ATCC 29213	49.6%	100%	58.9%	100%	28.0%	53.9%

2.5. VBNC Forms Induction

The influence of compounds **1** and **2** on VBNC *S. aureus* cells induction was tested. Both strains exhibited an about 2log reduction of viable cells when exposed to 2× MIC of compound **2** (Figure 3). However only *S. aureus* 10850 showed a significant gap (corresponding to the VBNC amount) between total viable and culturable cells. Compound **1** did not seem to induce VBNC forms, as demonstrated by the lack of discrepancy between culture and qPCR/flow cytometry (Figure 3).

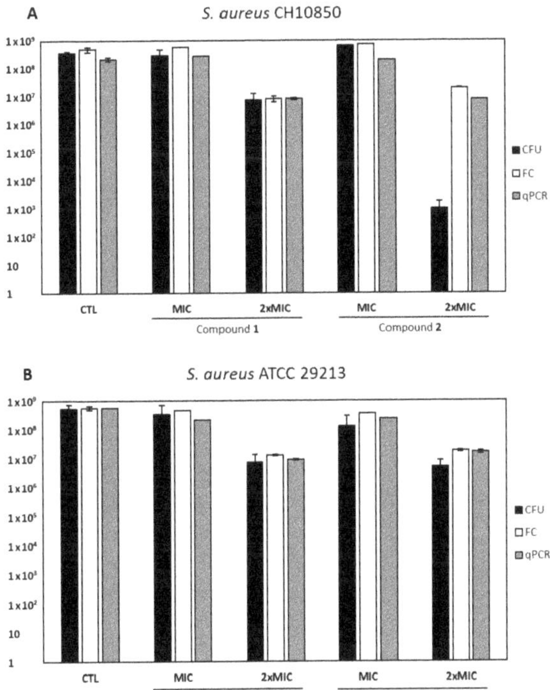

Figure 3. Effects of compounds **1** and **2** on *S. aureus* cell populations in in vitro biofilms. Biofilms of both *S. aureus* CH10850 (**A**) and *S. aureus* ATCC 29213 (**B**), in the absence (CTL) or presence of the two compounds at concentrations 1× and 2× MIC were developed in vitro and the amount of total viable and culturable cells counted after 24 h incubation at 37 °C. The amount of culturable cells was determined by plate count (CFU) and that of total (i.e., culturable and non-culturable) viable cells by qPCR and/or flow cytometry (FC).

3. Discussion

Considering the wide spread of antimicrobial resistance, the search for new and more efficient therapeutic approaches against MRSA strains represents a priority. Several studies have evidenced that molecules based on indole scaffolds can be used against different bacterial species [18,24,25]. Recently, it has been demonstrated that selected marine-alkaloid-derived molecules possess adjuvant activity against multi drug resistant bacteria [19], thus opening new challenges in this field of investigation. Therefore, we tested the fluorinated bisindole **2**, as well as the lead natural compound **1**, toward two different—one susceptible and one resistant to methicillin—strains of *S. aureus* (MSSA ATCC 39213 and MRSA CH 10850, respectively). Both compounds showed an antimicrobial activity, although compound **1** resulted in being more efficient than its fluorinated analogue compound **2** (MIC, 2 vs. 32 μg/mL in the MRSA and MIC 2 vs. 16 μg/mL in the MSSA). Indeed, bisindoles are known to act as antimicrobials by two possible mechanisms of action: the positive charge on the nitrogen at physiological pH makes them cationic surfactants able to destabilize the cytoplasmic membrane. Moreover, they could possibly inhibit the bacterial pyruvate kinase as reported for similar bisindoles [26,27].

Our compounds were tested as possible adjuvants. Both restored oxacillin activity against the MRSA strain. Indeed, while tested in association with oxacillin, an additive effect (ΣFICs = 1.0) and a synergistic one (ΣFICs = 0.5), for compounds **1** and **2**, respectively, was found. Interestingly, the fluorine substitution in compound **2** was able to increase the adjuvant property of the natural compound. In detail, the fluorinated bisindole **2** was able to decrease the oxacillin MIC by 256-fold (from 256 to 1 µg/mL) for the clinical MRSA strain studied here. To the best of our knowledge, an adjuvant property of this class of bisindoles has never been reported, hence, the data herein observed provide novel information and stress the potential use of these class of natural compounds, together with their synthetic analogues, to contrast antibiotic resistance. Based on these encouraging results, the toxicological aspect of both compounds was also investigated by assessing their ability to lyse human erythrocytes. Neither of the two compounds exhibited hemolytic activity, thus adding important information on their safety. Although a different behavior after plasma incubation was observed (i.e., 16-fold MIC increase of compound **1**), it is of note that both compounds retained their antimicrobial activities even after 6 h of preincubation in blood plasma, demonstrating their chemical stability to this body fluid at physiologically relevant time intervals.

Microbial biofilms generate serious human health problems, including infectious diseases such as endocarditis, periodontitis and bacteremia [4]. The obtained results evidenced the ability of the tested compounds to inhibit *S. aureus* biofilm formation, reaching in most cases the complete biofilm formation inhibition at 2× MIC concentration, as previously reported for similar compounds [18]. Only for *S. aureus* CH 10850 a lower biofilm formation inhibition was evidenced, stressing the higher resistance of MRSA to antimicrobials [4]. Considering the important role of indole in bacteria, the presence of two units of indole in both compounds **1** and **2** may suggest that the observed antibiofilm activity derives from a modulation of indole-based signaling pathways [18,28]. Indeed, it was found that intracellular indole and its derivatives can cause a temporary repression of the *agr*-quorum sensing system in *S. aureus* [29]. The eradication activity of compounds **1** and **2** was afterward assessed on preformed biofilms of *S. aureus* CH 10850 MRSA and *S. aureus* ATCC 39213. As shown, the fluorinated bisindole **2** had a most pronounced disaggregating activity (>50% for both the examined strains) compared to the natural compound **1** (maximum 37.5%). This result could be related to the presence of two fluorine atoms in compound **2**, conferring a reduced lipophilicity compared to the two bromine atoms of the natural product **1**. To possibly confirm this hypothesis, we calculated an important physicochemical property related to the lipophilicity/hydrophilicity of a molecule, the octanol–water portion coefficient (logP). LogP values are high for lipophilic molecules and low for hydrophilic ones. Calculated LogP (cLogP) were 2.36 for compound **2** and 3.68 for compound **1** (by OSIRIS Property Explorer) [30]. From the cLogP values, fluorinated derivative **2** is more than 10-fold less lipophilic than the natural product **1**, and therefore more soluble in the hydrophilic biofilm polysaccharide matrix, resulting in a more effective disaggregating activity on preformed biofilms.

Biofilm development can also lead to the induction of dormant cells, persistent bacterial phenotypes that seem to be suitable adjuvant targets [11]. Between the two tested molecules, compound **2** exerted a VBNC inducing effect at a 2× MIC concentration. Moreover, the bacterial response resulted strain-specific, as *S. aureus* ATCC 29213 exhibited a little difference between total viable and culturable cells (0.5 log), whereas *S. aureus* CH 10850 (MRSA) showed a four-log difference between qPCR/flow cytometry and CFU counts. This seems to indicate a more powerful action of compound **2** against this specific strain, as already suggested by the checkerboard assay, resulting in synergy. Indeed, this compound caused the same reduction of total viable cells in both strains, but CH 10850 exhibited a deeper state of dormancy (i.e., higher amount of VBNC forms) than the ATCC 29213 strain. These data confirm the role of methicillin resistance in the development of VBNC forms as a survival strategy to stress conditions [31]. Furthermore, the expression of the *mecA* gene, conferring methicillin resistance through PBP2a synthesis, has been suggested to correlate with a greater stability of VBNC forms of the same strain (*S. aureus* CH10850) under unfavorable conditions [32]. It is thus conceivable that compound **2** may exert antibiofilm activity even if it can constitute a stress factor able to induce VBNC *S. aureus* forms, as previously described for different antimicrobial compounds [31].

To briefly summarize, natural bisindole alkaloid **1** and its fluorinated derivative **2** showed antimicrobial activity against the tested *S. aureus* strains (MIC ranging from 2 to 32 µg/mL). Surprisingly, compound **2** (at 16 µg/mL) reduced the MIC of oxacillin from 256 to 1 µg/mL (ΣFICs = 0.5) for the clinical

MRSA strain. Although both molecules inhibited biofilm formation of *S. aureus* strains, compound **2** showed greater eradicating activity on preformed biofilm compared to the natural alkaloid **1**. None of the tested molecules exerted a VBNC inducing effect at their MIC values. Moreover, both compounds exhibited no hemolytic activity and a good stability in plasma, indicating a non-toxic profile. Although the tested compounds as well as the number of bacterial strains were limited, these preliminary results encourage us to further examine this class of interesting alkaloids, and in particular the fluorinated bisindole **2**, for in vivo applications to restore antibiotic treatment against MRSA infections.

4. Materials and Methods

4.1. Chemistry

All organic solvents used in this study were purchased from Sigma–Aldrich (St. Louis, MO, USA), Alfa Aesar (Haverhill, MA, USA), or TCI (Tokyo, Japan). Prior to use, acetonitrile was dried with molecular sieves with an effective pore diameter of 4 Å. Column chromatography purifications were performed under "flash" conditions using Merck (Darmstadt, Germany) 230–400 mesh silica gel. Analytical thin-layer chromatography (TLC) was carried out on Merck silica gel plates (silica gel 60 F254), which were visualized by exposure to ultraviolet light and an aqueous solution of cerium ammonium molybdate (CAM). ESI-MS spectra were recorded with a Waters (Milford, MA, USA) Micromass ZQ spectrometer. ^{1}H NMR and ^{13}C NMR spectra were recorded on a Bruker (Billerica, MA, USA) AC 400 or 100, respectively, spectrometer and analyzed using the TopSpin 1.3 (2013) software package. Chemical shifts were measured by using the central peak of the solvent.

4.1.1. General Procedure for the Synthesis of Derivatives **1–2**

Diphenyl phosphate (0.02 mmol) was added to a solution of the appropriate indole derivative (0.4 mmol) and (trifluoroacetylamino)acetaldehyde dimethyl acetal (0.2 mmol) in anhydrous acetonitrile (0.2 mL), and the resulting mixture was stirred at 80 °C for 24 h in a sealed tube, monitoring the progress of the reaction by TLC and HPLC-MS. After cooling to room temperature, saturated aqueous $NaHCO_3$ (30 mL) and dichloromethane (30 mL) were added and the two phases were then separated. The aqueous solution was extracted with dichloromethane (3 × 20 mL). After drying over dry Na_2SO_4, the combined organic phases were concentrated in vacuum and the resulting crude product was utilized without further purification. A mixture of that crude trifluoroacetamide derivative and potassium carbonate (1 mmoL) in MeOH (1.87 mL) and H_2O (0.13 mL) was stirred and heated at reflux for 2 h. The MeOH was removed under reduced pressure and water was added (30 mL). The aqueous solution was extracted with dichloromethane (3 × 30 mL) and the resulting solution was dried with Na_2SO_4 and then concentrated in vacuum. The crude material was purified by flash chromatography on silica gel.

4.1.2. 2,2-Bis(6-bromo-1*H*-indol-3-yl)ethanamine (**1**)

The physicochemical data of compound **1** are in agreement with those that were reported [18].

4.1.3. 2,2-Bis(6-fluoro-1*H*-indol-3-yl)ethanamine (**2**)

Compound **2** was prepared employing 6-fluoro-1*H*-indole and was isolated by column chromatography (dichloromethane/methanol/ammonia, 95:4:1) as a white solid in 70% yield (two steps). TLC: Rf = 0.18 (silica gel; dichloromethane/methanol/triethylamine, 90:9:1; UV, CAM). MS (ESI): m/z 310 [M-H]$^-$. ^{1}H NMR (400 MHz, CD_3OD, 293 K): δ = 3.37–3.41 (m, 2H, $CHCH_2NH_2$), 4.52 (dd, 1H, $J_1 = J_2$ = 7.5 Hz, $CHCH_2NH_2$), 6.72 (ddd, 2H, J_{5-7} = 2.0 Hz, J_{5-4} = 9.0 Hz, J_{5-F} = 9.5 Hz, H5), 7.04 (dd, 2H, J_{7-5} = 2.0 Hz, J_{7-F} = 9.5 Hz, H7), 7.14 (d, 2H, J = 3.0 Hz, H2), 7.44 (dd, 2H, J_{4-F} = 5.0 Hz, J_{4-5} = 9.0 Hz, H4) ppm. ^{13}C NMR (100 MHz, CD_3OD, 293 K): δ = 37.2, 45.6, 96.7 (d, 2C, J = 26 Hz, C5), 106.5 (d, 2C, J = 24 Hz, C7), 116.4 (2C, C3), 119.5 (d, 2C, J = 10 Hz, C4), 122.3 (d, 2C, J = 3 Hz, C9), 123.6 (2C, C2), 137.0 (d, 2C, J = 12 Hz, C8), 159.7 (d, 2C, J = 233 Hz, C6) ppm. The main physicochemical data of compound **2** are in agreement with those published [18].

4.2. Bacterial Strains

S. aureus CH 10850 (MRSA) [33] and *S. aureus* ATCC 29213 (MSSA), belonging to the strain collection of the Department of Life and Environmental Sciences (DiSVA), Polytechnic University of Marche (Ancona, Italy), were used. All the strains were cultured in brain heart infusion (BHI) broth or agar (Oxoid, Basingstoke, UK), subcultured in mannitol salt agar (MSA; Oxoid) and stored at −80 °C in BHI broth supplemented with 20% glycerol.

4.3. Determination of Minimum Inhibitory Concentration (MIC)

The minimum inhibitory concentration (MIC) of each molecule was determined by the microdilution method [34], with minor modifications. Bacteria were grown for 6 h in BHI broth at 37 °C, then diluted in Mueller Hinton II (Oxoid) to obtain ca. 5×10^5 CFU/mL in 100 μL, in the presence of increasing concentrations (2–128 μg/mL) of each compound dissolved in molecular biology grade dimethyl sulfoxide (DMSO, Sigma). Positive and negative controls included MHB inoculated or not with bacterial suspensions, respectively. Preliminary assays were performed to exclude the possible bacteriostatic and/or bactericidal activity of the solvent (i.e., DMSO); in any case, the volume of DMSO never exceeded 5% (*v/v*) of the final total volume. Tetracycline was used as a reference antibiotic, for comparison. MIC was defined as the lowest concentration of compound able to inhibit bacterial growth after 24 h of incubation at 37 °C, as detected by the unaided eye. All the experiments were performed three times using independent cultures.

4.4. Checkerboard Assays

The synergy of the two compounds and oxacillin (Sigma–Aldrich, St. Louis, Missouri, USA) against MRSA *S. aureus* CH 10850 was evaluated by the checkerboard assays [35], performed using 2-fold increasing concentrations of both compound (from 8 to 0.125μg/mL for compound **1** and from 512 μg/mL to 8 μg/mL for compound **2**) and oxacillin (from 512 to 0.5 μg/mL). Since the two compounds were resuspended in DMSO, the upper limit of the concentrations range tested was determined considering a final concentration of 1% DMSO. The combinations of each compound and oxacillin were evaluated by fractional inhibitory concentration (FIC) index, interpreted as follows: ≤0.5, synergy; >0.5 and ≤1.0, additive; >1.0 and <4, indifferent and ≥4, antagonistic.

4.5. Hemolytic Activity

The hemolytic activity of both compounds was evaluated as described by Ghosh et al. [36]. Briefly, 4 mL of freshly drawn, heparinized human blood was diluted with 25 mL of phosphate buffered saline (PBS), pH 7.4. After washing three times in 25 mL of PBS, the pellet was resuspended in PBS to 20 vol %. A 100 μL amount of erythrocyte suspension was added to 100 μL of different concentrations of compounds **1** and **2,** respectively. PBS and 0.2% Triton X-100 were used as the negative and positive control, respectively. Each condition was tested in triplicate. After 1 h of incubation at 37 °C each well was centrifuged at 1200 × *g* for 15 min, the supernatant was diluted 1:3 in PBS and transferred to a new plate. The OD_{350} was determined using the Synergy HT microplate reader spectrophotometer (BioTek, Winooski, VT, USA). The hemolysis (%) was determined as follows:

$$[(A - A_0)/(A_{\text{total}} - A_0)] \times 100 \quad (1)$$

where A is the absorbance of the test well, A_0 the absorbance of the negative control, and A_{total} the absorbance of the positive control; the mean value of three replicates was recorded.

4.6. Plasma Stability Assay

S. aureus CH 10850 and *S. aureus* ATCC 29213 were grown for 6 h in BHI broth and diluted in Mueller Hinton II to obtain a final concentration of 1.5×10^6 CFU/mL. Fresh human blood was centrifuged at 3000 rpm for 5 min to separate the plasma from the cells. Three aliquots of compound **1** and compound **2** were dissolved in DMSO at a concentration of 128 and 1024 μg/mL, respectively and diluted 2-fold in plasma to reach the final concentration of 64 and 512 μg/mL. After incubation at 37 °C for 0, 3 and 6 h [36], 50 μL of each compound serially diluted 1:2 in MHB were added to a 96-well plate containing 50 μL of bacterial suspensions in MHB and incubated at 37 °C for 24 h. MIC values were determined as mentioned above [34]. No change in MIC values among the trials performed after different plasma-preincubation times was considered a proof of plasma-stability.

4.7. Biofilm Formation Inhibition

Biofilms were developed in 24-well polystyrene plates (VWR). *S. aureus* strains were grown in tryptic soy broth (TSB, VWR, Radnor, PA, USA) at 37 °C for 24 h. The bacterial concentration was adjusted to 5×10^6 CFU/mL, as previously described, and 100 μL of each bacterial suspension were inoculated in 24-well polystyrene plates supplemented with the corresponding amount of the selected compounds at their MIC and 2× MIC values. Two wells were inoculated with bacteria in TSB, as controls. After 24 h of incubation at 37 °C, the wells were washed with PBS to eliminate unattached cells and covered with 0.1% (*v/v*) crystal violet (CV) dissolved in H_2O for 15 min and then washed in PBS and air-dried. The remaining CV was dissolved in 85% ethanol for 15 min at room temperature and 200 μL from each well was transferred to a 96-well plate for spectrophotometric quantification at 570 nm (Multiscan Ex Microplate Reader, Thermo Scientific, Waltham, MA, USA). Each data point was averaged from at least 8 replicate wells. All assays were performed in triplicate using independent cultures.

4.8. Biofilm-Disrupting Activity

Biofilms of each *S. aureus* strain were prepared with the procedure described above. After 24 h of incubation at 37 °C, the biofilms were gently washed in PBS, covered with the right amount of each compound at its MIC value, and left in contact for 30 min. For each plate, two wells were treated with saline and used as negative controls. After treatment, the biomass was evaluated by CV staining as described above. All data were expressed as the mean of three independent experiments performed in duplicate.

4.9. VBNC Detection

To evaluate the induction of staphylococcal VBNC forms, *S. aureus* CH 10850 and *S. aureus* ATCC 29213 biofilms were developed in Petri dishes (Ø 35 mm) by inoculating $OD_{650} = 0.1$ cultures in BHI broth, alone or supplemented with either compound **1** or compound **2** at their MIC and 2× MIC concentrations, and incubated at 37 °C for 24 h. At the end of the incubation, biofilms were gently washed with 1 mL of PBS to remove planktonic bacteria, detached and then resuspended in 1 mL of PBS.

4.9.1. Culture-Based Detection of Staphylococci

To evaluate the amount of the culturable cells, ten-fold serial dilutions of each mechanically detached biofilm were performed. For all dilutions 100 μL were spread onto BHI agar plates, incubated at 37 °C for 24 h prior to the enumeration of CFU.

4.9.2. Flow Cytometry Detection of Staphylococci

The abundance of total viable staphylococci, both culturable and non-culturable, was determined by flow cytometry. Assays were performed using 200 μL of a 1:1000 dilution of detached *S. aureus* biofilms after live/dead staining (1× SYBR Green and 40 μg/mL propidium iodide), in a Guava Millipore cytometer, and analyzed by the GUAVASOFT 2.2.3 software. To discriminate bacterial cells from the background, a gate for cell detection in side scatter and green fluorescence (GRN) was applied, using both channels at 488 nm and a threshold value in the GRN channel; SYBR green and propidium iodide fluorescence were excited using a 488 nm laser and collected at 525/30 and 617/30 nm, respectively. To better detect signals, they were logarithmically (4 decades) amplified and, to increase statistical significance, the total number of particles analyzed was set to 20,000 events/replicate. All assays were run in duplicate.

4.9.3. qPCR Detection of Staphylococci

Total DNA was extracted from 1 mL of biofilm aliquots diluted 1:10 (0× and 1× MIC) or undiluted (2× MIC) in PBS. Aliquots were centrifuged at 16,000 × *g* for 7 min, resuspended in 1 mL of STE (Tris-HCl 10 mM, NaCl 100 mM EDTA 1 mM) buffer supplemented with sucrose 20%, lysozyme 2.5 mg/mL and lysostaphin 100 μg/mL, and incubated for 1 h at 37 °C. Then, each aliquot was centrifuged, resuspended in 100 μL of PBS and the DNA was extracted using the QiaAmp DNA kit (Qiagen, Venlo, The Netherlands) according to the manufacturer instructions; a final elution volume of 80 μL was used.

S. aureus abundance was determined by *nuc*-qPCR using a Qiagen's Rotor-GeneQ MDx thermocycler, 0.2 μM of each primer [31] 10 μL of 2 × Rotor-Gene SYBR Green PCR master mix (Qiagen), and 2 μL DNA. Cycling conditions were 95 °C for 5 min, followed by 35 cycles of 95 °C for 10 s, 60 °C for 10 s and 72 °C for 10 s. A melting curve was obtained by ramping the temperature from 59 to 95 °C (0.5 °C/10 s) and analyzed with Qiagen's Rotor-GeneQ MDx software. DNA of *S. aureus* CH10850 and RNase-free water were used as positive and negative controls, respectively.

The number of viable *S. aureus* cells was determined as previously described [37,38]. Considering that *nuc* is a single copy gene [39] the amount of amplified DNA (ng) was divided for the weight (2.38928×10^{-10} ng) of *nuc*; and divided by 2 (qPCR template volume) to obtain the number of *S. aureus* cells corresponding to 1 μL of DNA extract. Staphylococcal abundance/mL of the original sample was then calculated multiplying the number of bacterial cell/μL of DNA extract by 80 (undiluted samples) or 800 (1:10 diluted samples), respectively. Plate counts were compared with both qPCR and flow cytometry quantifications; any discrepancy > 0.5 log was considered to attest the presence of a VBNC *S. aureus* subpopulation.

5. Conclusions

Marine alkaloid **1** and its fluorinated derivative **2** showed antimicrobial activity and inhibited biofilm formation of *S. aureus* strains. Moreover, bisindole **2** showed greater disaggregating activity (up to 56% for the examined strains) on preformed biofilm compared to the natural alkaloid **1**. Interestingly, compound **2** enabled 256-fold reduction in the MIC of oxacillin for the clinical MRSA strain herein studied. These encouraging data for analogue **2**, together with the evidences of its safety and stability, could represent the first step toward validating the potential of employing this adjuvant to restore oxacillin efficacy against MRSA infections.

Author Contributions: Conceptualization, R.C., F.B., S.L. and B.C.; methodology, R.C., B.C. and S.L.; validation, F.B., S.L. and B.C.; formal analysis, F.B., S.L. and B.C.; investigation, R.C., G.M. and M.T.; resources, S.L. and B.C.; data curation, R.C., E.F., F.B. and B.C.; writing—original draft preparation, S.L. and B.C.; writing—review and editing, R.C., G.M., M.T., E.F. and F.B.; visualization, S.L.; supervision, S.L. and B.C.; project administration, S.L. and B.C.; funding acquisition, E.F., F.B. and S.L. All authors have read and agreed to the published version of the manuscript.

Funding: This research was partially supported by a grant from the Department of Biomolecular Sciences (DISB), University of Urbino Carlo Bo (DISB_FRANGIPANI_PROG_SIC_ALIMENTARE).

Conflicts of Interest: The authors declare no conflict of interest.

References

1. Calfee, D.P. Trends in community versus health care-acquired methicillin-resistant *Staphylococcus aureus* infections. *Curr. Infect. Dis. Rep.* **2017**, *19*, 48. [CrossRef]

2. Tverdek, F.P.; Crank, C.W.; Segreti, J. Antibiotic therapy of methicillin-resistant *Staphylococcus aureus* in critical care. *Crit. Care Clin.* **2008**, *24*, 249–260. [CrossRef] [PubMed]
3. DeLeo, F.R.; Chambers, H.F. Reemergence of antibiotic-resistant *Staphylococcus aureus* in the genomics era. *J. Clin. Investig.* **2009**, *119*, 2464–2474. [CrossRef] [PubMed]
4. McCarthy, H.; Rudkin, J.K.; Black, N.S.; Gallagher, L.; O'Neill, E.; O'Gara, J.P. Methicillin resistance and the biofilm phenotype in *Staphylococcus aureus*. *Front. Cell Infect. Microbiol.* **2015**, *5*, 1. [CrossRef] [PubMed]
5. Flemming, H.C.; Wingender, J. The biofilm matrix. *Nat. Rev. Microbiol.* **2010**, *8*, 623–633. [CrossRef] [PubMed]
6. Bergkessel, M.; Basta, D.W.; Newman, D.K. The physiology of growth arrest: Uniting molecular and environmental microbiology. *Nat. Rev. Microbiol.* **2016**, *14*, 549–562. [CrossRef] [PubMed]
7. Fisher, R.A.; Gollan, B.; Helaine, S. Persistent bacterial infections and persister cells. *Nat. Rev. Microbiol.* **2017**, *15*, 453–464. [CrossRef]
8. Flemming, H.C.; Wingender, J.; Szewzyk, U.; Steinberg, P.; Rice, S.A.; Kjelleberg, S. Biofilms: An emergent form of bacterial life. *Nat. Rev. Microbiol.* **2016**, *14*, 563–575. [CrossRef]
9. Stewart, P.S.; Franklin, M.J. Physiological heterogeneity in biofilms. *Nat. Rev. Microbiol.* **2008**, *6*, 199–210. [CrossRef]
10. Ayrapetyan, M.; Williams, T.; Oliver, J.D. Relationship between the viable but nonculturable state and antibiotic persister cells. *J. Bacteriol.* **2018**, *200*, e00249-18. [CrossRef]
11. Melander, R.J.; Melander, C. The challenge of overcoming antibiotic resistance: An adjuvant approach? *ACS Infect. Dis.* **2017**, *3*, 559–563. [CrossRef] [PubMed]
12. Melander, R.J.; Minvielle, M.J.; Melander, C. Controlling bacteria behaviour with indole-containing natural products and derivatives. *Tetrahedon* **2014**, *70*, 6363–6372. [CrossRef] [PubMed]
13. Kobayashi, J. Search for new bioactive marine natural products and application to drug development. *Chem. Pharm. Bull.* **2016**, *64*, 1079–1083. [CrossRef]
14. Mari, M.; Tassoni, A.; Lucarini, S.; Fanelli, M.; Piersanti, G.; Spadoni, G. Brønsted acid catalyzed bisindolization of α-amido acetals: Synthesis and anticancer activity of bis (indolyl) ethanamino derivatives. *Eur. J. Org. Chem.* **2014**, *18*, 3822–3830. [CrossRef]
15. Mantenuto, S.; Lucarini, S.; De Santi, M.; Piersanti, G.; Brandi, G.; Favi, G.; Mantellini, F. One-pot synthesis of biheterocycles based on indole and azole scaffolds using tryptamines and 1, 2-diaza-1, 3-dienes as building blocks. *Eur. J. Org. Chem.* **2016**, *19*, 3193–3199. [CrossRef]
16. Salucci, S.; Burattini, S.; Buontempo, F.; Orsini, E.; Furiassi, L.; Mari, M.; Lucarini, S.; Martelli, A.M.; Falcieri, E. Marine bisindole alkaloid: A potential apoptotic inducer in human cancer cells. *Eur. J. Histochem.* **2018**, *62*, 2881. [CrossRef] [PubMed]
17. Choppara, P.; Bethu, M.S.; Vara Prasad, Y.; Venkateswara Rao, J.; Uday Ranjan, T.J.; Siva Prasad, G.V.; Doradla, R.; Murthy, Y.L.N. Synthesis, characterization and cytotoxic investigations of novel bis(indole) analogues besides antimicrobial study. *Arab. J. Chem.* **2015**, *12*, 2721–2731. [CrossRef]
18. Campana, R.; Favi, G.; Baffone, W.; Lucarini, S. Marine alkaloid 2,2-bis(6-bromo-3-indolyl) ethylamine and its synthetic derivatives inhibit microbial biofilms formation and disaggregate developed biofilms. *Microrganisms* **2019**, *7*, 28. [CrossRef]
19. Hubble, V.B.; Hubbard, B.A.; Minrovic, B.M.; Melander, R.J.; Melander, C. Using small-molecule adjuvants to repurpose azithromycin for use against *Pseudomonas aeruginosa*. *ACS Infect. Dis.* **2019**, *5*, 141–151. [CrossRef]
20. Isanbor, C.; O'Hagan, D. Fluorine in medicinal chemistry: A review of anti-cancer agents. *J. Fluor. Chem.* **2006**, *127*, 303–319. [CrossRef]
21. Zhang, Q.; Teschers, C.S.; Callejo, R.; Yang, M.; Wang, M.; Silk, P.J.; Ryall, K.; Roscoe, L.E.; Cordes, D.B.; Slawin, A.M.Z.; et al. Fluorine in pheromones: Synthesis of fluorinated 12-dodecanolides as emerald ash borer pheromone mimetics. *Tetrahedron* **2019**, *75*, 2917–2922. [CrossRef]
22. Gillam, E.M.J.; Notley, L.M.; Cai, H.; De Voss, J.J.; Guengerich, F.P. Oxidation of indole by cytochrome P450 enzymes. *Biochemistry* **2000**, *45*, 13817–13824. [CrossRef] [PubMed]
23. Mor, M.; Silva, C.; Vacondio, F.; Plazzi, P.V.; Bertoni, S.; Spadoni, G.; Diamantini, G.; Bedini, A.; Tarzia, G.; Zusso, M.; et al. Indole-based analogs of melatonin: In Vitro antioxidant and cytoprotective activities. *J. Pineal Res.* **2004**, *36*, 95–102. [CrossRef] [PubMed]
24. Ciulla, M.G.; Kumar, K. The natural and synthetic indole weaponry against bacteria. *Tetrahedron Lett.* **2018**, *59*, 3223–3233. [CrossRef]

25. Campana, R.; Sisti, M.; Sabatini, L.; Lucarini, S. Marine bisindole alkaloid 2,2-bis(6-bromo-3-indolyl) ethylamine to control and prevent fungal growth on building material: A potential antifungal agent. *Appl. Microbiol. Biotechnol.* **2019**, *103*, 5607–5616. [CrossRef]
26. Zoraghi, R.; Worrall, L.; See, R.H.; Strangman, W.; Popplewell, W.L.; Gong, H.; Samaai, T.; Swayze, R.D.; Kaur, S.; Vuckovic, M.; et al. Methicillin-resistant *Staphylococcus aureus* (MRSA) pyruvate kinase as a target for bis-indole alkaloids with antibacterial activities. *J. Biol. Chem.* **2011**, *286*, 44716–44725. [CrossRef]
27. Veale, C.G.L.; Zoraghi, R.; Young, R.M.; Morrison, J.; Pretheeban, M.; Lobb, K.A.; Reiner, N.E.; Andrersen, R.J.; Davies-Coleman, M.T. Synthetic analogues of the marine bisindole deoxytopsentin: Potent selective inhibitors of MRSA pyruvate kinase. *J. Nat. Prod.* **2015**, *78*, 355–362. [CrossRef]
28. Worthington, R.J.; Richards, J.J.; Melander, C. Small molecule control of bacterial biofilms. *Org. Biomol. Chem.* **2012**, *10*, 7457–7474. [CrossRef]
29. Lee, J.H.; Cho, H.S.; Kim, Y.; Kim, J.A.; Banskota, S.; Cho, M.H.; Lee, J. Indole and 7-benzyloxyindole attenuate the virulence of *Staphylococcus aureus*. *Appl. Microbiol. Biotechnol.* **2013**, *97*, 4543–4552. [CrossRef]
30. Organic Chemistry Portal. Available online: https://www.organic-chemistry.org/prog/ (accessed on 22 November 2019).
31. Pasquaroli, S.; Zandri, G.; Vignaroli, C.; Vuotto, C.; Donelli, G.; Biavasco, F. Antibiotic pressure can induce the viable but non-culturable state in *Staphylococcus aureus* growing in biofilms. *J. Antimicrob. Chemother.* **2013**, *68*, 1812–1817. [CrossRef]
32. Pasquaroli, S.; Citterio, B.; Mangiaterra, G.; Biavasco, F.; Vignaroli, C. Influence of sublethal concentrations of vancomycin and quinupristin/dalfopristin on the persistence of viable but non-culturable *Staphylococcus aureus* growing in biofilms. *J. Antimicrob. Chemother.* **2018**, *73*, 3526–3529. [CrossRef]
33. Donelli, G.; Francolini, I.; Romoli, D.; Guaglianone, E.; Piozzi, A.; Ragunath, C.; Kaplan, J.B. Synergistic activity of Dispersin B and Cefamandole Nafate in inhibition of Staphylococcal biofilm growth on polyurethanes. *Antimicrob. Agents Chemother.* **2007**, *51*, 2733–2740. [CrossRef] [PubMed]
34. Clinical and Laboratory Standards Institute (CLSI). *M07-A10: Methods for Dilution Antimicrobial Susceptibility Tests for Bacteria that Grow Aerobically*, 10th ed.; Clinical and Laboratory Standards Institute: Wayne, PA, USA, 2017; p. 66.
35. El-Azizi, M. Novel microdilution method to assess double and triple antibiotic combination therapy in vitro. *Int. J. Microbiol.* **2016**, 4612021.
36. Ghosh, C.; Manjunath, G.B.; Akkapeddi, P.; Yarlagadda, V.; Hoque, J.; Uppu, D.S.; Konai, M.M.; Haldar, J. Small molecular antibacterial peptoid mimics: The simpler the better! *J. Med. Chem.* **2014**, *57*, 1428–1436. [CrossRef] [PubMed]
37. Di Cesare, A.; Luna, G.M.; Vignaroli, C.; Pasquaroli, S.; Tota, S.; Paroncini, P.; Biavasco, F. Aquaculture can promote the presence and spread of antibiotic-resistant Enterococci in marine sediments. *PLoS ONE* **2013**, *8*, e62838. [CrossRef]
38. Mangiaterra, G.; Amiri, M.; Di Cesare, A.; Pasquaroli, S.; Manso, E.; Cirilli, N.; Citterio, B.; Vignaroli, C.; Biavasco, F. Detection of viable but non-culturable *Pseudomonas aeruginosa* in cystic fibrosis by qPCR: A validation study. *BMC Infect. Dis.* **2018**, *18*, 701. [CrossRef]
39. Hein, I.; Lehner, A.; Rieck, P.; Klein, K.; Brandl, E.; Wagner, M. Comparison of different approaches to quantify *Staphylococcus aureus* cells by real-time quantitative PCR and application of this technique for examination of cheese. *Appl. Environ. Microbiol.* **2001**, *67*, 3122–3126. [CrossRef]

Article

Bactericidal and In Vitro Cytotoxicity of *Moringa oleifera* Seed Extract and Its Elemental Analysis Using Laser-Induced Breakdown Spectroscopy

Reem K. Aldakheel [1,2], Suriya Rehman [3,*], Munirah A. Almessiere [1], Firdos A. Khan [4], Mohammed A. Gondal [5,*], Ahmed Mostafa [6] and Abdulhadi Baykal [7]

1 Department of Biophysics, Institute for Research & Medical Consultations (IRMC), Imam Abdulrahman Bin Faisal University, Dammam 31441, Saudi Arabia; reem.k.dakheel@gmail.com (R.K.A.); malmessiere@iau.edu.sa (M.A.A.)
2 Department of Physics, College of Science, Imam Abdulrahman Bin Faisal University, Dammam 31441, Saudi Arabia
3 Department of Epidemic Diseases Research, Institute for Research & Medical Consultations (IRMC), Imam Abdulrahman Bin Faisal University, Dammam 31441, Saudi Arabia
4 Department of Stem Cell Research, Institute for Research & Medical Consultations (IRMC), Imam Abdulrahman Bin Faisal University, Dammam 31441, Saudi Arabia; fakhan@iau.edu.sa
5 Department of Physics, Laser Research Group, King Fahd University of Petroleum & Minerals, Box 372, Dhahran 31261, Saudi Arabia
6 Department of Pharmaceutical Chemistry, College of Clinical Pharmacy, Imam Abdulrahman Bin Faisal University, Dammam 31441, Saudi Arabia; ammostafa@iau.edu.sa
7 Department of Nanomedicine Research, Institute for Research & Medical Consultations (IRMC), Imam Abdulrahman Bin Faisal University, Dammam 31441, Saudi Arabia; abaykal@iau.edu.sa
* Correspondence: surrehman@iau.edu.sa or suriyamir@gmail.com (S.R.); magondal@kfupm.edu.sa (M.A.G.); Tel.: +96-655-6211252 (S.R.); +96-613-8602351 or +96-613-8603274 (M.A.G.); Fax: +96-613-8604281 (M.A.G.)

Received: 9 July 2020; Accepted: 7 August 2020; Published: 13 August 2020

Abstract: In the current study, we present the correlation between the capability of laser-induced breakdown spectroscopy (LIBS) to monitor the elemental compositions of plants and their biological effects. The selected plant, *Moringa oleifera*, is known to harbor various minerals and vitamins useful for human health and is a potential source for pharmaceutical interventions. From this standpoint, we assessed the antibacterial and in vitro cytotoxicity of the bioactive components present in *Moringa oleifera* seed (MOS) extract. Detailed elemental analyses of pellets of MOSs were performed via LIBS. Furthermore, the LIBS outcome was validated using gas chromatography–mass spectrometry (GC-MS). The LIBS signal was recorded, and the presence of the essential elements (Na, Ca, Se, K, Mg, Zn, P, S, Fe and Mn) in the MOSs were examined. The bactericidal efficacy of the alcoholic MOS extract was examined against *Escherichia coli (E. coli)* and *Staphylococcus aureus (S. aureus)* by agar well diffusion (AWD) assays and scanning electron microscopy (SEM), which depicted greater inhibition against Gram-positive bacteria. The validity and DNA nuclear morphology of human colorectal carcinoma cells (HCT-116) cells were evaluated via an MTT assay and DAPI staining. The MTT assay results manifested a profoundly inhibitory action of MOS extract on HCT116 cell growth. Additionally, MOS extracts produced inhibitory action in colon cancer cells (HCT-116), whereas no inhibitory action was seen using the same concentrations of MOS extract on HEK-293 cells (non-cancerous cells), suggesting that MOS extracts could be non-cytotoxic to normal cells. The antibacterial and anticancer potency of these MOS extracts could be due to the presence of various bioactive chemical complexes, such as ethyl ester and D-allose and hexadecenoic, oleic and palmitic acids, making them an ideal candidate for pharmaceutical research and applications.

Keywords: antibacterial; anticancer; GC-MS; LIBS; *Moringa oleifera*; seed extract

1. Introduction

Traditional herbalists throughout the globe have emphasized the role of plants used as remedies to treat different diseases, like inflammation and bacterial infections [1]. Enormous research is being carried out, where plants are being tested for active compounds that have antibacterial, antifungal and anticancer activities [2,3]. One such plant is *Moringa oleifera* Lam. (MOL), which originates mainly from Africa, Asia and South America [4,5]. The moringa leaves and fruits are considered as vegetables in the Philippines, Thailand, India and Pakistan [6]. Recently, sundry countries (Mexico, Caribbean islands, Hawaii and Cambodia) have begun planting it for its plentiful health benefits and nutritional value in addition to its medicinal importance [7,8]. *Moringa* fruits and leaves have various biological applications in addition to being enriched with vitamins, minerals, and proteins. The WHO recommends using MOL as a food because of its superior nutritional values for human health [6]. It is a well-known fact that *Moringa* plants (leaves and fruits) were grown as for skin sanitation, an energy source and also to relieve tension [8,9]. Innumerable reports have discovered that MOL has numerous significant merits, such as antioxidant, antimicrobial, anticancer, anti-inflammatory, antiulcer, antihypertensive, anti-urolithic, anti-asthmatic, antidiabetic, analgesic, anti-aging, diuretic, cardiovascular, hepatoprotective, hypoglycemic and immunomodulatory characteristics [10–13]. The hypotensive, antibacterial and anticancer efficacies of the MOL leaves and fruits are due to the presence of sundry distinctive chemical compounds (benzyl glucosinolate, niazimicin, benzyl iso-thiocyanate complexes and pterygospermin) in their structure [11]. Several methods have been improved to extract the initial contents from MOL plants for the production of food supplements and medicines (with natural organic components) and the determination of their other health benefits. Extraction procedures based on pressurized liquid, ultrasound, microwaves and supercritical fluid have been introduced [14–17].

Nevertheless, all extraction mechanisms experience several restrictions in terms of using a large amount of harmful organic solvents, sample production proceedings and cost [13]. To get better results, LIBS has emerged as an efficient approach for the identification and quantification of elemental compositions from different medicinal plant extractions. Over the past decade, the LIBS method has broadly been exploited for analyzing chemical elements existing in diverse kinds of specimens. This analytical method has distinguishing properties, like cost effectiveness, real-time measurement, sensitivity, rapidity and in situ elemental analysis [18–20]. Additionally, this type of analysis is easy, eco-friendly and is less complicated when preparing the sample. It is known that intensive sample treatments frequently produce erroneous results due to the presence of contaminants and loss-related effects. Thus, the LIBS technique can accurately disclose the identity of various medicinal plants in a scientific manner. In addition, most edible plants consist of proteins, vitamins and minerals, which possess immense benefits to health [21,22].

The purpose of the study is to analyze the presence of different elements in *Moringa oleifera* seed (MOS) extracts by using the LIBS approach and to examine the biological activities of MOS extracts by evaluating the antibacterial and anticancer potencies.

2. Results

2.1. *Qualitative Analysis of MOS Using LIBS*

The laser-induced breakdown spectroscopy (LIBS) spectra (Figure 1) of the MOS extract were recorded in the range of 200 to 800 nm. For the detection of the major elements, the spectra were recorded from different spots of the pellets and by scanning at a 50 nm wavelength range each time. In order to reduce the background noise and to enhance the LIBS signal intensity, LIBS parameters, including the delay times (delay between the incident laser pulse and the recording of spectra), the number of accumulations and laser energies, were optimized prior to the application of the LIBS setup for MOS sample analysis. The recorded LIBS spectra of the MOSs comprised various significant spectral peaks of the atomic and ionic lines (with varying intensities) related to the abundance of elements present in the tested MOS samples. Based on the NIST database, the recorded spectral lines were identified and classified in terms of the characteristic elements present in the MOSs, suggesting their significant role towards antibacterial and anticancer activities. Clearly, the measured LIBS spectra (Figure 1) exhibited the presence of vital elements such as Ca, K, Mg, P, S, Fe, Mn, Zn, Na and Se in MOS samples.

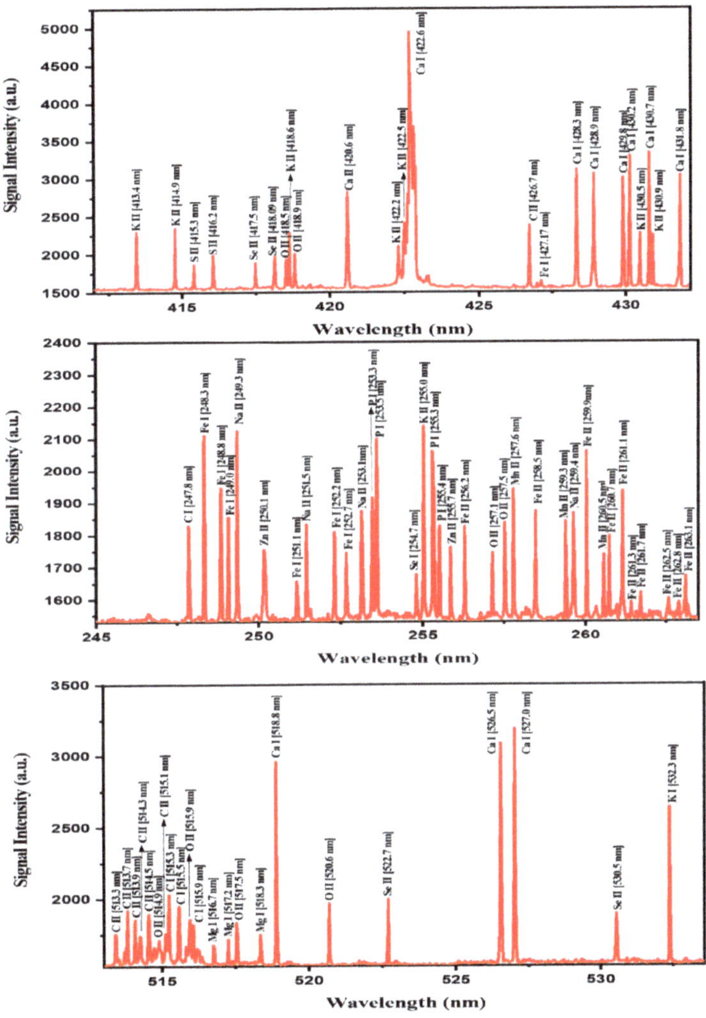

Figure 1. LIBS emission line spectra of various elements recorded in the wavelength range of 245–537 nm of MOS. The signature lines for different vital minerals present in MOSs are indicated in the figure.

Table 1 presents the LIBS signal intensities of the spectral transition lines corresponding to various detected elements in MOSs. In accordance with the Boltzmann distribution, the intensities of the LIBS spectral lines have a direct relationship with the elemental contents (concentration) present in MOSs [23,24]. This correlation was attained by considering the intensity ratio of the detected elemental lines with that of the C line taken as the reference (247.8 nm). The achieved intensity ratios of the elements Ca, K, Mg, P, S, Fe, Mn, Zn, Na and Se were in the ranges of 2.7–1.7, 1.2–1.4, 0.9, 1.1, 1.1, 1.1–0.9, 1.0, 0.9, 1.2 and 1.1, respectively, which were consistent with those reported in the literature [16].

Table 1. The detection of spectral lines of the different elements present in *MOS* by using our LIBS system.

Name of Element	Wavelength (nm)	Transition Configuration	LIBS Signal Intensity (arbitrary unit(a.u.))
Ca	422.6	$3p^6\,4s^2\,{}^1S_0 \rightarrow 3p^6\,4s\,4p\,{}^1P^{\circ}{}_1$	4960.9
	518.8	$3p^6\,4s\,4p\,{}^1P^{\circ}{}_1 \rightarrow 3p^6\,4s\,5d\,{}^1D_2$	2967.1
	527.0	$3p^6\,3d\,4s\,{}^3D_3 \rightarrow> 3p^6\,3d\,4p\,{}^3P^{\circ}{}_2$	3196.3
	248.3	$3d^6\,4s^2\,{}^5D_4 \rightarrow 3d^6\,({}^5D)\,4s\,4p\,({}^1P^{\circ})\,{}^5F^{\circ}{}_5$	2114.1
Fe	252.2	$3d^6\,4s^2\,{}^5D_4 \rightarrow 3d^6\,({}^5D)\,4s\,4p\,({}^1P^{\circ})\,{}^5D^{\circ}{}_4$	1810.1
	259. 9	$3d^6\,({}^5D)\,4s\,{}^6D_{9/2} \rightarrow 3d^6\,({}^5D)\,4p\,{}^6D^{\circ}{}_{9/2}$	2061.9
K	414.9	$3p^5\,3d\,{}^3P^{\circ}{}_0 \rightarrow 3p^5\,4p\,{}^3D_1$	2347.6
	430.5	$3p^5\,3d\,{}^3P^{\circ}{}_2 \rightarrow 3p^5\,4p\,{}^1D_2$	2270.9
	532.3	$3p^6\,4p\,{}^2P^{\circ}{}_{1/2} \rightarrow 3p^6\,8s\,{}^2S_{1/2}$	2639.7
Mg	518.3	$3s\,3p\,{}^3P^{\circ}{}_2 \rightarrow 3s\,4s\,{}^3S_1$	1753.5
Mn	259.3	$3d^5\,({}^6S)\,4s\,{}^7S_3 \rightarrow 3d^5\,({}^6S)\,4p\,({}^7P^{\circ}{}_3)$	1843.9
Na	249.3	$2s^2\,2p^5\,3s\,{}^1P^{\circ}{}_1 \rightarrow 2s^2\,2p^5\,3p\,{}^1S_0$	2128.4
	251.5	$2s^2\,2p^5\,3p\,{}^3S_1 \rightarrow 2s^2\,2p^5\,({}^2P^{\circ}{}_{1/2})\,3d\,{}^2[\rightarrow]^{\circ}{}_2$	1834.7
P	253.5	$3s^2\,3p^3\,{}^2P^{\circ}{}_{3/2} \rightarrow 3s^2\,3p^2\,({}^3P)\,4s\,{}^2P_{3/2}$	2100.0
	255.3	$3s^2\,3p^3\,{}^2P^{\circ}{}_{1/2} \rightarrow 3s^2\,3p^2\,({}^3P)\,4s\,{}^2P_{1/2}$	2032.1
S	416.2	$3s^2\,3p^2\,({}^3P)\,4p\,{}^4D^{\circ}{}_{7/2} \rightarrow 3s^2\,3p^2\,({}^3P)\,4d\,{}^4F_{9/2}$	1987.8
Se	418.09	$4s^2\,4p^2\,({}^3P)\,5p\,{}^4D^{\circ}{}_{7/2} \rightarrow 4s^2\,4p^2\,({}^3P)\,5d\,{}^4F_{9/2}$	1969.7
	522.7	$4s^2\,4p^2\,({}^3P)\,5s\,{}^4P_{5/2} \rightarrow 4s^2\,4p^2\,({}^3P)\,5p\,{}^4D^{\circ}{}_{7/2}$	2002.0
Zn	250.1	$3d^{10}\,4p\,{}^2P^{\circ}{}_{1/2} \rightarrow 3d^{10}\,5s\,{}^2S_{1/2}$	1751.4
	255.7	$3d^{10}\,4p\,{}^2P^{\circ}{}_{3/2} \rightarrow 3d^{10}\,5s\,{}^2S_{1/2}$	1762.3
C	426.7	$2s^2\,3d\,{}^2D_{5/2} \rightarrow 2s^2\,4f\,{}^2F^{\circ}{}_{7/2}$	2379.0
	247.8	$2s^2\,2p^2\,{}^1S_0 \rightarrow 2s^2\,2p\,3s\,{}^1P^{\circ}{}_1$	1831.6
O	418.9	$2s^2\,2p^2\,({}^1D)\,3p\,{}^2F^{\circ}{}_{7/2} \rightarrow 2s^2\,2p^2\,({}^1D)\,3d\,{}^2G_{9/2}$	2003.4

2.2. Volatile Content Analyses of MOS using GC-MS

The GC-MS analyses of the *Moringa oleifera* seed (MOS) showed the presence of 114 volatile complexes with diverse chemical groups: fatty acids, esters, ketones, alcohols, aldehydes, and hydrocarbons. Table 2 shows the retention times and Figure 2 displays the percentage composition of all the identified chemical compounds in MOS. The main types were oleic acid (22.53%), 2-3-di-hydroxy-propyl (13.48%), 9-octa-decenoic acid (Z), 2-3-di-hydroxy-propyl ester (11.35%), docosenamide (6.04%), ethyl oleate (6.03%), 1-3-propanediol, 2-ethyl-2-(hydroxyl-methyl) (5.52%), oleic anhydride (3.96%), 2-propanone and 1-1-dimethoxy (3.86%). These MOSs contain fatty acids and their ester derivatives (65.45%), alcohols (9.4%), nitrogen compounds (9.09%), ketones (5.34%) and aldehydes (2.88%). To form esters, fatty acids and alcohols in plants may undergo esterification.

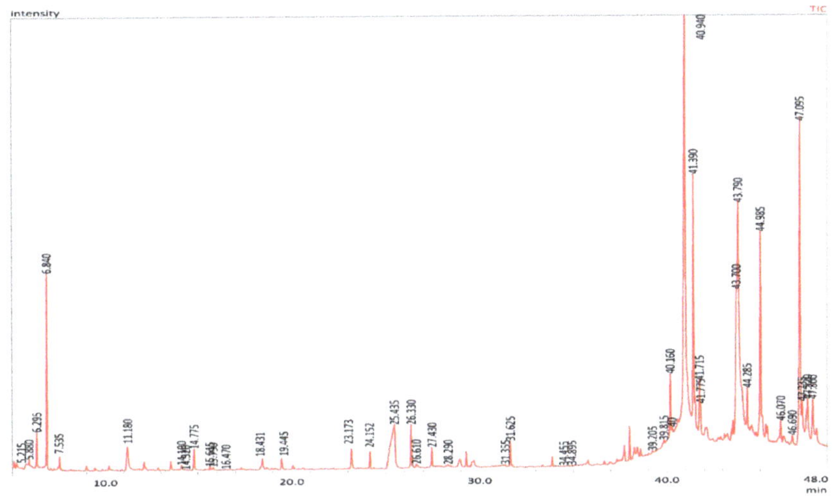

Figure 2. GC-MS chromatograph of MOS.

Table 2. Volatile compounds identified in MOS using GC-MS analysis.

No.	Compounds	RT	Peak area (%)
	Esters		
1	Propanoic acid, 2-oxo-, methyl ester	6.293	0.76
2	Acetic acid, ethoxyhydroxy-, ethyl ester	7.043	0.04
3	Diethoxymethyl acetate	12.827	0.09
4	2-Propenoic acid, 2-methyl-, 2-hydroxypropyl ester	13.134	0.02
5	6,9,12-Octadecatrienoic acid, phenylmethyl ester, (*Z*,*Z*,*Z*)-	17.797	0.01
6	Cyclopentanecarboxylic acid, 4-tridecyl ester	19.179	0.03
7	1-Cyclohexene-1-carboxylic acid, 2,6,6-trimethyl-, methyl ester	20.058	0.13
8	Hexanoic acid, 4-hexadecyl ester	28.291	0.07
9	Phthalic acid, diethyl ester	29.568	0.14
10	2-Propenoic acid, pentadecyl ester	31.625	0.59
11	Acetic acid, 3,7,11,15-tetramethyl-hexadecyl ester	33.627	0.02
12	Phthalic acid, dibutyl ester	35.692	0.07
13	Phthalic acid, diisobutyl ester	35.760	0.15
14	Palmitoleic acid, methyl ester	36.285	0.03
15	Pentadecanoic acid, 13-methyl-, methyl ester	36.642	0.09
16	9-Hexadecenoic acid, ethyl ester	37.652	0.15
17	Hexadecanoic acid, ethyl ester (Ethyl palmitate)	37.993	0.78
18	Heptadecanoic acid, ethyl ester	38.416	0.46
19	Propanoic acid, 3-mercapto-, dodecyl ester	38.588	0.25
20	9-Octadecenoic acid, methyl ester, (*E*)- (Methyl elaidate)	40.159	1.52
21	Oleic acid, methyl ester (Methyl oleate)	40.255	0.15
22	l-(+)-Ascorbic acid 2,6-dihexadecanoate	40.561	0.10
23	Ethyl oleate	41.389	6.03
24	Octadecanoic acid, ethyl ester (Ethyl stearate)	41.774	0.53
25	9-octadecenyl ester (Oleyl oleate)	43.516	0.34
26	2,3-dihydroxypropyl elaidate	43.794	13.48
27	9-Octadecenoic acid, 1,2,3-propanetriyl ester	44.287	0.90
28	Docosanoic acid, ethyl ester	45.273	0.37
29	Oleoyl chloride	46.066	0.65
30	9-Octadecenoic acid (*Z*)-, 2,3-dihydroxypropyl ester	47.099	11.35
31	Glycidol stearate	47.522	1.76
32	Hexadecanoic acid, 2-hydroxy-1-(hydroxymethyl)ethyl ester	47.803	1.57

Table 2. *Cont.*

No.	Compounds	RT	Peak area (%)
	Alcohols		
33	1,4-Cyclohexanediol, trans-	7.224	0.02
34	1,2-Propanediol, 3-methoxy-	7.430	0.05
35	Ethanol, 2,2-diethoxy-	7.535	0.31
36	1,2,4-Butanetriol	9.496	0.04
37	Glycerin	11.180	1.35
38	1-Butanol, 4-(ethylthio)-	11.742	0.02
39	Methoxyacetaldehyde diethyl acetal	11.883	0.02
40	1-Dodecanol	18.431	0.48
41	1-Tetradecanol	24.152	0.42
42	1,3-Propanediol, 2-ethyl-2-(hydroxymethyl)-	25.444	5.52
43	1-Tridecanol	26.331	1.06
44	3-Hexadecanol	31.348	0.05
45	3-Heptadecanol	35.561	0.06
	Aldehydes		
46	Butanal, 3-hydroxy-	8.375	0.00
47	Heptanal	10.006	0.02
48	Octanal	12.567	0.04
49	Nonanal	15.790	0.09
50	2-Methyl-oct-2-enedial	20.611	0.04
51	Undecanal	23.521	0.02
52	Tridecanal	34.455	0.01
53	10-Octadecenal	37.418	0.09
54	9-Octadecenamide	37.551	0.09
55	cis-9-Hexadecenal	38.241	0.55
56	cis-13-Octadecenal	47.228	1.92
	Ketones		
57	2-Propanone, 1,1-dimethoxy-	6.841	3.86
58	2-Butanone	8.151	0.03
59	Dihydroxyacetone	8.969	0.14
60	1,2-Cyclopentanedione	10.172	0.14
61	2-Propanone, 1-(1,3-dioxolan-2-yl)-	10.867	0.04
62	1,3-Dioxol-2-one,4,5-dimethyl-	13.528	0.28
63	2-Heptanol, 5-ethyl-	14.306	0.01
64	2-Methyl-4-octanone	16.466	0.03
65	2-Pentanone, 3,4-epoxy-	18.925	0.03
66	1-Oxa-spiro[4.5]deca-6,9-diene-2,8-dione	37.006	0.05
67	Z-11-Pentadecenol	39.209	0.05
68	Cyclopentadecanone, 2-hydroxy-	39.822	0.13
69	Cyclopentadecanone	43.430	0.21
70	2-Tetradecanone	46.691	0.34
	Acids		
71	Acetic acid, (acetyloxy)-	5.884	0.18
72	Butanoic acid, 3-hydroxy-	10.594	0.03
73	Octanoic acid	17.421	0.03
74	Nonanoic acid	20.363	0.02
75	n-Hexadecanoic acid (Palmitic acid)	37.305	0.05
76	Oleic acid	40.943	22.53
	Furans and lactones		
77	Furfural	7.624	0.02
78	2(5H)-Furanone	9.840	0.04
79	2-Hydroxy-gamma-butyrolactone	12.081	0.23

Table 2. *Cont.*

No.	Compounds	RT	Peak area (%)
80	2,5-Dimethyl-4-hydroxy-3(2H)-furanone	14.099	0.19
81	1,2-Ethanediol, 1-(2-furanyl)	19.096	0.03
82	5-Hydroxymethylfurfural	19.445	0.36
83	3-Deoxy-d-mannoic lactone	28.924	0.52
	Nitrogen-containing compounds		
84	*N,N*-Dimethylaminoethanol	5.216	0.12
85	1,3,5-Triazine-2,4,6-triamine	14.778	0.68
86	Acetic acid, 2-(*N*-methyl-*N*-phosphonatomethyl)amino-	15.617	0.11
87	1-Heptadecanamine	18.667	0.07
88	Nonanamide	23.034	0.02
89	Dodecanamide (Lauryl amide)	33.331	0.04
90	Tetradecanamide (Myristic amide)	37.717	0.44
91	Hexadecanamide (Palmitic amide)	41.718	1.26
92	Docosenamide	44.987	6.04
93	Nonadecanamide	45.352	0.32
	Sulfur-containing compounds		
94	Sulfurous acid, cyclohexylmethyl hexadecyl ester	23.173	0.67
	Hydrocarbons		
95	1-Butene, 4,4-diethoxy-2-methyl-	9.425	0.07
96	1-Methyl-2-octylcyclopropane	12.171	0.07
97	2-Trifluoroacetoxytridecane	13.923	0.03
98	trans-2,3-Epoxynonane	23.642	0.03
99	2-Heptafluorobutyroxypentadecane	23.887	0.04
100	1,2-Epoxyundecane	24.016	0.02
101	Heptacosane	24.357	0.03
102	1-Heptadecene	29.262	0.33
103	Octadecane, 1,1′-[(1-methyl-1,2-ethanediyl)bis(oxy)]bis-	29.426	0.01
104	1-Nonadecene	33.847	0.22
	Pyrans		
105	Tetrahydro-4H-pyran-4-ol	17.074	0.03
106	4H-Pyran-4-one, 2,3-dihydro-3,5-dihydroxy-6-methyl-	17.198	0.02
	Others		
107	5,6-Dihydroxypiperazine-2,3-dione dioxime	8.320	0.02
108	Tetraethyl silicate	11.807	0.04
109	Silanediol, dimethyl-, diacetate	17.300	0.07
110	D-Allose	26.609	0.19
111	Phenol, 2,4-bis(1,1-dimethylethyl)-	27.430	0.51
112	Oxirane, hexadecyl-	34.898	0.02
113	Ethyl iso-allocholate	36.936	0.06
114	Oleic anhydride	43.700	3.96

RT: Retention time in minutes.

2.3. In vitro *Cytotoxic Activity of MOS*

The cell viability of the MOS extract-treated HCT-116 cells was evaluated using an MTT assay after 48 h. The percentage of cell viability of the MOS extract-treated HEK-293 and HCT-116 cells was determined. The MTT assay examined the percentage of cell viability. The number of viable cells present in the control group was compared with the MOS-treated cells. In Figure 3, we show that cancer cells (control group, without MOS treatment) showed 100% cell viability, whereas cancer cells treated with MOS extract showed a significant decrease, which suggests that MOS extract could induce a significant drop in the viability of cancerous cells compared to control ones (without treatment using the MOS extract).

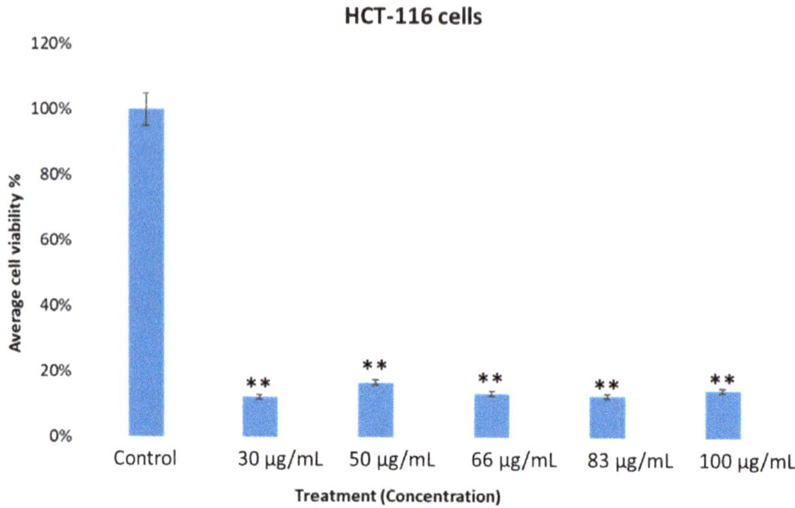

Figure 3. Effect of MOS extract on colon cancer cells (HCT-116) after treatment for 48 h with different concentrations. The average cell viability was calculated by MTT assay (** $p < 0.01$).

The average viability of the MOS extract-treated cancer cells at various concentrations showed quite encouraging outcomes, with $p < 0.01$. The specificity of the MOS extract in the concentration range of 30 to 100 µg/mL on normal cells was inspected using MTT assays after treatment for 48 h (Figure 4). When MOS extract was tested on normal cells (HEK-293), we found no inhibitory action.

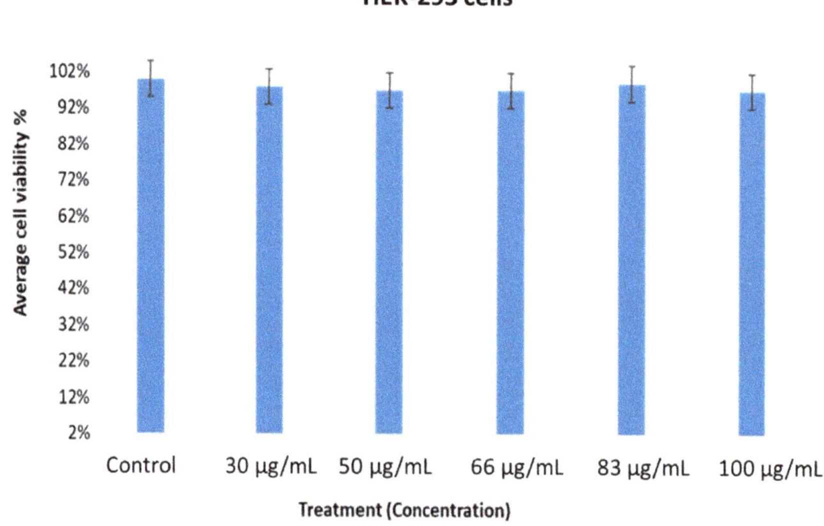

Figure 4. Effect of MOS extract on normal cells (HEK-293) after treatment for 48 h treated different concentrations. The average cell viability was calculated by MTT assay.

The MOS extract showed content-dependent specificity when the data were taken from three replications. A Student's *t*-test was used to understand the difference between the two treated groups and the results are presented as the mean ± standard deviation (SD).

2.4. Nuclear Breakdown of MOS Extract-Treated Cancerous Cells

Figure 5 illustrates the MOS extract-treated and untreated cell morphology that was imaged using confocal scanning microscopy (CSM). The MOS extract-treated cancer cells exhibited stronger inhibitory action (Figure 5A) than the control sample (Figure 5B). The CMS images of the nuclear cell morphology of both control (untreated) and MOS extract-treated (66 µg/mL) samples after stained by DAPI showed a substantial loss (nuclear disintegration) because of the treatments.

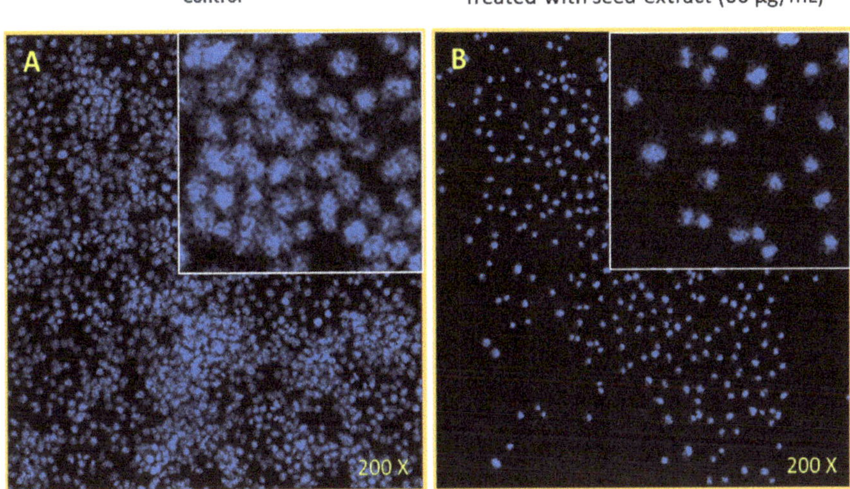

Figure 5. Confocal staining by DAPI. (**A**) The HCT-116 cells (non-treated) and (**B**) HCT-116 cells treated with MOS (66 µg/mL), 200× magnifications.

It was deduced that the MOS extract has strong anti-cancerous activities in colon cancer cells (as supported by the GC-MS analysis).

2.5. Antibacterial Efficacy of MOS Extract

The antibacterial potency of the MOS extract was assessed on Gram-negative and Gram-positive bacteria using the AWD method, wherein the inhibited areas due to antibacterial action around the inoculated wells were measured. This zone of inhibition was produced by the diffusion of the active chemical constituents present in the MOS extract. These results confirm the impact of the MOS extract on *S. aureus* and *E. coli*. Interestingly, the MOS extract was found to produce better antibacterial action in *S. aureus* bacteria than *E. coli*.

In Figure 6, the MOS extract shows concentration-dependent inhibition of *S. aureus* bacteria, with inhibition zones ranging from 18 to 24 mm. On the other hand, the inhibition zones of *E. coli* were found to range from 6 to 20 mm. The maximum and minimum inhibition zone diameters were acquired with the corresponding extract contents of 250 and 50 µg/mL (Figures 6 and 7). No inhibition was observed in *E.coli* with 50 µg/mL, whereas a higher concentration showed inhibition of the bacteria.

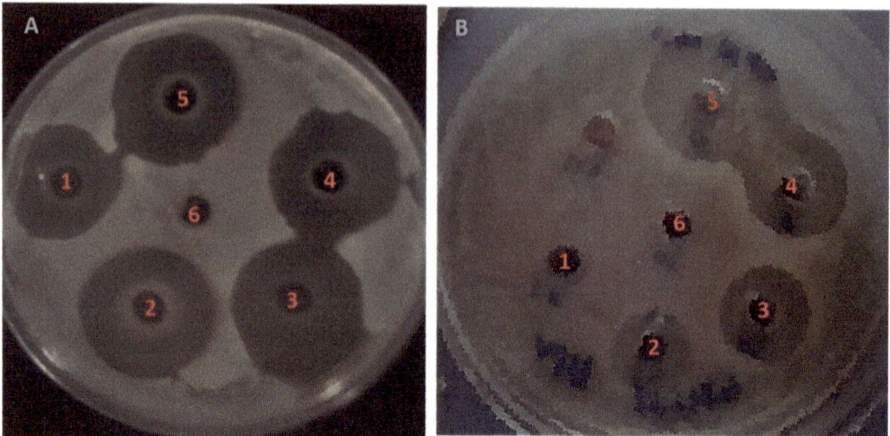

Figure 6. Agar well diffusion (AWD) plates showing the inhibition zones. (**A**) *S. aureus* and (**B**) *E. coli*. **1**: 50 µg/mL, **2**: 100 µg/mL, **3**: 150 µg/mL, **4**: 200 µg/mL, **5**: 250 µg/mL of MOS, **6**: Control (DH_2O).

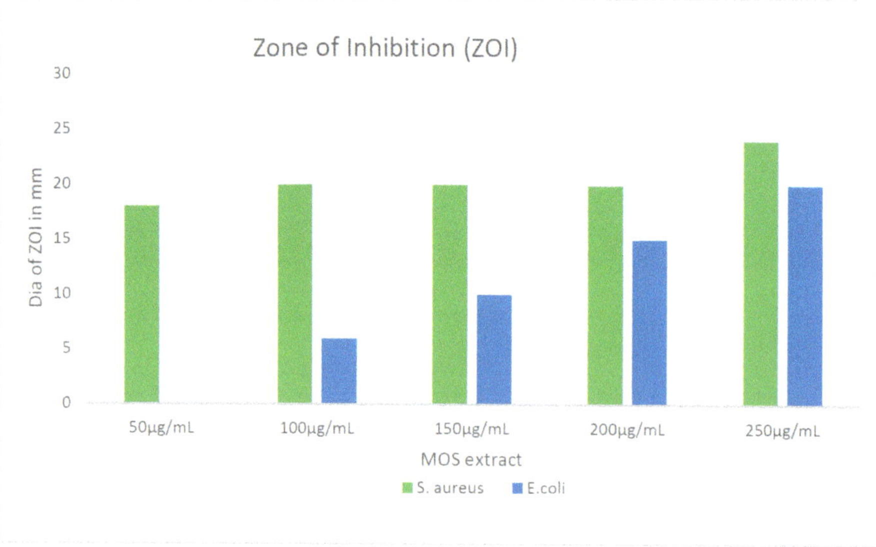

Figure 7. The graphical illustration of the zones of inhibition in millimeters (mm) examined in both *S. aureus* and *E. coli* after treatment using different concentrations of MOS extract. Data are the means ± SD of three different experiments.

The morphological changes in bacteria treated with MOS were studied using SEM. The untreated cells of *E. coli* were seen as regular rod-shaped cells with smooth cell surfaces (Figure 8a). The *E. coli* subjected to treatment with MOS appeared as damaged cells (Figure 8b) and the cell number also showed a reduction with a significant alteration of the cell wall and membrane. Similarly, the treatment of *S. aureus* with MOS extract also showed significant morphological changes in the structure and number of cells (Figure 8d), whereas the control *S. aureus* cells appeared as regular cocci with intact cell surfaces (Figure 8c). The damaged cells lost their cellular integrity, which led to the death of bacterial cells.

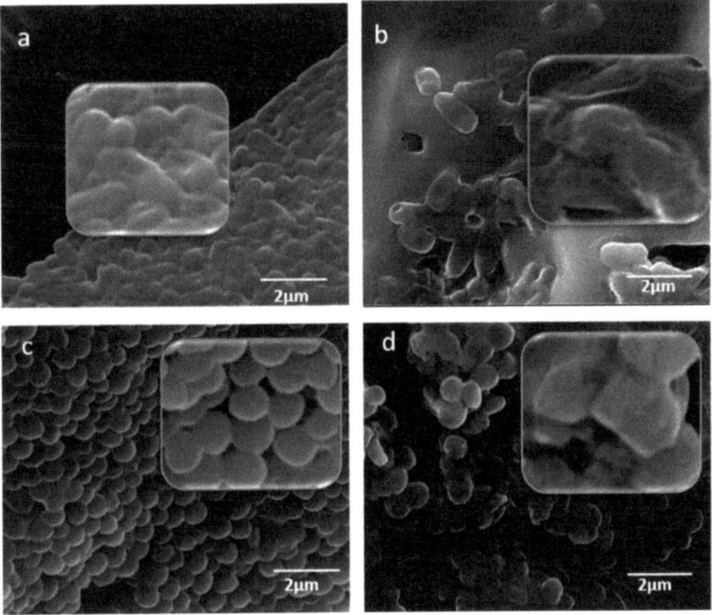

Figure 8. SEM micrographs for the study of morphogenesis by MOS extract. (**a**) *E. coli* cells (non-treated) and (**b**) treated *E. coli* cells, (**c**) *S. aureus* cells (non-treated) and (**d**) treated *S. aureus* cells.

3. Discussion

The LIBS and GC-MS techniques were utilized for the identification and quantitation of the elements existing in the MOS extract. For the first time, the anticancer and antibacterial activities of the MOS extract were assessed. Our results show the presence of diverse elements in MOSs, as confirmed by different methods. The GC-MS results of the MOS extract verified the existence of diverse anticancer and antimicrobial compounds. The spectra from different points on the surface of MOS pellets were examined by the LIBS method. The outcomes proved that MOSs are rich in different minerals that are useful to humans as food and medicine. Besides, the seeds have an influential role in antibacterial activity because of the availability of the following elements: K, Fe, P, Ca, Mg, S, Mn, Na, Se and Zn. It is clear that the amounts of Ca, K and Mg present in the MOSs were greater than the other elements. These results give evidence that MOSs are rich in different minerals, which are highly useful to humans as food supplements and medicine, indicating their remarkable impact in regulating the level of blood pressure, blood lipids, regulating the stomach function, protecting the liver, strengthening the bones, generating protein and enhancing the immunity of the human body [23–25]. Furthermore, the existence of Se in MOSs plays a vital role by protecting from fatal diseases like cancer, cardiovascular disease, cognitive decline, and thyroid disease. Moreover, these seeds exhibit a powerful antibacterial activity due to the presence of the detected elements, which is evidenced clearly by the antimicrobial studies conducted in this work.

The analysis of GC-MS of the MOSs showed the presence of 114 volatile complexes with diverse chemical groups: fatty acids, esters, ketones, alcohols, aldehydes, and hydrocarbons. To form esters, fatty acids and alcohols in plants may undergo esterification [19]. It has been reported that most of these compounds possess anticancer activities. In addition, palmitic acid shows selective cytotoxicity against human leukemic cells. The fatty acids also possess both antifungal and antibacterial activities. The omega-9 fatty acid was a detected primary complex (oleic acid) in the MOSs, which has numerous human health benefits (preventing ulcerative colitis and reducing blood pressure with remarkable antioxidant efficiency [26,27]). Therefore, GC-MS measurement proved the presence of numerous fatty acids and their related esters (cis-9-hexadecenoic acid (palmitoleic), oleic acid, octadecanoic (stearic) acids, n-hexadecenoic acid) and alcohols. It was reported that most of these compounds have anticancer activity. For example, D-allose inhibits cancer cell growth at G1 phase [17]. Palmitic acid has selective cytotoxicity against leukemic cells in humans. Fatty acids also have antifungal and antibacterial effects [28].

In the present study, the biological activities of MOS were evaluated. The cell viability of the MOS extract-treated (for 48 h) HCT-116 (cancer) cells was assessed via MTT assay. The viability status of the extract-treated HEK-293 (normal) and HCT-116 cells was determined. It was inferred that the MOS extract has strong anticancer activity in colon cancer cells. However, few reports done on MOS extract have shown that the enhanced anticancer activity of the extract correlated with the occurrence of high oleic acid and fatty acid contents [21]. Previously, it has been shown that *Moringa* plants (leaves and fruits) have been used for various applications, such as antioxidant, antimicrobial, anticancer, anti-inflammatory, antiulcer, antihypertensive, anti-urolithic, antidiabetic, anti-asthmatic, anti-aging, analgesic, diuretic, cardiovascular, hepatoprotective, hypoglycemic and immunomodulatory uses [22]. As per our knowledge, there is no information on whether *Moringa* leaf extract or seed extract cause any differential response in cancer cells. Nevertheless, it would be interesting to do a comparative study where the effects of *Moringa* seed and *Moringa* leaf extracts are examined in cancer cells. We have used HEK-293 (human embryonic kidney) cells as normal cells to compare the MOS anticancer activity to human colon cancer cell line HCT-116. The purpose was to check whether MOS extracts produce any cytotoxic effects in normal cells or not. In our studies, we have found that MOS extracts produce inhibitory action in colon cancer cells (HCT-116), whereas no inhibitory action was found using the same concentrations of MOS in HEK-293 cells (non-cancerous cells), which suggests that MOS extracts could be non-cytotoxic to normal cells.

MOS has 114 volatile complexes with diverse chemical groups: fatty acids, esters, ketones, alcohols, aldehydes and hydrocarbons. The hypotensive, antibacterial and anticancer efficacies of the MO leaves and fruits are due to presence of sundry distinctive chemical compounds (benzyl glucosinolate, niazimicin, benzyl iso-thiocyanate complexes and pterygospermin) in their structure [6]. Several methods have been improved to extract the initial contents from MOL plants for the production of food supplements medicines (with natural organic components) and the determination of their other health benefits. While we do not know the molecular mechanism of the anticancer activities of MOS extract in cancer cells, the role of the caspase signaling pathway cannot be ruled out in the process of programmed cell death. It would be interesting to study the different caspases, such as caspase-3 and caspase-9 in MOS extract-induced cell death.

The antimicrobial potency of the MOS extract was assessed in the Gram-negative and Gram-positive bacteria via AWD, wherein the inhibited areas around the inoculated wells were measured. Due to the occurrence of varying cell components, there is a discrepancy in the bactericidal action of the MOS extract for the two different types of test bacteria [29].

The study of morphogenesis using SEM showed that MOS extract affects the cell wall at the initial stage, and later penetrates and accumulates at the surface membrane. This leads to an interruption in the metabolic activities of bacterial cells and initiates cell death [30]. The present study illustrates that the MOS extract causes damage to the microbial cell surface, thereby causing significant antibacterial activity. The penetration of the extract into the bacterial cell alters the membrane integrity by structural alterations, the loss of membrane proteins, etc.

Natural compounds like fatty acids, esters, ketones, alcohols, aldehydes and hydrocarbons have been reported in a variety of plants [31]. The mechanisms of the antibacterial activities of such compounds is linked to their high affinity towards lipids due to their hydrophobic characteristics. Their antibacterial actions are evidently related to this lipophilic nature and to the bacterial membrane structure [32]. This leads the compounds to penetrate the cellular membrane of the microbial cell, which enhances the fluidity and permeability of membrane, alters the topology of membrane proteins and inflicts disruption in the respiration chain [33,34].

The present study indicated the presence of phenolic compounds and these are reported to disrupt the cell membrane, leading to the inhibition of cell metabolism causing the leakage of cellular content [35]. It has been found that phenolics inhibit processes associated with the cell membrane, for example, electron transport, ions, protein translocation, phosphorylation and other enzyme-dependent reactions. Therefore, the disturbed permeability of the cytoplasmic membrane may lead to cell death [36]. The interaction of plant bioactive compounds with bacterial cell membranes results in the inhibition of several Gram-positive and Gram-negative bacteria [37]. It has also been indicated that Gram-positive bacteria are more susceptible to the antibacterial action of natural compounds like fatty acids, esters, ketones, alcohols, aldehydes and hydrocarbons, compared to Gram-negative bacteria [38]. This is also concordant with the current study, owing to the fact that Gram-negative bacteria have an outer layer surrounding their cell wall, limiting the access of hydrophobic compounds.

Therefore, MOSs can serve as an antibacterial component that can be further studied and recommended as a potential antimicrobial and anticancer therapy. This natural source can be further evaluated and upgraded for pharmaceutical application.

4. Materials and Procedures

4.1. Seed Assembly and Extract Preparation

All the extracts used in this study were prepared from good quality MOSs procured from a place where they are naturalized, Dammam in Saudi Arabia (originally imported from India). The seeds, without a shell, were crushed to obtain fine powder, followed by compression to get pellets. Figure 9a–c shows different steps of sample production for detailed analysis with the LIBS technique. Diversified ratios of the MOS powder between 5–25 g was mingled with 220 mL of ethanol and were stirred for 5 h, then the MOS solutions were filtered and put into a rotary evaporator to get a dry powder. The samples (MOSs) were subjected to GC-MS and anticancer activity and antibacterial activity tests.

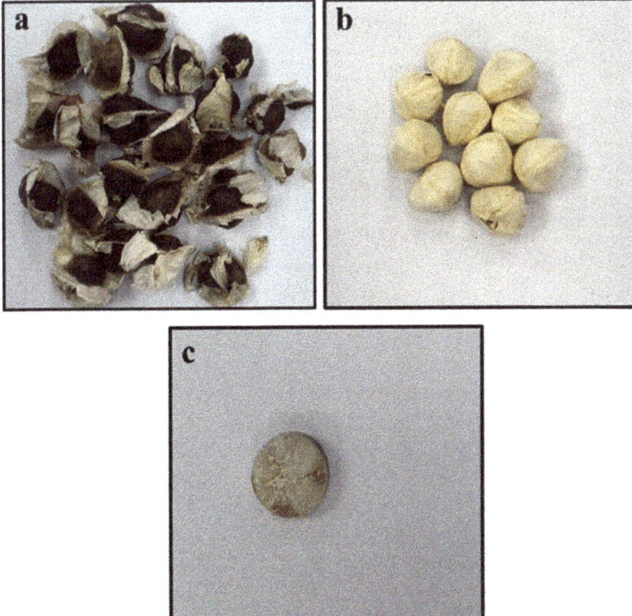

Figure 9. Pictorial view of the pelletization of *Moringa oleifera* seed samples, showing: (**a**) as-purchased *Moringa oleifera* seeds, (**b**) *Moringa oleifera* seeds without coat, (**c**) pelletized seed sample.

4.2. LIBS Setup

Figure 10 depicts the modified LIBS setup that was employed to determine the chemical constituents (elemental compositions) present in the MOS specimens. A quadrupled Q-switched Nd:YAG pulse laser (QUV-266-5) with an energy of 30 mJ, a pulse width of 8 ns, a 266 nm wavelength and a 20 Hz repetition rate was used with a UV convex lens of focal length 30 mm to focus the pulses onto the MOS pellets (which acted as the target) to ablate them. A plasma plume was generated when the target was ablated by the laser source. The emitted plasma was detected/collected using a fiber optic system positioned at 45° and the other end (500 mm) was connected to a spectrometer (Andor SR 500i-A) via grating with approximately 1200 lines per mm. An automated sample holder able to move across the plane was used to mount the pellet to avoid the formation of crusts on the surface of the target as a result of several laser pulses on the sample. The spectrum was recorded by an Intensified charge-coupled device (ICCD) camera (Model iStar 320T, 690 × 255 pixels with delay time setting at 300 ns) and data were transferred to an interfaced online computer (PC).

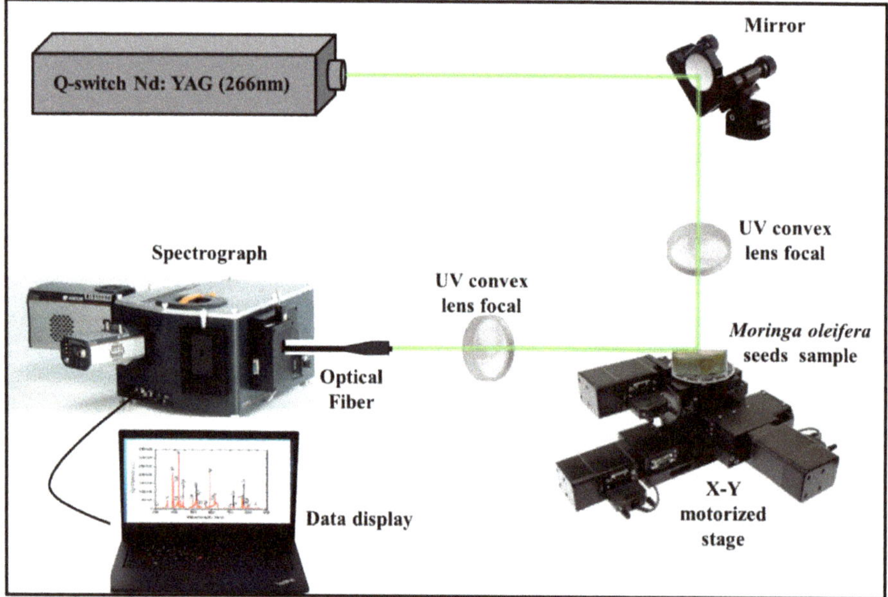

Figure 10. Schematic illustration of the LIBSsetup employed for the detection of vital elements present in MOS.

4.3. GC-MS Measurements

GC-MS (Shimadzu GC-2010 Plus) was applied for the analysis of ethanol MOS extracts that were provided by a split/splitless auto-injector (AOC-20i series) and coupled to a QP2010 Ultra single quadrupole instrument. The secession with GC was accomplished by using an Rxi-5MS fused silica capillary column (Restek, USA) with 30 m long, 0.25 mm wide and 100 μm thick films. The temperature was raised at a constant rate (5 °C/min) from the initial 60 °C (retained for 0.5 min) up to a final 280 °C (kept for 5 min). The temperature of the inlet was 270 °C (in the splitless mode) and helium (used as carrier gas of purity 99.99%) was blown through the MS transfer line (280 °C) at a 1 mL/min rate. The electron impact modes were used to operate the ion source (70 eV of energy at 250°C) and the mass spectrum (full scan) was recorded in the range of 33–550 *m*/*z*. The spectrum values were gained by regulating the GC-MS and via a GC-MS solution. The compounds of emanated volatiles were disclosed by means of the NIST-11 and WILEY-9 libraries and by calculating the relative area of each compound.

4.4. Anticancer Activity of MOS Extract

4.4.1. In Vitro Cell Viability and Cell Culture Assay

As already mentioned, the anticancer activities of the MOS extract were evaluated by treating HCT-116 and normal (HEK-293) human cell lines. The cells were obtained from Dr. Khaldoon M. Alsamman, College of Applied Medical Science, Imam Abdulrahman Bin Faisal University, Dammam, Saudi Arabia. First, cells were cultured in 96 well-plates according to the earlier specified method. Dulbecco's modified Eagle's medium (DMEM), supplemented with other reagents (selenium chloride, fetal bovine serum, L-glutamine and antibiotics), was added to the plates for cell growth [29,30].

The cells were treated with varying concentrations of MOS extracts (30–100 μg/mL) for 48 h, whereas in the control, no MOS extract was added. Equal concentrations of a solvent, dimethyl sulfoxide (DMSO) were used in both the control and MOS-treated samples. After 48 h, an MTT assay was performed for 4 h (Molecules, Wellington, New Zealand). Afterward, the growth medium was removed from the plates and (DMSO) was added to every well so the MTT formed formazan crystal. Then, the wells containing the cell cultures were checked at a wavelength of 570 nm via a micro-plate reader (Bio-Rad Labs., Boston, MA, USA). Finally, the recorded readings were analyzed through inbuilt software (Version 5.0, GraphPad Prism) and the statistical significance was studied via ANOVA tests.

4.4.2. Nuclear Staining via DAPI

The cells were treated with MOS extract (at 0.066 μg/mL) for 48 h and, in the control group, no MOS extract was added. The effects of MOS extract on cell nuclei were estimated with DAPI staining. Cold paraformaldehyde (4%) was used to pretreat the cells, which were then washed with 0.1% of Triton X-100 made from phosphate-buffered saline (PBS). At the same time, both the control and MOS- treated cells were stained using PBS with DAPI of 1.0 μg/mL concentration followed by rinsing with X-100 (0.1%). Cell morphologies of both cells (control and the treated) were analyzed *via* (CSM) (CSM-Zeiss, Frankfurt, Germany).

4.5. Antimicrobial Activity Assessment of MOS Extract

The effect of MOS extract on antibacterial efficacy at 50, 100, 150, 200 and 250 μg/mL against the *E. coli* ATCC35218 and *S. aureus* ATCC29213 strains was assessed using the AWD method. The inoculum was made from fresh bacteria grown overnight at 37 °C using nutrient broth (NB). This was followed by the preparation of inoculum to 0.5 McFarland standard and 100 μL of inoculum were disseminated over the surface of Mueller–Hinton agar (MHA) plates and dried in an aseptic environment. Next, the inoculated plates (6 mm) were punched using a sterile cork-borer. From the prepared extract, around 50 μL were added to the wells and further kept at 37 °C for an overnight incubation. Thereafter, the bacterial inhibition zone diameter over the entire well was recorded to evaluate the inhibition by the MOS extracts [29].

4.6. Antimicrobial Activity Assessment of MOS Extract Using SEM

The effect of MOS extract was additionally studied to investigate the structural damage caused to the selected bacteria by using SEM. The adjusted bacterial cell density, as described above, was treated with 250 μg/mL MOS for overnight incubation. The assay was carried out as per the protocol described by Rehman et.al [29].

4.7. Statistical Analysis

In the present study, cell viability data are presented as mean (±) standard deviation (SD), which were obtained from three independent experimental repeats. One-way ANOVA followed by

Dunnett's post hoc test with GraphPad Prism software version 5.0 (GraphPad Software, Inc., La Jolla, CA, USA) was done for the statistical analysis. $p < 0.05$ was considered to indicate a statistically significant difference.

5. Conclusions

The LIBS and GC-MS techniques were used to identify and quantify the elemental compositions of MOS extract. The antibacterial and antiproliferative effectiveness of the MOS extracts was evaluated. The LIBS spectra revealed the presence of various nutritional elements in the MOSs that are important for health. The GC-MS analysis reconfirmed the presence of several bioactive compounds in the MOS extract. The MTT assay and DAPI staining showed a significant impact of the MOS extract on the inhibition of the growth of the HCT-116 cells and the insignificant inhibitory action of the extract on the HEK-293 cells, indicating the excellent specificity of the extracts towards the cancer cells. The MOS extracts showed strong antibacterial activity in terms of the growth inhibition and morphogenic changes against *S. aureus* compared to *E. coli*, owing to their cell wall differences. Therefore, it is established that MOS extract can be a prospective antibacterial and anticancer agent for functional pharmaceutical formulations.

Author Contributions: M.A.G. conceptualized and designed the study; R.K.A., M.A.A., S.R. and F.A.K. carried out the experiments and prepared all the figures; M.A.A., S.R., A.M. and F.A.K. wrote the manuscript; S.R., M.A.A., M.A.G., F.A.K. and A.B. revised and finalized the manuscript. All authors have read and agreed to the published version of the manuscript.

Funding: This research received no external funding.

Conflicts of Interest: The authors declare no conflict of interest.

References

1. Shale, T.L.; Stirk, W.A.; van Staden, J. Screening of medicinal plants used in Lesotho for anti-bacterial and anti-inflammatory activity. *J. Ethnopharmacol.* **1999**. [CrossRef]
2. Macéé, S.R.H.; Truelstrup, H.L. Anti-bacterial activity of phenolic compounds against Streptococcus pyogenes. *Medicines* **2017**, *4*, 25. [CrossRef]
3. Nakhjavan, M.; Palethorpe, H.M.; Tomita, Y.; Smith, E.; Price, T.G.; Yool, A.J.; Pei, V.J.; Townsend, A.R.; Hardingham, G.H. Stereoselective anti-cancer activities of ginsenoside rg3 on triple negative breast cancer cell models. *Pharmaceuticals* **2019**, *12*, 117. [CrossRef]
4. Khor, K.Z.; Lim, V.; Moses, E.J.; Samad, N.A. The in Vitro and in Vivo Anticancer Properties of Moringa oleifera. *Evid.-based Complement. Altern. Med.* **2018**, *2018*, 14. [CrossRef]
5. Oduro, I.; Ellis, W.O.; Owusu, D. Nutritional potential of two leafy vegetables: Moringa oleifera and Ipomoea batatas leaves. *Sci. Res. Essays* **2008**, *3*, 57–60. [CrossRef]
6. Vergara-Jimenez, M.; Almatrafi, M.M.; Fernandez, M.L. Bioactive components in Moringa oleifera leaves protect against chronic disease. *Antioxidants* **2017**, *6*, 91. [CrossRef]
7. Natarajan, S.; Joshi, J.A. Characterisation of moringa (Moringa oleifera Lam.) genotypes for growth, pod and seed characters and seed oil using morphological and molecular markers. *Vegetos* **2015**, *28*, 64–71. [CrossRef]
8. Stohs, S.J.; Hartman, M.J. Review of the safety and efficacy of Moringa oleifera. *Phytother. Res.* **2015**, *29*, 796–804. [CrossRef]
9. Retta, N.; Awoke, T.; Mekonnen, Y. Comparison of total phenolic content, free radical scavenging potential and anti-hyperglycemic condition from leaves extract of Moringa stenopetala and Moringa oliefera. *Ethiop. J. Public Heal. Nutr.* **2019**, *1*, 20–27.
10. El-Hack, M.E.A.; Alagawany, M.; Elrys, A.S.; Desoky, E.-S.M.; Tolba, H.M.N.; Elnahal, A.S.M.; Elnesr, S.S.; Swelum, A.A. Effect of forage moringa oleifera l. (moringa) on animal health and nutrition and its beneficial applications in soil, plants and water purification. *Agriculture (Switzerland)* **2018**, *8*, 145. [CrossRef]
11. Ferreira, P.M.P.; Farias, D.F.; Oliveira, J.T.D.A.; Carvalho, A.D.F.U. Moringa oleifera: Bioactive compounds and nutritional potential. *Revista de Nutricao* **2008**, *21*. [CrossRef]
12. Anwar, F.; Zafar, S.N.; Rashid, U. Characterization of Moringa oleifera seed oil from drought and irrigated regions of Punjab, Pakistan. *Grasasy Aceites* **2006**. [CrossRef]

13. Silva, M.O.; Camacho, F.P.; Ferreira-Pinto, L.; Giufrida, W.M.; Vieira, A.M.S.; Visentaine, J.V.; Vedoy, D.R.L. Cardozo-Filho, L. Extraction and phase behaviour of Moringa oleifera seed oil using compressed propane. *Can. J. Chem. Eng.* **2016**, *94*, 2195–2201. [CrossRef]
14. Gondal, M.A.; Hussain, T.; Yamani, Z.H.; Baig, M.A. Detection of heavy metals in Arabian crude oil residue using laser induced breakdown spectroscopy. *Talanta* **2006**, *69*, 1072–1078. [CrossRef] [PubMed]
15. Sabsabi, M.; Cielo, P. Quantitative analysis of aluminum alloys by Laser-induced breakdown spectroscopy and plasma characterization. *Appl. Spectrosc.* **1995**, *49*. [CrossRef]
16. Zhao, S.; Zhang, D. An experimental investigation into the solubility of Moringa oleifera oil in supercritical carbon dioxide. *J. Food Eng.* **2014**. [CrossRef]
17. Yamaguchi, F.; Takata, M.; Kamitori, K.; Ninaka, M.; Dong, Y.; Tokuda, M. Rare sugar D-allose induces specific up-regulation of TXNIP and subsequent G1 cell cycle arrest in hepatocellular carcinoma cells by stabilization of p27kip1. *Int. J. Oncol.* **2008**, *32*, 377–385. [CrossRef]
18. Sharma, S.; Shukla, N.; Bharti, A.S.; Uttam, K.N. Simultaneous Multielemental Analysis of the Leaf of Moringa oleifera by Direct Current Arc Optical Emission Spectroscopy. *Natl. Acad. Sci. Lett.* **2018**, *41*, 65–68. [CrossRef]
19. Osuntokun, O.T.; Yusuf-Babatunde, M.A.; Fasila, O.O. Components and Bioactivity of Ipomoea batatas (L.) (Sweet Potato) Ethanolic Leaf Extract. *Asian J. Adv. Res. Rep.* **2020**, *10*, 10–26. [CrossRef]
20. Welch, R.; Tietje, A. Investigation of Moringa oleifera leaf extract and its cancer-selective antiproliferative properties. *J. South. Carolina Acad. Sci.* **2017**, *15*, 4.
21. Menendez, J.; Lupu, R. Mediterranean Dietary Traditions for the Molecular Treatment of Human Cancer: Anti-Oncogenic Actions of the Main Olive Oils Monounsaturated Fatty Acid Oleic Acid (18:1n-9). *Curr. Pharm. Biotechnol.* **2006**. [CrossRef]
22. Fahey, J. Moringa oleifera: A Review of the Medical Evidence for Its Nutritional, Therapeutic, and Prophylactic Properties. Part 1. *Trees Life J.* **2005**, *10*, 602–608. [CrossRef]
23. Zhong, J.; Wang, Y.; Yang, R.; Liu, X.; Yang, Q.; Qin, X. The application of ultrasound and microwave to increase oil extraction from *Moringa oleifera* seeds. *Ind. Crops Prod.* **2018**, *120*, 1–10. [CrossRef]
24. Rai, A.; Mohanty, B.; Bhargava, R. Experimental modeling and simulation of supercritical fluid extraction of *Moringa oleifera* seed oil by carbon dioxide. *Chem. Eng. Commun.* **2017**, *204*, 957–964. [CrossRef]
25. Belo, Y.N.; Al-Hamimi, S.; Chimuka, L.; Turner, C. Ultrahigh-pressure supercritical fluid extraction and chromatography of *Moringa oleifera* and *Moringa peregrina* seed lipids. *Anal Bioanal Chem.* **2019**, *411*, 3685–3693. [CrossRef]
26. Almessiere, M.A.; Altuwiriqi, R.; Gondal, M.A.; AlDakheel, R.K.; Alotaibi, H.F. Qualitative and quantitative analysis of human nails to find correlation between nutrients and vitamin D deficiency using LIBS and ICP-AES. *Talanta* **2018**, *185*, 61–70. [CrossRef]
27. Mehta, S.; Rai, P.K.; Rai, N.K.; Rai, A.K.; Bicanic, D.; Watal, G. Role of Spectral Studies in Detection of Antibacterial Phytoelements and Phytochemicals of Moringa oleifera. *Food Biophys.* **2011**. [CrossRef]
28. Pasquini, C.; Cortez, J.; Silva, L.M.C.; Gonzaga, F.B. Laser induced breakdown spectroscopy. *J. Braz. Chem. Soc.* **2007**, *18*, 463–512. [CrossRef]
29. Rehman, S.; Asiri, S.M.; Khan, F.A.; Jermy, B.R.; Ravinayagam, V.; Alsalem, Z.; Jindan, R.A.; Qurashi, A. Biocompatible Tin Oxide Nanoparticles: Synthesis, Antibacterial, Anticandidal and Cytotoxic Activities. *ChemistrySelect* **2019**, *4*, 4013–4017. [CrossRef]
30. Rehman, S.; Asiri, S.M.; Khan, F.A.; Jermy, B.R.; Ravinayagam, V.; Alsalem, Z.; Jindan, R.A.; Qurashi, A. Anticandidal and In Vitro Anti-Proliferative Activity of Sonochemically synthesized Indium Tin Oxide Nanoparticles. *Sci. Rep.* **2020**. [CrossRef]
31. Chávez-González, M.L.; Rodríguez-Herrera, R.; Aguilar, C.N. Essential Oils: A Natural Alternative to Combat Antibiotics Resistance. A Natural Alternative to Combat Antibiotics Resistance. *Antibiot. Resist. Mech. New Antimicrob. Approaches* **2016**, *11*, 227–235.
32. Paduch, R.; Kandefer-Szerszeń, M.; Trytek, M.; Fiedurek, J. Terpenes: Substances useful in human healthcare. *Arch. Immunol. et Ther. Exp.* **2007**, *55*, 27–315. [CrossRef]
33. Bajpai, V.K.; Baek, K.H.; Kang, S.C. Control of Salmonella in foods by using essential oils: A review. *Food Res. Int.* **2012**, *45*, 722–734. [CrossRef]
34. Friedly, E.C.; Crandall, P.G.; Ricke, S.C.; Roman, M.; O'Bryan, C.; Chalova, V.I. In vitro antilisterial effects of citrus oil fractions in combination with organic acids. *J. Food Sci.* **2009**, *74*, M67–M72. [CrossRef]

35. Trombetta, D.; Castelli, F.; Sarpietro, M.G.; Venuti, V.; Cristani, M.; Daniele, C.; Saija, A.; Mazzanti, G.; Bisignano, G. Mechanisms of antibacterial action of three monoterpenes. *Antimicrob. Agents Chemother.* **2005**. [CrossRef]
36. Nithyanand, P.; Shafreen, R.M.B.; Muthamil, S.; Murugan, R.; Pandian, S.K. Essential oils from commercial and wild Patchouli modulate Group A Streptococcal biofilms. *Ind. Crops Prod.* **2015**, *69*, 1–492. [CrossRef]
37. Tajkarimi, M.M.; Ibrahim, S.A.; Cliver, D.O. Antimicrobial herb and spice compounds in food. *Food Control* **2010**, *21*, 1199–1218. [CrossRef]
38. Calo, J.R.; Crandall, P.G.; O'Bryan, C.A.; Ricke, S.C. Essential oils as antimicrobials in food systems—A review. *Food Control* **2015**, *54*, 111–119. [CrossRef]

Article

Products Derived from *Buchenavia tetraphylla* Leaves Have In Vitro Antioxidant Activity and Protect *Tenebrio molitor* Larvae against *Escherichia coli*-Induced Injury

Tiago Fonseca Silva [1], José Robson Neves Cavalcanti Filho [1,2], Mariana Mirelle Lima Barreto Fonsêca [1,3], Natalia Medeiros dos Santos [1], Ana Carolina Barbosa da Silva [4], Adrielle Zagmignan [5,6], Afonso Gomes Abreu [6], Ana Paula Sant'Anna da Silva [1], Vera Lúcia de Menezes Lima [1], Nicácio Henrique da Silva [1], Lívia Macedo Dutra [7], Jackson Roberto Guedes da Silva Almeida [7], Márcia Vanusa da Silva [1], Maria Tereza dos Santos Correia [1,†] and Luís Cláudio Nascimento da Silva [4,6,*,†]

1 Departamento de Bioquímica, Universidade Federal de Pernambuco, Recife/PE 50670-901, Brazil; tiagotfs89@gmail.com (T.F.S.); robsonncavalcanti@gmail.com (J.R.N.C.F.); marianamirelle@gmail.com (M.M.L.B.F.); natmmedeiros@gmail.com (N.M.d.S.); annapsantanna@hotmail.com (A.P.S.d.S.); lima.vera.ufpe@gmail.com (V.L.d.M.L.); nhsilva@uol.com.br (N.H.d.S.); marciavanusa@yahoo.com.br (M.V.d.S.); mtscorreia@gmail.com (M.T.d.S.C.)
2 Faculdade de Odontologia, Universidade de Pernambuco, Camaragibe/PE 54756-220, Brazil
3 Curso de Farmácia, Faculdade Pernambucana de Saúde, Recife/PE 51150-000, Brazil
4 Curso de Biomedicina, Universidade Ceuma, São Luís/MA 65075-120, Brazil; carolinabsilva07@gmail.com
5 Curso de Nutrição, Universidade Ceuma, São Luís/MA 65075-120, Brazil; adrielle.zagmignan@ceuma.br
6 Programas de Pós-graduação, Universidade Ceuma, São Luís/MA 65075-120, Brazil; afonso.abreu@ceuma.br
7 Núcleo de Estudos e Pesquisas de Plantas Medicinais, Universidade Federal do Vale do São Francisco, Petrolina/PE 56304-205, Brazil; liviamdutra@gmail.com (L.M.D.); jackson.guedes@univasf.edu.br (J.R.G.d.S.A.)
* Correspondence: luiscn.silva@ceuma.br
† These authors contributed equally to this work.

Received: 28 December 2019; Accepted: 29 February 2020; Published: 16 March 2020

Abstract: The relevance of oxidative stress in the pathogenesis of several diseases (including inflammatory disorders) has traditionally led to the search for new sources of antioxidant compounds. In this work, we report the selection of fractions with high antioxidant action from *B. tetraphylla* (BT) leaf extracts. *In vitro* methods (DPPH and ABTS assays; determination of phenolic and flavonoid contents) were used to select products derived from *B. tetraphylla* with high antioxidant action. Then, the samples with the highest potentials were evaluated in a model of injury based on the inoculation of a lethal dose of heat-inactivated *Escherichia coli* in *Tenebrio molitor* larvae. Due to its higher antioxidant properties, the methanolic extract (BTME) was chosen to be fractionated using Sephadex LH-20 column-based chromatography. Two fractions from BTME (BTFC and BTFD) were the most active fractions. Pre-treatment with these fractions protected larvae of *T. molitor* from the stress induced by inoculation of heat-inactivated *E. coli*. Similarly, BTFC and BTFD increased the lifespan of larvae infected with a lethal dose of enteroaggregative *E. coli* 042. NMR data indicated the presence of aliphatic compounds (terpenes, fatty acids, carbohydrates) and aromatic compounds (phenolic compounds). These findings suggested that products derived from *B. tetraphylla* leaves are promising candidates for the development of antioxidant and anti-infective agents able to treat oxidative-related dysfunctions.

Keywords: oxidative stress; natural products; medicinal plants; anti-infective agents; alternative infection models

1. Introduction

A substantial amount of evidence has indicated the key role of free radicals and reactive oxygen species (ROS) in the etiology of degenerative pathologies associated with aging (Parkinson's and Alzheimer's diseases), cancer, cardiovascular diseases, and diabetes [1–7]. Free radicals are highly reactive molecules characterized by having unpaired electrons in the last valence layer, thus becoming potent oxidizing agents [8,9]. These entities are produced as a result of normal cellular metabolism and play an important role in cell function and signaling; however, they can also damage important macromolecules (DNA, proteins, and lipids), thereby impairing cellular functions and leading to cell death [2,3,8].

Due to the reactivity of free radicals, organisms have developed an efficient antioxidant defense system formed by enzymes (such as superoxide dismutase and catalase) and proteins (glutathione reductase, thioredoxin) [8,9]. However, in many situations, this system cannot cope with the overproduction of reactive species, generating a so-called oxidative stress state, which is related to the clinical manifestations described above [1]. An alternative way of combatting the damage caused by free radicals is the use of exogenous substances collectively called antioxidants [4,8,10].

The antioxidants (natural or synthetic) can act through different mechanisms in the organism, such as: (i) direct neutralization of free radicals; (ii) expression of molecules from the host antioxidant defense systems; (iii) inhibition of oxidant enzymes (leading to reduction of free radicals' generation/propagation) [10–13]. These compounds can also be used in the food industry to maintain the physical-chemical quality of fruits, meat, and other foods [14–16]. Due to the side effects of synthetic antioxidants, those from natural sources are preferred; this context leads to a constant search for plant-derived compounds with this property [4,10–12,14–18].

In addition, antioxidant compounds have been related to the therapeutic properties of the species considered medicinal. In fact, substances with antioxidant actions have been detected in different products derived from plants (juices, teas, extracts, infusions) used in the treatment and prevention of diseases [10–12,14,17]. However, the antioxidant potential of some medicinal plants is still unexploited; and the neotropical species called *Buchenavia tetraphylla* (Aubl.) RA Howard (synonymy *Buchenavia capitata*; Combretoideae family) is a good example. This plant is distributed from Cuba (Central America) to southeastern Brazil (South America). *B. tetraphylla* is popularly known as "tanimbuca" in Brazil, where it has ethnomedicinal importance for communities in the northeast region [19]. It is also known to have a broad spectrum of antimicrobial activity, inhibiting bacteria, fungi, and virus [20–22]. Buchenavianine and two derivatives (O-demethylbuchenavianine, N,O-bis-(demethyl)buchenavianine) have been isolated from *B. tetraphylla*. These compounds are classified as flavoalkaloids with a piperidine moiety at carbon 8 [20,23].

The antioxidant potential of plant products has been traditionally characterized by in vitro methods; however, the biological relevance of these tests has been contested by several works [10,24,25]. In general, in vitro methods have been limited to sample prospection and compound isolation. Thus, the need to employ cell-based methods and in vivo models for a better understanding of the pharmacological action of a candidate as an antioxidant agent is evident [10]. In this work, in vitro methods were used to select products derived from *B. tetraphylla* with antioxidant action. Then, the samples with the highest potentials were evaluated in alternative models of stress based on *Tenebrio molitor* larvae inoculated with *Escherichia coli*.

2. Results

2.1. Comparison of the Antioxidant Activity of Extracts Obtained from B. tetraphylla Leaves

Initially, the phenolic and flavonoid contents were compared in different extracts of *B. tetraphylla* (BTHE: hexane extract; BTCE: chloroform extract; BTEE: ethyl acetate extract; BTME: methanolic extract) (Table 1). Among the extracts, BTME presented higher concentrations of both classes of compounds with values of 123.03 ± 1.51 mg of gallic acid equivalent (GAE) per mg of dry extract (mg GAE/mg) and 108.90 ± 0.07 mg of quercetin equivalent (QE) per mg of dry extract (mg QE/mg) ($p < 0.05$). A Pearson coefficient of 0.71 was found between the phenolic and flavonoid contents, indicating a strong correlation.

Table 1. Comparative analysis of total phenolic compounds, flavonoid content, and DPPH radical scavenging of the crude extracts from the leaves of *Buchenavia tetraphylla*.

	BTHE	BTCE	BTEE	BTME	Trolox
Yield (%)	5.30	14.48	10.45	14.44	-
Phenolic compounds content (mg GAE/mg)	9.45 ± 1.29 [a]	26.53 ± 0.50 [b]	116.65 ± 10.26 [c]	123.03 ± 1.23 [c]	-
Flavonoid content (mg QE/mg)	10.03 ± 0.14 [a]	14.28 ± 0.48 [b]	24.92 ± 0.45 [c]	108.89 ± 0.06 [d]	-
DPPH (EC_{50} µg/mL)	6826.45	3779.98	562.75	79.04	44.10

Legend: BTHE: hexane extract; BTCE: chloroform extract; BTEE: ethyl acetate extract; BTME: methanolic extract. In each row, the values with significant differences ($p < 0.05$) are indicated by different superscript letters ([a, b, c, d]). The results are expressed as the mean ± standard deviation calculated from three independent assays performed in triplicate ($n = 3$).

The antioxidant potential of the extracts and fractions was evaluated using the DPPH (2,2-diphenyl-1-picrylhydrazyl radical) and ABTS ((2,2-azino-bis (3-ethylbenzo-thiazoline-6-sulfonic acid) radical) methods. In the DPPH assay, the highest in vitro antioxidant activity was observed for BTME with an EC_{50} (half maximal effective concentration) of 79.04 µg/mL (Table 1) and higher scavenging action in almost all tested concentrations in relation to other extracts ($p < 0.001$). The EC_{50} found for Trolox was 44.10 µg/mL (positive control). Similarly, in the ABTS assay, a greater scavenging action was observed for BTME (approximately 100%; $p < 0.05$), followed by BTEE (approximately 50%) (Figure 1A). Strong correlations were also found between the levels of phenolic compounds and the scavenger actions towards DPPH (0.96) and ABTS radicals (0.89). The flavonoid contents were moderately correlated with DPPH scavenging (0.68) and strongly correlated with ABTS scavenging (0.95). The results for both antioxidant assays were strongly related (0.85).

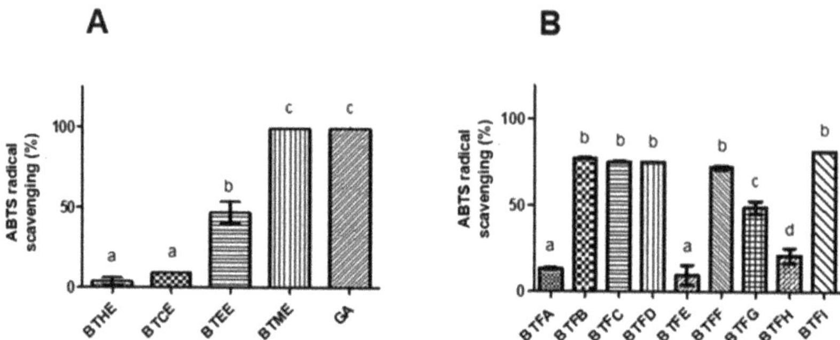

Figure 1. Comparative evaluation of antioxidant activity by the ABTS assay of the crude extracts (**A**) and fractions (**B**) of *Buchenavia tetraphylla* leaves. BTHE: hexane extract; BTCE: chloroform extract; BTEE: ethyl acetate extract; BTME: methanolic extract; BTFA: Fraction A; BTFB: Fraction B; BTFC: Fraction C; BTFD: Fraction D; BTFE: Fraction E; BTFF: Fraction F; BTFG: Fraction G; BTFH: Fraction H; BTFI: Fraction I. In each graph, values with significant differences ($p < 0.05$) are indicated by different superscript letters ($^{a, b, c}$). The results are expressed as the mean ± standard deviation calculated from three independent assays performed in triplicate ($n = 3$).

2.2. Comparison of Phenolic and Flavonoid Content and Antioxidant Activity of Fractions Obtained from Methanolic Extract of B. tetraphylla Leaves

Since a higher level of antioxidant activity was observed in BTME, it was submitted to fractionation using Sephadex LH-20 column chromatography. A total of 9 non-repetitive fractions were obtained (BTFA to BTFI). Among them, the highest phenolic content was detected in BTFD (168.99 ± 2.22 μg GAE/μg), followed by BTFC (156.02 ± 4.51 μg GAE/μg), BTFG (127.62 ± 19.11 μg GAE/μg), and BTFI (110.15 ± 0.78 μg GAE/μg) (Table 2). Almost the same pattern was observed for the flavonoid content; in this case, BTFC had the highest values (68.26 ± 2.87 μg QE/μg) ($p < 0.0001$), followed by BTFD (56.01 ± 5.54 μg QE/μg), BTFG (45.27 ± 4.13 μg QE/μg), and BTFI (39.29 ± 2.89 μg QE/μg) (Table 2). A strong correlation (Pearson coefficient of 0.88) was observed among the concentration of phenolic and flavonoid compounds in the fractions obtained from BTME. Following, the antioxidant action of each fraction was investigated, and BTFC showed the highest activity against the DPPH radical (EC_{50}: 50.41 μg/mL), followed by BTFD (EC_{50}: 237.7641 μg/mL), BTFG (EC_{50}: 294.38 μg/mL), and BTFI (EC_{50}: 376.25 μg/mL). On the other hand, the fractions BTFB, BTFC, BTFD, BTFF, and BTFI scavenged approximately 80% of the ABTS radical ($p < 0.05$), and no statistical differences were observed between them (Figure 1B).

Table 2. Comparative analysis of total phenolic compounds and flavonoid content and DPPH radical scavenging of the crude extracts from leaves of *Buchenavia tetraphylla*.

	BTFA	BTFB	BTFC	BTFD	BTFE	BTFF	BTFG	BTFH	BTFI
Yields (%)	0.10	0.29	1.22	2.53	2.04	0.96	0.46	0.56	0.98
Phenolic contents (μg GAE/μg)	49.44 ± 1.86 [a]	107.20 ± 7.23 [b]	155.67 ± 3.40 [c]	168.98 ± 1.81 [c]	49.80 ± 6.52 [a]	72.83 ± 1.13 [d]	127.62 ± 15.60 [e]	55.61 ± 3.01 [a]	110.10 ± 0.62 [b]
Flavonoid contents (μg QE/μg)	16.80 ± 1.76 [a]	12.65 ± 0.76 [a]	68.27 ± 2.35 [b]	56.01 ± 4.52 [c]	14.01 ± 1.72 [a]	4.28 ± 0.11 [d]	45.27 ± 3.37 [e]	0.86 ± 0.21 [d]	39.29 ± 2.36 [e]
DPPH (EC_{50} μg/mL)	2480.22	562.75	50.41	237.76	4132.98	2355.09	294.38	2578.18	376.25

Legend: BTFA-BTFI: Fractions obtained from *Buchenavia tetraphylla* methanolic extract. In each row, the values with significant differences ($p < 0.05$) are indicated by different superscript letters ([a, b, c, d, e]). The results are expressed as the mean ± standard deviation calculated from three independent assays performed in triplicate ($n = 3$).

2.3. Evaluation of the Hemolytic Effects of Extracts and Fractions from Buchenavia tetraphylla Leaves

Further, the hemolytic potential of each extract or fraction was evaluated using human erythrocytes (Figures 2 and 3). BTHE, BTCE, and BTEE induced toxic effects when tested at the highest concentrations (500 µg/mL and 1000 µg/mL) (Figure 2A–C). In contrast, it was observed that the BTME and its fractions did not induce significant hemolytic activity, even at the highest tested concentrations (Figures 2D and 3).

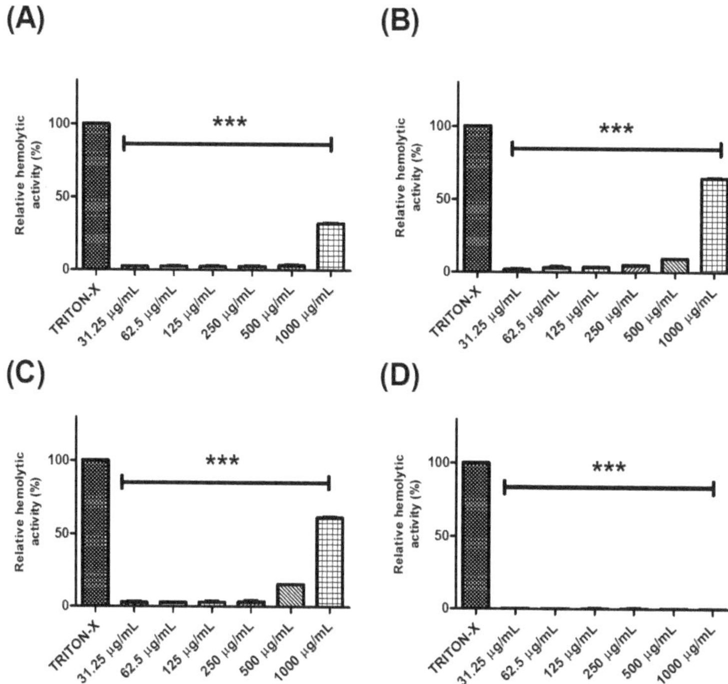

Figure 2. Hemolytic activity of the crude extracts of *Buchenavia tetraphylla* leaves. (**A**) BTHE (hexane extract); (**B**) BTCE (chloroform extract); (**C**) BTEE (ethyl acetate extract); (**D**) BTME (methanolic extract). *** Significant differences in relation to triton-X ($p < 0.0001$). The results are expressed as the mean ± standard deviation calculated from three independent assays performed in quadruplicate ($n = 4$).

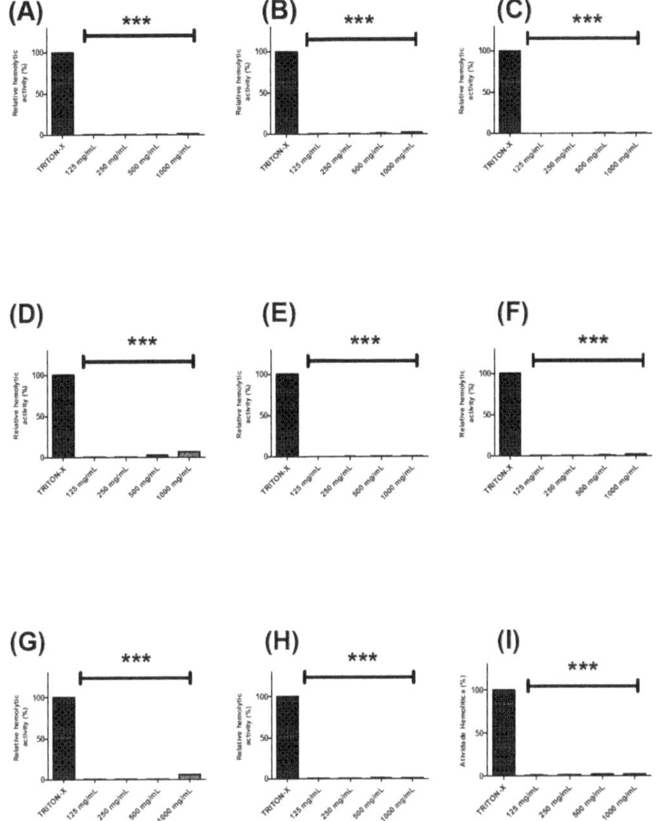

Figure 3. Hemolytic activity of the fractions obtained from the methanolic extract of *Buchenavia tetraphylla* leaves. (**A**) BTFA (Fraction A); (**B**) BTFB (Fraction B); (**C**) BTFC (Fraction C); (**D**) BTFD (Fraction D); (**E**) BTFE (Fraction E); (**F**) BTFF (Fraction F); (**G**) BTFG (Fraction G); (**H**) BTFH (Fraction H); (**I**) BTFI (Fraction I). *** Significant differences in relation to triton-X ($p < 0.0001$). The results are expressed as the mean ± standard deviation calculated from three independent assays performed in quadruplicate ($n = 4$).

2.4. Effects of Extracts and Fractions from Buchenavia tetraphylla Leaves on the Survival of Tenebrio molitor larvae Submitted to Stress Induced by Heat-Killed Escherichia coli

Based on the results presented in the above activities, we decided to evaluate the effects of the methanolic extract and its most active fractions (BTFC and BTFD) in a model of stress induced by heat-killed *E. coli* OP50 in *T. molitor* larvae. The heat treatment was used to ensure that larval death was not induced by bacterial growth; it was caused by the components present in the bacteria, such as lipopolysaccharide (LPS). Additionally, we used the nonpathogenic *E. coli* OP50 strain [26]. In this sense, at this point, we did not evaluate the antimicrobial action of the extract/fractions, but the ability of these to inhibit stress pathways induced by the presence of the bacteria. First, we evaluated the effects of several concentrations (measured at OD_{600} (optical density at 600 nm)) of heat-killed *E. coli* OP50 (data not shown). The best results were obtained with the suspension at an OD600 of 0.7; this dose induced the death of 50% of the larvae after 15 h and of 90% after 30 h. These larvae presented typical myelination points related to stress induction in this organism. The pre-treatment with BTME at 10 mg/kg was able to inhibit animal death, with survival rates of 100%, 70%, and 50% after 15 h, 30 h, and 60 h of infection, respectively. The concentration of 20 mg/kg showed no significant protective action (Figure 4A).

The results obtained with BTFD were even more significant. The group treated with 20 mg/kg exhibited viability rates of 90% and 80% after 45 h and 60 h, respectively. Similar results were obtained with BTFD at 10 mg/kg, where viability remained at 80% after 45 h and 70% after 60 h (Figure 4C). Regarding BTFC, the best protective action was observed for the dose of 10 mg/kg. In this case, survival rates remained at 70% up to 45 h and ended at 60% (60 h). In the case of the 20 mg/kg dose, after 15 h, 80% of the larvae remained viable, and this rate progressively decreased to 50% (30 h), 40% (45 h), and reached 30% after 60 h (Figure 4B).

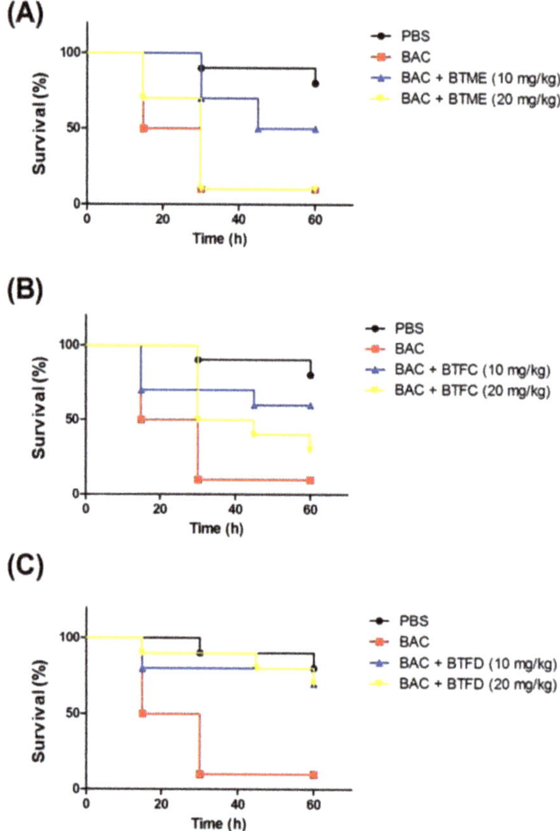

Figure 4. Effects of methanolic extract of *Buchenavia tetraphylla* leaves (BTME) and its fractions (BTFC and BTFD) on the survival of *Tenebrio molitor* larvae challenged with heat-killed *Escherichia coli* OP50. (**A**) Effects of BTME on the survival of *T. molitor* larvae challenged with heat-killed *E. coli* OP50; (**B**) effects of BTFC on the survival of *T. molitor* larvae challenged with heat-killed *E. coli* OP50; (**C**) effects of BTFD on the survival of *T. molitor* larvae challenged with heat-killed *E. coli* OP50. The larvae (n = 10/group) were pre-treated with each sample (at 10 mg/kg or 20 mg/kg) 12 h prior to inoculation of heat-killed bacteria. Larvae treated with phosphate-saline buffer (PBS) or *E. coli* OP50 (BAC) were used as negative or positive controls, respectively. In this set of assays, larvae survival was recorded each 12 h. The experiment was repeated three times.

2.5. *Effects of Fractions from Buchenavia tetraphylla Leaves in the Lifespan of Tenebrio molitor larvae Infected by Enteroaggregative Escherichia coli*

We also evaluated the efficacy of BTFC and BTFD in a model of infection provoked by Enteroaggregative *E. coli* 042. The larvae were treated with each fraction 2 h after the infection. *E. coli* 042 killed most of the larvae in less than 24 h (median survival of one day). In contrast, the group treated with *B. tetraphylla* had median survivals of two days (Figure 5). It is important to highlight that these fractions did not have antimicrobial activity towards *E. coli*.

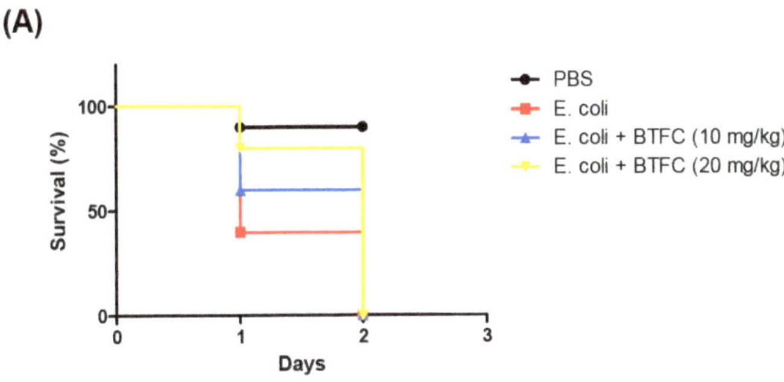

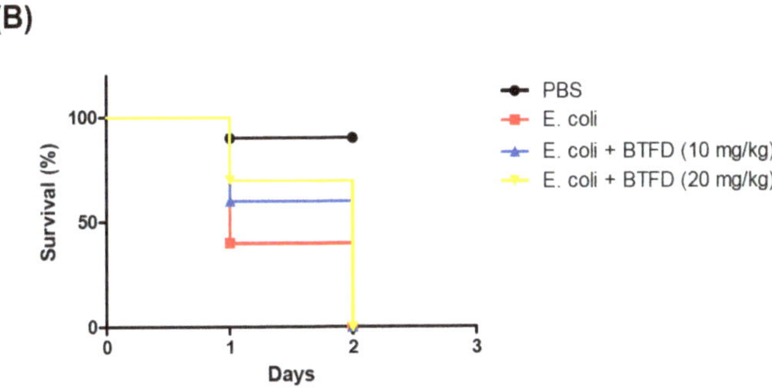

Figure 5. Effects of fractions from methanolic extract of *Buchenavia tetraphylla* leaves (BTFC and BTFD) on the survival of *Tenebrio molitor* larvae challenged with *Escherichia coli* 042. (**A**) Effects of BTFC on the survival of *Tenebrio molitor* larvae challenged with *E. coli* 042. (**B**) Effects of BTFD on the survival of *Tenebrio molitor* larvae challenged with *E. coli* 042. The larvae *(n = 10/group)* received a lethal dose of EAEC 042 and after 2 h were treated with fraction BTFC and BTFD (at 10 mg/kg or 20 mg/kg). Larvae treated with phosphate-saline buffer (PBS) or EAEC 042 were used as negative or positive controls, respectively. In this set of assays, larvae survival was recorded each 24 h. The experiment was repeated three times.

2.6. *Nuclear Magnetic Resonance Analysis of Buchenavia tetraphylla Leaves*

The ^{1}H NMR analysis and two-dimensional experiments revealed similar profiles for BTFC and BTFD. The intense overlapping made difficult the identification of metabolites; however, it was possible to suggest the presence of some classes of compounds. The fractions showed signals in spectral regions related to aliphatic and aromatic compounds (Figure 6). The signals at δ 0.50–5.90 ppm attributed to an aliphatic compound (such as terpenoids, fatty acids, carbohydrates) were more expressive in comparison to the signals at δ 6.00–8.50 ppm related to aromatic compounds (such as phenolic compounds). This was observed mainly in the BTFD fraction.

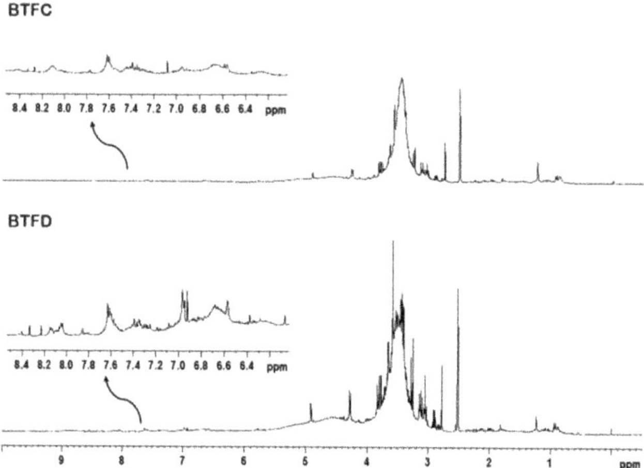

Figure 6. Representative ^{1}H NMR spectrum of the active fractions (BTFC and BTFD) obtained from the methanolic extract of *Buchenavia tetraphylla* leaves.

3. Discussion

The role of oxidative stress in the development of severe clinical conditions has promoted the search for new antioxidant agents from natural products [1,2,10,24,25]. Herein, we report the selection of high antioxidant fractions from *B. tetraphylla* leaf extracts using in vitro and in vivo approaches. Products derived from this plant have been shown as potential candidates for the development of new antimicrobial agents [20–22]; however, their antioxidant actions have not been properly addressed. The in vitro results obtained in this study showed that the methanolic extract (BTME) had a higher antioxidant activity, and this effect was correlated with its higher phenolic and flavonoid content. Among the fractions obtained from this extract, the best potentials were found for BTFC and BTFD (these fractions also exhibited greater levels of phenolic compounds). Furthermore, these agents were not toxic towards human erythrocytes.

These samples were selected to be evaluated in a model of stress induced by heat-killed *E. coli* in *T. molitor* larvae. This insect has been used as a model organism in studies of microbial pathogenesis and drug development (antimicrobial, antivirulence, and immunomodulator agents) [27–29]. Several factors have supported the use of this animal. First, *T. molitor* is susceptible to pathogens such as *Candida albicans*, *E. coli*, and *Staphylococcus aureus*, which are able to persist within the infected insect, and cause changes in tissues, hemolymph, and phagocytes [30–32]. Second, the immune system of this insect has some known signaling pathways, such as the Toll pathway, the prophenoloxidase cascade, and the autophagy pathway [33–36].

The antioxidant defense system of *T. molitor* is composed of several antioxidant and detoxifying enzymes such as superoxide dismutases, peroxidases, catalases, as well as tyrosinase, acetylcholinesterase, carboxylesterase, and glutathione S-transferase [37–39]. Previous studies have shown that during infection, the *T. molitor* larvae overproduce reactive species in response to the pathogen presence, leading to increased activity of several antioxidants and detoxifying enzymes that are correlated with larvae death [37,38,40].

The ability of *T. molitor* to produce reactive species in response to deleterious stimulus makes this insect a potential model for the study of antioxidant substances. Despite these advantages, no study has exploited *T. molitor* larvae for the screening of plant-derived antioxidant compounds. The protective action of the agent is evidenced by the increased survival of the treated larvae compared to untreated larvae. This approach using larvae could bring more information than those traditionally used for antioxidant prospecting, based on the chemical interaction of compounds and without biological relevance [10,24].

Since *T. molitor* is a multicellular organism, this approach also presents advantages over those that use cells as some insights into the toxicity of the antioxidant agent can also be assessed. Furthermore, the use of this insect offers some advantages in relation to *Caenorhabditis elegans*, the invertebrate organism traditionally used for in vivo antioxidant evaluation [41,42], such as ease of handling and direct inoculation of the compound agent in the larvae. In our in vivo model of stress induced by heat-killed *E. coli* (OP50 strain), BTFD induced higher protective effects than BTME and BTFC. However, in our assays using the live EAEC strain, the fractions had similar results (both increasing the larvae median survival in one day). It is important to highlight that BTFC and BTFD did no display antimicrobial activity *in vitro* (data not shown), suggesting that their protective effects are related to the antioxidant properties. In this sense, we showed the efficacy of two fractions rich in antioxidants to reduce the deleterious effects of *E. coli*-induced injury in *T. molitor* larvae.

Previous works reported the predominant presence of flavonoids and alkaloids in *B. tetraphylla* and other species from the same genus [20,21,43,44]. In this present research, the most bioactive fractions (BTFC and BTFD) showed a similar chemical composition with the presence of aliphatic (terpenes, fatty acids, carbohydrates) and aromatic compounds (phenolic compounds). In general, the pharmacological potentials of some terpenes are associated with their antioxidant action [45,46]. These studies were reviewed by Gonzalez-Burgos and Gomez-Serranillos [45], who highlighted some structural features involved in the antioxidant action of each type of terpene.

Among the classes of compounds detected, phenolic compounds are the most usually related to antioxidant activity. The high antioxidant abilities of phenolic compounds are related to their phenolic hydroxyl groups that can donate hydrogen atom or transfer electrons, resulting in the scavenging of harmful free radicals (such as hydroxyl radicals). The aromatic groups present in the phenolic acids can also delocalize the unpaired electron [47,48]. According to Dai and Mumper [47], the flavonoids are able to perform electron transfer (mainly due to the ortho-dihydroxy structure on the B ring) and electron delocalization (the 2,3-double bond with a 4-oxo function in the C ring, which relocates from the B ring). The authors also highlighted the essential role of 3- and 5-hydroxyl groups with the 4-oxo function in A and C rings and 3-hydroxyl groups.

4. Material and Methods

4.1. Collection and Extract Preparation

Leaves of *B. tetraphylla* were collected in November 2013, in Catimbau National Park (Pernambuco, Brazil). The samples were processed according to the taxonomic techniques, identified, and deposited in the Herbarium of Agronomic Institute of Pernambuco (voucher: IPA 80349). The leaves of *B. tetraphylla* were subjected to drying at room temperature and then pulverized using a Macsalab mill (Model 200 LAB). This material was stored in a closed, dark container until used.

For extraction, 25 g of the powder were mixed with 100 mL of the first solvent (hexane; for BTHE) on a rotary shaker table (125 rpm) at 25 °C. After 72 h, the sample was filtered, and the extracted liquid was dried in a rotary evaporator (45 rpm) at 50 °C. The remaining leaf residue was further extracted with 100 mL of chloroform (BTCE), and the above procedure was repeated completely, subsequently performed with ethyl acetate (BTEE), and finally, with methanol (BTME).

4.2. Fractionation of the Methanolic Extract

The methanolic extract (2 g) was fractionated by chromatography using a column (80 cm × 2.5 cm) incorporated with Sephadex LH-20 (GE Healthcare®, Chicago, IL, USA; 60 cm high). The systems of eluents were based on the combination of methanol and ethyl acetate (as shown below). A total of 120 fractions (7 mL each) were obtained and analyzed by fluorescent black light (25W; 127V; Empalux®, Curitiba, Brazil) and thin layer chromatography (POLYGRAM® SIL G60/UV$_{254}$; 20 × 20 cm; 0.20 mm;

Macherey-Nagel®, Düren, Germany). After these procedures, fractions with similar phytochemical profiles were gathered, resulting in 9 different fractions: BTFA (Fractions 1–4), BTFB (Fractions 5–11), BTFC (Fractions 12–18), BTFD (Fractions 19–26), BTFE (Fractions 27–36), BTFF (Fractions 37–54), BTFG (Fractions 55–61), BTFH (Fractions 62–69), and BTFI (Fractions 70–120). The elution systems (methanol:ethyl acetate; *v*/*v*) for obtaining each fraction were: 7:3 for BTFA; 6:4 for BTFB, BTFC, BTFD, BTFE, BTFF; 5:5 for BTFG; 6:4 for BTFH and BTFI.

4.3. Total Phenolic Content

Obtaining the total phenolic compounds in crude extracts and fractions was performed using the Folin-Ciocalteu reagent [49]. Samples of each extract/fraction (200 μL at 1000 μg/mL) were mixed with 1.0 mL of Folin-Ciocalteu reagent, and 800 μL of 20% sodium carbonate were added after 3 min. The mixture was incubated at room temperature, protected from light, and allowed to stand for 2 h. The absorbance of the mixture was measured at 765 nm (GeneQuant 1300, GE Healthcare). The total phenolic content was calculated in mg of gallic acid equivalent (GAE) per mg of dry extract using a calibration curve obtained with gallic acid ($y = 0.0043x + 0.0153$; $R^2 = 0.9932$). The results were expressed as mean ± standard deviation calculated from three independent assays performed in triplicate ($n = 3$).

4.4. Flavonoid Content

For flavonoid content, 100 μL (at 1000 μg/mL) of each sample were mixed with 100 μL of the reagent solution (2 g of aluminum chloride diluted in 2% ethanol solution). The mixture was incubated at room temperature and protected from light, and after 60 min, the absorbance was measured at 420 nm [50]. The content of flavonoids was calculated in mg of quercetin equivalent (QE) per mg of dry extract using a calibration curve constructed with standard quercetin solution ($y = 0.004x + 0.0121$; $R^2 = 0.993$). The results are expressed as the mean ± standard deviation calculated from three independent assays performed in triplicate ($n = 3$).

4.5. DPPH Assay

An aliquot of 250 μL of 1 mM DPPH solution (2,2-diphenyl-1-picrylhydrazyl; Sigma-Aldrich) was added to 40 μL of different sample concentrations (31.25–1000 μg/mL) and homogenized. After 30 min, the absorbance was measured at 517 nm [51]. Trolox was used as the control compound. The DPPH sequestering activity was calculated using the formula below. The results were expressed as the mean ± standard deviation calculated from three independent assays performed in triplicate ($n = 3$).

$$\text{DPPH scavenging (\%)} = (\text{Ac} - \text{As})/\text{Ac} \times 100$$

where: Ac = absorbance control; As = sample absorbance

4.6. ABTS Assay

The ABTS (2,2-azino-bis (3-ethylbenzo-thiazoline-6-sulfonic acid)) radical cation was prepared 16 h prior to the assay by mixing 5 mL of the stock solution (7 mM) with 88 μL of the 140 mM potassium persulfate solution. Aliquots (20 μL) of each extract/fraction and 2 mL of the ABTS radical were mixed, and the absorbance of the solutions was monitored at 734 nm after 6, 15, 30, 45, 60, and 120 min, respectively [52]. Gallic acid was used as the positive control. The ABTS scavenging was calculated using the formula below. The results were expressed as the mean ± standard deviation calculated from three independent assays performed in triplicate ($n = 3$).

$$\text{ABTS scavenging (\%)} = (\text{Ac} - \text{Aa})/\text{Ac} \times 100$$

where Ac (control absorbance) and Aa (sample absorbance).

4.7. Hemolytic Activity

Blood (5–10 mL) was obtained from healthy, non-smoker volunteers by venipuncture, after signing a free informed consent form. Human erythrocytes from citrated blood were immediately isolated by centrifugation at 1500 rpm for 10 min. After removal of the plasma, the erythrocytes were washed three times with phosphate-buffered saline (PBS; pH 7.4), and then, a suspension of 1% erythrocytes was prepared as the same buffer. Following, an aliquot of 1.1 mL of erythrocyte suspension was mixed with 0.4 mL of each extract/fraction (concentration range: 125 to 1000 μg/mL). The negative control and positive control received 0.4 mL of PBS and Triton X, respectively. After 60 min of incubation at room temperature, the cells were centrifuged, and the supernatant was used to measure the absorbance of hemoglobin released at 540 nm [21]. The hemolytic activity was expressed in relation to the action of Triton X-100 and calculated using the formula below. The results were expressed as the mean ± standard deviation calculated from three independent assays performed in quadruplicate ($n = 4$).

$$\text{Hemolytic activity (\%)} = [(\text{Aa} - \text{Ab}) \times 100]/(\text{Ac} - \text{Ab})$$

where: Aa is sample absorbance; Ab is the absorbance of the negative control; and Ac is the absorbance of the positive control.

4.8. Toxicity Model Using Heat-Killed E. coli

Larvae of *T. molitor* (~100 mg) were randomly allocated into groups ($n = 10$). After anesthesia and disinfection, 10 μL of the most active samples (methanolic extract or fractions C and D; at 10 mg/kg or 20 mg/kg) were injected in the ventral membrane between the second and third abdominal segments (tail to the head) [29]. One hour after the sample inoculation, the larvae received 10 μL of heat-killed *E. coli* OP50 (optical density at 600 nm: 0.7). The viability of the larvae was evaluated after 15, 30, 45, and 60 h (by evaluation the lack of movement after mechanical stimulus). Larvae inoculated with the microorganism and treated with PBS were used as the negative control; while larvae that received two doses of PBS were the positive control. The experiment was performed in three independent assays.

4.9. Infection Model Using Enteroaggregative E. coli

Larvae ($n = 10$) were prepared as described above and infected with 10 μL of enteroaggregative *E. coli* (EAEC) 042 (optical density at 600 nm: 0.1). After two hours, each animal received 10 μL of BTFC or BTFD (at 10 mg/kg or 20 mg/kg). Larvae inoculated with *E. coli* 042 and treated with PBS were used as the negative control; while larvae that received two doses of PBS were the positive control. The experiment was repeated three independent assays.

4.10. Nuclear Magnetic Resonance Analysis

The chemical characterization of the most active fractions (BTFC and BTFD) was performed by nuclear magnetic resonance (NMR) analysis. 1D and 2D NMR data were acquired at 298 K in DMSO-d_6 on a Bruker AVANCE III 400 NMR spectrometer operating at 9.4 T, observing ^{1}H and ^{13}C at 400 and 100 MHz, respectively. The NMR spectrometer was equipped with a 5 mm direct detection probe (BBO) with a *z*-gradient. One-bond (^{1}H-^{13}C HSQC) and long-range (^{1}H-^{13}C HMBC) NMR correlation experiments were optimized for average coupling constant $^1J_{(C,H)}$ and $^{LR}J_{(C,H)}$ of 140 and 8 Hz, respectively. All ^{1}H and ^{13}C NMR chemical shifts (δ) were given in ppm related to the TMS signal at 0.00 as an internal reference and the coupling constants (J) in Hz.

4.11. Statistical Analysis

The results were expressed as the mean ± standard deviation (SD). Statistical significance was determined by one-way ANOVA or two-way ANOVA followed by Tukey and Bonferroni tests. A p-value < 0.05 was considered statistically significant. Determination of EC_{50} (half maximal effective concentration) was performed by linear regression. Correlations were assessed using the Pearson coefficient. The larvae survival assays were analyzed using the Kaplan–Meier method to calculate survival fractions, and the log-rank test was used to compare survival curves.

5. Conclusions

In this study, the use of in vitro antioxidant assays allowed the selection of fractions from a methanolic extract with a high activity and low toxicity. The fractions (BTFC and BTFD) were able to extend the lifespan of *T. molitor* larvae submitted to stress induced by heat-killed *E. coli* significantly. The therapeutic treatment with these fractions had also positive effects on the infection induced by the pathogenic strain *E. coli* 042 (EAEC). ^{1}H NMR data indicated the presence of aliphatic (terpenes, fatty acids, carbohydrates) and aromatic compounds (phenolic compounds). These findings suggested that products derived from *B. tetraphylla* leaves are a promising candidate for the development of antioxidant agents able to treat the oxidative-related dysfunctions.

Author Contributions: T.F.S., M.V.d.S., M.T.d.S.C., and L.C.N.d.S. conceived of the study and performed the study design. T.F.S., V.L.d.M.L., N.H.d.S., J.R.G.d.S.A., M.V.d.S., M.T.d.S.C., and L.C.N.d.S. provided the reagents and equipment for all assays. J.R.N.C.F., M.M.L.B.F., N.M.d.S., and A.P.S.d.S. prepared the extracts and performed the in vivo and in vitro antioxidant assays. A.C.B.d.S., A.Z., and A.G.A. performed the anti-infective assays with *T. molitor*. J.R.G.d.S.A. and L.M.D. performed the chemical characterization of the extracts. All authors interpreted and discussed the results. T.F.S., J.R.G.d.S.A., L.M.D., A.C.B.d.S., M.T.d.S.C., and L.C.N.d.S. drafted and revised the manuscript. All authors approved the final version of this manuscript.

Funding: This work was funded by Fundação de Amparo à Ciência e Tecnologia de Pernambuco (FACEPE) Fundação de Amparo à Pesquisa e Desenvolvimento Científico do Maranhão (FAPEMA; BEPP-02241/18), Coordenação de Aperfeiçoamento de Pessoal de Nível Superior (CAPES), and Conselho Nacional de Desenvolvimento Científico e Tecnológico (CNPq).

Acknowledgments: The authors thank the support given by the Laboratory of Natural Products and the Laboratory of Molecular Biology, Department of Biochemistry, of the Federal University of Pernambuco (UFPE;Universidade Federal de Oer).

Conflicts of Interest: The authors declare no conflict of interest.

References

1. Poprac, P.; Jomova, K.; Simunkova, M.; Kollar, V.; Rhodes, C.J.; Valko, M. Targeting Free Radicals in Oxidative Stress-Related Human Diseases. *Trends Pharmacol. Sci.* **2017**, *38*, 592–607. [CrossRef] [PubMed]
2. Vallejo, M.J.; Salazar, L.; Grijalva, M. Oxidative Stress Modulation and ROS-Mediated Toxicity in Cancer: A review on in vitro models for plant-derived compounds. *Oxid. Med. Cell. Longev.* **2017**, *2017*, 4586068. [CrossRef] [PubMed]
3. Chikara, S.; Nagaprashantha, L.D.; Singhal, J.; Horne, D.; Awasthi, S.; Singhal, S.S. Oxidative stress and dietary phytochemicals: Role in cancer chemoprevention and treatment. *Cancer Lett.* **2018**, *413*, 122–134. [CrossRef] [PubMed]
4. De Andrade Teles, R.B.; Diniz, T.C.; Costa Pinto, T.C.; de Oliveira Junior, R.G.; Gama, E.S.M.; de Lavor, E.M.; Fernandes, A.W.C.; de Oliveira, A.P.; de Almeida Ribeiro, F.P.R.; da Silva, A.A.M.; et al. Flavonoids as therapeutic agents in alzheimer's and parkinson's diseases: A systematic review of preclinical evidences. *Oxid. Med. Cell. Longev.* **2018**, *2018*, 7043213. [CrossRef] [PubMed]
5. Reddy, P.H. Mitochondrial dysfunction and oxidative stress in asthma: Implications for mitochondria-targeted antioxidant therapeutics. *Pharmaceuticals* **2011**, *4*, 429–456. [CrossRef] [PubMed]
6. Sillar, J.R.; Germon, Z.P.; DeIuliis, G.N.; Dun, M.D. The role of reactive oxygen species in acute myeloid leukaemia. *Int. J. Mol. Sci.* **2019**, *20*, 6003. [CrossRef]
7. Teixeira, J.P.; de Castro, A.A.; Soares, F.V.; da Cunha, E.F.F.; Ramalho, T.C. Future therapeutic perspectives into the alzheimer's disease targeting the oxidative stress hypothesis. *Molecules* **2019**, *24*, 4410. [CrossRef]
8. Valko, M.; Leibfritz, D.; Moncol, J.; Cronin, M.T.; Mazur, M.; Telser, J. Free radicals and antioxidants in normal physiological functions and human disease. *Int. J. Biochem. Cell Biol.* **2007**, *39*, 44–84. [CrossRef]
9. Sies, H. Oxidative stress: A concept in redox biology and medicine. *Redox Biol.* **2015**, *4*, 180–183. [CrossRef]

10. Nascimento da Silva, L.C.; Bezerra Filho, C.M.; Paula, R.A.; Silva, E.S.C.S.; Oliveira de Souza, L.I.; Silva, M.V.; Correia, M.T.; Figueiredo, R.C. In vitro cell-based assays for evaluation of antioxidant potential of plant-derived products. *Free Radic. Res.* **2016**, *50*, 801–812. [CrossRef]
11. Halliwell, B. Free radicals and antioxidants: Updating a personal view. *Nutr. Rev.* **2012**, *70*, 257–265. [CrossRef] [PubMed]
12. Da Silva, L.C.; Alves, N.M.; de Castro, M.C.; Higino, T.M.; da Cunha, C.R.; Pereira, V.R.; da Paz, N.V.; Coelho, L.C.; Correia, M.T.; de Figueiredo, R.C. pCramoll and rCramoll as new preventive agents against the oxidative dysfunction induced by hydrogen peroxide. *Oxid. Med. Cell. Longev.* **2015**, *2015*, 520872. [PubMed]
13. Cacciatore, I.; Fornasari, E.; Baldassarre, L.; Cornacchia, C.; Fulle, S.; Di Filippo, E.S.; Pietrangelo, T.; Pinnen, F. A potent (R)-alpha-bis-lipoyl derivative containing 8-hydroxyquinoline scaffold: Synthesis and biological evaluation of its neuroprotective capabilities in sh-sy5y human neuroblastoma cells. *Pharmaceuticals* **2013**, *6*, 54–69. [CrossRef] [PubMed]
14. Marranzano, M.; Rosa, R.L.; Malaguarnera, M.; Palmeri, R.; Tessitori, M.; Barbera, A.C. Polyphenols: Plant sources and food industry applications. *Curr. Pharm. Des.* **2018**, *24*, 4125–4130. [CrossRef] [PubMed]
15. Ribeiro, J.S.; Santos, M.; Silva, L.K.R.; Pereira, L.C.L.; Santos, I.A.; da Silva Lannes, S.C.; da Silva, M.V. Natural antioxidants used in meat products: A brief review. *Meat Sci.* **2019**, *148*, 181–188. [CrossRef] [PubMed]
16. Lourenco, S.C.; Moldao-Martins, M.; Alves, V.D. Antioxidants of natural plant origins: From sources to food industry applications. *Molecules* **2019**, *24*, 4132. [CrossRef]
17. da Silva, L.C.; da Silva, C.A., Jr.; de Souza, R.M.; Jose Macedo, A.; da Silva, M.V.; dos Santos Correia, M.T. Comparative analysis of the antioxidant and DNA protection capacities of Anadenanthera colubrina, Libidibia ferrea and Pityrocarpa moniliformis fruits. *Food Chem. Toxicol.* **2011**, *49*, 2222–2228. [CrossRef]
18. Teles Fujishima, M.A.; Silva, N.; Ramos, R.D.S.; Batista Ferreira, E.F.; Santos, K.; Silva, C.; Silva, J.O.D.; Campos Rosa, J.M.; Santos, C. An Antioxidant Potential, Quantum-Chemical and Molecular Docking Study of the Major Chemical Constituents Present in the Leaves of Curatella americana Linn. *Pharmaceuticals* **2018**, *11*, 72. [CrossRef]
19. Agra, M.F.; Baracho, G.S.; Nurit, K.; Basilio, I.J.; Coelho, V.P. Medicinal and poisonous diversity of the flora of "Cariri Paraibano", Brazil. *J. Ethnopharmacol.* **2007**, *111*, 383–395. [CrossRef]
20. Beutler, J.A.; Cardellina, J.H., 2nd; McMahon, J.B.; Boyd, M.R.; Cragg, G.M. Anti-HIV and cytotoxic alkaloids from *Buchenavia capitata*. *J. Nat. Prod.* **1992**, *55*, 207–213. [CrossRef]
21. De Oliveira, Y.L.; Nascimento da Silva, L.C.; da Silva, A.G.; Macedo, A.J.; de Araujo, J.M.; Correia, M.T.; da Silva, M.V. Antimicrobial activity and phytochemical screening of *Buchenavia tetraphylla* (Aubl.) R. A. Howard (Combretaceae: Combretoideae). *Sci. World J.* **2012**, *2012*, 849302. [CrossRef] [PubMed]
22. Cavalcanti Filho, J.R.; Silva, T.F.; Nobre, W.Q.; Oliveira de Souza, L.I.; Silva, E.S.F.C.S.; Figueiredo, R.C.; de Gusmao, N.B.; Silva, M.V.; Nascimento da Silva, L.C.; Correia, M.T. Antimicrobial activity of *Buchenavia tetraphylla* against *Candida albicans* strains isolated from vaginal secretions. *Pharm. Biol.* **2017**, *55*, 1521–1527. [CrossRef] [PubMed]
23. Blair, L.M.; Calvert, M.B.; Sperry, J. Flavoalkaloids-Isolation, Biological Activity, and Total Synthesis. *Alkaloids Chem. Biol.* **2017**, *77*, 85–115. [PubMed]
24. Becker, K.; Schroecksnadel, S.; Gostner, J.; Zaknun, C.; Schennach, H.; Uberall, F.; Fuchs, D. Comparison of in vitro tests for antioxidant and immunomodulatory capacities of compounds. *Phytomedicine* **2014**, *21*, 164–171. [CrossRef]
25. Pellegrini, N.; Vitaglione, P.; Granato, D.; Fogliano, V. Twenty-five years of total antioxidant capacity measurement of foods and biological fluids: Merits and limitations. *J. Sci. Food Agric.* **2018**. [CrossRef]
26. Yun, E.J.; Lee, S.H.; Kim, S.; Kim, S.H.; Kim, K.H. Global profiling of metabolic response of Caenorhabditis elegans against Escherichia coli O157: H7. *Process. Biochem.* **2017**, *53*, 36–43. [CrossRef]
27. Canteri de Souza, P.; Custodio Caloni, C.; Wilson, D.; Sergio Almeida, R. An Invertebrate Host to Study Fungal Infections, Mycotoxins and Antifungal Drugs: *Tenebrio molitor*. *J. Fungi* **2018**, *4*, 125. [CrossRef]
28. Guo, Z.; Pfohl, K.; Karlovsky, P.; Dehne, H.W.; Altincicek, B. Dissemination of Fusarium proliferatum by mealworm beetle *Tenebrio molitor*. *PLoS ONE* **2018**, *13*, e0204602. [CrossRef]
29. Czarniewska, E.; Urbanski, A.; Chowanski, S.; Kuczer, M. The long-term immunological effects of alloferon and its analogues in the mealworm *Tenebrio molitor*. *Insect Sci.* **2018**, *25*, 429–438. [CrossRef]
30. McGonigle, J.E.; Purves, J.; Rolff, J. Intracellular survival of Staphylococcus aureus during persistent infection in the insect Tenebrio molitor. *Dev. Comp. Immunol.* **2016**, *59*, 34–38. [CrossRef]

31. Seong, J.H.; Jo, Y.H.; Seo, G.W.; Park, S.; Park, K.B.; Cho, J.H.; Ko, H.J.; Kim, C.E.; Patnaik, B.B.; Jun, S.A.; et al. Molecular Cloning and Effects of Tm14-3-3zeta-Silencing on Larval Survivability Against, *E. coli* and *C. albicans* in *Tenebrio molitor*. *Genes* **2018**, *9*, 330. [CrossRef] [PubMed]
32. De Souza, P.C.; Morey, A.T.; Castanheira, G.M.; Bocate, K.P.; Panagio, L.A.; Ito, F.A.; Furlaneto, M.C.; Yamada-Ogatta, S.F.; Costa, I.N.; Mora-Montes, H.M.; et al. *Tenebrio molitor* (Coleoptera: Tenebrionidae) as an alternative host to study fungal infections. *J. Microbiol. Methods* **2015**, *118*, 182–186. [CrossRef]
33. Johnston, P.R.; Makarova, O.; Rolff, J. Inducible defenses stay up late: Temporal patterns of immune gene expression in Tenebrio molitor. *G3* **2013**, *4*, 947–955. [CrossRef]
34. Park, S.J.; Kim, S.K.; So, Y.I.; Park, H.Y.; Li, X.H.; Yeom, D.H.; Lee, M.N.; Lee, B.L.; Lee, J.H. Protease IV, a quorum sensing-dependent protease of *Pseudomonas aeruginosa* modulates insect innate immunity. *Mol. Microbiol.* **2014**, *94*, 1298–1314. [CrossRef] [PubMed]
35. Tindwa, H.; Jo, Y.H.; Patnaik, B.B.; Lee, Y.S.; Kang, S.S.; Han, Y.S. Molecular cloning and characterization of autophagy-related gene *TmATG8* in *Listeria*-invaded hemocytes of Tenebrio molitor. *Dev. Comp. Immunol.* **2015**, *51*, 88–98. [CrossRef] [PubMed]
36. Yang, Y.T.; Lee, M.R.; Lee, S.J.; Kim, S.; Nai, Y.S.; Kim, J.S. *Tenebrio molitor* Gram-negative-binding protein 3 (*TmGNBP3*) is essential for inducing downstream antifungal Tenecin 1 gene expression against infection with *Beauveria bassiana* JEF-007. *Insect Sci.* **2018**, *25*, 969–977. [CrossRef]
37. Zhu, J.Y.; Ze, S.Z.; Stanley, D.W.; Yang, B. Parasitization by *Scleroderma guani* influences expression of superoxide dismutase genes in *Tenebrio molitor*. *Arch. Insect Biochem. Physiol.* **2014**, *87*, 40–52. [CrossRef]
38. Li, X.; Liu, Q.; Lewis, E.E.; Tarasco, E. Activity changes of antioxidant and detoxifying enzymes in *Tenebrio molitor* (Coleoptera: Tenebrionidae) larvae infected by the entomopathogenic nematode *Heterorhabditis beicherriana* (Rhabditida: Heterorhabditidae). *Parasitol. Res.* **2016**, *115*, 4485–4494. [CrossRef]
39. Gulevsky, A.K.; Relina, L.I.; Grishchenkova, Y.A. Variations of the antioxidant system during development of the cold-tolerant beetle, *Tenebrio molitor*. *Cryo Lett.* **2006**, *27*, 283–290.
40. Medina-Gomez, H.; Farriols, M.; Santos, F.; Gonzalez-Hernandez, A.; Torres-Guzman, J.C.; Lanz, H.; Contreras-Garduno, J. Pathogen-produced catalase affects immune priming: A potential pathogen strategy. *Microb. Pathog.* **2018**, *125*, 93–95. [CrossRef]
41. Peixoto, H.; Roxo, M.; Silva, E.; Valente, K.; Braun, M.; Wang, X.; Wink, M. Bark Extract of the Amazonian Tree *Endopleura uchi* (Humiriaceae) Extends Lifespan and Enhances Stress Resistance in *Caenorhabditis elegans*. *Molecules* **2019**, *24*, 915. [CrossRef] [PubMed]
42. Rohn, I.; Raschke, S.; Aschner, M.; Tuck, S.; Kuehnelt, D.; Kipp, A.; Schwerdtle, T.; Bornhorst, J. Treatment of *Caenorhabditis elegans* with small selenium species enhances antioxidant defense systems. *Mol. Nutr. Food Res.* **2019**, *63*, e1801304. [CrossRef] [PubMed]
43. Teodoro, G.R.; Brighenti, F.L.; Delbem, A.C.; Delbem, A.C.; Khouri, S.; Gontijo, A.V.; Pascoal, A.C.; Salvador, M.J.; Koga-Ito, C.Y. Antifungal activity of extracts and isolated compounds from *Buchenavia tomentosa* on *Candida albicans* and non-albicans. *Future Microbiol.* **2015**, *10*, 917–927. [CrossRef] [PubMed]
44. Teodoro, G.R.; Gontijo, A.V.L.; Salvador, M.J.; Tanaka, M.H.; Brighenti, F.L.; Delbem, A.C.B.; Delbem, A.C.B.; Koga-Ito, C.Y. Effects of acetone fraction from *Buchenavia tomentosa* aqueous extract and gallic acid on *Candida albicans* biofilms and virulence factors. *Front. Microbiol.* **2018**, *9*, 647. [CrossRef]
45. Gonzalez-Burgos, E.; Gomez-Serranillos, M.P. Terpene compounds in nature: A review of their potential antioxidant activity. *Curr. Med. Chem.* **2012**, *19*, 5319–5341. [CrossRef] [PubMed]
46. Xu, D.P.; Li, Y.; Meng, X.; Zhou, T.; Zhou, Y.; Zheng, J.; Zhang, J.J.; Li, H.B. Natural antioxidants in foods and medicinal plants: Extraction, assessment and resources. *Int. J. Mol. Sci.* **2017**, *18*, 96. [CrossRef]
47. Dai, J.; Mumper, R.J. Plant phenolics: Extraction, analysis and their antioxidant and anticancer properties. *Molecules* **2010**, *15*, 7313–7352. [CrossRef]
48. Castaneda-Arriaga, R.; Perez-Gonzalez, A.; Reina, M.; Alvarez-Idaboy, J.R.; Galano, A. Comprehensive investigation of the antioxidant and pro-oxidant effects of phenolic compounds: A double-edged sword in the context of oxidative stress? *J. Phys. Chem. B* **2018**, *122*, 6198–6214. [CrossRef]
49. Singleton, V.L.; Rossi, J.A. Colorimetry of total phenolics with phosphomolybdic-phosphotungstic acid reagents. *Am. J. Enol. Viticult.* **1965**, *16*, 144–158.
50. Woisky, R.G.; Salatino, A. Analysis of propolis: Some parameters and procedures for chemical quality control. *J. Apicult. Res.* **1998**, *37*, 99–105. [CrossRef]

51. Brand-Williams, W.; Cuvelier, M.E.; Berset, C. Use of a free radical method to evaluate antioxidant activity. *LWT Food Sci. Technol.* **1995**, *28*, 25–30. [CrossRef]
52. Re, R.; Pellegrini, N.; Proteggente, A.; Pannala, A.; Yang, M.; Rice-Evans, C. Antioxidant activity applying an improved ABTS radical cation decolorization assay. *Free Radic. Biol. Med.* **1999**, *26*, 1231–1237. [CrossRef]

Article

In Vitro Assessment of Antimicrobial, Antioxidant, and Cytotoxic Properties of Saccharin–Tetrazolyl and –Thiadiazolyl Derivatives: The Simple Dependence of the pH Value on Antimicrobial Activity

Luís M. T. Frija [1,*], Epole Ntungwe [2], Przemysław Sitarek [3], Joana M. Andrade [2], Monika Toma [4], Tomasz Śliwiński [4], Lília Cabral [5], M. Lurdes S. Cristiano [5], Patrícia Rijo [2,6,*] and Armando J. L. Pombeiro [1]

1 Centro de Química Estrutural (CQE), Instituto Superior Técnico, Universidade de Lisboa, Av. Rovisco Pais, 1049-001 Lisboa, Portugal; pombeiro@ist.utl.pt

2 CBIOS—Research Center for Health Sciences & Technologies, ULusófona de Humanidades e Tecnologias, Campo Grande 376, 1749-024 Lisboa, Portugal; ntungweepolengolle@yahoo.com (E.N.); p5319@ulusofona.pt (J.M.A.)

3 Department of Biology and Pharmaceutical Botany, Medical University of Lodz, Muszyńskiego Street 1, 90-151 Łódź, Poland; przemyslaw.sitarek@umed.lodz.pl

4 Laboratory of Medical Genetics, Faculty of Biology and Environmental Protection, University of Lodz, 90-151 Lodz, Poland; monika.toma@biol.uni.lodz.pl (M.T.); tomasz.sliwinski@biol.uni.lodz.pl (T.Ś.)

5 Department of Chemistry and Pharmacy (FCT) and Center of Marine Sciences (CCMar), Universidade do Algarve, P-8005-039 Faro, Portugal; liliacabral80@gmail.com (L.C.); mcristi@ualg.pt (M.L.S.C.)

6 iMed.ULisboa - Research Institute for Medicines, Faculdade de Farmácia, Universidade de Lisboa, Av. Prof. Gama Pinto, 1649-003 Lisboa, Portugal

* Correspondence: luisfrija@tecnico.ulisboa.pt (L.M.T.F.); patricia.rijo@ulusofona.pt (P.R.)

Received: 4 October 2019; Accepted: 7 November 2019; Published: 12 November 2019

Abstract: The antimicrobial, antioxidant, and cytotoxic activities of a series of saccharin–tetrazolyl and –thiadiazolyl analogs were examined. The assessment of the antimicrobial properties of the referred-to molecules was completed through an evaluation of minimum inhibitory concentration (MIC) and minimum bactericidal concentration (MBC) values against Gram-positive and Gram-negative bacteria and yeasts. Scrutiny of the MIC and MBC values of the compounds at pH 4.0, 7.0, and 9.0 against four Gram-positive strains revealed high values for both the MIC and MBC at pH 4.0 (ranging from 0.98 to 125 μg/mL) and moderate values at pH 7.0 and 9.0, exposing strong antimicrobial activities in an acidic medium. An antioxidant activity analysis of the molecules was performed by using the DPPH (2,2-diphenyl-1-picrylhydrazyl) method, which showed high activity for the TSMT (*N*-(1-methyl-2*H*-tetrazol-5-yl)-*N*-(1,1-dioxo-1,2-benzisothiazol-3-yl) amine, **7**) derivative (90.29% compared to a butylated hydroxytoluene positive control of 61.96%). Besides, the general toxicity of the saccharin analogs was evaluated in an *Artemia salina* model, which displayed insignificant toxicity values. In turn, upon an assessment of cell viability, all of the compounds were found to be nontoxic in range concentrations of 0–100 μg/mL in H7PX glioma cells. The tested molecules have inspiring antimicrobial and antioxidant properties that represent potential core structures in the design of new drugs for the treatment of infectious diseases.

Keywords: saccharin; tetrazole; 1,3,4-thiadiazole; H7PX glioma cells; antimicrobial screening; antioxidant capacity

1. Introduction

In 1878, 1,2-benzisothiazole-3-one 1,1-dioxide (**1**, Figure 1), commercially known as saccharin, was discovered accidentally by Fahlberg during an investigation of the oxidation of *o*-toluenesulfonamide [1,2]: it was published by Remsen and Fahlberg one year later [3]. For more than a century, saccharin has been commonly used as a noncaloric artificial sweetener in the form of its water-soluble salts (mainly sodium, ammonium, and calcium), and it is the principal sweetening component of diabetic diets. For about three decades (since reports on carcinogenicity in laboratory animals were published), the debate on its toxicity to humans has not reached a consensus [4–6]. Numerous *N*-substituted derivatives of saccharin have been assessed for in vitro biological activity [7–10]. For example, first-row transition metal saccharinates as well as dioxovanadium(VI), dioxouranium(VI), and cerium(IV) saccharinates have been classified as protease inhibitors, and several metal(II) saccharinates have displayed superoxide dismutase-like activity [11]. Besides, structure–activity relationship studies have shown that the saccharin scaffold is an effective element for the development of inhibitors of human leukocyte elastase (HLE), cathepsin G (Cat G), and proteinase 3 (PR3), as well as antimycobacterium and central nervous system agents [9,12–14]. Recently, different saccharin-based antagonists have been recognized for their interferon-signaling pathways, showing antitumor activity through the inhibition of cancer-related isoforms in humans [15]. It should also be noted that the first non-benzoannelated 4-amino-2,3-dihydroisothiazole 1,1-dioxide, which lacks a 3-oxo group, has been described and shows anti-HIV-1 activity. Additionally, saccharin and isothiazolyl derivatives have been used in agriculture as herbicides, fungicides, and pesticides [16].

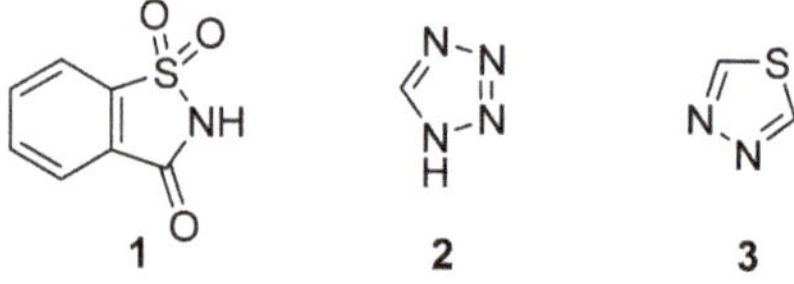

Figure 1. Structures of 1,2-benzisothiazole-3-one 1,1-dioxide (**1**, saccharin), 1*H*-tetrazole (**2**), and 1,3,4-thiadiazole (**3**).

Tetrazole (CN_4H_2) and its derivatives have attracted much attention as well due to their practical applications. The tetrazolic acid fragment $-CN_4H$ has acidity similar to the carboxylic acid group $-CO_2H$ and is almost allosteric with it, but it is metabolically more stable at the physiologic pH [17]. Hence, synthetic methodologies leading to the replacement of $-CO_2H$ groups by $-CN_4H$ groups in biologically active molecules are of major relevance [18]. Indeed, the number of patent claims and publications related to medicinal uses of tetrazolyl derivatives continues to grow rapidly and cover a wide range of applications: tetrazoles have been found, for instance, in compounds with antihypertensive, antiasthmatic, antitubercular, antimalarial, and antibiotic activity [19–22]. Several tetrazole derivatives have shown potential as anticonvulsants and anticancer and anti-HIV-1 drugs [23–25]. Tetrazoles have also had important applications in agriculture as plant growth regulators, herbicides, fungicides [26], and stabilizers in photography and photoimaging [27]. Due to the high enthalpy of formation, tetrazole decomposition results in the liberation of two nitrogen molecules and a significant amount of energy. Therefore, several tetrazole derivatives have been explored as explosives, propellant components for missiles, and gas generators for airbags (applicable to the automobile industry) [28]. In addition, various tetrazole-based compounds have good coordination properties and are able to form stable complexes with several metal ions. This ability is successfully used in analytical chemistry for the removal of heavy metal ions from liquids and in chemical systems formulated for metal protection against corrosion [29]. Many physical, chemical, physicochemical, and biological properties of tetrazoles are closely related to their ability to behave as acids and bases. In the tetrazole ring, the four nitrogen atoms connected in succession are able to be involved in proteolytic processes. This heterocyclic system is unusual in structure and unique in terms of acid–base characteristics.

In line with what is mentioned above for saccharin and tetrazole derivatives, the 1,3,4-thiadiazole scaffold represents an important class of core structures that are of great interest mainly because of their various biological activities and respective therapeutic applications. The 1,3,4-thiadiazole ring is a very weak base, possesses relatively high aromaticity, and is moderately stable in aqueous acid solutions although it is vulnerable to ring cleavage with an aqueous base [30]. Besides, this heterocyclic ring is very electron-deficient due to the electron-withdrawing effect of the nitrogen atoms and is relatively inert toward electrophilic substitution but susceptible to nucleophilic attack, whereas when substitutions are introduced into the 2′ or 5′ position of this ring, it is highly activated and readily reacts to produce varied derivatives [31]. To some extent, these specific properties lead to the application of 1,3,4-thiadiazole derivatives in pharmaceutical, agricultural, and material chemistry. Therefore, several 1,3,4-thiadiazole-based compounds display a broad spectrum of biological activities, such as antimicrobial [32], antituberculosis [33], antioxidant [34], anti-inflammatory [35], anticonvulsant [36], antidepressant, anxiolytic [37], antihypertensive [38], anticancer [39], and antifungal activity [40]. The most prominent thiadiazole derivative is possibly the acetazolamide [*N*-(5-sulfamoyl-1,3,4-thiadiazol-2-yl)acetamide], a very well-known carbonic anhydrase inhibitor that is used in the treatment of glaucoma [41], high-altitude illness [42], epileptic seizures [43], idiopathic intracranial hypertension [44], hemiplegic migraine [45], cystinuria [46], obstructive sleep apnea [47], and congenital myasthenic syndromes [48].

The search for new molecular entities, whether natural or synthetic, with relevant pharmacological activities for effective applications in medical practices continues to be a hot topic in health science as well as in science as a whole. This permanent search is well documented in the immense scientific literature and is mainly supported by the fact that many pathogenic organisms are able to develop mechanisms of resistance to high-activity medicines in their early lives [49–54]. In this context, and given our prior interest in the synthesis, reactivity, and bioactivity of tetrazole and thiazole derivatives, our collaborative interest was piqued by the great potential of these heterocycles for medical applications. Herein, we report on the antimicrobial, antioxidant, and cytotoxic activities of four mixed-azole compounds of benzisothiazole–tetrazolyl and benzisothiazole–thiadiazolyl.

2. Results and Discussion

2.1. Chemistry

The 3-chloro-1,2-benzisothiazole 1,1-dioxide (**2**), one of the strategic building blocks for the synthesis of the studied molecules, was first prepared through the halogenation of saccharin (**1**), as previously described (Scheme 1 (A)) [55]. The three saccharin–tetrazolyl analogs (TS (**5**), 2MTS (**6**), and TSMT (**7**)) were synthetized through a combination of the amino-tetrazoles **3**, **4a**, and **4b** with **2**, following a nucleophilic substitution reaction of the chloride anion by the amine functionality (Scheme 1 (B)). Similarly, compound **9** (MTSB) was prepared by coupling **2** and 5-methyl-1,3,4-thiadiazole-2-thiol (**8**) (Scheme 1 (C)). All of the reactions proceeded smoothly under experimental protocols originally developed by us [56–59], affording crystalline products in reasonable to very good yields.

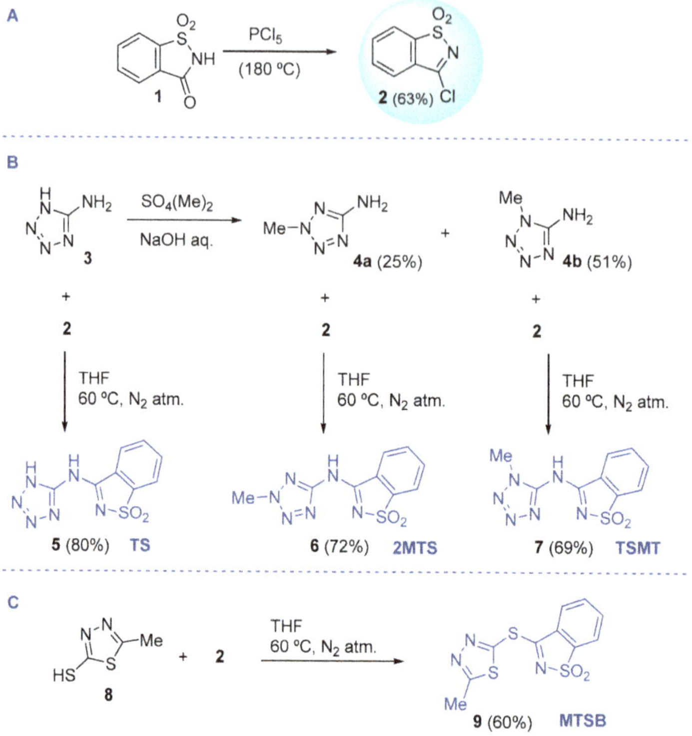

Scheme 1. Synthesis of saccharin–tetrazolyl (TS, 2MTS, and TSMT) and saccharin–thiadiazolyl (MTSB) derivatives.

2.2. Biological Assays

2.2.1. Antimicrobial Activity

The assessment of the antibacterial properties of compounds TSMT, MTSB, TS, and 2MTS was determined through minimum inhibitory concentration (MIC) and minimum bactericidal concentration (MBC) values against Gram-positive (*Staphylococcus aureus* and *Enterococcus faecalis*), Gram-negative (*Pseudomonas aeruginosa* and *Escherichia coli*), and yeast (*Saccharomyces cerevisiae* and *Candida albicans*) strains obtained from the American Type Culture Collection (ATCC) (Table 1). The MIC value corresponding to the lowest concentration at which no visible growth was observed was assessed by the microdilution method [60]. For MBC evaluation, the bacterial suspension in the wells was homogenized, serially diluted, spread in triplicate on appropriate medium, and incubated at 37 °C. All compounds and respective positive controls were tested at the same concentration of 500 μg/mL.

Table 1. Minimum inhibitory concentration (MIC) and minimum bactericidal concentration (MBC) values of TSMT, MTSB, TS, and 2MTS (obtained through the microdilution method against Gram-positive bacteria, Gram-negative bacteria, and yeast strains (in μM)).

	E. faecalis		*S. aureus*		*P. aeruginosa*		*E. coli*		*S. cerevisiae*		*C. albicans*	
Sample	**MIC**	**MBC**	**MIC**	**MBC**	**MIC**	**MBC**	**MIC**	**MBC**	**MIC**	**MBC**	**MIC**	**MBC**
TSMT	236.5	1892.1	118.3	1892.1	118.3	946.0	236.5	473.0	473.0	946.0	236.5	1892.1
MTSB	219.0	1752.1	109.5	1752.1	109.5	876.1	219.0	438.0	438.0	876.1	219.0	1752.1
TS	249.8	1998.1	124.9	1998.1	124.9	999.0	249.8	499.5	499.5	999.0	249.8	1998.1
2MTS	236.5	1892.1	118.3	1892.1	118.3	946.0	236.5	473.0	473.0	946.0	236.5	1892.1
Positivecontrol	1.95	nt	1.95	nt	0.977	nt	0.488	nt	15.6	nt	7.81	nt
	VAN		VAN		NOR		NOR		NYS		NYS	

VAN: vancomycin; NOR: norfloxacin; NYS: nystatin; nt: not tested. Data are the median values of at least three replicates.

The results of the antimicrobial assay showed that the compounds tested were bacteriostatic. This was concluded because the MBC values were much higher than the MIC values (see Table 1). The compounds were more active against the Gram-negative *P. aeruginosa* and the Gram-positive *S. aureus* bacteria. However, it seemed that the MIC and MBC values were more similar against *E. coli*. Additionally, antimicrobial tests were performed at different pH values (pH 4.0, 7.0, and 9.0), also using the microdilution method (see Tables 2–4). The microorganisms used in these tests were chosen based on the initial results from the antimicrobial screening of Gram-positive bacteria and comprised four Gram-positive strains, namely *S. aureus* CIP6538, *E. faecalis* ATCC 29212, methicillin-resistant *S. aureus* CIP106760 (MRSA), and vancomycin-resistant *E. faecalis* ATCC51299 (VRE).

Table 2. MIC and MBC values of TSMT, MTSB, TS, and 2MTS (obtained through the microdilution method against Gram-positive bacteria at pH 4.0).

	S. aureus CIP6538		*E. faecalis ATCC51299 (VRE)*		*S. aureus CIP106760 (MRSA)*		*E. faecalis 29212*	
Sample	**MIC**	**MBC**	**MIC**	**MBC**	**MIC**	**MBC**	**MIC**	**MBC**
TSMT	473.0	1892.1	236.5	1892.1	236.5	473.0	236.5	1892.1
MTSB	13.7	109.5	13.7	109.5	3.42	1.71	3.42	27.4
TS	62.5	499.5	31.2	249.8	3.90	1.95	7.80	62.5
2MTS	29.6	236.5	14.8	118.3	3.70	1.85	473.0	1892.1
Positive control	1.95	nt	1.95	nt	0.977	nt	0.488	nt
	VAN		VAN		NOR		NOR	

VAN: vancomycin; NOR: norfloxacin; NYS: nystatin; nt: not tested. Data are the median values of at least three replicates.

Table 3. MIC and MBC values of TSMT, MTSB, TS, and 2MTS (obtained through the microdilution method against Gram-positive bacteria at pH 7.0).

	S. aureus CIP6538		*E. faecalis ATCC51299 (VRE)*		*S. aureus CIP106760 (MRSA)*		*E. faecalis 29212*	
Sample	**MIC**	**MBC**	**MIC**	**MBC**	**MIC**	**MBC**	**MIC**	**MBC**
TSMT	473.0	1892.1	236.5	1892.1	473.0	1892.1	946.0	1892.1
MTSB	438.0	1752.1	219.0	1752.1	438.0	1752.1	876.1	1752.1
TS	499.5	1998.1	249.8	1998.1	499.5	1998.1	999.0	1998.1
2MTS	473.0	1892.1	236.5	1892.1	473.0	1892.1	946.0	1892.1
Positive control	1.95	nt	1.95	nt	0.977	nt	0.488	nt
	VAN		VAN		NOR		NOR	

VAN: vancomycin; NOR: norfloxacin; NYS: nystatin; nt: not tested. Data are the median values of at least three replicates.

Table 4. MIC and MBC values of TSMT, MTSB, TS, and 2MTS (obtained through the microdilution method against Gram-positive bacteria at pH 9.0).

	S. aureus CIP6538		*E. faecalis ATCC51299 (VRE)*		*S. aureus CIP106760 (MRSA)*		*E. faecalis 29212*	
Sample	**MIC**	**MBC**	**MIC**	**MBC**	**MIC**	**MBC**	**MIC**	**MBC**
TSMT	946.0	1892.1	473.0	1892.1	473.0	1892.1	473.0	1892.1
MTSB	438.0	1752.1	219.0	1752.1	438.0	1752.1	438.0	1752.1
TS	499.5	1998.1	499.5	1998.1	999.0	1998.1	499.5	1998.1
2MTS	473.0	1892.1	473.0	1892.1	473.0	1892.1	473.0	1892.1
Positive control	1.95	nt	1.95	nt	0.977	nt	0.488	nt
	VAN		VAN		NOR		NOR	

VAN: vancomycin; NOR: norfloxacin; NYS: nystatin; nt: not tested. Data are the median values of at least three replicates.

The results of the antimicrobial assays at the different pH values showed that lowering the medium pH to 4.0 had a positive effect on the compounds MTSB, TS, and 2MTS against the methicillin-resistant *S. aureus* CIP106760 (MRSA). Overall, it was attested that MTSB at pH 4.0 was the most active derivative against all Gram-positive strains. It should be noted that MTSB was the sole compound comprising the thiadiazole function, and presumably, the different activity must be correlated with the type of heterocycle ring. The results for pH 7.0 and 9.0 did not provide better results than those previously obtained.

In terms of pH-dependent antimicrobial mechanisms, it should be emphasized that several antimicrobial peptides (AMPs) have increasingly been reported as potent antibiotics that utilize pH-dependent antimicrobial mechanisms [61]. Some of these antibiotics display high pH optima related to their antimicrobial activity and show activity against microbes that present low pH optima, which reflects the acidic pH generally found at their sites of action, namely the skin. This effect should be comparable to our compounds and could be the explanation for the high antimicrobial activity of our compounds at low pH. Several pH-dependent AMPs and other antimicrobial proteins have been developed for medical purposes and have successfully gone through clinical trials, namely kappacins, LL-37, histatins, lactoferrin, and their derivatives. The major examples of the therapeutic applications of these antimicrobial compounds include wound healing as well as the treatment of multiple infections. Generally, such applications involve topical administration, a source of novel biologically active agents that could aid in the fulfilment of the urgent need for alternatives to conventional antibiotics, helping to avert a return to the pre-antibiotic era: our compounds could be a key development.

2.2.2. Antioxidant Activity

The antioxidant activity of the studied compounds was evaluated using a DPPH assay, which evaluates the potential of test samples to quench DPPH radicals via hydrogen-donating ability. The antioxidant agents convert DPPH into a stable diamagnetic molecule, 1-1diphenyl-2-picryl hydrazine, through electron or hydrogen transfers. A color change from purple to yellow indicates the increasing radical scavenging activity of the test compounds. Herein, it was observed that TSMT and MTSB derivatives possessed a high free radical scavenging ability (see Figure 2). These two molecules had the highest reducing power, indicating that they were good electron/hydrogen donors and could prevent oxidative stress.

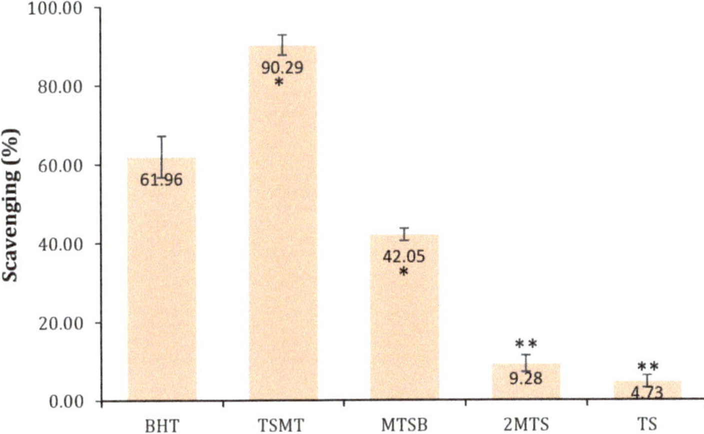

Figure 2. Antioxidant activity tested at a concentration of 10 μg/mL. The mean value ± SD was calculated from three independent experiments and compared to butylated hydroxytoluene (BHT) (* $p < 0.05$; ** $p < 0.001$).

2.2.3. Brine Shrimp Lethality Bioassay (General Toxicity)

The *Artemia salina* test is a known, simple, fast, and low-cost test and was used in this investigation to check the general toxicity of the compounds. As a general rule, it was observed that all of the compounds exhibited low toxicity in the *Artemia salina* model (Figure 3). Nevertheless, it should be emphasized that the MTSB and TSMT derivatives presented with higher toxicity, making them potential lead therapeutic agents, and thus they must further be tested using different cell- and microorganism-based assays.

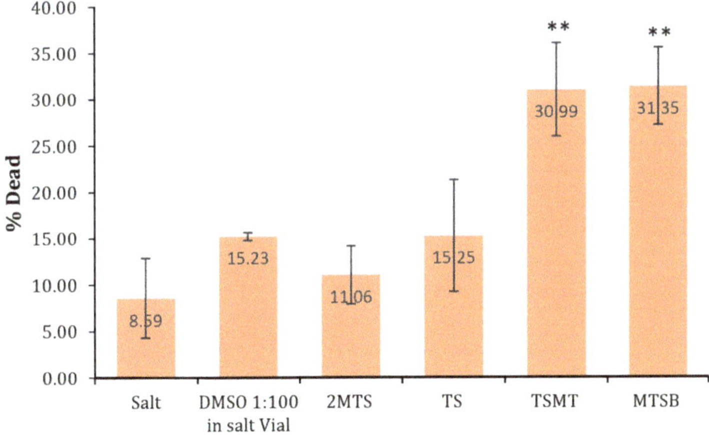

Figure 3. General toxicity screening at a concentration of 10 ppm using the *Artemia salina* test. The mean value ± SD was calculated from three independent experiments and compared to salt (** $p < 0.001$).

2.2.4. Cell Viability after Treatment with MTSB, TS, TSMS, and 2MTS in H7PX Glioma Cells (IV Grade)

In the course of this study, H7PX cells were treated with a concentration range of 0–100 μg/mL of the four compounds (MTSB, TS, TSMS, and 2MTS) over 24 h, after which the percentage of cell viability was determined by an MTT assay. It was demonstrated that none of the tested compounds reduced the viability of human H7PX cells in the stated range of concentration (0–100 μg/mL) after 24 h (Figure 4). None of the compounds' IC_{50} values were reached. Similar effects were observed with 48 h of incubation (data not shown).

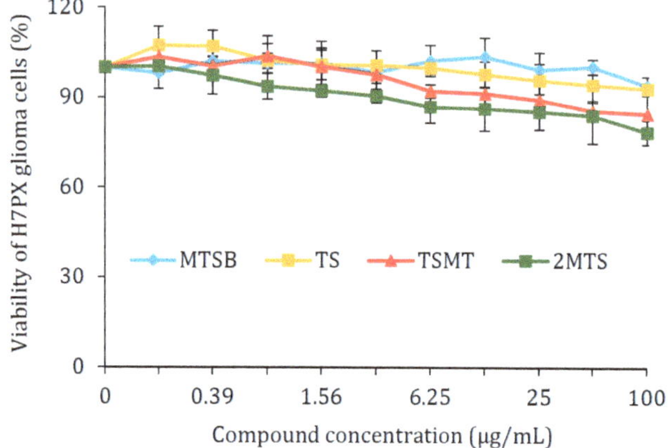

Figure 4. Effect of MTSB, TS, TSMT, and 2MTS treatment on the viability of H7PX glioma cells after 24 h of incubation (data are reported as means ± SD of three determinations).

3. Experimental Section

3.1. Chemistry

3.1.1. General

Unless indicated otherwise, solvents and starting materials were obtained from Sigma. All chemicals used were of reagent grade without further purification before use. Column chromatography was performed using silica gel 60 MN, and aluminum-backed silica gel Merck 60 F254 plates were used for analytical thin-layer chromatography (TLC). Melting points were recorded and are uncorrected. ^{1}H and ^{13}C NMR spectra were recorded at room temperature on a Bruker Avance II 400 (UltraShield™ Magnet, Billerica, MA, USA) spectrometer operating at 400 MHz (^{1}H) and 101 MHz (^{13}C). The chemical shifts are reported in ppm using TMS (tetramethylsilane) as an internal standard. Carbon, hydrogen, and nitrogen elemental analyses were carried out by the Microanalytical Service of the Instituto Superior Técnico—University of Lisbon. FT-IR spectra (4000–400 cm^{-1}) were recorded on a VERTEX 70 (Bruker, Billerica, MA, USA) spectrometer using KBr pellets. Mass spectra were obtained on a VG 7070E mass spectrometer through electron ionization (EI) at 70 eV.

3.1.2. Synthetic Protocols

The synthesis of 3-chloro-1,2-benzisothiazole-1,1-dioxide (**2**), 2-methyl-(2*H*)-tetrazole-5-amine (**4a**), 1-methyl-(1*H*)-tetrazole-5-amine (**4b**), *N*-(1*H*-tetrazol-5-yl)-*N*-(1,1-dioxo-1,2-benzisothiazol-3-yl) amine (**5**) (TS), *N*-(2-methyl-2*H*-tetrazol-5-yl)-*N*-(1,1-dioxo-1,2-benzisothiazol-3-yl) amine (**6**) (2MTS), *N*-(1-methyl-2*H*-tetrazol-5-yl)-*N*-(1,1-dioxo-1,2-benzisothiazol-3-yl) amine (**7**) (TSMT), and 3-[(5-methyl-1,3,4-thiadiazol-2-yl)sulfanyl]-1,2-benzisothiazole 1,1-dioxide (**9**) (MTSB) was carried out as previously described [55,56,58,59].

3.2. Biologic Activities

3.2.1. Antioxidant Activity (DPPH Method)

The antioxidant activity of all the compounds was measured by the DPPH method, as described by Rijo et al. [62]. Accordingly, a mixture containing 10 μL of sample and 990 μL of DPPH solution (0.002% in methanol) was incubated for 30 min at room temperature followed by absorbance measurements at 517 nm against the corresponding blank sample. The antioxidant activity of each compound was calculated using Equation (1). *AA* denotes the antioxidant activity, A_{DPPH} is the absorption of DPPH against the blank, and A_{sample} represents the absorption of the compound or control against the blank. All tests were carried out in triplicate at a sample concentration of 10 μg/mL. The reference standard used for this procedure was butylated hydroxytoluene (BHT) in the same conditions as the samples:

$$\mathrm{AA} = \frac{\mathrm{A_{DPPH}} - \mathrm{A_{Sample}}}{\mathrm{A_{DPPH}}} \times 100\%, \tag{1}$$

3.2.2. Brine Shrimp Lethality Bioassay (General Toxicity)

The general toxicity of the compounds was evaluated by the use of a test of lethality to *Artemia salina* (brine shrimp) [63]. Concentrations of 10 ppm of each sample were tested. The number of dead larvae was recorded after 24 h and was used to calculate the lethal concentration (%) according to Equation (2) (where $Total_{A.\ salina}$ = the total number of larvae in the assay and $Alive_{A.\ salina}$ = the number of alive *A. salina* larvae in the assay):

$$\text{Letal concentration} = \frac{Total_{A.\ salina} - Alive_{A.\ salina}}{Total_{A.\ salina}} \times 100\%, \tag{2}$$

3.2.3. Cell Culture

An H7PX primary glioblastoma cell line was cultivated in DMEM (Biowest, Nuaollé, France) supplemented with 10% FBS (Euroclone, Pero, Italy), 100 IU/mL penicillin (Sigma-Aldrich, Saint Louis, MO, USA), 100 μg/mL streptomycin (Sigma-Aldrich, Saint Louis, MO, USA) and 50 μg/mL gentamicin (Biowest, Nuaollé, France) in a humidified atmosphere (5% CO_2, 37 °C).

3.2.4. In Vitro Cell Viability by MTT Assay

An MTT (3-(4,5-dimethylthiazol-2-yl)-2,5-diphenyl tetrazolium bromide) assay was employed to measure the viability of H7PX cells (glioma cells in IV grade derived from patient) treated with different concentrations of TS (**5**), 2MTS (**6**), TSMT (**7**), or MTSB (**9**). Cells were seeded at 1×10^4 cells per well in 96-well culture plates and were left overnight before treatments for attachment. Subsequently, the cells were incubated for 24 h with all compounds over a range of concentrations: 0.0 (control), 0.39, 0.78, 1.56, 3.13, 6.25, 12.5, 25, 50, and 100 μg/mL. Following this, the cells were incubated with 0.5 mg/mL of MTT at 37 °C for 1.5 h. After that, the MTT was carefully removed, and DMSO (100 μL) was added to each well and vortexed at low speed for 5 min to fully dissolve the formazan crystals. Absorbance was measured at 570 nm with a reference at 630 nm using a Bio-Tek Synergy HT Microplate Reader (Bio-Tek Instruments, Winooski, VT, USA). All experiments were repeated in triplicate. Cell viability was expressed as a percentage relative to the untreated cells, which was defined as 100%. The study was approved by the Ethical Commission of the Medical University of Lodz, and informed consent was obtained from the patients (Nr. RNN/194/12/KE).

3.2.5. Antimicrobial Activity

Microorganisms and growth conditions: The strains used in this study comprised *Staphylococcus aureus* (ATCC 25,923 and CIP 106760), *Enterococcus faecalis* (ATCC 51,299 and ATCC29212), *Escherichia coli* ATCC 25922, *Pseudomonas aeruginosa* ATCC 27,853, and the yeasts *Candida albicans* ATCC 10,231 and *Saccharomyces cerevisiae* ATCC 2601. All bacteria were grown at 37 °C in Mueller–Hinton broth (Biokar Diagnostics, Beauvais, France), and the yeasts were grown in Sabouraud dextrose agar (Biokar Diagnostics, Allone, France).

Microdilution method (MIC determination): The minimum inhibitory concentrations (MICs) of the compounds (dissolved in the respective pH aqueous solutions) were evaluated using a twofold serial broth microdilution assay (CLSI, 2011) in Mueller–Hinton broth (MHB, Biokar Diagnostics, Beauvais, France). Overnight cultures were diluted in MHB with increasing concentrations of each compound (in μM). Vancomycin (VAN), norfloxacin (NOR), and nystatin (NYS) were used as positive controls for Gram-positive bacteria, Gram-negative bacteria, and yeasts, respectively. The negative control for the aqueous solutions at different pH values showed no inhibition growth. The cultures were incubated for 24 h at 37 °C, and Optical Density at 620 nm was measured using a Microplate Reader (Thermo Scientific Multiskan FC, Loughborough, UK). Assays were carried out in triplicate for each tested microorganism.

Minimum bactericidal concentration (MBC) assessment: To define the minimum bactericidal concentration (MBC) for each set of wells in the MIC determination, a loopful of agar was collected from the wells without any growth and inoculated on sterile Mueller–Hilton medium broth (for bacteria) through streaking. Plates inoculated with bacteria were incubated at 37 °C for 24 h. After incubation, the lowest concentration was noted as the MBC (for bacteria) at which no visible growth was observed.

3.2.6. Statistical Analysis

The values in this study are expressed as means ± SD. The Shapiro–Wilk test was used for verification of the normality of the data. Statistical differences were determined by one-way ANOVA. The results were analyzed using STATISTICA 12.0 software (StatSoft, Tulsa, OK, USA). Differences of $p < 0.05$ were considered statistically significant.

4. Conclusions

The antimicrobial, antioxidant, and cytotoxic activities of three saccharin–tetrazolyl (TS, TSMT, and 2MTS) derivatives and one saccharin–thiadiazolyl (MTSB) derivative were addressed throughout this investigation. The antimicrobial activity of the synthesized compounds was evaluated against a series of Gram-positive and Gram-negative bacteria and yeast strains. An evaluation of the MIC and MBC values of the four derivatives was completed at pH 4.0, 7.0, and 9.0 against four Gram-positive strains (*S. aureus*, *E. faecalis*, *S. aureus* (MRSA), and *E. faecalis* (VRE)), showing high values for the MIC and MBC at pH 4.0 (ranging from 3.42 to 473.0 μM). It was attested that the derivative MTSB, the sole compound comprising the thiadiazole function, was the most active against all of the considered Gram-positive strains at pH 4.0.

In addition, the antioxidant activity of the compounds (calculated by using the DPPH method) was the highest value for TSMT (90.29% compared to the BHT positive control of 61.96%). Finally, we demonstrated for the first time that the TS, TSMT, 2MTS, and MTSB compounds did not show in vitro cytotoxic effects on H7PX glioma cells.

The present study exposed the influence of the pH of the medium on the antimicrobial activity of the TS, TSMT, 2MTS, and MTSB compounds, which is similar to well-described antimicrobial peptide antibiotics. Therefore, the use of these kinds of molecules to produce new antibiotics should be considered in the future, although further studies are needed to confirm this hypothesis.

Author Contributions: L.M.T.F. and P.R. designed the study, analyzed the data, and wrote the paper; E.N., P.S., J.M.A., M.T., and T.Ś. performed the biological assays; L.M.T.F. and L.C. performed the synthesis and characterization of the compounds; M.L.S.C. and A.J.L.P. were responsible for supervision and reviewing the draft.

Funding: This work was partially supported by the Foundation for Science and Technology (FCT), Portugal ((UID/QUI/00100/2019) and (UID/MULTI/04326/2019 – CCMAR)). L.M.T.F. expresses gratitude to FCT for the post doc fellowship (SFRH/BPD/99851/2014) and work contract n° IST-ID/115/2018.

Conflicts of Interest: The authors declare no conflicts of interest.

References

1. Schulze, B.; Illgen, K. Isothiazol-1,1-dioxide—Vom Süßstoff zum chiralen Auxiliar in der stereoselektiven Synthese. *J. Prakt. Chem.* **1997**, *339*, 1–14. [CrossRef]
2. Ellis, J.W.J. Overview of sweeteners. *Chem. Educ.* **1995**, *72*, 671–675. [CrossRef]
3. Remsen, I.; Fahlberg, C. On the oxidation of substitution products of aromatic hydrocarbons. IV. On the oxidation of orthotoluenesulphonamide. *J. Am. Chem. Soc.* **1879**, *1*, 426–438. [CrossRef]
4. Price, M.J.; Biava, G.C.; Oser, L.B.; Vogin, E.E.; Steinfeld, J.; Ley, L.H. Bladder tumors in rats fed cyclohexylamine or high doses of a mixture of cyclamate and saccharin. *Science* **1970**, *167*, 1131–1132. [CrossRef] [PubMed]
5. Masui, T.; Mann, M.A.; Borgeson, D.C.; Garland, M.E.; Okamura, T.; Fujii, H.; Pelling, C.J.; Cohen, M.S. Sequencing analysis of HA-RAS, KI-RAS, and N-RAS genes in rat urinary-bladder tumors induced by N-[4-(5-Nitro-2-furyl)-2-thiazolyl]formamide (FANFT) and sodium saccharin. *Terat. Carcin. Mutagen.* **1993**, *13*, 225–233. [CrossRef] [PubMed]
6. Garland, M.E.; Sakata, T.; Fisher, M.J.; Masui, T.; Cohen, M.S. Influences of Diet and Strain on the Proliferative Effect on the Rat Urinary Bladder Induced by Sodium Saccharin. *Cancer Res.* **1989**, *49*, 3789–3794. [PubMed]
7. Groutas, W.C.; Houser-Archield, N.; Chong, L.S.; Venkataraman, R.; Epp, J.B.; Huang, H.; McClenahan, J.J. Efficient inhibition of human-leukocyte elastase and cathepsin-G by saccharin derivatives. *J. Med. Chem.* **1993**, *36*, 3178–3181. [CrossRef] [PubMed]
8. Groutas, W.C.; Chong, L.S.; Venkataraman, R.; Kuang, R.; Epp, J.B.; Houser-Archield, N.; Huang, H.; Hoydal, R.J. Amino acid-derivative phthalimide and saccharin derivatives as inhibitors of human leukocyte elastase, cathepsin G, and proteinase 3. *Arch. Biochem. Biophys.* **1996**, *332*, 335–340. [CrossRef] [PubMed]
9. Groutas, W.C.; Epp, J.B.; Venkataraman, R.; Kuang, R.; Truong, T.M.; McClenahan, J.J.; Prakash, O. Design, synthesis, and in vitro inhibitory activity toward human leukocyte elastase, cathepsin G, and proteinase 3 of saccharin-derived sulfones and congeners. *Bioorg. Med. Chem.* **1996**, *4*, 1393–1400. [CrossRef]
10. Elghamry, I.; Youssef, M.M.; Al-Omair, M.A.; Elsawy, H. Synthesis, antimicrobial, DNA cleavage and antioxidant activities of tricyclic sultams derived from saccharin. *Eur. J. Med. Chem.* **2017**, *139*, 107–113. [CrossRef] [PubMed]
11. Apella, M.C.; Totaro, R.; Baran, E.J. Determination of superoxide dismutase-like activity in some divalent metal saccharinates. *Biol. Trace Elem. Res.* **1993**, *37*, 293–299. [CrossRef] [PubMed]
12. Guenther, U.; Wrigge, H.; Theuerkauf, N.; Boettcher, M.F.; Wensing, G.; Zinserling, J.; Putensen, C.; Hoeft, A. Repinotan, a selective 5-HT1A-R-agonist, antagonizes morphine-induced ventilator depression in anesthetized rats. *Anesth. Analg.* **2010**, *111*, 901–907. [PubMed]
13. Malinka, W.; Ryng, S.; Sieklucka-Dziuba, M.; Rajtar, G.; Gownial, A.; Kleinrok, Z. 2-Substituted-3-oxoisothiazolo[5,4-b]pyridines as potential central nervous system and antimycobacterial agents. *Farmaco* **1998**, *53*, 504–512. [CrossRef]
14. Malinka, W.; Ryng, S.; Sieklucka-Dziuba, M.; Rajtar, G.; Gownial, A.; Kleinrok, Z. Synthesis and preliminary screening of derivatives of 2-(4-arylpiperazine-1-ylalkyl)-3-oxoisothiazolo[5,4,b]pyridines as CNS and antimycobacterial agents. *Pharmazie* **2000**, *55*, 416–425. [PubMed]
15. Csakai, A.; Smith, C.; Davis, E.; Martinko, A.; Coulp, S.; Yin, H. Saccharin derivatives as inhibitors of interferon-mediated inflammation. *J. Med. Chem.* **2014**, *57*, 5348–5355. [CrossRef] [PubMed]
16. Fischer, R.; Kretschik, O.; Schenke, T.; Schenkel, R.; Wiedemann, J.; Erdelen, C.; Loesel, P.; Drewes, M.W.; Feucht, D.; Andersch, W.W. *Ger. Offen.* DE 1999244668. 1999. *Chem. Abstr.* **2001**, *134*, 4932.
17. Singh, H.; Chawla, A.S.; Kapoor, V.K.; Paul, D.; Malhotra, R.K. Medicinal chemistry of tetrazoles. *Prog. Med. Chem.* **1980**, *17*, 151–183. [PubMed]
18. Noda, K.; Saad, Y.; Kinoshita, A.; Boyle, T.P.; Graham, R.M.; Husain, A.; Karnik, S.S. Tetrazole and carboxylate receptor antagonists bind to the same subsite by different mechanisms. *J. Biol. Chem.* **1995**, *270*, 2284–2289. [CrossRef] [PubMed]
19. Mavromoustakos, T.; Kolocouris, A.; Zervou, M.; Roumelioti, P.; Matsoukas, J.; Weisemann, R. An effort to understand the molecular basis of hypertension through the study of conformational analysis of Losartan and Sarmesin using a combination of nuclear magnetic resonance spectroscopy and theoretical calculations. *J. Med. Chem.* **1999**, *42*, 1714–1722. [CrossRef] [PubMed]

20. Toney, J.H.; Fitzgerald, P.M.D.; Grover-Sharma, N.; Olson, S.H.; May, W.J.; Sundelof, J.G.; Vanderwall, D.E.; Cleary, K.A.; Grant, S.K.; Wu, J.K.; et al. Antibiotic sensitization using biphenyl tetrazoles as potent inhibitors of *Bacteroides fragilis* metallo-β-lactamase. *Chem. Biol.* **1998**, *5*, 185–196. [CrossRef]
21. Gao, C.; Chang, L.; Xu, Z.; Yan, X.-F.; Ding, C.; Zhao, F.; Wu, X.; Feng, L.-S. Recent advances of tetrazole derivatives as potential anti-tubercular and anti-malarial agents. *Eur. J. Med. Chem.* **2019**, *163*, 404. [CrossRef] [PubMed]
22. Hashimoto, Y.; Ohashi, R.; Kurosawa, Y.; Minami, K.; Kaji, H.; Hayashida, K.; Narita, H.; Murata, S. Pharmacologic Profile of TA-606, a Novel Angiotensin II-Receptor Antagonist in the Rat. *J. Cardiovasc. Pharmacol.* **1998**, *31*, 568–575. [CrossRef] [PubMed]
23. Desarro, A.; Ammendola, D.; Zappala, M.; Grasso, S.; Desarro, G.B. Relationship between Structure and Convulsant Properties of Some b-Lactam Antibiotics following Intracerebroventricular Microinjection in Rats. *Antimicrob. Agents Chemother.* **1995**, *39*, 232–237. [CrossRef] [PubMed]
24. Tamura, Y.; Watanabe, F.; Nakatani, T.; Yasui, K.; Fuji, M.; Komurasaki, T.; Tsuzuki, H.; Maekawa, R.; Yoshioka, T.; Kawada, K.; et al. Highly Selective and Orally Active Inhibitors of Type IV Collagenase (MMP-9 and MMP-2): N-Sulfonylamino Acid Derivatives. *J. Med. Chem.* **1998**, *41*, 640–649. [CrossRef] [PubMed]
25. Abell, A.D.; Foulds, G.J. Synthesis of a cis-conformationally restricted peptide bond isostere and its application to the inhibition of the HIV-1 protease. *J. Chem. Soc. Perkin Trans.* **1997**, *1*, 2475–2482. [CrossRef]
26. Sandmann, G.; Schneider, C.; Boger, P. A New Non-Radioactive Assay of Phytoene Desaturase to Evaluate Bleaching Herbicides. *Z. Naturforsch. C* **1996**, *51*, 534–538. [CrossRef] [PubMed]
27. Koldobskii, G.I.; Ostrovskii, V.A.; Poplavskii, V.S. Advances in tetrazole chemistry. *Khim. Geterotsikl. Soedin.* **1981**, *10*, 1299.
28. Ostrovskii, V.A.; Pevzner, M.S.; Kofman, T.P.; Shcherbinin, M.B.; Tselinskii, I.V. *Targets in Heterocyclic Systems. Chemistry and Properties*; Attanasi, O.A., Spinelli, D., Eds.; Societa Chimica Italiana: Rome, Italy, 1999; Volume 3, p. 467.
29. Moore, D.S.; Robinson, S.D. Catenated Nitrogen Ligands Part II. Transition Metal Derivatives of Triazoles, Tetrazoles, Pentazoles, and Hexazine. *Adv. Inorg. Chem.* **1988**, *32*, 171–239.
30. Balaban, A.T.; Oniciu, D.C.; Katritzky, A.R. Aromaticity as a cornerstone of heterocyclic chemistry. *Chem. Rev.* **2004**, *104*, 2777–2812. [CrossRef] [PubMed]
31. Hu, Y.; Li, C.-Y.; Wang, X.-M.; Yang, Y.-H.; Zhu, H.-L. 1,3,4-Thiadiazole: Synthesis, reactions, and applications in medicinal, agricultural, and materials chemistry. *Chem. Rev.* **2014**, *114*, 5572–5610. [CrossRef] [PubMed]
32. Almajan, G.L.; Barbuceanu, S.F.; Bancescu, G.; Saramet, I.; Saramet, G.; Draghici, C. Synthesis and antimicrobial evaluation of some fused heterocyclic [1,2,4]triazolo[3,4-b][1,3,4]thiadiazole derivatives. *Eur. J. Med. Chem.* **2010**, *45*, 6139–6146. [CrossRef] [PubMed]
33. Kolavi, G.; Hegde, V.; Khazi, I.; Gadad, P. Synthesis and evaluation of antitubercular activity of imidazo[2,1-b][1,3,4]thiadiazole derivatives. *Bioorg. Med. Chem.* **2006**, *14*, 3069–3080. [CrossRef] [PubMed]
34. Khan, I.; Ali, S.; Hameed, S.; Rama, N.H.; Hussain, M.T.; Wadood, A.; Uddin, R.; Ul-Haq, Z.; Khan, A.; Ali, S.; et al. Synthesis, antioxidant activities and urease inhibition of some new 1,2,4-triazole and 1,3,4-thiadiazole derivatives. *Eur. J. Med. Chem.* **2010**, *45*, 5200–5207. [CrossRef] [PubMed]
35. Hafez, H.N.; Hegab, M.I.; Ahmed-Farag, I.S.; El-Gazzar, A.B.A. A facile regioselective synthesis of novel *spiro*-thioxanthene and *spiro*-xanthene-9′,2-[1,3,4]thiadiazole derivatives as potential analgesic and anti-inflammatory agents. *Bioorg. Med. Chem. Lett.* **2008**, *18*, 4538–4543. [CrossRef] [PubMed]
36. Jatav, V.; Mishra, P.; Kashaw, S.; Stables, J.P. CNS depressant and anticonvulsant activities of some novel 3-[5-substituted 1,3,4-thiadiazole-2-yl]-2-styryl quinazoline-4(3*H*)-ones. *Eur. J. Med. Chem.* **2008**, *43*, 1945–1954. [CrossRef] [PubMed]
37. Clerici, F.; Pocar, D.; Guido, M.; Loche, A.; Perlini, V.; Brufani, M. Synthesis of 2-amino-5-sulfanyl-1,3,4-thiadiazole derivatives and evaluation of their antidepressant and anxiolytic activity. *J. Med. Chem.* **2001**, *44*, 931–936. [CrossRef] [PubMed]
38. Hasui, T.; Matsunaga, N.; Ora, T.; Ohyabu, N.; Nishigaki, N.; Imura, Y.; Igata, Y.; Matsui, H.; Motoyaji, T.; Tanaka, T.; et al. Identification of benzoxazin-3-one derivatives as novel, potent, and selective nonsteroidal mineralocorticoid receptor antagonists. *J. Med. Chem.* **2011**, *54*, 8616–8631. [CrossRef] [PubMed]
39. Noolvi, M.N.; Patel, H.M.; Singh, N.; Gadad, A.K.; Cameotra, S.S.; Badiger, A. Synthesis and anticancer evaluation of novel 2-cyclopropylimidazo[2,1-b][1,3,4]-thiadiazole derivatives. *Eur. J. Med. Chem.* **2011**, *46*, 4411–4418. [CrossRef] [PubMed]

40. Liu, X.H.; Shi, Y.X.; Ma, Y.; Zhang, C.Y.; Dong, W.L.; Pan, L.; Wang, B.L.; Li, B.J.; Li, Z.M. Synthesis, antifungal activities and 3D-QSAR study of *N*-(5-substituted-1,3,4-thiadiazol-2-yl)cyclopropanecarboxamides. *Eur. J. Med. Chem.* **2009**, *44*, 2782–2786. [CrossRef] [PubMed]
41. Kaur, I.P.; Smitha, R.; Aggarwal, D.; Kapil, M. Acetazolamide: Future perspective in topical glaucoma therapeutics. *Int. J. Pharm.* **2002**, *248*, 1–14. [CrossRef]
42. Luks, A.M.; McIntosh, S.E.; Grissom, C.K.; Auerbach, P.S.; Rodway, G.W.; Schoene, R.B.; Zafren, K.; Hackett, P.H. Wilderness Medical Society consensus guidelines for the prevention and treatment of acute altitude illness. *Wilderness Environ. Med.* **2010**, *21*, 146–155. [CrossRef] [PubMed]
43. Wolf, P. Acute drug administration in epilepsy: A review. *CNS Neurosci. Ther.* **2011**, *17*, 442–448. [CrossRef] [PubMed]
44. Rangwala, L.M.; Liu, G.T. Pediatric idiopathic intracranial hypertension. *Surv. Ophthalmol.* **2007**, *52*, 597–617. [CrossRef] [PubMed]
45. Russell, M.B.; Ducros, A. Sporadic and familial hemiplegic migraine: Pathophysiological mechanisms, clinical characteristics, diagnosis, and management. *Lancet Neurol.* **2011**, *10*, 457–470. [CrossRef]
46. Tiselius, H.G. New horizons in the management of patients with cystinuria. *Curr. Opin. Urol.* **2010**, *20*, 169–173. [CrossRef] [PubMed]
47. Jalandhara, N.B.; Patel, A.; Arora, R.R.; Jalandhara, P. Obstructive sleep apnea: A cardiopulmonary perspective and medical therapeutics. *Am. J. Ther.* **2009**, *16*, 257–263. [CrossRef] [PubMed]
48. Schara, U.; Lochmuller, H. Therapeutic strategies in congenital myasthenic syndromes. *Neurotherapeutics* **2008**, *5*, 542–547. [CrossRef] [PubMed]
49. Superti-Furga, G.; Cochran, J.; Crews, C.M.; Frye, S.; Neubauer, G.; Prinjha, R.; Shokat, K. Where is the Future of Drug Discovery for Cancer? *Cell* **2017**, *168*, 564–565.
50. Gulçin, I. Antioxidant activity of food constituents: An overview. *Arch. Toxicol.* **2012**, *86*, 345–391. [CrossRef] [PubMed]
51. Walker, B.; Barrett, S.; Polasky, S.; Galaz, V.; Folke, C.; Engstrom, G.; Ackerman, F.; Arrow, K.; Carpenter, S.; Chopra, K.; et al. Environment. Looming global-scale failures and missing institutions. *Science* **2009**, *325*, 1345–1346. [CrossRef] [PubMed]
52. D'Costa, V.M.; King, C.E.; Kalan, L.; Morar, M.; Sung, W.W.; Schwarz, C.; Froese, D.; Zazula, G.; Calmels, F.; Debruyne, R.; et al. Antibiotic resistance is ancient. *Nature* **2011**, *477*, 457. [CrossRef] [PubMed]
53. Andersson, D.I.; Hughes, D. Antibiotic resistance and its cost: Is it possible to reverse resistance? *Nat. Rev. Microbiol.* **2010**, *8*, 260. [CrossRef] [PubMed]
54. Cassir, N.; Rolain, J.M.; Brouqui, P. A new strategy to fight antimicrobial resistance: The revival of old antibiotics. *Front. Microbiol.* **2014**, *5*, 551. [CrossRef] [PubMed]
55. Brigas, A.F.; Fonseca, C.S.C.; Johnstone, R.A.W. Preparation of 3-Chloro-1,2-Benzisothiazole 1,1-Dioxide (Pseudo-Saccharyl Chloride). *J. Chem. Res.* **2002**, *6*, 299–300. [CrossRef]
56. Frija, L.M.T.; Alegria, E.C.B.A.; Sutradhar, M.; Cristiano, M.L.S.; Ismael, A.; Kopylovich, M.N.; Pombeiro, A.J.L. Copper(II) and cobalt(II) tetrazole-saccharinate complexes as effective catalysts for oxidation of secondary alcohols. *J. Mol. Catal. A Chem.* **2016**, *425*, 283–290. [CrossRef]
57. Frija, L.M.T.; Fausto, R.; Loureiro, R.M.S.S.; Cristiano, M.L.S. Synthesis and structure of novel benzisothiazole-tetrazolyl derivatives for potential application as nitrogen ligands. *J. Mol. Catal. A Chem.* **2009**, *305*, 142–146. [CrossRef]
58. Ismael, A.; Paixão, J.A.; Fausto, R.; Cristiano, M.L.S. Molecular structure of nitrogen-linked methyltetrazole-saccharinates. *J. Mol. Struct.* **2012**, *1023*, 128–142. [CrossRef]
59. Cabral, L.; Brás, E.; Henriques, M.; Marques, C.; Frija, L.M.T.; Barreira, L.; Paixão, J.A.; Fausto, R.; Cristiano, M.L.S. Synthesis, Structure, and Cytotoxicity of a New Sulphanyl-Bridged Thiadiazolyl-Saccharinate Conjugate: The Relevance of S···N Interaction. *Chem. Eur. J.* **2018**, *24*, 3251–3262. [CrossRef] [PubMed]
60. Rijo, P.; Duarte, A.; Francisco, A.P.; Semedo-Lemsaddek, T.; Simões, M.F. In vitro antimicrobial activity of royleanone derivatives against Gram-positive bacterial pathogens. *Phytother. Res.* **2014**, *8*, 76–81. [CrossRef] [PubMed]
61. Malik, E.; Dennison, S.R.; Harris, F.; Phoenix, D.A. pH Dependent Antimicrobial Peptides and Proteins, Their Mechanisms of Action and Potential as Therapeutic Agents. *Pharmaceuticals* **2016**, *9*, 67. [CrossRef] [PubMed]

62. Rijo, P.; Falé, P.L.; Serralheiro, M.L.; Simões, M.F.; Gomes, A.; Reis, C. Optimization of medicinal plant extraction methods and their encapsulation through extrusion technology. *Measurement* **2014**, *58*, 249–255. [CrossRef]
63. Alanís-Garza, B.A.; González-González, G.M.; Salazar-Aranda, R.; Waksman de Torres, N.; Rivas-Galindo, V.M. Screening of antifungal activity of plants from the northeast of Mexico. *J. Ethnopharmacol.* **2007**, *114*, 468–471. [CrossRef] [PubMed]

MDPI
St. Alban-Anlage 66
4052 Basel
Switzerland
Tel. +41 61 683 77 34
Fax +41 61 302 89 18
www.mdpi.com

Pharmaceuticals Editorial Office
E-mail: pharmaceuticals@mdpi.com
www.mdpi.com/journal/pharmaceuticals

www.ingramcontent.com/pod-product-compliance
Lightning Source LLC
LaVergne TN
LVHW071613170726
843515LV00009B/2381
9783036528618